RECENT TRENDS FOR ENHANCING THE DIVERSITY AND QUALITY OF SOYBEAN PRODUCTS

Edited by **Dora Krezhova**

Recent Trends for Enhancing the Diversity and Quality of Soybean Products
Edited by Dora Krezhova

Published by InTech
Janeza Trdine 9, 51000 Rijeka, Croatia

Publishing Process Manager Vedran Greblo

Technical Editor Teodora Smiljanic

Cover Designer InTech Design Team

Additional hard copies can be obtained from orders@intechopen.com

Recent Trends for Enhancing the Diversity and Quality of Soybean Products,
Edited by Dora Krezhova
p. cm.
ISBN 978-953-307-533-4

Meet the editor

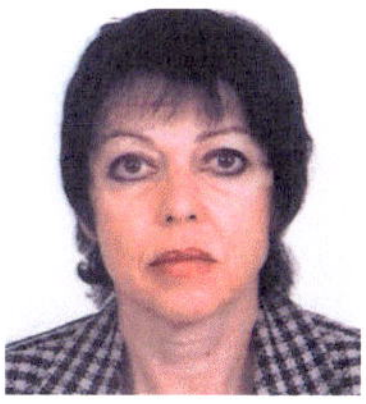

Professor Dora Krezhova was born in Sofia and obtained master degree in telecommunication from the Technical University of Sofia, Bulgaria. Since graduating, after a six-year position of research fellow at the Institute of computing techniques she moved to the Solar-Terrestrial Research Institute. Presently, she is Associate Professor at the Space and Solar-Terrestrial Research Institute at the Bulgarian academy of sciences. Krezhova published more than 187 research papers. She has contributed to the design of more than 10 scientific devices and systems and her PhD thesis and researches have made use of multichannel spectrometers for solving problems in remote sensing of the Earth and planets. For participation in the Bulgarian space programmes for remote sensing of the Earth by means of air and space-born equipment she was awarded with honorary badges, diplomas and medals. Her scientific contributions come from accurate determination of incident solar radiation and studies of spectral reflectance characteristics of natural formations, fluorescence of plants, and the impact of abiotic and biotic stress factors on plant physiology. She has a special interest in data processing and development of new mathematical models for description of the details of spectral characteristics and classification of natural formations and processes.

Contents

Preface

Soybean is one of the most important and valuable agricultural crops. Last decade it has risen to the top-traded commodities, with a multitude of uses. The high protein content of seeds is of primary importance for human food and animal feed. Fat-free soybean meal is a main low-cost source of protein for animal feeds and most pre-packaged meals, as well as a good source of protein for the human diet. Soybean is the leading oilseed crop produced and consumed worldwide. The soy vegetable oil is another valuable product of processing the soybean crop and a biofuel feedstock. This book presents new aspects and technologies for the applicability of soybean and soybean products in industry (human food, livestock feed, oil and biodiesel production, textile, medicine) as well as for future uses of some soybean sub-products. The contributions are organized in two sections considering soybean in aspects of food, nutrition and health and modern processing technologies. Each of the sections covers a wide range of topics. The authors are from many countries all over the world and this clearly shows that the soybean research and applications are of global significance. I am confident that book chapters contain useful information in many respects for broad audience and present ideas in help to researchers for advancing their fields.

Prof. Dora Krezhova
Space and Solar-Terrestrial
Research Institute at the
Bulgarian Academy of Sciences
Bulgaria

Part 1

Food, Nutrition and Health

1

Soybean in the European Union, Status and Perspective

Yves Bertheau[1] and John Davison[2]
[1]Inra SPE, route de Saint Cyr, 78 026 Versailles cedex,
[2]Inra, route de Saint Cyr, 78 026 Versailles cedex(retired)
France

1. Introduction

Originating from China, soybean is currently the most important agricultural commodity traded around the world, both in terms of volumes and money. This crop also shows the most important changes over the last decades by the predominance of genetically modified (GM) crops, dominated by herbicide tolerance traits, and its worldwide cultivation. Due to its important protein content and the increasing demand for proteins in relation with the intensification of livestock production, the soybean surfaces have dramatically increased in several South American countries, such as Brazil, Argentina, Paraguay, Uruguay and Bolivia when compared to the surfaces of soybean in the USA, and at a less extent in Canada (James, 2011). China is still the main (non-GM) soybean producer but the main exporters of (GM) soybean are the USA, Brazil and Argentina. China which was, until the 1930s, the main worldwide exporter but was dethroned in the 1950s by USA exporting soybean as basis of feedstuffs and China is now the main importer with ca 60% of US soy.
The success of GM soybean can be explained by the ease of cropping due to GM trait in countries with large fields, particularly for farmers for whom weed management and soil erosion have always been an issue.
This segmentation of market between food and feed use is still prevalent into the international trade: Asia mostly cultivates and uses non-GM soybean as a food component while other regions of the world mostly use GM and non-GM soybean as feed component (Birthal et al., 2010). However, since a few years, this trend is changing with the improvement of the living standards of Asian countries such as China, which now imports huge quantities of soybean for livestock feeding due to an increased demand for meat. Currently, the soybean daily price is at its second highest peak after the 2007/2008 peak. Altogether, the soybean daily price increased of only 83% over the 3 last decades due to the current prices' peak (IndexMundi, International Monetary Fund. April 20, 2011[1]).
Despite the fact that several other sources of protein are available for livestock, the flexibility of soybean in feedstuff preparation, particularly for pork and poultry productions, drives the international trade. Its use in bovine production, meat or milk, can be more easily replaced by alternative protein sources, or simply be replaced by pasturage.

---

[1] http://www.indexmundi.com/commodities/?commodity=soybeans&months=360

Soybean has been first introduced in Europe during the XVIIth century as high-class food, however despite several scientific and popular reports during the XIXth century soybean was rarely cropped in Western Europe. The first massive importation of soybean in Europe started by the beginning of the XXth century for oil and meal production, declined during the 2 world wars, but with increasing imports between them. Since the 1950s, Europe dramatically increased its importations of soybean due to a new animal production scheme with highly concentrated livestock production. However, new dossiers in the pipeline of GMO approvals now consider cultivation in the EU. We thus examine in more depth this forthcoming issue in environmental surveillance.

Due to this increasing part of GM soybean in the international market and consumers' reluctance of several countries to accept these products, a new segmentation of the market appeared between GM and non-GM soybean linked to labeling of GM food, and feed in some countries, with an exemption of labeling below a threshold of fortuitous or technically unavoidable presence, ranging from 0.9% in the EU-27[2] and Russia to 3-5% in Korea, Taiwan and Japan. Generally speaking, the labeling thresholds are representative of the countries' dependence on feed and food imports.

2. GMO in the EU

The development of green biotechnology dates back to the 1970s and led to the development of Genetically Modified Organisms (GMOs) in the 1980s. On May 21, 1994, the genetically engineered FlavrSavr™ tomato was the first crop approved by the U.S. Food and Drug Administration for commercialization. Due to the controversy over GMOs, which started with the first arrivals in Europe of soybeans cargoes in 1996, and to its poor organoleptic qualities, this tomato was withdrawn from the market in 1998 (Bruening and Lyons, 2000).

2.1 European consumers

Today, around 148 million hectares of GM plants are grown and traded around the world annually, among which ca. 71% of GM soy according to the ISAAA lobbyist (James, 2011). Despite this development, the European public's perception of GM crops is still very negative as demonstrated by the recent results of the Eurobarometer surveys (Bonny, 2008; de Cheveigné, 2004; Gaskell et al., 2006; Gaskell et al., 2010; TNS Opinion & Social, 2010). However, this consumers' reluctance seems less pronounced in the eastern part of the EU-27 (Consumerchoice Consortium, 2008).

After several scandals in the 1990s' such as BSE, dioxin contaminations, the reluctance of consumers' and citizens to embrace GMOs has been considered by the EC and European Union Member States (EU-MS) which have implemented a legal framework enabling consumers to maintain their freedom of choice through both food and feed labeling (European Commission, 1997, 2000a, b, 2001, 2002d, 2003c, d). In counterpart, the freedom of choice of producers to cultivate GM or non-GM crops is considered through a set of coexistence rules to be implemented by EU-MS, according to the European principle of subsidiarity as recommended by the EC (European Commission, 2003a, 2010a). However, despite numerous requests from NGOs and consumers associations, animals reared with GM feed do not require labeling.

---

[2] Current European Union with 27 Member States

2.2 GMO approvals in the EU

In order to re-assure the European public on food safety and more particularly the question of GMOs, the European Community has developed a series of regulations (Table 1) to ensure GMO safety, detection, traceability and labeling.

Food safety assessment is the responsibility of the European Food Safety Authority (EFSA) which cooperates with EU-MS national advisory committees and covers food additives, animal welfare, plant health, allergies, mycotoxins, biological hazards, chemical and biological contaminants. It also assesses the safety of GMOs (seed, food, feed, and derivatives). EFSA is an independent scientific body providing advice on all aspects of food safety, and a positive EFSA assessment is necessary for authorization to place food on the European market. GMO dossiers can be notified to the European Commission either under the 2001/18 directive or the (EC) regulation 1829/2003. Although not implied in its name, EFSA also provides advice on GMO environmental issues.

Once a positive EFSA assessment has been obtained, and once validated GMO detection methods and control sample and reference materials are available (all being provided by the applicant company), the application is then sent to the EC. On the basis of the opinion of EFSA, in some instances amended on the basis of national advisory agencies and committees, the EC drafts a proposal for granting or refusing the authorization, which it submits to the Section on GM Food and Feed of the Standing Committee on the Food Chain and Animal Health. If this Standing Committee accepts the proposal, it is finally adopted by the EC. Otherwise, it is passed on to the Council of Ministers which has a time limit of 3 months to reach a qualified majority for, or against, the proposal. In the absence of such a decision (which is frequently the case), the EC adopts the proposal. Over the last years, all GMO approvals in the EU were accepted on that scheme basis with approvals for a renewable 10 years period.

In contrast to several claims against the "lengthy and costly" approval European procedure, it should be noted that the notifiers often use dossiers of previous approvals in third countries, such as USA, with thus very few changes and thus very low costs of compliance with the EU approval procedure. Secondly, the European theoretical approval duration is *per se* not very long; however dossiers are in numerous instances incomplete. In this case the clock of approval is stopped each time details are requested from the notifiers. Due to these several stop-and-go steps in such an approval, the effective duration of European approval may be rather long. The European procedure of safety assessment of GMOs is currently under review, for instance on the statistics to be used in comparing animal cohorts, the guidelines about environmental impact assessment, or the more important use of the "substantial equivalence" concept in the comparisons between GM and conventional plants.

Despite the relatively rather strict European approval procedure, several EU-MS introduced national bans on GMO, be these for import and transformation such as Austria, for baby food in Italy or for cultivation as in France, Austria or Bulgaria.

However, there are currently ca. 50 GMOs in the pipeline of approval or approved for import and transformation, including several stacked GMOs and a few modified flowers. For soybean, 11 transformation events or stacked genes are in the European approval process with 2 GM soybean as fully approved and the first approved one (MON GTS 40-3-2) in the renewal process.

Due to the rather long European approval process, several reports outlined the possible shortage of soybean for the feed industry due to these "asynchronous approvals" (DG AGRI European Commission, 2007; Stein and Rodriguez-Cerezo, 2009, 2010a). The EC recently issued a proposal of modification of 2001/18 directive to allow Low Level Presence of EU unapproved GMOs, as also discussed in the *Codex Alimentarius* instance, for GMO already approved in a third country and whose dossiers are already under EFSA discussion for at least 3 months.

- **Directive 1990/219/EEC** covered the contained use of genetically modified organisms. Directive 1990/220/EEC was modified by Directive 98/81/EEC.
- **Directive 1990/220/EEC** covered the notification for a deliberate release and of the placing on the market of GMOs. Directive 1990/220/EEC was repealed by Directive 2001/18/EEC.
- **Regulation (EC) 258/1997** concerning novel foods and novel food ingredients, not heavily used in the EU before 1997 and establishing a compulsory labeling for these novel foods and ingredients, such as GMO, irradiated food, etc. Part of the current revision of food and feed legislation.
- **Regulation (EC) 1139/1998** laid down the compulsory indication on the labeling of foods and food ingredients produced from genetically modified soya (*Glycine max* L.) covered by Commission Decision 1996/281/EC and genetically modified maize (*Zea mays* L.) covered by Commission Decision 1997/98/EC, of particulars other than those provided for in Directive 1979/112/EEC.
- **Regulation (EC) 49/2000** amended the 1139/1998 EC regulation and established a 1% labeling threshold which was further decreased to 0.9% by regulation 1829/2003.
- **Regulation (EC) 50/2000** establishing a mandatory labeling of additives and flavorings that have been genetically modified or have been produced from genetically modified organisms.
- **Directive 2001/18/EEC** covers the deliberate release of GMOs in the environment (field trials and cultivation), in the absence of specific containment measures. It also regulates commercialization (importation, processing and transformation) of GMOs into industrial products. Finally, the Directive requests post-commercialization, case specific and general, surveillance plans on unforeseen effects of GMO on both health and environment.
- **Regulation (EC) 178/2002** resulted in the creation of EFSA and in a general obligation for traceability of at least one step forwards and one step backwards in the food chain.
- **Regulation (EC) 1946/2003** is concerned with the trans-boundary movement, and accompanying documentation, for LMOs (living modified organisms) destined for deliberate release, or for food and feed or for immediate processing, under the terms of the Cartagena Protocol on Biosafety.
- **Regulation (EC) 1829/2003** covers mainly the commercialization of food and feed. It facilitates GMO detection by obliging the providers of GMO plants to disclose methods for their detection (Regulation (EC) 1981/2006 provides for a fee to be paid by the applicant to the CRL for this service). These methods are then verified and validated by the EURL-GMFF, hosted by the DG JRC laboratory of Ispra (Italy) with the support of the ENGL, before being made public. This regulation imposes labeling for authorized GMOs above a threshold of 0.9%. Labeling is not required for conventional or organic food and feed containing the adventitious, or technically unavoidable, presence of authorized GMOs at levels less than 0.9%. Unauthorized GMO are not permitted entry in the EU, even at levels less that 0.9% (the so-called "zero tolerance").

- **Regulation (EC) 1830/2003** concerns the traceability and labeling of genetically modified organisms and the traceability of food and feed products produced from genetically modified organisms and amending Directive 2001/18/EC. It imposes a specific traceability requirement on GMOs, over and above that of the general traceability regulation 178/2002. Traceability archives must be kept for five years.
- **Regulation (EC) 65/2004** establishes a system for the development and assignment of unique identifiers for genetically modified organisms.
- **Regulation (EC) 882/2004** on official controls performed to ensure the verification of compliance with feed and food law, animal health and animal welfare rules.
- **Regulation (EC) 1981/2006** establishing a financial contribution on a flat-rate basis in order to contribute to supporting the costs incurred by the EURL-GMFF in the methods' validations.

Table 1. summarized overview of the European legislative frame on GMO.

After the commercial withdrawal of Event 176 maize, only 2 GM plants are currently approved for cultivation in the EU, namely the MON 810 maize and the Amflora® potato. However several other GMOs are in the pipeline for cultivation approvals, such as Bt11 maize or GTS 40-3-2 soybean. In this later case, Romania, which was cultivating GTS 40-3-2 soybean before its entrance in 2007 in the EU, is pushing hard for this approval.
In December 2008, the European council of ministries in charge of Environment asked for a reform of the EFSA approval process and for integrating socio-economic factors into the approval considerations. So far, only the French High Council of Biotechnologies integrates such considerations into its advice system through its Economic, Ethical and Social Committee[3] due to a recent law (République Française, 2008).

2.3 Labeling and traceability

According to (EC) 178/2002 regulation, traceability is mandatory in the EU for all food items, one step forward - one step backward, with additional specific requirements for GM products such as keeping traceability document for at least 5 years as described below and in Table 1.
The operation of GMO food control systems (e.g. detection, labeling and traceability methods) are not within EFSA's remit, and remain the responsibility of the EC, through the European Reference Laboratory for Genetically Modified Food and Feed (EURL-GMFF), and the Competent Authorities (CA) of individual EU-MS. It should be noted, since it is a source of frequent miscomprehension, that EC traceability and labeling regulations are not concerned with GMO safety, risk evaluation or risk management, since food that does not have a positive EFSA assessment does not reach the market. Traceability data on food and feed, including GMOs, may serve, however, to enable the re-call of products from the supermarkets in the case of unforeseen mishaps, such as the accidental or deliberate contamination of food chains. Traceability is a non-discriminatory and inexpensive requirement since most of the companies already have quality assurance protocols in place and since numerous analyses are routinely carried-out for multiple purposes, including vitamin or toxins contents. Quality assurance procedures offer several advantages to the companies such as specific market niches, efficient low-cost withdrawal of products and easier implementation of control procedures for future mandatory requirements (e.g.

[3] Comité économique, éthique et social (CEES)

traceability and labeling of allergens in food and feed). Fees incurred by the EURL-GMFF, for validating the detection methods by inter-laboratory trials, are on a flat-rate covered by a financial contribution of the notifying companies while a new network of National Reference Laboratories has been established beside the European Network of GMOs Laboratories (ENGL) (European Commission, 2004d, 2006a). The EC released several reports on traceability experience in the EU-MS (European Commission, 2006d, 2008b).

So far there is no European obligation of labeling animals, or their derived products, reared on GM feed despite several requests of NGOs[4] and consumers associations. However, this possibility of animal and derived products labeling has been recently introduced in Austria and Germany with a threshold of 0.9%. More generally speaking, GMO-free labeling has been introduced at 0.9% in Germany and Austria, while France is currently considering a definition of GMO-free products at 0.1% according to the recommendation of the Economic, Ethical and Social Committee[5] of its High Council of Biotechnology[6] (http://agriculture.gouv.fr/remise-de-l-avis-du-haut-conseil). GMO-free labeling is currently used by several German companies for e.g. milk which according to some claims increased their market shares (http://www.bund.net/bundnet/themen_und_projekte/gentechnik/verbraucherinnenschutz/kennzeichnung/nutzende_unternehmen/).

Since the first commercialization of GMOs in third countries, the EU has been facing a great number of alerts (Davison and Bertheau, 2007, 2008). In most of the cases, these alerts resulted from a misappropriate segregation of approved GMO (Starlink™ maize), or the seeds' commercialization of unapproved GMOs (US Bt10 instead of Bt11 maize), or the release of unapproved GMOs as in the case of US LL601 rice or Chinese Bt63 and Kefeng 6 rice. While the issues of presence of unapproved GMO in domestic markets were previously considered as an issue for countries with labeling policies, the recent increase of GMOs from emerging countries led to the reaction of the US agencies (APHIS News release, 2010; GAO, 2008). In several ways, it appears that USA will move toward a more surveying attitude similar to the EU (Davison, 2010).

The issue of domestic unapproved GMO in local market is the basis of the current work of *Codex Alimentarius* on the Low Level Presence (Codex Alimentarius, 2003). Asynchronous approvals of GMOs has been recently taken into consideration by the EC which proposed a 0.1% presence in feedstuffs of EU unapproved GMOs after a revision of EU legislative frame (Aramyan et al., 2009a; Reuter, 2010; Stein and Rodriguez-Cerezo, 2010b).

However, the recent results of the European research project Co-Extra (www.coextra.eu) show that supply chains operators already use a contractual threshold of ca. 0.1% for the 0.9% European labeling threshold (European Commission, 2010b). Together with the possible labeling of animals reared with GMOs, between 0.9 and 0.1%, and below 0.1%, such a situation will drastically impact on the availability of "GMO-free" products. The reaction of consumers toward this new European proposal remains currently unknown as *ex ante* studies appear very difficult for providing accurate results.

Generally speaking, the European traceability greatly improved over the last decade but with still several issues such as the sowing of EU unapproved Bt11 in France a few years ago. In these circumstances, the recent approval of a GM potato, specifically destined for

[4] Non Governmental Organization

[5] Comité économique, éthique et social

[6] Haut Conseil des Biotechnologies (HCB)

industrial uses, may lead to the same issue of inappropriate segregation as the US Starlink™ maize (Miller, 2010).

3. European GMO coexistence issues

3.1 General overview

The freedom to producers to either produce GM, conventional or organic crops is the counterpart and the necessary basis of consumers' freedom to choose, or not, GMO into their food. Accordingly, the EC released in 2003 and updated in 2010 a recommendation on the coexistence of GM, conventional and organic farming (European Commission, 2003a, 2010a). Practical implementation and rules is the responsibility of the EU-MS according to the European subsidiarity principle. In parallel, several European regions declared themselves as GMO-free (http://www.gmofree-euregions.net:8080/servlet/ae5Ogm) despite the fact that some do not have the administrative legality for such a positioning. The European Commission reported on implementation of coexistence rules in EU-MS which is far from being implemented in a harmonized way in all EU-MS (European Commission, 2006b, 2009e). COEX-NET is a network established to facilitate the exchange of information on coexistence issues between EU-MS CA.

Different national (French ANR-OGM, British Farm Scale Evaluation, German BMBF project, etc.) and European (SIGMEA, Transcontainer, Co-Extra) research programs were launched the last decade to establish the scientific bases of coexistence. To aid national Competent Authorities, the EC has recently created a new 'co-existence bureau' specific for co-existence issues, at JRC-IPTS[7], Seville, Spain, which recently released its first document on maize crops coexistence (Czarnak-Kłos and Rodríguez-Cerezo, 2010). If most of the current work focused on maize, currently the only plant sown, several other crops have been studied from a coexistence viewpoint, as for instance oilseed rape and sugar beet, two crops for which ferals and crossing with wild relatives are important in the EU (Colbach, 2009; Colbach et al., 2009; Darmency et al., 2009; Darmency et al., 2007; Gruber et al., 2008).

Up to now 2 trends can be distinguished in the European coexistence schemes, a flexible coexistence frame and one based on dedicated production areas, be these GMO or GMO-free. Until now, coexistence research has mostly focused on flexible coexistence, that is to say, the individual choice of a farmer, with a minimum of *ex-ante* duties (such as isolation distances, buffer and/or discard zones) together with an information system, including, for instance, a public register of GM crops to provide information to non-GM growing neighbors coupled with some *ex-post* economic solutions such as compensation schemes for economic losses (Demont et al., 2010; Demont et al., 2009; Desquilbet and Bullock, 2010; Devos et al., 2009; Messéan et al., 2006; Messéan et al., 2009). Current EU best practice guidelines and companies' stewardships for coexistence measures in maize give effective measures for the European 0.9% threshold by requesting separation distances, buffer and discard zones, and staggered flowering times, but without taking into consideration the threshold of 0.1%used by companies due to measurements and sampling uncertainties (Bartsch et al., 2009; Bock et al., 2002; European Commission, 2010b).

Generally speaking, the proposed flexible coexistence solutions are based on the assumption that individual choices should prevail, but information systems need to be available to allow other producers to know what is being produced and where, such as those already

---

[7] European Commission's Joint Research Centre, Institute for Prospective and Technological Studies

deployed in Portugal. However, at the same time territory / landscape multi-functionality is requested by the EU, and there is a growing call from consumers, and society as a whole, for both more sustainable production and so called quality-oriented produce as shown by the current yearly increase of ca. 20% of organic and other signs-of-quality based farming (Laisney, 2011). This leads to a conflict as GM produce is not seen as organic, even if it can be produced without, or with less pesticide, although it can definitely be more sustainable with yield increases over conventional farming (Cardwell, 2003; Grossman, 2003; Laurent et al., 2010; Marsden, 2008). Moreover, territory organization, with Natura 2000 areas (protected environmental areas, for example), is not taken into account by the flexible coexistence scheme while their domino impact is highly recognized (Demont et al., 2008). This dichotomy needs to be addressed. Finally, the landscape is highly structured by downstream supply chains (Coléno, 2008; Hannachi et al., 2009; Le Bail et al., 2010; Petit, 2009).

Due to the several requests of EU-MS to take into consideration socio-economic aspects into GMO approval, the 2010 updated EC recommendation considers more favorably the possibility of GMO-free areas. However the EC would not accept that requests by EU-MS be based on scientific or environmental grounds which are already assessed by EFSA. This last restriction is currently actively fought at the European Parliament. As a first demonstration of EU policy change, the Portuguese Madeira archipelago was established as the first GMO-free area, though cultivation of maize is relatively scarce in Madeira (Kanter, 2010).

On the opposite side, dissemination over long distance of pollen as well as the practical effect of the contractual threshold of operators militates in favor of dedicated production areas (Brunet et al., 2011; European Commission, 2010b). However, the research work on technical, economic and societal issues raised by this solution are drastically missing and the subject of strong opposition (DEFRA, 2006; Devos, 2008; Dobbs, 2011; European Economic and Social Committee, 2011; Jank et al., 2006; Sabalza et al. 2011).

As soybean is mostly an autogamous plant, numerous issues raised by pollen dissemination should not hamper the soybean cultivation in the EU. Similarly, the absence of out-crossing to wild-relatives, ferals and volunteers should facilitate the cultivation of GM soybean into a coexistence frame. However, the predominant herbicide trait would probably cause the same problems of resistant weeds as observed in the USA (Brasher, 2010; Cerdeira and Duke, 2006, 2007; Roberson, 2010). Transportation of GMO was identified as the cause of several incidental releases in the EU and third countries, including the growth of GM plants around harbors (Kawata et al., 2009; Kim et al., 2006; Lee et al., 2009).

Thus, due to its biological properties and despite the different structure of European farms and territories, the coexistence in the EU of GM and non-GM soybeans should be one of the easiest to implement; as it is in several third countries exporting non-GM soybean despite important GM soybean cropping.

4. Surveillance plans

Post-market release monitoring of GMOs approved both for import and cultivation (European Commission, 2001, 2003c) is one of several requests included in European GMO approval. According to the pre-market risk assessment (RA), this monitoring can be divided into case-specific monitoring (CSM), which covers any identified risk, and general surveillance (GS) for all risks that might not have been identified during the RA. EFSA published a series of documents about RA, CSM, and GS (Bartsch et al., 2006a; EFSA GMO

Panel, 2004, 2006a, b, 2010). Several guidance documents and reports on implementation of the monitoring were then published (Bartsch et al., 2006b; Bartsch et al., 2007; EU working group, 2003; European Commission, 2002a, b, c, 2004b, 2008b, 2009a).
Monitoring of the GMOs post-market release should include both health and environmental effects, should be carried out by the GMO consent holders, i.e. the companies having received a grant for a commercial release of a GMO, and may be supported by additional independent actions of EU-MS. The CSM and GS shall cover both GMO and non-GMO cultivated areas (EFSA, 2008; EFSA GMO Panel, 2004, 2006a).
Up to now, most of the surveillance activities of GMOs approved for import and processing have been delegated by the consent holders to European professional unions of importers, transporters, and processors, namely COCERAL, UNISTOCK and FEDIOL. However, the content of agreements between consent holders and such unions remains unknown. Due to the lack of precision, in particular about the methodology used for monitoring imported GMOs, the accuracy of such monitoring plans remains undetermined for the EFSA, GMO national advisory committees, and CA in charge of GMOs (Beissner et al., 2006).
As noted above, health and environmental monitoring, which also means animal health, of predictable and unexpected effects of GMO cultivation is mandatory in the EU on both GMO and non-GMO cropped areas (European Commission, 2001, 2002a, b, 2003c).

4.1 Specific surveillance

In the EU, several GMO CSM protocols have been pursued by notifiers, scientists, and enforcement authorities. A decade after the first GMO cultivation in the EU, a number of guidelines, conceptual frameworks, and reports are available for GMO CSM (Bartsch et al., 2007; Bontemps et al., 2004; Bourguet, 2004; Chaufaux et al., 2002; EU working group, 2003; European Commission, 2009b; Monsanto Co., 2006, 2009a, b). The consent holders[8] published results of insect resistance monitoring, but only from GMO cropped areas despite the European rules (Monsanto Co., 2006, 2009a, b).

4.2 General surveillance

This part of the chapter focuses on the monitoring activities of unexpected effects of GMO cultivation, i.e. general surveillance.
GMO GS frameworks proposed by GMO consent holders in relation to EFSA guidelines include literature survey, development and /or use of existing monitoring surveillance, and specific trials as necessary (EFSA GMO panel, 2006b, 2010).
General surveillance is designed to detect unanticipated effects on general safeguarded subjects such as natural resources, which must not be adversely affected by human activities like GMO cultivation. Monitoring has to be appropriate for detecting direct and indirect effects, immediate and long-term effects, as well as unforeseen effects. In its 2006 opinion, the EFSA GMO panel outlined that: "*general surveillance cannot be hypothesis driven, but should, when possible, make use of existing monitoring systems in addition to more focused monitoring systems (e.g. farm questionnaires). Data quality, management and statistical analysis are of high importance in the design of general surveillance plans and comparison should be made with baseline data*" (EFSA GMO panel, 2006b). A public consultation on the 2010 version of EFSA GMO panel opinion on GS is currently ongoing.

---

[8] Notifiers having received European approval for GMO import and/or cultivation.

The 2010 draft version of EFSA guidelines shows a drastic change of paradigm in the principle of environment GS and still does not establish guidelines for health effects surveillance. This draft version particularly outlines the importance of baselines, use and assessment of indicators after field trials, less oriented biodiversity studies without *a priori*, etc. This difference between 2006 and 2010 version may represent both the change into the EFSA GMO panel composition as well as an attempt of EU-MS, of their enforcement agencies and of the EC to master and retrieve the leadership in a scientific, but also highly political, issue.

For several years now, important scientific conceptual and practical works have indeed been developed in several EU-MS along with reports from national committees in charge of GMO approvals (ACRE, 2004; Breckling and Reuter, 2006; Garcia-Alonso et al., 2006; Graef et al., 2005; Monkemeyer et al., 2006; Wilhelm et al., 2009; Wolt et al., 2010; Zughart et al., 2008).

Most of these scientific works focused on environmental effects, while the effects on human health are roughly "delegated" by the consent holders to national health monitoring networks (Bakshi, 2003; Cellini et al., 2004; Covelli and Hohots, 2003; D'Agnolo, 2005; EFSA GMO panel, 2006b; Filip et al., 2004; Hepburn et al., 2008; Wal et al., 2003). To provide an example of EU-MS, in France, the "Sentinelles" network, ANSES[9], and InVS[10] might form parts of such a general surveillance plan on human health in application of the WHO and European rules, directives, and regulations. Animal health is relevant to the OIE[11] and European rules, directives, and regulations. As for GMO CSM and GS, the French Ministry of Agriculture (DGAl directorate) is in charge of animal health surveillance. However, no GMO-related GS activities in human and animal health are clearly identified in the European activity reports.

Indeed, GS of human and animal health is also particularly important given that GMOs not dedicated to food and feed purposes will rapidly arrive on the market as exemplified by the recent European approval of Amflora® potato for cultivation. For this kind of split approval, we must remember the first such issue raised by the incorrect segregation between food and feed/ industry storage facilities of the USA-approved Starlink™ maize (Alderborn et al., 2010; Beckie et al., 2010; Miller, 2010). Despite the past European experience of segregating crops dedicated to industrial uses as part of a specific derogatory cultivar list, the additional recent request Modena GM potato cultivation approval in the EU can lead us expect that more and more GMO dedicated to industrial use will enter the food chain and raise new controversies about human health.

From a decade of GMO cultivation in the EU, several remarks can be made about environmental GS reported by consent holders, scientists, and enforcement authorities.

- The consent holders include a literature survey and questionnaires to GMO cropping farmers and collaboration with existing networks in their environmental GS, as was done in Germany after the German Competent Authorities (CA) approval. However, in that latter case, a great deal of imprecision remains about the content of agreements with existing networks, the network's possible training, and the surveyed locations, i.e. representativeness and accuracy of the GS, particularly in non-GMO areas. Moreover, no statistics are provided which might alert the CA to start more in depth monitoring.

[9] Agence Nationale de Sécurité Sanitaire
[10] Institut National de Veille Sanitaire
[11] Office International des Epizooties

- Despite the positive EFSA assessment of the consent holders' monitoring reports, we can observe that, to date, the consent holders do not include non-GMO areas that are in practice delegated to the responsibility of the EU-MS (Alacalde et al., 2007; Lecoq et al., 2007; Monsanto Co., 2006, 2009a, b; Tinland, 2008; Tinland et al., 2007; Wandelt, 2007; Windels et al., 2009). Accordingly, several EU-MS have already, or are planning to, launched GS research projects and networks even in those EU-MS with bans on GMOs (Bartsch et al., 2009; Breckling and Reuter, 2006; de Jong, 2010; Gathmann, 2009; Gathmann and Bartsch, 2006; Pascher et al., 2009). This survey of non-GM cropped areas is however of utmost importance as recently shown in China (Lu et al., 2010). Unfortunately, we can again observe that private interests and benefits are supported by public funding when general goods are concerned, as usual in the "Tragedy of Commons" frame (Hardin, 1968; Hardin, 1998).

Most current environmental GS plans focus on changes in *ex ante* baseline and / or biodiversity assessments, sometimes along with a general approach looking at the effect of agricultural practices (Hintermann et al., 2002; Monkemeyer et al., 2006; Sanvido et al., 2007a; Sanvido et al., 2007b, 2009a; Sanvido et al., 2009b; Schmidt et al., 2009). However, the conceptual framework for environmental GS is far from being both a consensus and a reality. This situation motivated the European Commission to launch a call for proposals in 2010 (KBBE.2011.3.5-01) that address environmental GS and possible standardization (ACRE, 2004; Beismann et al., 2007; EFSA GMO panel, 2006b; Finck et al., 2006; Ostergard et al., 2009; Pascher et al., 2009; Sanvido et al., 2005; Schiemann, 2007; Seitz et al., 2010; Wilhelm et al., 2003; Wilhelm et al., 2009; Wilhelm and Schiemann, 2006).

The main conclusion that can be drawn from the current situation is that, despite the mandatory involvement of GMO consent holders into GMO GS and the monitoring of non-GMO areas, the main effort appears to be supplied by the EU-MS.

However, a number of environmental monitoring procedures are already in place in the EU, several of which partly embrace - episodically or on a longer term - biodiversity, GMO CSM, "epidemio-surveillance", or more general effects of agricultural practices on agro-environment. In several instances, these monitoring schemes are carried out by citizens in a so-called participatory science. These trends might be correlated with another trend for observing territory from societal and economic viewpoints (Barzman et al., 2005; Bodiguel, 2003; Cardwell and Bodiguel, 2005; Henle et al., 2008). Networks of citizens and/or scientists, as well as enforcement authorities already working on these issues, all have in common (i) a need for long-term studies, (ii) different demands on space and changes over time, (iii) different indicators which (iv) generally have to be reported to national CA and EC, sometimes according to international treaties. But up until now, results appear fragmented, collated in different databases generally without quality assessment or direct connection through a unique Web-based portal or automatic novelty detection capacity (Haggett, 2008).

Nonetheless the environmental liability directive and the right of European citizens to have access to environmental information reinforce the need for gathering these fragmented data (Cardwell, 2010; Ebert and Lahnstein, 2008; European Commission, 1985, 2003b, 2004a, c, 2006c, 2007b).

In the case that GM soybean cultivation would be allowed in the future, there are thus numerous issues that should be fulfilled by consent holders, particularly for herbicide tolerant crops whose uncontrolled use in third countries leads to numerous herbicide resistant weeds and costly companies' based eradication programs (Adams, 2011; Brasher, 2010; Cerdeira and Duke, 2006, 2007).

5. Soybean in the EU

5.1 A rapid historic overview of the last decades

In the Dillon round of GATT negotiations (1960-62), the EEC[12] negotiated zero duties on soybeans and several other agricultural products. At the time, soybean was of little importance in international trade with ca 4 MT traded in 1961. Furthermore, there were no varieties of soy available at this time that could be grown in this EU-6. Thus the EU-6 had no producers to protect and found in their interest to keep borders open to soybeans and their products. At that time, pasturages, cereal and some domestic protein rich crops provided most of the necessary feed. That period was the beginning of a drastic change into the European livestock production.

However, the high European internal costs of feed grains forced livestock producers to substitute cheaper soybean meal. In addition to the competition between European feed grains and imported soybean meal, soybean oil competed with domestic vegetable oils such as sunflower oil, olive oil or rapeseed oil, and, when used in margarine, competes with butter. To compete with cheap soybean oil, local oils were subsidized, which was attacked under GATT in 1987. In 1973, a shortage in US soybean exports impacted most of the current EU-MS, including France which was considered as the least soybean dependent EU-MS (Berlan et al., 1977; Hasha, 2002). As maize and soybean compete for both feeding and surface, this kind of shortage is expected to come back with the growing use of maize for bio-ethanol production (Headey, 2011). Several EU-MS attempted to reduce their growing dependence from soybean by national protein plans - but up to now unsuccessfully.

The European cropping of soybean was, up to 2007, restricted to some EU-MS and aimed at food or a few feed specialties (e.g. organic) with most of the production being based in Italy. Among the several reasons why EU is not a soybean producer we can distinguish a relatively unfavorable climate with cool spring and drought early summer, with a Northern predominance, in the EU compared to third countries producing soybean and a relatively high population density with rather small farms and fields. However, several soy varieties are cultivated in Canada and thus soy cropping in the EU-27 would now be possible after appropriate selection of cultivars, provided the seed companies could find some benefit in that selection. This would be probably the case after GM soybean approval for European cultivation.

The EU was in 2007 still under construction and two new countries coming from the implosion of the former soviet bloc entered the EU-25. At its entrance into the EU, Romania officially stopped cropping GM soybean and came back to old varieties of non-GM soybean whose cropping was also not favored by the current European Common Agricultural Policy (CAP; Badea and Pamfil, 2009; Dinu et al., 2010). However, the interest in GM soybean was declining from 1996 to 2002 (Brookes, 2005).

5.2 Soybean use in the EU

As most of the crops choices made by farmers over the last decades, soybean cultivation is highly linked to the several changes in CAP. Among those more related to soy cultivation (i.e. linked to oilseed and protein rich crops) we can notice the changes due the Blair House agreement, in 1992, for duty free soybean importation and the Berlin agreement, in 1999, for decreasing aids to oilseeds and open widely the European market to global trade.

[12] Economic European Communities or EU-6

	soybean imports	Soybean exports	Soybean crushing	Soybean oil and fats production	Soybean oil and fats imports	Soybean oil and fats exports	Soybean oil and fats consumption	Soybean meals production	Soybean meals imports	Soybean meals exports	Compound feedstuffs production (FEDIOL)	*Compound feedstuffs production (FEFAC)*	*Compound feedstuffs production in 2009 (FEFAC)*	*Compound feedstuffs turnover' (FEFAC) in 2008, M€*
	NL (4,004)	NL (1,079)	DE (3,364)	DE (643)	FR (547)	NL (633)	DE (591)	DE (2,676)	NL (5,236)	NL (4,769)	FR (22,424)	*FR (22,679)*	*SP (21,549)*	*FR (8,450)*
	DE (3,485)	BE (188)	SP (3,026)	NL (572)	DE (249)	SP (323)	FR (556)	SP (2,388)	FR (4,399)	DE (1,491)	DE (21,825)	*SP (22,117)*	*FR (21,236)*	*SP (6,804)*
	SP (3,279)	SL (84)	NL (2,898)	SP (549)	NL (223)	DE (301)	IT (415)	NL (2,257)	DE (3,573)	BE (782)	SP (19,220)	*DE (21,825)*	*DE (20,829)*	*IT (6,500)*
	IT (1,640)	DE (47)	PT (1,170)	IT (259)	UK (187)	BE (88)	SP (317)	IT (1,232)	SP (3,385)	SL (697)	NL (14,507)	*NL (14,643)*	*NL (14,108)*	*DE (6,290)*
	PT (1,190)	IT (39)	GB (559)	PT (199)	BE (185)	PT (68)	UK (279)	PT (930)	UK (2,216)	PT (247)	UK (14,222)	*IT (14,349)*	*IT (13,860)*	*NL (4,750)*

With an important increase since 2006 due to the food crisis.

Table 2. 2008 figures of soybeans, oil fats and meals' import, production and processing (sources: Fediol, 2011 and FEFAC, 2011). BE: Belgium; DE: Germany; FR: France; IT: Italy; NL: The Netherlands; PT: Portugal; SP: Spain; SL: Slovenia; UK: United Kingdom.

In 2006, the EU imported mostly soy beans and also meal. Crushing capacities have been developed in the Netherlands and Germany. As The Netherlands, Germany and Belgium have important harbors to where the most important part of EU imports of soybeans and meals are discharged, a large part of their imports are re-exported either as beans or more generally as meals and oils. Indeed, The Netherlands are currently a net exporter of soy meal toward other EU-MS. However, other European oilseeds compete with soybean. Generally speaking the proportion of rapeseed crushing is in constant increase over the last years, rapeseed crushing having overtaken soy crushing in 2005 due to the use of oilseed rape in European bio-fuels production. The table 2 provides the figures in 2008 of the soy beans, oil, fats and meals production imports and exports for the first top 5 EU-MS.
The table 3 provides a figure of soy beans and meals imports in the EU from 1980 to 2008. As it can be seen, European soy beans and meals imports are relatively stable, when compared to prices and yearly weather variations, since 2004, i.e. with the integration of 10 and then 2 new Member States into the EU in 2004 and 2007, respectively.

	2008 EU-27	2007 EU-27	2006 EU-25	2005 EU-25	2004 EU-25	2003 EU-15	2002 EU-15	2001 EU-15	2000 EU-15	1990 EU-12	1980 EU-9
Soy beans	15,298	15,064	14,127	14,670	14,732	17,353	18,239	17,922	14,779	13,301	11,760
Soy meals	23,227	24,321	23,405	23,029	22,632	20,352	19,605	17,870	15,840	10,471	7,226

Table 3. Imports (× 1,000 T) of soy beans and meals in the EU (source: Fediol, 2011).

5.2.1 Non-food non-feed use

Compared to the other uses of soybean the use of soybean, with or without chemical changes, in printer inks, as antifoam agent to bio-fuels and cosmetics are currently rather anecdotic (Gelder et al., 2008; Roebroeck, 2002). For instance most of the European sources of bio-fuel rely on oilseed rape. This part will probably increase with the new CAP reform favoring sustainable and environment friendly agriculture and supporting renewable, low carbon emitting energies sources.

5.2.2 Food use

Only a few percent of soybean is used for food purposes (Gelder et al., 2008). Lecithin and oil are the main products used in food, the latter being also used in margarine production, together with some specialties such as some kind of yogurts, vegetarian steaks, or the usual Asian specialties such as Tofu, nato (Roebroeck, 2002). Soy milk is mostly imported from Canada, by some worldwide companies.

5.2.3 Feed use

Soybean is mostly imported in the EU-27 for compound feedstuffs production (Popp, 2008). With the end, in the 1990's, of European intervention on cereals, which were used with some soy meal for compound feedstuff production, together with the ban of meat-and-bone meal from most of the feedstuffs (in fact meat-and-bone meal, despite being mostly destroyed, continues to be used in feeding short-living animals such as chickens and fish), the European feedstuff industry was looking for another source of cheap source of protein.

However, despite the end of European aids on cereals the proportion of cereals into compound feed increased since 1995, while the proportion of meals of all origins fluctuated between 30 Mt and 40 Mt.

Compound feed consumption in the EU-27 represents ca. 147.6 Mt, a quantity thus similar to the US consumption of 149 Mt, with a nearly constant percentage of the global consumption over the past decade. This compound feed consumption was accompanied by an increasing production of pig meat and poultry to be compared to a constant beef and veal production, a difference which is mostly due to the entrance of new EU-MS in 2004 and 2007 (FEFAC, 2010; European Commission DG-Agri, 2010*)*. However, the increase in European meat production is parallel to a general trend of decrease (beef, veal, pig) or stagnation (poultry) of European meat consumption. In 2009, the proportion of compound feed for animal rearing again decreased to ca. 30% of the total feedstuffs quantity which corresponds to a general change into the European livestock production schemes.

The soybean meal is used for all animal feeding, particularly since the ban in 2001 of meat-and-bone meals due to the mad-cow / BSE disease, with an exception for organic production or some animals growing under specific signs-of-quality. This important source of protein cannot be fully replaced by fish meal which was another reason for increasing the imports of protein rich commodities. Due to its high content of protein and relative poorness in fat, the soy meal is relatively difficult to replace in poultry, piglets and calves feeding. Alternative sources of protein such as sweet lupine, field pea or rapeseed meal are generally less palatable until the animals reach maturity. This explains the figure of compound feed mostly used for poultry and pig production (FEFAC, 2011). The issue of protein source is of less importance for mature animals and more particularly for cattle.

The origin of imported soybean may depend on EU-MS, for instance France mostly imports soy from Brazil while The Netherlands and Portugal are the top 2 importers of the US exports of soybean (Dahl and House, 2008). Up to 2008, EU was the first destination of exported soybean from USA, Brazil and Argentina. The European protein crops imports represented in 2009 ca. 20 Mha cultivated outside Europe.

However, the development of GM crops in the 3 main exporting countries definitely impeded exports, particularly in the US. Brazil, up to now, took into account the undesirable effect of asynchronous approvals of GM crops on its exports toward the EU (Aramyan et al., 2009a; Aramyan et al., 2009b; Boshnakova et al., 2009; DG AGRI European Commission, 2007; Dobrescu et al., 2009; Konduru, 2008; Stein and Rodriguez-Cerezo, 2009; Stein and Rodriguez-Cerezo, 2010b).

5.2.4 The animal labeling issue

As the animals fed with GMOs do not have to be labeled in the EU, most of the feedstuffs in the EU-27 is produced from GM soybean. However several NGO and consumers associations are requesting such animal labeling, a request supported by polls and experimental auctions studies (Kontoleon and Yabe, 2006; Noussair et al., 2004). The EU organic farming threshold of labeling is also of 0.9% (European Commission, 2007a, 2008a). However, this EU threshold can be superseded by national measures. More generally speaking the EU has numerous signs-of-quality, based, not only on brands as in third countries, but mostly on EC-approved processes or origins. The consumers' reluctance was thus taken into account in production procedures of most of these quality signs by eliminating GMO use into feedstuffs.

Germany and Austria recently introduced a legislative frame for GMO-free labeling. In these countries the GMO-free threshold complies with the European 0.9% threshold of fortuitous or technically unavoidable presence of GMO. This labeling is applicable to both vegetal produce and animals reared with "GMO-free" feed.
In another hand, several French producers, such as *Poulets de Loué*, or retailers, such as Carrefour, or quality signs producers such as *Comté* cheese, used non-GM (Identity Preserved, at 0.9%) soybean since the beginning of the XXIst century but without the possibility of retrieving profits of their efforts (Milanesi, 2008, 2009). In 2009, the French *Conseil de la Consommation* as well as the *Haut Conseil des Biotechnologies* (HCB) released recommendations for the creation of a GMO-free supply chain at 0.1%, with, in the latter advice, animal labeling according to 2 thresholds: below 0.1% and between 0.1 and 0.9%. Despite the fact that the French decree related to GMO-free labeling is so far not published, several producers and retailers took this opportunity, and the further policy change of the French Repression of Fraud services, to label their animals as being reared with less than 0.9%. If the latest HCB recommendation is followed up by the French government, 2 kinds of GMO-free animal labeling would thus prevail in France: "reared with GMO-free feedstuffs below 0.9%" and "reared with GMO-free feedstuffs below 0.1%". The HCB also requested into its recommendation that the French government should precede the implementation by an *ex ante* socio-economic analysis of the viability of such a GMO-free supply chain at 0.1%. A feasibility study on this request for a *ex ante* socio-economic analysis is currently ongoing.
However, the availability of non-GM (including GMO content below 0.9%, IP, and "GMO-free" at 0.1% or "hard IP") soybean is far from being sustainable. Up to 2008, Brazil was the most important exporter of non-GM and GMO-free soy toward EU with negotiated premiums. But the breeding of new soy GM varieties more appropriate to Brazilian climate induced a new increase of GM surfaces in Brazil, particularly in Matto-Grosso with the largest farms and fields (Fok, 2010). Despite the fact that Parana state dedicated a whole harbor to non-GM soy, this state also moved, toward GM soy, particularly for the most weedy fields and by the farmers the less experienced into weed management. The main source of European non-GM soy could thus disappear unless operators facilitate the maintenance of non-GM cropping.
One of the first issues, for maintaining the interest of non-GM cropping, is the rather low level of premium (ca 1/4th of the final one paid by final buyers) received by the Brazilian farmers. The second is that this non-GMO related premium is not discerned by the buyers, such as cooperatives, from other premiums, all premiums being thus provided into a non transparent package of several premiums. The incentive of producing non-GM soy is thus rather low in Brazil despite the fact that the tech fees imposed by the traits' providers may be ca 40% of the seeds prices (Bonny, 2009; Fok, 2010). According to ABRANGE, a Brazilian association of non-GM farmers, Brazil would be however currently providing 53% of non-GM soybean while India and China would be providing 18 and 17%, respectively (Milanesi, 2011). However, these claims are not in line with the observation of the 2009 increase of non-GM soybean in USA, after a decade-long decrease of non-GM soybean areas, due to both more incentive premiums for non-GM beans and increased production costs of GMOs (prices of seeds and herbicide) (Milanesi, 2011).
With premiums, long-term contracts are the second driving force for farmers for maintaining what several authors call market niches (Foster, 2007, 2010). Long-term contracts have thus been established by European producers, such as *Poulets de Loué*, with or

without the support of the European GMO-Free regions and Brazilian producers such as the Brazilian ABRANGE association. Generally speaking, the European GMO-free regions' network supports their producers into their search for long-term supply of "cheap" non-GM soy.

The premiums ranged from an average of 16 US$ in 2004 to ca. 70 US$ in 2009 for US farmers (Foster, 2010; Milanesi, 2011). While it is generally difficult to determine the premiums fluctuations over the year due to the confidentiality of the contracts, the changes observed into the non-GM soybean market of Tokyo show a general trend of a 10% premium over the GM soybean, with of course important peaks up to 40% in 2008 due to both the food crisis and an decrease of US 2007/08 soybean cropped surface in that year (Foster, 2010; Headey, 2011). Compared to the "only" 83% of soybean price increase over 30 years, such premiums could thus be very incentive, particularly when linked to long-term contracts.

The increase of price of compound feed due to this non-GM soy would be of less than 3% (Gryson and Eeckhout, 2011). Since feed cost represents ca. 77% of price of chickens, this would induce a final small increase of some cents per chickens' kilo (Milanesi, 2008, 2009).

Most of the European imports of both GM and non-GM soy are through the main commodities traders namely ADM, Bunge, Cargill and Louis Dreyfus (Green and Hervé, 2006). However, as these traders advertised they were facing a shortage of non-GM soy, several new SMEs, such as Solteam, are currently developing their own import network to provide European feed producers with non-GM commodities. With the growing surfaces of GM soybean in Brazil, alternative sources of non-GM soy are actively looked for beside long-term contracts and premiums use for sustaining the availability of American non-GM soy.

However, the main forthcoming issue might be the availability of low cost non-GM soybean varieties both in third countries and EU-MS (Milanesi, 2011; Then and Stolze, 2010). As currently observed, the availability of non-GM seeds is decreasing with a few new varieties being commercially released. Accordingly, old non-GM seeds cannot compete with new GM varieties what can explain, together with a lack of support of oilseeds by the European CAP, the dramatic decrease in yields and total production observed in Romania at its entrance in 2007 in the EU-27 (Dinu et al., 2010). Despite the fact that new but small plant breeder and seeds sellers (such as eMerge a Cargill subsidiary) are appearing, their ability to access to soybean germplams is questionable as private sector is focusing on GM varieties and public research is focused on germplasms (Heisey et al., 2005; Heisey et al., 2001; Naeve et al., 2010; Orf, 2004). The availability by the big seeds companies of non-GM soy varieties will highly depend on their forward or backward breeding strategies (Milanesi, 2011).

It may worth noting that while traders such as Cargill are developing such seeds companies which will help them to maintain the non-GM flows towards several importing countries: competing on global commodities trade does not mean, for such companies, excluding higher added value market niches. In the meantime, participatory breeding of non-GM soy varieties is also developing as observed for numerous other crops (Bellon and Morris, 2002; Desclaux et al., 2008; Smale, 1998). As noted by several authors, this availability of several kind of varieties is necessary for developing the segmented markets requested by farmers, retailers and consumers (Elbehri, 2007).

5.3 Soybean cultivation in the EU-27

As noticed by a recent motion of the European parliament, protein rich crop production occupies only 3% of EU-27 arable land and supplies only ca 30% of protein crops consumed as animal feed (LMC International, 2009a, b; Häusling, 2011). Table 4 provides figures on some EU-MS surfaces of soy cultivation (Eurostat, 2011).

With the *Agenda 2000* CAP reform, aid to European farmers became decoupled, i.e. aid were no longer received for oilseeds production, nor related to yields. There is thus no more European intervention for buying, export subsidizing or other market support available for oilseeds in the EU-27. Moreover, agricultural aid is now rather linked to environment preservation and sustainable agriculture, the second pillar of the new CAP, together with social criteria (Krautgartner et al., 2010b). These drastic changes into the European CAP could lead to drastic changes during the next years in the European cultivations' schemes.

	2010	2009	2008	2007	2006	2005	2004	2003	2002	2001	2000	1999
Italy	159	134.7	107.8	130.3	177.9	152.3	150.4	152.1	152	233.5	252.6	246.5
Romania	65.2	48.8	49.9	133.2[13]	190.8	143.1	121.3	128.8	71.8	44.8	117.	99.8
France	50.9	43.7	21.8	32.4	45.3	57.4	58.6	80.7	74.8	120.9	77.7	98.2
Austria	34.4	25.3	18.4	20.2	25	21.4	17.9	15.5	14	16.3	15.5	18.5
Hungary	33.5	31.5	29	32.9	35.9	33.6	27.3[14]	30.3	25	20.6	22.2	32.2

Table 4. Surface (× 1,000 ha) of soybean cultivation in the top 5 EU-MS (Eurostat, 2011).

In the Western part of the EU-27, soy cultivation attempts to satisfy the needs of high added values supplies such as food and meat production under signs-of-quality. However, in spite of several "protein plans", soy is still not considered as an important European crop.

While Western Europe was poorly considering soybean cultivation yet started to import soybean since ca. the second decade of the XXth century, the former USSR developed a soybean breeding institute since the beginning of this last century. As a result of Russian research, several soybean varieties were developed for the former Soviet bloc. An important area of soybean cropping is thus done in the eastern part of Europe, around the southern Danube basin, in particular in Romania, but also in Bulgaria and Hungary. As another example of such Soviet soybean production, and thus of varieties adapted to the European climate, Ukraine and Russia were cropping in 2006 725,000 and 810,100 ha, respectively (Otiman et al., 2008). However, Romania was the only country to extensively grow GM soybean over ca 137,000 ha for feedstuff production but with yield per ha nearly 2/3rd of the ones of USA, Brazil and Argentine. While irrigation is important for increasing the yield, it is also highly probable that the GM varieties were not fully adapted to the European eastern conditions.

Since its entrance in the EU in 2007, Romania stopped cultivating of GM soybean, but started with the MON 810 GM maize (Badea and Pamfil, 2009). Since that time Romanian farmers claim that after a period of self-sufficiency in feed, they had to import again soybean for livestock (Otiman et al., 2008). However, the decrease in soybean cultivation is more probably linked to the absence of European subsidies to soybean cultivation and a return to old, less productive, non-GM soy varieties. Romanian farmers are thus among those pushing to force the European approval of GM soybean cultivation, whose dossier is currently in the European approval pipeline. This may also be explained by the existence, in several Central and Eastern European Countries (CEEC), of large, corporate, farms up to 20,000 ha, inherited from the reforms after the Soviet bloc implosion, which face the same

[13] Integration into the EU
[14] Integration into the EU

weed management issue as the large farms of third countries (Csaki and Lerman, 1997; Eurostat. European Commission, 2010; Pouliquen, 2001).
Coexistence between GM and non-GM soy would probably not be an issue as soy is mostly autogamous as soon as the European seeds' threshold for non-GM seeds is defined (Roebroeck, 2002). However, case specific monitoring will be an important and costly workload for herbicide tolerant soybean cropping in order to avoid the issues of herbicide resistant weeds observed in the USA and eradication programs paid by companies (Brasher, 2010; Owen et al., 2010; Owen, 2009; Roberson, 2010).
As observed by most of the scholars and policy makers, European farming is highly dependent on CAP (Carlier et al., 2010; Cavaillès, 2009). Accordingly, European soybean cropping is currently only driven by global market prices. As finally observed by a recent EC sponsored study, a soybean shortage, such as the last US one in 2007, and thus an increase into soybean prices might induce reallocation of European arable surfaces toward soybean cultivation and probably allow several EU-MS to become self-sufficient. In this way, soybean might be considered as an opportunistic crop by European farmers driven by global soy prices, particularly for its non-GM counterpart. However, the European farmers have to "internalize" soybean cultivation into their agricultural practices and productions particularly in the EU-MS of Western Europe where soybean is not a familiar crop. For instance, farmers of Alsace region in France recently introduced a maize / soybean rotation as a tool to fighting Western corn rootworm. This "internalization" of rotation with soybean into maize culture may be rapid as the French government recently issued a decree making rotation mandatory.
Several recent changes in the CAP have to be kept in mind. In particular due to substantial reductions of aid in several agricultural sectors, European farmers have a closer look on the impact of the global commodities market on their sales prices with an increased trend toward crops' futures markets. In the meantime, farmers need to have new considerations toward the environment; the reduction into available chemicals for pest fighting, the carbon footprint, the multi-functionality of agriculture, etc. (Commission, 2006; European Commission, 2009d; Kaditi and Swinnen, 2006).
It is thus predictable that soybean cropping will differ from East to West among the EU-27, with probably GM crops in the eastern part, which is comparable to the gradient of sensitivity to GMO issues as observed for consumers (Consumerchoice Consortium, 2008).

5.4 Alternatives to imported soybean

Two considerations structure the soy importation issue: the first considering feedstuff production with GM soy and the second considering the use of non-GM soybean. Indeed alternative protein-rich crops did not succeed in the previous national or European "protein plans" and are less palatable to poultry and young pigs.

5.4.1 The issue of asynchronous approvals

As previously said, EU generally takes more time for approving GMO than several third countries, in part due to incomplete dossiers but probably also because of the EU-MS unclear economic interest of GM crops (European Commission, 2011).
The import of GM soybean is thus affected by this approval status as reported by numerous reports (ADAS ltd (for DEFRA feed import project), 2008; Aramyan et al., 2009a; Aramyan et al., 2009b; DG AGRI European Commission, 2007; Nowicki et al., 2010; Stein and Rodriguez-

Cerezo, 2009; Stein and Rodriguez-Cerezo, 2010b; Stein and Rodríguez-Cerezo, 2009; Tallage, 2010). All these reports concluded there is no alternatives to imported soybeans and meals, and thus recommended establishing a specific Low Level Presence threshold for EU unapproved GMOs to avoid any shortage in soybean which could hamper European livestock competitiveness, a recommendation recently officially taken in consideration by the EC by establishing a LLP threshold for EU unapproved GMOs dedicated to feed.

However, how reliable are these converging predictions? It might be helpful to have some insights on some recent reports on such issues of feed shortage in the EU-27. Are there some biases in those studies which are almost all based on modeling of feed use?

Models are clearly needed for simplifying complex situation and decision taking. In that way, it is thus understandable that models are used for forecasting international trade and soybean use in feedstuff production. However the choice of model or postulates, such as linear regression and "general equilibrium" instead of alternative is not neutral. Besides this essential questioning, common to all modeling issues, we will just examine some contextual questions.

- The first observation we can make is that those studies were carried out with limited funding in short time; thus impeding long studies and collective, contradictory expertise. Another problem is the use of very recently developed models, used by the EC without having been clearly in depth validated, or used out their scope, e.g. to foresee future trends while developed to analyze the past (Harrell, 2001). Some of these models were "validated" by discarding some crops in some parts of the evaluation but taking them into account into other evaluation parts. As these crops are used as adjustment variables in substitution strategies for low-price compound feedstuffs production, the validity of such models is highly questionable. The same issue of validating models applies to models attempting to merge ecological and CAP issues. It is finally rather surprising to read in a report about the development of a model, that one of the main goals was to simplify the yearly feeding of the model due to some lack of personal in the corresponding European Commission service. Avoiding complexity and simplifying the life of European personal does not help make sound policy.
- The large use of modeling is the expression of a general disinterest of economists for empirical studies and a preference toward modeling. This disinterest of economist scholars or consultants is due to (i) the difficulties to retrieve accurate information from companies and interviews, (ii) the duration necessary for establishing their own databases, together with (iii) a higher ranking in peer-reviewed journals for models, rather than for empirical data. There is thus a fundamental lack of sound, scholarly-established empirical data, i.e. not provided only by the companies in charge of feed production, before founding policy on models.
- A "business-as-usual" trend, i.e. a relative poorness in investigated scenarios and generally speaking in perspectives and alternatives. All considered scenarios take as read the requests of feed producers, i.e. the need for soybean and more generally proteins imports; just as previously cereals and then meat-and-bone were supposed to supply all needed proteins. This "business-as-usual" trend may in great part be explained by the power of lobbies, some kind of blindness, i.e. lack of prospective. However, it is generally recognized that a mass market is always turning into a market of niches and that EU is among the largest provider of market niches (Anderson, 2006). Such models are thus *inter-alia* not referring to market differentiation, European landscape use, consumers welfare, region competitiveness and ecological issues as

requested by the second pillar of the new CAP (Hermans et al., 2010; Kissinger and Rees, 2010; Konefal and Busch, 2010; van Delden et al., 2010). As outlined by Konefal and Bush, maize and soybean market standardization also introduced a multiplicity of segmented markets which were not taken into these models.

- Short and over-simplified studies, as generally the policy makers need rapid results and the academic community is rather slow to mobilize for participating in such applied work. Thus calls for this type of studies are generally awarded by consultants' cabinets or by the few scientists having already worked on that issue and thus able to "reinvest" their initial work. Work is thus mostly desk-search with several biases, such as a more difficult access to the scientific literature, a large use of "Google" which highlights URL according to a Google ranking algorithm mostly based on the number of external links or sponsorship, thus a way of working which favors industry and lobbies reports and sites.
- Group "consanguinity": such a strong relationship between sponsors, for instance a technical officer in charge of supervising an European study, originating from a European institute, whose recent reports all biased in the same way the effect of EU unapproved GMOs on feed availability. This first type of consanguinity is then reinforced by the tenders who have been chosen after a call for tenders. In most of the instances, the scientists have published reports in the same way, e.g. the dramatic effect of EU unapproved GMOs on feed availability. Reinvesting initial work is clearly not the best way for sound prospective in comparison of a collective and contradictory expertise.
- Influence of working environment. Indeed several studies were carried out in an EU country highly depending on feedstuffs ingredients importations, and further re-exportations, which may hamper the independence of viewpoint of the scientists and criteria retained for the scenarios. This may also be linked to previous studies funded by the feed industry which may influence the viewpoints and future results. Again, it would be necessary to amplify the panel of viewpoints, e.g. with scientists from countries with different production schemes. By not taking into consideration the socio-economic context and history of some scientists, the EU is decreasing its chances to find a systemic and long-term solution.

Taken altogether, these considerations of the studies and modeling context show the limits of the available data and predictions. This militates for more scholar-driven, long-term, multidisciplinary and with people sharing different viewpoints about futures of agriculture into collective expertise using also different modeling and postulates bases. Rapid, biased studies for a very complex matter highly influenced by both uncontrolled events, (such as seasonal incidents, or policy controlled issues, such as the disappearance of fallows in the new CAP, further cultivable areas reallocations and integration of new players), are not the best conditions for forecasting the future of European agriculture needs.

The users, i.e. policy makers, should be aware of the strengths and weaknesses of models used by the technical officers, what is generally not the case in the reports provided to the policy makers or media (van Tongeren et al., 2001). To conclude, the over-simplified models, developed moreover under detrimental contextual influences, have drastic limitations for forecasting trade and supply chains trends but are routinely used and dramatically impact the European policy without sound "scientific" ground-bases. These observations together with other not reported in this paper show that the EC was in fact over the last years attempting to "scientifically" legitimate previous political decisions for "smoothing" global trade issues.

5.4.2 Perspectives

In spite of a careful survey of European scientific and grey literature on alternatives to GM soy, almost no one Western EU stakeholder involved into meat production is currently considering soybean cultivation in the EU as a solution. Beside some recommendations to come back to pasturage for cattle, the general trend in compound feedstuff production is a larger incorporation of European non-GM rapeseed meal as it can be seen in the statistics of Fediol as a by-product of European bio-fuel production. Substitution is thus generally retained in national schemes for non-GM soy use. However new Eastern EU-MS have a long tradition of soy cultivation with some very large farms which might, in the 'business-as-usual" trend benefit from GM, or non-GM, soy cultivation.

5.4.2.1 Domestic substitution to imported soybean

Two ways of substitution of imported soybean have to be considered: firstly, the European cultivation of soy, as this protein-rich feed is difficult to substitute in feedstuff of several young animals and, secondly, the replacement of soybean by some other protein-rich crop for adults or some young animals.

As observed in several studies the trend over the last decade to use low-price soybean and soy meal induced a clear disaffection of plant breeding companies for leguminous fodder crops (alfalfa, clover, etc.) and several protein-rich crops (field pea, sweet lupine, etc.) due to their small volumes and a constant decrease of cultivation over the last decade (European Parliament. Directorate-General for Internal Policies, 2010; LMC International, 2009a, b; Häusling, 2011). Moreover, public research programs on such European substitutes to soybean declined over the 3 last decades. As a example of such general decline, domestic leguminous crops to be incorporated into feed dropped in France from 11% in 1991 toward 2.5% in 2006, despite CAP subsidies of field pea, field bean and lupine (European Commission, 2009c). If some studies are currently ongoing, for instance on the use of lupine and pea for poultry, there is a lack of European research on substituting soybean by domestic protein-rich crops which however present the interest of currently being non-GM (Laudadio et al., 2009; Häusling, 2011).

Several changes in the CAP such as the "20-20-20 in 2020" objectives (reduction of 20% of emissions from 1990, 20% share of energy consumption from renewable sources and 20% improvement in energy efficiency by 2020) conducted to an increase of rapeseed oil production for bio-fuels and thus of alternative meal, at least for some livestock.

Under the Blair House agreement, oilseeds plantings were limited to an adjusted Maximum Guaranteed Area for producers benefiting from crop specific oilseeds payments. This limited the EU oilseeds production area and penalized overproduction till the 2003 renegotiation of Blair House agreement. Finally, with the *Agenda 2000*, the CAP relies on 2 pillars: the market and income policy (first pillar) and the sustainable development of rural areas (second pillar). Since 2010, the producers are free of the hectare limits set out by Blair House agreement. Additionally, the disappearance of European mandatory fallows is freeing new arable surfaces for, current or new, long-term or opportunistic cultivations.

As the 2003 CAP reform (linked to renegotiation of Blair House agreement) brings greater consideration to environmental integration we may expect several changes, in particular about soybean whose production in Brazil is criticized due to deforestation, an important use of chemicals and social impacts (Carlier et al., 2010). The recent reduced European interest for biofuels production in the EU due to a contrasted carbon footprint, as well as a possible effect on food prices could also free agricultural surfaces for soy opportunistic cultivation. However, studies on soybean use in cow feeding show that soy might have less environment impacting than rapeseed (Lehuger et al., 2009). The new CAP which embraces more environmental considerations may thus face new issues in the balance of environmental footprint and might let European prefer importing soy.

After a ban of about a decade, several lobbyists are pushing the reintroduction of meat-and-bone meal, probably the richest protein source. For instance in 2002, 220,000T were estimated equivalent to 503,000 T of soy meal. The European dependence onto imported soy could thus be dramatically decreased if meat-and-bone is safely re-incorporated into feedstuffs. However, the acceptance of European consumers of such a reintroduction is far from being obvious following the 1990s' mad-cow disease scandal.

CAP Health check in 2008 reduced again aids to cereals opening the opportunity to grow more oilseeds including soybean despite a previous EU support to protein-rich crops such as field pea, field bean and lupine.

Surprisingly, in all alternatives to soybean imports even though by NGO or organic farmers, no one proposed cultivation of soybean as protein sources alternatives, at least for conventional livestock (Billon et al., 2009; Confédération paysanne, 2002). However, production of European soybean showed in 2009 a 12.4% increase which demonstrates the opportunism of European farmers in front of high trade prices of the 2008/2009 food crisis (Krautgartner et al., 2010a). Such an alternative of soy cropping instead of imports should be more effectively considered in EU-MS, even though it looks difficult to dedicate ca. 20 Millions of European hectares to soy, the equivalent in surfaces of currently imported soy (see above).

The recent entrance of several new EU-MS, with a past of soybean crops and some very large corporate farms, could also accompany this trend of growing more soybean in the EU. Interviewed Spanish representatives agreed that Spain could grow soybean, be these GM or not, as soon as the prices would be of interest. Additionally, the fight against the Western corn rootworm *(Diabrotica virgifera virgifera)* in French maize monocultures was successful when introducing soybean into a newly implemented rotation, a choice of crop made in function of cropping practices and apparatus compatibility. Together with the environmental and economic interest of introducing a leguminous plant into rotations and the more general European request of reducing chemicals in cultivation, the interest of maize monoculture is questionable when we consider that the infected area[15] covers most of the Central and Eastern European Countries. Finally, the current trend of increase of petrol and thus of nitrogen fertilizer prices also favors reintroduction of leguminous crops into rotation.

In addition to an increasing part of local, pasture based and on-farm production for both "conventional" and under signs-of-quality meat production, the European soybean dependence might thus drastically decrease at least for bovine animals. Soybean imports would then mostly depend on intensive livestock production such as poultry and probably pig.

Altogether, these several observations show a balanced approach of European farmers toward the global market and an important dependence of EU farmers to CAP. As long as soybean and soy meal prices are low, there is no interest for European farmers to enter the

[15] http://extension.entm.purdue.edu/wcr/images/pdf/2010/EUROPEMap2010.pdf

very competitive commodities market. But all occasions can be taken to improve their niche markets of soy be this GM or non-GM.

5.4.2.2 Alternative sources of non-GM soy

With the growing trend to label animals reared on non-GM feed, availability of non-GM soy is of a growing interest for livestock producers. With the increased cropping of GM soy in exporting countries, alternative sources of non-GM soy are thus actively looked for by European importers.

Since the 2008 issue of Chinese organic soy meal spiked with melamine, The Peoples' Republic of China is no longer considered as a reliable source of non-GM soy despite recent claims of its interest for this country (Anonymous, 2010; Hansen et al., 2007; Takada, 2010). China is the most important importer of soybean and this expected to continue (Taylor and Koo, 2010). As numerous GM crops are under development in China together with a growing request of soy for livestock production, we may expect this country may rapidly cease to be a putative exporter of non-GM soybean. Indeed, India is currently a new source of non-GM soy for certain European traders and has been identified as such by US surveys (Ash, 2011).

This current relative shortage of non-GM (<0.9%) or GMO-free (<0.1%) soybean could be an opportunity for European soy producers, provided they find more incentives to grow soy. The new CAP trend considering more environmental issues might favor such changes into the European farmers practices. Integrating crops rotations with leguminous crops, for decreasing the use of costly fertilizers and fighting some pests such as the expending Western corn root worm, would be additional causes of such practices' changes with premiums and long-term contracts for non-GM productions,.

Beside a new consideration of soybean into crop rotation in Western EU or an increase of soy surfaces in EU-MS cultivating soy for market niches, the European soy status may also change by the integration in 2007 of Central and Eastern European countries such as Romania with a past and a future wish of soybean cultivation. The move of these countries, some cases having very large corporate farms, toward GM or non-GM soy cultivation will greatly depend on non-GM demand, premiums and long-term contracts as well as the volatility of GM soy commodities' prices.

6. Conclusion

Europe is so far highly dependent on protein importations for compound feedstuffs production particularly of soy for young animal production, poultry and pigs. However, several factors may lead to an increased soybean production in the EU over the next decade. Among the several reasons for such an increase are societal considerations such as carbon footprint of imported soybean, development of market niches - be these or not GMO-free due to animal labeling - entrance of new EU-MS with a past of soybean cropping as well as a general increase of American exports toward China inducing tensions on prices particularly for poultry and pig feeding.

Soybean cropping would however probably be considered as an opportunistic European crop due to e.g. rotation for fighting corn rootworm which is prevalent in Central and Eastern EU and extending into the Western area, continuous rises in nitrogen fertilizer prices, a persistent ban of meat-and-bones meal as well as an absence of alternative European protein-rich crops.

As the cultivations are rapidly adjusted to the market requests, there is thus not an issue of soybean supply *per se* but an issue of importation of soybean at the lowest prices for intensive livestock production. Such issue of competitiveness of European livestock drove the EC to introduce a "technical solution" to EU unapproved GMOs, i.e. a LLP threshold, for feed, which might impact food supply chains since segregating of food and feed is difficult.

At the same time, the need for a larger use of bio-fuels increased the production of oilseed rape in Europe which in counterpart decreased the imports of soybeans. It is currently difficult to determine what would be the future of such by-products of bio-fuels as the European policy bio-fuels appears to be changing due to new calculations of their carbon footprint and the need for "feeding the world". Such trend to develop domestic bio-fuels will probably impact European soy imports and cultivations.

After several shortages in the 1970s and 2000s, the current increases in feed and food commodities' prices after the 2007-2008 food crisis militates for an European alimentary sovereignty due in particular to the impact of the increasing living standards of emerging countries and thus of protein-rich feedstuffs.

Environmental, sustainability and social criteria newly incorporated into the European agricultural aids frame will probably push domestic oilseed production, including soybean and jeopardize oilseeds imports. The main driver of European livestock production and soy imports will also depend on the possible extension among EU-MS of the labeling of animals reared with GMO-free feed.

By the different past histories of Western and Eastern parts of the EU-27, it is also to be expected that the soy cropping strategies, i.e. the choice between GM and non-GM soy cultivation, will differ between the two. The lowest sensitivity of Eastern consumers to GM food and cultivation could facilitate the implementation of GM soy in the Eastern part of the EU, while non-GM soy might develop in the Western part of the EU. Such search for a European "sovereignty" is in line with the development of numerous markets niches, a usual counterpart to a more and more standardized and global trade.

Generally speaking, the recent initiative of the EC for establishing a LLP threshold for feed did not take into consideration the change of paradigm i.e. the European move from an "economy of offer" toward an "economy of demand" nor the difficulties to segregate food and feed commodities.

7. Note added in proof

As this chapter was in proof reading, EFSA published its final version of PMEM (EFSA GMO Panel, 2011) and the EC published the LLP related regulation (European Commission, 2011).

8. References

ACRE (2004). ACRE Guidance Note 16. Guidance on best practice in the design of post-market monitoring plans in submissions to the Advisory Committee on Releases to the environment. Guidance for applicants seeking permission to release genetically modified crops into the environment (under Directive 2001/18/EC), 21 p.

Adams, S. (2011). Dow AgriSciences, MU researcher develop a way to control "superweed". Available at: http://munews.missouri.edu/news-releases/2011/0121-dow-agrisciences-mu-researcher-develop-a-way-to-control-%E2%80%9Csuperweed%E2%80%9D/

ADAS Ltd. (for DEFRA feed import project) (2008). What is the potential to replace imported soya, miaze and maize by-products with other feeds in livestock diets?, ADAS. Ltd, ed. (DEFRA, UK), 30 p.

Alacalde, E., Amijee, F., Blache, G., Bremer, C., Fernandez, S., Garcia-Alonso, M., Holt, K., Legris, G., Novillo, C., Schlotter, P., *et al.* (2007). Insect resistance monitoring for Bt maize cultivation in the EU: Proposal from the industry IRM working group. *Journal Fur Verbraucherschutz Und Lebensmittelsicherheit-Journal of Consumer Protection and Food Safety, 2*, pp. 47-49.

Alderborn, A., Sundstrom, J., Soeria-Atmadja, D., Sandberg, M., Andersson, H.C., and Hammerling, U. (2010). Genetically modified plants for non-food or non-feed purposes: Straightforward screening for their appearance in food and feed. *Food and Chemical Toxicology 48*, pp. 453-464.

Anderson, C. (2006). The long tail: why the future of business is selling less of more (Random House / Hyperion).

Anonymous (2010). Soybeans to go non-GM. In *China Business News*. Available at: http://cnbusinessnews.com/soybeans-to-go-non-gm/

APHIS News release (2010). USDA announces availability of biotechnology quality management system audit tandard and evaluation of comments. Available at: http://www.aphis.usda.gov/biotechnology/news_bqms.shtml

Aramyan, L.H., van Wagenberg, C.P.A., and Backus, G.B.C. (2009a). Impact of unapproved genetically modified soybean on the EU feed industry (Wageningen University, NL), 14 p.

Aramyan, L.H., Wagenberg, C.P.A.v., and Backus, G.B.C. (2009b). EU policy on GM soy. Tolerance threshold and asynchronic approval. Report 2009-052, 46 p.

Ash, M. (2011). Oil Crops Outlook 2011. In A report from the Economic Research Service, U. ERS, ed., 16 p.

Badea, E.M., and Pamfil, D. (2009). The status of agricultural biotechnology and biosafety in Romania. *Bulletin of University of Agricultural Sciences and Veterinary Medicine Cluj-Napoca Animal Science and Biotechnologies*, 66, pp. 8-15.

Bakshi, A. (2003). Potential adverse health effects of genetically modified crops. Journal of Toxicology and Environmental Health-Part B-Critical Reviews *6*, pp. 211-225.

Bartsch, D., Bigler, F., Castanera, P., Gathmann, A., Gielkens, M., Hartley, S., Lheureux, K., Renckens, S., Schiemann, J., Sweet, J., *et al.* (2006a). Concepts for general surveillance of genetically modified (GM) plants: the EFSA position. Journal fur Verbraucherschutz und Lebensmittelsicherheit *1*, pp. 15-20.

Bartsch, D., Bigler, F., Castanera, P., Gathmann, A., Gielkens, M., Hartley, S., Lheureux, K., Renckens, S., Schiemann, J., Sweet, J., *et al.* (2006b). The EFSA opinion on post-market environmental monitoring of GM plants. The 9th International Symposium on the Biosafety of Genetically Modified Organisms, Jeju Island, Korea, 24-29 September, 2006: biosafety research and environmental risk assessment, pp. 159-167.

Bartsch, D., Buhk, H.-J., Engel, K.-H., Ewen, C., Flachowsky, G., Gathmann, A., Heinze, P., Koziolek, C., Leggewie, G., Meisner, A., *et al.* (2009). BEETLE. Long-term effects of genetically modified crops on health and the environment (including biodiversity): Prioritisation of potential risks and delimitation of uncertainties. Executive summary and main report (Berlin, DE), 133 p.

Bartsch, D., Gathmann, A., Hartley, S., Hendriksen, N.B., Hails, R., Lheureux, K., Kiss, J., Mesdagh, S., Neemann, G., Perry, J., *et al.* (2007). First EFSA experiences with monitoring plans. Journal Fur Verbraucherschutz Und Lebensmittelsicherheit-Journal of Consumer Protection and Food Safety 2, pp. 33-36.

Barzman, M., Caron, P., Passouant, M., and Tonneau, J.P. (2005). Observatoire Agriculture et Territoires. Etude pour la définition d'une méthode de mise en place d'observatoires, 64 p. Available at : http://agriculture.gouv.fr/IMG/pdf/observatoire_rapport_final.pdf

Beckie, H.J., Hall, L.M., Simard, M.J., Leeson, J.Y., and Willenborg, C.J. (2010). A framework for postrelease environmental monitoring of second-generation crops with novel traits. *Crop Science*, 50, pp. 1587-1604.

Beismann, H., Finck, M., and Seitz, H. (2007). Standardisation of methods for GMO Monitoring on a European level. *Journal Fur Verbraucherschutz Und Lebensmittelsicherheit-Journal of Consumer Protection and Food Safety*, *2*, pp. 76-78.

Beissner, L., Wilhelm, R., and Schiemann, J. (2006). Current research activities to develop and test questionnaires as a tool for the General Surveillance of important crop plants. *Journal fur Verbraucherschutz und Lebensmittelsicherheit-Journal of Consumer Protection and Food Safety*, 1, pp. 95-97.

Bellon, M.R., and Morris, M.L. (2002). Linking global and local approaches to agricultural technology development: the role of participatory plant breeding research in the CGIAR. In *Economics working paper* 02- 03, 30 p.

Berlan, J.-P., Bertrand, J.-P., and Lebas, L. (1977). The growth of the American 'soybean complex'. *European Review of Agricultural Economics*, 4, pp. 395-416.

Billon, A., Neyroumande, E., and Deshayes, C. (2009). Vers plus d'indépendance en soja d'importation pour l'alimentation animale en Europe - cas de la France, W.W.F. ENESAD, ed. (Dijon, France, ENESAD, WWF France), 49 p. Available at: http://www.wwf.fr/content/download/2687/20819/version/1/file/rapport+substitution+au+soja+complet+2.pdf

Birthal, P.S., Rao, P.P., Nigam, S.N., Bantilan, M.C.S., and Bhagavatula, S. (2010). Groundnut and soybean economies of Asia. Facts, trends and outlook, I.I.C.R.I.f.t.S.-A. Tropics), ed. (Patancheru 502 324, Andra Pradesh, India, ICRISAT), 92 p. Available at: http://dspace.icrisat.ac.in/bitstream/10731/3648/1/rso-gn-sbean-asia.pdf

Bock, A.K., Lheureux, K., Libeau-Dulos, M., Nilsagard, H., and Rodriguez-Cerezo, E. (2002). Scenarios for co-existence of genetically modified, conventional and organic crops in European agriculture. EUR 20394EN, xi + 133 p. Available at: http://ftp.jrc.es/EURdoc/eur20394en.pdf

Bodiguel, L. (2003). A way to establish the multifunctional agriculture: the notion of territory. *Economie Rurale*, pp. 61-75.

Bonny, S. (2008). How have opinions about GMOs changed over time? The situation in the European Union and the USA. *CAB Reviews: Perspectives in Agriculture, Veterinary Science, Nutrition and Natural Resources*, 3, pp. 1-17.

Bonny, S. (2009). Taking stock of the world market in transgenic seeds. In 13th ICABR conference *The emerging bio-economy* (Ravello, Italy).

Bontemps, A., Bourguet, D., Pelozuelo, L., Bethenod, M.T., and Ponsard, S. (2004). Managing the evolution of Bacillus thuringiensis resistance in natural populations of the European corn borer, Ostrinia nubilalis: host plant, host race and pherotype of adult males at aggregation sites. *Proceedings of the Royal Society of London Series B-Biological Sciences*, 271, pp. 2179-2185.

Boshnakova, M., Dobrescu, M., Flach, B., Henard, M.-C., Krautgartner, R., Lieberz, S., and the group of FAS oilseeds specialists in the EU. (2009). EU-27. Oilseeds crop update - U.S. soybean exports to EU threatened. In EU-27, U.F.A. Service, ed. (Vienna, Austria, USDA GAIN), 11 p. Available at: http://gain.fas.usda.gov/Recent%20GAIN%20Publications/Oilseeds%20Crop%20Update%20%20U.S.%20Soybean%20Exports%20to%20EU%20Threatened_Vienna_EU-27_8-12-2009.pdf

Bourguet, D. (2004). Resistance to Bacillus thuringiensis toxins in the European corn borer: what chance for Bt maize? *Physiological Entomology*, 29, pp. 251-256.

Brasher, P. (2010). New weed strategies needed, scientists say. *CheckBiotech*. Available at: http://checkbiotech.org/node/29440/edit

Breckling, B., and Reuter, H. (2006). General surveillance of genetically modified organisms - the importance of expected and unexpected environmental effects. *Journal für Verbraucherschutz und Lebensmittelsicherheit*, 1, pp. 72-74.

Brookes G. (2005). The farm-level impact of herbicide-tolerant soybeans in Romania. *AgBioForum*, 8, 5, 235-241.

Bruening, G., and Lyons, J.M. (2000). The case of the FLAVR SAVR® tomato. *California Agriculture*, 54, pp. 6-7.

Brunet, Y., Dupont, S., Delage, S., Garrigou, D., Guyon, D., Dayau, S., Tulet, P., Pinty, J.-P., Lac, C., Escobar, J., *et al.* (2011). Long-distance pollen flow in large fragmented landscapes. In GM and non-GM supply chains coexistence and traceability, Y. Bertheau, ed. (Wiley Publishing), *In Press.*

Cardwell, M. (2003). Multifunctionality of agriculture: a European community perspective. In Agriculture and international trade: law, policy and the WTO, Cardwell, M. N. Grossman and M. R., Rodgers, C. P. (eds), CPL Press. pp. 131-164.

Cardwell, M. (2010). Public participation in the regulation of genetically modified organisms: a matter of substance or form? *Environmental Law Review*, 12, pp. 12-25.

Cardwell, M., and Bodiguel, L. (2005). Agriculture definition trend for advanced agriculture - comparative European Union/Great Britain/France approach. *Revue du Marche Commun et de l'Union Européenne*, 490, pp. 456-466.

Carlier, L., Rotar, I., and Vidican, R. (2010). The EU policy regulates and controls the farming practices. *Bulletin UASVM Agriculture*, 67, 1-11.

Cavaillès, E. (2009). La relance des légumineuses dans le cadre d'un plan protéine : quels bénéfices environnementaux ? In Etudes et documents, Commissariat général au développement durable - Service de l'économie, de l'évaluation et de l'intégration du développement durable, ed. (Paris, France, Commissariat général au développement durable - Service de l'économie, de l'évaluation et de l'intégration du développement durable), 44 p. Available at : http://www.developpement-durable.gouv.fr/IMG/pdf/E_D15.pdf

Cellini, F., Chesson, A., Colquhoun, I., Constable, A., Davies, H.V., Engel, K.H., Gatehouse, A.M.R., Karenlampi, S., Kok, E.J., Leguay, J.J., *et al.* (2004). Unintended effects and their detection in genetically modified crops. *Food and Chemical Toxicology*, 42, 1089-1125.

Cerdeira, A.L., and Duke, S.O. (2006). The current status and environmental impacts of glyphosate-resistant crops: A review. *Journal of Environmental Quality*, 35, pp. 1633-1658.

Cerdeira, A.L., and Duke, S.O. (2007). Environmental impacts of transgenic herbicide-resistant crops. *CAB Reviews: Perspectives in Agriculture, Veterinary Science, Nutrition and Natural Resources*, 2, 033, 14 p.

Chaufaux, J., Micoud, A., Delos, M., Naibo, B., Bombarde, F., Eychennes, N., Pagliari, C., Marque, G., and Bourguet, D. (2002). Transgenic maize and non target insects: what effects? *Phytoma*, pp. 13-16.

Codex Alimentarius (2003). Guideline for the conduct of food safety assessment of foods derived from recombinant-DNA plants. CAC/GL 45-2003. Annex 3: food safety assessment in situations of low-level presence of recombinant-DNA plant material in food. Available at: http://www.codexalimentarius.net/download/standards/10021/CXG_045e.pdf

Colbach, N. (2009). How to model and simulate the effects of cropping systems on population dynamics and gene flow at the landscape level: example of oilseed rape volunteers and their role for co-existence of GM and non-GM crops. *Environmental Science and Pollution Research*, 16, pp. 348-360.

Colbach, N., Monod, H., and Lavigne, C. (2009). A simulation study of the medium-term effects of field patterns on cross-pollination rates in oilseed rape (Brassica napus L.). *Ecological Modelling*, 220, pp. 662-672.

Coléno, F.C. (2008). Simulation and evaluation of GM and non-GM segregation management strategies among European grain merchants. Journal of Food Engineering *88*, 306-314.

Commission, E. (2006). Regulation (EC) No 1907/2006 of the European parliament and of the Council of 18 December 2006 concerning the Registration, Evaluation, Authorisation and Restriction of Chemicals (REACH), establishing a European Chemicals Agency, amending Directive 1999/45/EC and repealing Council Regulation (EEC) No 793/93 and Commission Regulation (EC) No 1488/94 as well as Council Directive 76/769/EEC and Commission Directives 91/155/EEC, 93/67/EEC, 93/105/EC and 2000/21/EC. *Official Jounal of the European Union*, L 396, pp. 1-849.

Confédération paysanne (2002). Un plan protéine pour l'Europe (Bagnolet, France), 26 p. Available at : http://www.confederationpaysanne.fr/images/imagesFCK/file/02/plan_proteines.pdf

Consumerchoice Consortium (2008). Do European consumers buy GM foods? European Commission: Framework 6, Project no. 518435. "Consumerchoice". Final report., 346 p. Available at: http://www.kcl.ac.uk/schools/biohealth/research/nutritional/consumerchoice/downloads.html

Covelli, N., and Hohots, V. (2003). The health regulation of biotech foods under the WTO agreements. *Journal of International Economic Law*, 6, pp. 773-795.

Csaki, C., and Lerman, Z. (1997). Land reform and farm restructuring in East Central Europe and CIS in the 1990s: Expectations and achievements after the first five years. *European Review of Agricultural Economics*, 24, pp. 428-452.

Czarnak-Kłos, M., and Rodríguez-Cerezo, E. (2010). European Coexistence Bureau (ECoB). Best Practice Documents for coexistence of genetically modified crops with conventional and organic farming. 1. Maize crop production. EUR 24509 EN. In JRC Scientific and Technical Reports, DG-JRC-IPTS, ed. (Seville, SP, DG-JRC-IPTS), 72 p. Available at: http://ecob.jrc.ec.europa.eu/documents/Maize.pdf

D'Agnolo, G. (2005). GMO: Human health risk assessment. *Veterinary Research Communications*, 29, pp. 7-11.

Dahl, E., and House, L. (2008). United States - European Union agricultural trade flows. In 110th EAAE Seminar 'System Dynamics and Innovation in Food Networks' (Innsbruck-Igls, Austria), 9 p. Available at: http://ageconsearch.umn.edu/bitstream/49841/2/Dahl-House.pdf

Darmency, H., Klein, E.K., De Garanbe, T.G., Gouyon, P.H., Richard-Molard, M., and Muchembled, C. (2009). Pollen dispersal in sugar beet production fields. *Theoretical and Applied Genetics*, 118, pp. 1083-1092.

Darmency, H., Vigouroux, Y., De Garambe, T.G., Richard-Molard, M., and Muchembled, C. (2007). Transgene escape in sugar beet production fields: data from six years farm scale monitoring. *Environmental Biosafety Research*, 6, pp. 197-206.

Davison, J. (2010). GM plants: science, politics and EC regulations. *Plant Science*, 178, pp. 94-98.

Davison, J., and Bertheau, Y. (2007). EU regulations on the traceability and detection of GMOs: difficulties in interpretation, implementation and compliance. *CAB Reviews: Perspectives in Agriculture, Veterinary Science, Nutrition and Natural Resources*, 2, 14 p.

Davison, J., and Bertheau, Y. (2008). The theory and practice of European traceability regulations for GM food and feed. *Cereal Foods World*, *53*, pp. 186-196.

de Cheveigné, S. (2004). Quand l'Europe mesure les représentations de la science : une analyse critique des Eurobaromètres. In *Sciences, Médias et Société*, J.L. Marec, and I. Babou, eds. (Lyon, F, Ecole normale supérieure. Lettres et Sciences Humaines. Laboratoire JE 2419. Communication, culture et société.), pp. 45-55.

de Jong, T.J. (2010). General surveillance of genetically modified plants in the EC and the need for controls. *Journal Fur Verbraucherschutz Und Lebensmittelsicherheit-Journal of Consumer Protection and Food Safety*, 5, pp. 181-183.

DEFRA (2006). Consultation on proposals for managing the coexistence of GM, conventional and organic crops (DEFRA, UK), 92 p. Available at: http://www.defra.gov.uk/environment/quality/gm/crops/documents/gmcoexist-condoc.pdf

Demont, M., Daems, W., Dillen, K., Mathijs, E., Sausse, C., and Tollens, E. (2008). Regulating coexistence in Europe: beware of the domino-effect! *Ecological Economics*, 64, pp. 683-689.

Demont, M., Devos, Y., and Sanvido, O. (2010). Towards flexible coexistence regulations for GM crops in the EU. *Eurochoices*, 9, pp. 18-24.

Demont, M., Dillen, K., Daems, W., Sausse, C., Tollens, E., and Mathijs, E. (2009). On the proportionality of EU spatial ex ante coexistence regulations. *Food Policy*, 34, pp. 508-518.

Desclaux, D., Nolot, J., Chiffoleau, Y., Gozé, E., and Leclerc, C. (2008). Changes in the concept of genotype × environment interactions to fit agriculture diversification and decentralized participatory plant breeding: pluridisciplinary point of view. *Euphytica*, 163, pp. 533-546.

Desquilbet, M., and Bullock, D.S. (2010). On the proportionality of EU spatial ex ante coexistence regulations: a comment. *Food Policy*, 35, pp. 87-90.

Devos, Y., Demont, M., Dillen, K., Reheul, D., Kaiser, M., and Sanvido, O. (2009). Coexistence of genetically modified (GM) and non-GM crops in the European Union. A review. *Agronomy for Sustainable Development*, 29, pp. 11-30.

Devos, Y.A. (2008). Transgenic crops: a kaleidoscope impact analysis. PhD Thesis. Department of plant production, faculty of bioscience engineering (Ghent, BE, Ghent University), 342 p. Available at: http://www.criticalphilosophy.ugent.be/index.php?id=2&type=file

DG AGRI European Commission (2007). Economic impact of unapproved GMOs on EU feed imports and livestock production (Brussels, B, European Commission). 11 p. Available at: http://ec.europa.eu/agriculture/envir/gmo/economic_impactGMOs_en.pdf

Dinu, T., Alecu, I.N., and Stoian, E. (2010). Assessing the economic impact and the traceability costs in the case of banning the cultivation of GM soybean in Romania. In *10th International Symposium* " Prospects of Agriculture and Rural Areas Development in the context of Global Climate Change ", Session "Management, Marketing, Accounting, Financial Analysis, Finance", M. Draghici, and M. Berca, eds. (20-21 May 2010 , Bucharest, Romania, RAWEX COMS Publishing House in co-editing with DO-MINOR Publishing House), pp. 62-67.

Dobbs, M. (2011). Legalising general prohibitions on cultivation of genetically modified organisms. *German Law Journal*, 11, pp. 1347-1372.

Dobrescu, M., Henard, M.-C., Krautgartner, R., Lieberz, S., and the group of FAS oilseeds specialists in the EU (2009). EU-27. Soybean imports from the United States still impeded. In EU-27, USDA GAIN, ed. (Vienna, Austria), 12 p. Available at: http://gain.fas.usda.gov/Recent%20GAIN%20Publications/EU-27%20Soybean%20Imports%20from%20the%20United%20States%20Still%20Impeded_Vienna_EU-27_11-3-2009.pdf

Ebert, I., and Lahnstein, C. (2008). GMO liability: options for insurers. In *Economic Loss Caused by Genetically Modified Organisms*, B.A. Koch, ed. (Springer Vienna), pp. 577-581.

EFSA (2008). Environmental risk assessment of genetically modified plants - Challenges and approaches. Paper presented at: EFSA Scientific Colloquium Summary Report (Parma, Italy, EFSA). Available at: http://www.efsa.europa.eu/fr/colloquiagmoera/publication/colloquiagmoera.pdf

EFSA GMO Panel (2004). Guidance document of the scientific panel on genetically modified organisms for the risk assessment of genetically modified plants and derived food and feed. *The EFSA Journal*, 99, 1-94.

EFSA GMO Panel (2006a). Guidance document of the scientific panel on genetically modified organisms for the risk assessment of genetically modified microorganisms and their derived products intended for food and feed use. *The EFSA Journal*, 374, pp. 1-115. Available at: http://www.efsa.europa.eu/fr/scdocs/scdoc/374.htm

EFSA GMO panel (2006b). Opinion of the scientific panel on genetically modified organisms on the post market environmental monitoring (PMEM) of genetically modified plants (Question No EFSA-Q-2004-061). *The EFSA Journal*, 319, 1-27.

EFSA GMO Panel (2010). Scientific opinion. Guidance on the environmental risk assessment of genetically modified plants. *The EFSA Journal*, 8, 1-111.

EFSA GMO Panel. (2011) Guidance on the Post-Market Environmental Monitoring (PMEM) of genetically modified plants. EFSA Journal 9:2316.

Elbehri, A. (2007). The changing face of the U.S. grain system differentiation and identity preservation trends. *Economic Research Report - Economic Research Service, USDA*, 39 p. Available at: http://www.ers.usda.gov/publications/err35/err35.pdf

EU working group (2003). Appendix 1: EU working group on insect resistance management: harmonised insect resistance management (IRM) plan for cultivation of Bt maize in the EU, 28 p. Available at: http://ec.europa.eu/food/food/biotechnology/docs/IRM_plan.pdf

European Commission (1985). Council Directive 85/374/EEC of 25 July 1985 on the approximation of the laws, regulations and administrative provisions of the Member States concerning liability for defective products *Official Jounal of the European Communities*, L 210, pp. 29-33.

European Commission. (2011) Commission Regulation (EU) No 619/2011 of 24 June 2011 laying down the methods of sampling and analysis for the official control of feed as regards presence of genetically modified material for which an authorisation procedure is pending or the authorisation of which has expired (Text with EEA relevance). Official Jounal of the European Union L 166:9-15.

European Commission (1997). Regulation (EC) No 258/97 of the European Parliament and of the Council of 27 January 1997 concerning novel foods and novel food ingredients. *Official Journal of the European Communities*, L 043, pp. 1-6.

European Commission (2000a). Commission regulation (EC) No 49/2000 of 10 January 2000 amending Council Regulation (EC) No 1139/98 concerning the compulsory indication on the labelling of certain foodstuffs produced from genetically modified organisms of particulars other than those provided for in Directive 79/112/EEC. *Official Journal of the European Communities*, L 6, pp. 13-14.

European Commission (2000b). Regulation (EC) No 50/2000 of 10 January 2000 on the labelling of foodstuffs and food ingredients containing additives and flavourings that have been genetically modified or have been produced from genetically modified organisms. *Official Journal of the European Communities*, L 6, 15-17.

European Commission (2001). Directive 2001/18/EC of the European Parliament and the Council of 12 March 2001 on the deliberate release into the environment of genetically modified organisms and repealing Council Directive 90/220/EEC. *Official Journal of the European Communities*, L 106, 1-38.

European Commission (2002a). Council decision 2002/811/EC of 3 October 2002 establishing guidance notes supplementing Annex VII to Directive 2001/18/EC of the European Parliament and of the Council on the deliberate release into the environment of genetically modified organisms and repealing Council Directive 90/220/EEC. *Official Journal of the European Communities*, L 280, pp. 27-36.

European Commission (2002b). Council decision 2002/812/EC of 3 October 2002 establishing pursuant to Directive 2001/18/EC of the European Parliament and of

the Council the summary information format relating to the placing on the market of genetically modified organisms as or in products. *Official Journal of the European Communities*, L 280, pp. 37-61.
European Commission (2002c). Council decision 2002/813/EC of 3 October 2002 establishing, pursuant to Directive 2001/18/EC of the European Parliament and of the Council, the summary notification information format for notifications concerning the deliberate release into the environment of genetically modified organisms for purposes other than for placing on the market. *Official Journal of the European Communities*, L 280, pp. 62-83.
European Commission (2002d). Regulation (EC) No 178/2002 of the European Parliament and of the Council of 28 January2002 laying down the general principles and requirements of food law, establishing the European Food Safety Authority and laying down procedures in matters of food safety. *Official Journal of the European Communities*, L 31, pp. 1-24.
European Commission (2003a). Commission Recommendation of 23 July 2003 on guidelines for the development of national strategies and best practices to ensure the coexistence of genetically modified crops with conventional and organic farming (notified under document number C(2003) 2624). (2003/556/CE). *Official Journal of the European Communities*, L 189, pp. 36-47.
European Commission (2003b). Directive 2003/4/EC of the European Parliament and of the Council of 28 January 2003 on public access to environmental information and repealing Council Directive 90/313/EEC. *Official Jounal of the European Union*, L 41, pp. 26-32.
European Commission (2003c). Regulation (EC) No 1829/2003 of the European Parliament and of the Council of 22 September 2003 on genetically modified food and feed. *Official Journal of the European Union*, L 268, pp. 1-23.
European Commission (2003d). Regulation (EC) No 1830/2003 of the European Parliament and of the Council of 22 September 2003 concerning the traceability and labelling of genetically modified organisms and the traceability of food and feed products produced from genetically modified organisms and amending Directive 2001/18/EC. *Official Journal of the European Union*, L 268, pp. 24-28.
European Commission (2004a). Commission decision No 2004/204/EC of 23 February 2004 laying down detailed arrangements for the operation of the registers for recording information on genetic modifications in GMOs, provided for in Directive 2001/18/EC of the European Parliament and of the Council. *Official Journal of the European Union*, L 065, pp. 20-22.
European Commission (2004b). Commission recommendation No 2004/787 of 4 october 2004 on technical guidance for sampling and detection of genetically modified organisms as or in products in the context of Regulation (EC) No 1830/2003. *Official Journal of the European Union*, L 348, pp. 18-26.
European Commission (2004c). Directive 2004/35/CE of the European Parliament and of the Council of 21 April 2004 on environmental liability with regard to the prevention and remedying of environmental damage. *Official Journal of the European Union*, L 143, pp. 56-75.
European Commission (2004d). Regulation (EC) No 882/2004 of the European Parliament and of the Council of 29 April 2004 on official controls performed to ensure the

verification of compliance with feed and food law, animal health and animal welfare rules. *Official Jounal of the European Union*, L 165, pp. 1-141.

European Commission (2006a). Commission Regulation (EC) No 1981/2006 of 22 December 2006 on detailed rules for the implementation of Article 32 of Regulation (EC) No 1829/2003 of the European Parliament and of the Council as regards the Community reference laboratory for genetically modified organisms (Text with EEA relevance), *Official Journal of the European Union*, L 368, pp. 99-103.

European Commission (2006b). Communication from the Commission to the Council and the European Parliament. Report on the implementation of national measures on the coexistence of genetically modified crops with conventional and organic farming. {SEC(2006) 313}. COM/2006/0104 final, 10 p. Availabel at: http://ec.europa.eu/agriculture/coexistence/com104_en.pdf

European Commission (2006c). Regulation (EC) No 1367/2006 of the European Parliament and of the Council of 6 September 2006 on the application of the provisions of the Aarhus Convention on Access to Information, Public Participation in Decision-making and Access to Justice in Environmental Matters to Community institutions and bodies. *Official Jounal of the European Union*, L 264, pp. 13-19.

European Commission (2006d). Report from the Commission to the Council and the European Parliament on the implementation of Regulation (EC) No 1830/2003 concerning the traceability and labelling of genetically modified organisms and the traceability of food and feed products produced from genetically modified organisms and amending Directive 2001/18/EC. 13 p. Available at: http://eur-lex.europa.eu/LexUriServ/LexUriServ.do?uri=COM:2006:0197:FIN:EN:PDF

European Commission (2007a). Council Regulation (EC) No 834/2007 of 28 June 2007 on organic production and labelling of organic products and repealing Regulation (EEC) No 2092/91. *Official Jounal of the European Union*, L 189, pp. 1-23.

European Commission (2007b). Directive 2007/2/EC of the European Parliament and of the Council of 14 March 2007 establishing an Infrastructure for Spatial Information in the European Community (INSPIRE). *Official Jounal of the European Union*, L 108, pp. 1-14.

European Commission (2008a). Commission Regulation (EC) No 889/2008 of 5 September 2008 laying down detailed rules for the implementation of Council Regulation (EC) No 834/2007 on organic production and labelling of organic products with regard to organic production, labelling and control. *Official Jounal of the European Union*, L 250, pp. 1-84.

European Commission (2008b). Report from the Commission to the Council and the European Parliament on the implementation of Regulation (EC) No 1830/2003 concerning the traceability and labelling of genetically modified organisms and the traceability of food and feed products produced from genetically modified organisms and amending Directive 2001/18/EC. COM (2008) 560 final. 11 p. Available at: http://eur-lex.europa.eu/LexUriServ/LexUriServ.do?uri=COM:2008:0560:FIN:EN:PDF

European Commission (2009a). Commission decision No 2009/770/EC of 13 October 2009 establishing standard reporting formats for presenting the monitoring results of the deliberate release into the environment of genetically modified organisms, as or in products, for the purpose of placing on the market, pursuant to Directive 2001/18/EC of the European Parliament and of the Council. *Official Journal of the European Union*, L 275, pp. 9-27.

European Commission (2009b). Commission Decision of 13 October 2009 establishing standard reporting formats for presenting the monitoring results of the deliberate release into the environment of genetically modified organisms, as or in products, for the purpose of placing on the market, pursuant to Directive 2001/18/EC of the European Parliament and of the Council. (notified under document C(2009) 7680). *Official Journal of the European Union*, L 975, pp. 9-27.

European Commission (2009c). Council regulation (EC) No 73/2009 of 19 January 2009 establishing common rules for direct support schemes for farmers under the common agricultural policy and establishing certain support schemes for farmers, amending Regulations (EC) No 1290/2005, (EC) No 247/2006, (EC) No 378/2007 and repealing Regulation (EC) No 1782/2003. *Official Jounal of the European Union*, L 30, pp. 16-99.

European Commission (2009d). Regulation (EC) No 1107/2009 of the Euopean Parliamne and of the Council of 21 October 2009 concerning the placing of plant protection products on the market and repealing Council Directives 79/117/EEC and 91/414/EEC. *Official Jounal of the European Union*, L 309, pp. 1-50.

European Commission (2009e). Report from the Commission to the Council and the European Parliament on the coexistence of genetically modified crops with conventional and organic farming (SEK(2009) 408). European Commission, ed. 12 p. Available at: http://eur-lex.europa.eu/LexUriServ/LexUriServ.do?uri=COM:2009:0153:FIN:EN:PDF

European Commission (2010a). Commission Recommendation of 13 July 2010 on guidelines for the development of national co-existence measures to avoid the unintended presence of GMOs in conventional and organic crops (2010/C 200/01). *Official Journal of the European Union*, C 200, pp. 1-5.

European Commission (2010b). A decade of EU-funded GMO research (2001 - 2010). EUR 24473, European Commission, ed. (Brussels, Belgium, European Commission), 264 p. Available at: http://ec.europa.eu/research/biosociety/pdf/a_decade_of_eu-funded_gmo_research.pdf

European Commission (2011). Report from the Commission to the European parliament and the Council on socio-economic implications of GMO cultivation on the basis of Member States contributions, as requested by the Conclusions of the Environment Council of December 2008. , D.-S. European Commission, ed. (Brussels, Belgium), 11 p. Available at: http://ec.europa.eu/food/food/biotechnology/reports_studies/docs/socio_economic_report_GMO_fr.pdf

European Economic and Social Committee (2011). Opinion of the European Economic and Social Committee on the 'Proposal for a Regulation of the European Parliament and of the Council amending Directive 2001/18/EC as regards the possibility for the Member States to restrict or prohibit the cultivation of GMOs in their territory' COM(2010) 375 final – 2010/0208 (COD). (2011/C 54/16). *Official Jounal of the European Union*, C 54, pp. 51-57.

European Parliament. Directorate-General for Internal Policies (2010). The evaluations of the impact of reforms that have affected the sector and the needs of European livestock system. In Policy Department Structural and cohesion policies B, E.P.D.f.I. Policies, ed. (Brussels, Belgium), 31 p. http://www.europarl.europa.eu/activities/committees/studies/download.do?language=en&file=32695

Eurostat. European Commission (2010). Agricultural statistics. Main results – 2008–09. In Eurostat pocketbooks, E. Commission, ed. (Eurostat), 186 p. Available at: http://epp.eurostat.ec.europa.eu/cache/ITY_OFFPUB/KS-ED-10-001/EN/KS-ED-10-001-EN.PDF

Filip, L., Miere, D., and Indrei, L.L. (2004). Genetically modified foods. Advantages and human health risks. *Revista medico-chirurgicala a Societatii de Medici si Naturalisti din Iasi*, 108, pp. 838-842.

Finck, M., Seitz, H., and Beismann, H. (2006). Concepts for general surveillance: VDI proposals standardisation and harmonisation in the field of GMO-monitoring. *Journal fur Verbraucherschutz und Lebensmittelsicherheit*, 1, pp. 11-14.

Fok, M. (2010). Un état de coexistence du soja transgénique et conventionnel au Paraná (Brésil). *Economie rurale*, 320, 53-68.

Foster, M. (2007). Challenges for agricultural markets, coexistence, segmentation of grain markets, costs and opportunities. ABARE CONFERENCE PAPER 07.9. Paper presented at: GMCC07 Third International Conference on Coexistence between Genetically Modified (GM) and non-GM based Agricultural Supply Chains. Seville, Spain, DG-JRC-IPTS (ed.). Available at: http://www.abare.gov.au/publications_html/conference/conference_07/agric_markets.pdf

Foster, M. (2010). Evidence of price premiums for non-GM grains in world markets. In Australian Agricultural and Resource Economics Society (Adelaide, Australia, ABARE, Australia), 16 p. Available at: http://www.abare.gov.au/publications_html/conference/conference_10/AARES_4.pdf

GAO (2008). Agencies are proposing changes to improve oversight, but could take additional steps to enhance coordination and monitoring. In Report to the Committee on Agriculture, Nutrition, and Forestry, US Senate (U.S. Government Accountability Office), 109 p. Available at: http://www.gao.gov/products/GAO-09-60

Garcia-Alonso, M., Jacobs, E., Raybould, A., Nickson, T.E., Sowig, P., Willekens, H., Van der Kouwe, P., Layton, R., Amijee, F., Fuentes, A.M., *et al.* (2006). A tiered system for assessing the risk of genetically modified plants to non- target organisms. *Environmental Biosafety Research*, 5, pp. 57-65.

Gaskell, G., Allansdottir, A., Allum, N., Corchero, C., Fischler, C., Hampel, J., Jackson, J., Kronberger, N., Mejlgaard, N., Revuelta, G., *et al.* (2006). Europeans and Biotechnology in 2005: Patterns and Trends. Eurobarometer 64.3. A report to the European Commission's Directorate-General for Research. In Eurobarometer, European Commission, ed. (Brussels, Belgium), 85 p. Available at: http://www.cibpt.org/docs/2006-jul-eurobarometro-bio-2nd-ed.pdf

Gaskell, G., Stares, S., Allansdottir, A., Allum, N., Castro, P., Esmer, Y., Fischler, C., Jackson, J., Kronberger, N., Hampel, J., *et al.* (2010). Europeans and biotechnology in 2010. Winds of change? A report to the European Commission's Directorate-General for Research. EUR 24537 EN. In Research*eu Studies and reports (Brussels, Belgium, European Commission), 176 p. Available at: http://ec.europa.eu/public_opinion/archives/ebs/ebs_341_winds_en.pdf

Gathmann, A. (2009). National implementation plan for MON810 monitoring in Germany - A way forward to improve General Surveillance? Journal fur Verbraucherschutz und Lebensmittelsicherheit *3*, 50-50.

Gathmann, A., and Bartsch, D. (2006). National coordination of GMO monitoring - a concept for Germany. *Journal fur Verbraucherschutz und Lebensmittelsicherheit*, 1, pp. 45-48.

Gelder, J.W.v., Kammeraat, K., and Kroes, H. (2008). Soy consumption for feed and fuel in the European Union. A research paper prepared for Milieudefensie (Friends of the Earth Netherlands), 22 p. http://www.foeeurope.org/agrofuels/FFE/Profundo%20report%20final.pdf

Graef, F., Zughart, W., Hommel, B., Heinrich, U., Stachow, U., and Werner, A. (2005). Methodological scheme for designing the monitoring of genetically modified crops at the regional scale. *Environmental Monitoring and Assessment*, 111, 1-26.

Green, R., and Hervé, S. (2006). IP - Traceability and grains traders: ADM, Bunge, Cargill, Dreyfus. *INRA Les cahiers d'ALISS*, 03, 86 p. Available at: http://www.prodinra.inra.fr/prodinra/pinra/data/2007/11/PROD200752c85e26_20071115125832834.pdf

Grossman, M.R. (2003). Multifunctionality and non-trade concerns. In Agriculture and international trade: law, policy and the WTO, Michael N Cardwell, Margaret R Grossman and Christopher Rodgers (eds), CABI. pp. 85-129.

Gruber, S., Colbach, N., Barbottin, A., and Pekrun, C. (2008). Post-harvest gene escape and approaches for minimizing it. *CAB Reviews: Perspectives in Agriculture, Veterinary Science, Nutrition and Natural Resources*, 3, 17 p.

Gryson, N., and Eeckhout, M. (2011). Co-existence and traceability in supply chains: a case study on Belgian compound feed. In GM and non-GM supply chains: their coexistence and traceability, Y. Bertheau, ed. (London, UK, Wiley Ltd.). In Press.

Haggett, S. (2008). Towards a multipurpose neural network approach to novelty detection (University of Kent), 236 p. Available at: http://kar.kent.ac.uk/24133/1/Towards_Detection.pdf

Hannachi, M., Coléno, F.C., and Assens, C. (2009). Collective strategies and coordination for the management of coexistence: the case studies of Alsace and western South of France. Paper presented at: Fourth International Conference on Coexistence between Genetically Modified (GM) and non-GM based Agricultural Supply Chains (Melbourne, AU). Available at: http://assens.perso.neuf.fr/GMO2009.pdf

Hansen, J., Lin, W., Tuan, F., Marchant, M.A., Kalaitzandonakes, N., Zhong, F., and Song, B. (2007). Commercialization of herbicide-tolerant soybeans in China: perverse domestic and international trade effects. In American Agricultural Economics Association Annual Meeting (Portland, OR, USA), 33 p. Available at: http://ageconsearch.umn.edu/bitstream/9906/1/sp07ha03.pdf

Hardin, G. (1968). Tragedy of commons. *Science*, 162, 1243-1248.

Hardin, G. (1998). Extensions of "The Tragedy of the Commons". *Science*, 280, 682-683.

Harrell, F.E. (2001). Regression modeling strategies. With applications to linear models, logistic regression, and survival analysis. Springer verlag.

Hasha, G. (2002). Livestock feeding and feed imports in the European Union—A decade of change. FDS-0602-01. In *Electronic Outlook Report from the Economic Research Service*, ERS, USDA, ed., 28 p. Available at: http://m.usda.mannlib.cornell.edu/usda/ers/FDS/2000s/2002/FDS-07-03-2002_Special_Report.pdf

Häusling M., Rapporteur of the Committee on Agriculture and Rural Development of the European Parliament (2011). The EU protein deficit: what solution for a long-standing problem? (2010/2111(INI)). In Report, European Parliament, ed. Available at: http://www.europarl.europa.eu/sides/getDoc.do?pubRef=-//EP//NONSGML+REPORT+A7-2011-0026+0+DOC+PDF+V0//EN&language=EN

Headey, D. (2011). Rethinking the global food crisis: the role of trade shocks. *Food Policy*, 36, pp. 136-146.

Heisey, P.W., King, J.L., and Rubenstein, K.D. (2005). Patterns of public-sector and private-sector patenting in agricultural biotechnology. *AgBioForum*, 8, 73.

Heisey, P.W., Srinivasan, C.S., and Thirtle, C. (2001). Public sector plant breeding in a privatizing world. In *Agriculture Information Bulletin*, E.R.S. Resource economics Division, USDA, ed. (Washington, DC, USA, Resource economics Division, Economic research Service, USDA). Available at: http://www.ers.usda.gov/publications/aib772/aib772.pdf

Henle, K., Alard, D., Clitherow, J., Cobb, P., Firbank, L., Kull, T., McCracken, D., Moritz, R.F.A., Niemelä, J., Rebane, M., *et al.* (2008). Identifying and managing the conflicts between agriculture and biodiversity conservation in Europe - A review. *Agriculture Ecosystems & Environment*, 124, pp. 60-71.

Hepburn, P., Howlett, J., Boeing, H., Cockburn, A., Constable, A., Davi, A., de Jong, N., Moseley, B., Oberdorfer, R., Robertson, C., *et al.* (2008). The application of post-market monitoring to novel foods. *Food and Chemical Toxicology*, 46, pp. 9-33.

Hintermann, U., Weber, D., Zangger, A., and Schmill, J. (2002). Biodiversity monitoring in Switzerland BDM - interim report. Environmental series. No. 342, Forest and Landscape, Swiss Agency for the Environment, Berne, Switzerland, ed. Available at: www.umwelt-schweiz.ch/buwal/shop/files/pdf/phpamJ0T6.pdf

James, C. (2011). Global status of commercialized biotech / GM crops: 2010. In ISAAA Brief, ISAAA, ed. (Ithaca, NY, USA).

Jank, B., Rath, J., and Gaugitsch, H. (2006). Co-existence of agricultural production systems. *Trends in Biotechnology*, 24, pp. 198-200.

Kaditi, E., and Swinnen, J.F.M. (2006). Trade Agreements, Multifunctionality and EU Agriculture. The Centre for European Policy Studies, Brussels, Belgium. 323 p.

Kanter, J. (2010). E.U. signals big shift on genetically modified crops. In *The New-York Times* (New-York, USA). Available at: http://www.nytimes.com/2010/05/10/business/energy-environment/10green.html?_r=1&src=busln

Kawata, M., Murakami, K., and Ishikawa, T. (2009). Dispersal and persistence of genetically modified oilseed rape around Japanese harbors. *Environmental Science and Pollution Research*, 16, pp. 120-126.

Kim, C.-G., Yi, H., Park, S., Yeon, J.E., Kim, D.Y., Kim, D.I., Lee, K.-H., Lee, T.C., Paek, I.S., Yoon, W.K., *et al.* (2006). Monitoring the occurrence of genetically modified soybean and maize around cultivated fields and at a grain receiving port in Korea. *Journal of Plant Biology*, 49, pp. 218-223.

Konduru, S.P. (2008). Three essays on the potential economic impacts of biotech crops in the presence of asynchronous regulatrory approval. PhD Thesis. Faculty of the Graduate School, Agricultural Economics (Columbia, U.S.A., University of Missouri-Columbia), 145 p. Available at: https://mospace.umsystem.edu/xmlui/handle/10355/6642

Konefal, J. and Busch, L. (2010). Markets of multitudes: how biotechnologies are standardising and differentiating corn and soybeans. *Sociologia Ruralis*, 50, 4, pp. 409-427

Kontoleon, A., and Yabe, M. (2006). Market segmentation analysis of preferences for GM derived animal foods in the UK. *Journal of Agricultural & Food Industrial Organization*, 4, Article 8.

Krautgartner, R., Henard, M.-C., Lieberz, S., Boshnakova, M., and the group of FAS oilseeds specialists in the EU (2010a). EU-27. Oilseeds - Increased domestic soybean & soybean meal production. In EU-27, USDA GAIN, ed. (Vienna, Austria), 14 p. Available at: http://gain.fas.usda.gov/Recent%20GAIN%20Publications/Oilseeds%20%20Increased%20Domestic%20Soybean%20and%20Soybean%20Meal%20Production_Vienna_EU-27_11-30-2010.pdf

Krautgartner, R., Henard, M.-C., Lieberz, S., Dobrescu, M., Flach, B., Wideback, A., Guerrero, M., Bendz, K., and the group of FAS oilseeds specialists in the EU (2010b). EU-27. Oilseeds and products annual. 2010. In EU-27, USDA GAIN, ed. (Berlin, Germany), 33 p. Available at : http://gain.fas.usda.gov/Recent%20GAIN%20Publications/Oilseeds%20and%20Products%20Annual_Berlin_EU-27_4-19-2010.pdf

Laisney, C. (2011). L'évolution de l'alimentation en France. Tendances émergentes et ruptures possibles. Futuribles, pp. 5-23.

Laudadio, V., Tufarelli, V., Dario, M., Cazzato, E., and Modugno, G.d. (2009). Evaluation of dehulled and micronized lupin (*Lupinus albus* L.) and pea (*Pisum sativum* L.) seed meal as an alternative protein source in diets for pre-laying hens. World Poultry Science Association (WPSA), 2nd Mediterranean Summit of WPSA, Antalya, Turkey, 4-7 October 2009, pp. 495-498.

Laurent, C., Berriet-Solliec, M., Kirsch, M., Labarthe, P., and Trouvé, A. (2010). Multifunctionality of agriculture, public policies and scientific evidences: some critical issues of contemporary controversies. *Applied Studies in Agribusiness and Commerce – APSTRACT*, pp. 53-58.

Le Bail, M., Lecroart, B., Gauffreteau, A., Angevin , F., and Messéan, A. (2010). Effect of the structural variables of landscapes on the risks of spatial dissemination between GM and non-GM maize. *European Journal of Agronomy*, 33, pp. 12-23.

Lecoq, E., Holt, K., Janssens, J., Legris, G., Pleysier, A., Tinland, B., and Wandelt, C. (2007). General surveillance: roles and responsibilities, the industry view. *Journal Fur Verbraucherschutz Und Lebensmittelsicherheit-Journal of Consumer Protection and Food Safety*, 2, pp. 25-28.

Lee, B., Kim, C.G., Park, J.Y., Park, K.W., Kim, H.J., Yi, H., Jeong, S.C., Yoon, W.K., and Kim, H.M. (2009). Monitoring the occurrence of genetically modified soybean and maize in cultivated fields and along the transportation routes of the Incheon Port in South Korea. *Food Control*, 20, pp. 250-254.

Lehuger, S., Gabrielle, B., and Gagnaire, N. (2009). Environmental impact of the substitution of imported soybean meal with locally-produced rapeseed meal in dairy cow feed. *Journal of Cleaner Production,* 17, pp. 616-624.

LMC International (2009a). Evaluation of measures applied under the Common Agricultural Policy to the protein crop sector. Case study monographs (Report to the European Commission), LMC International, ed., 231 p. Available at: http://ec.europa.eu/agriculture/eval/reports/protein_crops/case_studies_en.pdf

LMC International (2009b). Evaluation of measures applied under the Common Agricultural Policy to the protein crop sector. Main Report (Report to the European Commission.), L. International, ed., pp. vIII, ii, 169. Available at: http://ec.europa.eu/agriculture/eval/reports/protein_crops/fulltext_en.pdf

Lu, Y., Wu, K., Jiang, Y., Xia, B., Li, P., Feng, H., Wyckhuys, K.A.G., and Guo, Y. (2010). Mirid bug outbreaks in multiple crops correlated with wide-scale adoption of Bt cotton in China. *Science,* 328, pp. 1151-1154.

Marsden, T. (2008). Agri-food contestations in rural space: GM in its regulatory context. *Geoforum,* 39, pp. 191-203.

Messéan, A., Angevin, F., Gomez-Barbero, M., Menrad, K., and Rodriguez-Cerezo, E. (2006). New case studies on the coexistence of GM and non-GM crops in European agriculture. EUR22102 EN (Seville, Spain, DG-JRC-IPTS), 116 p. Available at: http://www.jrc.es/home/pages/eur22102enfinal.pdf

Messéan, A., Squire, G., Perry, J., Angevin, F., Gomez, M., Townend, P., Sausse, C., Breckling, B., Langrell, S., Dzeroski, S., et al. (2009). Sustainable introduction of GM crops into european agriculture: a summary report of the FP6 SIGMEA research project. *OCL - Oleagineux, Corps Gras, Lipides*, 16, pp. 37-51.

Milanesi, J. (2008). Analyse des coûts induits sur les filières agricoles par les mises en culture d'organismes génétiquement modifiés (OGM) - Etude sur le maïs, le soja et le poulet Label Rouge, C.d.R.e.G. (CREG). ed. (Pau, France, Université de Pau et des Pays de l'Adour), 123 p. Availabel at :

Milanesi, J. (2009). Quel avenir pour les filières animales « sans OGM » en France? Illustration par le poulet Label Rouge. In 3èmes journées de recherches en sciences sociales INRA SFER CIRAD (Montpellier, France). Available at : http://hal.archives-ouvertes.fr/hal-00521222/en/

Milanesi, J. (2011). Current and future availability of non-genetically modified soybean seeds in the U.S., Brazil and Argentina. In GM and non-GM supply chains: their coexistence and traceability, Y. Bertheau, ed. (Wiley Publishing Ltd.), *In Press.*

Miller, H.I. (2010). Split approvals and hot potatoes. *Nature Biotechnology,* 28, pp. 552-553.

Monkemeyer, W., Schmidt, K., Beissner, L., Schiemann, J., and Wilhelm, R. (2006). A critical examination of the potentials of existing German networks for GMO monitoring. *Journal fur Verbraucherschutz und Lebensmittelsicherheit,* 1, 67-71.

Monsanto Co. (2006). Monitoring report. Mon810 cultivation. Czech republic, France, Germany, Portugal and Spain. 2005, 236 p. Available at: https://yieldgard.eu/en-us/YieldGardLibraryGrower/2005%20YieldGard%20Monitoring%20Report.pdf

Monsanto Co. (2009a). Annual monitoring report on the cultivation of MON 810 in 2008. Czech Republic, Germany, Portugal, Slovakia, Poland, Romania and Spain, 388 p. Available at: https://yieldgard.eu/en-us/YieldGardLibraryGrower/2008%20YieldGard%20Monitoring%20Report.pdf
Monsanto Co. (2009b). Monitoring Report. MON 810 cultivation. Spain. 2003-2004, 57 p. Available at: https://yieldgard.eu/en-us/YieldGardLibraryGrower/20032004%20YieldGard%20Monitoring%20Report.pdf
Naeve, S.L., Orf, J.H., and Miller-Garvin, J. (2010). 2010 analysis of the U.S. non-GMO food soybean variety pipeline. In Japan Soy Food Summit (Tokyo, Japan). Available at: http://soybase.org/meeting_presentations/soybean_breeders_workshop/SBW_2011/Naeve.pdf
Noussair, C., Robin, S., and Ruffieux, B. (2004). Revealing consumers' willingness-to-pay: A comparison of the BDM mechanism and the Vickrey auction. *Journal of Economic Psychology*, 25, pp. 725-741.
Nowicki, P., Aramyan, L., Baltussen, W., Dvortsin, L., Jongeneel, R., Domínguez, I.P., Wagenberg, C.v., Kalaitzandonakes, N., Kaufman, J., Miller, D., *et al.* (2010). Study on the implications of asynchronous GMO approvals for EU imports of animal feed products. Final report (Contract N° 30-CE-0317175/00-74). Executed on behalf of Directorate-General for Agriculture and Rural Development, European Commission. Available at: http://ec.europa.eu/agriculture/analysis/external/asynchronous-gmo-approvals/full-text_en.pdf
Orf, J.H. (2004). Overview of recent genetic improvement in public and private breeding programs in the USA. pp. 220-227.
Ostergard, H., Finckh, M.R., Fontaine, L., Goldringer, I., Hoad, S.P., Kristensen, K., van Bueren, E.T.L., Mascher, F., Munk, L., and Wolfe, M.S. (2009). Time for a shift in crop production: embracing complexity through diversity at all levels. *Journal of the Science of Food and Agriculture*, 89, pp. 1439-1445.
Otiman, I.P., Badea, E.M., and Buzdugan, L. (2008). Roundup Ready soybean, a Romanian story. *Bulletin of University of Agricultural Sciences and Veterinary Medicine Cluj-Napoca Animal Science and Biotechnologies*, 65, pp. 352-357.
Owen, M., Dixon, P., Shaw, D., Weller, S., Young, B., Wilson, R., and Jordan, D. (2010). Sustainability of glyphosate-based weed management: the benchmark study. *ISB News Report August 2010*, 3.
Owen, M.D.K. (2009). Herbicide-tolerant genetically modified crops: resistance management. In Environmental impact of genetically modified crops, Ferry, N. and Gatehouse, A. M. R. (eds). *CABI*. pp. 115-164.
Pascher, K., Moser, D., Dullinger, S., Sachslehner, L., Gros, P., Traxler, A., Sauberer, N., Frank, T., and Grabherr, G. (2009). Establishment of an Austrian monitoring design to identify potential ecological effects of genetically modified plants. *In* Fourth International Conference on Coexistence between Genetically Modified (GM) and Non-GM based Agricultural Supply Chains (GMCC09) (Melbourne, Australia,

GMCC09), 11 p. Available at: http://www.gmcc-09.com/wp-content/uploads/poster_03_82-pascher_poster.pdf

Petit, S. (2009). The dimensions of land use change in rural landscapes: Lessons learnt from the GB Countryside Surveys. *Journal of Environmental Management*, 90, pp. 2851-2856.

Popp, J. (2008). The future of GM crops. *Hungarian Agricultural Research*, 2/3, pp. 17-20.

Pouliquen, A. (2001). Competitiveness and farm incomes in the CEES agri-food sectors. Implication before and after accession for EU markets and policies., 95 p. Available at: http://ec.europa.eu/agriculture/publi/reports/ceeccomp/fullrep_en.pdf

République Française (2008). Loi n° 2008-595 du 25 juin 2008 relative aux organismes génétiquement modifiés. NOR: DEVX0771876L. *Journal Officiel de la République française*, pp. 10218-10228.

Reuter (2010). E.U. considers easing rules on biotech crops in animal feed. In *The New-York Times, October 8, 2010*. Available at: http://www.nytimes.com/2010/10/09/business/global/09gmo.html?_r=3&ref=genetic_engineering

Roberson, R. (2010). Super weeds put USDA on hotseat, *CheckBiotech*, 2 p.

Roebroeck, L. (2002). Factsheet soybean, T.N. Biotechnology and Consumer Foundation, ed. (The Hague, The Netherlands), 44 p. Available at: http://www.voedingscentrum.nl/resources2008/Factsheet_sojapdf.pdf

Sabalza M., Miralpeix B., Twyman R.M., Capell T. and Christou P. (2011). EU legitimizes GM crops exclusion zones. Nature Biotechnology, 29, 4, pp. 315-317.

Sanvido, O., Aviron, S., Romeis, J., and Bigler, F. (2007a). Challenges and perspectives in decision-making during post-market environmental monitoring of genetically modified crops. *Journal Fur Verbraucherschutz Und Lebensmittelsicherheit-Journal of Consumer Protection and Food Safety*, 2, pp. 37-40.

Sanvido, O., Romeis, J., and Bigler, F. (2007b). Ecological impacts of genetically modified crops: Ten years of field research and commercial cultivation. *Advances in Biochemical Engineering/Biotechnology*, 107, pp. 235-278.

Sanvido, O., Romeis, J., and Bigler, F. (2009a). An approach for post-market monitoring of potential environmental effects of Bt-maize expressing Cry1Ab on natural enemies. *Journal of Applied Entomology*, 133, pp. 236-248.

Sanvido, O., Romeis, J., and Bigler, F. (2009b). Monitoring or Surveillance? Balancing between theoretical frameworks and practical experiences. *Journal fur Verbraucherschutz und Lebensmittelsicherheit*, 3, pp. 4-7.

Sanvido, O., Widmer, F., Winzeler, M., and Bigler, F. (2005). A conceptual framework for the design of environmental post-market monitoring of genetically modified plants. *Environmental Biosafety Research*, 4, 13-27.

Schiemann, J. (2007). Proceedings of the international workshop "Post Market Environmental Monitoring of Genetically Modified Plants: Harmonisation and Standardisation - a Practical Approach". 26-27 April, 2007, Berlin, Germany. *Journal fur Verbraucherschutz und Lebensmittelsicherheit*, 2, 91 p.

Schmidt, K., Mönkemeyer, W., Böttinger, P., Wilhelm, R., and Schiemann, J. (2009). Use of existing networks for post-market monitoring? *Journal fur Verbraucherschutz und Lebensmittelsicherheit*, 3, 13-13.

Seitz, H., Zughart, W., Finck, M., Beismann, H., Berhorn, F., and Eikmann, T. (2010). Standardization of methods for monitoring the environmental effects of genetically modified plants, development of VDI guidelines: a final report an R & D project (FKZ 804 67 010) on behalf of the Federal Office for Nature Conservation. BfN - Skripten (Bundesamt fur Naturschutz), 68 p. Available at: http://www.bfn.de/fileadmin/MDB/documents/service/Skript267.pdf

Smale, M., ed. (1998). Farmers, gene banks and crop breeding: economic analyses of diversity in wheat, maize, and rice (Springer Verlag). 288 p.

Stein, A.J., and Rodriguez-Cerezo, E. (2009). The global pipeline of new GM crops. EUR 23486 EN. In JRC scientific and technical reports, JRC-IPTS, ed. (Seville, SP, JRC-IPTS), 114 p. Available at: http://ftp.jrc.es/EURdoc/report_GMOpipeline_online_preprint.pdf

Stein, A.J., and Rodriguez-Cerezo, E. (2010a). International trade and the global pipeline of new GM crops. *Nature Biotechnology*, 28, pp. 23-25.

Stein, A.J., and Rodriguez-Cerezo, E. (2010b). Low-level presence of new GM crops: an issue on the rise for countries where they lack approval. *AgBioForum*, 13, pp. 173-182.

Stein, A.J., and Rodríguez-Cerezo, E. (2009). What can data on GMO field release applications in the USA tell us about the commercialisation of new GM crops? JRC52545. In JRC Technical notes, JRC-IPTS, ed., 16 p. Available at: http://ftp.jrc.es/EURdoc/JRC52545.pdf

Takada, A. (2010). China to compete with U.S., Canada in Asian market for soybean foods. In Bloomberg News. Available at: http://www.bloomberg.com/news/2010-11-19/china-s-non-modified-soybeans-to-compete-with-u-s-canada-in-asian-market.html

Tallage, Agri market forecasting (2010). Study on modelling of feed consumption in the European Union. Tender No. AGRI-2008-EVAL-09, Tallage, ed. (European Commission, DG-AGRI), 73 p. Available at: http://ec.europa.eu/agriculture/analysis/external/feed/fulltext_en.pdf

Taylor, R.D., and Koo, W.W. (2010). 2010 outlook of the U.S. and world corn and soybean industries, 2009-2019, C.f.A.P.a.T. Studies, ed. (Fargo, North Dakota, USA, Department of Agribusiness and Applied Economics. North Dakota State University), 35 p. Available at: https://ageconsearch.umn.edu/bitstream/92003/2/AgReport665%20-%20Corn_and_soybean.pdf

Then, C., and Stolze, M. (2010). Economic impacts of labelling thresholds for the adventitious presence of genetically engineered organisms in conventional and organic seed. Seed purity: costs, benefits and risk management strategies for maintaining markets free from genetically engineered plants (IFOAM EU Group), 60 p. Available at: http://www.ifoam.org/about_ifoam/around_world/eu_group-new/positions/publications/pdf/IFOAMEU_GMO-freeSeedStudy.pdf

Tinland, B. (2008). Implementation of post-market monitoring of insect-protected maize MON 810 in the UE. AFPP - 8ème Conference Internationale sur les Ravageurs en Agriculture, Montpellier SupAgro, France, 22-23 Octobre 2008, 165. Available at : http://www.cabdirect.org/abstracts/20093081604.html

Tinland, B., Delzenne, P., and Pleysier, A. (2007). Implementation of a post-market monitoring for insect-protected maize MON 810 in the EU. *Journal Fur*

Verbraucherschutz Und Lebensmittelsicherheit-Journal of Consumer Protection and Food Safety, 2, 7-10.

TNS Opinion & Social (2010). Eurobaromètre 72. L'opinion publique dans l'Union Européenne (Brussels, B).

van Tongeren, F., van Meijl, H., and Surry, Y. (2001). Global models applied to agricultural and trade policies: a review and assessment. *Agricultural Economics*, 26, pp. 149-172.

Wal, J.M., Hepburn, P.A., Lea, L.J., and Crevel, R.W.R. (2003). Post-market surveillance of GM foods: applicability and limitations of schemes used with pharmaceuticals and some non-GM novel foods. *Regulatory Toxicology and Pharmacology*, 38, pp. 98-104.

Wandelt, C. (2007). Implementation of general surveillance for amflora potato cultivation - Data management. *Journal Für Verbraucherschutz Und Lebensmittelsicherheit-Journal of Consumer Protection and Food Safety*, 2, pp. 70-71.

Wilhelm, R., Beissner, L., and Schiemann, J. (2003). Concept for the realisation of a GMO-monitoring in Germany. *Nachrichtenblatt des Deutschen Pflanzenschutzdienstes*, 55, pp. 258-272.

Wilhelm, R., Sanvido, O., Castanera, P., Schmidt, K., and Schiemann, J. (2009). Monitoring the commercial cultivation of Bt maize in Europe - conclusions and recommendations for future monitoring practice. Environmental Biosafety Research, 8, 4, pp. 219-225.

Wilhelm, R., and Schiemann, J. (2006). Does the baseline concept provide appropriate tools for decision making? *Journal fur Verbraucherschutz und Lebensmittelsicherheit*, 1, pp. 75-77.

Windels, P., Alcalde, E., Lecoq, E., Legris, G., Pleysier, A., Tinland, B., and Wandelt, C. (2009). General surveillance for import and processing: the EuropaBio approach. *Journal fur Verbraucherschutz und Lebensmittelsicherheit*, 3, pp. 14-16.

Wolt, J.D., Keese, P., Raybould, A., Fitzpatrick, J.W., Burachik, M., Gray, A., Olin, S.S., Schiemann, J., Sears, M., and Wu, F. (2010). Problem formulation in the environmental risk assessment for genetically modified plants. Transgenic Research, 19, pp. 425-436.

Zughart, W., Benzler, A., Berhorn, F., Sukopp, U., and Graef, F. (2008). Determining indicators, methods and sites for monitoring potential adverse effects of genetically modified plants to the environment: the legal and conceptional framework for implementation. *Euphytica*, 164, pp. 845-852.

2

Storage of Soybeans and Its Effects on Quality of Soybean Sub-Products

Ernandes Rodrigues de Alencar[1] and Lêda Rita D'Antonino Faroni[2]
[1]*Faculdade de Agronomia e Medicina Veterinária, Universidade de Brasília, Brasília, Distrito Federal, Brasil*
[2]*Departamento de Engenharia Agrícola, Universidade Federal de Viçosa, Viçosa, Minas Gerais, Brasil*

1. Introduction

World soybean production in the 2009/2010 harvest was roughly 260 million tons, and the major producers were the United States, Brazil and Argentina, producing 91.4, 69.0 and 57.0 million tons, respectively (USDA, 2011). Given the significant world production of soybeans, quality is essential for the sectors involved in production and/or processing of this commodity. Quality is an important parameter for commercialization and processing of the grains and can affect the value of the product and its derivatives.

During post-harvest stages, soybeans are subjected to qualitative and quantitative losses due to several external factors. These factors may be physical, such as temperature and humidity; chemical, such as oxygen supply; and biological, such as bacteria, fungi, insects and rodents (BROOKER et al., 1992). According to BAILEY (1974), secure storage retains the qualitative and quantitative aspects of the grains, creating conditions unfavorable to the development of insects, rodents and microorganisms. Grain storage in the natural environment of tropical regions, according to ABBA & LOVATO (1999), presents major problems as a result of the temperature and relative humidity when compared with temperate or cold climates.

Soybeans are composed of roughly 20% lipids and are susceptible to qualitative deterioration processes via degradation of these compounds when stored improperly and can result in serious damage to the food industry. According to NARAYAN et al. (1988a), physical, chemical and biochemical alterations may occur in soybeans, depending on conditions and storage time. The qualitative changes of soybeans during storage contribute to the loss of oil and meal quality (ORTHOEFER, 1978), as well as other derivatives such as tofu and soymilk (NARAYAN et al. 1988b; LIU, 1997; HOU & CHANG, 1998; KONG et al., 2008).

2. Soybean storage

The objective of storage is to preserve the characteristics that grains present after harvest, therefore it is possible to obtain and market sub-products with satisfactory quality. Vitality of the grains can be preserved and the grinding quality and nutritive properties of the food can be maintained (BROOKER et al., 1992).

2.1 Soybean quality during storage

The pursuit of quality grain and sub-products should be a priority for producers, processors and for distributors of these products. According to BROOKER et al. (1992), the main characteristics that determine soybean quality are: low and uniform moisture content; low percentage of foreign material, discoloration, susceptibility to breakage, damage by heat (internal cracks), insect and fungal damage, elevated values of density, oils and protein concentration, and seed viability. Some factors can affect these characteristics such as the environmental conditions during grain formation on plants, season and harvesting system, drying system, techniques of storage, transport and characteristics of the species, and the variety.

The grain mass is an ecological system in which deterioration is the result of interaction between physical, chemical and biological variables (internal and external). The rate of deterioration during storage depends on the rates of change of these variables, which are directly affected by temperature and water content, and also by their inter-relationship with the grain and the storage structure (SINHA & MUIR, 1973). Insects, mites, rodents and fungi are the main biological factors responsible for qualitative and quantitative losses in stored grains, where development of these organisms is influenced by environmental factors such as temperature and relative humidity (PADIN et al., 2002).

2.2 Principal variations which affect quality of stored grains

Among the many variables that affect the storability of grains and their sub-products, moisture content and temperature are highlighted, associated with the storage time. Moisture content can be considered the most important factor on the quality of stored grain. ACASIO (1997) suggested that grains with moisture contents greater than 13% w.b. must be dried to reduce risk of deterioration in the form of dry matter loss by respiration, fungi attack, spontaneous heat production and reduction of germination percentage. Table 1 shows periods of safe storage for soybean with different moisture contents (BARRE, 1976 as cited in ACASIO, 1997).

Moisture content (%) w.b.	Safe storage period
10.0 - 11.0	4 years
10.0 -12.5	1-3 years
12.5 - 14.0	6-9 months
14.0 - 15.0	6 months

Source: BARRE (1976) as cited in ACASIO (1997)

Table 1. Safe storage period for soybeans.

Another determinant variable in the quality of stored products is temperature. When it comes to storage of soybeans, temperature not only affects the development of fungi but can promote chemical changes such as hydrolytic and oxidative rancidity. This physical variable also affects the development and reproduction of insect pests, where the optimum temperature for most species is between 27 and 35 °C. Soybeans with water contents between 14 and 14.3% w.b. and maintained at a temperature of between 5 and 8 °C can be stored for two years without development of fungi, while grain stored at 30 °C can be infected by fungi within a few weeks and severely damaged after six months of storage (ACASIO, 1997). It is emphasized that decision making must take into consideration the ideal conditions for grain storage, analyzing the combination of the variables of moisture and temperature, and not each one separately.

2.3 Qualitative parameters of soybeans and alterations during storage

2.3.1 Bulk density

Bulk density of grains is defined as the ratio between mass and volume (kg m^{-3}). This parameter generally increases with the decrease in moisture content of the product, except for coffee, paddy rice and barley (SILVA et al. 2009). It is emphasized that this trend depends on the percentage of damaged grains, initial moisture content, temperature reached during drying, final moisture content and the grain variety (BROOKER et al., 1992). Bulk density can be used as a qualitative indicator and the decrease in its value during storage may be associated with quality losses.

ALENCAR et al. (2009) studied the effect of different combinations of moisture contents and temperatures on the quality of stored soybeans (Fig. 1). The authors observed that the bulk density remained almost constant at different combinations of moisture content and temperature, except for grains stored with 12.8 and 14.8% moisture content and temperature of 30 and 40 °C, respectively. According to the authors, the decrease in bulk density of stored grain with 12.8% moisture content at the temperature of 30 °C (Fig. 1A), was due to infestation by *Plodia interpunctela* and *Sitotroga cerealella* whose optimum conditions of temperature and relative humidity are 30 °C and 75%, respectively (MBATA & OSUJI, 1983, MASON, 2006, HANSEN et al., 2004). On the other hand, the decrease in grains stored at 14.8% moisture content and temperature of 40 °C (Fig. 1B), was attributed to the development of fungi, where a high incidence (87%) of *Aspergillus glaucus* was verified.

2.3.2 Germination

Germination can be defined as the appearance of the first visible signs of growth or root protrusion, and is affected by several factors, including attack by insects, fungal infection, temperature, moisture and damage to the grains or seeds (BLACK, 1970, as cited in AL-YAHYA, 2001). The germination percentage has been used as an indicator of deterioration in different types of grains during storage.

HUMMIL et al. (1954) studied the qualitative deterioration of wheat grain stored with different moisture contents and inoculated or not with fungi at different temperatures. These authors observed a rapid deterioration of the wheat grains stored at 18% w.b. They verified that the process was more pronounced at a temperature of 35 °C. KARUNAKARAN et al. (2001) stored wheat with moisture contents between 15 and 19% w.b. at different temperatures in order to verify the time of safe storage, using the germination percentage as a quality standard. Results obtained for the water content of 17% w.b. at temperatures of 25, 30 and 35 °C were equal to 15, 7 and 5 days, respectively. Qualitative deterioration of soybeans stored with initial moisture contents between 9.8 and 13.8% w.b. in tropical conditions (30 °C and 82% RH) was simulated by Locher & Bucheli (1998). These authors confirmed a marked decrease in the germination percentage between 5 and 9 months of storage, where this behavior was more pronounced in seeds with greater initial moisture contents. Bhattacharya & Raha (2002) studied alterations in soybeans stored with moisture content of 14.0%, in the presence of different fungi species. The germination percentage of soybeans after 10 months of storage was zero. GUNGADURDOSS (2003), when studying the viability of soybean seeds under different storage conditions concluded that temperature

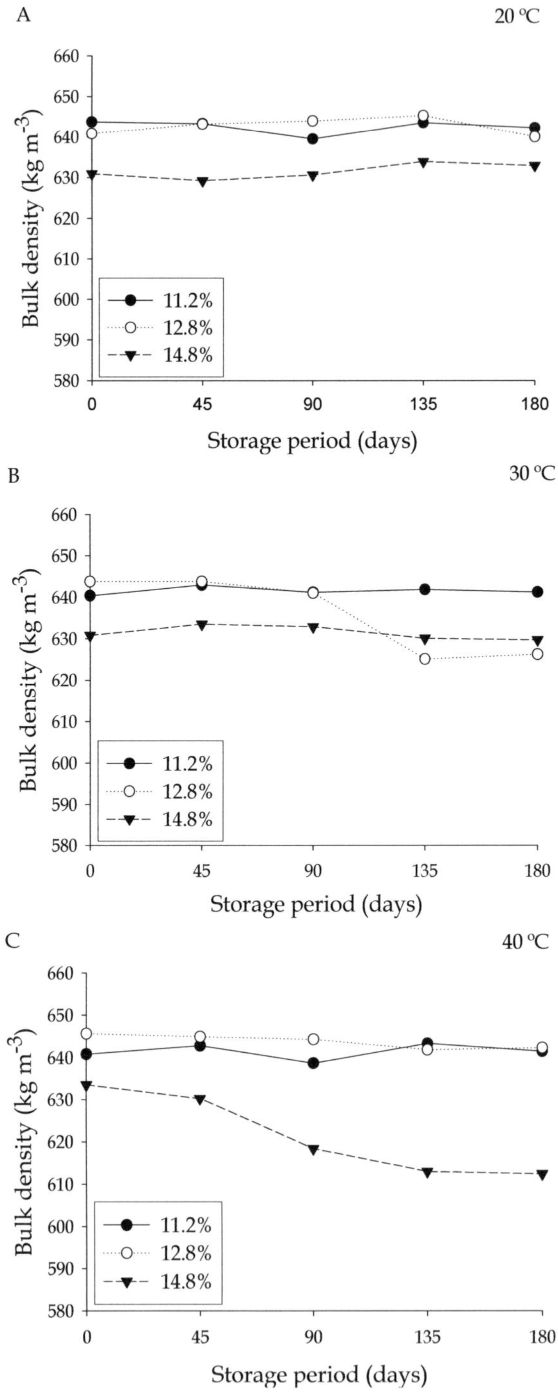

Source: ALENCAR et al. (2009)

Fig. 1. Average measurements of bulk density for soybeans stored with moisture contents of 11.2, 12.8 and 14.8% (w.b.) at the temperatures of 20, 30 and 40 °C.

was the predominant factor in maintaining the viability of soybean seeds. The effect of different combinations of moisture content and temperature on germination percentage of soybeans was evaluated by ALENCAR et al. (2008), during 180 days of storage (Fig. 2). It was verified that there was a decrease in the percentage of germinated grains, where this trend was less pronounced in grains stored with 11.2 and 12.8% moisture contents and temperatures of 20 °C (Fig. 2B).

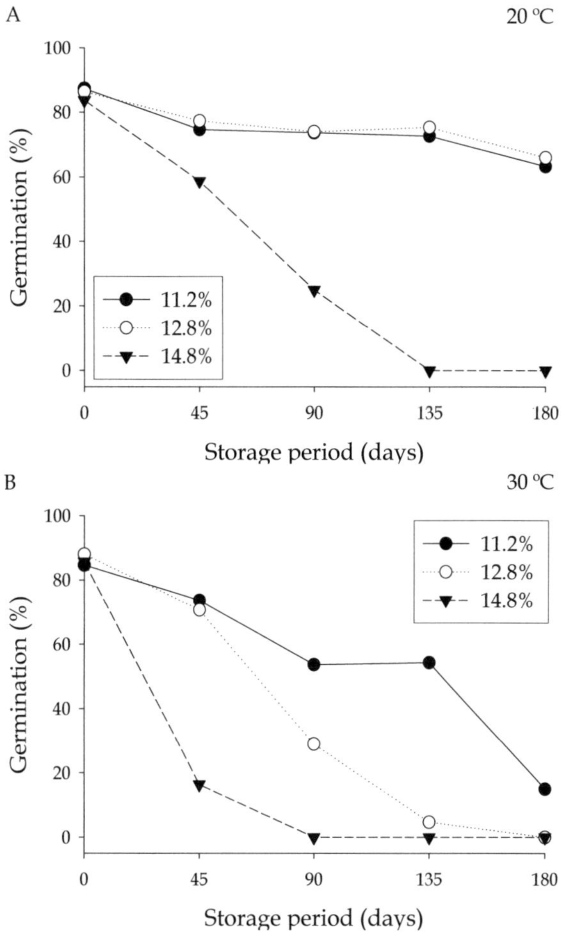

Source: ALENCAR et al. (2008)

Fig. 2. Average values of germination percentage of the soybeans with moisture contents of 11.2, 12.8 and 14.8% (w.b.) at the temperature of 20 and 30 °C, along the period of storage.

2.3.3 Electrical conductivity

According to SANTOS et al. (2004), the deterioration of grain is considered all and any degenerative change after the grain has reached its maximum quality, as evidenced by genetic damage, loss of integrity of system membranes, selective reduction of capacity, lipid peroxidation, leaching of solutes, changes in respiratory activity of the grains and seeds, changes in enzyme activity and protein synthesis, the inability to maintain the electrochemical gradient and loss of cellular compartmentalization and accumulation of toxic substances. Membrane damage is the initial event of degenerative changes in grains and seeds (DELOUCHE, 2006). According to HESLEHURST (1988), determination of electrical conductivity can be used to evaluate vigor, since the value of the conductivity is related to the amount of ions leached into solution, which is directly associated with cell membrane integrity. Poorly structured membranes and damaged cells are usually associated with the deterioration process of grains and seeds. Losses in germination and vigor in aged grains and seeds, according to LIN (1990), are correlated with increased electrolyte leaching, which increases with the decrease of membrane phospholipids. The lowest values corresponding to the lower release of exudates, indicate a high physiological potential (greater vigor), revealing a lower intensity of disorganization of cell membrane systems (VIEIRA et al., 2002).

ALENCAR et al. (2008) used electrical conductivity as a qualitative parameter for soybeans stored with moisture contents of 11.2, 12.8 and 14.8%, at temperatures of 20, 30 and 40 °C for 180 days (Table 2 and Fig. 3). In general, the authors observed a tendency for increased electrical conductivity during storage, where this trend was more pronounced as the moisture content and temperature were increased. It is reinforced that for the soybeans stored with 11.2 and 12.8% moisture content and temperature of 20 °C, electrical conductivity remained almost constant.

Temperature (ºC)	Moisture content (%)	**Regression equation adjusted**	R^2
	11.2	$\hat{y} = 153.4$	
20	12.8	$\hat{y} = 147.1$	
	14.8	$\hat{y} = 149.4 - 0.2029X + 0.0087X^2$	0.89
	11.2	$\hat{y} = 145.4 - 0.0403X$	0.91
30	12.8	$\hat{y} = 145.5 - 0.1477 + 0.0045X^2$	0.87
	14.8	$\hat{y} = 143.7 + 2.8360X - 0.0056X^2$	0.94
	11.2	$\hat{y} = 159.6 - 0.4581X + 0.0053X^2$	0.97
40	12.8	$\hat{y} = 175.5 + 3.768X - 1.002X^2$	0.88
	14.8	$\hat{y} = 194.2 + 4.403X - 1.530X^2$	0.84

Source: ALENCAR et al. (2008)

Table 2. Regression equations adjusted for electrical conductivity of the solution containing the soybeans with moisture contents of 11.2, 12.8 and 14.8% (w.b.) at temperatures of 20, 30 and 40 ºC along the storage period (X), and respective coefficients of determination.

2.3.4 Color of the grains

Appearance of the grains is considered a critical and decisive factor in the commercialization process. The color of soybeans, according to SINCLAIR (1992), has been used as an indicator of quality, and discoloration is indicative of physical or chemical alterations, presence of metabolites or other unfavorable characteristics. According to this author, changes in color of the soybeans are caused mainly by microorganisms, although changes in climatic conditions can enhance or affect color of the grain, but is not the main cause of the problem. In the United States upper limits are established for the classification of soybeans with distinct colors of yellow that is predominant, but may be green, black,

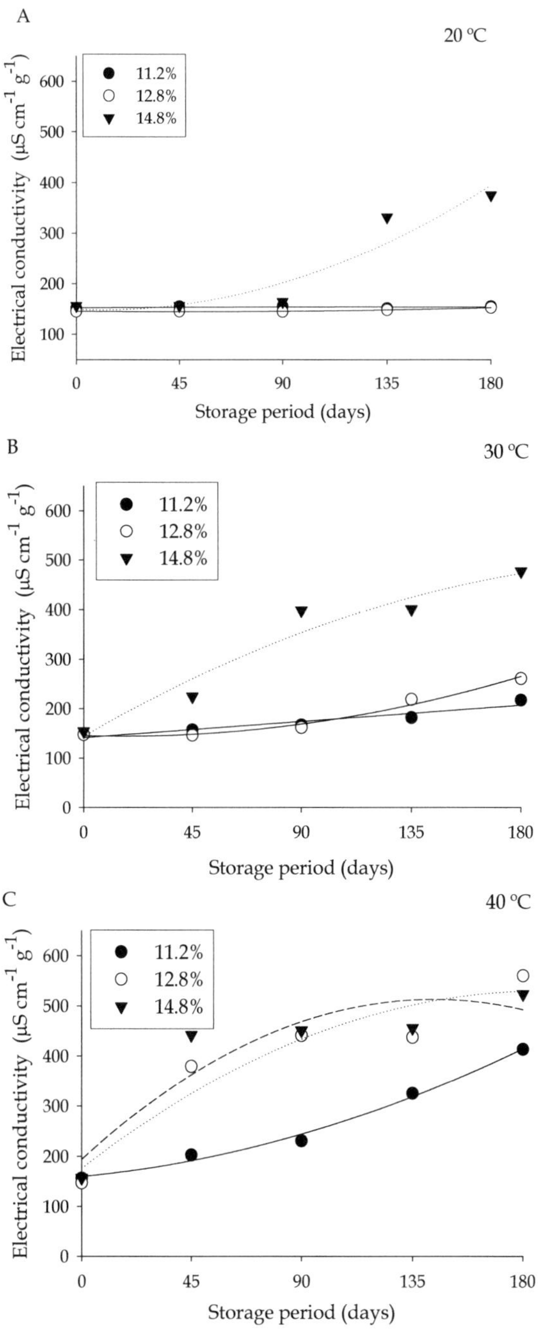

Source: ALENCAR et al. (2008)

Fig. 3. Regression curves of electrical conductivity (µS cm^{-1} g^{-1}) of the solution which contains soybeans stored with moisture contents of 11.2, 12.8 and 14.8% w.b. at the temperatures of 20, 30 and 40 °C.

brown or bicolored (USDA, 2006). The percentage limits of grain characterized as other colors for soybeans in types 1, 2, 3 and 4 are 1.0, 2.0, 5.0 and 10.0%, respectively. These values indicate that product is of very poor quality.

The darkening of soybeans, according to SAIO et al. (1980), is an important qualitative indicator of deterioration during storage, and for LIU (1997), the variation in color characterizes the aging of the grains and is associated with qualitative changes such as reducing the germination percentage. HOU & CHANG (2004) evaluated alterations in the color of soybeans stored with 5.4% moisture content in different conditions of temperature and relative humidity. These authors observed a significant darkening, according to variation of the Hunter coordinates, for the soybeans only when stored at 30 °C and 84% relative humidity (Table 3). ALENCAR et al. (2009) studied the change in color of the soybeans stored with moisture contents of 11.2, 12.8 and 14.8% at temperatures of 20, 30 and 40 °C for 180 days. The authors evaluated the color difference (ΔE), from values of Hunter L, a and b coordinates, and found a significant increase, where this increase is more pronounced when grains are stored with elevated moisture content and under increased temperature (Fig. 4 and Table 4). This trend of increasing color difference is directly related to the increase in the percentage of damaged grains, which are considered by Brazilian law as serious defects. The damaged grains are defined as grains or pieces of grain that present visible damage and have an accentuated dark brown color, affecting the cotyledon (MAPA, 2007). Alterations in the color of soybeans can also be viewed from the aspect of soybean flour obtained from these grains (Fig. 5).

Period (month)	Hunter L value	**Hunter a value**	Hunter b value
0	51.04±0.51[a]	4.05±0.04[g]	15.58±0.09[a]
1	49.56±0.55[c]	4.22±0.15[f]	15.33±0.11[a]
2	50.34±0.37[b]	4.58±0.06[e]	15.38±0.11[a]
3	48.67±0.40[d]	4.83±0.16[d]	14.59±0.25 [b]
4	48.38±0.09[d]	5.19±0.06[b]	14.76±0.03 [b]
5	45.14±0.44[e]	5.53±0.07[a]	12.87±0.13 [c]
6	43.37±0.18[f]	4.81±0.07[d]	11.19±0.13 [e]
7	45.37±0.59[e]	5.05±0.15[c]	12.53±0.22 [d]
8	41.99±0.55[g]	4.28±0.09[f]	10.57±0.28 [f]
9	38.97±0.94[h]	3.87±0.18[h]	8.75±0.58 [g]

Values followed by the same letter in the column are not statistically different at 5% probability
Source: HOU & CHANG (2004)

Table 3. Color of soybeans stored at 30 °C and 84% relative humidity.

2.4 Soybean oil

Soybean oil has emerged as one of the products obtained from processing of the grain, being one of the major products of this nature in the world market. It is most utilized to prepare food for humans and pets. Because of its properties it is suitable for a wide range of applications including use in margarines, salad oil, mayonnaise, and other food products (MORETTO & FETT, 1998). Virtually all soybean oil is extracted by solvent and commercial extraction techniques have remained unaltered since the early nineteenth century (ERICKSON & WIEDERMANN, 2006). Table 5 shows the main components of crude and refined soy oil, according to ERICKSON & WIEDERMANN (2006).

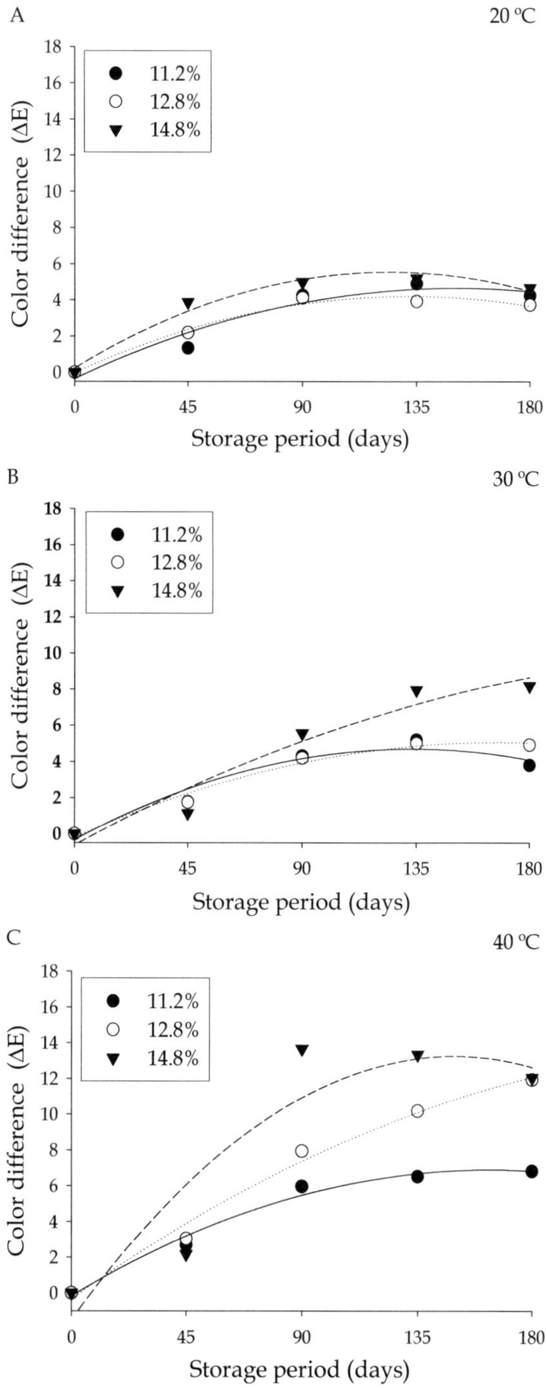

Source: ALENCAR et al. (2009)

Fig. 4. Regression curves of the color difference of the soybeans stored with moisture contents of 11.2. 12.8 and 14.8% (w.b.) at the temperatures of 20, 30 and 40 °C.

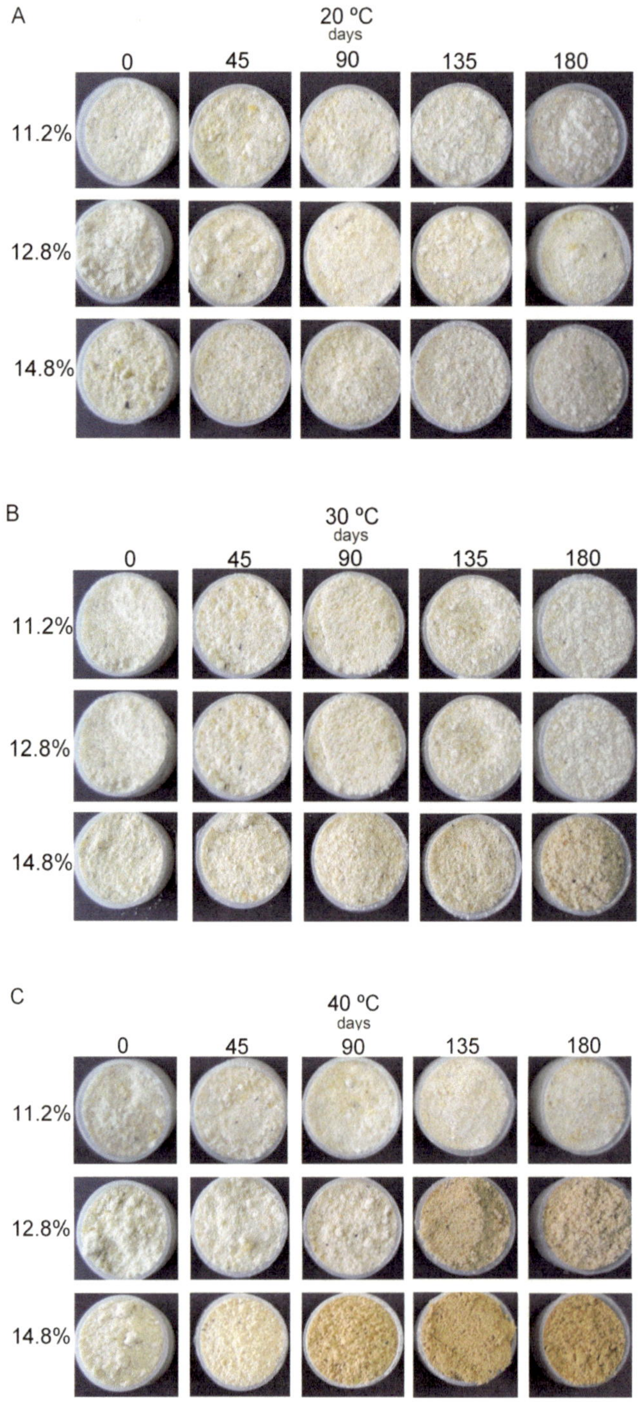

Source: ALENCAR (2006)

Fig. 5. Visual aspect of whole soybean flour obtained from soybeans with moisture contents of 11.2, 12.8 and 14.8% (w.b.) at the temperatures of 20, 30 and 40 °C, during storage.

Temperature (°C)	Moisture content (%)	**Regression equation adjusted**	R^2
20	11.2	$\hat{y} = -0.358 + 0.0663X - 0.0002X^2$	0.91
	12.8	$\hat{y} = -0.031 + 0.0639X - 0.0002X^2$	0.78
	14.8	$\hat{y} = 0.227 + 0.0852X - 0.0003X^2$	0.78
30	11.2	$\hat{y} = -0.329 + 0.0753X - 0.0003X^2$	0.86
	12.8	$\hat{y} = -0.202 + 0.0627X - 0.0002X^2$	0.91
	14.8	$\hat{y} = -0.619 + 0.0759X - 0.0001X^2$	0.92
40	11.2	$\hat{y} = -0.169 + 0.0864X - 0.0003X^2$	0.95
	12.8	$\hat{y} = -0.330 + 0.1020X - 0.0002X^2$	0.96
	14.8	$\hat{y} = -1.488 + 0.1972X - 0.0007X^2$	0.82

Source: ALENCAR et al. (2009)

Table 4. Regression equations adjusted for the color difference of the soybeans with moisture contents of 11.2, 12.8 and 14.8% (w.b.) at 20, 30 and 40 °C during the storage period (X), and respective coefficients of determination.

Component	Crude oil	Refined oil
Triglycerides (%)	95.0 - 97.0	99.0
Phosphatides (%)	1.5 - 2.5	0.003 - 0.045
Unsaponifiable matter (%)	1.6	0.3
Sterols (%)	0.33	0.13
Tocopherols (%)	0.15 - 0.21	0.11 - 0.18
Hydrocarbons (squalene) (%)	0.014	0.01
Free fatty acids (%)	0.3 - 0.7	< 0.05
Trace metals		
Iron (ppm)	1.0 - 3.0	0.1 - 0.3
Copper (ppm)	0.03 - 0.05	0.02 - 0.06

Source: ERICKSON & WIEDERMANN (1989)

Table 5. Principal components of crude and refined soy oil.

2.4.1 Qualitative parameters of oils and alterations resulting from storage conditions of the grains

In all stages of oil and fat processing various analysis are needed for quality control. In the refining process, for example, determining the percentage of free fatty acids is necessary in the neutralization step, or as a qualitative indicator (O'BRIEN, 2004). Other widely used analyses as quantitative indices of oils and fats are: peroxide value, iodine value, color, saponification number, water content and others.

2.4.1.1 Free fatty acids

During the storage of grains, the lipid fraction is slowly hydrolyzed by water at high temperature (physical process) or by natural lipolytic enzymes or those produced by bacteria and/or fungi, contributing to the hydrolytic rancidity of the product (ARAÚJO, 2004). Increase in the content of free fatty acids from lipids occurs by the action of lipase and phospholipase enzymes present in the soybeans or produced by the associated microflora, which contribute to the breaking of ester linkages of triglycerides (ZADERNOWSKI et al.,

1999). Thus, the percentage of free fatty acids is an important indicator of quality throughout the processing of oils and fats. O'BRIEN (2004) stated that hydrolytic rancidity can affect taste, odor and other characteristics of oil. This author stresses that vegetable oils may present relatively high contents of free fatty acids if the grains or seeds present damages due to procedures in the field or incorrect storage practices, being that high values of free fatty acids can cause excessive losses in refining. WILSON et al. (1995) claimed the refining losses between 1 and 1.5% are considered normal; however, such losses may reach 4% or more for greater levels of free fatty acids.

Many authors have related the increase in free fatty acid percentages to storage conditions. The variation in percent free fatty acids in crude oil extracted from soybeans stored with different moisture contents was observed by FRANKEL et al. (1987). Soybeans stored with 13% (w.b.) resulted in lower increases in the percentage of free fatty acids when compared with the values obtained by grains stored with 16 and 20% (w.b.) moisture content. With regards to the crude oil extracted from the soybeans stored at 13% (w.b.), it was verified that the increase in the free fatty acid percentage was from 0.2 to 1.25% after 49 days of storage; in the crude oil obtained from the grains stored with moisture contents of 16 and 20% w.b., increase was from 0.5 to 2.0% after 27 days and from 0.6 to 2.3% after 28 days, respectively. NARAYAN et al. (1988a) verified the increase in free fatty acid percentage in soybeans stored at different temperature conditions (between 16 and 40 °C) and relative humidity (between 50 and 90%), obtaining average values equal to 0.69, 4.32, 5.37 and 9.85% after 12, 24, 36 and 108 months of storage. ALENCAR et al. (2010) evaluated the effect of different combinations of temperature and moisture content on the percentage of free fatty acids of crude oil extracted from soybeans stored for 180 days. The authors adopted the grain moisture contents of 11.2, 12.8 and 14.8% and temperatures of 20, 30 and 40 °C (Fig. 6 and Table 6), and generally observed a significant increase in free fatty acid content of crude oil, except for the grains with moisture content of 11.2% at 20 °C. The increasing trend in the percentage of free fatty acids was more pronounced as water content and temperature increased, and the greatest percentage of free fatty acids from crude oil was 12.5% for grain stored at 14 8% moisture content after 180 days.

Temperature (°C)	Moisture content (%)	**Regression equation adjusted**	R^2
	11.2	$\hat{y} = 0.41$	
20	12.8	$\hat{y} = 0.370 + 0.0012X$	0.85
	14.8	$\hat{y} = 0.438 + 0.0069X$	0.80
	11.2	$\hat{y} = 0.352 + 0.025X$	0.82
30	12.8	$\hat{y} = 0.332 + 0.035X$	0.72
	14.8	$\hat{y} = 0.307+ 0.0251X$	0.86
	11.2	$\hat{y} = 0.277 + 0.0066X$	0.96
40	12.8	$\hat{y} = 0.440 + 0.0121X$	0.87
	14.8	$\hat{y} = -0.294 + 0.0692X$	0.84

Source: ALENCAR et al. (2010)

Table 6. Regression equations adjusted for free fatty acids of oil extracted from soybeans with moisture contents of 11.2, 12.8 and 14.8% (w.b.) at the temperatures of 20, 30 and 40 °C along the storage period (X), and respective coefficients of determination.

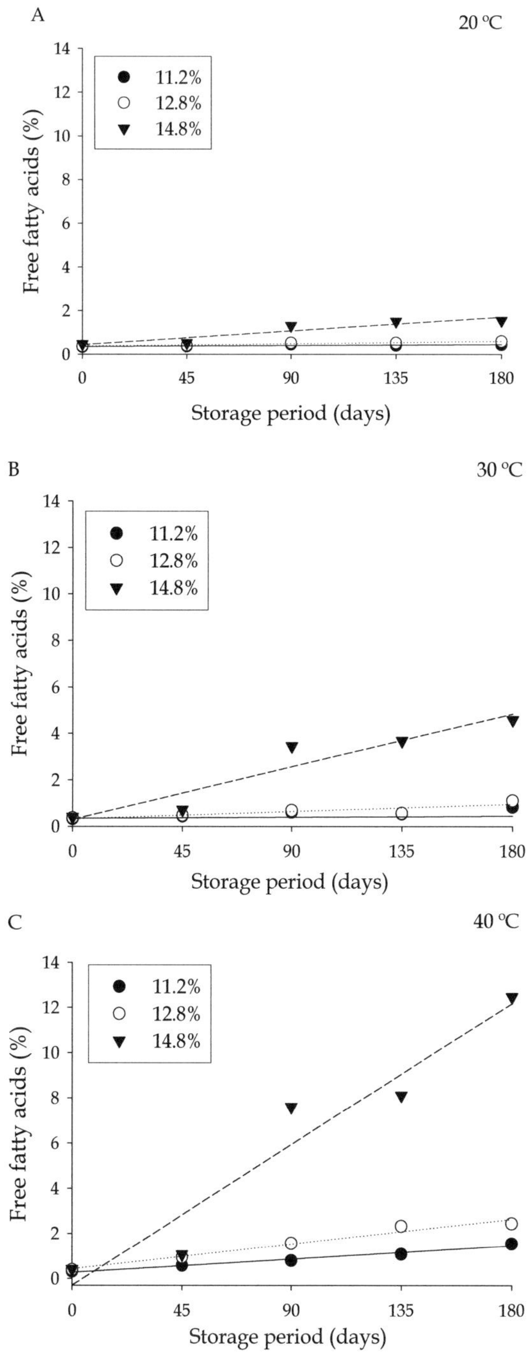

Source: ALENCAR et al. (2010)

Fig. 6. Regression curves of the percentage of free fatty acids (%) of crude oil extracted from soybeans stored with moisture contents of 11.2, 12.8 and 14.8% (w.b.) at temperatures of 20, 30 and 40 °C.

2.4.1.2 Peroxide index

Lipid oxidation is a spontaneous and inevitable phenomenon, according to SILVA et al. (1999), with direct implications on the market value of either the fatty bodies, or of all the products formulated from them, and peroxidation is the main cause of deterioration of fatty bodies (lipid materials and greases). It is the main cause of deterioration of oils and fats, and the hydroperoxides formed from the reaction between oxygen and unsaturated fatty acids are the primary products. Although these compounds do not exhibit taste or odor, they are rapidly decomposed even at room temperature into aldehydes, ketones, alcohols, hydrocarbons, esters, lactones and furans, causing unpleasant taste and odor in oils and fats (O'BRIEN, 2004; EYS et al., 2006). Other consequences of lipid oxidation in foods are changes in nutritional value, functionality, and also in the integrity and safety of the product via the formation of potentially toxic polymer compounds (SILVA et al., 1999; ARAÚJO, 2004; NAZ et al., 2004; RAMALHO & JORGE, 2006). According to HOU & CHANG (2004), the appearance of off-flavors (unpleasant aroma and taste) in soybean products can be partially attributed to lipid peroxidation.
One of the methods used to determine the degree of oxidation in fats and oils is the peroxide index. The peroxide index (PI) is a measure of oxidation or rancidity in its initial phase, as shown in Table 7 and measures the concentration of substances (in terms of milliequivalents of peroxide per thousand grams of sample) which oxidize potassium iodide to iodine is widely used in determining the quality of oils and fats, showing good correlation with taste (O'BRIEN, 2004).

Range	Degree of oxidation
<1	Freshness
$1< PI<5$	Low oxidation
$5<PI<10$	Moderate oxidation
$10<PI<20$	High oxidation
>20	Poor flavor

Source: O'BRIEN (2004)

Table 7. Classification of the degree of oxidation of soybean oil in accordance with the peroxide index (PI, meq kg^{-1}).

Works are encountered in literature that report the effect of different soybean storage conditions on the peroxide index of crude oil. NARAYAN et al. (1988a) studied the evolution of the peroxide index of crude oil extracted from soybeans stored at different temperatures and humidities. Average values observed for the peroxide index were 18, 40, 65 and 98 meq kg^{-1} after 12, 24, 36 and 108 months. ALENCAR et al. (2010) evaluated the peroxide value in crude oil obtained from soybeans stored with moisture contents of 11.2, 12.8 and 14.8%, at the temperatures of 20, 30 and 40 °C for 180 days. The authors verified an increase for all combinations of water content and temperature, where the highest values were obtained as the water content and temperature were increased (Table 8).

2.4.1.3 Color of the oil

The color and appearance of oils and fats, according to O'BRIEN (2004), are not monitored only due to the visual character, but also because they are related to the cost of processing and quality of the final product. Most oils present a reddish-yellow color as the result of the

Temperature (°C)	Moisture content (%)	Storage period (days) 0	45	90	135	180
	11.2	1.51	2.17	2.87	2.81	2.68
20	12.8	1.45	2.09	3.29	3.57	2.76
	14.8	1.52	3.58	4.33	3.62	7.84
	11.2	1.34	2.57	2.80	3.14	2.79
30	12.8	1.52	2.66	2.75	3.52	2.96
	14.8	1.48	4.25	3.51	5.89	8.09
	11.2	1.87	2.73	3.64	6.60	7.64
40	12.8	1.48	2.58	8.37	9.85	14.54
	14.8	1.30	4.47	11.77	13.88	14.76

Source: ALENCAR (2006)

Table 8. Average values of the peroxide index of crude oil extracted from soybeans stored at 20, 30 and 40 °C and moisture contents of 11.2, 12.8 and 14.8% (w.b.) during storage.

presence of carotenoids and chlorophyll. However, some crude oils may present a relatively high pigmentation due to damage of the raw material in the field, storage or processing failures; alterations in color indicate qualitative deterioration of the oil.
Alterations in color of the crude oil obtained from soybeans stored under different conditions were evaluated by ALENCAR et al. (2010). In this study different combinations of water content (11.2, 12.8 and 14.8%) and temperature (20, 30 and 40 °C) were obtained for the grains stored for 180 days, and the qualitative photometric index of the oil was analyzed. The authors observed a significant increase in the photometric color index for all combinations of water content and temperature, as for the temperature of 30 °C (Fig. 7 and Table 9). WILSON et al. (1995) associated an increase in the photometric color index to the percentage of grains damaged by fungi. It is emphasized that the degumming of crude oil extracted from seriously damaged grains is hampered and the refined oil is darker than that obtained from healthy kernels, as well as greater losses in refining (LIST et al., 1977).

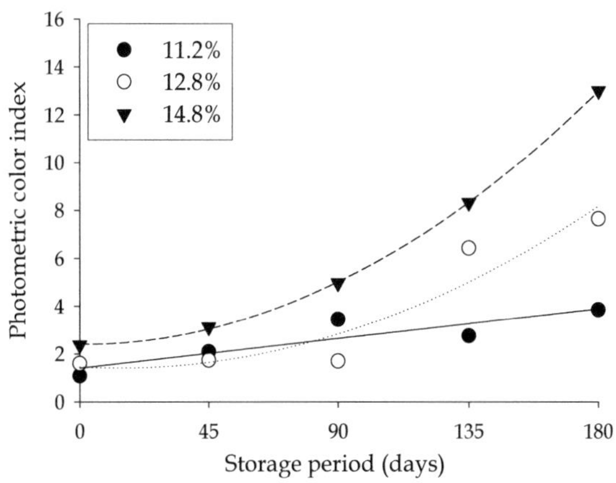

Source: ALENCAR et al. (2010)

Fig. 7. Regression curve of the photometric color index of crude oil extracted from soybeans stored with moisture contents of 11.2, 12.8 and 14.8% (w.b.) at the temperature of 30 °C.

Temperature (°C)	Moisture content (%)	**Regression equation adjusted**	R^2
30	11.2	$\hat{y} = 1.41 + 0.0138X$	0.46
	12.8	$\hat{y} = 1.46 - 0.0067X + 0.00024X^2$	0.74
	14.8	$\hat{y} = 2.40 - 0.0007X + 0.00033X^2$	0.85

Source: ALENCAR et al. (2010)

Table 9. Regression equations adjusted for the photometric color index of the oil extracted from the soybeans with moisture contents of 11.2, 12.8 and 14.8% (w.b.) at the temperature of 30 °C during the storage period (X), and respective coefficients of determination.

2.5 Effects of different storage conditions on the quality of other sub-products derived from soybeans

The quality of soybeans can also influence the qualitative parameters of other sub-products, including soymilk and tofu. SAIO et al. (1980) evaluated qualitative parameters of soymilk and tofu made from soybeans stored with a moisture content of 10.61%, and different combinations of temperature (15, 25 and 35 °C) and relative humidity (60 and 90%) during 180 days. The authors verified, resulting from the storage conditions adopted, significant changes in color and pH of the soymilk and hardness of tofu. The physicochemical quality of tofu, obtained from soybeans stored at temperatures of 3-4, 20 and 30 °C and relative humidities of 86.0, 57.0 and 84.0%, respectively, was assessed by HOU & CHANG (2004). For grains stored at 30 °C and relative humidity of 84.0%, the authors observed a reduction in yield (512g/100g of grains at time zero to 71g/100 g of grains after 7 months of storage) and alterations in texture with increasing hardness and color (Table 10). ACHOURI et al. (2008) evaluated the quality of the soymilk obtained from soybeans stored for 10 months at 18 °C and 50% relative humidity. Under these storage conditions the authors observed no significant change in the water uptake factor and pH of the soymilk, but there was significant variation in color and total volatiles. KONG et al. (2008) evaluated the physicochemical quality of soymilk and tofu made from soybeans stored with moisture contents between 6 and 14% in different combinations of temperature (40 to 50 °C) and humidity (55 to 80%). For the soymilk a decrease in pH and protein content was verified, this tendency being more accentuated as temperature and relative humidity increased. Reduction in the pH of the soymilk was observed for the soybeans stored under temperatures between 22 and 50 °C and relative humidity between 55.0 and 80.0%. It is highlighted that the protein content of the soymilk reduced by 24.0% in grains stored for 10 months at 40 °C. With regards to tofu, KONG et al. (2008) observed a significant reduction in yield for grains stored at 30 and 40 °C, as well as alterations in texture and color of the product. Also according to these authors, there is a strong relationship between the color of the grains and tofu, with respect to the Hunter (L, a and b) coordinates, as shown in Table 11.

3. Conclusion

The combination of high grain moisture and temperature during soybean storage accelerates the deterioration process of the sub-products of soybean. Proposed preventive measures of post-harvest handling to reduce risks of quality loss in soybean grains and sub-products are: store soybean grains with moisture content up to 15% (w.b.) at 20 °C without

Period (month)	Hunter L value	**Hunter a value**	Hunter b value
0	87.13±0.37[a]	-0.50±0.17[e]	13.57±0.26[b]
1	87.30±0.05[a]	-0.31±0.03[e]	13.57±0.09[b]
2	85.20±0.18[b]	0.50±0.25[d]	13.36±0.09[b]
3	85.13±0.12[b]	0.71±0.10[d]	13.95±0.13[b]
4	84.09±0.41[c]	1.06±0.17[c]	13.11±0.23[b]
5	83.60±0.41[d]	1.04±0.04c	12.63±0.25[c]
6	82.30±0.22[e]	1.71±0.06[b]	12.97±0.23[bc]
7	70.14±0.49[f]	4.56±0.12[a]	15.90±0.04[a]

Values followed by the same letter in the column do not differ statistically at 5% probability
Source: HOU & CHANG (2004)

Table 10. Tofu color obtained from soybeans stored at 30 °C and 84% relative humidity.

Coordinate	**Adjusted equations**	R^2
L	$\hat{y} = 19.275 + 1.1084X$	0.623
a	$\hat{y} = 9.0894 + 1.3217X$	0.546
b	$\hat{y} = 9.929 + 0.4453X$	0.125

Source: KONG et al. (2008)

Table 11. Regression equations which relate the Hunter L, a and b coordinates of tofu (y) and soybean grains (X).

risk of deterioration up to 180 days; in regions with temperatures around 30 °C, store soybean with moisture content up to 13% (w.b.); do not store soybean with moisture content above 11% (w.b.) in regions where the grain mass temperature can reach 40 °C with the risk of accelerating deterioration of grains and sub-products.

4. Acknowledgements

Part of the data presented in this chapter was obtained during Master's research of the first author, as part of the graduate program in Agricultural Engineering of the Universidade Federal de Viçosa (UFV).

5. References

Abba, E.J. & Lovato, A. (1999). Effect of seed storage temperature and relative humidity on maize (*Zea mays* L.) seed viability and vigour. *Seed Science and Technology*, Vol.27, No.1, (April 1999), pp. 101-114, ISSN 0251-0952

Acasio, A. (September 2010). *Handling and storage of soybeans and soybean meal*, 21.09.2010, Available from http://www.feedmachinery.com

Achouri, A.; Boye, J.I. & Zamani, Y. (2008). Soybean variety and storage effects on soymilk flavour and quality. *International Journal of Food Science and Technology*, Vol.43, No.1, (January 2008), pp. 82-90, ISSN 1365-2621

Alencar, E.R. (2006). *Efeitos das condições de armazenagem sobre a qualidade da soja (Glycine max (L.) Merrill) e do óleo bruto*. Dissertação, Mestrado em Engenharia Agrícola, Universidade Federal de Viçosa, Viçosa, Brasil

Alencar, E.R.; Faroni, L.R.D; Lacerda Filho, A.F.; Garcia, L.F.& Meneghitti, M.R. (2008). Qualidade fisiológica dos grãos de soja em função das condições de armazenamento, *Engenharia na Agricultura*, Vol 16, No.3, (April 2008), pp. 155-166, ISSN 1414-3984

Alencar, E.R.; Faroni, L.R.D; Lacerda Filho, A.F.; Peternelli, L.A. & Costa, A.R. (2009). Qualidade dos grãos de soja armazenados em diferentes condições, *Revista Brasileira de Engenharia Agrícola e Ambiental*, Vol.13, No.5, (September 2009), pp. 606-613, 2009, ISSN 1807-1929

Alencar, E.R.; Faroni, L.R.D.; Peternelli, L.A.; Silva, M.T.C. & Costa, A.R. (2010). Influence of soybean storage conditions on crude oil quality, *Revista Brasileira de Engenharia Agrícola e Ambiental*, Vol.14, No.3, (March 2010), pp. 303-308, ISSN 1807-1929

Al-Yahya, S.A. (2001). Effect of storage conditions on germination in wheat. *Journal Agronomy & Crop Science*, Vol.186, No.4, (June 2001), pp. 273-279, ISSN 0931-2250

Araújo, J.M.A. (2004). *Química de Alimentos: Teoria e Prática*, Editora UFV, ISBN 978-85-7269-351-6, Viçosa, Brasil

Bailey, J.E. (1974). Whole grain storage, In: *Storage of cereal grains and their products*, C.M. Christensen, (Ed.), 333-360, AACC, ISBN 0-913250-05-8, St. Paul, United States

Bhattacharya, K. & Raha, S. (2002). Deteriorative changes of maize, groundnut and soybean seeds by fungi in storage. *Mycopathologia*, Vol.155, No.3, (November 2002), pp. 135-141, ISSN 0301-486X

Brooker, D.B.; Bakker-Arkema, F.W. & Hall, C.W. (1992). *Drying and storage of grains and oilseeds*, Springer, ISBN 0442205155, New York, United States

Delouche, J. (June 2006). *Germinação, deterioração e vigor da semente*, 26.06.2006, Available from http://www.seednews.inf.br/portugues/seed66/artigocapa66.shtml

Erickson, D.R. & Wiedermann, L.H. (April 2006). *Soybean oil: modern processing and utilization*, 03.04.2006, Available from http://www.asaim-europe.org/Backup/pdf/sboprocess.pdf

Eys, J.E.; Offner, A. & Bach, A. (June 2006). *Manual of quality analyses for soybean products in the feed industry*, 12.06.2006, Available from http://www.asaim-europe.org/Backup/Library/library_e.htm

Frankel, E.N.; Nash, A.M. & Snyder, J.M. (1987). A methodology study to evaluate quality of soybeans stored at different moisture levels. *Journal of the American Oil Chemists' Society*, Vol.64, No.7, (July 1987), pp. 987-992, ISSN 0003-021X

Gungadurdoss, M. (2003). Improvement of seed viability of vegetable soybean (*Glycine max* (L) Merrill). In: Food and Agricultural Research Council, 2003, Réduit, Mauritius. Proceedings...: Réduit, Mauritius: Lalouette, J.A., BACHRAZ (Eds.), 117-123.

Hansen, L.S.; Skovgard, H. & Hell, K. (2004). Life table study of *Sitotroga cerealella* (Lepidoptera: Gelichiidae), a strain from West Africa. *Journal of Economic Entomology*, Vol.97, No.4, (August 2004), pp. 1484-1490, ISSN 1938-291X

Heslehurst, M.R. (1988). Quantifying initial quality and vigour of wheat seeds using regression analysis of conductivity and germination data from aged seeds. *Seed Science and Technology*, Vol.16, No.1, (April 1999), pp. 75-85, ISSN 0251-0952

Hou, H.J. & Chang, K.C. (1998). Yield and quality of soft tofu as affected by soybean physical damage and storage. *Journal of Agricultural and Food Chemistry*, Vol.46, No.12, (November 1998), pp. 4798-4805, ISSN 0003-021X

Hou, H.J. & Chang, K.C. (2004). Storage conditions affect soybean color, chemical composition and tofu qualities. *Journal of Food Processing and Preservation,* Vol.28, No.6, pp. 473-488, (December 2004), ISSN 0145-8892

Hummil, B. C.W.; Cuendit, L. S.; Christensen, C.M. & Geddes, W.F. (1954). Grain storage studies XIII; comparative changes in respiration, viability, and chemical composition of mold-free and mold-contaminated wheat upon storage. *Cereal Chemistry,* Vol.31, No.2, (March 1954), pp. 143-150, ISSN 0009-0352

Karunakaran, C.; Muir, W.E.; Jayas, D.S.; White, N.D.G. & Abramson, D. (2001). Safe storage time of high moisture wheat. *Journal of Stored Products Research,* Vol.37, No.3, (July 2001), pp. 303-312, ISSN 0022-474X

Kong, F.; Chang, S.K.C.; Liu, Z. & Wilson, L.A. Changes of soybean quality during storage as related to soymilk and tofu making. *Journal of Food Science,* Vol. 73, No.3, (April 2008), pp. S134-S144, ISSN 1750-3841

Lin, S.S. (1990). Alterações na lixiviação eletrolítica, germinação e vigor da semente de feijão envelhecida sob alta umidade relativa do ar e alta temperatura. *Revista Brasileira Fisiologia Vegetal,* Vol.2, No.2, (May 1990), pp. 1-6, ISSN 1806-9355

List, G.R.; Evans, C.D.; Warner, K.; Beal, R.E.; Kwolek, W.F.; Black, L.T. & Moulton, K.J. (1977). Quality of oil from damaged soybeans. *Journal of the American Oil Chemists' Society,* Vol.54, No.1, (January 1977), pp. 8-14, ISSN 0003-021X

Liu, K. (1997). *Soybeans: chemistry, technology and utilization,* Chapman & Hall, ISBN 0834212994, New York, United States

Locher, R. & Bucheli, P. (1998). Comparison of soluble sugar degradation in soybean seed under simulated tropical storage conditions. *Crop Science,* Vol.38, No.5, (September 1998), pp. 1229-1235, 0011-183X

MAPA. Ministério da Agricultura, Pecuária e Abastecimento. (2007). *Instrução Normativa Nº 11, de 15 de maio de 2007,* Brasília, Brasil

Mason, L.J. (May 2006). *Indianmeal moth Plodia interpunctella (Hubner),* 30.05.2006, Available from http://www.entm.purdue.edu/entomology/ext/targets/e-series/EseriesPDF/E-223.htm

Mbata, G.N. & Osuji, F.N.C. (1983). Some aspects of the biology of *Plodia interpunctella* (Hübner) (Lepidoptera: Pyralidae), a pest of stored groundnuts in Nigeria. *Journal of Stored Products Research,* Vol.19, No.3, (July 1983), pp. 141-151, ISSN 0022-474X

Moretto, E. & Fett, R. (1998). *Tecnologia de oleos e gorduras vegetais,* Varela, ISBN 85-85519-41-X, São Paulo, Brasil

Narayan, R.; Chauhan, G.S. & Verma, N.S. (1988a). Changes in the quality of soybean during storage. Part 1 - Effect of storage on some physico-chemical properties of soybean. *Food Chemistry,* Vol.27, No.1, (January 1988), pp. 12-23, ISSN 0308-8146

Narayan, R.; Chauhan, G.S. & Verma, N.S. (1988b). Changes in the quality of soybean during storage. Part 2 - Effect of soybean storage on the sensory qualities of the products made therefrom. *Food Chemistry,* Vol.30, No.3, (December 1988), pp. 181-190, ISSN 0308-8146

Naz, S.; Sheikh, H.; Siddiqi, R. & Sayeed, S.A. (2004). Oxidative stability of olive, corn and soybean oil under different conditions. *Food Chemistry,* Vol.88, No.2, , (November 2004), pp. 253-259, ISSN 0308-8146

O'Brien, R.D. (2004). *Fats and Oils Formulating and Processing for Applications*, CRC Press, ISBN 0849315999, Boca Raton, United States

Orthoefer, F.T. (1978). Processing and utilization. In: *Soybean: physiology, agronomy and utilization*, A.G. Norman, (Ed.)., 219-246, Academic Press, ISBN 0-12-521160-0, New York, United States

Padín, S.; Bello, G.D. & Fabrizio, M. (2002). Grain loss caused by *Tribolium castaneum, Sitophilus oryzae* and *Acanthoscelides obtectus* in stored durum wheat and beans treated with *Beauveria bassiana. Journal of Stored Products Reseach,* Vol.38, No.1, (January 2002), pp. 69-74, ISSN 0022-474X

Ramalho, V.C. & Jorge, N. (2006). Antioxidantes utilizados em óleos, gorduras e alimentos gordurosos. *Química Nova,* Vol.29, No.4, (July 2006), pp. 755-760, ISSN 0100-4042

Saio, K.; Nikkuni, I.; Ando, Y.; Osturu, M.; Terauchi, Y. & Kito, M. (1980). Soybean quality changes during model storage studies. *Cereal Chemistry,* Vol.57, No.2, (March 1980), pp. 77-82, ISSN 0009-0352

Santos, C.M.R.; Menezes, N.L. & Villela, F.A. (2004). Alterações fisiológicas e bioquímicas em sementes de feijão envelhecidas artificialmente. *Revista Brasileira de Sementes,* Vol.26, No.1, (July 2004), pp. 110-119, ISSN 0101-3122

Silva, F.A.M.; Borges, M.F.M. & Ferreira, M.A. (1999). Métodos para avaliação do grau de oxidação lipídica e da capacidade antioxidante. *Química Nova,* Vol.22, No.1, (January 2006), pp. 94-103, ISSN 0100-4042

Silva, J.S.; Berbert, P.A.; Rufato, S. & Afonso, A.D.L. (2008). Indicadores da qualidade dos grãos. In: *Secagem e Armazenagem de Produtos Agrícolas,* J.S. Silva. (Org.), 63-108, ISBN 9788562032004, Aprenda Fácil, Viçosa, Brasil

Sinclair, J.B. (1995). Reevaluation of grading standards and discounts for fungus-damaged soybean seeds. *Journal of the American Oil Chemists' Society,* Vol.72, No.12, (December 1995), pp. 1415-1419, ISSN 0003-021X

Sinha, R. N. & Muir, W.E. (1973). *Grain storage: part of a system,* Avi Pub. Co., ISBN 087055123X, Westport, United States

USDA. (May 2006). *U.S. soybean inspection,* 31.05.2006, Available from http://www.usda.gov/gipsa/reference-library/brochures/soyinspection.pdf

USDA. (March 2011). *U.S. soybean inspection,* 14.03.2011, Available from http://www.usda.gov/oce/commodity/wasde/latest.pdf

Vieira, R.D.; Penariol, A.L.; Perecin, D. & Panobianco, M. (2002). Condutividade elétrica e teor de água inicial das sementes de soja. *Pesquisa Agropecuária Brasileira,* Vol.37, No.9, (September 2002), pp. 1333-1338, ISSN: 0100-204X

Wilson, R.F.; Novitzky, W.P. & Fenner, G.P. (1995). Effect of fungal damage on seed composition and quality of soybeans. *Journal of the American Oil Chemists' Society,* Vol.72, No.12, (December 1995), pp. 1425-1429, ISSN 0003-021X

Zadernowski R.; Nowak-Polakowska H. & Rashed, A.A. (1999). The influence of heat treatament on the activity of lipo and hydrophilic components of oat grain. *Journal of Food Processing and Preservation,* Vol.23, No.3, pp. 177-191, (September 1999), ISSN 0145-8892

3

High Pressure Treatments of Soybean and Soybean Products

Kashif Ghafoor[1], Fahad Y.I. AL-Juhaimi[1] and Jiyong Park[2]
[1]*Department of Food and Nutrition Sciences, King Saud University, Riyadh,*
[2]*Department of Biotechnology, Yonsei University, Seoul 120-749,*
[1]*Saudi Arabia*
[2]*South Korea*

1. Introduction

Soybean (*Glycine max*) is an industrial crop extensively cultivated for its oil and protein content. The global demand for soybean has increased dramatically over the last few years. Since the application of high hydrostatic pressure (HHP) to different food systems in the late 1800s (Bridgman, 1914), many researchers today are again applying this promising technology to the processing of foods. There is increasing worldwide interest in the use of HHP because of the advantages of this technology over other methods of processing and preservation. HHP offers homogeneity of treatment at every point in the product due to the fact the applied pressure is instantaneously and uniformly distributed within the HHP chamber. Therefore, processing time is not a function of sample size. Important advantages in using this technology are (1) significant or total inactivation of microorganisms (Knorr, 1993), and (2) better functional and nutritional retention of ingredients in the processed products, with improved food quality parameters (Hayashi, 1989). In addition, there is significant energy economy in comparison to thermal stabilization techniques, because once the desired pressure is reached, it can be maintained without the need for further energy input. Recently, processing of foods with HHP and low to moderate temperatures (less than 70 °C) was introduced as an alternative to thermal preservation. However, it was not until the late 1980s that researchers began investigating ways to commercialize high pressure treatment of foods (Hayashi, 1989).

The main uses of soybeans can be categorized into three groups: industrial, human food and livestock feed. Soybeans for human consumption are processed in many forms. Of major importance in Asian countries are soy foods such as tofu, soy sauce, miso, soy sprouts, and soymilk. These soy foods were recently introduced to the American market as were soy flour and soymilk previously. Soy flour is often mixed with other flours to increase the protein content, and soymilk provides an alternative source of protein for people allergic to the protein in cow's milk (Riaz, 1988). In addition to its versatility, the soybean is a commodity of unique chemical composition. On a mean dry matter basis, soybeans contain about 40% protein and 20% oil. Soybeans contains the highest protein content among food crops, and are second highest with respect to oil content among all the food legumes. Thus, the composition of soy products range in protein content from about 40% for full-fat flours to 95% or more for protein isolates (Wolf & Cowan, 1975).

2. High pressure processing of soybean

Due to the great variety of foods obtained from soybeans, different processing methods are required. Traditional methods include two types of processes: (1) fermentation, which uses fungi to produce fermented products; and (2) soaking and grinding of the soybeans to make bean curd and soymilk. During fermentation, protein is digested into peptides and later into amino acids for increasing digestibility of protein by the human body. Soaking and grinding are usually combined with thermal treatment to inactivate biologically active compounds such as trypsin inhibitors, lipoxygenase, and hemagglutinins, while increasing digestibility of proteins. In addition, thermal treatment (continued steaming) helps to diminish the characteristic beany flavor of raw soy products due to the volatilization of monocarbonyl compounds, which results from oxidation of fatty acids by the enzyme lipoxygenase. However, excessive heating may destroy certain amino acids that are sensitive to heat such as lysine, with losses of possibly more than 50% (Estrada-Giron et al., 2005).

Until recently, little research was being done on the effects of HHP on soybean grains and their sub-products. This is because of the recent interest in the use of HHP as a potential technology to improve the quality of cereals and textured products. These studies include reduction in the microbial population of soymilk curd, commonly known as tofu, to obtain a product with longer shelf life and to avoid secondary contamination. The solubilization of protein from whole soybean grains subjected to different treatments of pressure, time, and temperature was also reported. Additional information is available in a more extensive context about the effects of this technology on the inactivation of pure soybean lipoxygenase and lipoxygenase from some legumes (Estrada-Giron et al., 2005).

2.1 Microbial inactivation

The effects of high hydrostatic pressure on microbial inactivation depend on several factors such as type of microorganism, extent and duration of the high pressure treatment, temperature, and composition of suspension media or food. Therefore, suitable pressure treatment should be applied taking into account these factors to assure microbial inactivation of pathogenic, spoilage, and vegetative cells present in foods. Prestamo et al. (2000) reported that the microbial population of tofu pressurized at 400 MPa and 5 °C for 5, 30, and 45 min decreases from an initial microorganism count of 5.54×10^4 cfu/g to 0.31, 1.56, or 2.38 log units, respectively. Prestamo et al. (2000) also postulated that the effectiveness of HHP treatment to reduce microbial population at 400 MPa largely depends on the exposure time (Fig. 1). In the same study, after HHP treatment of tofu, psycrotrophs were reduced 2 log units from an initial population of 1×10^3 cfu/g. Mesophilic microorganisms were reduced 1 log unit from an initial number of 1.6×10^3 cfu/g, whereas yeast and molds decreased from an initial population of 2.64×10^3 cfu/g to 1×10^2 cfu/g. Other microorganisms such as *Pseudomonadaceae*, *Salmonella*, and Gram-negative bacteria (confirmed before HHP treatment) were not detected after HHP treatment of tofu. *Yersenia enterocolitica* and *Listeria monocytogenes*, which are more resistant to high pressure, were not found before and after HHP treatment. *Hafnia halvei* and *Bacillus cereus* remained active after high pressure treatment of tofu.

In addition to temperature and the extent and duration of high pressure treatment, a factor that significantly influences the effectiveness of HHP treatment on the inactivation and consequently the reduction in microbial population is the medium composition in which microorganisms are dispersed.

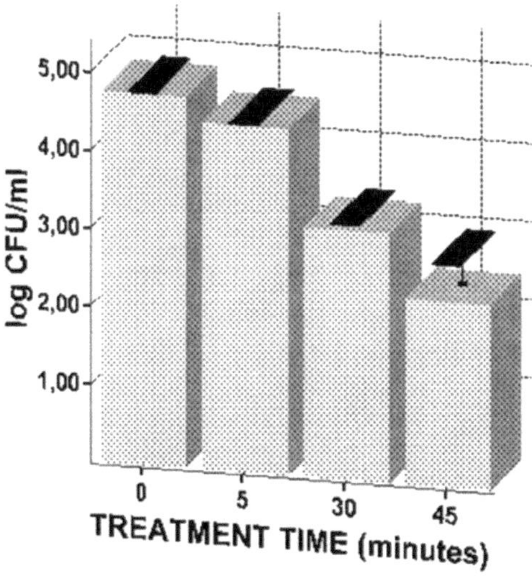

Fig. 1. Viable aerobic mesophilic population in tofu after treatment at 400 MPa and 5 °C for 5, 30, and 45 min (from Prestamo et al., 2000).

Food constituents such as sucrose, fructose, glucose, and salts affect the baro-resistance of microorganisms present in food (Oxen & Knorr, 1993). This effect is often observed since food constituents appear to protect microorganisms from the effects of high pressure. Therefore, a non-nutritive solution can reduce the microorganism's baro-tolerance. The presence of microorganisms such as *Hafnia halvei* and *Bacillus cereus* that remained active after HHP treatment could explain the baro-protective effect that food components exert over the extent on microbial reduction (Prestamo et al., 2000).

2.2 Proteins

Unlike allergenic proteins in cereals such as rice, soybeans contain a large number of proteins with important functional properties (Wolf & Cowan, 1975). Eighty percent of the proteins in soybeans are glycinin and β- and γ-conglycinin, which are globular salt-soluble proteins. On the basis of their sedimentation constants at pH 7.6 and ionic strength buffer of 0.5, the globulins are characterized as 11S or glycinin and 7S or β- and γ-conglycinin (Fukushima, 1991), with other less abundant globulins including 2S or α-conglycinin, 9S globulins, and 15S globulins. Functional properties associated with these kinds of soybean proteins:

1. Hydration properties such as swelling, solubility, and viscosity.
2. Protein–protein interactions resulting in precipitation and gelling.
3. Interfacial properties identified as surface tension and related foam/emulsion stability (Utsumi et al., 1998).

Therefore, the method of processing intact soybeans is important since the retention of proteins in the soybean seed is of special interest because of the high-quality vegetable protein, which also contains most of the essential amino acids (Steinke et al., 1992). When soybeans are immersed in hot water at 50–60 °C for 1 h, a considerable amount of protein solubilized from the soybean seeds is released to the surrounding water (Asano et al., 1989). Later studies identified these solubilized proteins as 7S globulins, which accounted for about 3% of the total protein in mature soybean seeds (Hirano et al., 1992).

In soybean seeds immersed in distilled water and treated at 300 MPa and 20 °C for 0–180 min, the solubilized proteins accounted for 0.5–2.5% of the total seed proteins. No apparent changes in shape, color, and size between treated and untreated soybeans were reported. The solubility of protein in surrounding water increased with increasing pressure, reaching a maximum value at 400 MPa (Fig. 2). Similar to what occurs when heat treatment is applied, SDS-PAGE patterns of high pressure treated seeds exhibited solubilization of 7S globulin, consisting of 27 and 16 KDa bands with staining intensity increasing as pressure increased to 400 MPa. The increase in staining intensity is indicative of the amount of release protein; thus, at higher intensity larger amounts of protein are released. At 700 MPa, 11S glycinin and 2S β-conglycinin also increased their staining intensity (Omi et al., 1996).

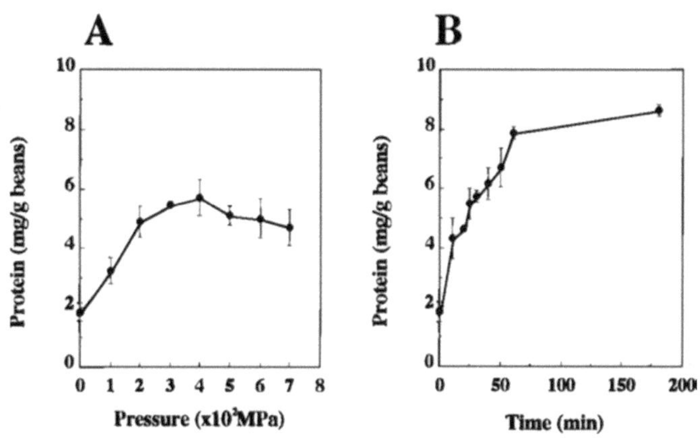

Fig. 2. Effect of high-pressure treatment on the release of proteins from soybean seeds. Water-immersed soybean seeds were pressurized at (A) 0-700 MPa and 20 °C for 25 min and at (B) 300 MPa and 20 °C for 0-180 min (From Omi et al., 1996).

2.3 Enzymes

At present, research regarding the inactivation of enzymes in intact grains and their sub-products is scarce. However, it is well known that high pressure modifies the activities of a whole range of unwanted food enzymes, which can result in a reduction in food quality or cause spoilage during storage. Recent investigations have reported the effect of combined pressure and temperature on soybean lipoxygenase. Lipoxygenase is one of the main anti-nutritional factors in soybean processing, which is also known to occur in other legume seeds, some cereal grains, and oil seeds. At least three types of lipoxygenase are well identified in soybeans as lipoxygenase I, II, and III. These enzymes catalyze the oxidation of unsaturated fatty acids in the presence of molecular oxygen. The presence of lipoxygenase can have detrimental effects on foods, for example:

1. Degradation of the essential fatty acids linoleic, linolenic, and arachidonic acid to yield fatty acid hydroperoxides.
2. Degradation of formed hydroperoxides, resulting in the formation of volatile compounds such as aldehydes, ketones, and alcohols, which cause the development of off-flavors.
3. Production of free radicals that can damage other compounds, including vitamins and proteins (Whitaker, 1972).

In soybean products, off-flavor development is highly dependent on the action of lipoxygenase since subsequent decomposition of the resulting hydroperoxides yields especially rancid flavor and beany aroma. Nevertheless, lipoxygenase is sensitive to heat and is destroyed at 82 °C when processed for 15 min (Baker & Mustakas, 1972). It is well known that thermal processing methods reduce considerably or completely inactivate unwanted enzyme activity, which limits or largely determines the conditions of storage needed to extend shelf life of food products. Although the behavior of enzymes under the influence of heat has been extensively studied, the effects of HHP treatment on enzyme inactivation are not clearly understood.

P/T treatment (MPa/°C)	Total time (min)	Cycling time	Activity retention
350/40	40	1×40	0.709
350/40	40	4×10	0.314
400/35	40	4×10	0.324
400/40	40	4×10	0.300
450/40	40	1×40	0.481
475/10	40	1×40	0.652
475/25	60	4×15	0.160
475/30	60	4×15	0.373
500/15	30	1×30	0.367
500/15	30	3×10	0.018
525/25	20	1×20	0.099
525/25	20	2×10	0.110

Adapted from Ludikhuyze et al. (1998a).

Table 1. Influence of multi-cycling on the inactivation of lipoxygenase in Tris-HCl at pH 9

Thermal inactivation of enzymes at atmospheric pressure occurs in the temperature range 60–70 °C. In contrast, pressure-temperature inactivation occurs in the pressure range 50-650 MPa at temperatures between 10 and 64 °C. Also, depending on the objectives of the research, pressure treatment may be applied in a single cycle or multi-cycles. Multi-cycling is the multiple application of pressure alone or in combination with temperature for the same total treatment time but with various numbers of cycles. Ludikhuyze et al. (1998a) reported the multi-cycling application of pressure to inactivate lipoxygenase. These authors found that in the pressure range 350-525 MPa and thermal treatment at 10-40 °C, the use of multi-cycles exerted an additional inactivation effect on lipoxygenase, compared to single cycle treatments (Table 1). Furthermore, temperature treatments at 10 °C caused an enhanced inactivation of lipoxygenase because the temperature inside the vessel dropped below zero upon depressurization.

In crude green bean extract, irreversible lipoxygenase inactivation was reported in the temperature range 55-70 °C at ambient pressure, whereas at room temperature, pressures around 500 MPa were required to inactivate lipoxygenase. High pressure treatment at 200 MPa and 50 °C resulted in 10% inactivation, while at least 50%, lipoxygenase inactivation occurred at pressures greater than 500 MPa and thermal treatment between 10 and 30 °C (Indrawati et al., 1999). The effect of HHP on enzyme inactivation in food systems is different compared to its effects on pure components dissolved in buffer solutions. As an example, solutions of commercial soybean lipoxygenase type I (100 mg/ml) dissolved in 0.2 M citrate-phosphate (pH range of 4.0-9.0) and 0.2 M Tris buffer (pH range of 6.0-9.0) were subjected to pressures of 0.1, 200, 400, and 600 MPa for 20 min. Under these conditions, lipoxygenase in citrate-phosphate buffer lost more than 80% of its activity at alkaline pH, whereas it was completely inactivated at acidic conditions and pressure treatment of 400 and 600 MPa (Tangwongchai et al., 2000). In Tris buffer, lipoxygenase activity was significantly inactivated at pH 9.0 and 400 MPa and lost all activity at 600 MPa and all pH values. Similar results were observed by Seyderhelm et al. (1996) who reported that lipoxygenase in Tris buffer pH 7.0 was completely inactivated at 600 MPa and temperatures 45 and 50 °C for 10 min and 5–10 min, respectively.

3. Effect of HHP on technological properties of tofu

Tofu with 0, 2.5 or 5% trehalose was pressurized at 100–686 MPa and approximately -20 °C for 60 min to determine changes in temperature and sensory evaluation of high-pressure-frozen tofu as affected by trehalose.

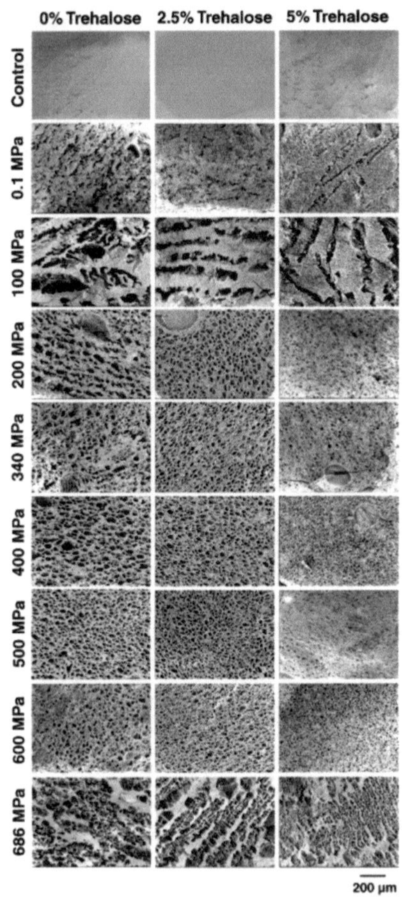

Fig. 3. Cryo-scanning electron micrographs of tofu frozen at high pressure. Tofu was pressurized at 100–686 MPa and approximately −20 °C for 60 min. After decompression, it was left in the pressure medium for 30 min then thawed at 20 °C. Control: unfrozen tofu (From Fuchigami et al., 2002).

Tofu froze during pressurization at 100 or 686 MPa; conversely, tofu did not freeze between 200 and 600 MPa and -20 °C, but it froze rapidly when the pressure was released. It was found that tofu frozen at 0.1, 100 or 686 MPa had larger ice crystals and was firmer (less like unfrozen tofu) than tofu frozen at 200–600 MPa. In the sensory evaluation, results showed that mouth feel (texture. of tofu frozen at 400 MPa) was more like the control when 2.5% trehalose was added (Fuchigami et al., 2002).

The micro structure of the tofu gel network high pressure frozen at 686 MPa was compared with untreated tofu (Fig. 3). Tofu (0% trehalose) frozen at 0.1–500 MPa maintained a comparatively coarse network (data not shown), but tofu gel frozen at 686 MPa was compressed. Compression of the protein gel network might have occurred above 600–686 MPa; however, the gel network in tofu frozen at 686 MPa became coarse with the addition of trehalose. This indicates that trehalose with high-pressure-freezing appears to protect against compression (effects of concentration of protein and coagulants on frozen tofu) (Fuchigami et al., 2002).

4. High pressure inactivation of soybean lipoxygenase

The high pressure inactivation of lipoxygenase in soy milk and crude soybean extract was studied in the pressure range 0.1–650 MPa with temperature varying from 5 to 60 °C.

T (°C)	Soy milk	Crude soybean extract
63	0.55 ± 0.02[a] r^2 = 0.993	0.68 ± 0.03 r^2 = 0.995
65	1.35 ± 0.05 r^2 = 0.995	1.54 ± 0.05 r^2 = 0.996
67	3.57 ± 0.09 r^2 = 0.998	4.31 ± 0.12 r^2 = 0.997
69	13.25 ± 0.50 r^2 = 0.994	14.43 ± 0.42 r^2 = 0.997
71	47.72 ± 4.64 r^2 = 0.972	53.40 ± 2.74 r^2 = 0.987
E_a (kJ/mol)	538.78 ± 29.04 r^2 = 0.991	526.94 ± 29.54 r^2 = 0.991

[a] Standard error. (From Wang et al., 2008).

Table 2. Estimated inactivation rate constants ($\times 10^{-2}$ min^{-1}) for the isothermal inactivation of lipoxygenase in soy milk and in crude soybean extract

For both systems, the isobaric–isothermal inactivation of lipoxygenase was irreversible and followed a first-order reaction at all pressure–temperature combinations tested. In the entire pressure–temperature area studied, the lipoxygenase inactivation rate constants increased with increasing pressure at constant temperature for both systems; the rate constants were somewhat smaller in soy milk system than in crude soybean extract. At constant elevated pressure, lipoxygenase exhibited the greatest stability around 20 °C in both systems, indicating that the Arrhenius equation was not valid over the entire temperature range. For both systems, the temperature dependence of the lipoxygenase inactivation rate constants at high temperature decreased with increasing pressure, while the highest sensitivity of the lipoxygenase inactivation rate constants to pressure was observed at about 30 °C. The pressure–temperature dependence of the lipoxygenase inactivation rate constants was successfully described either using an empirical mathematical model or using a thermodynamic kinetic model for both systems. On a kinetic basis, neither the reaction order of inactivation nor the pressure and temperature sensitivities of the inactivation rate constants were influenced by the different levels of food complexity between the two systems (Wang et al., 2008).

In 63 to 71 °C temperature range, isothermal inactivation of soybean lipoxygenase followed first-order kinetics, allowing inactivation rate constants (*k*) to be determined from plots of the natural logarithm of relative residual activity, as a function of inactivation time. The estimated *k* values, together with standard errors and regression coefficients, are summarized in Table 2. Over the entire temperature domain studied, lipoxygenase was less thermostable in crude soybean extract than in soy milk and the temperature sensitivity of the rate constants for lipoxygenase inactivation in both systems could be estimated using the Arrhenius relation (Wang et al., 2008). First-order kinetics for thermal inactivation of soybean lipoxygenase has been frequently reported in the literature (Indrawati et al., 1999; Ludikhuyze et al., 1998b). Ludikhuyze et al. (1998a, 1998c) investigated the thermal inactivation kinetics of commercial soybean lipoxygenase in Tris–HCl buffer (0.01 M, pH 9) at two different concentrations (0.4 and 5 mg/ml) over the temperature range 60–70 °C (Table 3).

T (°C)	Lipoxygenase concentration	
	0.4 mg/ml[b]	5 mg/ml[c]
60		2.09 ± 0.13 r^2 = 0.978
62	2.02 ± 0.09[a] r^2 = 0.987	4.86 ± 0.21 r^2 = 0.991
64	4.94 ± 0.16 r^2 = 0.993	10.8 ± 0.61 r^2 = 0.984
66	9.18 ± 0.32 r^2 = 0.992	29.1 ± 3.37 r^2 = 0.949
68	15.5 ± 0.52 r^2 = 0.992	
E_a (kJ/mol)	319.8 ± 27.3 r^2 = 0.986	408.2 ± 14.7 r^2 = 0.997

[a] Standard error. [b] Ludikhuyze et al. (1998c). [c] Ludikhuyze et al. (1998a). (From Wang et al., 2008).

Table 3. Inactivation rate constants ($\times 10^{-2}$ min^{-1}) for the isothermal inactivation of commercial soybean lipoxygenase in 0.01 M, pH 9 Tris–HCl buffer

Lipoxygenase in soy milk or in crude soybean extract exhibited a higher thermal stability with the corresponding smaller inactivation rate constants. The two activation energy values derived from the plots of the natural logarithm of inactivation rate constants, as a function of the reciprocal of the absolute temperature were larger, pointing to higher temperature sensitivity of the k values. Likewise, kinetic inactivation of lipoxygenase from many different sources, such as green peas, green beans, potatoes, asparagus, wheat germ, and germinated barley, have also been studied (Bhirud & Sosulski, 1993; Ganthavorn et al., 1991; Guenes & Bayindirli, 1993; Hugues et al., 1994; Indrawati et al., 1999; Park et al., 1988; Svensson & Eriksson, 1974). Indrawati et al. (1999) reported that thermal inactivation of lipoxygenase in green bean juice could be described by a two-fraction first-order inactivation model, referring to the existence of two fractions (isozymes) with different thermal stability. However, in their study, this phenomenon was not observed.

Van Loey et al. (1999) studies soybean lipoxygenase inactivation [0.4 mg/mL in Tris-HCl buffer (0.01 M, pH 9)] quantitatively under constant pressure (up to 650 MPa) and temperature (-15 to 68 °C) conditions and kinetically characterized by rate constants, activation energies, and activation volumes. The irreversible lipoxygenase inactivation followed a first-order reaction at all pressure-temperature combinations tested. In the entire pressure-temperature area studied, LOX inactivation rate constants increased with increasing pressure at constant temperature. On the contrary, at constant pressure, the inactivation rate constants showed a minimum around 30 °C and could be increased by either a temperature increase or decrease. On the basis of the calculated rate constants at 102 pressure temperature combinations, an iso-rate contour diagram was constructed as a function of pressure and temperature. The pressure-temperature dependence of the LOX inactivation rate constants was described successfully using a modified kinetic model (Van Loey et al., 1999)

5. Immunoreactivity and nutritional quality of soybean products

Penas et al. (2011) reported that sprouts obtained from HHP-treated soybean seeds demonstrated an important reduction in immune-reactivity. Furthermore, they were a good source of proteins and essential amino acids, with Met and Cys corresponding to the limiting amino acids, as indicated by the chemical score (CS), and a high essential amino acid index (EAAI) (Table 4). These results suggested that HHP could constitute an important technological approach for the industrial production of hypoallergenic and nutritive soybean sprouts.

The HHP treatment of raw seeds (PRS) produced lower Gly and Cys levels than raw seeds (RS), while no significant differences ($P \leq 0.05$) were observed for total EAA (Table 4). The germination process resulted in a significant decrease in Glu, Trp and Cys, while no changes were observed in the other amino acids, compared to RS. The total EAA content showed a 4% reduction compared to RS. Pomeranz et al. (1977) found only minor differences in the amino acid composition of germinated and ungerminated soybean, while Mostafa et al. (1987) observed a marked increase in the relative contents of both EAA and NEAA after germination. The levels of sulphur amino acids in germinated soybean seeds remained almost constant, whereas Asp increased compared to raw seeds. Discrepancies between the data reported by other authors and those reported in the present work could be attributed to differences in the germination conditions and seed varieties.

The application of HHP treatment to seeds prior to germination (GPS) led to a reduction in Glu and Ala as NEAA and Trp, Met and Cys in EAA in comparison with GS, and also Pro, and Ile compared to PRS. GPS showed similar statistical ($P \leq 0.05$) values of total EAA content to GS (32 and 34 g/100 g protein, respectively), whilst significant ($P \leq 0.05$) differences were found compared to RS and PRS (34 g/100 g protein) (Table 4).

In another study, the effects of HPP on soybean cotyledon as a cellular biological material were investigated from the viewpoints of the cell structure and enzyme reaction system (Ueno et al., 2010). Damage to cell structure was evaluated by measuring dielectric properties using the Cole−Cole arc, the radius of which decreased as pressure level increased. Results suggested that cell structure was damaged by HPP. The distribution of free amino acids was measured after HPP (200 MPa) of soybean soaked in water or sodium glutamate (Glu) solution. HPP resulted in high accumulation of free amino acids in water-soaked soybean, due to proteolysis. HPP of soybean in Glu solution caused higher accumulation of γ-aminobutyric acid, suggesting that both proteolysis and specific Glu metabolism were accelerated by HPP. They concluded that HPP partially degraded cell structure and accelerated biochemical reactions by allowing enzyme activities to remain. These events were described as "high-pressure induced transformation" of soybean.

	RS	PRS	GS	GPS	Whole egg protein [B]
Protein (% d.w.)	45.0[a]	44.3[a]	47.3[b]	48.6[c]	
		Non-essential amino acids			
Glu	15.9[c]	15.7[c]	14.8[b]	14.0[a]	
Asp	7.25[a]	7.26[a]	7.09[a]	6.80[a]	
Arg	5.44[ab]	5.60[b]	5.05[a]	5.27[ab]	
Pro	4.26[b]	4.25[b]	3.79[ab]	3.71[ab]	
Ala	4.05[b]	3.97[ab]	4.04[b]	3.71[a]	
Ser	3.87[a]	3.94[a]	3.92[a]	3.68[a]	
Gly	3.50[b]	3.07[a]	3.15[ab]	3.11[a]	
		Essential amino acids			
Leu	6.10[a]	6.18[a]	6.52[a]	5.68[a]	8.6
Lys	4.32[a]	4.40[a]	4.18[a]	4.13[a]	7.0
Val	4.19[a]	4.26[a]	4.26[a]	4.11[a]	6.6
Phe	4.13[bc]	4.17[c]	3.70[ab]	3.90[abc]	9.3 (Phe+Tyr)
Ile	4.12[b]	4.22[b]	3.78[ab]	3.75[ab]	5.4
Tyr	2.91[c]	2.67[abc]	2.81[bc]	2.57[ab]	
Trp	2.26[c]	2.29[c]	1.99[b]	1.71[a]	1.7
Thr	2.89[b]	2.52[ab]	2.50[ab]	2.32[a]	4.7
His	2.03[a]	1.93[a]	1.96[a]	1.85[a]	2.2
Met	1.32[b]	1.36[b]	1.32[b]	1.16[a]	5.7 (Met+Cys)
Cys	1.24[c]	1.08[b]	1.06[b]	0.94[a]	
Total EAA	35.5[b]	35.1[b]	34.1[ab]	32.1[a]	
CS	45.0	42.8	41.8	37.0	
EAAI	72.4	70.9	68.5	65.6	

[A] Data are the mean of three independent results. Different superscripts in the same row mean statistically significant differences ($P \leq 0.05$). PRS: HHP-treated soybean seeds; RS: raw soybean seeds; GPS: germinated PRS, HHP-treated soybean seeds; GS: germinated soybean seed. EAA: essential amino acid; CS: chemical score; EAAI: essential amino acid index. [B] (FAO/WHO/UNU, 1985). (From Penas et al., 2011).

Table 4. Effect of HHP and/or germination on the total protein and amino acids (g/100 g protein) content of soybean seeds and sprouts during germination.[A]

6. Inactivation of soymilk trypsin inhibitors

Protease inhibitors (PIs) are generally considered the main anti-nutritional factors in soybeans. Soybean PIs belong to a broad class of proteins that inhibit proteolytic enzymes, such as trypsin and chymotrypsin. Both compounds are important animal digestive enzymes for splitting proteins to render dipeptides and tripeptides (Scheider, 1983). However, the specificity of these inhibitors is not necessarily restricted to trypsin and chymotrypsin but also to elastase and serine proteases for which serine constitutes the active site. Nevertheless, the literature reports two main types of soybean PIs, specifically called trypsin inhibitors (TIs). The Kunitz soybean inhibitor, with a molecular weight of 20,000 and two disulfide bridges, exhibits specificity to inhibit trypsin. The Bowman-Birk inhibitor, on the other hand, with a molecular weight ranging from 6000 to 10,000 and seven disulfide bonds, exhibits specificity to inhibit chymotrypsin (Liener, 1994).

Residual trypsin was measured in soymilk subjected to selected pressures, temperatures and holding times. Treatment combination at higher pressures and temperatures, for selected holding times resulted in an increased inhibition rate of trypsin inhibitors in soymilk. It was not possible to obtain inactivation rate parameters for treatments at 550 MPa and 80 °C because the data did not fit a first order kinetics model. However, a clear increase of residual trypsin was observed as treatment times increased. Soaking of soybeans in sodium bicarbonate solution, prior to preparation of soymilk, resulted in smaller inhibition rates of trypsin at the working selected pressures, combined with thermal treatment and holding times, than in soybeans soaked in distilled water. The use of sodium bicarbonate, as soaking medium of soybeans, did not result in a significant increase in the percentage of residual trypsin in soymilk treated at 550 MPa and 80 °C for the selected holding times (Fig. 4).

Ven et al. (2005) also evaluated HPP as an alternative for the inactivation of TIs in soy milk and also studied the effect of HPP on in whole soybeans and soy milk. For complete lipoxygenase inactivation either very high pressures (800 MPa) or a combined temperature/pressure treatment (60 °C/600 MPa) was needed. Pressure inactivation of TIs was possible only in combination with elevated temperatures. For TIs inactivation, three process parameters, temperature, time, and pressure, were optimized using experimental design and response surface methodology. A 90% TIs inactivation with treatment times of <2 min can be reached at temperatures between 77 and 90 °C and pressures between 750 and 525 MPa.

7. Conclusions

High hydrostatic pressure (HHP) processing is an innovative technology for processing of soybean which is an important food from nutritional point of view. HHP enables the inactivation of pathogenic bacteria at ambient temperatures. It also showed an increase in protein solubility and staining intensity due to the release of more protein after application of HHP. It also favored the inactivation of quality deteriorating enzymes such as lipoxygenase at room temperatures. Treatment of soybean with HHP improved the bio-availability of nutrients such as amino acids and the reduction of immune-reactivity. HHP also favored the activity of proteases, probably by reducing the activity of their inhibitors. It can be inferred that soybean and its products which are valuable food commodities can be effectively processed using this innovative processing technology, however more research needs to be done on HHP optimization and its effects on various physicochemical properties of soybean and different soy-foods.

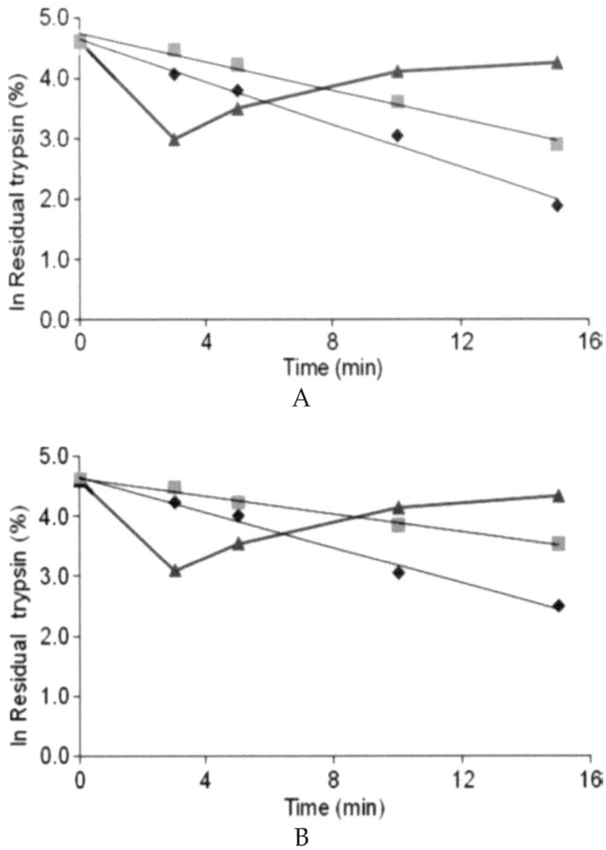

Fig. 4. Residual trypsin in HHP-processed (550 MPa) soymilk from soybeans previously soaked in water (A) and sodium bicarbonate solution (0.5%) (B). Temperatures: 19 (♦), 65 (■), and 80 °C (▲) (From Guerrero-Beltrán et al., 2009).

8. References

Asano, M., Okubo, K., & Yamauchi, F. (1989). Effect of immersing temperature on the behavior of exuding compounds from soybean. *Journal of the Japanese Society for Food Science and Technology*, 36, pp. 636–642, ISSN: 1344-6606

Baker, E. C., & Mustakas, G. C. (1972). Heat inactivation of trypsin inhibitor, lipoxygenase and urease in soybean: Effect of acid and base additive. *Journal of the American Oil Chemists Society*, 50, pp. 137–141, ISSN: 0003-021X

Bhirud, P. R., & Sosulski, F. W. (1993). Thermal inactivation kinetics of wheat germ lipoxygenase. *Journal of Food Science*, 58, pp. 1095–1098, ISSN: 1750-3841

Bridgman, P. W. (1914). The coagulation of albumen by pressure. *Journal of Biological Chemistry*, 19, pp. 511–512, ISSN: 0021-9258

Estrada-Giron, Y., Swanson, B. G., & Barbosa-Canovas, G. V. (2005). Advances in the use of high hydrostatic pressure for processing cereal grains and legumes. *Trends in Food Science and Technology*, 16, pp. 194–203, ISSN: 0924-2244

Fuchigami, M., Ogawa, N., & Teramoto, A. (2002). Trehalose and hydrostatic pressure effects on the structure and sensory properties of frozen tofu (soybean curd). *Innovative Food Science and Emerging Technologies*, 3, pp. 139-147, ISSN: 1466-8564

Fukushima, D. (1991). Recent progress of soybean protein foods: chemistry, technology and nutrition. *Food Review International*, 7, pp. 323–351, ISSN: 8755-9129

Ganthavorn, C., Nagel, C. W., & Powers, J. R. (1991). Thermal inactivation of asparagus lipoxygenase and peroxidase. *Journal of Food Science*, 56, pp. 47–49, 79, ISSN: 1750-3841

Guenes, B., & Bayindirli, A. (1993). Peroxidase and lipoxygenase inactivation during blanching of green beans, green peas and carrots. *LWT-Food Science and Technology*, 26, pp. 406–410, ISSN: 0023-6438

Guerrero-Beltrán, J. A., Estrada-Girón, Y., Swanson, B. G., & Barbosa-Cánovas, G. V. (2009). Pressure and temperature combination for inactivation of soymilk trypsin inhibitors. *Food Chemistry*, 116, pp. 676-679, ISSN: 0308-8146

Hayashi, R. (1989). Application of food processing and preservation: philosophy and development. In: *Engineering and Food*, W. E. L. Spiess, & H. Schuber, pp. (815–826), Elsevier Applied Science, ISBN: 1851664475, London

Hirano, H., Kagawa, H., & Okubo, K. (1992). Characterization of proteins released from legume seeds in hot water. *Phytochemistry*, 31, pp. 731–735, ISSN: 0031-9422

Hugues, M., Boivin, P., Gauillard, F., Nicolas, J., Thiry, J. M., & Richard-Forget, F. (1994). Two lipoxygenases from germinated barley - Heat and kilning stability. *Journal of Food Science*, 59, pp. 885–889, ISSN: 1750-3841

Indrawati, A., Van Loey, A. M., Ludikhuyze, L. R., & Hendrickx, M. (1999). Single, combined, or sequential action of pressure and temperature on lipoxygenase in green beans (*Phaseolus vulgaris* L.): A kinetic inactivation study. *Biotechnology Progress*, 15, pp. 273–277, ISSN: 1520-6033

Ludikhuyze, L. R., Indrawati, Van den Broeck, I., Weemaes, C. A., & Hendrickx, M. E. (1998a). High pressure and thermal denaturation kinetics of soybean lipoxygenase: A study based on gel electrophoresis. *LWT-Food Science and Technology*, 31, pp. 680–686, ISSN: 0023-6438

Ludikhuyze, L. R., Indrawati, Van den Broeck, I., Weemaes, C. A., & Hendrickx, M. A. (1998b). Effect of combined pressure and temperature on soybean lipoxygenase. 2. Modelling inactivation kinetics under static and dynamic conditions. *Journal of Agricultural and Food Chemistry*, 46, pp. 4081–4086, ISSN: 0021-8561

Ludikhuyze, L. R., Indrawati, Van den Broeck, I., Weemaes, C. A, & Hendrickx M. A. (1998c). Effect of combined pressure and temperature on soybean lipoxygenase. 1. Influence of extrinsic and intrinsic factors on isobaric–isothermal inactivation kinetics. *Journal of Agricultural and Food Chemistry*, 46, pp. 4074–4080, ISSN: 0021-8561

Mostafa, M. M., Rahma, E. H., & Rady, A. H. (1987). Chemical and nutritional changes in soybean during germination. *Food Chemistry*, 23, pp. 257–275, ISSN: 0308-8146

Omi, Y., Kato, T., Ishida, K., Kato, H., & Matsuda, T. (1996). Pressure-induced release of basic 7S globulin from cotyledon dermal tissue of soybean seeds. *Journal of Agricultural and Food Chemistry*, 44, pp. 3763–3767, ISSN: 0021-8561

Oxen, P., & Knorr, D. (1993). Baroprotective effect of high solute concentration against inactivation of Rhodotorula rubra. *LWT-Food Science and Technology*, 26, pp. 220–223, ISSN: 0023-6438

Park, K. H., Kim, Y. M., & Lee, C. W. (1988). Thermal inactivation kinetics of potato tuber lipoxygenase. *Journal of Agricultural and Food Chemistry*, 36, pp. 1012–1015, ISSN: 0021-8561

Penas, E., Gomez, R., Frias, J., Baeza, M. L., & Vidal-Valverde, C. (2011). High hydrostatic pressure effects on immunoreactivity and nutritional quality of soybean products. *Food Chemistry*, 125, pp. 423–429, ISSN: 0308-8146

Pomeranz, Y., Shogren, M. D., & Finney, K. F. (1977). Flour from germinated soybeans in high protein bread. *Journal of Food Science*, 42, pp. 824–828, ISSN: 1750-3841

Prestamo, G., Lesmes, M., Otero, L., & Arroyo, G. (2000). Soybean vegetable protein (tofu) preserved with high pressure. *Journal of Agricultural and Food Chemistry*, 48, pp. 2943–2947, ISSN: 0021-8561

Riaz, M. N. (1988). Soybeans as functional foods. *Cereal Foods World*, 44, pp. 88–92, ISSN: 0146-6283

Seyderhelm, I., Boluslawski, S., Michaelis, G., & Knorr, D. (1996). Pressure induced inactivation of selected food enzymes. *Journal of Food Science*, 61, pp. 308–310, ISSN: 1750-3841

Steinke, F. H., Waggle, D. H., & Volgarev, M. N. (1992). *New Protein Foods in Human Health: Nutrition, Prevention and Therapy*. CRC Press, Inc., ISBN: 0849369045, Florida

Svensson, S. G., & Eriksson, C. E. (1974). Thermal inactivation of lipoxygenase from peas (*Pisum sativum* L.). III. Activation energy obtained from single heat treatment experiments. *LWT-Food Science and Technology*, 7, pp. 142–144, ISSN: 0023-6438

Tangwongchai, R., Ledward, D. A., & Ames, J. M. (2000). Effect of high-pressure treatment on lipoxygenase activity. *Journal of Agricultural and Food Chemistry*, 48, pp. 2896–2902, ISSN: 0021-8561

Ueno, S., Shigematsu, T., Watanabe, T., Nakajima, K., Murakami, M., Hayashi, M., & Fujii, T. (2010). Generation of free amino acids and γ-aminobutyric acid in water-soaked soybean by high-hydrostatic pressure processing. *Journal of Agricultural and Food Chemistry*, 58, pp. 1208–1213, ISSN: 0021-8561

Utsumi, S., Matsuda, Y., & Mori, T. (1998). Structure-functions relationships of soy proteins. In: *Functional food*, G. Mazza (Ed.), pp. (46-51), Technomic Publishing Company Inc., ISBN: 1566764874 , Lancaster

Van Loey, I. A. M., Ludikhuyze, L. R., & Hendrickx, M. E. (1999). Soybean lipoxygenase inactivation by pressure at subzero and elevated temperatures. *Journal of Agricultural and Food Chemistry*, 1999, 47, pp. 2468–2474, ISSN: 0021-8561

Ven, C., Matser, & Berg, R. W. (2005). Inactivation of soybean trypsin inhibitors and lipoxygenase by high-pressure processing. *Journal of Agricultural and Food Chemistry*, 2005, 53, pp. 1087–1092, ISSN: 0021-8561

Wang, R., Zhou, X., & Chen, Z. (2008). High pressure inactivation of lipoxygenase in soy milk and crude soybean extract. *Food Chemistry*, 106, 603–611, ISSN: 0308-8146

Whitaker, J. R. (1972). *Principles of Enzymology for the Food Sciences*. Marcel and Dekker, ISBN: 0824791487, New York

Wolf, W. J., & Cowan, J. C. (1975). *Soybeans as a Food Source*. CRC Press, Inc., ISBN: 0878191127, Cleveland

4

The Main Components Content, Rheology Properties and Lipid Profile of Wheat-Soybean Flour

Nada Nikolić and Miodrag Lazić
University of Niš, Faculty of Technology, Leskovac
Serbia

1. Introduction

Soybean (*Glycine max* (L.) Merr.) is a species of legume, native to a Eastern Asia and an important global crop, today. Soybean is rich in high quality proteins, contains essential amino acids, similar to those found in meat, minerals such as Fe, Zn, Cu, Mn, Ca and Mg as well as phytic acid. The bulk of seed soybean proteins contains albumins and globulins as major components, but there are minor, undesirable components such as inhibitors of trypsyn and chymotrypsin, and sugar-binding lecitins. The inhibitors and lecitins are generally inactivated by heat treatment. A new immunochemical methods can be used for quantitative detection of soybean proteins and production of healthful foods (Brandon & Friedman, 2002). The soybean lipid contains a significant amount of unsaturated acids: α-linolenic acid, known as omega-3 acid, linoleic, γ-linolenic and arachidonic acid, known as omega-6, and oleic acids known as omega-9 acid and are very important in human nutrition (Liu, 1997). The soybean lipid also contains saturated acids: palmitic and stearic acid (Bressani, 1972; Olguin et al., 2003; Bond et al., 2005), as well as tocopherols (Ortega-Garcia et.al., 2004, Yoshida et al., 2006). These soybean components make the products with soybean have higher nutritional value.

1.1 Soybean in bread making industry

In the bread making industry, the soybean is used with the aim to increase the bread protein value and decrease carbohydrate value. In ordinary white bread protein content ranges from 8 to 9% and by including soybean, the protein content can be made up to 16% (Ribotta et al. 2010) and at the same time, the dough and bread are richer with lipoxygenase enzymes preparation which can make the dough physical properties better. The fact is that bread made with soybean costs less and it is especially important in countries where wheat is not a major domestic crop. Bread and products with higher protein content and lower carbohydrate content are more suitable for use in some diets than bread and products formulations currently used (Mohamed et al., 2006). As the main protein component in white bread is gluten, a component which causes celiac disease, the usage of soy is useful for decreasing the gluten content in bread. A portion of 0.3 to 5% of soybean flour portion is usually added (Auerman, 1979), but rational addition for increasing bread protein value is 20-30% of soybean flour. Besides whole soybean flour, different soybean products can also be used: defatted soy flour (Mashayekh et al., 2008), physically modified soy flour (Maforimbo et al., 2008), soy flour and durum wheat flour mixture (Sabanis & Tzia, 2009), commercial soy protein isolate (Roccia et al., 2009), diffrerent kinds of soy protein powder (Qian et al., 2006) and part of soy seeds such as a hulls (Anjum et a., 2006). Based on these investigations results, different bread formulations are defined, and soy is used in portion up to 20%. When soy flour in wheat flour was to a level of 10% and in durum wheat flour up to 20%, the produced bread was without any negative effects in quality attributes such as colour, hardness and flavour, promising nutritious and healthy alternative to consumers (Sabanis & Tzia, 2009). By investigating the effect of defatted soy flour on sensory and rheological properties of wheat bread, Mashayekh et al. 2008, concluded that adding 3 and 7% defatted soy flour gives as good a loaf of bread as the 100% wheat bread and acceptable consumer attribute with rheological and sensory characteristics. Adding small quantity of soy protein powder of 3% to wheat flour did not change the sensory properties of bread and a large quantity of soy flour adding, exceeding 7%, can lead to stickiness and leguminous flavour (Roccia et al., 2009). The results of investigation (Anjum et al., 2006) of soy hulls usage showed the content of 4.5% soy hulls combined with wheat flour is acceptable and suitable level by the consumers.

1.2 Wheat-soybean bread manufacturing

The manufacture of bread from flour without gluten represents considerable technological difficulties (Jong et al., 1968; Schober et al., 2003) because gluten is the most important structure forming protein for making bread (Gujural et al., 2003; Moore et al,. 2004) and by using appropriate soy-wheat flour mixture these difficulties can be avoided. When soy was added to wheat, the soy globulins interact with wheat gluten proteins forming aggregates of high molecular weight. As reduction-reoxidation treatment facilitated the interaction of glutenin subunits and soy proteins (11S subunits), interaction probably occurs through the oxidation of SH groups (Maforimbo, 2008). Investigations of the changes in glutenin macro polymer content, protein composition and free sulfhydryl content showed that active soy flour decreased glutenin macro polymer content due to gluten depolymerization and glutenin macro polymer content increased by inactive soy flour because soy proteins became insoluble and precipitated together. Soy proteins were associated to wheat protein through physical interaction and covalent and non-covalent bonds during mixing and resting and these interactions produced large and medium-size polymers. This increased solubility of insoluble gluten proteins, producing a weakening of the gluten network (Perez et al., 2008) and decreasing availability of water to build up in gluten network (Roccia et all. 2009). Physicochemical status of soy protein in the product had a great influence on how wheat-soy proteins will interact (Perez et al., 2008). Incorporation of soy proteins changes the rheolocical and bread properties. The investigations showed that adding soy protein powder depresses loaf volume, gives poor crumb characteristics and decreases acceptability by consumers (Qian et al., 2006.) Ribotta et al. 2010. tested different additive combinations for improved bread quality obtained from soy-wheat flour in ratio of 90:10 w/w and found that the combination with transglutaminase showed a major improving effect on dough rheological properties and crumb uniformity.

1.3 Aims of investigations

Dough rheological properties have great relevance in predicting the mixing behaviour, sheeting and baking performance (Dobraszczyk & Morgenstain, 2003) and supplementation wheat flour with other changed these properties. Ribotta et al. 2005, presented data about effect of soybean addition on farinographic properties such as water absorption (WA), dough development time (DT), dough stability (Dst) and dough degree of softening (DSf). The present work has been undertaken with the objective to investigate the effect of whole soybean seed flour (»full-fat soy flour«) portion of 3 to 30% on dough farinographic, but extensographic and amylographic properties, too. Based on the composition of wheat and soybean flour lipids, the aim is to obtain lipid composition in wheat-soybean flour mixtures, and compare it to wheat flour only, with an emphasis on content of total saturated fatty acids (TS), total monounsaturated fatty acids (TMUS), total polyunsaturated fatty acids (TPUS) and total unsaturated fatty acids (TU). In order to value wheat-soybean flour mixtures, they were classified into groups by using statistical analysis and Euclidean distances and the correlations between content of some lipid components and rheological properties were found.

2. Experimental

2.1 Soybean seed

The whole soybean seeds (*Glycine max* L.) cultivars ZP Lana, grown in Serbia in summer of 2006 were used. The seeds were purchased in local store „Green Apple" (Leskovac, Serbia) and milled. The particle size was determined by method of insemination via riddles with gaps size from 0.315 to 3.15 mm. The overall particle size was determined using equations:

$$100 / d_{sr} = \sum \Delta d_i / d_i \tag{1}$$

where Δd_i is weight of fraction with appropriate particle size in %, and

$$d_i = (d_{i-1} + d_{i+1}) / 2 \tag{2}$$

where d_{i-1} is bottom riddle gap size and d_{i+1} is upper riddle gap size.

2.2 Chemicals

Chemicals used for oil extraction were high quality chemicals (Centrohem, Serbia) and for HPLC and GC analysis they were analytical grade (Riedel-de Haën, Honeywell Specialty Chemicals, Germany).

2.3 Wheat and soybean flour and flour mixtures

The wheat flour, Kikinda Mill, Serbia (WF) was bought from the local market. The soybean flour (SF) as »full-fat soy flour« was obtained by soybean seeds milling (IKA Model M120), to an overall particle size (d_{sr}) of 0.4 mm. Quantities of 291, 285, 270, 240 and 210 g of wheat flour and 9, 15, 30, 60 and 90 g of soybean flour, respectively, were used to make flour mixture with 3, 5, 10, 20 and 30 % (w/w) soybean flour portion, without adding additives.

2.4 Flour analyses

Flour protein content was determined by the Kjeldahl method (Nx5.95). The moisture content was determined by Scaltec SMO 01 (Scaltec instruments, Germany) instruments: flour (5 g) was put into the disk plate analyzer, dried at 110°C to a constant weight, and the moisture content was read out on the display. The ash content was determined by staking at 800°C during 5 h. For gluten content determination, the dough was prepared by adding a sodium chloride solution first; wet gluten was isolated by dough washing and weighed. The starch content was determined by polarimetry according to Grossfeld's method and the total carbohydrates content according to Luff-Schoorl's method (Trajković et al., 1983). The values for samples are from triplicate analysis and followed by standard deviation.

2.5 The wheat and soybean flour lipid content

The wheat or soybean flour (50g) were put into Erlenmeyer flask, 500 ml of trichloroethylene was added and extracted for 30 minutes, under reflux and by mixing (200 min^{-1}) at approximately 88°C. The extract was separated by using Buchner funnel under weak vacuum. The plant material was extracted three more times by the same method; the extracts were mixed together and eluted by water in the separation funnel (3 x 10 ml). The eluted extracts' volume was recorded and an aliquot (3 ml) was taken for the dry residue determination test. This test was performed by drying at 110°C to a constant weight and the dry residue content was read out on the analyzer display (Scaltec SMO 01, Scaltec instruments, Germany). The lipid content in wheat and soybean flour was calculated based on average value of three measurements. The remnant of lipid extracts, after dry residue determination test, is evaporated under vacuum and obtained residue was used for HPLC and GC analysis.

2.6 Rheology measurements

The Brabender farinograph (Brabender Model 8 10 101, Duisburg, Germany) according to ISO 5530-1 test procedure, was used for water absorption values (WA value in ml/100g), development time (DT in minutes), dough stability (DSt in minutes), degree of softening (DSf in BU) and farinograph quality number (QN) determination. For extensograph measurement, the Brabender extensograph (Brabender, Model 8600-01, Duisburg, Germany) and test procedure ISO 5530-2 were used. The samples were prepared from wheat flour and wheat-soybean flour mixtures, distilled water and salt, and data for energy (E in cm^2), resistance (R in EU), extensibility (Ex in mm) and ration number (R/Ex) were recorded on extensograph curve. To obtain amylograph data, such as gelatinization temperature (T_{max} in °C) and gelatization maximum (η_{max} in AU), the amylograph (Brabender Model PT 100, Duisburg, Germany) and ISO 7973 test procedure were used.

2.7 HPLC analysis

For HPLC analysis, Holčapek et al., (1999) modified HPLC method and the Agilent 1100 High Performance Liquid Chromatograph, a Zorbax Eclipse XDB-C18 column: 4.4 m x 150 mm x 5 μm, Agilent technologies, Wilmington, USA and an UV/ViS detector were used. The flow rate of binary solvent mixture (methanol, solvent A, and 2-propanol/*n*-hexane, 5:4 by volume, solvent B) was 1 ml/min, with a linear gradient (from 100% A to 40% A+ 60% B in 15 min). The column temperature was held constant at 40°C. The components were detected at 205 nm. The free fatty acids (FFA), methyl esters (ME), monoacylglycerols (MAG), diacylglycerols (DAG) and triacylglycerols (TAG) were identified by comparing the retention times of the lipid components with those of standards. The samples of the reaction mixture were dissolved into a mixture of 2-propanol:*n*-hexane, 5:4 v/v and filtered through 0.45 μm Millipore filters.

2.8 GC analysis

For GC analysis, fatty acids methyl esters were prepared. The lipid were alkaline hydrolyzed and methylated by methanol and BF_3 as catalysts. The final fatty acids methyl esters concentration was about 8 mg/ml in heptane. For obtaining a methyl esters GC spectra, the HP 5890 SERIES II GAS-CHROMATOGRAPH, HP with FID detector and 3396 A HP integrator was used. Column was ULTRA 2 (25m x 0.32mm x 0.52 μm) (Agilent Technologies, Wilington, USA), injector temperature of 320°C, and injector volume of 0.4 μl. The carrier gas was He at a constant flow rate of 1 ml/min. The flame ionization detector was at 350°C and split ratio was 1:20. Oven temperature was initially 120°C and was maintained at 120°C, for 1 min, then increased by 15°C/min until 200°C, increased by 3 °C/min until 240°C, increased by 8°C/min until 300°C and maintained at 300°C for 15 min. The fatty acids were identified by comparison of retention times of the lipid components with those of standards.

2.9 Energetic value

Based on total carbohydrates (CHC), protein (PC) and lipid content (LC), the energetic value (EV) of wheat flour and white-soybean flour mixtures was calculated as:

$$EV = (CHC + PC)\cdot 17 + LC \cdot 37 \tag{3}$$

2.9.1 Statistical analysis

STATISTICA, version 5.0 software was used to perform the statistical analysis: the means and standard deviations, the correlation coefficients and cluster analysis. The means and standard deviations were obtained by Descriptive Statistics, marking the Median & Quartiles and Confirm Limits for Means. In order to classify wheat flour and wheat-soybean mixtures, the cluster analysis and the Euclidean method with the complete linkage was used.

3. Results and discussion

3.1 The main components content

The results of wheat and soybean flour moisture, starch, protein, ash, lipid, gluten and carbohydrates content are showed in Table 1. Data are presented as means of three determinations with standard deviation. Based on these data, considering the soybean flour portion in mixtures and compared to wheat flour, it is evident the moisture (from 12.8 to 11.6%), the starch (from 76.6 to 56.8%), gluten (23.9 to 16.7%) and carbohydrates content (from 78.8 to 62.8%) decreased, while the protein (from 8.6 to 20.0%), ash (from 0.48 to 2.08%) and lipid (1.2 to 7.2%) content increased with increasing soybean flour portion in mixtures.

Taking into account the protein, carbohydrates and lipid content, based on formula (3) for energetic value determination, it was obtained that the soybean flour increased energetic value in flour mixtures (from 1530 obtained for wheat flour, to 1544, 1554, 1579, 1625, 1674 kJ/100g, when soy flour portion was 3, 5 10, 20 and 30%, respectively) and the maximal increasing of 9.4% was in mixture with soybean flour portion of 30%.

Content (g/100g)	Moisture	Starch	Protein	Ash	Lipid	Gluten	Carbo-hydrates
Wheat flour	12.8±0.6	76.6±1.2	8.6±0.34	0.48±0.04	1.2±0.05	23.9±0.4	78.8±2.2
Soybean flour	8.7±0.6	10.7±0.6	46.7±0.6	5.80±0.6	21.2±0.6	-	25.4±0.6
3%	12.7	74.6	9.7	0.64	1.8	23.2	77.2
5%	12.6	73.3	10.5	0.75	2.2	22.7	76.1
10%	12.4	69.9	12.4	1.01	3.2	21.5	73.5
20%	11.9	63.4	16.2	1.54	5.2	19.1	68.1
30%	11.6	56.8	20.0	2.08	7.2	16.7	62.8

Table 1. The main components content in wheat and soybean flour and their mixtures

3.2 Rheology properties

Farinograph, extensograph and amylograph data of flours and flour mixtures with different portions of soybean flour are given in Table 2. The results showed the farinograph data depended on the soybean flour portion in mixtures. The water absorption increased from 53.4 to 64.2% with increasing soybean flour portion in mixtures. It is well known, that the main dough component in wheat flour responsible for water absorption is gluten. The soy flour is without gluten but had even more than 5 times higher protein content than wheat flour (46.7 to 8.6 g/100g). The higher absorption ability could be due to soy protein components, such as globulins, which interacted with gluten protein in the composite dough (Maforimbo et al., 2008). The same effect of soy flour on water absorption value was reported by Ribotta et al., 2005, when heat-treated full-fat flour, enzyme-active defatted flour and soy protein isolates were used for wheat flour substitution in portion from 3 to 12% and enzyme-active full-fat flour in portion from 5 to 12%. The differences in dough development time and dough stability among flour mixtures with different soybean flour portions, ranged from 1.3 to 7.3 min and 0.7 to 2.9 min, respectively and both values for mixtures were higher than for wheat flour where value for dough development time was 1 min and for dough stability was 0.1 min. The effect of delaying dough development time and dough stability by addition the mostly of investigated soy flours was obtained in Ribotta et al. (2005) experiments, too.

Degree of dough softening was increased with increasing soybean flour portion in mixtures, from 45 to 66 BU and all values were lower than value for wheat flour, i.e. than 90 BU. The same effect of soybean flour on this farinographic characteristic was obtained by Ribotta et al. (2005), for all investigated samples, except for soy protein isolate Samsory 90 HI and portion of 5, 10 and 12%, when dough softening value was higher. According to appropriate triangle area on farinograph curves, the quality number, known as Honkocy number could be in the range from 0 to100, and quality groups are A_1 (area of 0-1.4 cm^2), A_2 (area of 1.5-5.5 cm^2), B_1 (area of 5.6-12.1 cm^2), B_2 (area of 12.2-17.9 cm^2), C_1 (area of 18.0-27.4 cm^2) and C_2 (area of 27.5-50.0 cm^2) (Djaković, 1980). The soybean flour addition in all investigated portions had positive influence on quality number and quality group which was B1 instead B2, which was for wheat flour.

Data obtained on extensograph, showed that dough with soybean flour portion of 20 and 30% had lower values for energy and dough resistance in comparison to dough made of wheat flour only. The extensibility of dough with soybean flour ranged from 125 to 89 EU and those values were lower than extensibility of dough with wheat flour only. The ration number, R/Ex, varied depending on soybean flour portion and ranged from 2.76 to 3.42. Based on curve of volume versus ratio number (Djaković, 1980), obtained for round bread, the bread volume was predicted to be in range from 576 to 557 cm^3 and it was lower than bread volume obtained from wheat flour only which was 581 cm^3.

Farinograph data						
	Wheat flour	3%	5%	10%	20%	30%
WA (ml/100g)	53.4±1.4	53.3±1.4	53.8±1.3	53.9±1.6	61.9±1.6	64.2±1.7
DT (min)	1±0.1	1.3±0.1	1.3±0.1	6.5±0.3	6.8±0.3	7.3±0.4
DSt (min)	0.1±0.1	0.7±0.2	0.8±0.2	1.8±0.4	2.3±0.5	2.9±0.5
DSf (BU)	90±1.6	45±1.4	50±1.4	65±1.7	65±1.7	66±1.5
QN	52.8±1.4	57.7±1.4	66.7±1.4	61.7±1.4	64.2±1.4	68.8±1.4
Group	B2	B1	B1	B1	B1	B1
Extensograph data						
E (cm^2)	67.8±1.3	70.5±1.5	74.2±1.5	79.0±1.6	57.1±1.4	43.4±1.2
R (EU)	315±5.5	345±6.0	350±6.0	355±5.5	345±5.5	305±5.5
Ex (EU)	126±3.0	125±3.0	123±3.0	117±2.5	101±2.5	89±2.0
R/Ex	2.50±0.2	2.76±0.3	2.84±0.4	3.03±0.4	3.41±0.4	3.42±0.4
V (cm^3)	581±10	576±10	574±10	566±10	559±10	557±10
Amylograph data						
T_{max} (°C)	81.2±0.5	81.5±0.5	82.0±0.5	85.0±0.5	86.5±0.5	88.8±0.5
η_{max} (AU)	630±20	315±15	250±10	180±10	120±10	90±5

Table 2. Rheological properties of wheat flour (WF) and wheat–soy bean flour mixtures

By amylograph data, dough with soybean flour had higher gelatinization temperature (in range 81.5 to 88.8°C) than dough with wheat flour only (81.2°C), and this value was higher when soybean flour portion was higher. Based on Stevenson et al. (2006) results, gelatinization temperature for wheat-soybean flour mixtures value could be lower. They found the gelatinization temperature of soybean starches were lower compared to wheat starch, due to the short amylopectin branch-chains and positive relationship between gelatinization temperature and amylopectin branch-chains. The reason why our gelatinization temperature values were higher maybe a different behaviour of soy starch in combination with wheat starch, which was present in wheat-soybean flour mixtures. The maximal pasta viscosity decreased from 315 to 90 AU when soybean flour portion was increased. The lowest pasta viscosity was when the soybean flour portion was 30% (w/w) and it was seven times lower than maximal pasta viscosity value for wheat flour only. The low peak viscosity of soybean starch could be due to short amylopectin branch-chain which has been correlated in wheat starches (Sasaki & Matsuki, 1998; Shibananuma et al., 1996).

3.3 Lipid profile

The lipid profile of wheat flour and soybean flour obtained by HPLC analysis and based on these results the lipid profile of wheat-soybean flour mixtures, is presented in Table 3. The content of components was determined by measuring the peak area at 1.76 min for free fatty acids, peak area at 2.15 min for methyl esters, peaks area in the range of 3.44-4.58 min for monoacylglycerols, peaks area in the range of 5.28-8.68 min for diacylglycerols and peaks area in the range of 10.91-15.81 min for triacylglycerols (Holčapek et al., 1999).

Flour	FFA (g/100g)	ME (g/100g)	MAG (g/100g)	DAG (g/100g)	TAG (g/100g)
Wheat	11.9±0.4	23.2±1.3	2.5±0.2	12.2±0.9	50.2±1.5
Soybean	25.9±1.2	1.6±0.3	0.8±0.1	2.3±0.4	69.4±1.6
3%	12.3	22.5	2.4	11.9	50.8
5 %	12.6	22.1	2.4	11.7	51.2
10 %	13.3	21.0	2.3	11.2	52.1
20 %	14.7	18.9	2.2	10.2	54.0
30 %	16.1	16.7	2.0	9.2	55.9

Table 3. The lipid profile of wheat and soybean flour and their mixtures obtained by HPLC

Fatty Acid content in g/100g	Wheat flour	Soybean flour	3%	5%	10%	20%	30%
Myristic ($C_{14:0}$)	0	0.11±0.01	<0.035	<0.035	<0.035	<0.035	<0.035
Palmitic ($C_{16:0}$)	19.45±0.45	10.35±0.6	19.17	18.99	18.54	17.63	16.71
Linoleic ($C_{18:2}$)	57.91±0.72	55.23±1.2	57.83	57.77	57.63	57.37	57.11
Oleic ($C_{18:1}$)	20.23±0.21	13.07±0.8	19.20	19.87	19.52	18.79	18.08
Linolenic ($C_{18:3}$)	0	13.46±0.9	0.40	0.67	1.34	2.69	4.04
Stearic ($C_{16:0}$)	1.36±0.14	5.23±0.8	1.48	1.56	1.76	2.18	2.58
Nonadecanoic ($C_{19:0}$)	0	0.52±0.09	<0.053	<0.053	<0.053	0.10	0.16
Arachidic ($C_{20:0}$)	0	0.60±0.1	<0.061	<0.061	<0.061	0.12	0.18
Behenic ($C_{22:0}$)	0.26±0.06	0.48±0.06	0.27	0.27	0.28	0.30	0.33
ND RT 25.96	0	0.17±0.04	<0.052	<0.052	<0.052	<0.052	<0.052
Lignocerinic ($C_{24:0}$)	0	0.55±0.01	<0.056	<0.056	<0.056	0.11	0.17
Phthalic acid	0.68±0.08	0	<0.069	<0.069	<0.069	0.14	0.20
TU	78.14	81.46	77.43	78.31	78.49	78.85	79.23
TMU	20.23	13.07	19.20	19.87	19.52	18.79	18.08
TPU	57.91	68.39	58.23	58.44	58.97	60.06	61.15
TS	21.07	17.84	20.92	20.82	20.58	20.44	20.13
TU/TS	3.71	4.58	3.70	3.76	3.81	3.86	3.94

Table 4. Fatty acids composition of wheat and soybean flour and their mixtures obtained by GC

The lipid from soybean flour had higher free fatty acids content (25.9 g/100g lipid) and triacylglycerols content (69.4 g/100g lipid) than lipid from wheat flour (11.9 and 50.2 g/100g lipid, respectively), while the contents of methyl esters, monoacylglycerols and diacylglycerols were lower than appropriate contents in wheat flour, even 14.5, 3.1 and 5.3 times, respectively. The content of free fatty acids and triacylglycerols in mixtures had the same changing tendency as soybean flour portion: it increased when the soybean flour portion in flour mixtures increased, while the contents of monoacylglycerols and diacylglycerols had opposite changing tendency: they decreased when the soybean flour portion in flour mixture increased. The properties of dough such as water absorption, development time, dough stability and gelatinization temperature had the same dependency on soybean flour portion as the content of free fatty acids and triacylglycerols and these components seemed to have a proper influence on the mentioned dough properties. Further, the content of monoacylglycerols and diacylglycerols had a proper influence on dough energy, resistance, extensibility and gelatization maximum. Some of these dependences were confirmed by determination of correlations coefficients.

The fatty acids composition of flours obtained by GC analysis and based on these results, the fatty acids composition in flour mixtures is presented in Table 4. The lipid contained: linoleic ($C_{18:2}$), α-linolenic ($C_{18:3}$), oleic ($C_{18:1}$), palmitic ($C_{16:0}$), stearic ($C_{18:0}$), arachidonic ($C_{20:4}$), behenic ($C_{22:0}$), nonadecanoic ($C_{19:0}$), γ-linolenic, lignocerinic, myristic ($C_{14:0}$) and several non-determined components (ND). GC analysis showed the lipid from wheat flour contained 78.14 g/100g of the total unsaturated fatty acids, consisting of linoleic (57.91 g/100g) and oleic acid (20.23 g/100g). The total polyunsaturated fatty acids content in lipid was 57.91 g/100g and it was from linoleic acid. The monounsaturated fatty acids content was 20.23 g/100g, from oleic acid, while the total saturated fatty acids content was 21.07 g/100g where the main fatty acids were palmitic and stearic acid with the content of 19.45 and 1.36 g/100g, respectively. The ratio of total unsaturated to total saturated fatty acids content was 3.7.

The lipid from soybean flour contained 81.46 g/100g of the total unsaturated fatty acids, composed of linoleic, oleic and linolenic acid with content of 55.23, 13.07 and 13.46 g/100g of lipid, respectively. The palmitic and stearic acid were the main saturated fatty acids with content of 10.35 and 5.23 g/100g, respectively, while the content of other detected saturated fatty acid was 0.69 g/100g. The myristic, linolenic, nonadecanoic, arachidic and lignocerinic fatty acids were detected only in soybean flour, while phthalic acid was detected only in wheat flour. Based on this composition and content of fatty acids, the ratio of total unsaturated to total saturated content in soybean flour was 4.58 and it was higher than ratio in wheat flour. Such fatty acids composition had influence on fatty acids composition in wheat-soybean flour mixtures: the content of total saturated fatty acids decreased when soybean flour portion in flour mixture increased, and all values were lower than the content of total saturated fatty acids in wheat flour; the content of total unsaturated fatty acids increased when soybean flour portion in flour mixture increased and all values (except in mixture with portion of soybean flour of 3%) were higher than the content of total saturated fatty acids in wheat flour. Finally, the ratio of total unsaturated to total saturated fatty acids content was higher in flour mixtures (except in mixture with portion of soybean flour of 3%) than in wheat flour.

3.4 Statistical analysis data

The correlation coefficients between the rheological properties (water absorption, dough stability, Ex, gelatinization temperature and gelatization maximum) and the content of some lipid components (free fatty acids, monoacylglycerols, diacylglycerols, triacylglycerols, palmitic (Pal), linoleic (Lin) and oleic (Ole) acid) in wheat flour and wheat-soybean flour mixtures are presented in Table 5.

	WA	DSt	Ex	T_{max}	μ_{max}	FFA	MAG	DAG	TAG	Pal	Lin
DSt	**0.88**										
Ex	**-0.94**	-0.84									
T_{max}	**0.91**	**0.98**	**-0.90**								
μ_{max}	-0.64	**-0.86**	0.67	-0.78							
FFA	**0.96**	**0.96**	**-0.93**	**0.98**	-0.77						
MAG	-0.75	-0.78	**0.91**	-0.83	0.71	-0.83					
DAG	-0.81	-0.81	**0.94**	**-0.86**	0.73	**-0.86**	**-0.97**				
TAG	**0.94**	**0.94**	**-0.85**	**0.94**	-0.72	**0.97**	-0.69	-0.72			
Pal	**-0.89**	**-0.89**	**0.97**	**-0.93**	0.78	**-0.94**	**0.95**	**0.98**	-0.84		
Lin	-0.45	-0.47	0.71	-0.53	0.51	-0.51	**0.85**	**0.88**	-0.30	0.77	
Ole	**-0.85**	**-0.85**	**0.90**	**-0.85**	0.78	**-0.88**	**0.88**	0.84	-0.81	**0.92**	0.65

Table 5. Correlation coefficients between rheological properties and lipid components content (correlations are significant at $p< 0.05$, N=12)

The sample size was twelve (N=12 (6x2): wheat flour and five wheat-soybean flour mixtures with minimal and maximal obtained value). As there were many correlations, only the one which had absolute value 0.85 and above 0.85 were taken into consideration. There were 54.5% of correlations, among which 30.3% were proper, and 24.2% were opposite correlations. The correlations can be divided into three groups: correlations between rheological properties, between rheological properties and lipid components content and between lipid components content. In the first group, there are the correlations where high water absorption value is associated with high dough stability and gelatinization temperature and low extensibility, high dough stability is associated with high gelatinization temperature and low gelatization maximum, and correlation between low gelatization maximum and high extensibility. In the second group of correlations, the high content of free fatty acids caused high water absorption, dough stability, and gelatinization temperature but low extensibility as well as the high monoacylglycerols and diacylglycerols content was proper correlated with extensibility. Also, when triacylglycerols content was higher, the water absorption, dough stability and gelatinization temperature were higher and extensibility was lower. The palmitic and oleic acid had opposite effect on the rheological properties, such as water absorption, dough stability and gelatinization temperature and proper effect on extensibility. The linolenic acid content was not associated with any rheological properties. Among lipid components content, there were proper correlations between free fatty acids and triacylglycerols content, between monoacylglycerols and diacylglycerols contents on one side and palmitic, linoleic and oleic acid content on the other side. It can mean that monoacylglycerols and diacylglycerols are mainly consists of palmitic, linoleic and oleic acids. The higher oleic acid content was associated with the higher palmitic acid content and it was only correlation among fatty acids.

By cluster analysis, based on multiple variables, wheat and wheat-soybean flour mixtures were classified into groups. Number of variables was six: wheat and five wheat-soybean flour mixtures (3, 5, 10, 20 and 30% w/w of soybean flour portions); number of cases i.e. parameters were eight: water absorption, dough stability, free fatty acids, diacylglycerols, triacylglycerols, palmitic, linolenic and oleic acid content. Linkage distances were obtained and presented by dendrogram in Fig. 1.

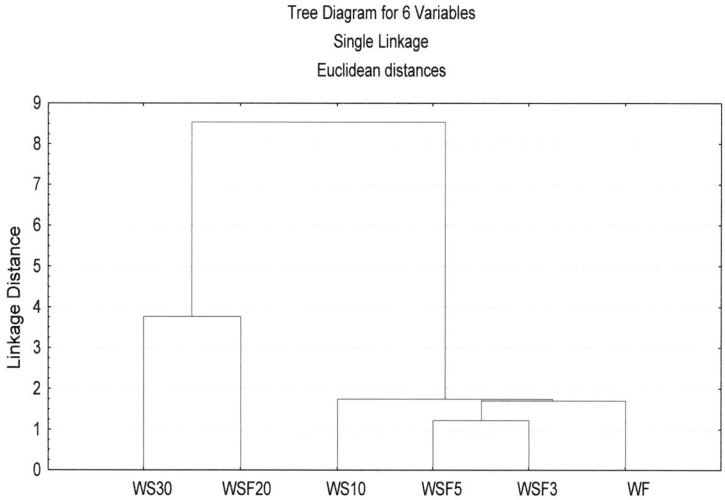

Fig. 1. Dendrogram obtained for wheat flour and wheat-soybean flour mixtures

The cluster analysis based on flour rheological and lipid characteristics, shows the linkage distance between wheat flour and flour mixtures increases when soybean flour portion in mixtures increases. The mixtures with soybean flour portion of 3 and 5% (w/w) are joined with wheat flour at the same distance level of 1.6 and make the first group. The mixtures with soybean flour portion of 10 % (w/w) are joined with wheat flour at the distance level of 1.7 and could be added to the first group. The mixture with soybean flour portion of 20 and 30% (w/w) is joined with wheat flour at distance level of 8.6 and make the second group. This means that soybean flour portion of 30% could be used to more enrich dough with soybean protein and the main rheological properties remain satisfactory as at portion of 20%. This provides a possibility of soybean flour being included in portions even higher than 30%, so future work could include this investigations as well as investigations to examine what happens with lipid components during dough mixing and baking: does their content and composition stay the same as in flour mixture or there occur changes.

4. Conclusion

The soybean flour addition increased the protein content up to 20.0%, the ash content up to 2.08% and lipid content up to 7.2%, while decreased starch, gluten and carbohydrates content for 19.8, 7.2 and 16 g/100 g flour mixture, respectively. Dough rheological properties and lipid profile depend on soybean flour portion. The soybean flour addition had positive influence on the quality number and group and extended duration of dough stability. The dough water absorption and the degree of softening increased with increasing soybean flour portion. The dough with soybean flour portion of 20 and 30% had lower values for energy in comparison to dough made of wheat flour only and it could be economically important. Values of gelatinization temperature for dough with soybean flour were higher than dough with wheat flour, maybe due to a specific behaviour of soy starch in combination with wheat starch. The maximal pasta viscosity decreased when soybean flour portion increased, even seven times when the soybean flour portion was 30%, probably due to short amylopectin branch-chain in soybean starch. All the wheat-soybean flour mixtures had higher free fatty acids and triacylglycerols content than wheat flour. Wheat-soybean flour mixtures had higher content of stearic and behenic acid compared to wheat flour, had almost the same content of linoleic acid as wheat flour and contained linolenic acid which was absent in wheat flour. The ratio of total unsaturated to total saturated fatty acids content was higher in flour mixtures than in wheat flour. The rheological properties of dough, such as water absorption and stability, had the same dependency on soybean flour portion in mixtures as the content of free fatty acids, while the content of triacylglycerols had the same dependency as water absorption and dough development time. In the same way, the content of monoacylglycerols and diacylglycerols had influence on dough extensibility and all these dependences were confirmed by statistical analysis with positive correlation coefficient value, higher than 0.85. The cluster analysis showed that the mixtures with soybean flour portions of 20 and 30% (w/w) were joined with wheat flour at the same distance level, so the soybean flour portion of 30% could be used to enrich even more dough with soybean protein and the main rheological properties remain satisfactory as with portion of 20%.

5. Acknowledgement

This work was supported under the project No.OI 172047 by the Ministry of Science of the Republic of Serbia.

6. References

Anjum, F.M.; Khan, M.I.; Butt, M.S.; Hussain, S. & Abrar, M. (2006) Functional properties of soy hulls supplemented wheat flour. *Nutrition and Food Science*, Vol.36, No.2, pp. 82-89, ISSN 0034-6659

Auerman, L.J., (1979). *Technology of bread production* (Ed Petar Spasic), Faculty of Technology, Novi Sad, Serbia

Bond, B.; Fernandez, D.R.; VanderJagt, D.J.; Williams, M.; Huang, Y.S.; Chuang, L.; Millson M.; Andrews, R. & Glew, R.H. (2005). Fatty acid, amino acid and trace mineral analysis of three complementary foods from Jos Nigeria. *Journal of Food Composition and Analysis,* Vol.18, No.7, (November, 2005), pp. 675-681, ISSN 0889-1575

Brandon, D. & Friedman, M. (2002). Immunoassays of soy proteins. *Journal of Agricultural and Food Chemistry*, Vol.50, No.22, (September, 2002), pp. 6635-6642, ISSN 1520-5118

Bressani, R. (1975). Nutritional contribution of soy protein to food systems. *Journal of the American Oil Chemists' Society,* Vol.52, No.4, (April, 1975) , pp. 254-262, ISSN 0003-021x

Djaković, Lj. (1980). *Wheat flour*, (Third edition), Faculty of Technology, Novi Sad, Serbia

Dobraszczyk, B.J. & Morgenstern, M.P. (2003). Rheology and bread making process. *Journal of Cereal Science,* Vol.38, No.3, (November, 2003), pp. 229-245, ISSN 07335210

Gujral, H.S.; Guardiola, I.; Carbonell, J.V. & Rosell, C.M. (2003). Effect of cyclodextrinase on dough rheology and bread quality from rice flour. *Journal of Agricultural and Food Chemistry,* Vol.51, No.13, (May, 2003), pp. 3814-3818, ISSN 0021-85661

Holčapek, M.; Jandera, P.; Fisher, J. & Prokeš, B. (1999). Analytical monitoring of biodiesel by high-performance liquid chromatography with various detection methods. *Journal of Chromatography A*, Vol.858, No.1, (October, 1999), pp. 13-31, ISSN 1873-3778

Jong, G.; Slim, T. & Creve, H. (1968). Bread without gluten. *Baker's Digest*, Vol.42, pp. 24-27, ISSN 0191-6114.

Liu, K. (1997). *Soybean: chemistry, technology and utilization*, Springer, ISBN 0-8342-1299-4

Mashayekh, M.; Mahmoodi, M.R. & Entezari, M.H. (2008). Effect of fortification of defatted soy flour on sensory and rheological properties of wheat bread. *International Journal of Food Science and Technology*, Vol.43, No.9, (September, 2008), pp. 1693-1698, ISSN 1365-2621

Maforimbo, E.; Skurray, G.; Uthaykumaran, S. & Wringly, C. (2008). Incorporation of soy proteins into the wheat-gluten matrix during dough mixing. *Journal of Food Science*, Vol.47, No. 2, (March, 2008), pp. 380-385 ISSN 0022-1147

Mohamed, A.A.; Shogren, R.L. & Sessa, D.J. (2006). Low carbohydrates bread: Formulation, processing and sensory quality. *Food Chemistry*, Vol.99, No.4, pp. 686-692, ISSN 0308-8146

Moore, M.; Tilman, T.S.; Dockery, P. & Arendt, E.K. (2004). Textural Comparisons of Gluten-Free and Wheat-Based Dough, Batters and Breads. *Cereal Chemistry*, Vol.81, No.5, (September/October, 2004), pp. 567-575, ISSN 0009-0352.

Olguin, M.C.; Hisano, N.; D'Ottavio, E.A.; Zingale, M.I.; Revelant, G.C. & Calderari, S.A. (2003). Nutritional and antinutritional aspects of an Argentinean soy flour assessed on weanling rats. *Journal of Food Composition and Analysis*, Vol.16, No.4, (August, 2003), pp. 441-449, ISSN: 0889-1575

Ortega-Garcia, J.O; Medina-Juarez L.A.; Gamez-Meza, N. & Noriega-Rodrigez, J.A. (2005). Optimisation of bleaching conditions for soybean oil using response surface methodology. *Food Science Technology International*, Vol.11, No.6, (December, 2006), pp. 443-450, ISSN 1532-1738

Qian, H.; Zhou, H. & Gu, J. (2006). Effect of different kinds of soy protein powder on bread-making properties. *Nongye Gongcheng Xuebao/Transactions of the Chinese Society of Agricultural Engineering*, Vol.22, No.10, (October, 2006), pp. 233-236, ISSN 10026819

Pérez, G.T.; Ribotta, P.D.; Steffolani, M.E. & Leon, A.E. (2008). Effect of soybean proteins on gluten depolymerization during mixing and resting. *Journal of the Science of Food and Agriculture*, Vol.88, No.3, (February, 2008), pp. 455-463, ISSN 0022-5142

Ribotta, P.D.; Arnulphi, S.A. & León, A.E. (2005). Effect of soybean addition on rheological properties quality of wheat flour. *Journal of the Science of Food and Agriculture*, Vol.85, No.11, (August, 2005), pp. 1889-1896, ISSN 1097-0010

Ribotta, P.D.; Pérez, G.T.; Aňo'n, M.C. & Leon, A.E. (2010). Optimization of additive combination for improved soy-wheat bread quality. *Food and Bioprocess Technology*, Vol.3, No.3, (June, 2010), pp. 395-405, ISSN 1935-5149

Roccia, P.; Ribotta, P.D.; Pérez, G.T. & León, A.E. (2009). Influence of soy protein on rheological properties and water retention capacity of wheat gluten. *LWT Food Science and Technology*, Vol.42, No.1, pp. 358-362, ISSN 1096-1127

Sabanis, D. & Tzia, C. (2009). Effect of rice, corn and soy flour addition on characteristics of bread produced from different wheat cultivars. *Food and Bioprocess Technology*, Vol.2, No.1, (March, 2009), pp. 68-79, ISSN 1935-5149

Sasaki, T. & Matsuki, J. (1998). Effect of wheat structure on swelling power. *Cereal Chemistry*, Vol.75, No.4, (May/June, 1998), pp. 385-391, ISSN 0009-0352

Schober, T.J.; O'Brien, C.M.; McCarthy, D.; Dardnedde, A. & Arendt, E.K. (2003). Influence of gluten-free flour mixes and fat powders on the quality of gluten-free biscuits. *European Food Research and Technology*, Vol.216, No.5, (May, 2003), pp. 369-376, ISSN 1438-2385

Shibanuma, Y.; Takeda, Y. & Hizukuri, S. (1996). Molecular and pasting properties of some wheat starches. *Carbohydrate Polymers*, Vol.29, No.3, (March, 1996), pp. 253-261, ISSN 0144-8617.

Stevenson, D.G.; Russell, K.D. & Jay-lin, J. & George, E.I. (2006). Structures Functional Properties from seeds of three soybeans (*Glycine max* (L.) Merr.) Varietes. *Starch*, Vol.58, No.10, (October, 2006), pp. 509-519, ISSN 1521-379x

Trajković, J.; Baras, J.; Mirić S. & Šiler, S. (1983). *Analysis of viands*, Technology and Metallurgy, Beograd, Serbia

Yoshida, H.; Kanerei, S.; Yuka, T. & Mizushina, Y. (2006). Regional characterization of tocopherols and distribution of fatty acids within soybean seeds (*Glycine max* L.). *Journal of Food Lipid*, Vol.13, No.1, (March, 2006), pp. 12-26, ISSN 1745-4522

5

Lipids, Nutrition and Development

Juliana M.C. Borba, Maria Surama P. da Silva
and Ana Paula Rocha de Melo
Federal University of Pernambuco
Brazil

1. Introduction

Lipids are classified as simple, compound and derived based on the hydrolysis, which result in breaking the fatty acids off, leaving free fatty acids and a glycerol, using up three water molecules. Simple lipids are esters of fatty acids with various types of alcohol. They are distinguished into fats and oils. Compound lipids contain an inorganic or organic group in addition to fatty acids and glycerol. They include phospholipids, glicolipids and lipoproteins. Finally, derived lipids are obtained by hydrolysis of simple and complex lipids. These lipids contain glycerol and other alcohols. They correspond to steroid hormones, ketone bodies, hydrocarbons, fatty acids, fatty alcohols, mono and diglycerides, terpenes and carotenoids. Sometimes they are present as waste products of metabolism. Lipids also can be classified, depending on its solubility or function, as polar or apolar compounds and as structural or reserve substances, respectively (Basso, 2007).

The main source of body energy comes from the triglycerides. These compounds are esters formed by one molecule of glycerol and three molecules of fatty acids. Fatty acids are carboxylic acids that usually have in its structure an unbranched carbon chain and one carboxyl. According to the saturation of the carbon chain, they can be classified as saturated, monounsaturated and polyunsaturated fatty acids (Basso, 2007).

In general, all mammals are able to synthesize saturated and monounsaturated fatty acids, but this ability is limited to polyunsaturated fatty acids (PUFAs), without them the organisms could not function properly. For this reason, these compounds are considered "essential" fatty acids. Thus, these fatty acids must be supplied by the diet. Linoleic acid 18:2 (n-6), a member of the n-6 family of fatty acids, was identified as the first "essential" fatty acid, whereas α-linolenic acid, 18:3 (n-3) represents the other essential fatty acid. These two essential fatty acids are the only sources for important longer chain fatty acids and physiological synthesis of complex lipids (Yehuda et al., 1999). Linoleic acid [18:2 (n-6)] is usually found in large quantities in soybean, corn, canola and safflower oil while α – linolenic [18:3 (n-3)] is easily found in green leafy vegetables, linseed and marine fish oil (Takahashi, 2005).

The main source of dietary lipids is obtained through the intake of triglycerides which can be found as fats or oils. The concept of fat or oil is based on the consistency and on the fatty acid present in the triglyceride molecule. At room temperature, oils are liquid because are constituted of tiacylglycerols containing a high proportion of mono and/or polyunsaturated fatty acids. These come from the vegetable sources such as soybean, corn, sunflower, olive or canola oil or from animal source such as fish oil. On the other hand, fats are solid or pasty

at room temperature and contain a large proportion of saturated fatty acids and/or unsaturated with trans double bonds. Fats can be from animal source such as butter, beef tallow or pig and from vegetable source such as cocoa butter and hydrogenated fats. Trans fats are produced naturally at the rumen or from an industrial process adding hydrogen to unsaturated fatty acids found in vegetable oils (partial or total hydrogenation) (Basso, 2007). After hydrogenation, vegetable oils may be converted from liquids to solids, resulting in margarines and shortenings which have excellent culinary properties even though have detrimental health effects. The partial hydrogenation increases the melt pointing of the fats at a room temperature and the degree of hydrogenation controls the final consistency of the manufactured products. These fats are regularly found in foods such as ice cream, chocolates, biscuits, cookies, cakes, mass and margarines (Basso, 2007; Kinsella et al., 1981).
Changes in the structure of the lipids molecules have metabolic and nutritional repercussions (Kummerow, 2009), since, produce a significant loss of essential fatty acids (Martin et al., 2004). According to Martin et al. (2004), human nutrition has been moulded substantially along the modernization and industrialization process. Because the property of hydrogenated fats in increasing validity of food products and making them look like more crispy and less oily, food industry and "fast food"restaurants has been intensifying the use of trans fatty acids. On the other hand, since 1990´s it has been a substantial interest of the scientif community in investigating the adverse effects over health caused by the long lasting intake of trans fatty acids.

1.1 Effects of early and prolonged intake of hydrogenated fat for health

Once considered a problem only in high-income countries, overweight and obesity are now dramatically present in high rates in low- and middle-income countries, particularly in urban areas. World Health Organization predicts that by 2015, approximately 2.3 billion adults will be overweight and more than 700 million will be obese. According to World Health Organization/Food and Agriculture Organization (2003), obesity and overweight is caused by an energy imbalance between calories consumed and calories expended. Among several factors related to the present development of overweight and obesity, an increased intake of foods high in fat and sugar but low in vitamins, minerals and fiber is one of the worldwide problems of this century (WHO, 2003; Remig et al., 2010). Fats are important nutrients in diet and have wide chemical properties that drive diverse metabolic effects.
Although optimal dietary fat quantity has been keenly pursued over past decades, attention has recently centered on the value of dietary fat quality (Gillingham et al., 2011). Recent population studies have shown the important role of monounsaturated and polyunsaturated fats as key nutrients in preventing chronic diseases in modern societies. In addition, sufficient intakes of polyunsaturated fats during childhood are required for optimal growth and development (Carrillo-Fernandez et al., 2011). Consumption of a healthy diet, containing adequate rates of omega6:omega3 throughout the life is important to maintain cardiovascular and possibly also cognitive and immune health (WHO/FAO, 2003). In addition to PUFA, dietary monounsaturated also promote benefits to health in terms of blood lipid profile, blood pressure, insulin sensitivity and glycemic control. Due to existing and emerging research on health attributes of monounsaturated rich diets, and to the low prevalence of chronic disease in populations consuming monounsaturated rich Mediterranean diets, recommendations have been made to replace saturated fat acids with unsaturated fats (Gillingham et al., 2011).

Although several studies have been shown the importance for health in replacing saturated for unsaturated fats in diets, currently, food products consumed in a typical Western diet contain a significant proportion of fats that have been industrially altered. It is estimated that 2% to 8% of energy needs in the typical Western diet come from chemically modified lipid products (Craig-Schmidt, 2006). Trans fatty acids frequently present in partially hydrogenated vegetable oils have been consumed on large scale all over the world. In terms of consumption, trans fatty acids have long been used in food manufacturing for reasons that were described above. Increasing epidemiologic and biochemical evidence suggest that high consumption of trans fats is related to many metabolic alterations and also induce a significant risk factor for cardiovascular disturbs (de Oliveira et al, 2011). Recently, Mozaffarian et al. (2006) showed that a 2% absolute increase in energy intake from trans fat has been associated with a 23% increase in cardiovascular risk. In general, these cardiovascular disturbs are related to some of known mechanisms such as reduction of c-HDL concentration, increase of low density lipoprotein and triglycerides; disturbance in prostaglandin balance and they may also promote insulin resistance (Castro-Martínez et al., 2010). Since , the American Heart Association as well as the World Health Organization recommend limiting trans fats to <1% energy and many others health institutions in the United States of America all recommend limiting dietary trans-fat intake from industrial sources as much as possible.

The presence of an inflammatory process in the arteries seems to be another risk factor in heart disease, and studies show that hydrogenated trans fats increase the inflammation in the arteries and promote endothelial dysfunction (Sun et al., 2007; Lopez-Garcia et al., 2005). Endothelial cells are responsible for the regulation of local vascular tone by means of releasing relaxing and contracting factors synthesized mainly from arachidonic acid. Trans fat inhibits COX-2, an enzyme which converts arachidonic acid to prostacyclin that is needed to prevent blood clots in the coronary arteries (Kummerow, 2009).

In addition to the relation between high rates of trans fatty acids in diet and increased risk for developing cardiovascular disease, a detrimental relationship was found between trans fatty acids intake and depression risk. Rising secular trends in the incidence of depressive disorders have been paralleled by a dramatic change in the sources of fat in the Western diet. This change mainly consists in the replacement of polyunsaturated or monounsaturated fatty acids by saturated fats and trans-unsaturated fats (Pawels & Volterrani, 2008). These findings suggest that cardiovascular disease and depression may share some common nutritional determinants related to subtypes of fat intake (Sanchez-Villegas et al., 2011).

It seems that the exposure to trans fatty acids in utero has negative consequences early in life. How much the in utero environment dictates birth weight and the programming of long-term obesity related disorders is still unclear, especially when compared with that of early neonatal growth rate. The placental transfer of trans fatty acids is still contradictory, both in human and animals (Haggarty, 2010). However, experimental studies have demonstrated that placenta is not completely impermeable to these compounds, since a number of trans fatty acids cross this barrier and accumulates in the liver and in the total body lipids of the fetus. However, despite that the myelinogenesis process is not finished in fetus, the amount of trans isomers transferred to fetal brain was negligible in all studies. At least in animals, this finding suggests the brain might be protected from the trans fatty acids accumulation, but no data have yet been reported for human newborns (Haggarty, 2010).

1.2 The role of lipids on the development of nervous system

Among the various organic systems, the nervous system plays the main role in controlling several physiological processes. During development, this system presents a rapid growth spurt period or a vulnerable period which corresponds to the highest rate of cellular migration and differentiation, neurogenesis, synaptogenesis, myelinization and maturation of neurotransmitter pathways. Depending on the animal species, this critical period of intense neural development occurs at different time points early in life (Dobbing, 1968). In human, for example, it occurs during the last trimester of gestation until the second year of life, while in rats it corresponds the lactation period. During this period, the nutrition is one of the essential environmental factors to a normal development because it provides nutrients without them the neurodevelopment would be impaired (Walker, 2005).

The classical studies about malnutrition show that nutritional deficiency in macro or micronutrients has deleterious effects on the brain (Winick & Rosso, 1969). In rats, malnutrition induces functional and developmental failures, as well as reduction in brain size (Morgane et al., 1992;1993). In children, malnutrition showed influence both short and long term problems of cognition and behavior (Grantham-McGregor & Baker-Henningham, 2005; Benton, 2008).

Essential fatty acids are important constituents of structural lipids in nervous membranes cell and signaling molecules, as such, are involved in many brain functions. Around 30–40% of the total phospholipids in these structures are docosahexaenoic acid molecules (Young et al., 2000), which appear to be specifically concentrated in membranes surrounding synapses (Carlson, 2001). Changes in the quantity and quality of the dietary fatty acids are often associated with developmental and functional alterations in the nervous system.

At the cellular level, an α-linolenic deficient diet can induce less complex patterns of dendritic branching (Wainwright, 2002), smaller neurite growth in hippocampal neurons (Calderon & Kim 2004) and reduced neuronal soma size in some brain regions (Ahmad et al., 2002). Modification in the fatty acid composition of rat brain cell membranes of neurons, astrocytes, oligodendrocytes, and of subcellular fractions, such as myelin and synaptosomes, are also induced by a diet with reduced levels of n-3 fatty acids (Bourre et al., 1984). It has been shown that essential fatty acids imbalance as well as specific fatty acid deficiencies in the maternal diet can affect the neuromotor development of pups, including the ability to respond to environmental stimulation (Lamptey & Walker, 1976; Wainwright, 2002; Anselmo et al., 2006).

Fats and oils as they exist in nature must be processed before they are suitable for human consumption (González et al., 2007). On the other hand, only a few studies have described the effects of the ingestion of trans fatty acids early in life on the rat brain development. In this study we investigated the replacement of soybean oil in the diet by partially hydrogenated vegetable oil, rich in trans fatty acids, from the beginning of gestation through lactation on reflex ontogeny.

2. Methods

2.1 Animals and diets

Female pregnant *Wistar* rats weighing 200-250 g were obtained from the colony of Department of Nutrition of the Federal University of Pernambuco. Twenty-four hours after the birth of the entire mothers´ nestling, born on the same day, were contained in a large group and randomly each litter was culled to six males and two female pups. The litters were also randomly assigned to isocaloric diets containing as lipid source 7% soybean oil (control group-C; n=32) or 7% of hydrogenated vegetal fat (experimental group-E; n=39), since gestation until weaning (21 days old). Both diets (table 1) were formulated based on recommendations of the American Institute of Nutrition-AIN-93 (Reeves et al., 1993). The animals were kept on a 12:12-h light -dark photoperiod at 24o C temperature during the whole period. The animals were maintained according to recommendations from the National Institute of Health (USA) and approved by the Ethics Committee for Use Animal, Federal University of Pernambuco (CEUA, protocol.23076.020339/2010-24). After delivery, from P1 until 21d, the pups were weighted at 1, 7, 14 and 21days of age. Indicators of somatic maturation and reflex ontogeny were studied daily from P1 to P21 between 07:00 am and 09:00 am. Daily it was observed the occurrence of the reflex responses, being considered the consolidation day to be the first one, of a series of three consecutive days where the reply was verified. For each reflex it was established a maximum observation time of 10s according to the experimental model established by Smart & Dobbing (1971).

Ingredients	**Control diet (g/100g)**	**Experimental diet (g/100g)**
Casein	20.0	20.0
Cellulose	5.0	5.0
Corn starch	52.95	53.49
Sucrose	10.0	10.0
Soyabean oil	7.0	-
Hydrogenated vegetable fat	–	6.46
Vitamin mix 1	1.0	1.0
Mineral mix 2	3.5	3.5
L-Cystine	0.3	0.3
Choline Bitartrate	0.25	0.25

Table 1. Composition of the diets. Vitamin mixture (Rhoster Ind.Com. LTDA. SP. Brazil) containing (m%): folic acid (20); niacin (300); biotin (2); calcium pantothenate 160; pyridoxine 70; riboflavin 60; thiamine chloride 60; vitamin B_{12} 0.25; vitamin K_1 7.5. Additionally containing (UI%): vitamin A 40.000; vitamin D_3 10.000; vitamin E 750. 2 Mineral mixture (Rhoster Ind. Com. LTDA. SP. Brazil) containing (m%): $CaHP0_4$ (38); K_2HP0_4 (24); $CaCO_3$ (18.1); NaF (0.1); NaCl(7.0); MgO (2.0); $MgS0_4$ $7H_20$ (9.0); $FeS0_4$ $7H_20$ (0.7); $ZnS0_4$ H_20 (0.5); $MnSO^+$ H_20 (0.5); $CuS0_4$ $5H_20$ (0.1); $Al_2(S0_4)_3K_2S0_4$ $24H_20$ (0.02); Na_2SeO_3 $5H_20$ (0.001); KCl (0.008).

2.2 Indicators of somatic maturation

The following indicators of somatic maturation, as illustrated in figure 1, were analyzed in order to study whether the replacement of soybean oil by vegetable hydrogenated fat in the diet influenced the development of physical features of the rat pups.

2.3 Indicators of reflex ontogeny

The following indicators of the reflex ontogeny investigated at the present study are described and illustrated below (figure 2):

Palmar Grasp (PG)- This reflex is present at birth and consists of a dorso flexion of the digits ('grasping') in response to the stimulation of the hand-palm with a small metallic stick. The

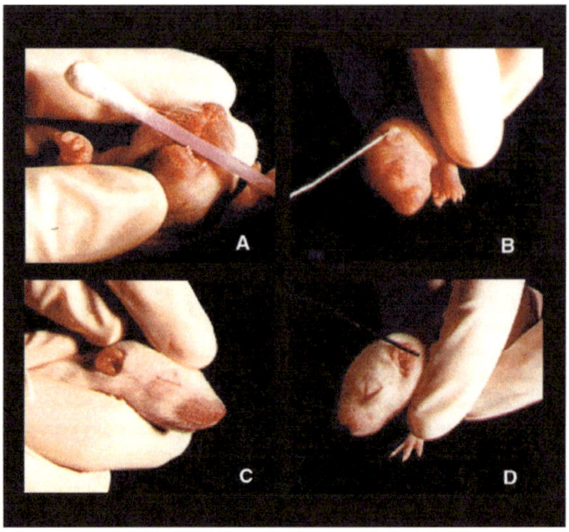

Fig. 1. Indicators of somatic maturation. A) Eruption of the Upper Incisors (EUI) and Eruption of the Lower Incisors (ELI); B) Ear Unfolding (EU); C) Eye Opening (EO); D) Auditory Conduit Opening (ACO).

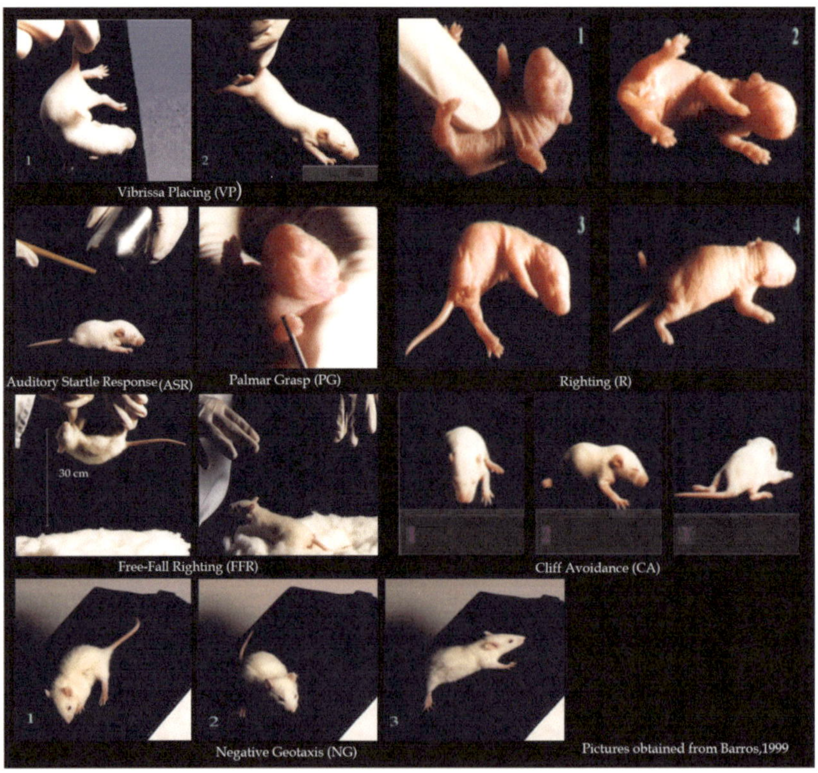

Fig. 2. The figure represents the indicators of reflex ontogeny observed during lactation period.

expected response is the disappearance of the palmar grasping response, as the organism matures.

Righting (R)- The newborn is placed on its back on a flat surface and the expected mature response is to turn over on the ventral surface, resting in the normal position with the four feet on the ground.

Vibrissa Placing (VP)- The pup is held by the tail, with the head facing the edge of a table and the vibrissae just touching the vertical surface of the table. The expected response is to lift the head and to extend the fore legs in direction of the table.

Cliff Avoidance (CA) - The newborn is placed on the edge of a 'cliff' (for instance, on the edge of a table), with the forepaws and face just over the edge. The expected response is to move away from the cliff, to avoid dropping.

Negative Geotaxis (NG)- The pup is placed on an inclined ramp (45° slope) with its head pointing to the ground. The expected mature response is to turn around and crawl up the slope.

Free-Fall Righting (FFR)- The pup is held with the back downwards 35cm above a cotton pad and dropped. The expected response is to turn in mid-air to land on its four paws.

Auditory Startle Response (ASR)- The newborn is exposed suddenly to a loud, sharp noise. The expected response is a prompt extension of the head and the limbs, followed by withdrawal of the limbs and a crouching posture.

2.4 Statistical analysis

The body weight was analyzed by Student's t-test. Results are presented as means±standard error of the mean (SEM). Differences were significant when $p < 0.05$. Results of the somatic maturation and reflex ontogeny were evaluated by Mann-Whitney test. Results are expressed as median ± interquartile. Statistical significance was set at $p < 0.05$.

3. Results

3.1 Body weight

The figure 3 shows that pups fed an experimental diet did not exhibited significant differences in the body weight at the 1st, 7th, 14th and 21th day, during the lactation period when compared to the controls. (C = 6.7 g ± 0.12; 16.3 g ± 0.49; 29.6 g ± 1.18; 46.4 g ±1.52; E = 6.6 g ± 0.11; 16.5 g ± 0.22; 30.3 g ± 0.48; 47 g ± 0.86).

3.2 Indicators of somatic maturation

The effect of the experimental diet on somatic development is shown in Figure 4, where the results are expressed in median (min. - max.) and compared with the control group. There was no difference between the number of days for ear unfolding, eye opening and the eruptions of superior and inferior incisors. However, the opening of the external auditory canal was significantly delayed in the experimental group as compared to the control (C: 12-1; E: 14-2.5; $p<0.05$).

3.3 Indicators of reflex ontogeny

As can be seen in Figure 5, the development of the early reflexes which appear in the first postnatal week such as righting, cliff avoidance and vibrissa placing did not differ between control and experimental pups. In the second postnatal week, the negative geotaxis was the only reflex which was delayed in the experimental group when compared to the control (C: 11-3; E: 9-1.5).

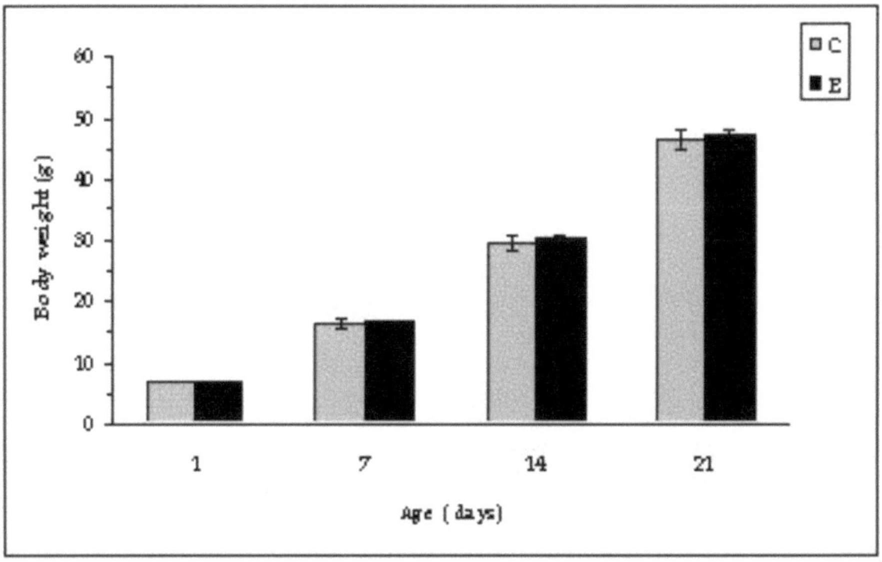

Fig. 3. Body weight in control (C) and experimental (E) offsprings from the 1st , 7th, 14th and the 21st day. Each bar represents the mean ± SEM. C = control group fed a diet containing 7% of soybean oil (n=32); E = experimental group fed a diet containing 7% of hydrogenated vegetal fat (n=39).

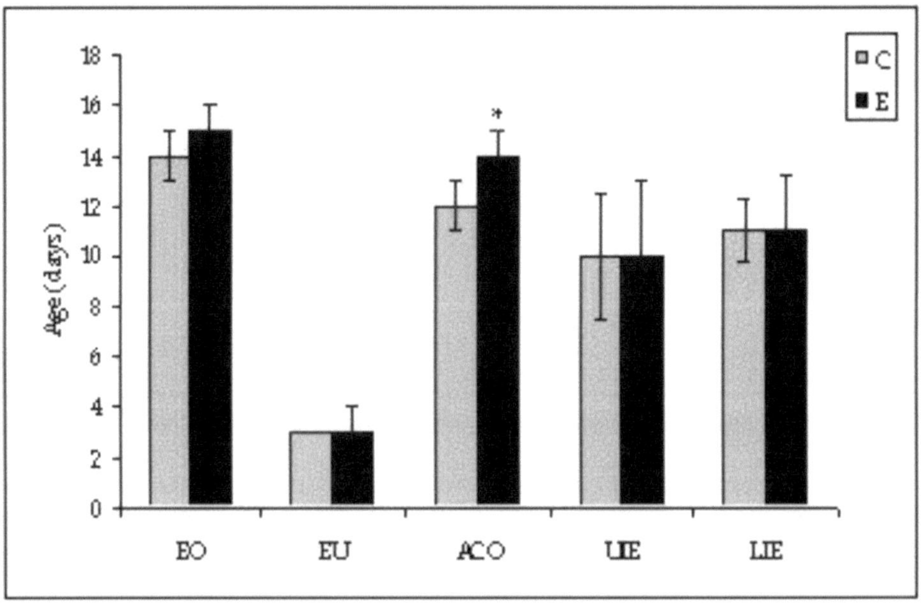

Fig. 4. Somatic maturation indicators of suckling rats fed the control diet (C; n=32) and the experimental diet (E; n=39). Each bar represents the median ± interquartile. EO = Eye Opening; EU = Ear Unfolding; ACO = Auditory Conduit Opening; UIE = Upper Incisors Eruption; LIE = Low Incisors Eruption. * $p < 0.05$ vs C group (Mann-Whitney Test).

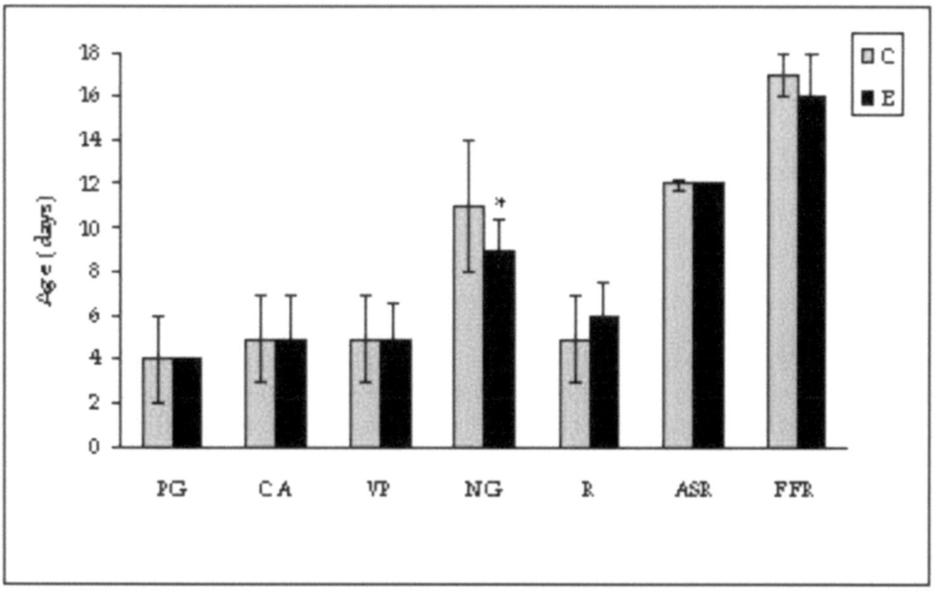

Fig. 5. Reflex ontogeny. Each bar represents the median + inertquartile. C = control group (n=32); E = experimental group (n=39). PG = Palmar Grasp; CA = Cliff Avoidance; VP = Vibrissa Placing; NG = Negative Geotaxis; R = Righting; ASR = Auditory Startle Response; FFR = Free-fall Righting. * $p< 0.05$ vs C group (Mann-Whitney Test).

4. Discussion

The currently hydrogenation of vegetable oils produces trans fatty acids which can be found in most manufactured products. The high consumption of these foods is related with the increase of health problems. When consumed, trans fat acids can be found in plasma and in the maternal milk (Carlson et al., 1997) and its concentration varies with daily mother intake. Circulating trans fat acids are also carried and incorporated into tissues such as the brain, liver, adipose tissue and spleen and its levels on these tissues depend on the amount ingested (Larqué et al., 2001).
In this study, we observed that offsprings fed a hydrogenated fat based diet did not exhibited differences in body weight from P0 until P21 when compared to the controls. One could explain this finding by the fact of the experimental drawing, herein used, kept the same proportion of lipids in the both diets, only replacing soybean oil for hydrogenated vegetal fat in the experimental diet. Although this dietary treatment has been offered during a period of an intense growth and body development, it did not compromise the body weight gain of the pups over the lactation period. These data are similar to those found in a previous study of our laboratory when rats fed a diet containing 5%coconut-oil or soybean-oil during pregnancy and lactation did not show significant changes in body weight gain from P0 to P21 when compared to the control group (Borba et al., 2010). Santillán et al., (2010) did not find differences in the body weight of mice fed a commercial diet enriched with soybean or sunflower oil diet over gestation and lactation when compared with those fed a commercial diet. On the other hand, mice known to be prone to obesity and insulin resistance when consumed a high fat diet during pregnancy and lactation exhibited a

growth delay at the first week of life, but accelerated the growth in the subsequent two weeks (Kavanagh, et al., 2010). It is known that the fat content variation in human milk is clearly the result of different dietary, metabolic and physiologic controls (German, 2011). In humans, the fatty acid composition in maternal diet and in breastmilk during lactation may not affect the infant body composition in the early postpartum period but may be a factor in the development of childhood overweight later in life (Anderson et al., 2010).

The somatic maturation and reflex development from the 2nd over the 21st postnatal days of life are good indicators to understand how environmental factors can influence the functional maturity of the brain development (Smart & Dobbing, 1971; Gramsbergen, 1998).

Among the somatic parameters herein investigated, the opening of the external auditory canal was delayed in the experimental group when compared to the control group. However, we did not observe this effect on the auditory startle response. These results suggest that the consumption of trans fat during the critical period of development did not influence the function of the auditory system since the hearing sensation was preserved. It has been known since the 19th century that hearing may occur through bone conduction; however the way how this physiologic pathway works is not completely understood. Some factors can contribute to the bone conduction, such as: sound radiated into the external ear canal, middle ear ossicle inertia, inertia of the cochlear fluids, compression of the cochlear walls and pressure transmission from the cerebrospinal. Of these five, inertia of the cochlear fluid seems the most important. The efficiency of the bone conduction is largely dependent on the skull bone where the skull acts as a rigid body at low frequencies and incorporates different types of wave transmission at higher frequencies (Stenfelt & Goode, 2005).

Regarding the others somatic maturation indicators no differences were observed between the group fed a trans fat based diet and the group fed a soybean oil based diet. When the reflex ontogeny was analyzed, we observed that there was a significant delayed in the negative geotaxis of the pups fed an experimental diet. These data suggested that there was a negative effect possibly induced by the lack of any essential fatty acid or by the trans fat *per se* on the development of the motor and the cerebellar system. The negative geotaxis reflex is stimulated by the abnormal position of the head and the body which are under control of the vestibular and postural systems (Adams et. al, 1985). This reflex requires a sequence of organized motor events (Ramirez & Spear, 2010) but this only occurs if the motor system is maturated. In rats, spinal cord descending projections develop relatively early. Projections from vestibulospinal and reticulospinal origin reach the cervical levels of the spinal cord at 13rd or 14rd embrionary day. Around the same period, motoneurons in the ventral horn of thoracic and lumbosacral spinal cord segments start developing and two days thereafter, their axons invade the muscle mass of the caudal limb bud (Altman & Bayer, 1984; 1997; Gramsbergen, 1993).

The development of cerebellum occurs in the postnatal period, reaching its peak of development at the end of the first week (Smart & Dobbing, 1971). This period results of a number of events including: neuronal and glial proliferation, outgrowth of axons and dendrites, establishment of synaptical contacts, as well as myelination (Altman & Bayer, 1997). This late development makes the cerebellum a structure particularly vulnerable to insufficient supply of nutrients or to side and possible beneficial effects of pharmacological treatments (Gramsbergen, 2003). It has been shown that a restriction of daily food intake to dams delays the motor development and behavior associated with a disturbed cerebellar development of the offspring (Gramsbergen, 2003). On the other hand, Collucia et al. (2009) showed that omega-3 supplementation during gestation and lactation improved motor coordination in juvenile-adults rats.

One of the problems with the process of hydrogenation is the fact that it possibly produces a loss of essential fatty acids of the original vegetable oils. Hill et al. (1982) showed that rats fed a diet containing as a lipid source partially hydrogenated soybean oil showed a reduction of essential fatty acids levels in the liver and heart. It is evident that partially hydrogenated fats have excellent culinary properties, but from a nutritional point of view, the consumption of trans fatty acids represents a loss of essential fatty acids intake that may have a hazardous impact on health. This study is the first evidence that the consumption of hydrogenated vegetable fat during the critical period of development may compromise some parameters of the reflex and somatic development of rat pups.

5. Conclusion

Although the maternal intake of a diet containing trans fatty acid in replacement of soybean oil have not changed the body weight in the early postpartum period of the pups, it influenced negatively both somatic and reflex development. Recently, there is an increase in the level of interest in fatty acids and lipids. This interest is not limited to brain biochemistry, but also to the effects of levels and ratios of fatty acids on physiological and behavioral aspects. For these reasons, more research is warranted regarding the influence of maternal dietary on the fatty acid composition of the breast milk and their effects on body composition, the development of overweight and behavior changes later in life of rat pups.

6. Acknowledgements

This publication was made possible in part by support from the Universidade Federal de Pernambuco, Pró-Reitoria para Assuntos de Pesquisa e Pós-Graduação (PROPESQ). The authors gratefully acknowledge the invaluable assistance of Dr. Edeones França for the animal care.

7. References

Adams, J., Buelke-Sam, J., Kimmel, C.A., Nelson, C.J., Reiter, L.W., Sobotka, T.J., Tilson, H.A. & Nelson, B.K. (1985). Collaborative Behavioral Teratology Study: protocol design and testing procedures. *Neurobehav. Toxicol. Teratol.*, Vol.. 7, No. 6 (November-December 1985), pp. 579-586, ISSN 0275-1380

Ahmad, A., Moriguchi, T. & Salem N. (2002). Decrease in neuron size in docosahexaenoic acid-deficient brain. *Pediatr. Neurol.*,Vol. 26, No. 3, (March 2002), pp. 210-8, ISSN 0022-3166

Altman, J. & Bayer, S.A. (1984). The development of the rat spinal cord. *Adv. Anat. Embryol. Cell. Biol.*, Vol. 85, (1984), pp. 1-164, ISSN 0022-3077

Altman, J. & Bayer, S.A. (1997). Development of the cerebellar system in relation to its evolution, structure and functions, In: *The Central Nervous System, Structure and Function*, Brodal, P. (Ed.), pp. 1-783, Oxford University Press, ISBN 084-93-9490-2, New York, USA

Anderson, A.K., McDougald, D.M. & Steiner-Asiedu, M. (2010). Dietary trans fatty acid intake and maternal and infant adiposity. *European Journal of Clinical Nutrition*, Vol. 64, (November 2010) pp.1308-1315, ISSN 0954-3007

Anselmo, C.W., Santos, A.A., Freire, C.M., Ferreira, L.M., Cabral Filho, J.E., Catanho, M.T. & Medeiros, Mdo C. (2006). Influence of a 60 Hz, 3 micro T, electromagnetic field on the reflex maturation of Wistar rats offspring from mothers fed a regional basic diet during pregnancy. *Nutr. Neurosci.*, Vol. 9, No. 5-6, (October-December 2006), pp. 201-6, ISSN 1028-415X

Basso, R. (2007). Bioquímica e Metabolismo dos Lípides, In: *Tratado de Alimentação, Nutrição & Dietoterapia*, Silva, S.M.C.S. & Mura, J.D.P. (Ed.), pp. 55-73, Roca, ISBN 978-85-7241-678-8, São Paulo, Brazil

Benton, D. (2008). Sucrose and behavioral problems. *Crit. Rev. Food. Sci. Nutr.*, Vol. 48, No. 5, (May 2008), pp. 385-401, ISSN 1040-8398

Borba, J.M., Rocha-de-Melo, A.P., dos Santos, A.A., da Costa, B.L., da Silva, R.P., Passos, P.P. & Guedes, R.C. (2010). Essential fatty acid deficiency reduces cortical spreading depression propagation in rats: a two-generation study. *Nutr. Neurosci.*,Vol. 13, No. 3, (June 2010), pp. 144-50, ISSN 1028-415X

Bourre, J.M., Pascal, G., Durand, G., Masson, M., Dumont, O. & Piciotti, M. (1984). Alterations in the fatty acid composition of rat brain cells (neurons, astrocytes, and oligodendrocytes) and of subcellular fractions (myelin and synaptosomes) induced by a diet devoid of n-3 fatty acids. *J. Neurochem.*, Vol. 43, No. 2, (August 1984), pp. 342-8, ISSN 0022-3042

Calderon, F. & Kim, H.Y. (2004). Docosahexaenoic acid promotes neurite growth in hippocampal neurons. *J. Neurochem.*, Vol.90, No. 4, (August 2004), pp. 979-88, ISSN 0022-3042

Carlson, S.E., Clandinin, M.T., Cook, H.W., Emken, E.A. & Filer, L.J. Jr. (1997). Trans Fatty acids: infant and fetal development. *Am. J. Clin. Nutr.*, Vol.66, No. 3, (September 1997), pp.715S-36S, ISSN 0002-9165

Carlson, S.E. (2001). Docosahexaenoic acid and arachidonic acid in infant development. *Semin. Neonatol.*, Vol. 6, No. 5, (October 2001), pp. 437-449, ISSN 1744-165X

Carrillo-Fernández, L., Dalmau- Serra, J., Martínez Álvarez, J.R., Solà Alberich, R. & Pérez Jiménez, F. (2011). Dietary fats and cardiovascular health. *Anales de Pediatría (Barc)*, Vol.74, No. 03, (February 2011), e1-e16, ISSN 1695-4033

Castro-Martínez, M.G., Bolado-García, V.E., Landa-Anell, M.V., Liceaga-Cravioto, M.G., Soto-González, J. & López-Alvarenga, J.C. (2010). Dietary trans fatty acids and its metabolic implications. *Gac. Med. Mex.*, Vol. 146, No. 4, (July-August 2010), pp. 281-288, ISSN 0016-3813

Craig-Schmidt, M.C. (2006). World-wide consumption of trans fatty acids. *Atherosclerosis Supplements*, Vol. 7, pp. 1–4, ISSN 1567-5688

Coluccia, A., Borracci, P., Renna, G., Giustino, A., Latronico, T., Riccio, P. & Carratu, M.S. (2009). Developmental omega-3 supplementation improves motor skills in juvenile-adult rats. *Int. J. Devl. Neuroscience*, Vol. 27, (May 2009), pp. 599–605, ISSN 0736-5748

de Oliveira, J.L., Oyama, L.M., Hachul, A.C., Biz, C., Ribeiro, E.B., Oller do Nascimento, C.M. & Pisani, L.P. (2011). Hydrogenated fat intake during pregnancy and lactation caused increase in TRAF-6 and reduced AdipoR1 in white adipose tissue, but not in muscle of 21 days old offspring rats. *Lipids Health Dis.*, Vol. 10, No. 22, (January 2011), pp. 1-9, ISSN 1476-511X

Dobbing, J. (1968). The development of the blood-brain barrier. *Prog. Brain. Res.* Vol. 29, pp. 417-27, ISSN 0079-6123

German, J. B. (2011). Dietary lipids from an evolutionary perspective: sources, structures and function. *Maternal & Child Nutrition*, Vol. 7, No. 2, (2011), pp. 2-16, ISSN 1740-8709

Gillingham, L.G., Harris-Janz, S. & Jones, P.J. (2011). Dietary Monounsaturated Fatty Acids Are Protective Against Metabolic Syndrome and Cardiovascular Disease Risk Factors. *Lipids*, Vol. 46, No. 3, (March 2011), pp. 209-28, ISSN 0024-4201

González, C., Resa, J.M., Concha, R.G. & Goenaga, J.M. (2007). Enthalpies of mixing and heat capacities of mixtures containing acetates and ketones with corn oil at 25°C. *J. Food Eng.*, Vol., 79, pp. 1104-1109, ISSN 0260-8774

Gramsbergen, A. (1993). Posture and locomotion in the rat: independent or interdependent development? *Neurosci. Biobehav. Rev.*, Vol. 22, No. 4, (July 1998), pp. 547-53, ISSN 0149-7634

Gramsbergen A. (2003). Clumsiness and disturbed cerebellar development: insights from animal experiments. *Neural Plast.*,Vol. 10, No. 1-2, (2003), pp. 129-40, ISSN 1687-5443

Grantham-McGregor S & Baker-Henningham H. (2005). Review of the evidence linking protein and energy to mental development. *Public. Health. Nutr.*, Vol. 8, No. 7A, (October 2005), pp. 1191-201, ISSN 1368-9800

Haggarty, P. (2010). Fatty Acid Supply to the Human Fetus. *Annal Review of Nutrition*, Vol. 30, (May 2010), pp. 237-255, ISSN 0199-9885

Hill, E.G., Johnson, S.B., Lawson, L.D., Mahfouz, M.M. & Holman, R.T.(1982). Perturbation of the metabolism of essential fatty acids by dietary partially hydrogenated vegetable oil. *Proc. Natl. Acad. Sci.*, Vol. 79, No. 4, (February 1982), pp. 953-7, ISSN 0027-8424

Lamptey, M.S. & Walker, B.L. (1976). A possible essential role for dietary linolenic acid in the development of the young rat. *Journal of Nutrition*, Vol. 106, No. 1, (January1976), pp. 86-93, ISSN 0022-3166

Larqué, E., Zamora, S. & Gil, A. (2001). Dietary trans fatty acids in early life: a review. *Early Hum Dev.*, Vol. 65, Suppl 2, (November 2001), pp. S31-S41, ISSN 0378-3782

Lopez-Garcia, E., Schulze, M. B., Meigs, J. B., Manson, J.A. E. , Rifai, N., Meir, J. S., Willett , W. C. & Hu, F. B. (2005). Consumption of Trans Fatty Acids Is Related to Plasma Biomarkers of Inflammation and Endothelial Dysfunction. *Journal of Nutrition*, Vol. 135, No. 3, (March 2005), pp. 562-566, ISSN 0022-3166

Kavanagh, K., Sajadian, S., Jenkins, K.A., Wilson, M.D., Carr ,J.J., Wagner, J.D. & Rudel, L.L.(2010). Neonatal and fetal exposure to trans-fatty acids retards early growth and adiposity while adversely affecting glucose in mice. *Nutr. Res.*,Vol. 30, No. 6, (June 2010), pp. 418-26, ISSN 0271-5317

Kinsella, J.E., Bruckner, G., Mai, J. & Shimp, J. (1981). Metabolism of trans fatty acids with emphasis on the effects of trans, trans-octadecadienoate on lipid composition, essential fatty acid, and prostaglandins: an overview. *The American Journal of Clinical Nutrition*, Vol.34, (October 1981), pp. 2307-2318, ISSN 0002-9165

Kummerow, F.A. (2009). The negative effects of hydrogenated trans fats and what to do about them. *Atherosclerosis*, Vol. 205, No. 2, (March 2009), pp. 458–465, ISSN 0021-9150

Martin, C. A., Matshushita, M. & Souza, N. E. (2004). Ácidos graxos trans: implicações nutricionais e fontes na dieta. *Revista de Nutrição*, Vol. 17, No. 3, (July-September 2004), pp. 361-368, ISSN 1415-5273

Morgane, P.J., Austin-LaFrance, R., Bronzino, J., Tonkiss, J., & Galler, J.R. (1992). Malnutrition and development central nervous system. In: The Vulnerable Brain and Environmental Risks, Issacson, R.L., Jensen, K.F. (Ed.), pp. 2-42, Springer Us, ISBN 0306441489, New York, USA

Morgane, P.J., Austin-LaFrance, R., Bronzino, J., Tonkiss, J., Díaz-Cintra, S., Cintra, L., Kemper, T. & Galler, J.R. (1993). Prenatal malnutrition and development of the brain. *Neurosci. Biobehav. Rev.*,Vol. 17, No. 1, pp. 91-128, ISSN 0149-7634

Mozaffarian, D., Katan, M.B., Ascherio, A., Stampfer, M.J. & Willett, W.C. (2006). Trans Fatty Acids and Cardiovascular Disease. *N. Engl. J. Med.*, Vol. 354, (April 2006), pp. 1601-1613, ISSN 0028-4793

Pawels, E.K. & Volterrani, D. (2008). Fatty acid facts, Part I. Essential fatty acids as treatment for depression, or food for mood? *Drug News Perspect,* Vol. 21, No. 8, (October 2008), pp. 446–451, ISSN 0214-0934

Ramirez, R.L. & Spear, L.P. (2010). Ontogeny of ethanol-induced motor impairment following acute ethanol: assessment via the negative geotaxis reflex in adolescent and adult rats. *Pharmacol Biochem. Behav.,* Vol. 95, No.2, (April 2010), pp. 242-248, ISSN 0091-3057

Reeves, P. G., Nielsen, F. H. & Fahey, G. C. (1993). AIN-93 Purified Diets for Laboratory Rodents: Final Report of the American Institute of Nutrition. *Journal of Nutrition,* Vol. 123, pp. 1939-1951, ISSN 0022-3166

Remig, V., Franklin, B., Margolis, S., Kostas, G., Nece, T. & Street , J. C. (2010). Trans Fats in America: A Review of Their Use, Consumption, Health Implications, and Regulation. *Journal of the American Dietetic Association,* Vol. 110, No. 4 , (April 2010), pp. 585-592, ISSN 0002-8223

Sánchez-Villegas, A., Verberne, L., De Irala, J., Ruíz-Canela, M., Toledo, E., Serra-Majem, L. & Martínez-González, M.A. (2011). Dietary Fat Intake and the Risk of Depression: The SUN Project. *PLoS One,* Vol. 26, No. 1, (January 2011), e16268, ISSN 1932-6203

Santillán, M.E., Vincenti, L.M., Martini, A.C., de Cuneo, M.F., Ruiz, R.D., Mangeaud, A. & Stutz, G. (2010). Developmental and neurobehavioral effects of perinatal exposure to diets with different omega-6:omega-3 ratios in mice. *Nutrition,* Vol. 26, No. 4, (April 2010), pp. 423-31, ISSN 0899-9007

Smart, J.L. & Dobbing, J. (1971). Vulnerability of developing brain. II. Effects of early nutritional deprivation on reflex ontogeny and development of behavior in the rat. *Brain Research,* Vol. 28, pp. 85-95, ISSN 0006-8993

Stenfelt, S. & Goode, R.L. (2005). Bone-conducted sound: physiological and clinical aspects. *Otol. Neurotol.,*Vol. 26, No. 6, (November 2005), pp. 1245-61, ISSN 1531-7129

Sun, Q., Ma, J., Campos, H. & Hu, F.B. (2007). Plasma and erythrocyte biomarkers of dairy fat intake and risk of ischemic heart disease. *Am. J. Clin. Nutr.,* Vol. 86, No.4, (May 2007), pp. 929-937, ISSN 0002-9165

Takahashi, N.S. (October 2005). Importância dos ácidos graxos essenciais. 04.12.2010, Available from: <ftp://ftp.sp.gov.br/ftppesca/acidos_graxos.pdf>

Wainwright, P.E. (2002). Dietary essential fatty acids and brain function: a developmental perspective on mechanisms. *Proc Nutr Soc.,* Vol. 61, No. 1, (February 2002), pp. 61-9, ISSN 0029-6651

Walker, C.D. (2005). Nutritional aspects modulating brain development and the responses to stress in early neonatal life. *Prog. Neuropsychopharmacol. Biol. Psychiatry,* Vol. 29, No. 8, (December 2005), pp. 1249-63, ISSN 0278-5844

WHO/FAO (2003). Diet, Nutrition and the Prevention of Chronic Diseases: Report of a Joint WHO/FAO Expert Consultation. *WHO Technical Report Series,* No. 916, World Health Organization: Geneva

WHO. (2003). Diet, nutrition and the prevention of chronic diseases. In: Organization, WH, editor. WHO Technical Report Series. Geneva

Winick, M. & Rosso, P. (1969). The effect of severe early malnutrition on cellular growth of human brain. *Pediatric Research,* Vol. 3, (1969), pp. 181-184, ISSN 0031-3998

Yehuda S., Rabinovitz S. & Mostofsky D. I.(1999). Essential Fatty Acids Are Mediators of Brain. *Journal of Neuroscience Research,* Vol. 56, (1999), pp. 565–570, ISSN 0360-4012

Young, C., Gean, P.W., Chiou, L.C. & Shen, Y.Z. (2000). Docosahexaenoic acid inhibits synaptic transmission and epileptiform activity in the rat hippocampus. *Synapse,* Vol. 37, No. 2, (August 2000), pp. 90-94, ISSN 0887-4476

6

Inhibition of Soybean Lipoxygenases – Structural and Activity Models for the Lipoxygenase Isoenzymes Family

Veronica Sanda Chedea[1] and Mitsuo Jisaka[2]
[1]*Laboratory of Animal Biology, National Research Development Institute for Animal Biology and Nutrition (IBNA),*
[2]*Faculty of Life Sciences and Biotechnology, Shimane University,*
[1]*Romania*
[2]*Japan*

1. Introduction

Lipoxygenases (EC 1.13.11.12, linoleate:oxygen, oxidoreductases, LOXs) which are widely found in plants, fungi, and animals, are a large monomeric protein family with non-heme, non-sulphur, iron cofactor containing dioxygenases that catalyze the oxidation of polyunsaturated fatty acids (PUFA) as substrate with at least one 1Z, 4Z-pentadiene moiety such as linoleic, linolenic and arachidonic acid to yield hydroperoxides (Gardner, 1991).

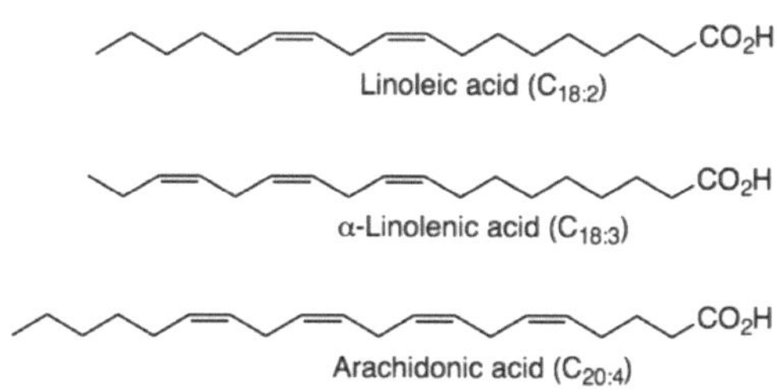

Fig. 1. Lipoxygenase substrates, linoleic, α- linolenic and arachidonic acid.

Theorell et al. (1947) succeeded in crystallizing and characterizing lipoxygenase (LOX) from soybeans and since then among plant LOXs, soybean lipoxygenase isozyme 1 (LOX-1) can be regarded as the mechanistic paradigm for these nonheme iron dioxygenases (Coffa et al., 2005; Minor et al., 1996; Fiorucci et al., 2008).

Designing agents to modulate activities of the variety of so closely homologous enzymes, such as different LOXs, require an intimate knowledge of their 3D structures, as well as information about metabolism of the potential xeno- or endobiotics. So far only the structures of soybean isozymes LOX-1 and LOX-3 have been determined for native enzymes, and several structures of their and rabbit 15-LOX (from reticulocytes) molecular complexes with inhibitors are known. Due to lack of sufficiently purified human enzymes most of the structural research has been done on soybean LOX (Skrzypczak-Jankun et al., 2003).

Understanding the mechanism of inhibition of LOXs can have profound effect in the development of many anti-cancer and anti-inflammatory drugs. On the basis of the available LOX data it was suggested that a combination of LOX modulators might be needed to shift the balance of LOX activities from procarcinogenic to anticancerogenic as a novel strategy for cancer chemoprevention (Skrzypczak-Jankun et al., 2003).
The aim of the present study is to present knowledge on different lipoxygenases having the soybean lipoxygenases as a structural and activity template for their inhibition by natural antioxidant compounds as theoretical approach for food biochemistry and medical applications.

2. Lipoxygenase structure and activity

The three-dimensional structure of soybean lipoxygenase-1 has been determined to 2.3 Å resolution by single crystal X-ray diffraction methods (Boyington et al., 1993). It is a two-domain, single-chain prolate ellipsoid of dimensions 90 x 65 x 60 Å with a molecular mass of 95 kDa. The 839 residues are organised in two domains: one 146 residue N-terminal domain (domain I), and a major, 693 residue C-terminal domain (domain II) (Prigge et al., 1997). Overall, the three-dimensional structure of lipoxygenase-1 shows a helical content of 38.0% and a β-sheet content of 13.9%. The structure of another crystal form of soybean lipoxygenase-1 determined to 1.4 Å resolution (Minor et al., 1996) showed very similar results. The structure of lipoxygenase-3, another soybean lipoxygenase isozyme (Skrzypczak-Jankun, 1997) shows that the lipoxygenase-3 isozyme is very similar in structure despite significant differences in sequence: 857 residues vs 839, deletions at 7 positions, insertions at 25 positions, and substitutions at 224 residues (72% identity).
Soybean seed isoenzymes are 94–97 kDa monomeric proteins with distinct isoelectric points ranging from about 5.7 to 6.4, and can be distinguished by optimum pH, substrate specificity, product formation and stability (Siedow, 1991; Mack et al., 1987). LOX-1 is the smallest in size (838 amino acids; 94 kDa), exhibits maximal activity at pH 9.0 and converts linoleic acid preferentially into the 13-hydroperoxide derivative. LOX-2 is characterized by a larger size (865 amino acids; 97 kDa), by a peak of activity at pH 6.8, and forms equal amounts of the 13-and 9-hydroperoxide compounds (Loiseau et al., 2001). LOX-2 oxygenates the esterified unsaturated fatty acid moieties in membranes in contrast to LOX-1 which only uses free fatty acids as substrates (Maccarrone et al., 1994). LOX-3 (857 amino acids; 96.5 kDa) exhibits its maximal activity over a broad pH range centred around pH 7.0 and displays a moderate preference for producing a 9-hydroperoxide product. It is the most active isoenzyme with respect to both carotenoid cooxidation and production of oxodienoic acids (Ramadoss, 1978).

3. Lipoxygenase reaction

The initial step of LOX reaction is removal of a hydrogen atom from a methylene unit between double bonds in substrate fatty acids (Fig. 2A). The resulting carbon radical is stabilized by electron delocalization through the double bonds. Then, a molecular oxygen is added to the carbon atom at +2 or –2 position from the original radical carbon, forming a peroxy radical as well as a conjugated *trans,cis*-diene chromophore. The peroxy radical is then hydrogenated to form a hydroperoxide. The initial hydrogen removal and the following oxygen addition occur in opposite (or antarafacial) sides related to the plane

formed by the 1Z,4Z-pentadiene unit. In most LOX reactions, particularly those in plants, the resulting hydroperoxy groups are in *S*-configuration, while one mammalian LOX and some marine invertebrate LOXs produce *R*-hydroperoxides. Even in the reactions of such "*R*-LOXs", the antarafacial rule of hydrogen removal and oxygen addition is conserved.
In cases of plant LOXs, including soybean LOXs, the usual substrates are C18-polyunsaturated fatty acids (linoleic and α-linolenic acids), and the products are their 9*S*- or 13*S*-hydroperoxides (Fig. 2B). Most plant LOXs react with either one of the regio-specificity, while some with both. Therefore, based on the regio-specificty, plant LOXs are classified into 9-LOXs, 13-LOXs, or 9/13-LOXs.

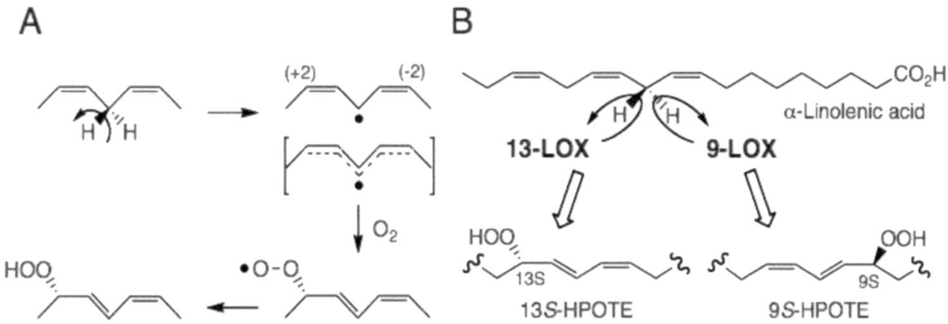

Fig. 2. LOX reaction showing the principal steps of LOX reaction (Panel A), and the actual reactions of plant LOXs and α-linolenic acid (Panel B). HPOTE: hydroperoxyoctadecatrienoic acid.

4. Biological and metabolic functions

4.1 In plants

Lipid peroxidation is common to all biological systems, both appearing in developmentally and environmentally regulated processes of plants (Feussner & Wasternack, 2002). The hydroperoxy polyunsaturated fatty acids, synthesized by the action of various highly specialised forms of lipoxygenases, are substrates of at least seven different enzyme families (Feussner & Wasternack, 2002). Signaling compounds such as jasmonates, antimicrobial and antifungal compounds such as leaf aldehydes or divinyl ethers, and a plant-specific blend of volatiles including leaf alcohols are among the numerous products. Thus, the lipoxygenase pathway becomes an initial step in the interaction of plants with pathogens, insects, or abiotic stress and at distinct stages of development (Feussner & Wasternack, 2002).

4.2 In humans

Besides polyunsaturated fatty acids, H_2O_2, fatty acid hydroperoxides, and synthetic organic hydroperoxides support the lipoxygenase-catalyzed xenobiotic oxidation the major reactions documented thus far including oxidation, epoxidation, hydroxylation, sulfoxidation, desulfuration, dearylation, and N-dealkylation (Kulkarni, 2001). It is noteworthy that lipoxygenases are also capable of glutathione conjugation of certain xenobiotics (Kulkarni, 2001). Available data suggest that lipoxygenases contribute to in vivo metabolism of endobiotics and xenobiotics in mammals (Kulkarni, 2001).

Recent reviews describe the role of lipoxygenase in cancer (Bhattacharya et al., 2009; Pidgeon et al., 2007; Moreno, 2009), inflammation (Duroudier et al., 2009; Hersberger, 2010) and vascular biology (Chawengsub et al., 2009; Mochizuki & Kwon, 2008) and for an extensive presentation of the role of eicosanoids in prevention and management of diseases the reader is referred to the review of Szefel et al. (2011).

5. Interaction of lipoxygenase with inhibitors as theoretical approach for food industry and medical applications

In terms of the structure and function, LOXs are unique, because their metal cofactor is a single ion bound by the side chains of the surrounding amino acids and the carboxylic group of the C-terminus, and their inhibitors bind to or near the Fe co-factor (Skrzypczak-Jankun et al., 2007). Lipoxygenases are inhibited by a large number of chemicals, some of which also serve as co-substrates (Kulkarni, 2001).

5.1 Importance of lipoxygenase inhibition for food industry

Besides their physiological role, plant lipoxygenases are of significant importance to the food industry, since these enzymes have been implicated in the generation of the flavour and aroma in many plant products. For instance, they are responsible for the undesirable 'beany', 'green' and 'grassy' flavours produced during processing and storage of protein products derived from legume seeds (Fukushima, 1994; Robinson et al., 1995) and the development of the stale flavour in beer during storage (Kobayashi et al., 1993). Lipoxygenases also play an important role in the baking industry. They are quite effective as bleaching agents, increase mixing tolerance and improve dough rheology (Nicolas & Potus, 1994; Larreta-Garde, 1995; Cumbee et al., 1997; Borelli[b] et al., 1999).

Freshly refined soybean oil is practically odourless and bland, but "green, grassy, fishy" off-flavors may develop quickly if the oil is heated or stored under conditions that expose it to light and oxygen or by contamination with pro-oxidant metals such as copper and iron (Berk, 1992). "Beany" flavour is the principal inconvenience of traditional soymilk and its products (e.g., tofu) and is caused by some ketones and aldehydes, particularly hexanals and heptanals, produced through LOX catalyzed oxidation (Berk, 1992).

Fish lipids are susceptible to oxidation owing to the high levels of polyunsaturated fatty acids (PUFA), even in frozen storage, and this can affect the flavour, texture, taste, aroma and shelf life of fish (Ke & Ackman, 1976). Since the direct interaction between oxygen and highly unsaturated lipids is kinetically hindered (Kanner et al., 1987), the enzymatic initiation of oxidation by enzymes such as lipoxygenase, peroxidases and microsomal enzymes has been gaining favour.

Green tea glazing was shown to improve the storage quality of frozen bonito fillets (Lin & Lin, 2005). In addition, hot water tea extract was shown to suppress the pro-oxidant activities of the dark meat and skin of blue sprat (Seto et al., 2005). Banerjee (2006) proposes that the improvement in the shelf life of fish by green tea polyphenols is at least in part due to inhibition of LOX resulting in delaying oxidation of fish lipids and because of that impregnation of muscle fillets in tea extract by itself or in combination with other natural inhibitors may improve the shelf-life and storage quality of fish fillets.

Besides its function of oxidizing the polyunsaturated fatty acids (linoleic, linolenic and arachidonic), the enzyme may also catalyse the co-oxidation of carotenoids, resulting in the loss of natural colorants and essential nutrients (Robinson et al., 1995). LOX have been implicated in the generation of the flavour and aroma in many plant products, in the decolourisation of pigments and in the potential of compromising the anti-oxidant status (Casey, 1999). In pasta the involvement of LOX in colour loss is demonstrated by positive correlation between the decrease of β-carotene content after pastification and LOX activities in semolina. In addition to this, the hydroperoxidation and bleaching activities of LOX are highly correlated demonstrating that the bleaching might be ascribable to a co-oxidative action by LOX (Borrelli[a] et al., 1999).

During pasta processing in which the maximal pigment degradation by LOX activity occurs (Borrelli[b] et al., 1999), it is shown that externally added β-carotene can act as inhibitor of the LOX-catalysed linoleate hydroperoxidation and an inverse relation between the % of carotenoid loss and the initial carotenoid content in semolina from durum varieties, showing similar LOX activity, was found (Trono et al., 1999).

The complete characterisation of lipoxygenase from pea seeds (*Pisum sativum var. Telephone L.*) gives possibility to avoid destructive influence during food processing and storage (Szymanowska et al., 2009) by the action of this enzyme.

5.2 LOX inhibition in cancer

Molecular studies of the well-known relationship between polyunsaturated fatty acid metabolism and carcinogenesis have revealed novel molecular targets for cancer chemoprevention and treatment (Lipkin et al., 1999; Willett, 1997; Klurfeld & Bull, 1997; Guthrie & Carroll, 1999).

The role of lipoxygenase in the development and progression of cancer is complex due to the variety of lipoxygenase genes that have been identified in humans, in addition to different profiles of lipoxygenase observed between studies on human tumor biopsies and experimentally induced animal tumor models (Pidgeon et al., 2007). The literature emerging on the role of lipoxygenases in tumor growth, for the most part, suggests that distinct lipoxygenase isoforms, whose expression are lost during the progression of cancer, may exhibit anti-tumor activity, while other isoforms may exert pro-tumorigenic effects and are preferentially expressed during the development of various cancers.

The involvement of 5-lipoxygenase and 12-lipoxygenase in human cancer progression is now supported by a growing body of literature. The involvement of 15-lipoxygenase-1 in colorectal cancer involves its implication in carcinogenesis having pro-carcinogenic as well as anti-carcinogenic roles (Bhattacharya et al., 2009). The co-localization of these enzymes and the similarities of their bioactions on cancer cell growth suggest that the simultaneous inhibition of these enzymes may represent novel and promising therapeutic approaches in selected cancer types (Pidgeon et al., 2007). Therefore, when targeting the regulation of arachidonic acid metabolism, blocking 5-lipoxygenase, 12-lipoxygenase and 15-lipoxygenase-1 without altering the expression of the anti-carcinogenic 15-lipoxygenase-2 may be the most effective, however at present no drug recapitulates these capabilities (Pidgeon et al., 2007).

5.3 Mechanisms of lipoxygenase inhibition

In general, lipoxygenase inhibitors can bind covalently to iron or form the molecular complexes blocking access to iron (Skrzypczak-Jankun et al., 2007). It was pointed out by Walther et al., that a course of inhibition, by the drug ebselen, (noncompetitive vs competitive) and its reversibility depend on the oxidation state of iron, i.e. whether the enzyme is catalytically silent with Fe^{2+} when it binds covalently, causing irreversible inhibition or preoxidized and active with Fe^{3+} in the presence of the fatty acid substrate (Walther et al., 1999). In both cases the enzyme's performance can be illustrated by a classic Lineweaver-Burk plot. Many inhibitors do not follow such a linear relation between velocity and the inhibitor's concentration showing a hyperbolic curve instead as observed by Skrzypczak-Jankun et al. (2002) for polyphenolic inhibitors (curcumin, quercetin, epigallocatechin gallate and epigallocatechin) interacting with soybean lipoxygenase-3. In general, the kinetic data are seldom reported (Skrzypczak-Jankun et al., 2007). Xenobiotic oxidation by soy lipoxygenase has been investigated and described, while human enzymes lack such thorough studies (Skrzypczak-Jankun et al., 2007). The in vivo susceptibility of lipoxygenases' inhibitors may depend not only on the source of lipoxygenase and its isozyme (Pham et al., 1998; Schewe et al., 1986) but also on the oxidation state of iron and the competition between peroxidase and co-oxidase activities of enzyme (Borbulevych et al., 2004).

The first mode to inhibit the lipoxygenase would be a direct reduction of iron to its inactive form. For soybean lipoxygenase, it has been demonstrated that nordihydroguaiaretic acid rapidly reduces the active ferric species of the enzyme to its inactive ferrous form, thus causing interruption of the catalytic cycle (Kemal et al., 1987). For the polyphenol inhibition of lipoxygenase it was firstly suggested that this molecules strongly complex of the ferric iron moiety of the lipoxygenase, thus preventing its reduction *via* the catalytic cycle as proposed for the action of 4-nitrocatechol on the soybean lipoxygenase-1 (Spaapen et al., 1980). The second observation that complexation of the flavanols with Fe^{3+} did not abolish the inhibitory effect may rule out a direct complexation of the iron moiety in ferric lipoxygenase by these catechol compounds. The X-ray analysis shows 4-nitrocatechol near iron with partial occupancy, blocking access to Fe but not covalently bound to it (Skrzypczak-Jankun et al., 2004). If a similar mode of action holds for the interaction of flavanols with mammalian lipoxygenases, the corresponding iron polyphenol complexes may retain their lipoxygenase-inhibitory effect.

A third conceivable mode of action of polyphenols is the effective reduction of hydroperoxides that are essential activators of lipoxygenase *via* conversion of the enzymatically silent ferrous species to the active ferric form. The observation of Schewe et al. (2001) that lowering of the hydroperoxide tone by glutathione plus glutathione peroxidase did not modulate the inhibitory effects of flavanols on 15-lipoxygenase-1 does not support the latter possibility.

In case of carotenoids, more specifically, β-carotene, the lipoxygenase was inhibited by keeping it in the inactive form of Fe(II) (Serpen & Gökmen, 2006). These authors suggest that β-carotene reacts with linoleyl radical (L•) at the beginning of the chain reaction, so it prevents the accumulation of conjugated diene forms (LOO•, LOO− and LOOH). Since L• transforms back to its original form of LH, the enzyme cannot complete the chain reaction and thus remains in the inactive Fe(II) form, which is not capable of catalyzing linoleic acid hydroperoxidation (Serpen & Gökmen, 2006).

Wu et al. (1999) have reported that β-carotene scavenges the linoleyl peroxy radical (LOO•) by a hydrogen transfer mechanism and the oxidation of β-carotene occurs during this action. In these conditions, it is absolutely clear that the amount of inactivated enzyme depends on the concentration of β-carotene present in the medium (Serpen & Gökmen, 2006).

According to Mahesha et al. (2007) the lipoxygenase inhibition by isoflavones follows the next mechanism: an electron donated by isoflavones is accepted by the ferric form (Fe^{3+}) of lipoxygenase, which is reduced to resting ferrous form (Fe^{2+}), thus inhibiting lipoxygenase.

Genistein is neither consumed nor does state change during the course of the reaction of lipoxygenase (Mahesha et al., 2007), while quercetin entrapped within lipoxygenase undergoes degradation (Borbulevych et al., 2004).

5.4 Inhibition of soybean lipoxygenase by different classes of polyphenols

Gillmor et al. (1997) obtained the structure of the rabbit reticulocyte enzyme as a complex with the inhibitor RS75091. Located in one of the hydrophobic channels of the enzyme, the inhibitor was found to be close but not biding the iron atom of the catalytic situs. These observations provided the first indications of how the native enzyme can interact with potential ligands (Pham et al., 1998).

Natural flavonoids don't affect only the lipoxygenase oxidation of its classical substrates but also the co-oxidation of xenobiotics by this enzyme. Epigallocatechin-gallate, quercetin and rutin proved to reduce the co-oxidation rate of guaiacol, benzidine, paraphenylenediamine and dimethoxybenzidine by soybean lipoxygenase-1 (Hu et al., 2006). This data suggest that flavonoids may have anticarcinogenic and antitoxic effect through inhibition of oxidative activation generated by lipoxygenase (Hu et al., 2006). Green tea polyphenols have potent free radical quenching and antioxidant activities (Wiseman et al., 1997) and have structural features that may specifically interfere with the arachidonic acid cascade, including the lipoxygenase pathway (Hong & Yang, 2003; Hussain et al., 2005). In addition, with growing concerns regarding the safety of synthetic antioxidants such as BHT and BHA, alternative mechanisms of antioxidant protection by the use of natural antioxidants have been in review over the years (Barlow, 1990).

Polyphenols, mainly flavonoids and phenolic acids, are abundant in a number of dietary sources such as certain cocoas, tea, wine, fruits and vegetables. More than 8000 different flavonoids of natural origin are known (Schewe & Sies, 2003). The flavonoids exist in nature as aglycons (free form) or conjugated (with O-glucosides or methylated). The aglycons can be subdivided in different subclasses (flavanols, flavanones, flavones, izoflavones, flavonols, anthocyanidines, aurones, chalcones) in function of how the B ring from their structure is linked to the heterocycle C, of the oxidation state and of the functional groups linked to the C ring (Beecher, 2003).

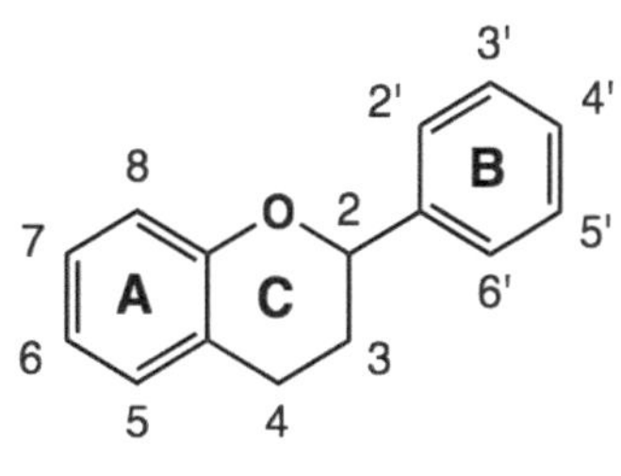

Fig. 3. The basic structure of flavonoids.

The basic structure of flavonoids is represented by the flavan nucleus containing 15 carbons structured in 2 benzene rings, named A and B and linked by a C_3 unit, which together with an oxygen atom forms the γ-pyronic or γ-pyranic ring, named the C ring as shown in Fig.3.

A number of *in vitro* and *in vivo* studies as well as clinical trials suggest beneficial effects of flavonoids for health, counteracting the development of cardiovascular diseases, cancer and obesity. Bors et al. (1990) were the first to claim three partial structures contributing to the radical-scavenging activity of flavonoids: (a) an o-dihydroxyl structure in the B ring (catechol structure) as a radical target site providing good electron delocalization and stabilization of the phenoxy radical; (b) a 2,3-double bond with conjugation to the 4-oxo group which is necessary for delocalization of an unpaired electron from the B ring, (c) hydroxyl groups at the 3-and 5-positions, which are necessary for enhancement of radical scavenging activity, increasing the delocalisation of electrons across the flavonoid scaffold. The catechol group is essential for the radical-scavenging activity of flavan-3-ols and flavanones lacking 2,3-double bonds (Bors et al., 1990).

In parallel to the free radical-scavenging properties the following structural features were found to enhance the inhibitory potency: (i) presence of a catechol arrangement in the B or A ring, (ii) a carbonyl group together with a 2,3-double bond in the C ring (Schewe & Sies, 2003). Other structural features were opposite to the free radical scavenging potencies: (i) presence of a 3-OH group in the C ring diminished than reinforced the inhibition of lipoxygenases, (ii) in the absence of a catechol arrangement there was an inverse correlation to the total number of OH groups in the flavonoid molecule (Schewe & Sies, 2003). Although either reducing or ferric iron chelating properties are prerequisites for a lipoxygenase- inhibitory compound, and both of them are also inherent to flavonoids, the inhibitory effects cannot be ascribed solely to one of these mechanisms (Schewe & Sies, 2003). The inhibition of lipoxygenases by flavonoids appears to be of more complex nature (Schewe & Sies, 2003).

Inhibitors studies of lipoxygenase from pea showed that phenolic antioxidant components were effective and can be used to protect food lipids against oxidation (Szymanowska et al., 2009). The conducted research proved that activity of lipoxygenase from pea seeds could be effectively inhibit by some phenolic compounds. The most effective inhibitor is caffeic acid (about 57% of inhibition). Flavonoids like catechin and quercetin considerably inhibit the lipoxygenase activity. Inhibitors used for investigation in this study were placed in the following order: caffeic acid > quercetin > catechin > benzoic acid > ferulic acid > kaempferol (Szymanowska et al., 2009).

5.5 Lipoxygenase inhibition by quercetin

Quercetin is the most abundant among the flavonoid molecules and can be found in the fruits, vegetables, seeds, nuts, and flowers of many plants. Its documented impact on human health includes cardiovascular protection, anticancer, antiviral, anti-inflammatory activities, antiulcer effects and cataract prevention. Like other flavonoids, quercetin appears to combine both lipoxygenase-inhibitory activities and free radical-scavenging properties in one agent and thus belongs to a family of very effective natural antioxidants (Sadik et al., 2003). Quercetin is a flavonol that can be easily oxidized in an aqueous environment, and in the presence of iron and hydroxyl free radicals (Borbulevych et al., 2004).

The inhibition of rabbit 15-lipoxygenase-1 and of soybean lipoxygenase-1 by quercetin was studied in detail (Sadik et al., 2003). Quercetin modulates the time course of the lipoxygenase reaction in a complex manner by exerting three distinct effects: (i) prolongation of the kinetic lag period, (ii) instant decrease in the initial rate after the lag phase being overcome, (iii) time-dependent inactivation of the enzyme during reaction, but not in the absence of substrate (Schewe & Sies, 2003). The literature data obviously indicate that quercetin represents one of the most potent inhibitors of different LOXs (Schneider & Bucar[a], 2005; Schneider & Bucar[b], 2005).

Structural analysis reveals that quercetin entrapped within LOX undergoes degradation and the resulting compound has been identified by X-ray analysis as protocatechuic acid (3,4-dihydroxybenzoic acid) positioned near the iron site (Borbulevych et al., 2004).

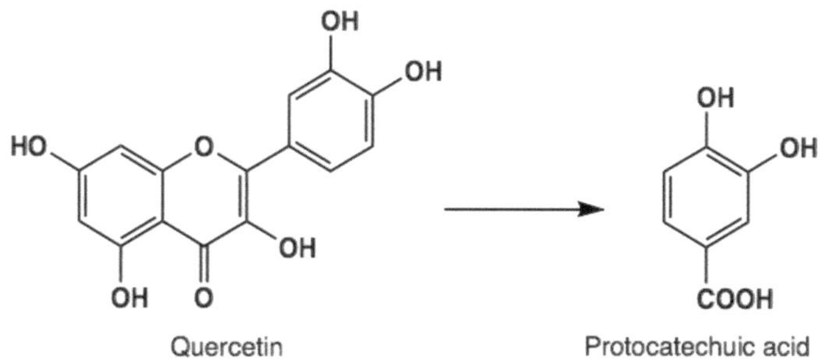

Fig. 4. Product of quercetin degradation by soybean LOX-3 (Borbulevych et al., 2004).

We demonstrated that pH values may influence the molecular interactions between soybean LOX-1 and quercetin, and especially the alcaline pH favours the ionic display of quercetin in order to interact with LOX better (Chedea et al., 2006).

Quercetin inhibited the 12 (S)-hydroxytetraenoic acid production at concentrations below those necessary for growth inhibition in colorectal cancer cells overexpressing the enzyme 12(S)-lipoxygenase with an IC_{50} of 1μM (Bednar et al., 2007). The finding that LOX can turn different compounds into simple catechol derivatives (with one aromatic ring only) might be of importance as an additional small piece of a "jigsaw puzzle" in the much bigger picture of drug metabolism (Borbulevych et al., 2004). Their interactions with LOX can be more complicated than simply blocking the access to the enzyme's active site. The studies on LOX and quercetin contribute to the understanding of biocatalytic properties of this enzyme and its role in the metabolism of this popular (as a medicinal remedy) flavonol and possibly other, similar compounds (Borbulevych et al., 2004). Acting both as a substrate and a source of inhibition, quercetin seems to play an antinomic role (Fiorucci et al., 2008). But this could be explained as quercetin, one of the most representative flavonoids, is a highly functionalized substrate and can thus be activated and degraded following several ways (Fiorucci et al., 2008).

5.6 Inhibition of soybean lipoxygenase by epigallocatechin gallate

Flavanols (or flavan-3-ols or catechins) are a class of flavonoids that include the catechins and the catechin gallates. Catechins are described as colorless, astringent, water-soluble polyphenols found in many fruits and grains, such as coffee, red grapes, prunes and raisins. Their main source however comes from a beverage made from tea leaves of *Camellia sinensis*. (-)-Epigallocatechin gallate (EGCG) together with other galloylated catechins constitute more than 90% of the total catechin content in green tea (Lekli et al., 2010). Laboratory studies strongly indicate that tea inhibits certain cancers, and there is a multitude of evidence confirming the antioncogenic properties of the individual catechins. For instance: EGCG alone shows anticancer effectiveness against carcinogen-induced skin, lung, forestomach, esophagus, duodeum, liver and colon tumors in rodents. It was found to cause apoptosis and/or cell cycle arrest in human carcinoma cells of skin and prostate cancers (Zimeri & Tong, 1999). Catechins have also known inhibitory activity toward dioxygenases with a potential to be utilized in disease prevention and treatment.

The study of Banerjee (2006) shows that green tea polyphenols are very potent inhibitors of mackerel muscle LOX, with EGCG (epigallocatechin gallate) as the most effective inhibitor (IC_{50} 0.13 nM) followed by ECG (epicatechin gallate) (IC_{50} 0.8 nM), EC (epicatechin) (IC_{50} 6.0 nM), EGC (epigallocatechin) (IC_{50} 9.0 nM) and C (catechin) (IC_{50} 22.4 nM). Chocolate and cocoa are also sources of catechins. Epigallocatechin gallate isolated from the seeds of *Theobroma cacao* had the best inhibitory activity on rabit 15-lipoxygenase-1, with an IC_{50}=4M, epicatechin gallate had IC_{50}=5M and epicatechin an IC_{50}=60M (Schewe et al., 2002). A better inhibition of epicatechin (IC_{50} approx. 15M) was registered in the case of recombinant human platelet 12-lipoxygenase.

Obtained from X-ray analysis, the 3D structure of the resulting complex of (-)-epigallocatechin gallate (EGCG) interacting with soybean lipoxygenase-3 reveals the inhibitor depicting (-)-epigallocatechin that lacks the galloyl moiety (Skrzypczak-Jankun et al., 2003). The A-ring is near the iron co-factor, attached by the hydrogen bond to the C-terminus of the enzyme, and the B-ring hydroxyl groups participate in the hydrogen bonds and the van der Waals interactions formed by the surrounding amino acids and water molecules (Skrzypczak-Jancun et al., 2003).

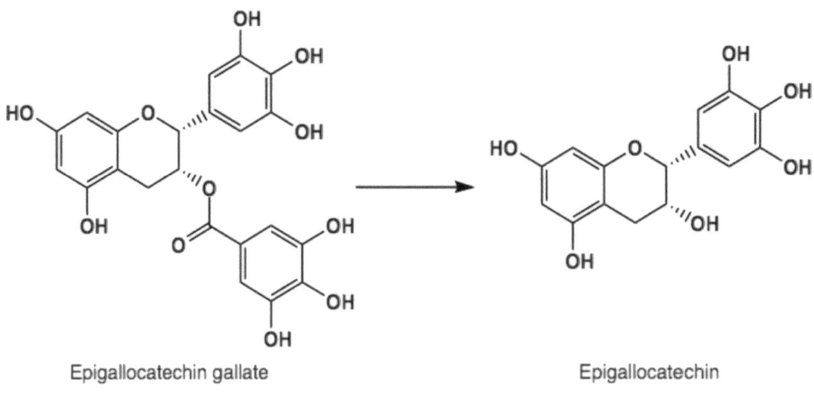

Fig. 5. Lipoxygenase-3 in complex with epigallocatechin gallate as an inhibitor determines the degradation of natural flavonoid to epigallocatechin (Skrzypczak-Jancun et al., 2003).

X-ray analysis of soybean lipoxygenase-3 crystals soaked with EGCG shows the molecular complex of LOX-3 with (-)epigallocatechin molecule.

5.7 Inhibitory effects of soybean isoflavones on lipoxygenase activity

Soybeans are important sources of isoflavone levels (Song et al., 1998), present as 12 derivatives, including free genistin, daidzin, glycitin and their acetyl, malonyl or glycosilated forms. Isoflavones are composed of 2 benzene rings (A and B) linked through a heterocyclic pyrane C ring. The position of the B ring discriminate flavonoid flavones (C2-position) from isoflavones (C3-position).

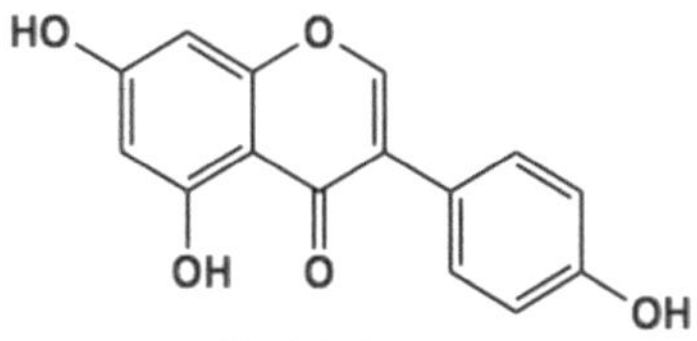

Genistein

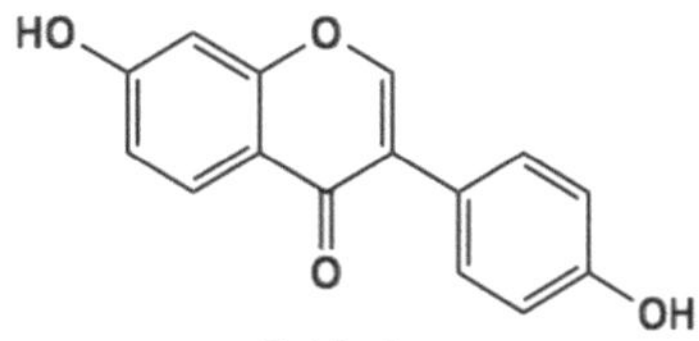

Daidzein

Fig. 6. Dietary isoflavones inhibitors of lipoxygenase.

The impact of dietary isoflavones, daidzein and genistein, on the health of adults and infants is well documented, an increasing interest for these compounds being registered due to their biological effects including: estrogen-like activity, prevention of breast (Warri et al., 2008), prostate (Matsumura et al., 2008) and colon cancer (Mac Donald et al., 2005), antioxidant activity (Malencić et al., 2007; Sakthivelu et al., 2008), prevention of menopausal symptoms and osteoporosis (Ma et al., 2008), and heart disease (Xiao, 2008).

In the work of Mahesha et al. (2007) the inhibition of soy lipoxygenase-1 and 5-lipoxygenase from human polymorph nuclear lymphocyte by isoflavones, genistein and daidzein as glycosilated and unglycosilated compounds was studied. Soybean isoflavones inhibit LOX either as aglycons, or as glucosides. Isoflavones exert combined dual actions as inhibitors: they compete with the hydroperoxide formation to prevent the generation of LOX active ferric state (1) and also are capable of reducing the ferric enzyme to its inactive ferrous form (2) (Mahesha et al., 2007).

Vicaş et al. (2011) showed that genistein was almost twice more potent inhibitor than daidzein at similar concentration with concentration that induces 50% soybean lipoxygenase-1 inhibition values of 5.33 mM versus 11.53 mM. Genistein and daidzein proved to be noncompetitive with inhibition constants K_i of 33.65 and 43.45 mM, respectively (Vicaş et al., 2011). The inhibitory efficiency of the genistein and daidzein depended both on their concentration and on the substrate's concentration (Vicaş et al., 2011).

5.8 Inhibition of soybean lipoxygenase by carotenoids

There are over 600 fully characterized, naturally occurring molecular species belonging to the class of carotenoids. In humans, some carotenoids (the provitamin A carotenoids: α-carotene β-carotene, γ-carotene and the xanthophyll, β-cryptoxanthin) are best known for converting enzymatically into vitamin A; diseases resulting from vitamin A deficiency remain among the most significant nutritional challenges worldwide. Also, the role that carotenoids play in protecting those tissues that are the most heavily exposed to light (e.g. photo protection of the skin, protection of the central retina) is perhaps most evident, while other potential roles for carotenoids in the prevention of chronic diseases (cancer, cardiovascular disease) are still being investigated. Because carotenoids are widely consumed and their consumption is a modifiable health behaviour (via diets or supplements), health benefits for chronic disease prevention, if real, could be very significant for public health (Mayne, 2010). Carotenoids are isoprenoid molecules which contain a polyene chains, with or without cyclisation at the ends.

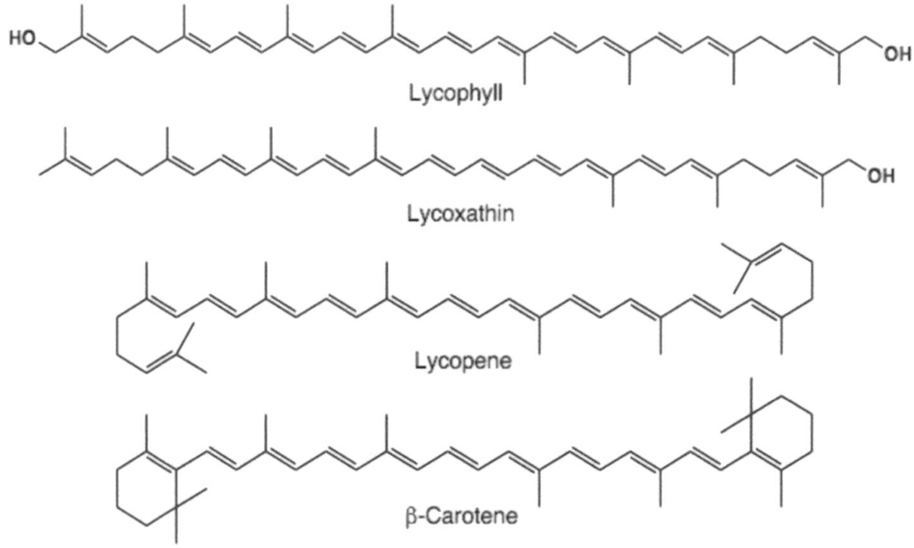

Fig. 7. Carotenoids serving as lipoxygenase co-substrates

The existence of an enzyme "carotene oxidase" in soybeans, which catalyzes the oxidative destruction of carotene was reported by Bohn and Haas in 1928 (Bohn & Haas, 1928). Four years later, Andre and Hou found that soybeans contained an enzyme, lipoxygenase (linoleate oxygen oxidoreductase), which they termed "*lipoxidase*", catalyzing the peroxidation of certain unsaturated fatty acids (Andre & Hou, 1932).

In 1940 the observation that "lipoxydase" is identical to "carotene oxidase" was published (Sumner & Sumner, 1940). These early findings of lipoxygenase peroxidizing the unsaturated fats and bleaches the carotene were reported as the result of studies on the oxidation of crystalline carotene or carotene dissolved in unsaturated oil. Surprisingly it was found that the carotene oxidase had an almost negligible bleaching action upon the crystalline carotene. On the contrary, when one employs carotene dissolved in a small quantity of fat, the bleaching is extremely rapid. With excessive quantities of fat, the rate of bleaching of the carotene diminishes, and it was concluded that the effect of added fat upon the rate of bleaching of carotene is probably due to a coupled oxidation (Sumner & Sumner, 1940).

Studying the soya-lipoxygenase-catalyzed degradation of carotenoids from tomato Biacs and Daood (2000) found that β-carotene was the most sensitive component, followed by lycoxanthin and lycopene. Their results also implied that β-carotene can actively perform its antioxidant function during the course of lipid oxidation. It seems that oxidative degradation and, accordingly, antioxidant activity of each carotenoid depends on the rate of its interaction with the peroxyl radical produced through the lipoxygenase pathway (Biacs & Daood, 2000) and thus is able to inhibit lipoxygenase. The inhibition of the hydroperoxide formation by carotenoids has been attributed to their lipid peroxyl radical-trapping ability (Burton & Ingold, 1984).

In vitro, lycopene is a substrate of soybean lipoxygenase. The presence of this enzyme also significantly increased the production of lycopene oxidative metabolites (dos Anjos Ferreira et al., 2004; Biacs & Daood, 2000). It was reported that during the co-oxidation of β-carotene

by LOX-mediated hydroperoxidation reactions, inhibition of LOX activity takes place also (Lomnitski et al., 1993; Trono et al., 1999; Pastore et al., 2000). The activity of soybean lipoxygenase-1 was inhibited by β-carotene which breaks the chain reaction at the beginning stage of linoleic acid hydroperoxidation (Serpen & Gökmen, 2006). Besides soybean lipoxygenase (Ikedioby & Snyder, 1977; Hildebrand & Hymowitz, 1982) carotene oxidation during lipoxygenase-mediated linoleic acid oxidation has been reported in various studies for the enzymes extracted from potato (Aziz et al., 1999), pea (Yoon & Klein, 1979; Gökmen et al., 2002), wheat (Pastore et al., 2000), olive (Jaren-Galan et al., 1999) and pepper (Jaren-Galan & Minguez-Mosquera, 1997). Soybean lipoxygenase-1 and recombinant pea lipoxygenase-2 and lipoxygenase-3, oxidizing β-carotene, yield apocarotenal, epoxycarotenal, apocarotenone and epoxycarotenone (Wu et al., 1999).
Through molecular modeling Hazai et al. (2006) predicted that lycopene and lycophyll bind with high affinity in the superficial cleft at the interface of the β-barrel and the catalytic domain of 5-LOX (the "cleavage site") suggesting potential direct competitive inhibition of 5-LOX activity by these molecules after in vivo supplementation, particularly in the case of the dial metabolite.

5.9 Quinone and semiquinone formation during the lipoxygenase inhibition reaction

In his excellent review from 2001, Kulkarni presents the studies up to that date indicating the semiquinone and quinone formation in different lipoxygenase catalyzed reactions of xenobiotics oxidation. Diethylstilbestrol (DES) is a human transplacental carcinogen. DES-quinone, one of the metabolites of DES, binds to DNA and is presumed to be the ultimate toxicant. Although DES-quinone formation by human tissue lipoxygenase has yet to be examined, soybean lipoxygenase has been shown to initiate one-electron oxidation to DES semiquinone in the presence of H_2O_2 (Nunez-Delicado et al., 1997). Subsequent dismutation of two molecules of DES semiquinone yields one molecule each of DES-quinone and DES.
Although phenol is oxidized slowly by different lipoxygenase isoenzymes, potato 5-lipoxygenase (Cucurou et al., 1991) and soybean lipoxygenase-1 (Cucurou et al., 1991; Mansuy et al., 1988), substituted phenols and catechols undergo extensive one-electron oxidation and yield the corresponding reactive phenoxyl radicals or semiquinones. These free radicals polymerize to yield a mixture of complex metabolites.
However, it seems that quercetin may act, in most of cases, after being metabolically activated (Metodiewa et al., 1999), and despite a constant increase of knowledge on both positive and negative biological effects of this natural product, it remains often unclear which activated form should play a role in a given process (Fiorucci et al., 2007). Indeed, semiquinone and quinone forms of quercetin, deriving from the abstraction of respectively one or two H•, are involved in many oxidative processes (Metodiewa et al., 1999; Gliszczynska-Swiglo et al., 2003; Hirakawa et al., 2002). For instance, quercetin reduces peroxyl radicals involved in lipid peroxidation, and through this reaction, a semiquinone species is produced, which then undergoes a disproportionation to generate a quinone form (Fiorucci et al., 2007). Three semiquinone forms for quercetin have been considered by Fiorucci et al. (2008) in order to study the quercetin binding to lipoxygenase-3 by molecular modeling simulations. In the case of lipoxygenase–catechol complexes, the formation of the catechol-iron(III) complex of soybean lipoxygenase 1 gradually results in reduction of the cofactor and release of the semiquinone but no evidence of quinone formation in the UV-visible spectra of samples of the native enzyme treated with catechol was obtained (Spaapen et al., 1980; Nelson, 1988; Pham et al., 1998).

Besides their antioxidant properties, catechins have been described to display pro-oxidant activity having the potential to oxidize the quinones or semiquinones resulting in redox cycling and reactive oxygen species production as well as in thiol, DNA and protein alkylation (Galati & O'Brien, 2004; van der Woude et al., 2006).
Our previous study shows that the oxidation products of catechins are formed within the cellular matrix but also in the extracellular medium (Chedea et al., 2010). We have demonstrated by UV-Vis spectroscopy, that the quinones are involved in the modulation of lipoxygenase activity in the presence of catechins within the cells (Chedea et al., 2010). This conclusion is in agreement with that of Sadik et al. (2003) and Banerjee (2006). An irreversible covalent modification of soybean LOX by flavonoids has been suggested by Sadik et al. (2003) whereby during the formation of fatty acid peroxyl radical in the LOX pathway, the flavonoids are co-oxidized to a semi-quinone or quinone, which in turn may bind to sulfhydryl or amino groups of the enzyme causing inhibition (Banerjee, 2006).

6. Conclusions

The knowledge presented in this study addressed the lipoxygenase pathway inhibition by antioxidant polyphenols at two levels: human diet and human health or to a larger extent human disease prevention and treatment. Review articles (Jachak, 2006; Schneider & Bucar[a], 2005; Schneider & Bucar[b], 2005) summarize natural products with inhibitory properties toward LOX (Skrzypczak-Jankun et al., 2007). Natural remedies almost never consist of a single ingredient and usually are a mixture of many in proper proportions, with a synergistic effect of their simultaneous action being absolutely necessary for beneficial medicinal results. Thus, one should proceed with caution, since the action of a selected single compound may not be the same (Skrzypczak-Jankun et al., 2007). The X-ray studies of soybean complexes with quercetin, curcumin, EGCG, EGC indicated conversion of these inhibitors into their metabolites (Skrzypczak-Jankun et al., 2007), which is not surprising considering the co-oxidative activity of LOXs (Kulkarni, 2001). As already presented a question arises concerning the lipoxygenase inhibion: "What is really inhibiting LOX, a given chemical or its LOX metabolite?" (Skrzypczak-Jankun et al., 2007). The results presented so far indicate a complex mode of inhibition involving the inhibitor itself but also its reaction product with lipoxygenase.
The lipoxygenase researcher faces the next antinomy: despite the difference in the number of amino acids between plant and mammalian LOXs, these proteins are amazingly similar in topology with high similarities in the active site of these enzymes. It is believed that all LOXs follow the same catalytic mechanism; however, it is probably the vicinity of the iron site that determines the regio and stereospecificity of the particular enzyme (Skrzypczak-Jankun et al., 2003). In this contradictorily state of facts soybean lipoxygenases stands as a control point in terms of structure, activity and thus inhibition.

7. Aknowledgements

V.S. Chedea is a Japan Society for the Promotion of Science (JSPS) postdoctoral fellow. The authors wish to thank Dr. Ewa Skrzypczak-Jankun for the enlightening discussions during this manuscript writing and Ms. Nana Henmi for helping with the structures drawing.

8. References

André, E. & Hou, K.-W. (1932). La présence d'une oxydase des lipides ou lipoxydase dans graine de soja, glycine soja lieb, *Comptes Rendus*, Vol.194, pp. 645–647.

Aziz, S., Wu, Z. & Robinson, D.S. (1999). Potato lipoxygenase catalyzed co-oxidation of β-carotene. *Food Chemistry* Vol.64, pp. 227–230.

Banerjee, S. (2006). Inhibition of mackerel (*Scomber scombrus*) muscle lipoxygenase by green tea polyphenols, *Food Research International*, Vol.39, pp. 486–491.

Barlow, S. M. (1990). Toxicological aspects of antioxidants used as food additives, In B. J. F. Hudson (Ed.). *Food antioxidants*, London: Elsevier Applied Science, pp. 253–308.

Bednar, W., Holzmann, K. & Marian, B. (2007). Assessing 12(S)-lipoxygenase inhibitory activity using colorectal cancer cells overexpressing the enzyme, *Food and Chemical Toxicology*, Vol.45, pp. 508-514.

Beecher, G.R. (2003). Overview of dietary flavonoids: Nomenclature, occurrence and intake, *Journal of Nutrition*, Vol. 133, pp. 3248S–3254S.

Bergström[a], S., Krabisch, L., Samuelsson, B., & Sjövall, J. (1962). Preparation of prostaglandin F from prostaglandin E, *Acta Chemica Scandinavica*, Vol.16, pp.969-974.

Bergström[b], S., Dressler F., Ryhage R., Samuelsson B. & Sjövall, J. (1962). The isolation of two further prostaglandins from sheep prostate glands, *Arkiv foer Kemi*, Vol.19, pp. 563-567.

Berk, Z. (1992). Technology of production of edible flours and protein products from soybeans, In: *FAO Agricultural Services Bulletin*. Vol. 97. Date of access: 19.01.2010, Available from: http://www.fao.org/docrep/t0532e/t0532e01.htm.

Bhattacharya, S., Mathew, G., Jayne, D.G., Pelengaris, S. & Khan, M. (2009). 15-Lipoxygenase-1 in colorectal cancer: A review, *Tumor Biology*, Vol. 30, pp. 185–199.

Biacs, P.A. & Daood, H.G. (2000). Lipoxygenase-catalysed degradation of carotenoids from tomato in the presence of antioxidant vitamins. *Biochemical Society Transactions*, Vol. 28, pp. 839.

Bohn, R.M. & Haas, L.W. (1928). Chemistry and methods of enzymes, in *Chemistry and methods of enzymes*, ed by Sumner JB and Somers GF, Academic Press, New York.

Borbulevych, O.Y., Jankun, J., Selman, S. & Skrzypezak-Jankun, E. (2004). Lypoxygenase interactions with natural flavonoid, quercetin, reveal a complex with protocatechuic acid in its X-Ray structure at 2.1 Å resolution. *Proteins, Structure, Function and Bioinformatics*, Vol. 54, pp.13-19.

Borrelli[a], G.M., Troccoli, A., Fares, C., Trono, D., De Leonardis, A.M., Padalino, L., Pastore, D., Del Giudice, L. & Di Fonzo, N. (1999). Lipoxygenase in durum wheat: What is the role in pasta colour?, *Options Mediterraneennes*, pp. 497-499, Date of access 12.02.2011, Available from: http://ressources.ciheam.org/om/pdf/a40/00600082.pdf

Borrelli[b], G.M., Troccoli, A., Di Fonzo, N. & Fares, C. (1999). Durum wheat lipoxygenase activity and other quality parameters that affect pasta colour, *Cereal Chemistry*, Vol. 76, pp. 335-340.

Bors, W., Heller, W., Michael, C. & Saran, M. (1990). Flavonoids as antioxidants: Determination of radical-scavenging efficiencies. In L. Packer & A. N. Glazer (Eds.). *Methods in enzymology* (Vol. 186, pp. 343–355). San Diego: Academic Press.

Boyington, J. C., Gaffney, B. J. & Amzel L. M. (1993). Structure of soybean lipoxygenase-1. *Biochemical Society Transactions*, Vol. 21, pp. 744–748.

Burton, G.W. & Ingold, K.U. (1984). β-carotene: An unusual type of lipid antioxidant. *Science,* Vol. 224, pp. 569-573.

Casey, R. (1999). *Lipoxygenases in seed proteins*, pp. 685-708. Shewry, P.R. and Casey, R.eds. Kluwer Academic Publishers.

Chawengsub, Y., Gauthier, K. M. & Campbell, W. B. (2009). Role of arachidonic acid lipoxygenase metabolites in the regulation of vascular tone, *American Journal of Physiology- Heart Circulatory Physiology,* Vol. 297, pp. H495–H507, ISSN: 0363-6135, eISSN: 1522-1539.

Chedea, V. S., Vicaş, S.I., Oprea, C. & Socaciu, C. (2006). Lipoxygense kinetic in relation with lipoxygenase's inhibition by quercetin, *Proceedings of the XXXVI Annual Meeting of the European Society for New Methods in Agricultural Research (ESNA)*, 10 - 14 September 2006, Iasi-Romania, pp. 449-456, ISBN (10) 973-7921-81-X ; ISBN (13) 978-973-7921-81-9.

Chedea, V.S., Braicu, C. & Socaciu, C. (2010). Antioxidant/prooxidant activity of a polyphenolic grape seed extract, *Food Chemistry,* Vol. 121, pp. 132–139.

Coffa, G., Schneider, C. & Brash, A.R. (2005). A comprehensive model of positional and stereo control in lipoxygenases. *Biochemical and Biophysical Research Communications,* Vol. 338, pp. 87–92, ISSN: 0006-291X.

Cucurou, C., Battioni, J. P., Daniel, R. & Mansuy, D. (1991). Peroxidase-like activity of lipoxygenase: different substrate specificity of potato 5-lipoxygenase and soybean 15-lipoxygenase and particular affinity of vitamin E derivatives for the 5-lipoxygenase. *Biochimica et Biophysica Acta,* Vol. 1081, pp. 99–105.

Cumbee, B., Hildebrand, D.F. & Addo, K. (1997). Soybean flour lipoxygenase isozymes effects on wheat flour dough rheological and breadmaking properties. *Journal of Food Science,* Vol. 62, pp. 281–283, ISSN:1750-3841.

dos Anjos Ferreira, A. L., Yeum, K. J., Russell, R. M., Krinsky, N. I. & Tang, G. (2004). Enzymatic and oxidative metabolites of lycopene. *The Journal of Nutritional Biochemistry,* Vol. 15, pp. 493-502, ISSN: 0955-2863.

Duroudier, N.P., Tulah, A.S. & Sayers, I. (2009). Leukotriene pathway genetics and pharmacogenetics in allergy. *Allergy,* Vol. 64, pp. 823–839.

Feussner, I. & Wasternack, C. (2002). The Lipoxygenase Pathway. *Annual Review in Plant Biology,* Vol. 53, pp. 275-297.

Fiorucci, S., Golebiowski, J., Cabrol-Bass, D. & Antonczak, S. (2007). DFT Study of quercetin activated forms involved in antiradical, antioxidant, and prooxidant biological processes, *Journal of Agricultural and Food Chemistry,* Vol. 55, pp. 903-911.

Fiorucci, S., Golebiowski, J., Cabrol-Bass, D. & Antonczak, S. (2008). Molecular simulations enlighten the binding mode of quercetin to lipoxygenase-3. *Proteins,* Vol. 73, pp. 290–298.

Fukushima, D. (1994). Recent progress on biotechnology of soybean proteins and soybean protein food products. *Food Biotechnology,* Vol. 8, pp.83–135.

Galati, G. & O'Brien, P. J. (2004). Potential toxicity of flavonoids and other dietary phenolics: Significance for their chemopreventive and anticancer properties. *Free Radical Biology & Medicine,* Vol. 37, pp. 287-303, ISSN: 0891-5849.

Gardner, H.W. (1991). Recent investigations into the lipoxygenase pathway in plants. *Biochimica Biophysica Acta,* Vol. 1084, No. 3, pp. 221-239.

Gillmor, S. A., Villasenor, A., Fletterick, R., Sigal, E. & Browner, M.F. (1997). The structure of mammalian 15-Lipoxygenase reveals similarity to the lipases and the determinants of substrate specificity. *Nature Structural & Molecular Biology*, Vol. 4, pp. 1003–1009, ISSN: 1545-9993.

Gliszczynska-Swiglo, A., van der Woude, H., de Haan, L., Tyrakowska, B., Aatrs, J. M. M. J. G. & Rietjens, I. M. C. M. (2003). The role of quinone reductase (NQO1) and quinone chemistry in quercetin cytotoxicity. *Toxicology in Vitro*, Vol. 17, pp. 423-431, ISSN: 0887-2333.

Gökmen, V., Bahçeci, S. & Acar, J. (2002). Characterization of crude lipoxygenase extract from green pea using a modified spectrophotometric method. *European Food Research and Technology*, Vol. 215, pp. 42–45, ISSN:1438-2377.

Guthrie, N. & Carroll, K. K. (1999). Specific versus non-specific effects of dietary fat on carcinogenesis. *Progress in Lipid Research*, Vol. 38(3), pp. 261-271, ISSN: 0163-7827.

Hazai, E., Bikádi, Z., Zsila, F. & Lockwood, S. F. (2006). Molecular modeling of the non-covalent binding of the dietary tomato carotenoids lycopene and lycophyll, and selected oxidative metabolites with 5-lipoxygenase. *Bioorganic and Medicinal Chemistry*, Vol. 14, pp. 6859–6867.

Hersberger, M. (2010). Potential role of the lipoxygenase derived lipid mediators in atherosclerosis: leukotrienes, lipoxins and resolvins. *Clinical Chemistry and Laboratory Medicine*, Vol. 48(8), pp.1063–1073, ISSN: 1434-6621.

Hildebrand, D.F. & Hymowitz, T. (1982). Carotene and chlorophyll bleaching by soybeans with and without seed lipoxygenase-1. *Journal of Agricultural and Food Chemistry*, Vol. 30, pp. 705–708, ISSN: 0021-8561.

Hirakawa, K., Oikawa, S., Hiraku, Y., Hirosawa, I. & Kawanishi, S. (2002). Catechol and hydroquinone have different redox properties responsible for their differential DNA-damaging ability. *Chemical Research in Toxicology*, Vol. 15, pp.76-82, ISSN: 0893-228X.

Ho, C. T., Chen, Q., Shi, H. & Rosen, R. T. (1992). Antioxidative effect of polyphenol extract prepared from various Chinese teas. *Preventive Medicine*, Vol. 21, pp. 520-525, ISSN: 0091-7435.

Hong, J.G. & Yang, C.S. (2003). Effects of tea polyphenols on arachidonic acid metabolism in human colon. In F. Shahidi et al. (Eds.), ACS symposium series: Vol. 851. *Food factors in health promotion and disease prevention* (pp. 27–38). Washington, DC: American Chemical Society.

Hu, J., Huang, Y., Xiong, M., Luo, S. & Chen, Y. (2006). The effects of natural flavonoids on lipoxygenase-mediated oxidation of compounds with benzene ring structure-a new possible mechanism of flavonoid anti-chemical carcinogenesis and other toxicities. *International Journal of Toxicology*, Vol. 25, pp. 295-301.

Hussain, T., Gupta, S., Adhami, V. M. & Mukhtar, H. (2005). Green tea constituent epigallocatechin-3-gallate selectively inhibits COX-2 without affecting COX-1 expression in human prostate carcinoma cells, *International Journal of Cancer*, Vol. 113, pp. 660–669.

Ikediobi, C.O. & Snyder, H.E. (1977). Co-oxidation of β-carotene by an isoenzyme of soybean lipoxygenase. *Journal of Agricultural and Food Chemistry*, Vol. 25, pp. 124–127, ISSN: 0021-8561.

Jachak, S. M. (2006). Cyclooxygenase inhibitory natural products: current status. *Current Medicinal Chemistry*, Vol. 13, pp. 659-678, ISSN: 0929-8673.

Jaren-Galan, M. & Minguez-Mosquera, M.I. (1997). β-Carotene and capsanthin co-oxidation by lipoxygenase: kinetic and thermodynamic aspects of reaction. *Journal of Agricultural and Food Chemistry*, Vol. 45, pp. 4814–4820, ISSN: 0021-8561.

Jaren-Galan, M., Carmona-Ramon, C. & Minguez-Mosquera, M.I. (1999). Interaction between chloroplast pigments and lipoxygenase enzymatic extract of olives. *Journal of Agricultural and Food Chemistry*, Vol. 47, pp. 2671–2677, ISSN: 0021-8561.

Kanner, J., German, J. B. & Kinsella, J. E. (1987). Initiation of lipid peroxidation in biological systems. *CRC Critical Reviews in Food Science and Nutrition*, Vol. 25, pp. 317–364.

Ke, P. J. & Ackman, R. G. (1976). Metal catalyzed oxidation in mackerel skin and meat lipids. *Journal of the American Oil Chemists Society*, Vol. 53, pp. 636–640.

Kemal, C., Louis-Flamberg, P., Krupinski-Olsen, R. & Shorter, A.L. (1987). Reductive inactivation of soybean lipoxygenase 1 by catechols: a possible mechanism for regulation of lipoxygenase activity. *Biochemistry*, Vol. 26, No. 22, pp. 7064-7072.

Klurfeld, D.M. & Bull, A.W. (1997). Fatty acids and colon cancer in experimental models, *The American Journal of Clinical Nutrition* ,Vol. 66, pp. 1530-1538, ISSN: 0002-9165.

Kobayashi, N., Kaneda, H., Kano, Y. & Koshino, S. (1993). The production of linoleic and linolenic acid hydroperoxides during mashing. *Journal of Fermentation and Bioengineering*. Vol. 76, pp. 371–375.

Kulkarni, A.P. (2001). Lipoxygenase - a versatile biocatalyst for biotransformation of endobiotics and xenobiotics. *Cellular and Molecular Life Sciences*, Vol. 58, pp. 1805-1825.

Larreta-Garde, V. (1995). Lipoxygenase in making breads, cookies and crackers. *Oléagineux Corps Gras Lipides*, Vol. 2, pp. 363–365.

Lekli, I., Ray, D. & Das, D.K. (2010). Longevity nutrients resveratrol, wines and grapes. *Genes and Nutrition*, Vol. 5, pp. 55-60, ISSN: 1555-8932.

Lin, C. C. & Lin, C. S. (2005). Enhancement of the storage quality of frozen bonito fillets by glazing with tea extracts, *Food Control*, Vol. 16, pp. 169–175.

Lipkin, M., Yang, K., Edelmann, W., Xue, L., Fan, K., Risio, M., Newmark, H. & Kucherlapati R. (1999). Preclinical mouse models for cancer chemoprevention studies. *Annals of the New York Academy of Sciences*, Vol. 889, pp. 14-19.

Loiseau, J., Benoît, L. V., Macherel, M.-H. & Le Deunff, Y. (2001). Seed lipoxygenases: occurrence and functions. *Seed Science Research*, Vol. 11, pp. 199–211.

Lomnitski, L., Bar-Natan, R., Sklan, D. & Grossman, S. (1993). The interactions between β-carotene and lipoxygenase in plant and animal systems. *Biochimica Biophysica Acta*, Vol. 1167, pp. 331–338.

Ma, D.F., Qin, L.Q., Wang, P.Y. & Katoh, R. (2008). Soy isoflavone intake inhibits bone resorption and stimulates bone formation in menopausal women: meta-analysis of randomized controlled trials. *European Journal of Clinical Nutrition*, Vol. 62, pp. 155-161, ISSN: 0954-3007.

Maccarrone, M., van Aarie, P.G.M., Veldink, G.A. & Vliegenthart, J.F.G. (1994). In vitro oxygenation of soybean biomembranes by lipoxygenase-2. *Biochimica et Biophysica Acta*, Vol. 1190, pp. 164–169.

Macdonald, R.S., Guo, J., Copeland, J., Browning, J.D. JR., Sleper, D., Rottinghaus, G.E. & Berhow, M.A. (2005). Environmental influences on isoflavones and saponins in soybeans and their role in colon cancer. *Journal of Nutrition*, Vol. 135, pp. 1239-1242.

Mahesha, H.G., Singh, S.A. & Appu Rao, A.G. (2007). Inhibition of lipoxygenase by soy isoflavones: Evidence of isoflavones as redox inhibitors. *Archives of Biochemistry and Biophysics*, Vol. 461, pp. 176-185.

Malencic, D., Popovic, M. & Miladinovic, J. (2007). Phenolic content and antioxidant properties of soybean (*Glycine max* (L.) Merr.) seeds, *Molecules*, Vol. 12, pp. 576–581.

Mansuy, D., Cucurou, C., Biatry, B. & Battioni, J. P. (1988). Soybean lipoxygenase-catalyzed oxidations by linoleic acid hydroperoxide: different reducing substrates and dehydrogenation of phenidone and BW755C. *Biochemical and Biophysical Research Communications*, Vol. 151, pp. 339–346.

Matsumura, K., Tanaka, T. & Kawashima, H. (2008). Involvement of the estrogen receptor beta in genistein induced expression of p21 (waf1/cip1) in PC-3 prostate cancer cells. *Anticancer Research*, Vol. 28, pp. 709-714.

Mayne, S.T. (2010). Carotenoids a colorful and timely research field. In *Carotenoids, physical, chemical and biological functions and properties* edited by Landrum T., CRC Press Taylor and Francisc Group, USA.

Metodiewa, D., Jaiswal, A. K., Cenas, N., Dickancaite´, E. & Segura-Aguilar, J. (1999). Quercetin may act as a cytotoxic prooxidant after its metabolic activation to semiquinone and quinoidal product. *Free Radical Biology and Medicine*, Vol. 26, pp. 107-116.

Minor, W., Steczko, J., Stec, B., Otwinowski, Z., Bolin, J.T., Walter, R. & Axelrod, B. (1996). Crystal structure of soybean lipoxygenase L-1 at 1.4 ANG. resolution. *Biochemistry*, Vol. 35 pp. 10687-10701.

Mochizuki, N. & Kwon, Y.-G. (2008). 15-Lipoxygenase-1 in the vasculature: Expanding roles in angiogenesis, *Circulation Research*, Vol.102, pp. 143-145.

Moreno, J.J. (2009). New aspects of the role of hydroxyeicosatetraenoic acids in cell growth and cancer development. *Biochemical Pharmacology*, Vol. 77, pp. 1–10, ISSN: 0006-2952.

Nelson, M. J. (1988). Catecholate complexes of ferric soybean lipoxygenase 1. *Biochemistry*, Vol. 27, pp. 4273-4278.

Nicolas, J. & Potus, J. (1994). Enzymatic oxidation phenomena and coupled oxidations - Effects of lipoxygenase in bread making and of polyphenol oxidase in fruit technology. *Sciences des Aliments*, Vol. 14, pp. 627–642.

Nunez-Delicado, E., Sanchez-Ferrer, A. & Garcia-Carmona, F. (1997). Hydroperoxidative oxidation of diethylstilbestrol by lipoxygenase. *Archives of Biochemistry and Biophysics*, Vol. 348, pp. 411–414, ISSN: 0003-9861.

Pastore, D., Trono, D., Padalino, L., Simone, S., Valenti, D., Fronzo, N.D. & Passarella, S. (2000). Inhibition by *α*-tocopherol and L-ascorbate of linoleate hydroperoxidation and *β*-carotene bleaching activities in durum wheat semolina. *Journal of Cereal Science*, Vol. 31, pp. 41–54, ISSN: 0733-5210.

Pham, C., Jankun, J., Skrzypczak-Jankun, E., Flowers, R.A. & Funk, M.O. (1998). Structural and thermochemical characterization of lipoxygenase-catechol complexes. *Biochemistry*, Vol. 37, pp. 17952-17957.

Pidgeon, G. P., Lysaght, J., Krishnamoorthy, S., Reynolds, J. V., O'Byrne, K., Nie, D. & Honn, K. V. (2007). Lipoxygenase metabolism: roles in tumor progression and survival. *Cancer and Metastasis Reviews*, Vol. 26, pp. 503–524, ISSN: 0167-7659.

Prigge, S.T., Boyington, J.C., Faig M., Doctor, K.S., Gaffney, B.J. & Amzel, L.M. (1997). Structure and mechanism of lipoxygenases. *Biochemie*, Vol. 79, pp. 629-636.

Ramadoss, C.S., Pistorius, E.K. & Axelrod, B. (1978). Coupled oxidation of carotene by lipoxygenase requires two isoenzymes. *Archives of Biochemistry and Biophysics*, Vol. 190, pp. 549–552.

Robinson, D.S., Wu, Z.C., Domoney, C. & Casey, R. (1995). Lipoxygenases and the quality of foods. *Food Chemistry*, Vol. 54, pp. 33–43.

Sadik, C.D., Sies, H. & Schewe, T. (2003). Inhibition of 15-lipoxygenases by flavonoids: structure–activity relations and mode of action. *Biochemical Pharmacology*, Vol. 65, pp. 773-781.

Sakthivelu, G., A. Devi, M.K., Giridhar, P., Rajasekaran, T., Ravishankar, G.A., Nikolova, M.T., Angelov, G.B., Todorova, R.M. & Kosturkova, G.P. (2008). Isoflavone composition, phenol content, and antioxidant activity of soybean seeds from India and Bulgaria. *Journal of Agricultural and Food Chemistry*, Vol. 56, pp. 2090-2095, ISSN: 0021-8561.

Schewe, T. & Sies, H. (2003). Flavonoids as protectants against prooxidant enzymes, *Research monographs*, Institut für Physiologische Chemie I; Heinrich-Heine-Universität Düsseldorf- Date of access: 7.07.2008. Available from: http://www.uni-duesseldorf.de/WWW/MedFak/PhysiolChem/index.html.

Schewe, T., Kühn, H. & Rapoport, S. (1986). Positional specificity of lipoxygenases and their suitability for testing potential drugs. *Prostaglandins, Leukotrienes & Medicine*, Vol. 23, pp. 155-160, ISSN: 0952-3278.

Schewe, T., Kühn, H. & Sies H. (2002). Flavonoids of cocoa inhibit recombinant human 5-Lipoxygenase. *The Journal of Nutrition*, Vol. 56, pp. 1825-1829.

Schewe, T., Sadik, C., Klotz, L.-O., Yoshimoto, T., Kühn, H. & Sies, H. (2002). Polyphenols of cocoa: inhibition of mammalian 15-Lipoxygenase. *Biological Chemistry*, Vol. 382, pp. 1687- 1696.

Schneider, I. & Bucar[a] F. (2005). Lipoxygenase inhibitors from natural plant sources. Part I Medicinal plants with action on arachidonate 5-lipoxygenase. *Phytotherapy Research*, Vol. 19, pp. 81-102.

Schneider, I. and Bucar[b] F. (2005). Lipoxygenase inhibitors from natural plant sources. Part II Medicinal plants with action on arachidonate 12-lipoxygenase, 15-lipoxygenase and leukotriene receptor antagonists. *Phytotherapy Research*, Vol. 19, pp. 263-272.

Serpen, A. & Gökmen, V. (2006). A proposed mechanism for the inhibition of soybean lipoxygenase by β-carotene. *Journal of the Science of Food and Agriculture*, Vol. 86, pp. 401-406.

Seto, Y., Lin, C. C., Endo, Y. & Fujimoto, K. (2005). Retardation of lipid oxidation in blue sprat by hot water tea extracts, *Journal of the Science of Food and Agriculture*, Vol. 85, pp. 1119–1124.

Siedow, J.N. (1991). Plant lipoxygenases. Structure and function. *Annual Review of Plant Physiology and Plant Molecular Biology*, Vol. 42, pp. 145-188.

Skrzypczak-Jankun, E., Amzel, L.M., Kroa, B.A. & Funk, M.O. (1997). Structure of soybean lipoxygenase L3 and a comparison with its L1 isoenzyme. *Proteins*, Vol. 29, pp. 15-31.

Skrzypczak-Jankun, E., Zhou, K. & Jankun J. (2003). Inhibition of lipoxygenase by (-)-epigallocatechin gallate: X-ray analysis at 2.1 A reveals degradation of EGCG and shows soybean LOX-3 complex with EGC instead. *International Journal of Molecular Medicine*, Vol. 12, pp. 415-420.

Skrzypczak-Jankun, E., Chorostowska-Wynimko, J., Selman, S.H., & Jankun, J. (2007). Lipoxygenases - A challenging problem in enzyme inhibition and drug development. *Current Enzyme Inhibition*, Vol. 3, pp. 119-132.

Skrzypczak-Jankun, E., Zhou, K., McCabe, N.P., Kernstock, R., Selman, S.H., Funk, M.O., Jr. & Jankun, J. (2002). In *Natural Polyphenols in Complexes with Lipoxygenase - Structural Studies and their Relevance to Cancer*, pp. 205. 93rd Annual Meeting AACR, San Francisco, 2002; AACR: San Francisco.

Song, T.B., Buseman, K. & Murphy, G. (1998). Soy isoflavone analysis: quality control and a new internal standard. *American Journal of Clinical Nutrition*, Vol. 68, pp. 1474-1479.

Spaapen, L. J., Verhagen, J., Veldink, G. A. & Vliegenthart, J. F. (1980). Properties of a complex of Fe(III)-soybean lipoxygenase-1 and 4-nitrocatechol. *Biochimica and Biophysica Acta*, Vol. 617, No.1, pp. 132-140.

Sumner, J.B. & Sumner, R.J. (1940). The coupled oxidation of carotene and fat by carotene oxidase. *Journal of Biological Chemistry*, Vol. 134, pp. 531–533.

Szefel, J., Piotrowska, M., Kruszewski, W.J., Jankun, J., Lysiak-Szydlowska, W. & Skrzypczak-Jankun, E. (2011). Eicosanoids in prevention and management of diseases, *Current Molecular Medicine*, Vol. 11, pp. 13-25.

Szymanowska, U., Jakubczyk, A., Baraniak, B. & Kur, A. (2009). Characterisation of lipoxygenase from pea seeds (*Pisum sativum* var. Telephone L.). *Food Chemistry*, Vol. 116, pp. 906–910.

Theorell, H., Holman, R. T. & Åkeson, Å. (1947). Cristalline lipoxidase. *Acta Chemica Scandinavica*, Vol. 1, pp. 571-576.

Trono, D., Pastore, D. & Fonzo, N.D. (1999). Carotenoid dependent inhibition of durum wheat lipoxygenase. *Journal of Cereal Science*, Vol. 29, pp. 99–102.

van der Woude, Boersma, H., Alink, M. G., Vervoort, G. M., J. & Rietjens, I. M. C. M. (2006). Consequences of quercetin methylation for its covalent glutathione and DNA adduct formation. *Chemico-Biological Interactions*, Vol. 160, pp. 193-203.

Vicaş, S. I., Chedea, V. S. & Socaciu, C. (2011). Inhibitory effects of isoflavones on soybean lipoxygenase-1 activity, *Journal of Food Biochemistry*, Vol. 35, pp. 613-627.

Walther, M., Holzhutter, H.G., Kuban, R.J., Wiesner, R., Rathmann, J. & Kuhn, H. (1999). The inhibition of mammalian 15-lipoxygenases by the anti-inflammatory drug ebselen: dual-type mechanism involving covalent linkage and alteration of the iron ligand sphere *Molecular Pharmacology*, Vol. 56, pp. 196-203.

Warri, A., Saarinen, N.M. & Makela, S. (2008). The role of early life genistein exposures in modifying breast cancer risk. *British Journal of Cancer*, Vol. 98, pp. 1485-1493, ISSN 0007-0920.

Willett, W.C. (1997). Nutrition and cancer. *Salud Publica Mexican*, Vol. 39, No. 4, pp. 298-309.

Wu, Z., Robinson, D. S., Hughes, R. K., Casey, R., Hardy, D. & West, S. I. (1999). Co-oxidation of β-carotene catalysed by soybean and recombinant pea lipoxygenases. *Journal of Agricultural and Food Chemistry*, Vol. 47, pp. 4899–4906.

Xiao, C. W. (2008). Health effects of soy protein and isoflavones in humans. *Journal of Nutrition*, Vol. 138, No.6, pp. 1244 - 1249.

Yoon, S. & Klein, B.P. (1979). Some properties of pea lipoxygenase isoenzymes. *Journal of Agricultural and Food Chemistry*, Vol. 27, pp. 955–962, ISSN: 0021-8561.

Zimeri, J. & Tong, C.H. (1999). Degradation kinetics of (-)epigallocatechin gallate as a function of pH and dissolved oxygen in a liquid model system. *Journal of Food Science*, Vol. 64, pp. 753-758.

7

Soybean Phytoestrogens – Friends or Foes?

Branka Šošić-Jurjević, Branko Filipović and Milka Sekulić
*University of Belgrade, Institute for Biological Research „Siniša Stanković",
Serbia*

1. Introduction

Proper and balanced nutrition is very important in prevention and treatment of chronic diseases. Many individuals modify their diet and/or take different nutraceuticals expecting to attain optimum health, extend their lifespan and prevent diseases such as cardiovascular, cancer, osteoporosis, obesity, or diabetes type II.

Based on „Japanese phenomenon" (Adlercreutz, 1998), numerous advertisements suggest that soy-based diet, and its phytoestrogens (PE) in particular, provide protection against many chronic diseases and contribute to the long lifespan often observed in Asia. That is why soy and other phytoestrogen - rich plants became increasingly popular in the U.S. and western countries in the past 30 years. Furthermore, in these countries, PEs are often consumed in its purified form, as nutritional supplements, "designed" for special medical purposes. These supplements are freely available in pharmacies, health food shops, grocery shops and are usually consumed without medical control. There is a lack of awareness that uncontrolled consumption of natural PEs may be potentially harmful to human health. Even more concerning is that some people consume supplements in excess of suggested daily dosage (Wuttke et al., 2007).

The soybean (*Glycine max*), compared to other legumes, is richer in protein levels and quality, based on its digestibility and concentration of essential amino acids (Rand et al., 2003). It is also good source of fiber, certain vitamins and minerals, such as folate and potassium (Rochfort and Panozzo, 2007). It has very high antioxidant content, similar to fruits famous for their antioxidant activity (Galleano et al., 2010). Also, despite their high carbohydrate content, the glycemic load of soybeans is relatively low due to their low glycemic index. In addition, soy-food has high levels of iron in the form of ferritin (Lönnerdal et al., 2006). The concentration of calcium in soymilk is much lower than in cow milk, however, its absorption from soy milk is similar to that from cow milk (Reinwald and Weaver, 2010).

Besides the favorable nutritional attributes, soybean contains a number of biologically active components (saponins and lunasin, phytic acids, phytosterols, trypsin inhibitors, and peptides) including isoflavones genistein (G), daidzein (D) and glycitein (Gy). As soybean phytoestrogens, isoflavones are considered the most important in prevention and treatment of hormone-dependent cancers, cardiovascular diseases, osteoporosis, menopausal symptoms and other age-related diseases. In addition, some studies suggest that soy and its isoflavones affect body weight homeostasis.

Modern world is a controversy with ever-increasing obesity on one side, and a high percent of starving people around the globe, on the other side. Having that in mind, combined with

observed beneficial health and weight-lowering effects, high nutritional value makes soy probably one of the most strategically important plants.
However, aside from potential beneficial effects (still under intensive investigation and not fully proven), soybean phytoestrogens may also act as endocrine disruptors, by interfering with the function of reproductive system, as well as with other endocrine systems, namely thyroid and adrenal, and may, under some circumstances, increase cancer risk. This is why scientists are intensively trying to precisely evaluate potential benefits versus adverse effects of soy. Due to the importance, the researches are done both in vitro and in vivo, using different experimental approaches, animal models and various human studies. Results obtained so far are highly inconsistent and depend on experimental conditions, applied doses, animals and humans' age and sex, type of diet, presence of other PE sources in the diet, or other factors. Moreover, it remains unclear whether soy extracts, soy concentrate and purified isoflavones have identical effects. This is why the role of soy food in diet became a somewhat confusing topic in recent years. With approximately 2000 soy-related papers published annually, and half of it related to isoflavones (Messina, 2010), it is becoming extremely difficult to compare all of the available data.
Due to the many differences in the chemical composition of soy products, and the fact that two thirds of human population cannot produce equol (Setchell et al., 2002), the authors decided to primarily focus their attention on effects of purified genistein and daidzein. We will evaluate the latest findings, using clear statements from the literature, as well as our own results, focusing on major potential healthful effects while also considering adverse effects of purified soybean phytoestrogens. More important, the authors will try to analyze the data in order to evaluate whether the net beneficial /adverse effect for each targeted organ system depends on sex and age.

2. Structure, absorption and bioavailability of soybean phytoestrogens

Soybeans and its products are the most abundant source of isoflavones in the human diet. Isoflavones are normally taken up with food, absorbed in the gastrointestinal tract, and eliminated via urine. The absorption and bioavailability of isoflavones has been the subject of frequent debates among scientists. One of the main factor influencing the absorption and bioavailability is the chemical structure of the compound (D'Archivio et al., 2010). Structurally, soy isoflavones G, D and Gy are diphenolic compounds, which are present in soy and non-fermented soyfood isoflavones in its glycosylated forms, as glycones genistin, daidzein and glycitin (Setchell, 1999; Fig. 1).

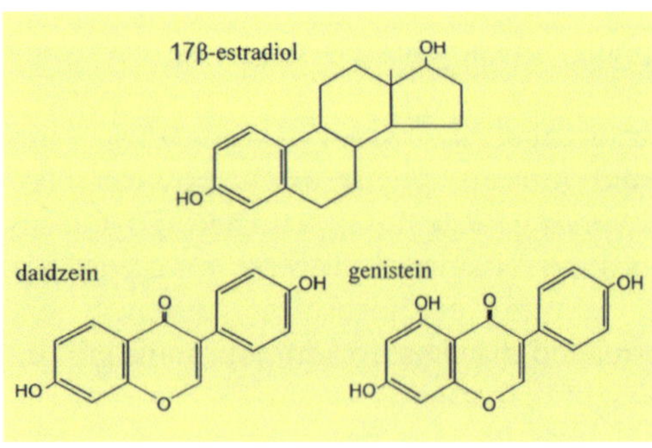

Fig. 1. Structure of soybean phytoestrogens and mammalian 17β – estradiol.

As a prerequisite for absorption, the sugar must be removed from the compound at some point during ingestion (Setchell et al., 2002). Soy isoflavone glycosides are hydrolyzed to their aglycones by lactase phloridizin hydrolase in the apical membrane of the lumen of the small intestine, as well as by bacterial intestinal glucosidases (Wilkinson et el., 2003). Aglycones undergo passive diffusion across the small and large intestinal brush border (Larkin et al., 2008). However, some authors claim that glycosides may be absorbed also through the active sodium–dependent glucose transporter (Gee et al., 2000). Results obtained when we examined effects of soy extract on fluidity of erythrocyte membrane, showed that genistein and isoflavone glucosides intercalate and increase the order and rigidity of the outer layer of cellular membrane. Therefore, isoflavone glucosides may be also transported across the cell membrane directly, via entropy-driven flip-flop (Ajdžanović et al., 2010, 2011). Biological significance of this mechanism is unclear.

The absorption and bioavailability of isoflavones depends to some extent on interaction with other food components (Birt et al., 2001). The assumption that isoflavones are absorbed more efficiently from fermented than from non-fermented soy foods was re-examined and then rejected (Maskarinec et al., 2008). Since intestinal microflora is capable of hydrolyzing the isoflavone glycosides from nonfermented soyfood, recommendations favoring fermented soyfood cannot be justified.

Genistein is stronger than daidzein in its agonistic activity for the ERs, as well as in its antioxidative potential. On the other hand, daidzein can be further metabolized into its bacterial metabolite equol, which has stronger estrogenic and antioxidative properties than both genistein and daidzein, or some other isoflavone metabolites (Mitchell et al., 1998). Although it appears that all animals produce equol following soy ingestion, in humans this is the case in approximately 30% of population (Lampe et al., 1998; Setchell et al., 2002). This is thought to be dependent on inter-individual variability in the presence of specific intestinal bacteria (Rowland et al., 2000). Besides the microflora composition, individual differences in gut transit time and redox potential of colon and genetic polymorphisms are likely to contribute to this great variability (Duffy et al., 2007). When evaluating the effects of age on equol production, it was demonstrated that during the first months of life, equol levels in plasma and urine were significantly lower than in adults, which may be due to the immature intestinal flora (Setchell et al. 2002). Lampe et al. (1998) detected no significant differences in the prevalence of equol production between genders.

3. Biological basis of soybean phytoestrogen actions

The estrogenic effect of the isoflavones was first recognized when examining impaired fertility in grazing animals (Bennetts et al., 1946). Three decades later, Setchell et al. (1987) established that isoflavone-rich soy was a factor in reduced fertility of cheetahs in North American zoos. Isoflavones were classified as phytoestrogens following in vivo and in vitro demonstration of their binding potency of isoflavones for estrogen receptors (ER), as well as for sex-hormone binding globuline (Kuiper et al., 1998).

Testosterone actions in numerous male tissues are mediated through its conversion to estrogen catalyzed by aromatase enzymes. Specific α and β ER are detected in different male and female tissues (Korach, 1994), but the ratio between ERα and ERβ is different (Rosen, 2005). This finding has finally changed the classical view of the estrogens as exclusively female hormones. ERβ is known to modulate ERα transcriptional activity acting as an activator at low concentrations of mammalian estrogen - estradiol 17β (E2) and as an inhibitor at high concentrations of E2. E2 has equal binding affinities for ERα and ERβ, while isoflavones have a higher potency for ERβ.

The molecular structures of genistein, daidzein and E2 are similar in many aspects. The intra-molecular distance between the hydroxyl groups at each end of the molecules is almost identical for both isoflavones and E2. These distances determine hydrogen bond interaction with amino acids of the ligand-binding site of the ER (Vaya and Tamir, 2004). Though molecular binding for ER between isoflavones and E2 are similar, both G and D binding potency for ERs is significantly lesser in comparison to E2. In addition, they bind with higher potency to estrogen receptor (ER) β in comparison to ERα (Kuiper et al., 1998). These features classify them as potential natural selective estrogen receptor modulators (Phyto SERMs). Thus, soybean isoflavones may exert estrogenic, antiestrogenic, or estrogen non-reactive biological actions, depending on their concentration and concentration of endogenous estrogen, tissue, and amount and type of estrogen receptors present in the tissue (Wuttke et al. 2007). Therefore, it is of importance to determine the estrogenic action of isoflavones compared with the effects of E2 (in females) and both testosterone and estradiol (in males) in each individual organ.

Phyto SERMs represent a new and very promising class of potential hormonal therapy agents. Major potential advantage of SERMs over estrogen analogue therapy is that it may demonstrate all of the favorable effects of estrogens. However, in order to declare isoflavones as safe, it needs to be demonstrated that they do not share the risks associated with estrogens used in hormone replacement therapy, osteoporosis treatment or in treatment of prostate carcinoma (Wuttke et al., 2007).

Besides their estrogenic activities, isoflavones also exhibit non-hormonal actions such as antioxidant effects. Antioxidant properties are one of the most important claims for food ingredients, dietary supplements and anticancer products. In addition, the free radical theory of aging continues to be among the most popular theories. Therefore, the antioxidant property of isoflavones offers an additional important mechanism through which they protect against age-related diseases. All soy isoflavones act as antioxidants, playing role in scavenging free radicals that can cause DNA damage and lipid peroxidation (Kruk et al., 2005) and activate antioxidant enzymes such as catalase, superoxide dismutase, glutathione peroxidase and glutathione reductase (Mitchell et al. 1998). The determining factors for isoflavone antioxidant activities are the absence of the 2, 3-double bond and the 4-oxo-group on the isoflavone nucleus and the position of the hydroxyl groups, with hydroxyl substitution being of utmost importance at the 4′ position, of moderate importance at the 5 position, and of little significance at the 7 position. That is why G has higher antioxidant capacity than daidzein and the reason why both have stronger antioxidant activity than their glycosides (Cherdshewasart and Sutjit, 2008).

Genistein in high concentration is a potent inhibitor of Tyr kinases (Akiyama et al., 1987), DNA topoisomerases I and II, and ribosomal S6 kinase, resulting in inhibition of cell growth. Tyrosine kinases are responsible for <1% of protein phosphorylation within cells, but they appear to phosphporylate many proteins required for regulation of cell functions. Genistein has been shown to induce cell cycle arrest and apoptosis in numerous cell lines, including ER (+) and ER (-). Many of the reported beneficial effects of isoflavones and particularly those on tumor growth may be attributed to this mechanism. However, some authors critycize this by underlying that the concentrations necessary for such inhibition in the tested cell systems or organs by far exceed the serum concentrations achieved by isoflavone ingestion alone (Jiménez and Montiel 2005; Wuttke et al., 2007).

Growing evidence shows that isoflavones may also modulate the activity/expression of steroidogenic enzymes. These enzymes are present in the adrenal glands and gonads but also in many tissues that have the ability to convert circulating precursors into active hormones (i.e. brain, liver, reproductive tracts, adipose tissue, skin and breast tissue). Genistein and daidzein were reported to inhibit the activity of 3β- hydroxysteroid dehydrogenases (HSD) purified from bovine adrenal microsomes (Wong & Keung, 1999). The same isoflavones were also shown to inhibit 3β-HSD type II in mitochondrial and microsomal preparations of the human adrenocortical H295R cell line, and subsequently a similar inhibition of the conversion of dehydroepiandrosterone (DHEA) to androstenedione by these isoflavones was observed in total membrane fractions of Sf9 insect cells in which human 3β-HSD had been over-expressed (Ohno et al., 2002). However, Mesiano et al. (1999) showed that genistein and daidzein specifically inhibited the activity of 21-hydroxylase (P450c21/CYP 21) in H295 cells but had no effect on other steroidogenic enzymes, including 3β-HSD.

4. Soybean phytoestrogens in prevention and therapy of cancer

The incidence of hormon-dependent cancers, namely breast and prostate, is lower in Asia than in western countries (Messina et al., 2006; Parkin, 2005). Migrants from Asia, who maintained their traditional diet, even when living in the West, had a lower risk of these diseases. However, shifting towards a more of a western diet increased the risk (Ziegler et al., 1993). Once SERM properties of soybean isoflavones were discovered, it was hipothesized that high soy dietary intake might be associated with low incidence of hormone-dependent cancers in Asian population, as well as with other putative health benefits (Setchell, 1999). That is why soyfood and its isoflavones in a form of dietary supplements or concentrated extracts have been increasingly used in the western populations in the recent years.

However, when Patisaul and Jefferson (2010) disscussed potential safety of infant soy formula, they stressed the essential difference between Asians (on a traditional „soy-reach" diet) and Caucasians (on a traditional „Western" diet) in exposure to soy over the lifespan. In Asia, soy consumption is high during entire lifespan, except for a brief breast-feeding period in early infancy. People in the West feed their babies soy infant formula, so the pattern is just the opposite - the highest intake of isoflavones occurs in the first year of life and then drop to near zero, with eventual increase later in advanced adult age. In relation to this, some authors support the opinion that lower incidence of breast cancer in Asian women is due to their continous exposure to soy from early life throuought their whole lifespan (Warri et al., 2008). Maskarinec et al. (2004) concluded that Caucasian women who ate more soy during their lifespan had denser breast tissue (a risk factor for breast cancer) than those who did not.

4.1 Effects on breast cancer

Overexposure to estrogen (early menarche, short duration of breastfeeding and low parity) is a major contributing factor in the development of breast cancer. As soybean isoflavones have a relatively high binding potency for ERs, a concern has been raised that high phytoestrogen intake may promote growth of estrogen-sensitive tumors or put breast cancer survivors at risk of reoccurrence (Helferich et al., 2008; Messina &Loprinzi, 2001).

The data about the role of isoflavones in prevention and therapy of breast cancer are controversial. Some authors proposed that genistein at low, physiologically relevant level, may stimulate ER-positive tumors due to their estrogenic properties, while at higher level, anti-cancer actions of isoflavones may be predominant (Duffy et al., 2007). Shu et al. (2009) also suggested dose-dependent effects of ingested soybean isoflavones: intake of low doses was associated with increased mortality rate and breast cancer recurrence, while intake of more than 40 mg per day appeared to have antiproliferative effects. These results were evident in women with both ER-positive and ER-negative breast cancer. The authors suggested that soy isoflavones protect against breast cancer by competing with estrogens in binding to the estrogen receptor. At the same time soy isoflavones increase the synthesis of sex hormone-binding globulin, lowering the biological availability of sex hormone, inhibit 17β-hydroxysteroid dehydrogenases (thus reducing estrogen synthesis), and increase clearence of steroids from the circulation (Taylor et al., 2009). However, Harris et al. (2004) showed that isoflavones inhibit sulfotransferase (enzymes that catalize estrogen inactivation in mammary gland) ten times more than sulfatase enzymes, which catalize local estrogen production. This may lead to increase in free estrogen levels in the tumor tissue, which in turn may stimulate tumor growth.

Recent epidemiological and clinical data were summarized in review article of Messina & Wood (2008) and the authors concluded that isoflavone intake have either a modest protective role or no effect on breast tissue density in pre and postmenopausal women and on breast proliferation in postmenopausal women with or without a history of breast cancer.

The results of animal studies are also controversial. Experiments on monkeys that examined effects of soy on mammary gland indicated to the possibility that proliferating effect of estradiol may be antagonized by isoflavone-rich soy protein diet (Jones et al., 2002; Wood et al., 2004).

A number of studies conducted in immunodeficient nu/nu or SCID mice strains demonstrated enhanced proliferation, or no effect of isoflavones on tumor development and progression (Allred et al. 2004; Hsieh et al., 1998). In addition, Heferich and co-workers (2008) implanted estrogen-dependent tumors into ovariectomized mice and found that dietary genistein was able to reduce the inhibitory effect of tamoxifen on tumor growth. However, prevention and inhibition of the progression of experimentally induced mammary tumors by isoflavones was also detected, as well as that post pubertal soy treatment before the induction of tumor had a slightly preventive effect (Pei et al., 2003; Sarkar et al., 2002). Results on Sprag–Dawley rats also proposed that pre-pubertal exposure to soybean isoflavones have highly significant tumor preventive effects (Gallo et al., 2001; Lamartiniere et al., 2002).

More recent review of the animal models used to investigate the health benefits of soy isoflavones concluded that results obtained in different animal models demonstrate minimal effects of isoflavones in breast and prostate cancer prevention (Cooke, 2006).

In vitro genistein inhibited proliferation of ER-positive and ER-negative breast cancer cells at high doses (>10M), but promote tumor growth at lower, more nutritionally relevant doses (Wang et al., 1996). Tamoxifen is the oldest and most-prescribed SERM for breast cancer treatment and it also have mixed effects depending on dose. The SERM-like activities of soy isoflavones makes dietary guidelines particularly difficult to be issued with confidence. Carcinogen-induced mammary cancers predominantly express ERα, and there are some indications that substances that activate mainly ERβ have an antiproliferative effect. In addition, it was reported that genistein may interact with tamoxifen, both synergistically and antagonistically (Shu et al., 2009; Taylor et al., 2009). The inhibition of proliferation in human breast cancer cell line with tamoxifen could be overridden by physiological concentration of genistein (Jones et al., 2002), which indicate that genistein may negate healing effect of tamoxifen on breast cancer patients.

Besides the ER-dependent mechanisms, high doses of genistein may inhibit tumor development and growth by other molecular mechanisms: by antiproliferative actions through inhibition of tyrosine kinase and DNA topoisomerase activities (Akiyama et al., 1987; Markovits et al., 1989), by induction of cell cycle arrest and apoptosis (Bektic et al., 2005), as well as by exerting anti-angiogenic actions (Fotsis et al., 1993).

4.2 Effects on prostate cancer

Lifelong exposure to isoflavones plays a role in the low incidence of prostate cancer observed in Asian males. However, the effects of soy consumption on existing prostate cancer may differ in relation to disease stage. Kurahashi et al. (2007) reported that soy isoflavones in the diet decreased the risk of localized prostate cancer, while soy-containing miso soup increased the risk of advanced prostate cancer. The obtained results may be due to loss of estrogen receptors in advanced tumors, or due to possible errors in food measurement and small sample of men with advanced prostate cancer.

Hamilton-Reeves et al. (2007) reported that soy protein isolate with or without isoflavones affected hormone receptor expression patterns in men at high risk for developing advanced prostate cancer. Intake of soy protein isolate with isoflavones significantly suppressed androgen receptor expression but did not alter estrogen receptor beta expression in prostate, while intake of soy protein isolate without isoflavnoes tended to suppress AR expression ($P = 0.09$). The authors concluded that soy protein isolate consumption may be beneficial in preventing prostate cancer, and hypothesized that soy isoflavones may attenuate but not prevent progression of latent prostate cancer.

Hussain et al. (2003) found that patients with prostate carcinoma consuming a soy-enriched diet had a statistically significant drop in prostate-specific antigen (PSA) levels, compared to the control group. However, more recent study of deVere White et al. (2010) demonstrated that higher amounts of aglycone isoflavones genistein and daidzein did not lower PSA levels in men with low-volume prostate cancer.

Osterweil (2007) observed a dose-dependent decrease in the risk of localized prostate cancer with isoflavone consumption. Men with higher intake of isoflavones had a decreased risk of prostate cancer compared to those with lower intake of isoflavones.

Few animal studies have been conducted to investigate the role of soy isoflavones on prostate cancer development and progression. Genistein markedly inhibited prostate tumor metastasis in mice (Lakshman et al., 2008). Isoflavone-containing diets retarded the development of prostate cancer in rats (Pollard & Suckow, 2006). In contrast to this, Naik et al. (1994) showed that genistein added to the drinking water or intraperitonealy injected have no effect on the growth of the subcutaneously implanted MAT-LyLu prostate carcinoma in rats.

Zhou et al. (1999) in their in vitro studies found that dietary soy products may inhibit experimental prostate tumor growth through a combination of direct effect on tumor cells and indirect effects on tumor neovasculature. In addition, dietary phytoestrogens down-regulated androgen and estrogen receptor expression in adult male rats prostate (Lund et al., 2004). More recent in vitro studies demonstrated that phytoestrogens at high concentrations exert an anti-androgen effect through the interaction with AR (Mentor-

Marcelet al., 2001). In vitro tests also showed that soy isoflavone genistein induced apoptosis and inhibited growth of both androgen-sensitive and androgen-independent prostate cancer cells (Hussain et al., 2003).
Wuttke et al. (2010) in a recent review provided detailed analysis of both in vitro and animal experimental data and concluded that isoflavones may protect the prostate to make it less prone to develop cancer.
In conclusion, based on inconsistent evidence, it is apparent that the use of phytoestrogens as chemopreventive agents is still in its infancy, justifying a need for further research. Experimental studies based on nutritionally relevant doses are needed to clarify potential health benefits, as well as estrogenic, antiandrogenic and/or nonestrogenic isoflavone activities in the breast and prostate tumors.

5. Soybean phytoestrogens in prevention and therapy of cardiovascular diseases

Soy protein and isoflavones received great attention and provoked heated discussions due to their potential role in reducing risks of cardiovascular diseases. Following is a historical overview of the most relevant results and announcements related to clinical trials, as well as of animal and in vitro research, providing insight into potential mechanisms of isoflavone action.

5.1 Effects on serum lipid levels

Obesity is associated with disruption in lipid and sugar metabolism, and is a principal cause of chronic diseases, namely cardiovascular diseases, hypertension, atherosclerosis and type II diabetes mellitus. This makes obesity a major health problem, which has reached pandemic proportions. The treatment for obesity is lifestyle change, including diet restriction and exercise. However, pharmacological treatment is often necessary. Isoflavones are of particular interest as an alternative to statins or fibrates in potential lowering of serum lipid levels.
Epidemiologic studies demonstrated a reduced rate of mortality due to coronary hearth disease in Japanese postmenopausal women populations consuming a traditional Japanese diet. On the other side expatriate Japanese living in the US had higher blood pressure and cholesterol levels than the Japanese still living in Japan. Some authors proposed that detected differences are not of genetic origin but are due to diet rich in soy products, fish and fiber (Adlercreutz et al., 1998).
Anderson et al. (1995) published a meta-analysis that attracted widespread attention, demonstrating that intake of at least 25g of soy protein per day lowered total and low density lipoprotein (LDL) cholesterol. Lipid lowering potential of soy protein was also demonstrated in various animal studies (Greaves, et al., 1999; Potter, et al., 1995). This led to U.S. Food and Drug Administration (FDA) issuing a health claim for soy protein and coronary hearth disease (1999). FDA also claimed that the evidence did not support significant role of isoflavones in lipid-lowering effects of soy protein. Some more recent reports also demonstrated a significant reduction in plasma concentrations of total and LDL cholesterol in humans exposed to soy proteins (Greany et al., 2004; Teixeira et al., 2000).
Due to their estrogenic activity, isoflavones may be the bioactive component attributed to soy protein. This possibility was examined using different experimental approaches and

animal models. Some research studies highlighted a favorable hypolipidemic effect related to isoflavones, at least when consumed in combination with soy proteins. Removal of the isoflavone-containing fraction from soy protein resulted in a loss of its beneficial effect on the serum lipid profile and atherosclerosis progression in mice (Kirk et al., 1998), in golden Syrian hamsters (Lucas, et al., 2001), and in rhesus monkeys (Anthony et al. 1996). High isoflavone, combined with high soy protein intake leads to significantly decreased serum total and LDL cholesterol compared to low isoflavone intake. Some authors reported that ingested purified isoflavones exert lipid-lowering effects (Ae Park et al. 2006; Kojima et al. 2002; Sosić-Jurjević et al., 2007). However, others showed minimal or no effects of isolated isoflavones on blood lipid levels (Greaves et al., 1999; Molsiri et al., 2004).

Clinical trials also show diverse beneficial effects of isoflavone supplements on cardiovascular system. These discrepancies may be a result of different intestinal bacterial flora and hence bioavailability of soy isoflavone metabolites. Other reasons might be differences in dose-response effects (Hooper et al., 2008), sex and length of isoflavone supplementation (Zhan & Ho, 2005), limited number of subjects, or pre-existing metabolic status of subjects included in supplement trials (Villa et al., 2009).

In contrast to previously mentioned data, in 22 random trials, isolated soy protein combined with isoflavones, compared with milk or other proteins, decreased LDL cholesterol by approximately 3%. This reduction was small in comparison to amount of soy protein (average 50g per day) intake (Sacks et al., 2006). There was no detected benefit on level of HDL cholesterol, triglycerides or blood pressure. These authors concluded that soy food may be beneficial to cardiovascular health because of their high content of fiber, vitamins, high content of polyunsaturated fat, rather than and its isoflavone content. Recent review of the animal models used to investigate the health benefits of soy isoflavones also concluded that the efficiency of isoflavones in improving lipid profile is less than earlier research suggested (Cooke, 2006).

For this reason, American Hearth Association issued a discoursing statement, and warned that earlier research indicating clinically important favorable effects of soy products on low density lipoprotein (LDL) is not confirmed by most studies during the past 10 years. U.S. FDA announced its intent to reevaluate the data related to cardio protective effects of soy (2007).

More recent research demonstrated that the combined intervention of genistein and l-carnitine act synergistically in reducing serum lipid and LDL levels, as well as reducing body weight in mice and rats (Che et al., 2011; Yang et al., 2006). In addition, synergy portfolio diet, containing plant sterols, viscous fibers and soy protein reduced serum LDL cholesterol similar to traditional statin drugs (Jenkins et al., 2003). Therefore, soybean isoflavones, either as natural components of food or as nutritional supplements, in combination with other functional food may favorably alter indicators of cardiovascular disease risk.

Though positive effects on metabolism in humans have been widely debated, studies in rodents should help in identifying and evaluating the biologically relevant mechanisms involved in isoflavone actions.

ERs are important mediators of the action of estrogen on lipid metabolism both in males and females. Men with mutations in the aromatase gene (enzyme that converts androgens to estrogens) display truncal obesity, insulin resistance and hyperlipidemia (Carani et al., 1997). Due to structural similarities of isoflavones and E2, G and D might also directly influence the regulation of adipogenesis. However, it must be noticed that genistein preferably binds to ERβ, while ERα is predominantly found in liver. In ovariectomized mice, estradiol and genistein did not increase estrogen-responsive genes in the liver, and the authors suggested that the cholesterol-lowering ability of estrogen requires estrogen receptors (they postulated crosstalk between ERs and NF κβ) but not estrogen receptor-dependent gene transcription (Evans et al., 2001).

Isoflavones may have distinct influences on metabolism in males and females. Males have a different number and distribution of ERs compared to females. It is important to realize the impact of other hormones such as androgens and thyroid hormones on liver and other metabolic tissues. Using ovariectomized Wistar rats Molsiri et al. (2004) obtained no significant difference in serum lipid levels after s.c. genistein injections, while we detected lipid lowering effect of both G and D (similar to this obtained for testosterone-treated groups) in orchidectomized young and middle-aged adults, as well as in testis-intact middle-aged male rats (Sosić-Jurjević et al., 2007 and our unpublished data).

Aside from having estrogenic activity (Potter et al., 1995), both G and D exert „phytofibrate" and or „phytoglitazone"activity, and activate peroxisome proliferator-activated receptors (PPAR) α and γ (Mezei et al., 2006). PPARs bind a wide number of ligands and directly affect lipid metabolism by enhancing transcription of PPAR-regulated genes (Shen et al., 2006). Generally, PPARα controls the transcription of many genes involved in lipid catabolism, whereas PPARγ controls the expression of genes involved in adipocyte differentiation and insulin sensation. PPARα is important for β-oxidation and is mainly expressed in liver, kidney, heart, and muscle, where lipoprotein metabolism is important. PPARγ is mainly expressed in adipose tissues and is considered the master regulator of adipogenesis (Rosen, 2005; Ørgaard & Jensen, 2008).

Isoflavones may also affect lipid metabolism indirectly, via effect on thyroid function and/or thyroid hormone action in liver. T3 and its receptor (TR) play important role in regulation of energy homeostasis, metabolic processes and body weight. Hypothyroidism causes hypercholesterolaemia characterized by increased levels of LDL (Sasaki, et al., 2006).TRβ1 is the major TR in the liver while T3 action is mediated via TRα1 in the heart. TRβ1 agonist KB-141 lower cholesterol, increases metabolic rate and decreases body weight (Grover et al., 2005). Xiao et al. (2007) described that expression of the rat hepatic thyroid hormone receptor β1 is upregulated by isoflavones. In addition, E interplay with TH in regulation of different physiological functions including effects on growth, bone mass, and triglycerides. E can be viewed as a modulator whose response relies on interplay with T3 signaling mechanisms (DiPippo et al., 1995).

Many researchers have tried to link effects of soy intake on lipid metabolism with modulation of thyroid hormone levels. However, it is still difficult to demonstrate clear-cut effects on thyroid (this topic would be analyzed in more details in a subchapter related to endocrine disruptive potential of isoflavones). On the other hand, most researchers who examined lipid-lowering potential of isoflavones did not include in their research examining of the thyroid status, or deiodinase I enzyme activity in liver. When examining the effects of G and D that should mimic exposure to supplements (10mg/kg) in orchidectomized middle-aged male rats our research team obtained that both G and D decreased the serum total cholesterol and LDL levels similar to control testosterone treatment, and brought about an increase in serum triglycerides similar to that observed after control estradiol treatment (Sosić-Jurjević et al., 2007). Within the same animal model we detected significant decrease of serum thyroid hormones (Sosić-Jurjević et al., 2010). However, when we examined deiodinase I enzyme activity in liver of G and D treated rats, it was significantly increased (our unpublished data) in comparison to the control values. Therefore, the local production of T3 in liver was increased and the local increase of T3 might contribute to the detected decrease in total cholesterol and LDL levels.

5.2 Effects on atherosclerosis progression

Atherosclerosis is part of the normal aging process but its progression depends on a wide range of environmental and genetic factors (Davies et al., 2004). Generally, atherosclerosis refers to the formation and hardening of fatty plaques (atheromas) on the inner surface of the arteries. The arteries not only harden, they become narrow. Such narrowed vessels can be easily blocked by constriction or objects in the bloodstream. Atherosclerosis begins with injury to endothelial cells, exposing portions of the artery surface below the endothelium. Free radicals or other irritants could start the process, as well as high blood pressure. Platelets cluster around the injured endothelial cells and release prostaglandins, which cause the endothelial cells to proliferate. LDL-cholesterol particles release their fat into the areas made porous by prostaglandins. Macrophages swell themselves on oxidized LDL-cholesterol until they become "foam cells" that invade atheromas. The atheromas are hardened by fibrin, which forms scar tissue, and finally calcium patches.

The atheroprotective effects of soy-based diets have been partly attributed to the associated reduction in cholesterol levels in human studies (Jenkins et al., 2002). Similar findings have also been reported in nonhuman primates fed soy-based diets (Anthony et al., 1997; Register et al., 2005). Animal studies with rabbits and hamsters, which are considered a good non-primate model for studies of atherosclerosis, demonstrated that soy isoflavones reduce atherosclerotic lesion areas in the aortic arch by means of LDL reduction (Alexandersen et al., 2001; Lucas et al., 2003). In addition, atherosclerotic changes induced by a cholesterol rich diet were prevented by isoflavones in rabbits, hamsters and premenopausal monkeys (Adams, et al. 2005; Lucas et al., 2001). The intake of genistein and daidzein decreases LDL oxidation (Tikkanen, et al., 1998). Both genistein and daidzein have also been shown to protect human umbilical cord endothelial cells and bovine aortic endothelial cells from the atherogenic effect of oxidized LDL (Kapiotis et al., 1997).

However, numerous animal studies suggest that dietary soy inhibits atherosclerotic lesion development by mechanisms other than lowering serum cholesterol.

Isoflavones are reported to prevent lipid peroxidation by scavenging lipid-derived peroxyl radicals (Patel et al., 2001) and inhibit copper-dependent LDL oxidation (Kerry & Abbey, 1998). It is well known that oxidized LDL is more prone to induce atherosclerosis than unoxidized form. In addition, proteome analyses revealed protein targets that in response to soy isoflavones increase the anti-inflammatory response in blood mononuclear cells thereby contributing to the atherosclerosis-preventive activities of a soy-rich diet (Wenzel et al., 2008). Studies in apolipoprotein E knock-out mice showed that atherosclerotic lesions are reduced when fed a soy-containing diet despite unchanged serum lipid levels (Adams et al., 2002). Findings from a recent study in aged lipoprotein receptor knock-out mice has underscored the importance of oxidative stress coupled with a failure to up-regulate There is now compelling evidence that isoflavone supplementation have anti-inflammatory functions and hence can represent an effective therapeutic strategy to enhance Nrf2 activity to protect the aging vasculature (Adams et al., 2002; Mulvihill & Huff, 2010).

Vasodilatatory effects of isoflavones may be also related to their estrogenic actions. Both estrogen receptors α and β are expressed in the arteries (Christian et al., 2006). Estrogens have been shown to stimulate inducible NO synthase in endothelial cells and the increased

NO production causes relaxation of artherial myocites (Mahn et al., 2005). Research on the effect of genistein on plasma nitric oxide concentrations, endothelin-1 levels and endothelium-dependent vasodilatation in postmenopausal women revealed that genistein therapy improved flow-mediated endothelium-dependent vasodilatation in healthy postmenopausal women. This improvement is probably mediated by a direct effect of genistein on vascular function and could be the result of an increased ratio of nitric oxide to endothelin (Squadrito et al., 2002).

In conclusion, despite the fact that dietary soy products and isoflavones are heavily advertised for their hypolipidemic effect, their therapeutic potential is lesser than was previously hoped and depend on many factors related to inter-individual differences.

6. Soybean phytoestrogens in bone protection

Bone remodeling is a continuous process of bone resorption and bone formation for the purpose of maintaining normal bone mass. As a skeletal disease characterized by low bone mass and microarchitectural deterioration of bone tissues, osteoporosis is usually caused by a chronic imbalance in the bone remodeling cycle. This skeletal disorder occurs as part of the natural aging process and is associated with the rapid decline in ovarian function and subsequent reduction of circulating estrogen in women after the menopause and declining testosterone in middle-aged and older men. However, in contrast to postmenopausal osteoporosis in women, the age related bone loss in men is less well-defined. Observational studies have indicated that estrogen administration is important in bone remodeling. Thus, hormone replacement therapy (HRT) administered in a dose-dependent manner, not only significantly reduces bone loss, but also lowers the incidence of hip and vertebral fractures (Lindsay et al., 1976, 1984; Michaelsson et al., 1998). In the other hand, although HRT has a protective effect on bone tissue, it can increase the risk of breast, endometrial ovarian or prostate cancer developing (Davison & Davis, 2003; Loughlin & Richie, 1997; Nelson et al., 2002). For this reason, much attention has been paid to the examination of alternative therapeutic compounds that may have protective effects on bone, without adverse effects on other tissues. Epidemiological studies have demonstrated a low incidence of postmenopausal fractures and high bone mineral density (BMD) in Asian populations with a particularly soy-rich diet (Cooper et al., 1992; Lauderdale et al., 1997; Somekawa et al., 2002). Thus, phytoestrogens have been proposed as an alternative to conventional hormone therapy for preventing osteoporosis and have shown beneficial effects on bone health (Barnes, 2003; Morin, 2004).

Bone remodeling is regulated by the activity of two different cell lines. Osteoblasts stimulate bone formation and calcification, while osteoclasts promote bone resorption. It has been shown that isoflavones affect osteoblastic bone formation and osteoclastic bone resorption in vitro. The anabolic effects of genistein and daidzein on bone metabolism have been investigated in culture using femoral trabecular and cortical bone tissues obtained from elderly female rats (Gao & Yamaguchi, 1999; Yamaguchi & Gao, 1997, 1998). Genistein induced a significant increase in calcium content, alkaline phosphatase activity as a marker of osteoblasts, as well as DNA content, which is an index of bone cell numbers in bone tissues (Yamaguchi & Gao, 1997). In bone tissue culture medium daidzein significantly elevated bone components (Gao & Yamaguchi, 1999). Both genistein and daidzein increased newly synthesized protein content, alkaline phosphatase activity and DNA content in cultures of osteoblastic MC3T3-E1 cells (Sugumoto & Yamaguchi, 2000, 2000a; Yamaguchi & Sugumoto, 2000).

In addition to effects on osteoblasts, many authors have reported that isoflavones are efficacious in suppressing osteoclast activity in vitro. Genistein completely inhibited bone resorption and osteoclast-like multinucleated cells in culture with bone-resorbing factors (Gao & Yamaguchi, 1999a; Yamaguchi & Gao, 1998a). Also, daidzein inhibited the development of osteoclasts from cultures of porcine bone marrow and reduced bone resorption (Rassi et al., 2002).

While in vitro studies reveal possible actions of isoflavones on individual bone cells, in vivo studies provide insight into the effects of isoflavones on the intact system and coupling effects between osteoblasts and osteoclasts. Most of the animal bone studies investigating isoflavone action have been performed in rodents. Aged ovariectomized female and orchidectomized male rats represent a suitable model for simulating osteoporosis due to estrogen or androgen deficiency (Comelekoglu et al., 2007; Filipović et al., 2007; Pantelić et al., 2010; Turner, 2001; Vanderschueren et al., 1992). Using this animal model, supplementation with isoflavones has been shown to prevent bone loss (Fig. 2) induced by gonadal hormone deficiency (Filipović et al., 2010; Khalil et al., 2005; Lee et al., 2004; Om & Shim, 2007; Ren et al., 2007; Soung et al., 2006). In a randomized placebo controlled trial with estrogen and phytoestrogen on ovariectomized nonhuman primates, Ham et al. (2004) failed to show any efficacy of soy phytoestrogens in decreasing all indices of bone turnover as estrogen does, but soy phytoestrogens were able to increase bone volume, trabecular number and decrease trabecular separation, stressing the importance of phytoestrogens in postmenopausal osteoporosis prevention.

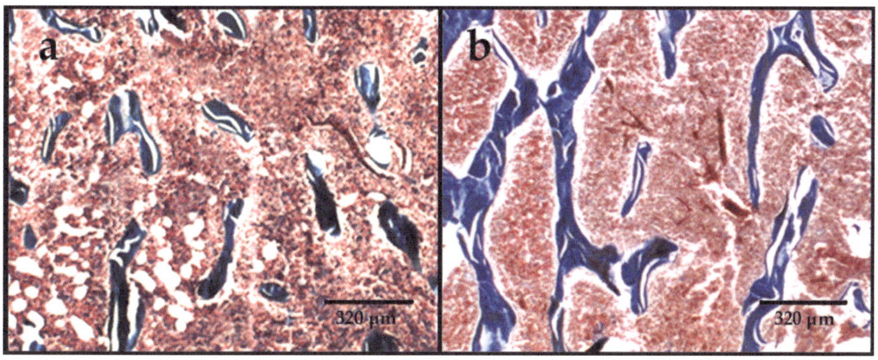

Fig. 2. Trabecular bone microarchitecture of the proximal tibia in control orchidectomized (a) and daidzein-treated orchidectomized (b) rat; azan staining method; unpublished image of Filipovic et al.

Phytoestrogens may elicit a bone sparing effect by both genomic and nongenomic mechanisms. They are able to interact with enzymes and receptors and, their stable structure and low molecular weight enables them to pass through cell membranes (Adlercreutz et al., 1998). The structural similarity of phytoestrogens to mammalian estrogens and their ability to bind to estrogen receptors (Setchell et al., 1999) suggests that the actions of phytoestrogens are mediated via estrogen receptors. ERα and ERβ have been detected in bone (Arts et al., 1997; Onoe et al., 1997). The relative binding affinity of phytoestrogens for ERβ is greater than that for ERα, and the protective effect of phytoestrogens on bone is probably produced through binding to estrogen receptors, particularly ERβ (Kuiper et al., 1998). In addition, phytoestrogens such as coumestrol, genistein and daidzein increase alkaline phosphatase activity in osteoblast-like cells (Kanno et al., 2004). Daidzein stimulates osteoblast differentiation, induces changes in the action of the cytoskeleton responsible for cell adhesion and motility and activates transcription factors associated with cell proliferation and differentiation (de Wilde et al., 2004, Ge et al., 2006; Jia et al., 2003). Also, isoflavones promote insulin-like growth factor-I (IGF-I) production which enhances osteoblastic activity (Ajrmandi et al., 2000)

Isoflavones inhibit bone resorption, via direct targeting of osteoclasts. They can decrease differentiation and increase apoptosis of osteoclasts or interfere with signaling pathways such as intracellular calcium, cAMP or protein kinase and protein tyrosine phosphatase (Gao & Yamaguchi, 2000; Sliwinski et al, 2005). Furthermore, osteoblasts are essential for in vitro osteoclastogenesis through cell-to-cell interactions of cytokines. Isoflavones regulate the expression and osteoblastic production of osteoclastogenesis-regulatory cytokines, such as interleukin-6 (IL-6), which stimulates osteoclast formation, and osteoprotegerin (OPG), which is identical to osteo-clastogenesis inhibitory factor, and the receptor activator of NF-κB ligand (Chen et al., 2002).

In addition to ERs, it has been shown that PPAR are new targets of phytoestrogens. PPAR directly influences osteogenesis and adipogenesis in a divergent way (Dang & Lowik, 2005). These authors suggested that biphasic dose-dependent effects of phytoestrogens are the result of concurrent activation of ERs and PPARs. Dominant ER-mediated effects that increase osteogenesis and decrease adipogenesis can only be seen at low concentrations of phytoestrogens, whereas dominant PPAR-mediated effects that decrease osteogenesis and increase adipogenesis are only evident at high concentrations.

Calcitonin (CT), a hormone secreted from thyroid C cells is known to inhibit osteoclast activity directly through its receptors (Nicholson et al., 1986). It was shown that synthesis and release of CT from thyroid C cells decreased after ovariectomy in rats, due to lack of estrogens (Filipović et al., 2002; Sakai et al., 2000). On the other hand, estrogen treatment had a stimulatory effect on CT secretion in ovariectomized rats (Filipović et al., 2003; Grauer et al., 1993). However, chronic Ca treatment of ovariectomized rats positively affected CT release without any significant changes in morphometric parameters of the C cells, suggesting an important role for estrogen in the regulation of CT synthesis (Filipović et al., 2005). Exogenous CT administration was reported to inhibit CT secretion in rats and therefore CT treatment probably suppresses C cell function due to a negative feedback (Sekulić et al., 2005). Recently, daidzein was found to stimulate CT secreting thyroid C cell activity in addition to increasing trabecular bone mass and decreasing bone turnover (Filipović et al., 2010). These results suggest that, besides direct action, daidzein may affect bone structure indirectly through enhancement of thyroid C cell activity.

Although animal studies demonstrate a clear skeletal benefit of phytoestrogens, clinical trials have given different results. Soy isoflavones were observed to retard bone loss in some (Huang et al., 2006; Newton et al., 2006), but not in other studies (Arjmandi et al., 2005; Brink et al., 2008). To date, only one study indicated that supplementation of intact soy protein providing 83 mg isoflavones daily might increase both hip and spine BMD in men (Newton et al., 2006). Also, the results of meta-analyses of soy foods and isoflavones extracted from soy protein have given conflicting results concerning the prevention of bone loss (Liu et al., 2009; Ma et al., 2008). The large heterogeneity in these conclusions might have arisen because many results were pooled from different individual studies, involving different treatment durations, different doses of soy isoflavone and study quality (Liu et al., 2009). These authors suggested that, because changes in bone mineral density (BMD) occur slowly over time, in short-term intervention studies this change may represent a transient remodeling rather than a long-term steady-state. In addition, a favorable effect on the spine BMD was achieved with large doses of isoflavones (≥ 80 mg/day, median 99 mg/day), but not with lower doses (< 80 mg/day, median 60 mg/day). Thus, the potential of soy isoflavones to prevent bone loss can be achieved by a dosage of 80 mg/day (Huang et al., 2006).

Finally, a wealth of supporting data from many in vitro mechanistic studies on bone cell lines and in vivo investigations using models of osteoporosis shows bone-sparing effects from phytoestrogens. These studies indicate that positive effects of phytoestrogens, as a SERM, may be achieved through estrogen receptors or other mechanisms. However, the results of clinical studies are more inconsistent. The different efficacy of phytoestrogen treatments, in studies involving either animal or human subjects, depended on dose, route and duration of administration. The data are, however, rather tantalizing because it is possible that soy isoflavones may offer the maximum benefit for prevention of osteoporosis. Therefore, it is necessary to perform large-scale clinical dietary intervention studies with phytoestrogens to determine their effects on bone tissue in humans.

7. Soybean phytoestrogens as potential endocrine disruptors

Endocrine systems of vertebrates have essential role in regulation of growth (including bone growth/remodeling), reproduction, stress, lactation, metabolism, energy balance, osmoregulation, and all other processes involved in maintaining homeostasis. Disruption in function of any endocrine system, involving either increased or decreased hormone secretion, result inevitably in disease, the effects of which may extend to many different organs and functions, and may even be life-threatening.

An endocrine-disrupting compound (EDC) is defined by the U.S. Environmental Protection Agency (EPA) as "an exogenous agent that interferes with synthesis, secretion, transport, metabolism, binding action, or elimination of natural blood-borne hormones that are present in the body and are responsible for homeostasis, reproduction, and developmental process." All hormone-sensitive physiological systems are vulnerable to EDCs, including brain and hypothalamic neuroendocrine systems; pituitary; thyroid; adrenal gland; cardiovascular system; mammary gland; adipose tissue; pancreas; ovary and uterus in females; and testes and prostate in males.

The exposure to such chemicals does not necessarily mean that disturbance of the relevant endocrine system will occur, as much depends on the level, duration and timing of exposure. However, even subtle changes, however small, in combination and/or under different conditions and/or in later generations might reduce the ability of humans (animals) to adapt. It may also happen that the magnitude of the disruption becomes evident only in presence of an additional stress factor.

It is beyond the scope of this chapter to discuss the potential interference of soy isoflavones with all endocrine organs; instead, the focus will be on three major endocrine axes that are affected by soybean phytoestrogens: pituitary - gonadal, -thyroid and - adrenocortical systems.

7.1 Effects on female reproductive system

Soybean isoflavones are ligands for both ERα and ERβ, despite the fact that their estrogenic potency is much lower than that of E2. Therefore, they can mimic and/or antagonize the mechanisms of E2 action and thus interfere with both endocrine and reproductive functions of the pituitary-gonadal axis. The rat uterotrophic assay is a widely used screening test for the detection of estrogenic, endocrine-disrupting chemicals. Genistein administration to ovariectomized rats induced a dose-dependent uterine growth and altered expression of estrogen-regulated genes (Diel et al., 2004). As E2 does not stimulate these uterine parameters in ERα KO mice (Couse & Korach, 2001), this test is considered as the proof for estrogenic action of phytoestrogens via ERα.

The first recognized health benefit of isoflavones was their potential to alleviate climacteric complaints, namely hot flushes and night sweats in perimenopausal women (Adlercreutz, 1998). Within the short period of time, numerous isoflavone and soy products became available in a form of food supplements and remedies. They were advertised as natural alternative to hormone replacement therapy, useful in prevention of climacteric symptoms. However, majority of recent placebo-controlled clinical trials support the opinion that isoflavone preparations are not superior to placebo, as placebo effect is 30% to 50% when dealing with psychosomatic climacteric complaints (Patisaul & Jefferson, 2010). Animal studies also demonstrated that only high doses of isoflavones were able to suppress overactivation of hypothalamic gonadotropin –release hormone pulse generator induced by estrogen deprivation (the major cause of hot flushes and other climacteric symptoms (Wuttke et al., 2007). It is important to stress that exposure to high doses of soy isoflavones (150mg/kg) is similar in biological effects to classical hormone replacement therapy. Therefore, their consumption bears a risk of increased proliferation of endometrial and mammary gland tissue with so far unpredictable risk of cancer development.

Multiple human studies demonstrated that exposure of premenopausal women to soybean isoflavones have a suppressive effect on pituitary-gonadal axis; consumption of isoflavones-rich soy food suppresses serum estrogen and progesterone levels and attenuate the preovulatory surge of follicle-stimulating hormone (FSH) and luteinizing hormone (LH) (Hooper et al., 2009; Nagata et al., 1998; Schmidt et al., 2006;). However, some researchers found no impact of isoflavones on female hormone levels (Maskarinec et al., 2002). Soybean phytoestrogens may also affect the women menstrual cycle, but findings are inconsistent. It was shown that a diet with soy protein delays menstruation and prolongs the follicular phase of the menstrual cycle (Cassidy et al., 1994). Other studies demonstrated increased or unchanged follicular phase length, decreased or unchanged midcycle LH and FSH, increased, decreased or unchanged estradiol, decreased dehidroepiandrosterone sulfate, and decreased or unhanged luteal phase progesterone in relation to isoflavone ingestion (Cassidy et al., 1994; Duncan et al., 1999). Therefore, women who try to become pregnant or have menstrual cycle irregularities should be cautious with consumption of isoflavone-enriched soy products or supplements.

Animal studies in rodents produced clear evidence of adverse effects of G on the female reproductive system following treatment during development (Chen et al., 2007; Kouki et al., 2003; National Toxicilogy Program, 2008). Studies that demonstrated clear evidence of developmental toxicity for G involved treatment during the period of lactation in rodents, as well as multigenerational studies that included exposure during gestation, lactation, and post-weaning. In adulthood, the effects of neonatal exposure to 50 mg G/kg bw/day were manifested as a lower number of live pups per litter (Padilla-Banks et al., 2006), a lower number of implantation sites and corpora lutea (Jefferson et al., 2005), and a higher incidence of histomorphological changes of the reproductive tract (i.e., cystic ovaries, progressive proliferative lesions of the oviduct, cystic endometrial hyperplasia, and uterine carcinoma) relative to control females (Newbold et al., 2001).

In addition, the reproductive performance of the neonatally-treated mice was tested during adulthood and there was a significant negative trend for the number of dams with litters. Because the effects were more pronounced in animals at 6 months of age than at 2 or 4 months of age, the authors suggested that reproductive senescence may occur earlier in these animals as a result of the neonatal G treatments (Jefferson et al., 2005). These authors explained that, although G-treated mice ovulate under exogenous hormonal influence, the ovulation rate was changed. The lower doses of G treatment enhanced ovulation rate, while the higher doses decreased this parameter. Ovulation of too many oocytes early in life may reduce the number of oocytes available for fertilization and lead to lower fertility rates later in life (McLachlan et al., 1982). The development of the ovary and ovarian follicles was altered following neonatal G treatment (Jefferson et al., 2002). Ovaries of G-treated mice contained multioocyte follicles (MOFs) at 19th postnatal day. This phenotype is a marker for altered development of the ovary, which lead to oocytes of poor quality (Jefferson et al., 2005). These oocytes are less potent, since the oocytes derived from single oocyte follicles were far more likely to be fertilized in vitro than oocytes derived from MOFs (Iguchi et al., 1990). In our laboratory, results obtained on the ovaries of immature rats treated with 50mg G/kg for three days (from 19th till 21th postnatal day) showed that G disturbed the follicular parenchyma-ovarian stroma ratio (Fig. 3), induced increase of total ovary volume (Medigović et al., 2009).

Data from experiments using DNA microarray analysis for examining the effects of genistein in the developing rat uterus indicate that genistein alters the expression of 6-8 times as many genes as does E2, most of which were down-regulated (Barnes, 2004).

Data are not consistent about onset of puberty and sexual maturation in rats and mice following exposure during gestation and lactation or continuous exposure to soy diet or supplements. An earlier onset of vaginal opening was observed in mice exposed directly to G during the period of lactation (Nikaido et al., 2004.) and in rats treated by sc injection as neonates with 10 mg G/kg bw/day (Bateman & Patisaul, 2008). However, other authors reported delay in vaginal opening (Anzalone et al., 1998).

Only a very small number of studies have been published on D and its estrogenic metabolite equol, and no studies have evaluated the effects of developmental exposure to glycitein. Detection of typical estrogenic effects in these studies are controversial. Kouki et al. (2003) reported no effect on estrous cyclicity in rats treated by sc injection with ~19 mg D/kg bw/day on PND1-5. In contrast, treatment with the same dose levels of G caused the predicted estrogenic effect in all of these studies. Similar to these authors, in our laboratory (unpublished data) no uterotrophic response was detected after subcutaneous injection of immature female rats with 50mg D/kg/day (treatment lasted from 19th postnatal till 21th postnatal day), though the same treatment with G caused predicted estrogenic response.

Isoflavones can pass from mother to fetus through placenta. However, this exposure is considerably lower than in infants fed with soy formula. Initially developed as an alternative to bovine milk formulas for babies with a milk allergy, use of soy infant formula became more popular among environmentally oriented population with vegetarian life style. A recent prospective study in human infants observed that female infants fed soy-based formulas exhibit estrogenized vaginal epithelium at times when their breast fed or cow milk- based formula fed peers did not (Bernbaum et al., 2008). Patisaul and Jefferson

(2010) concluded that further determination if soy infant formula have long-term reproductive health effects should be a public health imperative.

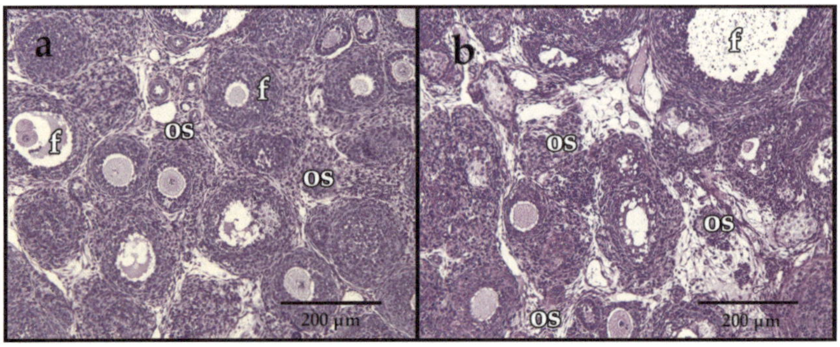

Fig. 3. Ovaries of 21 day old control (a) and genistein - treated rat; hematoxylin - eosin staining method; OS, ovarian stroma; f, follicular parenchyma; hematoxylin - eosin staining method; unpublished image of Medigović et al.

7.2 Effects on male reproductive system

Soy phytoestrogens, alone or in combination with some other EDC, may adversely affect androgen hormone production, spermatogenesis, sperm capacitation and fertility. Results of recent meta-analysis suggest that neither soy foods nor isoflavone supplements alter bioavailable T concentration in adult men (Hamilton-Reeves et al., 2010). However, Tanaka et al. (2009) reported that short-term administration of soy isoflavones decreased testosterone and dihydrotestosterone (DHT) and increased sex hormone-binding globulin levels.

Male reproductive system is particularly sensitive in prenatal stage and during early infancy, when disruption of the hormonal balance in favor of estrogens can lead to irreversible abnormalities in sex specific physiology and behavior in the adulthood (Patisaul & Jafferson, 2010).

Only few animal studies reported results on the developmental effects of exposure to soy infant formula. Their study designs were based on the same group of male marmosets treated during infancy, and assessed either as juveniles (Sharpe et al., 2002) or adults (Tan et al., 2006). The soy infant formula-fed male marmosets had significantly lower plasma testosterone levels than their cow milk formula-fed co-twins. Histopathological analysis on the testes of a subset of the co-twins revealed an increase in Leydig cell abundance per testes in the soy infant formula-fed marmosets compared to their cow milk formula–fed co-twin, in the absence of a significant change in testicular weight. A follow up study was conducted on the remaining animals when they were sexually mature (80 weeks of age or older). The males fed with soy infant formula as infants had significantly heavier testes and increased number of both Leydig and Sertoli cells per testicle compared to cow milk formula-fed controls. In addition, there was no significant onset of puberty, level of adult plasma testosterone, or fertility. The authors suggest that the increase in testes weight was likely due to an increase in testicular cell populations. Therefore, these results demonstrated permanent effects on testicular cell populations, but no obvious effects on reproductive function, namely fertility or permanent changes in testosterone levels of experimental animals.

Some studies on rats and mice demonstrated increased testicular weight when animals were treated with soy diet or isoflavone supplements during gestation and lactation or continuous exposure, similar to the effect described above in marmosets treated with soy infant formula during infancy (Akingbemi et al., 2007; McVey et al., 2004; Piotrowska et al., 2011; Ruhlen et al., 2008; Wisniewski et al., 2005). Other authors reported a decrease (Atanassova et al., 1999; Wisniewski et al., 2003) or no effect on testicular weight (Fielden et al., 2003; Kang et al., 2002).

Controversial results are found as to the effects of lifelong exposure of rodents to phytoestrogens on reproductive function, namely fertility or changes in testosterone levels. The litter size was not affected when male rats were exposed to dietary soy throughout life (Atasnassova et al., 1999). Also, chronic dietary exposure to G did not adversely affect spermatogenesis or seminal vesicle weight in rats (Delclos et al., 2001; Roberts et al., 2000). On the other hand, a few studies indicate negative effects of phytoestrogens on male reproductive success. Thus, a continuous exposure to low combined doses of G and vinclozolin affects male rats' reproductive health by inducing reproductive developmental anomalies, alterations in sperm production and quality, and fertility disorders (Eustache, 2009).

Exposure to G was found to induce hyperplasia of Leydig cells in mice (Lee et al., 2004b). The exposure to isoflavones during 5 weeks decreased the level of circulating testosterone, depending on the dose used (Weber et al., 2001). No significant differences in serum testosterone concentration was detected in rats receiving high doses of G and D from intrauterine life through sexual maturity (Piotrowska et al., 2011). In vitro investigation showed that G can promote the testosterone production of rat Leydig cells at a low concentration, but both D and G can inhibit it at a higher concentration (Zhu et al., 2009).

Effect of phytoestrogens on male reproduction system is a complex process that depends on developmental stage and time of exposure, applied dosage, and other factors. Together, these factors determine the potential risk for adverse consequences with long-lasting effects on male reproductive function. At present, the evidence is insufficient to determine whether soy products cause or do not cause adverse developmental effect on male reproductive system, due to the small number of studies, limitations in their experimental designs, and failure to detect adverse functional effects.

7.3 Effects on pituitary-thyroid axis

Goitrogenic effects of a soybean diet in animals were reported in 1933 (McCarrison, 1933). Similar to animals, goiter and hypothyroidism were reported in infants fed with adapted soy formula without adequate iodine supply (Van Wik et al., 1959). This effect was eliminated by supplementing commercial soy infant formulas with iodine, or by switching to cow milk (Chorazy et al., 1995). However, infants with congenital hypothyroidism that were fed with iodine supplemented diet still needed higher doses of L-thyroxine (Jabbar et al., 1997). In addition, the incidence rate of autoimmune thyroid disease was doubled in teenage children who consumed soy formula as infants (Fort et al., 1990). However, results of clinical studies with adults are not consistent: some authors suggest that isoflavones have a mild or no effect on thyroid function (Dillingham et al., 2007; Duncan et al., 1999), while others indicate that isoflavones suppress the thyroid function (Haselkorn et al., 2003; Ralli, 2003; Sathyapalan et al., 2011).

Rats provide a useful risk assessment model for various thyroid toxins (Choksi et al., 2003). However, compared to the human, rodent thyroid gland is more sensitive to adverse chemicals (Capen, 1997). Several investigators have reported induction of goiter in iodine-deficient rats maintained on a soybean diet (Ikeda et al., 2000; Kajiya et al., 2005; Kimura et al., 1976), although only in cases of iodine deficiency or presence of some other goitrogenic factor. Rats receiving low iodine diet that included 20% of defatted soybeans developed severe hypothyroidism, characterized by a reduction in serum thyroxin and an increase in serum TSH (Ikeda et al., 2000). In addition, a diet containing higher percentage of soy (40% of defatted soybeans) in combination with iodine deficiency induced the development of thyroid carcinoma in rats (Kimura et al., 1976).

Doerge and his associates demonstrated that genistein and daidzein inhibit the activity of thyroid peroxidase (TPO), the key enzyme in the synthesis of thyroid hormones (TH), both in vitro and in vivo (Divi et al., 1997; Chang & Doerge, 2000; Doerge et al., 2002). However, despite significant inactivation of this enzyme, serum thyroid hormone levels were unaffected by isoflavone treatments in young adult rats of both sexes. Most other authors, who performed their studies on young adult animals of both sexes, also reported that soy or isoflavones alone, in the absence of other goitrogenic stimulus, did not affect thyroid weights, histopathology and the serum levels of TSH and thyroid hormones (Chang & Doerge, 2000, Schmutzler et al., 2004). The authors suggested that soy could cause goiter, but only in animals or humans consuming diets marginally adequate in iodine, or who were predisposed to develop goiter, or exposed to additional goitrogenic compounds such as perchlorate, a potent inhibitor of the sodium-iodide-symporter (NIS) of thyrocytes.

Increasing evidence is available that set points of the HPT axis change during various life phases and tend to be less sensitive to negative feedback by thyroid hormones in aging individuals. However, the results on isoflavone effects in aged humans and rodents are scarce. In rodent models, we are the first who demonstrated that both genistein and daidzein induce micro-follicular changes in the thyroid tissue, including hypertrophy of Tg-immunopositive follicular epithelium and colloid depletion (Fig.4), and reduce the level of serum thyroid hormones in orchidectomized (Orx) middle-aged male rats, a model of andropause (Šošić-Jurjević et al., 2010). The concentration of total T4 in serum decreased more prominently than concentration of total T3 in serum in comparison to the corresponding control values. This reduction consequently led to a feedback stimulation of pituitary TSH cells, detected by the increase in cell volume and relative volume density of TSHβ-immunopositive cells per pituitary unit volume, as well as by the increased concentration of TSH in serum. Besides the TPO, there might be other molecular targets for isoflavone interference with the pituitary-thyroid axis.

Soy isoflavones may interfere with thyroid hormones at binding sites of serum distribution proteins such as transthyretin (TTR). In vitro analysis demonstrated that soy isoflavones are potent competitors for T4 binding to TTR in serum and cerebrospinal fluid (Radović et al., 2006). As an outcome of this interference, isoflavones may alter free thyroid hormone concentrations, resulting in altered availability and metabolism of thyroid hormones in target tissues (Köhrle, 2008; Radović et al., 2006). The role of serum binding proteins for thyroid hormone in thyroid homeostasis is not well understood. No single serum T4 - binding protein is essential for good health or for the maintenance of euthyroid state in humans (Robbins, 2000). There are a number of clinical situations in which serum binding proteins are elevated or reduced (even completely absent) and the thyroid state remain normal (Refetoff, 1989). In contrast, there is evidence that the role of serum binding proteins is to allow the equal distribution of hormone delivery to tissues (Mendel et al., 1987). In rats, TTR is a major serum transport protein of thyroid hormones. In humans TTR is produced in the choroid plexus and appears to be important for thyroid hormone action in the brain (Richardson et al., 2007). Thus, TTR may mediate transport of environmental chemicals into various compartments such as placenta (Meerts et al., 2002). Chemical binding to the TTR may not only decrease the availability of thyroid hormone to various tissues, it may also selectively target these chemicals for transport and uptake.

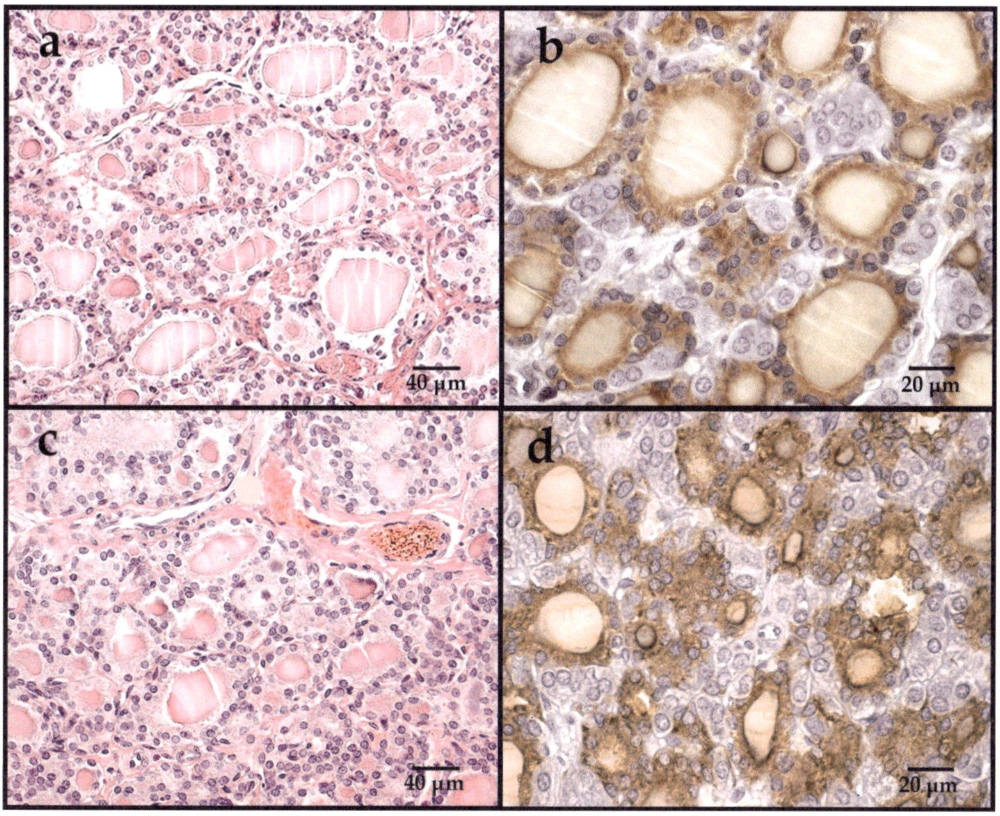

Fig. 4. Thyroid gland tissue of control orchidectomized (a and b) and daidzein-treated orchidectomized (c and d) rat; hematoxylin - eosin and immuno-staining for thyroglobulin; unpublished image of Šošić-Jurjević et al.

In order to accurately asses thyroid function it must be understood that deiodinase enzymes are essential control points of cellular thyroid activity that determine intracellular activation and deactivation of thyroid hormones. Apart from the hormone synthesis by the thyroid gland, deiodination pathways in liver and kidney are the main contributors to thyroid hormone metabolism, turnover and homeostasis. Enzyme 5'-deiodinase type I (5'DI) is the key enzyme in thyroid hormone activation and inactivation in extra thyroidal tissues. This enzyme catalyzes deiodination of the thyroid hormone precursor thyroxine (T4) to the biologically active triiodo-thyronine (T3), as well as the inactivation of T4 and T3 to „reverse" T3 and T2. It is expressed in different tissues, with highest expression rate found in rat liver, kidney, thyroid gland and pituitary (Bianco et al., 2002). It is regulated in a TH-dependent manner (Köhrle, 2002). In response to iodine deficiency or hypothyroidism, plasma TH values are reduced, TSH is increased and the organism tries to restore normal T3 levels by down-regulation of 5′DI in brain and liver, respectively. In addition, activity of 5′DI in different tissues seems to be sex- (Köhrle et al., 1995; Lisboa et al., 2001) and age- dependent (Corrêa da Costa et al., 2001). According to Corrêa da Costa et al. (2001) decreased serum T3 was detected only in old males, which was explained by a two-times-higher hepatic deiodination of T4 to T3, detected in aged females in comparison to males. Genistein was shown to increase hepatic 5′DI activity of about 33% in young adult female rats, but the detected increase was not statistically significant (Schmutzler et al., 2004). In our model – system (orchidectomized middle-aged rats) G significantly increased ($p<0.05$) 5′DI activity by 33% (unpublished data). However, neither 5′DI in thyroid nor pituitary 5′DII activity were affected by G or D treatment (unpublished data). These data indicate that although pituitary-thyroid axis in male rats is more vulnerable compared to the one in young adults, it still has great ability to compensate the adverse effects of isoflavones.

Isoflavones may also affect the thyroid function indirectly, via its estrogenic action. Estrogen receptors were located both in pituitary thyrotrophs and in thyroid follicular cells (González et al., 2008; Hampl et al., 1985). Donda et al. (1990) found that pituitary TSH cells in adult female rats have a higher density of T_3 and TRH receptors than in male rats, probably due to a modulatory effect of estradiol. Males are more prone to develop goitrogenesis in response to goitrogenic stimuli, probably due to higher TSH levels in comparison to females (Capen, 1997). It seems that estradiol make the TSH cells more sensitive to the negative feedback regulation with thyroid hormones (Ahlquist et al., 1987). In orchidectomized middle-aged rats we demonstrated that pharmacological doses of testosterone and estradiol disturbed the endocrine homeostasis of pituitary-thyroid axis, but in different directions. Testosterone acted stimulatory, probably through central stimulation of pituitary TSH cells, since both serum TSH and T4 levels were increased. Estradiol acted inhibitory and, though detected structural changes corresponded to centrally induced hypothyroidism, the level of TSH in serum was not significantly altered, suggesting that estradiol may interfere with TSH action within the thyrocytes (Sekulic et al., 2010).

Estrogen was also demonstrated to inhibit activity of thyroid follicular cells in the absence of TSH both in vitro and in vivo (Furlanetto et al., 2001; Vidal et al., 2001). Our previous research of a young adult and middle-aged rat menopause models indicated that chronic estradiol treatment modulated pituitary TSH cells and thyroid structure and decreased serum levels of thyroid hormones, with no significant changes in serum TSH level (Šošić-Jurjević et al., 2005, 2006). Genistein acted as estrogen agonist in an estrogen-responsive pituitary cell line (Stahl et al., 1998).

In conclusion, though there are multiple molecular targets for interference of isoflavones with pituitary-thyroid-peripheral network, this system has considerable capacity to compensate disturbances of its feedback mechanism. If thyroid function is impaired, the risk of developing hypothyroidism increases. Elderly population and individuals with thyroid dysfunction should be aware of potential risk when use isoflavone supplements.

7.4 Effects on pituitary-adrenocortical axis

Results concerning the potential effects of the soy phytoestrogens on pituitary-adrenocortical axis in humans are very limited. The animal studies and in vitro experiments demonstrated remarkable influence of isoflavones on morphology and function of adrenal cortex. The continuous administration of genistein (40mg/kg) to weanling rats resulted in greater total protein content in zona fasciculata (ZF) and zona reticularis (ZR) of adrenal cortex, and low serum corticosterone concentration (corticosterone is a major glucocorticoid hormone in rats; Ohno et al., 2003). Genistein administration to orchidectomized middle-aged rats, as a model of andropause, increased zona glomerulosa (ZG), ZF (Fig. 5) and ZR cell volumes, and decreased serum aldosterone and corticosterone concentrations ($p<0.05$), whereas serum DHEA concentration significantly increased (Ajdžanović et al., 2009a). Genistein and daidzein increased androgen and decreased glucocorticoid production (Mesiano et al., 1999) in human adrenocortical cells in a culture. Recent study on human adrenocortical H295R cell line demonstrated that daidzein and genistein strongly inhibited secretion of cortisol with IC50 values below 1 μM (Ohlsson et al., 2010).

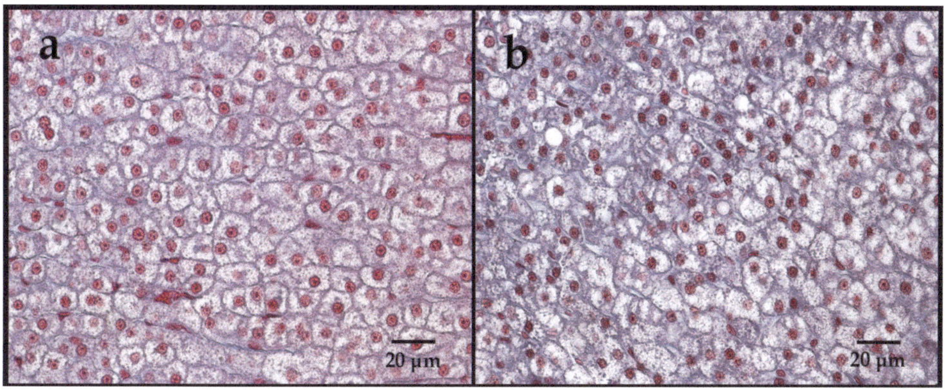

Fig. 5. Zona fasciculata of adrenal cortex in control orchidectomized (a) and genistein-treated orchidectomized (b) rat; azan staining method; unpublished image of Ajdžanović et al.

The isoflavones possess structural features similar to estradiol, which enables them to act via ERs (Lephart et al., 2004). Production of steroids in human fetal adrenocortical cells is modulated by estrogens (Fujieda et al., 1982, Mesiano & Jaffe, 1993; Voutilainen et al., 1979). It was shown that 17β-estradiol in high concentrations increased ACTH-stimulated androgen production and inhibited glucocorticoid synthesis in cultured human fetal adrenal cortical cells (Mesiano & Jaffe, 1993). Although these results indicate the influence of estrogens on the adrenocortical cells in vitro, their physiological significance is still unclear. Under physiological conditions the endogenous estrogen concentration does not reach 1 μM/L in nonpregnant adults. However, it is possible that dietary phytoestrogens, as estrogen-related compounds, could reach circulating levels high enough to exert estrogenic actions. Consuming the large amounts of soy-derived foods, for example in Japanese diet, circulating concentrations of phytoestrogens can reach higher levels (1-5 micromole/L) (Adlercreutz & Mazur, 1997).

Isoflavones may also affect activity or expression of steroidogenic enzymes, which seems to be the case for its action on rat adrenal cortex (Malendowicz et al., 2006; Mesiano et al., 1999; Ohno et al., 2003). Within the adrenals, steroids are produced through the action of five forms of cytochrome P450 and 3β-hydroxysteroid dehydrogenase (3βHSD) (Simpson & Waterman, 1992). Differential expression of these enzymes in the three adrenocortical zones

leads to the production of specific steroids within each zone (Suzuki et al., 2000). As a precursor of steroidogenesis, the glomerulosa cells use pregnenolone which can be metabolized by either 3βHSD or 17α-hydroxylase, 17, 20-lyase. The relative expression of these enzymes influences the synthesis of aldosterone and cortisol/corticosterone in ZG and ZF, as well as adrenal androgens in ZR (Conley & Bird, 1997). The major physiological regulators of adrenal aldosterone production are angiotenzin II (Ang II) and potassium. Ang II stimulates aldosteron production through the activation of multiple intracellular signaling pathways including a number of tyrosine kinases (Berk & Corson, 1997; Ishida et al., 1995). It was showed that genistein, as a potent inhibitor of various tyrosine kinases may inhibit aldosteron production (Akiyama et al., 1987; Dhar et al., 1990). Genistein and daidzein are also potent competitive inhibitors of human adrenocortical 3βHSD and cytochrome P450 21-hydroxylase, suppressing cortisol and stimulating DHEA production in vitro (Mesiano et al., 1999). Part of the inhibition of aldosteron production may result from an increase in 17α-hydroxylase, 17, 20-lyase activity, which removes the substrate from the pathway leading to aldosterone and directs it towards the synthesis of adrenal androgens (Sirianni et al., 2001).
Isoflavones could also affect adrenal function indirectly, by affecting pituitary ACTH cells. It was previously reported that estrogen replacement lowered the *proopiomelanocortin* (POMC) gene mRNA level and the ACTH response to repeated stressful stimuli in ovariectomized rats (Redei et al., 1994). A certain synergism between CRH (*corticotrophin releasing hormone*) and the various cytokines, namely IL-1, IL-2 and IL-6, has been shown to exist in stimulation of the pituitary ACTH secretion (Bateman et al., 1989; Besedowsky & del Ray, 1996). Genistein may interrupt the stimulatory effects of CRH and cytokines on POMC gene transcription and reduce the level of ACTH, through inhibition of tyrosine kinase phosphorylation cascades (Katahira et al., 1998), but the biological significance of this mechanism is still unclear. We treated orchidectomized middle-aged rats with different doses of genistein or daidzein (10 and 30mg/kg body weight); (Ajdžanović et al., 2009; Ajdžanović et al., 2010, Milošević et al., 2009), and detected similar decrease in pituitary ACTH cellular volume and plasma ACTH levels. Corticosterone levels were also decreased, supporting that some other mechanism, aside from feedback regulation, is involved in effect of isoflavones on pituitary ACTH cell regulation. Keeping in mind that aging is associated with augmented activity of the pituitary-adrenal axis and higher incidence of stress-related psychiatric disorders (Hatzinger et al. 2000), this decline might be considered beneficial at some point.

On the other hand, chronic treatment of weanling rats with genistein (40mg/kg body weight) elevated ACTH level, most probably due to decreased serum corticosterone level and thus release from a negative feedback regulation (Ohno et al., 2003). This finding is of importance since glucocorticoids have important "programming" effects during development. This means that alternations in the circulating levels of glucocorticoid hormones may affect the timing and set points of other endocrine axes (Manojlovic-Stojanoski et al., 2010), as well as brain development, memory and learning capabilities in adults (de Kloet et al., 1988).

Based on animal studies and in vitro research it may be concluded that soy isoflavones interfere with the function of pituitary-adrenocortical axis. This hormonal axis plays a major role in control of stress response and regulation of numerous body processes (digestion, metabolism of carbohydrates, protein and fat, attenuation of the inflammatory response, mood, emotions and sexuality). Therefore, the biological impact of this interference is high. Potential health risks for various age groups should be further assessed.

8. Conclusion

So, are soy isoflavones friends or foes? The answer is complex and may ultimately depend on age, sex, health status, quantity of intake, and even the composition of an individual's intestinal micro flora. In vitro and animal research, as well as human research including both clinical and epidemiologic data, suggests that isoflavone-containing products pose a risk to estrogen-sensitive breast cancer patients and in women at high risk of developing this disease. Results of animal and human studies suggest a modest benefit in prevention of prostate cancer. Exposure to isoflavones by feeding soy infant formula bears a risk of adverse effects on the long-term development of infants. Women who tend to get pregnant or have irregularities in menstrual cycle, as well as persons who are at risk of thyroid dysfunction, should avoid soy isoflavone supplements. The usage of soy protein (with or without isoflavones) seems to have a modest beneficial effect on cardiovascular system and protective role in prevention and treatment of osteoporosis. Research of potential synergy of isoflavones and drugs, and/or other functional food could be a new promising strategy in reducing risk of age-related diseases, improving life quality and expanding life span.

9. Acknowledgment

This work was supported by the Ministry of Education and Science of the Republic of Serbia, Grant No. 173009. Publishing of this chapter was in part financially supported by SOJAPROTEIN, member of Victoria Group, Soybean Processing Company - PLC, Bečej, Serbia. The authors express their sincere thanks to late Dr. Dana Brunner for her guidance and contribution, and to Mr. Kristijan Jurjevic for his valued assistance with English manuscript preparation.

10. References

Adams, MR., Golden, DL., Register, TC., Anthony, MS., Hodgin, JB, Maeda, N. & Williams, JK. (2002). The atheroprotective effect of dietary soy isoflavones in apolipoprotein E-/- mice requires the presence of estrogen receptor-alpha. *Arterioscler Thromb Vasc Biol,* Vol. 22, pp. 1859-1864

Adams, MR., Golden, DL., Williams, JK., Franke, AA., Register, TC. & Kaplan, JR. (2005). Soy protein containing isoflavones reduces the size of atherosclerotic plaques without affecting coronary artery reactivity in adult male monkeys. *J Nutr,* Vol. 135, pp. 2852-2856

Adlercreutz, H. & Mazur W. (1997). Phyto-oestrogens and Western diseases. *Ann Med,* Vol. 29, pp. 95-120

Adlercreutz, H. (1998), Evolution, nutrition, intestinal microflora, and prevention of cancer: a hypothesis. *Proc Soc Exp Biol Med,* Vol. 217, pp. 241-246

Adlercreutz, H., Höckerstedt, K., Bannwart, C., Bloigu, S., Hämäläinen, E., Fotsis, T. & Ollus, A. (1987). *J Steroid Biochem,* Vol.27, pp. 1135-1144

Ae Park, S., Choi, MS., Cho, SY., Seo, JS., Jung, UJ., Kim, MJ., Sung, MK., Park, YB. & Lee, MK.(2006). Genistein and daidzein modulate hepatic glucose and lipid regulating enzyme activities in C57BL/KsJ-db/db mice. *Life Sci,* Vol. 79, pp. 1207-1213

Ahlquist, JA., Franklyn, JA., Wood, DF., Balfour, NJ., Docherty, K., Sheppard, MC. & Ramsden, DB. (1987) Hormonal regulation of thyrotropin synthesis and secretion. *Horm Metab Res Suppl,* Vol. 17, pp. 86-89

Ajdžanović, V., Sosić-Jurjević, B., Filipović, B., Trifunović, S., Brkić, D., Sekulić, M. & Milošević, V. (2009). Genistein affects the morphology of pituitary ACTH cells and decreases circulating levels of ACTH and corticosterone in middle-aged rats. *Biol Res,* 42, pp. 13-23

Ajdžanović, V., Sosić-Jurjević, B., Filipović, B., Trifunović, S., Manojlović-Stojanoski, M., Sekulić M. & Milošević V. (2009) a. Genistein-induced histomorphometric and hormone secreting changes in the adrenal cortex in middle-aged rats. *Exp Biol Med,* 234, pp. 148-156

Ajdžanović, V., Spasojević, I., Filipović, B., Sosić-Jurjević, B., Sekulić, M. & Milosević V. (2010). Effects of genistein and daidzein on erythrocyte membrane fluidity: an electron paramagnetic resonance study. *Can J Physiol Pharmacol,* Vol. 88, pp. 497-500

Ajdžanović, V., Spasojević, I., Sosić-Jurjević, B., Filipović, B., Trifunović, S., Sekulić, M. & Milošević, V. (2011). The negative effect of soy extract on erithrocyte membrane fluidity: An electron paramagnetic resonance study. *J Membrane Biol,* Vol. 239, pp. 131-135

Akingbemi, BT., Braden, TD., Kemppainen, BW., Hancock, KD., Sherrill, JD., Cook, SJ., He, X. & Supko, JG. (2007). Exposure to phytoestrogens in the perinatal period affects androgen secretion by testicular Leydig cells in the adult rat. *Endocrinology,* Vol. 148, pp. 4475-4488

Akiyama, T., Ishida, J., Nakagawa, S., Ogawara, H., Watanabe, S., Itoh, N., Shibuya, M. & Fukami, Y. (1987). Genistein a specific inhibitor of tyrosine-specific protein kinases. *J Biol Chem,* 262, pp. 5592-5595.

Alexandersen, P., Haarbo, J., Breinholt, V. & Christiansen, C. (2001). Dietary phytoestrogens and estrogen inhibit experimental atherosclerosis. *Climacteric,* Vol. 4, pp. 151-159

Allred, CD., Allred, KF., Ju, YH., Goeppinger, TS., Doerge, DR. & Helferich, WG. (2004). Soy processing influences growth of estrogen-dependent breast cancer tumors. *Carcinogenesis,* Vol. 25. pp. 1649-1657

Anderson, JW., Johnstone, BM. & Cook-Newell, ME. (1995). Meta-analysis of the effects of soy protein intake on serum lipids. *N Engl J Med,* Vol. 333, pp. 276-282

Anthony, MS., Clarkson, TB., Bullock, BC. & Wagner, JD. (1997). Soy protein versus soy phytoestrogens in the prevention of diet-induced coronary artery atherosclerosis of male cynomolgus monkeys. *Arterioscler Thromb Vasc Biol,* Vol. 17, pp. 2524-2531

Anthony, MS., Clarkson, TB., Hughes, CL. Jr Morgan, TM. & Burke, GL. (1996). Soybean isoflavones improve cardiovascular risk factors without affecting the reproductive system of peripubertal rhesus monkeys. *J Nutr,* Vol. 126, pp. 43-50

Anzalone, CR., Lu, JK. & LaPolt, PS. (1998). Influences of age and reproductive status on ovarian ovulatory responsiveness to gonadotropin stimulation. *Proc Soc Exp Biol Med,* Vol. 217, pp. 455–460

Arjmandi, BH., Lucas, EA., Khalil, DA., Devareddy, L., Smith, BJ., McDonald, J., Arquitt, AB., Payton, ME. & Mason, C. (2005). One year soy protein supplementation has positive effects on bone formation markers but not bone density in postmenopausal women. *Nutr J,* Vol. 4, pp. 8

Arts, J., Kuiper, GG., Janssen, JM., Gustafsson, JA., Lowik, CW., Pols, HA. & van Leeuwen, JP. (1997). Differential expression of estrogen receptors alpha and beta mRNA during differentiation of human osteoblast SV-HFO cells. *Endocrinology,* Vol. 138, pp. 5067- 5070

Atanassova, N., McKinnell, C., Walker, M., Turner, KJ., Fisher, JS., Morley, M., Millar, MR., Groome, NP. & Sharpe, RM. (1999). Permanent effects of neonatal estrogen exposure in rats on reproductive hormone levels, Sertoli cell number, and the efficiency of spermatogenesis in adulthood. *Endocrinology,* Vol. 140, pp. 5364-5373

Barnes, S. (2003). Phyto-oestrogens and osteoporosis: what is a safe dose? *Br J Nutr,* Vol. 89, pp S101-S108

Barnes, S. (2004). Soy isoflavones--phytoestrogens and what else? *J Nutr,* Vol. 134, pp. 1225S-1228S

Bateman, A., Singh, A., Kral, T. & Solomon S. (1989). The immune hypothalamic-pituitary-adrenal axis. *Endocr Rew,* 10, pp. 92-112

Bateman, HL. & Patisaul, HB. (2008). Disrupted female reproductive physiology following neonatal exposure to phytoestrogens or estrogen specific ligands is associated with decreased GnRH activation and kisspeptin fiber density in the hypothalamus. *Neurotoxicology,* Vol. 29, pp. 988-997

Bektic, J., Guggenberger, R., Eder, IE., Pelzer, AE., Berger, AP., Bartsch, G. & Klocker, H. (2005). Molecular effects of the isoflavonoid genistein in prostate cancer. *Clin Prostate Cancer,* Vol. 4, pp. 124-129

Bennetts, HW., Underwood, EJ. & Shier, FL. (1946). A specific breeding problem of sheep on subterranean clover pastures in Western Australia. *Br Vet J,* Vol. 102, pp. 348-352

Berk BC & Corson MA. (1997). Angiotensin II signal transduction in vascular smooth muscle. *Circulation Research,* 80, pp. 607–616

Bernbaum, JC., Umbach, DM., Ragan, NB., Ballard, JL., Archer, JI., Schmidt-Davis, H. & Rogan, WJ. (2008). Pilot studies of estrogen-related physical findings in infants. *Environ Health Perspect,* Vol. 116, pp.416-20

Besedovsky, HO., & del Rey, A. (1996). Immune-neuro-endocrine interactions: facts and hypotheses. *Endocr Rev,* 17, pp. 64–102

Bianco, AC., Salvatore, D., Gereben, B., Berry, MJ. & Larsen, PR. (2002). Biochemistry, cellular and molecular biology, and physiological roles of the iodothyronine selenodeiodinases. *Endocr Rev,* Vol. 23, pp. 38-89

Birt, DF., Hendrich, S. & Wang, W. (2001). Dietary agents in cancer prevention: flavonoids and isoflavonoids. *Pharmacol Ther,* Vol. 90, pp.157-177

Brink, E., Coxam, V., Robins, S., Wahala, K., Cassidy, A. & Branca, F. (2008). Long-term consumption of isoflavone-enriched foods does not affect bone mineral density, bone metabolism, or hormonal status in early postmenopausal women: a randomized, double-blind, placebo controlled study. *Am J Clin Nutr,* Vol. 87, pp. 761–770

Capen C. (1997). Mehanicistic data and risk assesment of selected toxic end points of the thyroid gland. *Toxicologic Pathology,* Vol. 25, pp. 39-48

Carani, C., Qin, K., Simoni, M., Faustini-Fustini, M., Serpente, S., Boyd, J., Korach, KS. & Simpson, ER. (1997). Effect of testosterone and estradiol in a man with aromatase deficiency. *N Engl J Med,* Vol. 337, pp. 91-95

Cassidy, A., Bingham, S. & Setchell KDR. (1994). Biological effects of a diet of soy protein rich in isoflavones on the menstrual cycle of premenopausal women. *Am J Clin Nutr,* Vol. 60, pp. 333–340

Chang, HC. & Doerge, DR. (2000). Dietary genistein inactivates rat thyroid peroxidase in vivo without an apparent hypothyroid effect. *Toxicol Appl Pharmacol,* Vol. 168, pp. 244-252

Che, JH., Kwon, E., Kim, SH., You, JR., Kim, BH., Lee, SJ., Chung, JH. & Kang, BC. (2011). Acute and subchronic toxicity of FCD, a soybean extract combined with l-carnitine, in Sprague-Dawley rats. *Regul Toxicol Pharmacol,* Vol. 59, pp. 285-292

Chen, XW., Garner, SC. & Anderson, JJ. (2002), Isoflavones regulate interleukin-6 and osteoprotegerin synthesis during osteoblast cell differentiation via an estrogen-receptordependent pathway. *Biochem Biophys Res Commun,* Vol. 295, pp. 417–422

Chen, Y., Jefferson, WN., Newbold, RR., Padilla-Banks, E. & Pepling, ME. (2007). Estradiol, progesterone, and genistein inhibit oocyte nest breakdown and primordial follicle assembly in the neonatal mouse ovary in vitro and in vivo. *Endocrinology* Vol. 148, pp. 3580–3590

Cherdshewasart, W. & Sutjit, W. (2008). Correlation of antioxidant activity and major isoflavonoid contents of the phytoestrogen-rich Pueraria mirifica and Pueraria lobata tubers. *Phytomedicine,* Vol. 15, pp. 38-43

Choksi, NY., Jahnke, GD., St Hilaire, C. & Shelby, M. (2003) Role of thyroid hormones in human and laboratory animal reproductive health. *Birth Defects Res B Dev Reprod Toxicol,* Vol. 68, pp. 479-491

Chorazy, PA., Himelhoch, S., Hopwood, NJ., Greger, NG. & Postellon, DC. (1995). Persistent hypothyroidism in an infant receiving a soy formula: case report and review of the literature. *Pediatrics,* Vol. 96, pp. 148-50

Christian, RC., Liu, PY., Harrington, S., Ruan, M., Miller, VM. & Fitzpatrick, LA. (2006). Intimal estrogen receptor (ER)beta, but not ERalpha expression, is correlated with coronary calcification and atherosclerosis in pre- and postmenopausal women. *J Clin Endocrinol Metab,* Vol. 91, pp. 2713-2720

Comelekoglu, U., Bagis, S., Yalin, S., Ogenler, O., Yildiz, A., Sahin, NO., Oguz, I. & Hatungil, R. (2007). Biomechanical evaluation in osteoporosis: ovariectomized rat model. *Clin Rheumatol,* Vol. 26, pp. 380-384

Conley, AJ., & Bird, IM. (1997) The role of cytochrome P450 17α-hydroxylase and 3β-hydroxysteroid dehydrogenase in the integration of gonadal and adrenal steroidogenesis via the delta 5 and delta 4 pathways of steroidogenesis in mammals. *Biology of Reproduction,* Vol. 56, pp. 789–799

Cooke, GM. (2006). A review of the animal models used to investigate the health benefits of soy isoflavones. *J AOAC Int,* Vol. 89, pp. 1215-1227

Cooper, C., Campion, G. & Melton, LJD. (1992). Hip fractures in the elderly, a world-wide projection. *Osteoporosis International,* Vol. 2, pp. 285-289

Corrêa da Costa, V.M., Moreira, D.G. & Rosenthal, D. (2001) Thyroid function and aging: gender-related differences. *J Endocrinol,* Vol. 171, pp. 193-198

Couse, JF. & Korach, KS. (2001). Contrasting phenotypes in reproductive tissues of female estrogen receptor null mice. *Ann N Y Acad Sci,* Vol. 948, pp.1-8

Dang, ZC. & Lowik, C. (2005), Dose-dependent effects of phytoestrogens on bone. *Trends Endocrinol Metab,* Vol. 16, pp. 207– 213

D'Archivio, M., Filesi, C., Varì, R., Scazzocchio, B. & Masella, R. (2010). Bioavailability of the polyphenols: status and controversies. *Int J Mol Sci,* Vol. 11, pp. 1321-1342

Davies, JR., Rudd, JH. & Weissberg, PL. (2004). Molecular and metabolic imaging of atherosclerosis. *J Nucl Med,* Vol. 45, pp. 1898-1907

Davison, S. & Davis, SR. (2003). Hormone replacement therapy: current controversies. *Clin Endocrinology,* Vol. 58, pp. 249–261

De Kloet, ER., Rosenfeld, P., Van Eekelen, JA., Sutanto, W. & Levine, S. (1988). Stress, glucocorticoids and development. *Prog Brain Res,* Vol. 73, pp. 101-120

de Wilde, A., Heberden, C., Chaumaz, G., Bordat, C. & Lieberherr, M. (2006), Signaling networks from Gbeta1 subunit to transcription factors and actin remodeling via a membrane- located ERbeta-related protein in the rapid action of daidzein in osteoblasts. *J Cell Physiol,* Vol. 209, pp. 786–801

Delclos, KB., Bucci, TJ., Lomax, LG., Latendresse, JR., Warbritton, A., Weis, CC. & Newbold, RR. (2001). Effects of dietary genistein exposure during development on male and female CD (Sprague-Dawley) rats. *Reproductive Toxicology,* Vol. 15, pp. 647–663

deVere White, RW., Tsodikov, A., Stapp, EC., Soares, SE., Fujii, H. & Hackman, RM. (2010). Effects of a high dose, aglycone-rich soy extract on prostate-specific antigen and serum isoflavone concentrations in men with localized prostate cancer. *Nutr Cancer,* Vol. 62, pp. 1036-1043

Dhar, A., Paul, AK. & Shukla SD. (1990). Platelet-activating factor stimulation of tyrosine kinase and its relationship to phospholipase C in rabbit platelets: studies with genistein and monoclonal antibody to phosphotyrosine. *Mol Pharmacol,* Vol. 37, pp. 519-525

Diel, P., Schmidt, S., Vollmer, G., Janning, P., Upmeier, A., Michna, H., Bolt, HM. & Degen, GH. (2004). Comparative responses of three rat strains (DA/Han, Sprague-Dawley and Wistar) to treatment with environmental estrogens. *Arch Toxicol,* Vol. 78, pp. 183-193

Dillingham, BL., McVeigh, BL., Lampe, JW. & Duncan, AM. (2007). Soy protein isolates of varied isoflavone content do not influence serum thyroid hormones in healthy young men.*Thyroid,* Vol. 17, pp. 131-137

DiPippo, VA., Lindsay, R. & Powers, CA. (1995). Estradiol and tamoxifen interactions with thyroid hormone in the ovariectomized-thyroidectomized rat. *Endocrinology,* Vol. 136, pp. 1020-1033

Divi, RL., Chang, HC. & Doerge, DR. (1997). Anti-thyroid isoflavones from soybean: isolation, characterization, and mechanisms of action. *Biochem Pharmacol,* Vol. 54, pp. 1087-1096

Doerge, DR. & Chang, HC. (2002). Inactivation of thyroid peroxidase by soy isoflavones, in vitro and in vivo. *J Chromatogr B Analyt Technol Biomed Life Sc,* Vol. 777, pp. 269-279

Donda, A., Reymond, F., Rey, F. & Lemarchand-Beraud, T. (1990) Sex steroids modulate the pituitary parameters involved in the regulation of TSH secretion in the rat. *Acta Endocrinol (Copenh),* Vol. 122, pp. 577-584

Duffy, C., Perez, K. & Partridge, A. (2007). Implications of phytoestrogen intake for breast cancer. *CA Cancer J Clin,* Vol. 57, pp. 260-277

Duffy, C., Perez, K. & Partridge, A. (2007). Implications of phytoestrogen intake for breast cancer. *CA Cancer J Clin,* Vol. 57, pp. 260-277

Duncan, AM., Underhill. KE., Xu. X., Lavalleur. J., Phipps. WR. & Kurzer. MS. (1999). Modest hormonal effects of soy isoflavones in postmenopausal women. *J Clin Endocrinol Metab,* Vol. 84: 3479-3484

Eustache, F., Mondon, F., Canivenc-Lavier, MC., Lesaffre, C., Fulla, Y., Berges, R., Cravedi, J. P., Vaiman, D. & Auger, J. (2009). Chronic dietary exposure to a low-dose mixture of genistein and vinclozolin modifies the reproductive axis, testis transcriptome and fertility. *Environmental Health Perspectives*, Vol. 117, pp. 1272–1279

Evans, MJ., Eckert, A., Lai, K., Adelman, SJ. & Harnish, DC. (2001). Reciprocal antagonism between estrogen receptor and NF-kappaB activity in vivo. *Circ Res*, Vol. 89, pp. 823-830

Fielden, MR., Samy, SM., Chou, KC. & Zacharewski, TR. (2003). Effect of human dietary exposure levels of genistein during gestation and lactation on long-term reproductive development and sperm quality in mice. *Food and Chemical Toxicology*, Vol. 41, pp. 447–454

Filipović, B., Sosić-Jurjević, B., Ajdzanović, V., Brkić, D., Manojlović-Stojanoski, M., Milosević, V. & Sekulić M. (2010). Daidzein administration positively affects thyroid C cells and bone structure in orchidectomized middle-aged rats. *Osteoporos Int*, Vol. 21, pp. 1609-1616

Filipović, B., Šošić-Jurjević, B., Ajdžanović, V., Trifunović, S., Manojlović Stojanoski, M., Ristić, N., Nestorović, N., Milošević, V. & Sekulić M. (2007) The effect of orchidectomy on thyroid C cells and bone histomorphometry in middle-aged rats. *Histochem Cell Biol*, Vol. 128, pp. 153–159

Filipović, B., Šošić-Jurjević, B., Manojlović Stojanoski, M., Nestorović, N., Milošević V. & Sekulić M. (2005). The effect of chronic calcium treatment on thyroid C cells in ovariectomized rats. *Life Sci*, Vol. 77, pp. 121-129

Filipović, B., Šošić-Jurjević, B., Manojlović-Stojanoski, M., Nestorović, N., Milošević, V. & Sekulić, M. (2002). The effect of ovariectomy on thyroid C cells of adult rats. *Yugoslov Med Biohem*, Vol. 21, pp. 345-350

Filipović, B., Šošić-Jurjević, B., Nestorović, N., Manojlović Stojanoski, M., Kostić N., Milošević, V. & Sekulić M. (2003). The thyroid C cells of ovariectomized rats treated with estradiol. *Histochem Cell Biol*, Vol. 120, pp. 409-414

Food and Drug Administration, Health claims and qualified health claims; dietary lipids and cancer, soy protein and coronary heart disease, antioxidant vitamins and certain cancers, and selenium and certain cancers. Reevaluation. (2007). Available from http://www.fda.gov/OHRMS/DOCKETS/98fr/E7-24813.pdf

Fort, P., Moses, N., Fasano, M., Goldberg, T. & Lifshitz, F. (1990). Breast and soy-formula feedings in early infancy and the prevalence of autoimmune thyroid disease in children. *J Am Coll Nutr*, Vol. 9, pp. 164–167

Fotsis, T., Pepper, M., Adlercreutz, H., Fleischmann, G., Hase, T., Montesano, R. & Schweigerer, L. (1993). Genistein, a dietary-derived inhibitor of in vitro angiogenesis. *Proc Natl Acad Sci USA*, Vol. 90, pp. 2690-2694

Fujieda, K., Faiman, C., Reyes, FI. & Winter JSD. (1982) The control of steroidogenesis by human fetal adrenal cells in tissue culture. IV. The effects of exposure to placental steroids. *J Clin Endocrinol Metab*, Vol. 54, pp. 89 –94

Furlanetto, T.W., Nunes, R.B., Sopelsa, A.M.I., Maciel, R.M.B. (2001) Estradiol decreases iodide uptake by rat thyroid follicular FTRL-5 cells. Brazil J Biol Res 34: 259-263

Galleano, M., Pechanova, O. & Fraga, CG. (2010). Hypertension, nitric oxide, oxidants, and dietary plant polyphenols. *Curr Pharm Biotechnol*, Vol. 11, pp. 837-848

Gallo, D., Giacomelli, S., Cantelmo, F., Zannoni, GF., Ferrandina, G., Fruscella, E., Riva, A., Morazzoni, P., Bombardelli, E., Mancuso, S. & Scambia, G. (2001). Chemoprevention of DMBA-induced mammary cancer in rats by dietary soy. Breast *Cancer Res Treat,* Vol. 69, pp. 153-64

Gao YH. & Yamaguchi, M. (1999) a. Inhibitory effect of genistein on osteoclast-like cell formation in mouse marrow cultures. *Biochem Pharmacol,* Vol. 58, pp. 767-772

Gao, YH. & Yamaguchi M. (1999). Anabolic effect of daidzein on cortical bone in tissue culture: comparison with genistein effect. *Mol Cell Biochem,* Vol. 194, pp. 93-98

Gao, YH. & Yamaguchi, M. (2000). Suppressive effect of genistein on rat bone osteoclasts: involvement of protein kinase inhibition and protein tyrosine phosphatase activation. *Int J Mol Med,* Vol. 5, pp. 261–267

Ge, Y., Chen, D., Xie, L. & Zhang, R. (2006). Enhancing effect of daidzein on the differentiation and mineralization in mouse osteoblast-like MC3T3-E1 cells. *Yakugaku Zasshi,* 126, 651–656

Gee, JM., DuPont, MS., Day, AJ., Plumb, GW., Williamson, G. & Johnson, IT. (2000). Intestinal transport of quercetin glycosides in rats involves both deglycosylation and interaction with the hexose transport pathway. *J Nutr,* Vol. 130, pp. 2765-2771

González, M., Reyes, R., Damas, C., Alonso & R., Bello, AR. (2008). Oestrogen receptor alpha and beta in female rat pituitary cells: an immunochemical study. *Gen Comp Endocrinol,* Vol. 155, pp. 857-868

Greany, KA., Nettleton, JA., Wangen, KE., Thomas, W. & Kurzer, MS. (2004). Probiotic consumption does not enhance the cholesterol-lowering effect of soy in postmenopausal women. *J Nutr,* Vol. 134, pp. 3277-3283

Greaves, KA., Parks, JS., Williams, JK. & Wagner, JD. (1999). Intact dietary soy protein, but not adding an isoflavone-rich soy extract to casein, improves plasma lipids in ovariectomized cynomolgus monkeys. *J Nutr,* Vol. 129, pp. 1585-1592

Grover, GJ., Mellstrom, K. & Malm, J. (2005). Development of the thyroid hormone receptor beta-subtype agonist KB-141: a strategy for body weight reduction and lipid lowering with minimal cardiac side effects. *Cardiovasc Drug Rev,* Vol. 23, pp.133-148

Ham, KD., Carlson, C.S. (2004). Effects of estrogen replacement therapy on bone turnover in subchondral bone and epiphyseal metaphyseal cancellous bone of ovariectomized cynomolgus monkeys.*J Bone Miner Res, Vol.*19(5)pp. 823-829.

Hamilton-Reeves, JM., Rebello, SA., Thomas, W., Slaton, JW. & Kurzer, MS. (2007). Isoflavone-rich soy protein isolate suppresses androgen receptor expression without altering estrogen receptor-beta expression or serum hormonal profiles in men at high risk of prostate cancer. *J Nutr,* Vol. 137, pp. 1769-1775

Hamilton-Reeves, JM., Vazquez, G., Duval, SJ., Phipps, WR., Kurzer, MS. (2009). Messina, MJ. Clinical studies show no effects of soy protein or isoflavones on reproductive hormones in men: results of a meta-analysis. *Fertil Steril,* Vol.94, pp. 997-1007

Hampl, R., Nemec, J., Jeresova, J., Kimlova, I. & Starka, L. (1985) Estrogen receptors in human goitrous and neoplastic thyroid. *Endocrinol Exp,* Vol. 19, pp.227-230

Harris, RM., Wood, DM., Bottomley, L., Blagg, S., Owen, K., Hughes, PJ., Waring, RH.. & Kirk, CJ. (2004). Phytoestrogens are potent inhibitors of estrogen sulfation: implications for breast cancer risk and treatment. *J Clin Endocrinol Metab,* Vol. 89, pp. 1779-1787

Haselkorn, T., Stewartm SL. & Horn-Rossm, PL. (2003). Why are thyroid cancer rates so high in southeast asian women living in the United States? The bay area thyroid cancer study. *Cancer Epidemiol Biomarkers Prev,* Vol. 12, pp. 144-150

Hatzinger, M., Wotjak, CT., Naruo, T., Simchen, R., Keck, ME., Landgraf, R., Holsboer, F. & Neumann, ID. (2000). Endogenous vasopressin contributes to hypothalamic-pituitary-adrenocortical alterations in aged rats. *J Endocrinol,* Vol. 164, pp. 197-205

Hooper, L., Kroon, PA., Rimm, EB., Cohn, JS., Harvey, I., Le Cornu, KA., Ryder, JJ. & Hall, WL. (2008). Cassidy A.Flavonoids, flavonoid-rich foods, and cardiovascular risk: a meta-analysis of randomized controlled trials. *Am J Clin Nutr,* Vol. 88, pp. 38-50

Hooper, L., Ryder, JJ., Kurzer, MS., Lampe, JW., Messina, MJ., Phipps, WR. & Cassidy, A. (2009). Effects of soy protein and isoflavones on circulating hormone concentrations in pre- and post-menopausal women: a systematic review and meta-analysis. *Hum Reprod Update,* Vol. 15, pp. 423-440

Hsieh, CY., Santell, RC., Haslam, SZ. & Helferich, WG. (1998). Estrogenic effects of genistein on the growth of estrogen receptor-positive human breast cancer (MCF-7) cells in vitro and in vivo. *Cancer Res,* Vol. 58, pp. 3833-3838

Huang, HY., Yang, HP., Yang, HT., Yang, TC., Shieh, MJ. & Huang, SY. (2006). One-year soy isoflavone supplementation prevents early postmenopausal bone loss but without a dosedependent effect. *J Nutr Biochem,* Vol. 17, pp. 509–517

Hussain, M., Banerjee, M., Sarkar, FH., Djuric, Z., Pollak, MN., Doerge, D., Fontana, J., Chinni, S., Davis, J., Forman, J., Wood, DP. & Kucuk, O. (2003). Soy isoflavones in the treatment of prostate cancer. *Nutr Cancer,* Vol. 47, pp. 111-117

Iguchi, T., Fukazawa, Y., Uesugi, Y. & Takasugi, N. (1990). Polyovular follicles in mouse ovaries exposed neonatally to diethylstilbestrol in vivo and in vitro. *Biol Reprod,* Vol. 43, pp. 478–484

Ikeda, T., Nishikawa, A., Imazawa, T., Kimura, S. & Hirose, M. (2000). Dramatic synergism between excess soybean intake and iodine deficiency on the development of rat thyroid hyperplasia. *Carcinogenesis,* Vol. 21, pp. 707-713

Ishida, M., Marrero, MB., Schieffer, B., Ishida, T., Bernstein, KE. & Berk BC. (1995). Angiotensin II activates pp60c-src in vascular smooth muscle cells. *Circulation Research,* Vol. 77, pp. 1053–1059

Jabbar, MA., Larrea, J. & Shaw, RA. (1997). Abnormal thyroid function tests in infants with congenital hypothyroidism: the influence of soy-based formula. *J Am Coll Nutr,* Vol. 16, pp. 280-282

Jefferson, WN., Couse, JF., Padilla-Banks, E., Korach, KS. & Newbold, RR. (2002). Neonatal exposure to genistein induces estrogen receptor (ER)alpha expression and multioocyte follicles in the maturing mouse ovary: evidence for ERbeta-mediated and nonestrogenic actions. *Biol Reprod,* Vol. 67, pp.1285–1296

Jefferson, WN., Padilla-Banks, E. & Newbold, RR. (2005). Adverse effects on female development and reproduction in CD-1 mice following neonatal exposure to the phytoestrogen genistein at environmentally relevant doses. *Biol Reprod,* Vol. 73, pp. 798–806

Jenkins, DJ., Kendall, CW., Jackson, CJ., Connelly, PW., Parker, T., Faulkner, D., Vidgen, E., Cunnane, SC., Leiter, LA. & Josse, RG. (2002). Effects of high- and low-isoflavone soyfoods on blood lipids, oxidized LDL, homocysteine, and blood pressure in hyperlipidemic men and women. *Am J Clin Nutr,* Vol. 76, pp. 365-372

Jenkins, DJ., Kendall, CW., Marchie, A., Faulkner, D., Vidgen, E., Lapsley, KG., Trautwein, EA., Parker, TL., Josse, RG., Leiter, LA. & Connelly, PW. (2003). The effect of combining plant sterols, soy protein, viscous fibers, and almonds in treating hypercholesterolemia. Metabolism, Vol. 52, pp. 1478-1483

Jia, TL., Wang, HZ., Xie, LP., Wang, XY. & Zhang, RQ. (2003), Daidzein enhances osteoblast growth that may be mediated by increased bone morphogenetic protein (BMP) production. *Biochem Pharmacol,* Vol. 65, pp. 709–715

Jiménez, E. & Montiel, M. (2005). Activation of MAP kinase by muscarinic cholinergic receptors induces cell proliferation and protein synthesis in human breast cancer cells. *J Cell Physiol,* Vol. 204, pp. 678-686

Jones, JL., Daley, BJ., Enderson, BL., Zhou, JR. & Karlstad, MD. (2002). Genistein inhibits tamoxifen effects on cell proliferation and cell cycle arrest in T47D breast cancer cells. *Am. Surg,* Vol. 68, pp. 575-577; discussion 577-578

Kajiya, H., Takekoshi, S., Miyai, S., Ikeda, T., Kimura, S. & Osamura, RY. (2005). Dietary soybean enhances Pit-1 dependent pituitary hormone production in iodine deficient rats. *J Mol Histol,* Vol. 36, pp. 265-274

Kanemori, M. & Prygrocki, M. (2005). Results of breast conservation therapy from a single-institution community hospital in Hawaii with a predominantly Japanese population. *Int J Radiat Oncol Biol Phys,* Vol. 62, pp. 193-197

Kang, KS., Che, JH. & Lee, YS. (2002). Lack of adverse effects in the f1 offspring maternally exposed to genistein at human intake dose level. *Food and Chemical Toxicology,* Vol. 40, pp. 43–51

Kanno, S., Hirano, S. & Kayama, F. (2004), Effects of the phytoestrogen coumestrol on RANK-ligand-induced differentiation of osteoclasts. *Toxicology,* Vol. 203, pp. 211–220

Kapiotis, S., Hermann, M., Held, I., Seelos, C., Ehringer, H. & Gmeiner, BM. (1997). Genistein, the dietary-derived angiogenesis inhibitor, prevents LDL oxidation and protects endothelial cells from damage by atherogenic LDL. *Arterioscler Thromb Vasc Biol,* Vol. 17, pp. 2868-2874

Katahira, M., Iwasaki, Y., Aoki, Y., Oiso, Y. & Saito H. (1998). Cytokine regulation of the rat proopiomelanocortin gene expression in AtT-20 cells. *Endocrinology,* Vol.139, pp. 2414-2422

Kerry, N. & Abbey, M. (1998). The isoflavone genistein inhibits copper and peroxyl radical mediated low density lipoprotein oxidation in vitro. *Atherosclerosis,* Vol. 140, pp. 341-347

Khalil, DA., Lucas, EA., Smith, BJ., Soung, DY., Devareddy, L., Juma, S., Akhter, MP., Recker, R. & Arjmandi, BH. (2005). Soy isoflavones may protect against orchidectomy-induced bone loss in aged male rats. *Calcif Tissue Int.* Vol. 76, pp. 56–62

Kimura, S., Suwa, J., Ito, M. (1976). Sato, H. Development of malignant goiter by defatted soybean with iodine-free diet in rats. *Gann,* Vol. 67, pp. 763-765

Kirk, EA., Sutherland, P., Wang, SA., Chait, A. & LeBoeuf, RC. (1998). Dietary isoflavones reduce plasma cholesterol and atherosclerosis in C57BL/6 mice but not LDL receptor-deficient mice. *J Nutr,* Vol. 128, pp. 954-959

Köhrle J. (2002). Iodothyronine deiodinases. *Methods Enzymol,* Vol. 347, pp. 125-167

Köhrle, J. (2008). Environment and endocrinology: the case of thyroidology. *Ann Endocrinol (Paris),* Vol. 69, pp. 116-122

Köhrle, J., Schomburg, L., Drescher, S., Fekete, E. & Bauer, K. (1995). Rapid stimulation of type I 5'-deiodinase in rat pituitaries by 3,3',5-triiodo-L-thyronine. *Mol Cell Endocrinol,* Vol. 108, pp. 17-21

Kojima, T., Uesugi, T., Toda, T., Miura, Y. & Yagasaki, K. (2002). Hypolipidemic action of the soybean isoflavones genistein and genistin in glomerulonephritic rats. *Lipids,* Vol. 37, pp. 261-265

Korach, KS. (1994). Insights from the study of animals lacking functional estrogen receptor. *Science,* Vol. 266, pp. 1524-1527

Kouki, T., Kishitake, M., Okamoto, M., Oosuka, I., Takebe, M. & Yamanouchi, K. (2003). Effects of neonatal treatment with phytoestrogens, genistein and daidzein, on sex difference in female rat brain function: estrous cycle and lordosis. *Horm Behav,* Vol. 44, pp.140–145

Kruk, I., Aboul-Enein, HY., Michalska, T., Lichszteld, K. & Kładna, A. (2005). Scavenging of reactive oxygen species by the plant phenols genistein and oleuropein. *Luminescence,* Vol. 20, pp. 81-89

Kuiper, GG., Lemmen, JG., Carlsson, B., Corton, JC., Safe, SH., van der Saag, PT., van der Burg, B. & Gustafsson, JA. (1998). Interaction of estrogenic chemicals and phytoestrogens with estrogen receptor beta. *Endocrinology,* Vol. 139, pp. 4252–4263

Kurahashi, N., Iwasaki, M., Sasazuki, S., Otani, T., Inoue, M. & Tsugane, S. (2007). Japan Public Health Center-Based Prospective Study Group. Soy product and isoflavone consumption in relation to prostate cancer in Japanese men. *Cancer Epidemiol Biomarkers Prev,* Vol. 16, pp. 538-545

Lakshman, M., Xu, L., Ananthanarayanan, V., Cooper, J., Takimoto, CH., Helenowski, I., Pelling, JC. & Bergan, RC. (2008). Dietary genistein inhibits metastasis of human prostate cancer in mice. *Cancer Res,* Vol. 68, pp. 2024-2032

Lamartiniere, CA., Cotroneo, MS., Fritz, WA., Wang, J., Mentor-Marcel, R. & Elgavish, A. (2002). Genistein chemoprevention: timing and mechanisms of action in murine mammary and prostate. *J Nutr,* Vol. 132, pp. 552S-558S

Lampe, JW., Karr, SC., Hutchins, AM. & Slavin, JL. (1998). Urinary equol excretion with a soy challenge: influence of habitual diet. *Proc Soc Exp Biol Med,* Vol. 217, pp. 335-339

Larkin, T., Price, WE. & Astheimer, L. (2008). The key importance of soy isoflavone bioavailability to understanding health benefits. *Crit Rev Food Sci Nutr,* Vol. 48, pp. 538-552

Lauderdale, DS., Jacobsen, SJ., Furner, SE., Levy, PS., Brody, JA. & Goldberg, J. (1997). Hip fractures incidence among elderly Asian-American populations. *American Journal of Epidemiology,* Vol. 146, pp. 502-509

Lee, BJ., Jung, EY., Yun, YW., Kang, JK., Baek, IJ., Yon, JM., Lee, YB., Sohn, HS., Lee, JY., Kim, KS. & Nam, SY. (2004) a. Effects of exposure to genistein during pubertal development on the reproductive system of male mice. *J Reprod Dev,* Vol. 50, pp. 399–409

Lee, YB., Lee, HJ., Kim, KS., Lee, JY., Nam, SY., Cheon, SH. & Sohn HS. (2004). Evaluation of the preventive effect of isoflavone extract on bone loss in ovariectomized rats. *Bioscience, Biotechnology and Biochemistry,* Vol. 68, pp. 1040-1045

Lephart, ED., Setchell, KD., Handa, RJ.& Lund TD. (2004). Behavioral effects of endocrine-disrupting substances: phytoestrogens. *ILAR J*, Vol. 45, pp. 443-454

Lindsay, R., Hart, DM. & Clark, DM. (1984). The minimum effective dose of estrogen for prevention of postmenopausal bone loss. *Obstet Gynecol*, Vol. 63, pp. 759–763

Lindsay, R., Hart, DM., Aitken, JM., MacDonald, ED., Anderson, JB. & Clarke, AC. (1976). Long-term prevention of postmenopausal osteoporosis by oestrogen. *Lancet*, Vol. 1, pp. 1038–1041

Lisbôa, PC., Curty, FH., Moreira, RM., Oliveira, KJ. & Pazos-Moura, CC. (2001). Sex steroids modulate rat anterior pituitary and liver iodothyronine deiodinase activities. *Horm Metab Res*, Vol. 33, pp. 532-535

Liu, J., Ho, SC., Su, YX., Chen, WQ., Zhang, CX. & Chen, YM. (2009). Effect of long-term intervention of soy isoflavones on bone mineral density in women: a meta-analysis of randomized controlled trials. *Bone*, Vol. 44, pp. 948–953

Lönnerdal, B., Bryant, A., Liu, X. & Theil, EC. (2006). Iron absorption from soybean ferritin in nonanemic women. *Am J Clin Nutr*, Vol. 83, pp.103-107

Loughlin, K. & Richie, J. (1997). Prostate cancer after exogenous testosterone treatment for impotence. *J Urology*, Vol. 157, pp.1845

Lucas, EA., Khalil, DA., Daggy, BP. & Arjmandi, BH. (2001). Ethanol-extracted soy protein isolate does not modulate serum cholesterol in golden Syrian hamsters: a model of postmenopausal hypercholesterolemia. *J Nutr*, Vol. 131, pp. 211-214

Lucas, EA., Lightfoot, SA., Hammond, LJ., Devareddy, L., Khalil, DA., Daggy, BP., Soung do, Y. & Arjmandi, BH. (2003). Soy isoflavones prevent ovariectomy-induced atherosclerotic lesions in Golden Syrian hamster model of postmenopausal hyperlipidemia. *Menopause*, Vol. 10, pp. 314-321

Lund, TD., Munson, DJ., Adlercreutz, H., Handa, RJ. & Lephart, ED. (2004). Androgen receptor expression in the rat prostate is down-regulated by dietary phytoestrogens. *Reprod Biol Endocrinol*, Vol. 2, pp. 5

Ma, DF., Qin, LQ., Wang, PY. & Katoh, R. (2008). Soy isoflavone intake increases bone mineral density in the spine of menopausal women: meta-analysis of randomized controlled trials. *Clin Nutr*, Vol. 27, 57–64

Mackey, R. & Eden, J. (1998). Phytoestrogens and the menopause. *Climateric*, Vol. 1, pp. 302-308

Mahn, K., Borrás, C., Knock, GA., Taylor, P., Khan, IY., Sugden, D., Poston, L., Ward, JP., Sharpe, RM., Viña, J., Aaronson, PI. & Mann, GE. (2005). Dietary soy isoflavone induced increases in antioxidant and eNOS gene expression lead to improved endothelial function and reduced blood pressure in vivo. *FASEB J*, Vol. 19, pp. 1755-1757

Malendowicz, LK., Trejter, M., Rebuffat, P., Ziolkowska, A., Nussdorfer, GG. & Majchrzak, M. (2006). Effects of some endocrine disruptors on the secretory and proliferative activity of the regenerating rat adrenal cortex. *Int J Mol Med*, 18, pp. 197-200

Manojlović-Stojanoski, M., Nestorović, N., Ristić, N., Trifunović, S., Filipović, B., Sosić-Jurjević, B. & Sekulić, M. (2010). Unbiased stereological estimation of the rat fetal pituitary volume and of the total number and volume of TSH cells after maternal dexamethasone application. *Microsc Res Tech*, Vol. 73, pp. 1077-1085

Markovits, J., Linassier, C., Fossé, P., Couprie, J., Pierre, J., Jacquemin-Sablon, A., Saucier, JM., Le Pecq, JB. & Larsen, AK. (1989). Inhibitory effects of the tyrosine kinase inhibitor genistein on mammalian DNA topoisomerase II. *Cancer Res,* Vol. 49, pp. 5111-5117

Maskarinec, G., Franke, AA., Williams, AE., Hebshi, S., Oshiro, C., Murphy, S. & Stanczyk, FZ. (2004). Effects of a 2-year randomized soy intervention on sex hormone levels in premenopausal women. *Cancer Epidemiol Biomarkers Prev,* Vol. 13, pp. 1736-1744

Maskarinec, G., Watts, K., Kagihara, J., Hebshi, SM. & Franke, AA. (2008). Urinary isoflavonoid excretion is similar after consuming soya milk and miso soup in Japanese-American women. *Br J Nutr,* Vol. 100, pp. 424-429

Maskarinec, G., Williams, AE., Inouye, JS., Stanczyk, FZ. & Franke, AA. (2002). A randomized isoflavone intervention among premenopausal women. *Cancer Epidemiol Biomarkers Prev,* Vol. 11, pp. 195-201

McCarrison, R. (1993). A Paper on FOOD AND GOITRE. *Br Med J,* Vol. 14, pp. 671-675

McLachlan, JA., Newbold, RR., Shah, HC., Hogan, MD. & Dixon, RL. (1982). Reduced fertility in female mice exposed transplacentally to diethylstilbestrol (DES). *Fertil Steril,* Vol. 38, pp. 364–371

McVey, MJ., Cooke, GM., & Curran, IH. (2004). Increased serum and testicular androgen levels in F1 rats with lifetime exposure to soy isoflavones. *Reproductive toxicology (Elmsford, N.Y),* Vol. 18, pp. 677-685

Medigović, I., Ristić, N., Manojlović-Stojanoski, M., Šošić-Jurjević, B., Pantelić, J., Milošević, V., Nestorović, N. (2009). The effects of genistein and estradiol on the ovaries of immature rats: Stereological study. *Second Congress of Physiological Sciences of Serbia with International Participation,* p. 108, Kragujevac, Serbia, September 17-20, 2009

Meerts, IA., Assink, Y., Cenijn, PH., Van Den Berg, JH., Weijers, BM., Bergman, A., Koeman. JH. & Brouwer, A. (2002). Placental transfer of a hydroxylated polychlorinated biphenyl and effects on fetal and maternal thyroid hormone homeostasis in the rat. *Toxicol Sci,* Vol. 68, pp. 361–371

Mendel, CM., Weisiger, RA., Jones, AL. & Cavalieri, RR. (1987). Thyroid hormone-binding proteins in plasma facilitate uniform distribution of thyroxine within tissues: a perfused rat liver study. *Endocrinology,* Vol. 120, pp. 1742–1749

Mentor-Marcel, R., Lamartiniere, CA., Eltoum, IE., Greenberg, NM. & Elgavish, A. (2001). Genistein in the diet reduces the incidence of poorly differentiated prostatic adenocarcinoma in transgenic mice (TRAMP). *Cancer Res,* Vol. 61, pp. 6777-6782

Mesiano, S. & Jaffe, RB. (1993). Interaction of insulin-like growth factor-II and estradiol directs steroidogenesis in the human fetal adrenal toward dehydroepiandrosterone sulfate production. *J Clin Endocrinol Metab,* Vol. 77, pp.754 –758

Mesiano, S., Katz, SL., Lee, JY. & Jaffe RB. (1999). Phytoestrogens alter adrenocortical function: genistein and daidzein suppress glucocorticoid and stimulate androgen production by cultured adrenal cortical cells. *J Clin Endocrinol Metab,* Vol. 84, pp. 2443-2448

Messina, M. (2010). Insights gained from 20 years of soy research. J Nutr, Vol. 140, pp. 2289S-2295S

Messina, M., Nagata, C. & Wu, AH. (2006). Estimated Asian adult soy protein and isoflavone intakes. *Nutr Cancer,* Vol. 55, pp. 1-12

Messina, MJ. & Loprinzi, CL. (2001). Soy for breast cancer survivors: a critical review of the literature. *J Nutr,* Vol. 131, pp. 3095S-3108S

Messina, MJ. & Wood, CE. (2008). Soy isoflavones, estrogen therapy, and breast cancer risk: analysis and commentary. *Nutr J*, Vol. 7, pp. 17

Mezei, O., Li, Y., Mullen, E., Ross-Viola, JS. & Shay, NF. Dietary isoflavone supplementation modulates lipid metabolism via PPARalpha-dependent and -independent mechanisms. *Physiol Genomics*, Vol. 26, pp. 8-14

Michaelsson, K., Baron, JA., Farahmand, BY., Johnell, O., Magnusson, C., Persson, PG., Persson, I. & Ljunghall S. (1998). Hormone replacement therapy and risk of hip fracture: population based case-control study. The Swedish Hip Fracture Study Group. *BMJ*, Vol. 316, pp. 1858–1863

Milošević, V., Ajdžanović, V., Sosić-Jurjević, B., Filipović, B., Brkić, M., Nestorović, N. & Sekulić, M. (2009). Morphofunctional characteristics of ACTH cells in middle-aged male rats after treatment with genistein. *Gen Physiol Biophys*, 28, pp. 94-97

Mitchell, JH., Gardner, PT., McPhail, DB., Morrice, PC., Collins, AR. & Duthie, GG. (1998). Antioxidant efficacy of phytoestrogens in chemical and biological model systems. *Arch Biochem Biophys*, Vol. 360, pp. 142-148

Molsiri, K., Khemapech, S., Patumraj, S. & Siriviriyakul, P. (2004). Preventive mechanism of genistein on coronary endothelial dysfunction in ovariectomized rats: an isolated arrested heart model. *Clin Hemorheol Microcirc*, Vol. 31, pp. 59-66

Morin, S. (2004). Isoflavones and bone health. *Menopause*; Vol. 11, pp. 239- 241

Mulvihill, EE. & Huff, MW. (2010). Antiatherogenic properties of flavonoids: implications for cardiovascular health. *Can J Cardiol*, Vol. 26, pp. 17A-21A

Nagata, C., Takatsuka, N., Inaba, S., Kawakami, N. & Shimizu, H. (1998). Effect of soymilk consumption on serum estrogen concentrations in premenopausal Japanese women. *J Natl Cancer Inst*, Vol. 90, pp. 1830-1835

Naik, HR., Lehr, JE. & Pienta, KJ. (1994). An in vitro and in vivo study of antitumor effects of genistein on hormone refractory prostate cancer. *Anticancer Res*, Vol. 14, pp. 2617-2619

National Toxicology Program. (2008). NTP Multigenerational Reproductive Study of Genistein (CAS no. 446-72-0) in Sprague-Dawley Rats (Feed Study). Technical Report 539. Research Triangle Park, NC:National Toxicology Program

Nelson, HD., Humphrey, LL., Nygren, P., Teutsch, SM. & Allan, JD. (2002). Postmenopausal hormone replacement therapy: scientific review. *JAMA*. Vol. 288, pp. 872–881

Newbold, RR., Banks, EP., Bullock, B. & Jefferson, WN. (2001). Uterine adenocarcinoma in mice treated neonatally with genistein. *Cancer Res*, Vol. 61, pp. 4325–4328

Newton, KM., LaCroix, AZ., Levy, L., Li, SS., Qu, P., Potter, JD. & Lampe, JW. (2006) Soy protein and bone mineral density in older men and women: a randomized trial. *Maturitas*, Vol. 55, pp. 270–277

Nicholson, CG., Mosley, JM., Sexton, PM., Mendelsohn, FAO. & Martin, TJ. (1986). Abundant calcitonin receptors in isolated rat osteoclasts. *J Clin Invest*, Vol. 78, pp.355–360

Nikaido, Y., Yoshizawa, K., Danbara, N., Tsujita-Kyutoku, M., Yuri, T., Uehara, N. & Tsubura, A. (2004). Effects of maternal xenoestrogen exposure on development of the reproductive tract and mammary gland in female CD-1 mouse offspring. *Reprod Toxicol*, Vol. 18, pp. 803-811

Ohlsson, A., Ullerås, E., Cedergreen, N. & Oskarsson, A. (2010). Mixture effects of dietary flavonoids on steroid hormone synthesis in the human adrenocortical H295R cell line. *Food Chem Toxicol,* Vol. 48, pp. 3194-3200

Ohno, S., Nakajima, Y., Inoue, K., Nakazawa, H. & Nakajin, S. (2003). Genistein administration decreases serum corticosterone and testosterone levels in rats. *Life Sci,* Vol. 74, pp. 733-742

Ohno, S., Shinoda, S., Toyoshima, S., Nakazawa, H., Makino, T. & Nakajin, S. (2002). Effects of favonoid phytochemicals on cortisol production and on activities of steroidogenic enzymes in human adrenocortical H295R cells. *J Steroid Biochem Mol Biol,* Vol. 80, pp. 355–363

Om, AS. & Shim, JY. (2007). Effect of daidzein, a soy isoflavone, on bone metabolism in Cd-treated ovariectomized rats. *Acta Biochim Pol,* Vol. 54, pp. 641-646

Onoe, Y., Miyaura, C., Ohta, H., Nozawa, S. & Suda, T. (1997). Expression of estrogen receptor in rat bone. *Endocrinology,* Vol. 138, pp. 4509-4512

Ørgaard, A. & Jensen, L. (2008). The effects of soy isoflavones on obesity. *Exp Biol Med (Maywood),* Vol. 233, pp. 1066-1080

Osterweil, N. (2007). Soy as prostate cancer protection yields paradoxical results. Medpage Today March 16, 2007

Padilla-Banks, E., Jefferson, WN. & Newbold, RR. (2006). Neonatal exposure to the phytoestrogen genistein alters mammary gland growth and developmental programming of hormone receptor levels. *Endocrinology,* Vol. 147, pp. 4871-4882

Pantelić, J., Filipović, B., Sosić-Jurjević, B., Medigović, I. & Sekulić, M. (2010). Effects of testosterone and estradiol treatment on bone histomorphometry in orchidectomized middle-aged rats. *Proceedings of 4th Serbian Congress for Microscopy,* pp. 143-144, Belgrade, Serbia, October 11-12, 2010

Parkin, DM., Whelan, SL., Ferlay, J. & Storm, H. (2005). Cancer incidence in five continents, IARC Cancer Base, 2005; No. 7. Lyon:IARC Press

Patel, RP., Boersma, BJ., Crawford, JH., Hogg, N., Kirk, M., Kalyanaraman, B., Parks, DA., Barnes, S. & Darley-Usmar, V. 82001). Antioxidant mechanisms of isoflavones in lipid systems: paradoxical effects of peroxyl radical scavenging. *Free Radic Biol Med,* Vol. 31, pp. 1570-1581

Patisaul, HB. & Jefferson, W. (2010). The pros and cons of phytoestrogens. *Front Neuroendocrinol,* Vol. 31, pp. 400-419

Pei, RJ., Sato, M., Yuri, T., Danbara, N., Nikaido, Y. & Tsubura, A. (2003). Effect of prenatal and prepubertal genistein exposure on N-methyl-N-nitrosourea-induced mammary tumorigenesis in female Sprague-Dawley rats. *In Vivo,* Vol. 17, pp. 349-357

Piotrowska, K., Baranowska-Bosiacka, I., Marchlewicz, M., Gutowska, I., Noceń, I., Zawiślak, M., Chlubek, D. & Wiszniewska, B. (2011). Changes in male reproductive system and mineral metabolism induced by soy isoflavones administered to rats from prenatal life until sexual maturity. *Nutrition,* Vol 27, pp. 372-379

Pollard, M. & Suckow, M. (2006). Dietary prevention of hormone refractory prostate cancer in Lobund-Wistar rats: a review of studies in a relevant animal model. *Comp Med,* Vol. 56, pp. 461-467

Potter, SM. (1995). Overview of proposed mechanisms for the hypocholesterolemic effect of soy. *J Nutr,* Vol. 125, pp. 606S-611S

Radović, B., Mentrup, B., Köhrle, J. (2006). Genistein and other soya isoflavones are potent ligands for transthyretin in serum and cerebrospinal fluid. *Br J Nutr,* Vol. 95, pp. 1171-1176

Ralli, M. (2003). Soy and the Thyroid: Can This Miracle Food Be Unsafe? *Nutrition Noteworthy,* Vol. 6, pp. 6

Rand, WM., Pellett, PL. & Young, VR. (2003). Meta-analysis of nitrogen balance studies for estimating protein requirements in healthy adults. *Am J Clin Nutr,* Vol. 77, pp. 109-127

Rassi, CM., Lieberherr, M., Chaumaz, G., Pointillart, A. & Cournot, G. (2002). Down-regulation of osteoclast differentiation by daidzein via caspase 3. *J Bone Miner Res,* Vol. 17, pp. 630-638

Redei, E., Li, L., Halasz, I., McGivern, RF. & Aird, F. (1994). Fast glucocorticoid feedback inhibition of ACTH secretion in the ovariectomized rat: effect of chronic estrogen and progesterone. *Neuroendocrinology,* Vol. 60, pp. 113-123

Refetoff, S. (1989). Inherited thyroxine-binding globulin abnormalities in man. *Endocr Rev,* Vol. 10, pp. 275–293

Register, TC., Cann, JA., Kaplan, JR., Williams, JK., Adams, MR., Morgan, TM., Anthony, MS., Blair, RM., Wagner, JD. & Clarkson, TB. (2005). Effects of soy isoflavones and conjugated equine estrogens on inflammatory markers in atherosclerotic, ovariectomized monkeys. *J Clin Endocrinol Metab,* Vol. 90, pp. 1734-1740

Reinwald, S. & Weaver, CM. (2010). Soy components vs. whole soy: are we betting our bones on a long shot? *J Nutr,* Vol. 140, pp. 2312S-2317S

Ren, P., Ji, H., Shao, Q., Chen, X., Han, J. & Sun, Y. (2007).Protective effects of sodium daidzein sulfonate on trabecular bone in ovariectomized rats. *Pharmacology,* Vol. 79, pp. 129-136

Richardson, SJ., Lemkine, GF., Alfama, G., Hassani, Z. & Demeneix, BA. (2007). Cell division and apoptosis in the adult neural stem cell niche are differentially affected in transthyretin null mice. *Neurosci Lett,* Vol. 421, pp. 234–238

Robbins, J. (2000) Thyroid hormone transport proteins and the physiology of hormone binding. In: Braverman L, Utiger R, eds. Werner & Ingbar's The Thyroid: a fundamental and clinical text. 8th ed. Philadelphia: Lippincott, Williams and Wilkins, pp. 105–120

Roberts, D., Veeramachaneni, DN., Schlaff, WD. & Awoniyi, CA. (2000). Effects of chronic dietary exposure to genistein, a phytoestrogen, during various stages of development on reproductive hormones and spermatogenesis in rats. *Endocrine,* Vol. 13, pp. 281–286

Rochfort, S. & Panozzo, J. (2007). Phytochemicals for health, the role of pulses. *J Agric Food Chem,* Vol. 55, pp. 7981-7894

Rosen, ED. (2005). The transcriptional basis of adipocyte development. *Prostaglandins Leukot Essent Fatty Acids,* Vol. 73, pp. 31-34

Rowland, IR., Wiseman, H., Sanders, TA., Adlercreutz, H. & Bowey, EA. (2000). Interindividual variation in metabolism of soy isoflavones and lignans: influence of habitual diet on equol production by the gut microflora. *Nutr Cancer,* Vol. 36, pp. 27-32

Ruhlen, RL., Howdeshell, KL., Mao, J., Taylor, JA., Bronson, FH., Newbold, RR., Welshons, WV. & vom Saal, FS. (2008). Low phytoestrogen levels in feed increase fetal serum estradiol resulting in the "fetal estrogenization syndrome" and obesity in CD-1 mice. *Environ Health Perspect,* Vol. 116, pp. 322-328

Sacks, FM., Lichtenstein, A., Van Horn, L., Harris, W., Kris-Etherton, P. & Winston, M. (2006). Soy protein, isoflavones, and cardiovascular health: a summary of a statement for professionals from the american heart association nutrition committee. *Arterioscler Thromb Vasc Biol,* Vol. 26, pp. 1689-1992

Sarkar, FH. & Li, Y. (2002). Mechanisms of cancer chemoprevention by soy isoflavone genistein. *Cancer Metastasis Rev,* Vol. 21, pp. 265-280

Sasaki, S., Kawai, K., Honjo, Y. & Nakamura, H. (2006). Thyroid hormones and lipid metabolism. *Nippon Rinsho,* Vol. 64, pp. 2323-2329

Sathyapalan, T., Manuchehri, AM., Thatcher, NJ., Rigby, AS., Chapman, T., Kilpatrick, ES. & Atkin, SL. (2011). The Effect of Soy Phytoestrogen Supplementation on Thyroid Status and Cardiovascular Risk Markers in Patients with Subclinical Hypothyroidism: A Randomized, Double-Blind, Crossover Study. *J Clin Endocrinol Metab,* [Epub ahead of print]

Schmidt, S., Degen, GH., Seibel, J., Hertrampf, T., Vollmer, G. & Diel, P. (2006). Hormonal activity of combinations of genistein, bisphenol A and 17beta-estradiol in the female Wistar rat. *Arch Toxicol,* Vol. 80, pp. 839-845

Schmutzler, C., Hamann, I., Hofmann, PJ., Kovacs, G., Stemmler, L., Mentrup, B., Schomburg, L., Ambrugger, P., Grüters, A., Seidlova-Wuttke, D., Jarry, H., Wuttke, W. & Köhrle, J. (2004). Endocrine active compounds affect thyrotropin and thyroid hormone levels in serum as well as endpoints of thyroid hormone action in liver, heart and kidney. *Toxicology,* Vol. 205, pp. 95-102

Sekulić, M., Šošić-Jurjević, B., Filipović, B., Milošević, V., Nestorović, N. & Manojlović-Stojanoski, M. (2005). The effects of synthetic salmon calcitonin on thyroid C and follicular cells in adult female rats. *Folia Histochem Cytobiol.* Vol. 43, pp. 103-108

Sekulić, M., Sosić-Jurjević, B., Filipović, B., Pantelić, J., Nestorović, N., Manojlović-Stojanoski, M. & Milošević, V. (2010). Testosterone and estradiol differently affect thyroid structure and function in orchidectomizrd middle-aged rats. *Proceedings of 14th International Thyroid Congress,* p. 302 Paris, France, September 11-16, 2010

Setchell, KD., Brown, NM. & Lydeking-Olsen, E. (2002). The clinical importance of the metabolite equol-a clue to the effectiveness of soy and its isoflavones. *J Nutr,* Vol. 132, pp. 3577-3584

Setchell, KD., Gosselin, SJ., Welsh, MB., Johnston, JO., Balistreri, WF., Kramer, LW., Dresser, BL. & Tarr, MJ. (1987). Dietary estrogens--a probable cause of infertility and liver disease in captive cheetahs. *Gastroenterology,* Vol. 93, pp. 225-233

Sharpe, RM., Martin, B., Morris, K., Greig, I., McKinnell, C., McNeilly, A. S., Walker, M., Setchell, KD., Cassidy, A., Ingram, D., Sanders, K., Kolybaba, M., Lopez, D., Fitzpatrick, M., Fort, P., Moses, N., Fasano, M., Goldberg, T., Lifshitz, F., Chang, H. C., Doerge, DR., Bingham, S., Setchell, K., Herman-Giddens, ME., Slora, EJ., Wasserman, RC., Bourdony, CJ., Bhapkar, MV., Koch, GG., & Hasemeier, CM. (2002). Infant feeding with soy formula milk: effects on the testis and on blood testosterone levels in marmoset monkeys during the period of neonatal testicular activity. *Hum Reprod,* Vol. 17, pp. 1692-1703

Shen, P., Liu, MH., Ng, TY., Chan, YH. & Yong EL. (2006). Differential effects of isoflavones, from Astragalus membranaceus and Pueraria thomsonii, on the activation of

PPARalpha, PPARgamma, and adipocyte differentiation in vitro. *J Nutr,* Vol. 136, pp. 899-905

Shu, XO., Zheng, Y., Cai, H., Gu, K., Chen, Z., Zheng, W. & Lu, W. (2009). Soy food intake and breast cancer survival. *JAMA,* Vol. 302, pp. 2437-2443

Simpson, ER. & Waterman, MR. (1992). Regulation of expression of adrenocortical enzymes. In *The Adrenal Gland,* edn 2. Ed VHT James. New York: Raven Press

Sirianni, R., Sirianni, R., Carr, BR., Pezzi, V. & Rainey WE. (2001). A role for src tyrosine kinase in regulating adrenal aldosterone production. *J Mol Endocrinol,* 26, pp. 207-215

Sliwinski, L., Folwarczna, J., Janiec, W., Grynkiewicz, G. & Kuzyk, K. (2005). Differential effects of genistein, estradiol and raloxifene on rat osteoclasts in vitro. *Pharmacol Rep,* Vol. 57, pp. 352–359

Somekawa, Y., Chiguchi, M., Ishibashi, T. & Aso, T. (2001). Soy intake related to menopausal synptoms, serum lipids, and bone mineral density in postmenopausal Japanese women. *Obstetrics and Gynecology,* Vol. 97, pp.109-115

Sosić-Jurjević, B., Filipović, B., Ajdzanović, V., Brkić, D., Ristić, N., Stojanoski, MM., Nestorović, N., Trifunović, S. & Sekulić, M. (2007). Subcutaneously administrated genistein and daidzein decrease serum cholesterol and increase triglyceride levels in male middle-aged rats. *Exp Biol Med,* Vol. 232, pp. 1222-1227

Sosić-Jurjević, B., Filipović, B., Ajdžanović, V., Savin, S., Nestorović, N., Milošević, V., Sekulić, M. (2010). Suppressive effects of genistein and daidzein on pituitary-thyroid axis in orchidectomized middle-aged rats. *Exp Biol Med,* Vol. 255, pp. 590-598

Sosić-Jurjević, B., Filipović, B., Milosević, V., Nestorović, N., Negić, N. & Sekulić, M. (2006). Effects of ovariectomy and chronic estradiol administration on pituitary-thyroid axis in adult rats. *Life Sci,* Vol. 79, pp. 890-897

Sosić-Jurjević, B., Filipović, B., Milošević, V., Nestorović, N., Manojlović Stojanoski, M. & Sekulić, M. (2005) Chronic estradiol exposure modulates thyroid structure and decreases T4 and T3 serum levels in middle-aged female rats. *Horm Res,* Vol. 63, pp. 48-54

Soung, DY., Devareddy, L., Khalil, DA., Hooshmand, S., Patade, A., Lucas, EA. & Arjmandi, BH. (2006). Soy affects trabecular microarchitecture and favorably alters select bone-specific gene expressions in a male rat model of osteoporosis. *Calcif Tissue Int,* Vol. 78, pp. 385–391

Squadrito, F., Altavilla, D., Morabito, N., Crisafulli, A., D'Anna, R., Corrado, F., Ruggeri, P., Campo, GM., Calapai, G., Caputi, AP. & Squadrito, G. (2002). The effect of the phytoestrogen genistein on plasma nitric oxide concentrations, endothelin-1 levels and endothelium dependent vasodilation in postmenopausal women. *Atherosclerosis,* Vol. 163, pp. 339-347

Stahl, S., Chun, TY. & Gray, WG. (1998). Phytoestrogens act as estrogen agonists in an estrogen-responsive pituitary cell line. *Toxicol Appl Pharmacol,* Vol. 152, pp. 41-48

Sugimoto, E. & Yamaguchi, M. (2000) a. Stimulatory effect of Daidzein in osteoblastic MC3T3-E1 cells. *Biochem Pharmacol,* Vol. 59, pp. 471-475

Sugimoto, E. & Yamaguchi, M. (2000). Anabolic effect of genistein in osteoblastic MC3T3-E1 cells. *Int J Mol Med,* Vol. 5, pp. 515-520

Suzuki, T., Sasano, H., Tadeyama, J., Kaneko, C., Freije, WA., Carr, BR. & Rainey, WE. (2000). Developmental changes of steroidogenic enzymes in human postnatal adrenal cortex: immunohistochemical studies. *Clinical Endocrinology,* 53, pp. 739–747

Tan, KA., Walker, M., Morris, K., Greig, I., Mason, JI. & Sharpe, RM. (2006). Infant feeding with soy formula milk: effects on puberty progression, reproductive function and testicular cell numbers in marmoset monkeys in adulthood. *Hum Reprod,* Vol. 21, pp. 896-904

Tanaka, M., Fujimoto, K., Chihara, Y., Torimoto, K., Yoneda, T., Tanaka, N., Hirayama, A., Miyanaga, N., Akaza, H. & Hirao, Y. (2009). Isoflavone supplements stimulated the production of serum equol and decreased the serum dihydrotestosterone levels in healthy male volunteers. *Prostate Cancer Prostatic Dis,* Vol. 12, pp. 247–252

Taylor, CK., Levy, RM., Elliott, JC. & Burnett, BP. (2009). The effect of genistein aglycone on cancer and cancer risk: a review of in vitro, preclinical, and clinical studies. *Nutr Rev,* Vol. 67, pp. 398-415

Teixeira, SR., Potter, SM., Weigel, R., Hannum, S., Erdman, JW Jr. & Hasler, CM. (2000). Effects of feeding 4 levels of soy protein for 3 and 6 wk on blood lipids and apolipoproteins in moderately hypercholesterolemic men. *Am J Clin Nutr,* Vol. 71, pp. 1077-1084

Tikkanen, MJ., Wähälä, K., Ojala, S., Vihma, V. & Adlercreutz, H. (1998). Effect of soybean phytoestrogen intake on low density lipoprotein oxidation resistance. *Proc Natl Acad Sci U S A,* Vol. 95, pp. 3106-3110

Turner, AS. (2001). Animal models of osteoporosis--necessity and limitations. *Eur Cell Mater,* Vol. 1, pp. 66-81

Van Wyk, JJ., Arnold, MB., Wynn, J. & Pepper, F. (1959). The effects of a soybean product on thyroid function in humans. *Pediatrics,* Vol. 24, pp. 752-760

Vanderschueren, D., Van Herck, E., Suiker, AMH., Visser, WJ., Schot, LPC. & Bouillon, R. (1992). Bone and mineral metabolism in aged male rats: short- and long-term effects of androgen deficiency. *Endocrinology,* Vol. 130, pp. 2906–2916

Vaya, J. & Tamir, S. (2004). The relation between the chemical structure of flavonoids and their estrogen-like activities. *Curr Med Chem,* Vol. 11, pp. 1333-1343

Vidal, S., Cameselle-Teijeiro, J., Horvath, E., Kovacs, K. & Bartke, A. (2001) Effects of protracted estrogen administration on the thyroid of Ames dwarf mice. *Cell Tissue Res,* Vol. 304, pp. 51-58

Villa, P., Costantini. B., Suriano. R., Perri, C., Macrì, F., Ricciardi, L., Panunzi, S. & Lanzone, A. (2009). The differential effect of the phytoestrogen genistein on cardiovascular risk factors in postmenopausal women: relationship with the metabolic status. *J Clin Endocrinol Metab,* Vol. 94, pp. 552-558

Voutilainen, R., Kahri, AI. & Salmenpera, M. (1979) The effects of progesterone, pregnenolone, estriol, ACTH and hCG on steroid secretion of cultured human fetal adrenals. *J Steroid Biochem,* Vol. 10, pp. 695–700

Wang, TT., Sathyamoorthy, N. & Phang, JM. (1996). Molecular effects of genistein on estrogen receptor mediated pathways. *Carcinogenesis,* Vol. 17, pp. 271-275

Warri, A., Saarinen, NM., Makela, S. & Hilakivi-Clarke, L. (2008). The role of early life genistein exposures in modifying breast cancer risk. *Br J Cancer,* Vol. 98, pp. 1485-1493

Weber, KS., Setchell, KDR., Stocco, DM. & Lephard, ED. (2001). Dietary soydphytoestrogens decrease testosterone levels and prostate weight without altering LH, prostate 5a-reductase or testicular steroidogenic acute regulatory peptide levels in adult male Sprague-Downey rats. *J Endocrinol,* Vol. 170, pp. 591–599

Wenzel, U., Fuchs, D. & Daniel, H. (2008). Protective effects of soy-isoflavones in cardiovascular disease. Identification of molecular targets. *Hamostaseologie,* Vol. 28, pp. 85-88

Wilkinson, AP., Gee, JM., Dupont, MS., Needs, PW., Mellon, FA., Williamson, G. & Johnson, IT. (2003). Hydrolysis by lactase phlorizin hydrolase is the first step in the uptake of daidzein glucosides by rat small intestine in vitro. *Xenobiotica,* Vol. 33, pp. 255-264

Wisniewski, AB., Cernetich, A., Gearhart, JP. & Klein, SL. (2005). Perinatal exposure to genistein alters reproductive development and aggressive behavior in male mice. *Physiol Behav,* Vol. 84, pp. 327–334

Wisniewski, AB., Klein, SL., Lakshmanan, Y. & Gearhart, JP. (2003). Exposure to genistein during gestation and lactation demasculinizes the reproductive system in rats. *Journal of Urology,* Vol. 169, pp. 1582–1586

Wong, CK. & Keung, WM. (1999). Bovine adrenal 3β-hydrogenase E.C.1.1.1.145)/5-ene-4ene isomerase (E>C>5.3.3.5.1); characterization and its inhibition by iosoflavones. *J Steroid Biochem Mol Biol,* Vol. 71, pp. 191–202

Wood, CE., Register, TC., Anthony, MS., Kock, ND. & Cline, JM. (2004). Breast and uterine effects of soy isoflavones and conjugated equine estrogens in postmenopausal female monkeys. *J Clin Endocrinol Metab,* Vol. 89, pp. 3462-3468

Wuttke, W., Jarry, H. & Seidlova-Wuttke, D. (2007). Isoflavones - safe food additives or dangerous drugs? *Ageing Res Rev,* Vol. 6, pp. 150–188

Wuttke, W., Jarry, H. & Seidlova-Wuttke, D. (2010). Plant-derived alternative treatments for the aging male: facts and myths. *Aging Male,* Vol. 13, pp. 75-81

Xiao, CW., Mei, J., Huang, W., Wood, C., L'abbé, MR., Gilani, GS., Cooke, GM. & Curran, IH. Dietary soy protein isolate modifies hepatic retinoic acid receptor-beta proteins and inhibits their DNA binding activity in rats. *J Nutr,* Vol. 137, pp. 1-6

Yamaguchi, M. & Gao YH. (1998). Anabolic effect of genistein and genistin on bone metabolism in the femoral-metaphyseal tissues of elderly rats: the genistein effect is enhanced by zinc. *Mol Cell Biochem,* Vol. 178, pp. 377-382

Yamaguchi, M. & Gao, YH. (1997). Anabolic effect of genistein on bone metabolism in the femoral-metaphyseal tissues of elderly rats is inhibited by the anti-estrogen tamoxifen. *Res Exp Med (Berl),* Vol. 197, pp. 101-107

Yamaguchi, M. & Sugimoto, E. (2000). Stimulatory effect of genistein and daidzein on protein synthesis in osteoblastic MC3T3-E1 cells: activation of aminoacyl-tRNA synthetase. *Mol Cell Biochem,* Vol. 214, pp. 97-102

Yamaguchi, M. (2006). Regulatory mechanism of food factors in bone metabolism and prevention of osteoporosis. *Yakugaku Zasshi,* Vol. 126, pp. 1117-1

Yamaguchi, M.. & Gao YH. (1998) a. Inhibitory effect of genistein on bone resorption in tissue culture. *Biochem Pharmacol,* Vol. 55, pp. 71-76

Yang, JY., Lee, SJ., Park, HW. & Cha, YS. (2006). Effect of genistein with carnitine administration on lipid parameters and obesity in C57Bl/6J mice fed a high-fat diet. *J Med Food,* Vol. 9, pp. 459-467

Zhan, S. & Ho, SC. (2005). Meta-analysis of the effects of soy protein containing isoflavones on the lipid profile. *Am J Clin Nutr,* Vol. 81, 397-408

Zhou, JR., Gugger, ET., Tanaka, T., Guo, Y., Blackburn, GL. & Clinton, SK. (1999). Soybean phytochemicals inhibit the growth of transplantable human prostate carcinoma and tumor angiogenesis in mice. *J Nutr,* Vol. 129, pp. 1628-1635

Zhu, FR., Wang, YG., Chen, J., Hu, YX., Han, FS. & Wang HY. (2009). Effects of phytoestrogens on testosterone production of rat Leydig cells. *Zhonghua Nan Ke Xue,* Vol. 15, pp. 207-211

Ziegler, RG., Hoover, RN., Pike, MC., Hildesheim, A., Nomura, AM., West, DW., Wu-Williams, AH., Kolonel, LN., Horn-Ross, PL., Rosenthal, JF. & Hyer, MB. (1993). Migration patterns and breast cancer risk in Asian-American women. *J Natl Cancer Inst,* Vol. 85, pp. 1819-1827

8

Soybeans (*Glycine max*) and Soybean Products in Poultry and Swine Nutrition

Leilane Rocha Barros Dourado[1,2], Leonardo Augusto Fonseca Pascoal[1,3], Nilva Kazue Sakomura[1,4], Fernando Guilherme Perazzo Costa[1,3] and Daniel Biagiotti[1,2]

[1]*Department of Animal Science,*
[2]*Federal University of Piauí,*
[3]*Federal University of Paraíba and*
[4]*São Paulo State University*
Brazil

1. Introduction

Soy is a legume and has been successfully cultivated around the world. Today, the world's top producers of soy are the United States, Brazil, Argentina, China and India. According Brazilian Association of Vegetable Oil Industries (Abiove), the Brazil is responsible for some 28 percent of the world's soybean production, with the estimate of a production of 57 million tons. The Brazil is the world's second largest producer and exporter of soybeans, soybean meal and soybean oil. The soybean complex, which gathers the productive chain of soybean, soybean meal and soybean oil, is the main item in the country's Trade Balance. Other activity that involves the use of soy products (oil) is the production of biodiesel.

In fact, so much in the Brazil as in most of the countries of the World, the soy represents one of the largest oilseeds of the world and to main source of vegetable protein for the poultry and swine feeding.

2. The nutritional composition of the soybeans and soybean products used in the feeding of poultry and swine

Soybeans and soybean products are now used widely in animal feeding. The crop is grown as a source of protein and oil for the human market and for the animal feed market. Soybean meal is generally regarded as the best of plant protein source in terms of its nutritional value. Also, it has a complementary relationship with cereal grains in meeting the amino acids (AA) requirements of farm animals. Consequently, it is the standard to which other plant protein sources are compared (Blair, 2008).

Soybeans provide an excellent source of both energy and protein for poultry and swine. As with any ingredient, their usage rate depends upon economics, although in the case of soybeans such economics relate to the relative price of soybean meal and of supplemental fats. Soybeans contain about 38% crude protein, and around 20% oil (Leeson & Summers, 2008). However, soybeans contain compounds that inhibit the activity of the proteolytic enzyme trypsin. They also contain other antinutrients, including hemagglutinins or lectins, which contribute to reduce nutrient use. Nutritional composition of the soybeans and soybean products is affecting by percentage of anti-nutritional factors (ANFs), variety genetic, efficiency of the oil-extraction process and the amount of residual hulls present, the heat processing and other factors.

2.1 Anti-nutritional factors (ANFs) and nutritional quality

The nutritional quality of soybean products for poultry and swine feeding is determined not only by the quantity of nutrients (protein, amino acids, fat and others), but mainly by nutrients availability for the animals. According Durigan (1989), anti-nutritional factor whole substance is synthesized by normal plant metabolism, which may result from different mechanisms to reduce the efficiency of utilization of the diet. The ANFs present in soybean can to affect the nutrients availability for poultry and swine.

Most of the ANFs present in the raw soybeans, as protease inhibitors and lectins is heat-labile, but others as phytic acid and polysaccharides non starch (PNS) only decrease with enzyme addition in diet, because poultry and swine has no ability to produce enzymes to degrade the PNS.

From Liener (1994), soybeans contain some heat-labile protease inhibitors and hemagglutinins. Soy also contains factors that are relatively heat-stable, though of lesser significance, such as: Goitrogens: substances that cause goiters, an enlargement of the thyroid gland; Tannins: complex plant compounds that are often bitter or astringent; Flatus-producing oligosaccharides: carbohydrates of small molecular weight that cause flatulence (gas); Phytates: which bind minerals preventing absorption; Saponins and Antivitamins.

This ANFs of soybean can cause inhibition of growth, decreased feed efficiency, goitrogenic responses, pancreatic hypertrophy, hypoglycemia, and liver damage in nonruminant animals depending on species, age, size, sex, state of health and plane of nutrition (Palacios et al., 2004)

2.1.1 Proteases inhibitors

Proteases inhibitors are substances that ability has to inhibit the activity of certain digestive enzymes (Durigan, 1989). They are polypeptides of 181 and 71 amino acid residues, respectively, which form well-characterized stable enzyme inhibitor complexes with pancreatic trypsin on a one-to-one molar ratio. The content of soya bean trypsin inhibitors varies in different varieties of soya bean and germination process (Bau et al., 1997). These are known as the Kunitz inhibitor and the Bowman-Birk inhibitor which are active against trypsin, while the latter is also active against chymotrypsin (Liener, 1994), because Bowman (1944) identified a protein in soy can inhibit trypsin and chymotrypsin and subsequently purified by Birk et al (1961), called Bowman-Birk inhibitor. It is a heat-stable protein, due to its large number of sulfur bridges. Later Kunitz (1945) identified and crystallized other protein, Kunitz trypsin inhibitor, which strongly inhibited the activity of digestive enzyme trypsin.

These protease inhibitors interfere with the digestion of proteins, resulting in decreased animal growth. Protease inhibitors stimulate protein synthesis and enzyme secretion from the pancreas. Inhibition of proteolysis, the presence of undigested protein in the intestinal tract, and a decreased release of amino acids in raw soy diets induce a compensatory reaction in the pancreas and a general stimulatory effect on other endogenous secretions

(Rackis & Gumbmann, 1981). The compensatory effect of the pancreas is effective since that the urease activity of the soy is in up to 0,20 (Butolo, 2010). The effect 8of hypertrophy of pancreas followed by a stimulation of its secretory activity can also result in an endogenous loss of the pancreatic enzymes, trypsin and chymotrypsin which are rich in the sulphur-containing amino acids, and thus accentuating the deficiency of methionine, being the first limiting amino acid in soybean (Johri, 2005)

Coca-Sinova et al. (2008) had evaluated the coefficient of apparent ileal digestibility (%) of DM, N, energy, and amino acids (AA) of the diet with different soybean meal (SBM) origin in broilers of 21 d of Age. They observed that digestibility coefficients were higher for SBM contained lower levels of TIA - trypsin inhibitor activity (1,8 mg/g), when compared with SBM with higher levels of TIA (4,8mg/g).

2.1.2 Lectins (haemagglutinins)

Lectins are glycoproteins with the ability to bind carbohydrate-containing molecules on the epithelial cells of the intestinal mucosa, with the property of agglutinating the erythrocytes of higher animals (Liener, 2000).) The cells of the intestine in the presence of lectin, tend to collapse by reducing the absorption (Butolo, 2010). According Fasina et al. (2004), when lectins are ingested by animals, they can be degraded by intestinal digestive enzymes or survive intestinal digestion and bind to enterocytes on the brush border membrane (BBM). However if bind, lectins may cause antinutritional effects such as disruption of the intestinal microvilli, shortening or blunting of villi, impairment of nutrient digestion and absorption, increased endogenous nitrogen loss, bacterial proliferation, and increased intestinal weight and size (Pusztai, 1993 cited by Fasina et al., 2004).

2.1.3 Goitrogens

The soybean and its products have been considered goitrogenic in humans and animals (Doerge et al., 2002)., because the acidic methanolic extract of soybeans contains compounds that inhibit thyroid peroxidase-(TPO) catalyzed reactions essential to thyroid hormone synthesis (Divi et al., 1997). Pigs feeding goitrogens (0.075% 1-methyl-2-mercaptoimidazole or .5% potassium thiocyanate) produced symptoms of hypothyroidism in a relative short period of time, usually 3 to 4 weeks with pronounced growth depression, but when the goitrogens were withdrawn from the diet, there was a marked increase in growth rate. The effects of goitrogens are more common in humans, mainly infants (Shepard et al., 1960), than in pigs, because the soybeans used in feed for pigs is generally thermally processed. the goitrogens of soybean can removed by heat treatment (Liener, 1970; Zhenyu et al., 2000).

2.1.4 Tannins

Tannins are complex plant compounds that are often bitter or astringent, they are naturally-occurring plant polyphenols which combine with proteins and other polymers such as cellulose, hemicellulose and pectin, to form stable complexes (Mangan, 1988). Egounlety et al. (2003) observed that the hulls were much richer in tannins than the whole soybean (2. 31 x 1.52 mg catechin equivalent/g). Soaking soybean for 12-14 h reduced the tannin content by 54.6%. No tannin was detected in dehulled and cooked and in fermented Soybean.

In contrast to the position with ruminant animals where tannins in the diet may have considerable benefits, and in plants where tannins give partial protection against predators, in simple-stomached animals, including man, tannins in the diet are generally undesirable, because they present effects as decrease of protein digestibility and reduction of the animal growth(Mangan, 1988).

2.1.5 Saponins

Saponins are steroid or triterpenoid glycosides, common in a large number of plants and plant products that are important in human and animal nutrition (Francis et al., 2002). Saponins have long been known to cause lysis of the erthrocytes when given in vitro. The hemolytic activity of saponins, coupled with this cholesterol inhibition effect, has been extensively used as a means of detecting and quantifying saponins in plant material. Such saponins activity results from their affinity for membrane sterols (Yoshiki et al., 1998).
Most listings of soybean antinutritional factors in the past included saponins, although with little or no justification. Toxicity was attributed to them simply by analogy with saponins from other sources that, in deed, are toxic (Anderson et al., 1995). Soyabean saponins did not impair growth of chicks when added at five times the concentration in a normal soyabean-supplemented diet (Ishaaya et al., 1969 as cited in Francis et al., 2002). Acording Yoshiki et al. (1998) no hemolytic activity in soybean saponin was observed, while 80% methanolic soybean extract had strong activity. As a result of detailed studies, the hemolytic compounds in soybean are indentified as linoleic acid and lipoxin, wich are secondary metabolitic products from lipoxigenase.

2.1.6 Antivitamins

There exists a fairly large category of natural substances that interfere with the utilization of certain minerals and vitamins. As examples, isolated soya-bean protein has been shown to interfere with the availability of such minerals as zinc, manganese, copper and iron as well as vitamin D (Liner, 1970) or other diverse but ill-defined factors appear to increase the requirements for vitamins A, B12, D, and E (Liner, 1994).
Raw soybean contains an enzyme lypoxygenase which catalyses oxidation of carotene, the precursor of vitamin A and can be destroyed by heating soybeans for 15 min at atmospheric pressure. Autoclaving of soybean protein or supplementation with vitamin D3 for about 8-10 times can eliminate the rachitogenic activity (Johri, 2005). Fisher et al. (1969) reported anti-vitamin E activity of isolated soy protein for the chick.

2.1.7 Olygosaccharides and polysaccharides non starch

Soybean carbohydrates make up approximately 35% of soybean (SB) seed and 40% of soybean meal (SBM) dry matter (DM). Approximately half of these carbohydrates are nonstructural in nature, including low molecular weight sugars, oligosaccharides, and small amounts of starch, while the other half are structural polysaccharides, including a large amount of pectic polysaccharides (Karr-Lilienthal et al., 2005). The fibre component of the grain consists primarily of nonstarchpolysaccharides (NSP) which in cereals form part of the cell wall structure. In legumes, NSP also play a role as an energy storage material. The role of fibre in monogastric diets has attracted much attention in recent years, due to the facts that (a) the soluble NSP elicit anti-nutritive effects (Choct, 1997). The components of non-digestible carbohydrates of a feedstuff are NSP, consisting of water insoluble cellulose and water soluble gums, hemicelluloses, pectic substances and mucilages (Mekbungwan, 2007) .
Polysaccharides are plymers of monosaccharides joined through glycosidic linkages and are defined and classified in terms of the following structural (Choct, 1997). Choct et al. (1995)

showed that the addition of 40 g/kg NSP to a commercial broiler diet decreased the weight gain, feed efficiency and apparent metabolizable energy (AME) by 28.6, 27.0 and 21.2%, respectively.

According Smits & Annison (1996) the physicochemical properties of non-starch polysaccharides (NSPs) are responsible for their antinutritive activities in the poultry and swine. In particular, soluble viscous NSPs depress the digestibilities of protein, starch and fat. It is suggested that the gut microflora can mediate the antinutritive effects of soluble and viscous NSP. On the other hand, insoluble and non-viscous NSPs may have a beneficial effect. Hetland et al. (2004) reported that digestibility of starch is higher and digesta passage rate faster when a moderate level of insoluble fibre is present in the diet. The effect of insoluble fibre on gut functions stems from its ability to accumulate in the gizzard, which seems to regulate digesta passage rate and nutrient digestion in the intestine. NSP content of soybean meal is approximately 61 and 103 g kg-1 (dry matter basis) for soluble NSP and insoluble NSP, respectively (Bach Knudsen, 1997).

Choct et al. (2010) reported in review that non-digestible oligosaccharides can be fermented throughout all sections of the gastrointestinal tract including the large intestine, and the effects are most variation depends of specie. Soy oligosaccharides increased microbial activities as indicated by the increased volatile fat acids (VFA) contents and can cause intestinal disorder.

The α-galactoside family of oligosaccharides cause a reduction in metabolizable energy with reduced fiber digestion and quicker digesta transit time. Birds do not have an α-1:6 galactosidase enzyme in the intestinal mucosa (Leeson & Summers, 2008). Enzyme addition in diet for poultry and swine is most utilized for improve the nutritional value of soybean products. Zanella et al. (1999) found better body weight gain for broilers fed soybean meal (45% CP), extruded soybeans (38% CP) and roasted soybeans with addition of blend enzyme (xylanase, protease and amylase) ,when compared to without enzyme..

2.1.8 Phytate

Three terminologies, namely phytate, phytin and phytic acid, are used in the literature to describe the substrate for phytase enzymes. The most commonly used term, phytate, refers to the mixed salt of phytic acid (myo-inositol hexaphosphate; IP6). The term, phytin, specifically refers to the deposited complex of IP6 with potassium, magnesium and calcium as it occurs in plants, whereas phytic acid is the free form of IP6 (Selle & Ravindran, 2007). Historically, phytates have been considered solely as antinutrients because they are known as strong chelators of divalent minerals such as Ca^{2+}, Mg^{2+}, Zn^{2+} and Fe^{2+}. Moreover, phytates are also capable of binding with starch and proteins while preventing their assimilation through the digestive system (Afinah et al., 2010). The soybean meal present 0.34% of phytate the soybean has 0.34% of phytate which represents approximately 60% of the amount of total phosphorus (Kornegay, 2000). To improve the utilization of phosphorus and other nutrients complexed to phytate has been used in the enzyme phytase in diets for poultry and pigs. In Brazil more than 50% of poultry diets were formulated using phytase.

2.2 Soybean processing

Soybean processing is useful and necessary to destroy or remove undesirable constituents, make nutrients more accessible or to improve palatability. However, processing toward these ends also leads to changes in the composition of the various soybean materials compared with whole soybeans. These changes may be intentional, as in the case of heating to diminish trypsin inhibitor activity (Anderson et al.,1995). Although exist other options for reducing the antinutritional effects of soybean products, the cheaper and more efficient is heat-processing (Pusztai et al., 1997).

The most common procedures have involved a combination or extraction, cooking and fermentation. With soybeans, moist-heat treatment is particularly effective in reducing trypsin inhibitor activity below biological threshold levels, as determined by short-term animal bioassay. With present day manufacturing processes, residual trypsin inhibitor activity in edible-grade soy protein products is about 5-20% of the activity originally present in raw soybeans (Rackis & Gumbmann, 1981).

The heat necessary to destroy trypsin inhibitors and other hemagglutinins found in raw soybeans is dependent upon exposure time, and so high temperatures for a shorter time period are as effective as lower temperatures for longer times (Leeson & Summers, 2008).

In the 1930's, soybeans were mechanically processed using hydraulic or screw presses, which squeezed out the oil of the heated or cooked soybeans. In the late 1940's and early 1950's, most of the industry converted to the solvent-extraction process, which removes more oil from the soybean. Today, more than 99 percent of the U.S. processing capacity is using the solvent extraction process produced in large crushing facilities that produce meals of consistent high quality (Johnson & Smith, sd).

According Soybean Meal INFOcenter, the soybean products appear with the initial processes that included cleaned and dehulled and other three processes is used to separate the soybean oil from the protein meal. The first is solvent extraction, which is the one used most commonly around the world, uses hexane to leach or wash (extract) the oil from flaked oilseeds. This method reduces the level of oil in the extracted flakes to one percent or less. After this continuous pressing is performed at elevated temperatures, using a screw press to express the oil from ground and properly conditioned soybeans. The pressed cake is reduced to between 4 percent and 6 percent oil content by this method. At the end, dydraulic or batch pressing, this is an intermittent pressing operation carried out at elevated temperatures in a mechanical or hydraulic press after the soybeans have been rolled into flakes and properly conditioned by heat treatment. It is the oldest known method of processing oilseeds. According Butolo (2010), the industrialization flow for obtaining soybean oil, soybean meal or full fat soybean is divided into four distinct phases, which are shown in Figure 1.

A great processing variety exists, in table 1 sowed the composition of soybean feed ingredient products.

The quality of soya meal is the result of many factors, including bean variety, origin and storage. The various processing steps employed from the time the bean is received can affect the quality of the resulting meal and oil obtained. Heat treatment of the meal is essential to optimize its protein quality. The variables of moisture, temperature and time are interrelated and are important to achieve proper cooking conditions. The magnitude of these variables must be determined for each plant, preferably using a biological assay for evaluation. Many in vitro tests designed to measure protein quality in soya have been proposed and evaluated (Wright, 1968). Simple crude protein or amino acid assays provide information on the protein, but do not provide useful information on the quality of the protein. Chemists have used trypsin inhibitor analyses, urease activity, protein solubility in potassium hydroxide, protein solubility in water and dye binding methods to assay for protein quality (Johnson & Smith, nd).

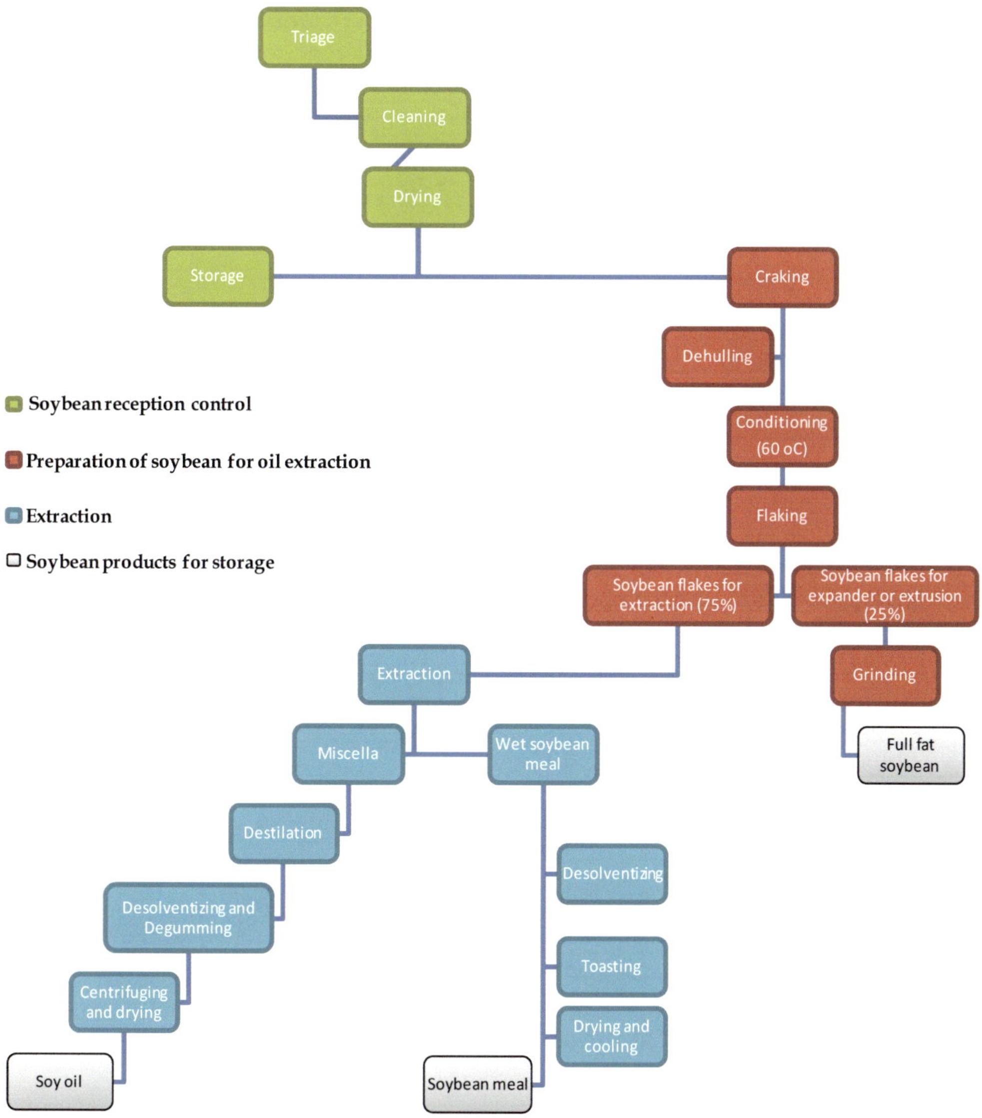

Fig. 1. The simple flowchart of soybean processing (Adapted of Butolo, 2010).

Inadequate heating fails to completely destroy the ANFs, which may have a detrimental impact on animal performance, while excessive heating reduces the availability of lysine via the Maillard reaction and possibly, to a lesser extent, of other amino acids (Caprita et al. 2010). While processed beans should be periodically tested for trypsin inhibitor or urease levels, a simple on-going test is to taste the beans. Under-heated beans have a characteristic 'nutty' taste, while over-heated beans have a much darker color and a burnt taste. The problem with overheating is potential destruction of lysine and other heat-sensitive amino acids (Leeson & Summers, 2008).

Nutrient profile	Full-fat extruded soybean	Soybean meal (48%)	Soybean protein concentrate	Soybean Hulls	Soy oil
Dry matter (%)	90.47	88.21	89.88	88.80	99.60
Gross energy (Kcal kg^{-1})	4938	4164	4495	3854	9333
Starch (%)	6.70	3.00	-	-	-
Fat (%)	17.64	1.40	0.43	2.86	99.60
Crude fibre (%)	6.24	4.27	2.64	33.00	-
Crude protein(%)	37.00	47.90	62.92	13.50	-
Arginine(%)	2.71	3.50	5.32	0.81	-
Lysine (%)	2.23	2.92	4.07	0.89	-
Methionine + cystine (%)	1.08	1.37	1.90	0.39	-
Threonine (%)	1.47	1.86	2.60	0.51	-
Triptophan (%)	0.47	0.64	0.87	0.14	-
Calcium (%)	0.23	0.31	0.27	0.49	-
Av. Phosphorus(%)	0.17	0.21	0.27	0.05	-
Potassium(%)	1.67	2.11	2.18	-	-
Poultry					
Metabolizable energy (Kcal kg^{-1})	3429	2302	2677	871	8790
Dig. Arginine(%)	2.54	3.31	5.12	0.64	-
Dig. Lysine (%)	2.02	2.70	3.75	0.59	-
Dig. Methionine + cystine (%)	0.93	1.21	1.67	0.21	-
Dig. Threonine (%)	1.29	1.66	2.29	0.25	-
Dig. Triptophan (%)	0.43	0.58	0.79	0.09	-
Swine					
Digestible energy (Kcal kg^{-1})	4250	3540	4035	2370	8600
Dig. Arginine(%)	2.52	3.31		0.68	-
Dig. Lysine (%)	1.99	2.66		0.53	-
Dig. Methionine + cystine (%)	0.93	1.23		0.26	-
Dig. Threonine (%)	1.28	1.62		0.31	-
Dig. Triptophan (%)	0.40	0.57		0.09	-

Table 1. Nutrient profile of soybean products for poultry and swine nutrition (From Rostagno et al., 2005).

The most widely adopted method more economical and faster is the measurement of urease activity (urease test or urease index), however difficult the correlation of the protein solubility in KOH. Levels of the enzyme urease are used as an indicator of trypsin inhibitor activity. Urease is much easier and cheap to measure than is trypsin inhibitor and both molecules show similar characteristics of heat sensitivity. Two analytical methods for urease test were described, the first is based on ph difference in which 200 mg of sample (soybean or soybean product) is incubated in 10.0 ml of phosphate buffered urea solution at 300C for 30 minutes, after which the increase in pH units (ΔpH) from pH 7.00 is recorded (Palić et al. 2008; Butolo, 2010). The second is based a simple colorimetric assay in which urea-phenol-red solution is brought to an amber color by using either 0.1 N HCl or 0.1 N NaOH. About 25 g of soybean meal is then added to 50 ml of indicator in a petri dish. After 5 minutes, the sample is viewed for the presence of red particles. If there are no red particles showing, the mixture should stand another 30 minutes, and again if no red color is seen, it suggests overheating of the meal. If up to 25% of the surface is covered in red particles, it is an indication of acceptable urease activity, while 25 - 50% coverage suggest need for more detailed analysis. Over 50% incidence of red colored particles suggests an under-heated meal (Leeson & Summers, 2008). But urease index is not useful to determine excessive heat treatment since additional heating has no effect on the urease index (Caprita et al. 2010).

The KOH protein solubility test is based on the solubility of soybean proteins in a dilute solution of potassium hydroxide. The procedure involves the incubation of a sample with a 0.2% KOH solution for 20 min at room temperature. Following this incubation, the sample is centrifuged and the supernatant is analyzed for the protein concentration. The solubility of the protein, expressed as a percentage, was calculatedby dividing the protein content of the KOH extracted solution by the protein content of the original soybean sample (Caprita et al. 2010). KOH protein solubility is a better indicator of overprocessing than underprocessing of soybeans (Batal et al., 2000). In table 2 are shown the levels of urease activity and protein solubility in potassium hydroxide acceptable in most soybeans processing.

Protein Dispersibility Index (PDI) can be used to measure protein quality. According Butolo (2010) to determine the protein dispersibility index should be mixed 8g of soybean meal with 150ml water, then it is centrifuged to 8.500ppm for 10 minutes, filter and determining the soluble nitrogen by the Kjeldahl method. Batal et al. (2000) indicated that PDI demonstrates more consistent response to heating of soyflakes than did urease index or protein solubility in KOH, because the urease index is not linear and that it rapidly falls from approximately 2.0 units of pH change to near zero as SBM is heated contributes to the difficulty in determining a precise maximum acceptable.

Degree of soybean processing	Urease test (pH change)	KOH Protein solubility (%)
Under-processed	> 0.20	90
Normal	-	85
Adequately processed	0.05 - 0.20	77-80
Over-processed	-	< 77

Table 2. Globally accepted relation between the degree of soybean processing for urease activity and protein solubility in potassium hydroxide (from Palić et al. 2008; Butolo, 2010).

Nitrogen Solubility Index (NSI) is other methods can be used for determines protein quality. It uses a slow stirring technique. Nitrogen is extracted from the ground flour by placing approximately 1.5 g into a 200 ml beaker and adding 75 ml of 0.5% KOH. The sample is stirred 20 minutes at 120 rpm. PDI and NSI are a more consistent and sensitive indicator for monitoring both underheating and over-heating of SBM (Caprita et al. 2010).

It is very important whether the assessment process quality of soybeans for the animal is unlikely to have decreased performance.

2.3 Storage and genetic variety

The storage and genetic variation are factors that can alter the nutritional composition of soybean as well as the performance of monogastric animals. Narayan et al., (1988) found that chemical characteristics, moisture content, fat, water-soluble nitrogen (WSN), nitrogen solubility index (NSI), sugars, trypsin inhibitor activity, available lysine, pigment and

lipoxygenase activity of seeds decreased during storage whereas non-protein nitrogen (NPN), extent of browning, free fatty acid (FFA) content and peroxide value are increased. According Cromwell et al. (2002) the rate and efficiency of weight gain, scanned backfat and longissimus area, and calculated carcass lean percentage were not different ($P > 0.05$) for pigs fed diets containing conventional or genetically modified, herbicide (glyphosate)-tolerant soybean. For poultry, Taylor et al. (2007) concluded that the diets containing soybean meal produced from genetically modified (GM) glyphosate-tolerant were nutritionally equivalent to diets containing soybean meal produced from the control and conventional reference soybean varieties when fed to broilers.
Other genetically modified soybeans were study for Palacios et al. (2004) that compared the growth performance of chicks and pigs fed diets containing modified soybeans: Kunitz trypsin inhibitor-free (KF), lectin-free (LF), lectin and Kunitz trypsin inhibitor-free (LFKF), conventional soybeans (CSB), and commercially obtained, dehulled, solvent-extracted soybean meal (SBM). They verify that Chicks fed diets containing any of the raw soybean varieties gained less weight than did chicks and among the raw soybean treatments, there was a greater effect on growth performance by removing both lectins and Kunitz trypsin inhibitor (LFKF), than by removing each antinutritional factor separately. Feeding raw soybeans to chicks decreased average daily gain (ADG) by 49% for CSB, 37% for KF, 38% for LF, and 27% for LFKF compared with the ADG achieved by chicks fed SBM. For to pigs deprecreased ADG by 78% for CSB, 60% for LF, and 35% for LFKF compared with the ADG achieved by pigs fed the same variety but extruded. This results and others (Brune et al., 2010; Becker-Ritt et al., 2004; Vasconcelos et al., 1997) prove that variability in the amounts of these components (proteases inhibitors and lectins) can be affect by cultivars differences.
As observed, there seems great potential for reduction in content of anti-nutrients within GM soybeans, as studies have shown that the isogenic variant lacking the Kunitz trypsin inhibitor and other soybean variants low in Kunitz trypsin inhibitor are nutritionally superior to conventional raw soybeans but not as good as commercial soybean meal. Others genetic improvements in reducing the phytate-bound phosphorous, and reduction or elimination of oligosaccharide carbohydrates are the most important economical traits that are being researched

3. Soybean products for poultry and swine nutrition

3.1 Soybean meal and full fat soybean

Soybean meal is the most popular source of supplemental protein in livestock feeds (Table 3). That popularity derives from its nutrient content, its relative freedom from intractable antinutritional factors, and other issues (Pettigrew et al.2002).
Many studies are conducted comparing the inclusion of soybean meal with other soy products or other protein source, and where the animals fed soybean meal has a better perform in most cases. In general, full-fat soybeans may replace soybean meal in swine and poultry diets with similar performance anticipated. The decision on which soybean product to use needs to be based on the product's composition, availability and unit costs. Bertol et al. (2001) observed that substitution of 50% of soybean meal by full-fat extruded soybeans, texturized soybean protein and concentrated soybean protein in the weaning diet, promoted better performance, with additional 1 to 2 kg of body weight gains per piglet at the end of the nursery phase.

Protein Source	Million Metric Tons	Percent
Soybean meal	152.1	66.5
Rapeseed meal	30.7	13.5
Cottonseed meal	14.4	6.3
Sunflower meal	12.2	5.3
Palm kernel meal	6.1	2.7
Peanut meal	5.9	2.6
Fish meal	5.3	2.3
Copra meal	1.9	0.8
TOTAL	228.6	

* Soy Stats (2010)

Table 3. World Protein Meal Consumption*.

Micronizing is the name given to a cooking process that uses infrared rays to cook cereals and pulses at lower temperatures and for shorter times than other heating methods. Gas burners are used to generate the infrared rays that are absorbed by the products. The raw materials are passed under the burners on variable speed belts to achieve the desired level of "cook". The product is then passed through a roller mill to create flakes. These flakes can be used whole or ground into a meal (MMfeeds). The increase available energy and improve digestibility are both achieve due to the gelatinisation of starch molecules during the cooking process. Trindade Neto et al. (2002) observed that pigs fed micronized soybean takes more days to reach 50 and 90 kg of body weight when compared with those fed soybean meal.

3.2 Soybean hulls

Soybean hulls, due to their high fiber contents, are known to be poorly digested by non-ruminant animals. Recent studies, however, suggest that the hulls have potential as an alternative feed ingredient for swine and poultry. The soybean hulls can be included up to 10 and 12% for growing or finishing pig diets, respectively, replacing the wheat bran on a weight basis without any adverse effects on palatability of diets and animal performances (Chee et al., 2005). However, Moreira et al. (2009) not recommend the use of soybean hulls to piglets due reducing daily feed intake and daily weight gain for the animals fed feed containing soybean hull (15% inclusion in the diet) compared to the control feed without soybean hull.

Esonu et al. (2005) studing laying hens, found that inclusion of up to 20% soybean hulls, improves the Feed cost/dozen eggs, and when cellulolytic enzyme supplementation at 30% dietary level of soybean hull meal in layer diet could not significantly affect the performance of laying hens.

Currently, it is very common the use of soybean hulls in programmers of feed restriction and welfare of breeders and laying hens.

3.3 Soy protein isolates

Soybean protein concentrate (or soy protein concentrate) is the product obtained by removing most of the oil and water-soluble non-protein constituents from selected, sound, cleaned, dehulled soybeans. The traded product among 650 to 900 g/kg CP on a moisture-free basis. Soybean protein isolate (or soy protein isolate) is the dried product obtained by removing most of the non-protein constituents from selected, sound, cleaned, dehulled soybeans. Both soy protein concentrate and isolate have the potential to be used in poultry diets as a source of protein and AA (Blair, 2008). In Table 4 are showed broilers performance when feed with different soybean products in diets.

Protein source	Weight gain (g)
Casein	364 b
Soybean meal	405 a
Soy protein concentrate	356 b
Soy protein isolate	366 b

Batal & Parson (2003)

Table 4. Effect of protein sources on weight gain of chicks(week 0-3).

3.4 Soy oil

Soy oil has found many food uses due to its excellent nutritional qualities, widespread availability, economic value and wide-use functionality. Soy oil is a highly concentrated source of feed energy. Its caloric value is the major reason for its increased use.

Gaiotto et al. (2000) evaluated performance of broilers fed diets containing 4% supplemental fat from the sources: soybean oil (SOY4), beef tallow (TAL4), acidulated soapstock (SOAP4), mixtures 2%:2% (SOAP2/TAL2), (SOAP2/SOY2) and (SOY2/TAL2), and confirmed the superiority of soybean oil relative to the other fat sources fed to broiler and demonstrated that the quality of acidulated soapstock and beef tallow may be improved when used in 1:1 mixtures with soybean oil.

For laying hens Costa et al. (2008) evaluated soy and canola oil, and they observed better results for those characteristics were obtained as soybean oil increased. However, the egg mass conversion was negatively influenced by increase of canola oil. The addition of soybean oil promoted better performance as compared to canola oil.

For swine, Mascarenhas et al. (2002) evaluated the effects from two lipid sources (soybean oil and coconut oil) on performance from 60 to 100 kg boars and they observed that diets with coconut oil as lipidic source showed the best results of weight gain.

4. Conclusion

The benefits of the use of soybean and soybean products can be observed in the nutrition of poultry and swine, but it is very important to know the factors that affect the composition of the same ingredients for that may be included in adequate amounts without reducing animal performance.

5. References

ABIOVE. (Dez 2010). Brazilian Association of Vegetable Oil Industries. In : *About ABIOVE*. 17.12.2010. Available from : http://www.abiove.com.br/english/menu_us.html, Access in: march 2011.

Afinah, S. ; Yazid, A. M. ; Anis Shobirin, M. H. & Shuhaimi, M. (2010). Phytase: application in food industry. *International Food Research Journal*, vol. 17, pp. 13-21 , ISSN: 1985-4668

Anderson, R.L. & Wolf, W.J. (1995). Compositional changes in trypsin inhibitors, phytic acid, saponins and isoflavones related to soybean processing. *Journal of Nutrition*, vol. 125, pp.581-588, ISSN 1541-6100

Bach Knudsen, K.E. (1997). Carbohydrate and lignin contents of plant materials used in animal feeding. *Animal Feed Science and Technology*, vol. 67, pp. 319-338, ISSN: 0377-8401

Batal, A. B. & Parsons, C. M. (2003). Utilization of Different Soy Products as Affected by Age in Chicks. *Poultry Science*, vol. 82, pp.454-462, ISSN 0032-5791

Bau, H.; Villaume, C. ; Nicolas, J.P. & Jean, L.M. (1997). Effect of Germination on Chemical Composition, Biochemical Constituents and Antinutritional Factors of Soya Bean (Glycine max) Seeds. Journal of Science and Food Agriculture, vol. 73, pp. 1-9, ISSN : 1097-0010

Becker-Ritt, A. B. ; Mulinari, F. ; Vasconcelos, I. V. & Carlini, C. R. (2004). Antinutritional and/or toxic factors in soybean (Glycine max (L) Merril) seeds: comparison of different cultivars adapted to the southern region of Brazil. *Journal of the Science of Food and Agriculture*, vol. 84, pp.263-270, ISSN : 1097-0010

Bertol, T. M. ; Mores, N. ; Ludke, J. V. & Franke, M. R. (2001). Proteínas da Soja Processadas de Diferentes Modos em Dietas para Desmame de Leitões. *Brazilian Journal of Animal Science*, vol.30, no.1, pp.150-157, ISSN 1806-9290

Birk, Y. & Gertheler, A. (1961).Effects of mild chemical and enzimatic treatments of soybean meal and soybean trypsin inhibitors on their nutritive and biochemical properties. *Journal of Nutrition*, vol.75, pp. 379-387, ISSN 1541-6100

Blair, R. (2008). *Nutrition And Feeding Of Organic Poultry*, CAB International, ISBN 978-1-84593-406-4, Oxfordshire, England.

Bowman, D.W. (1944). Fractions derived from soybeans and navybeans whice retard tryptic digestion of casein. Proc. Soc. Exp. Biol. Med., New York, v.57, pp.139-140,

Brune, M. F. S. S. ; Pinto, M. O. ; Peluzio, M. C. G. ; Moreira,M. A. & Barros, E. G. (2010). Avaliação bioquímico-nutricional de uma linhagem de soja livre do inibidor de tripsina Kunitz e de lectinas. *Ciência e Tecnologia de Alimentos*, vol. 30, No. 3, pp. 657-663, ISSN : 0101-2061

Butolo, J. E. (2010). *Qualidade de ingredientes na alimentação animal*. CBNA, ISBN 85-902473-1-7, Campinas-São Paulo- Brazil

Căpriţă, R. ; Căpriţă, A. & Creţescu, I. (2010). Protein Solubility as Quality Index for Processed Soybean. *Animal Science and Biotechnologies*, vol. 43, pp.375-378.

Chee, K.M.; Chun, K. S.; Huh, B. D.; Choi, J. H. ; Chung, M. K.; Lee, H. S.; Shin, I. S. & Whang, K. Y. (2005). Comparative feeding values of soybean hulls and wheat bran for growing and finishing swine. *Asian-Australian Journal Animal Science*, vol. 18, No. 6, pp. 861-867, ISSN 1011-2367

Choct, M.; Hughes, R.J.; Wang, J.; Bedford, M.R. ; Morgan, A.J. & Annison, G. (1995). Feed enzymes eliminate antinutritive effect of non-starch polysaccharides and modify fermentationin broilers. Proceeding of Australian Poultry Science Symposium 7, pp. 121-125, ISSN 1034-6260, Sydney : University of Sydney.

Choct, M.; Dersjant-Li, Y. ; McLeish, J. & Peisker, M. (2010). Soy oligosaccharides and soluble non-starch polysaccharides: a review of digestion, nutritive and anti-nutritive effects in pigs and poultry. *Asian-Australian Journal Animal Science* , Vol. 23, No. 10, pp. 1386 - 1398, ISSN 1011-2367

Coca-Sinova, A.; Valencia, D. G. ; Jiménez-Moreno, E. ; Lázaro, R. & Mateos. G. G. (2008). Apparent Ileal Digestibility of Energy, Nitrogen, and Amino Acids of Soybean Meals of Different Origin in Broilers. *Poultry Science*, vol. 87, pp.2613–2623, ISSN : 0032-5791

Costa, F. G. P. ; Souza, C. J. ; Goulart, C. C. ; Lima Neto, R. C. ; Costa, J. S. & Pereira, W. E. (2008). Desempenho e qualidade dos ovos de poedeiras semipesadas alimentadas com dietas contendo óleos de soja e canola. *Brazilian Journal of Animal Science*, vol.37, no.8, pp.1412-1418, ISSN 1806-9290

Cromwell, G. L. ; Lindemann, M. D. ; Randolph, J. H. ; Parker, G. R. ; Coffey, R. D. ; Laurent, K. M. ; Armstrong, C. L. ; Mikel, W. B. ; Stanisiewski, E. P. & Hartnell, G. F. (2002). Soybean meal from Roundup Ready or conventional soybeans in diets for growing-finishing swine. Journal Animal Science, vol.80, pp.708–715, ISSN 1525-3163

Doerge, D. R. & Sheehan, D. M. (2002). Goitrogenic and Estrogenic Activity of Soy Isoflavones. *Environmental Health Perspectives*, Vol. 110, No. 3, pp. 349-353, ISSN : 0091-6765

Durigan, J.F. (1989). Fatores antinutricionais em alimentos. Anais do Simpósio Interfase Nutrição Agricultura, p.155-225, Piracicaba, São Paulo, Brazil.

Egounlety, M. & Aworh, O.C. (2003). Effect of soaking, dehulling, cooking and fermentation with Rhizopus oligosporus on the oligosaccharides, trypsin inhibitor, phytic acid and tannins of soybean (Glycine max Merr.), cowpea (Vigna unguiculata L. Walp) and groundbean (Macrotyloma geocarpa Harms). *Journal of Food Engineering*, vol. 56, pp. 249–254, ISSN : 0260-8774

Esonu, B.O. ; Izukanne, R.O. & Inyang, O.A. (2005). Evaluation of Cellulolytic Enzyme Supplementation on Production Indices and Nutrient Utilization of Laying Hens Fed Soybean Hull Based Diets. International. *Journal of Poultry Science*, vol. 4, no. 4, pp. 213-216, ISSN 1682-8356

Fisher,H. ; Griminger, P. & Budowski, P. (1969). Anti-vitamin E activity of isolated soy protein for the chick. Zeitschrift für Ernährungswissenschaft (European Journal of Nutrition), Vol.9, No. 4. pp. 271-278 , ISSN : 1436-6215

Francis, G. ; Kerem, Z. ; Makkar, H. P. S. & Becker, K. (2002). The biological action of saponins in animal systems: a review. *British Journal of Nutrition*, vol. 88, pp. 587–605, ISSN : 1475-2662

Gaiotto, J. B. ; Menten, J.F.M. ; Racanicci, A.M.C. &Iafigliola, M.C. (2000). Óleo de Soja, Óleo Ácido de Soja e Sebo Bovino Como Fontes de Gordura em Rações de Frangos de Corte. *Brazilian Journal of Poultry Science*. vol.2 no.3, pp. 219-227, ISSN 1516-635X

Hetland, H.; Choct, M. & Svihus, B. (2004). Role of insoluble non-starch polysaccharides in poultry nutrition. *World's Poultry Science Journal*, vol.60, pp.415-422, ISSN 0043-9339

Ilka M Vasconcelos, I. M. ; Siebra, E. A. ; Maia, A. A. B. ; Moreira, R. A. ; Neto, A. F. ; Campelo, G. J. A. & Oliveira, J.T. A. (1997). Composition, Toxic and Antinutritional Factors of Newly Developed Cultivars of Brazilian Soybean (Glycine max). Journal of Science and Food Agriculture, vol. 75, pp. 419-426, ISSN : 1097-0010

Johnson, L.; Smith, K. (Jan 2011). Soybean Processing. In: *Fact Sheet*, 19.01.2011, Available from: http://www.soymeal.org ISSN : 1065-3309

Johri, T.S. (2005). Endogenous and exogenous feed toxins. In: Poultry Nutrition Research in India and its Perspective. 19.01.2011, Available from: http://www.fao.org/docrep/article/agrippa/659_en-10.htm

Karr-Lilienthala, L.K.; Kadzereb, C.T. ; Grieshopc C.M. & Fahey Jr, G.C. (2005). Chemical and nutritional properties of soybean carbohydrates as related to nonruminants: A review. *Livestock Production Science*, vol. 97, pp. 1 -12, ISSN 0301-6226

Kunitz, M. (1945). Crystallization of trypsin inibitor from soybeans. *Science*, v.101, p.668-669, ISSN 1095-9203

Leeson, S. & Summers, J. D. (2008). *Comercial Poultry Nutrition*, Nottingham University Press, ISBN 978-1-904761-78-5, Guelph, Ontario

Liener, I.E. (1994). Implications of antinutritional components in soybean foods.*Critical Reviews in Food Science and Nutrition*, vol. 34, no. 1, pp. 31-67. ISSN : 1549-7852

Mascarenhas, A. G. ; Donzele, J. L. ; Oliveira, R. F. M. ; Ferreira, A. S. ; Lopes, R. S. & Tavares. S. L. (2002). Fontes e Níveis de Energia Digestível em Rações para Suínos Machos Inteiros dos 60 aos 100 kg. *Brazilian Journal of Animal Science*, vol.31, no.3, pp.1403-1408, ISSN 1806-9290

Mangan, J. L. (1988). Nutritional effects of tannins In animal feeds. *Nutrition Research Reviews*, vol. 1, pp. 209-231, ISSN : 1475-2700

Mekbungwan, A. Application of tropical legumes for pig feed. (2007). *Animal Science Journal*, Vol. 78, pp. 342–350, ISSN 1525-3163

Moreira, I. ; Mourinho, F. L. ; Carvalho, P. L. O. ; Paiano, D. ; Piano, L. M. & Kuroda Junior, I. S. (2009). Avaliação nutricional da casca de soja com ou sem complexo enzimático na alimentação de leitões na fase inicia. *Brazilian Journal of Animal Science*, vol.38, no.12, pp.2408-2416, ISSN 1806-9290

Narayan, R.; Chauhan, G. S. & Verma, N. S. (1988). Changes in the quality of soybean during storage. Part 1—Effect of storage on some physico-chemical properties of soybean. *Food Chemistry*, Vol. 27, No. 1, pp. 13-23, ISSN 0308-8146

Palacios, M. F. ; Easter, R. A. ; Soltwedel, K. T. ; Parsons, C. M. ; Douglas, M. W. ; Hymowitz, T. & Pettigrew, J. E. (2004). Effect of soybean variety and processing on growth performance of young chicks and pigs. Journal of Animal Science, vol. 82, pp.1108-1114, ISSN : 1525-3163

Palić, D. V. ; Lević, J. D. ; Sredanović, S. A. & Đuragić, O. M. (2008). Quality control of full-fat soybean using urease activity: critical assessment of the method. *Acta Periodica Technologica*. vol. 39, pp. 47-53, ISSN 1450-7188

Rackis, J. J. & Gumbmann, M. R. (1981). Protease inhibitors: physiological properties and nutritional significance. In: Antinutrients and Natural Toxicants in Foods. pp. 203-237, Food & Nutrition Press, Westport, CT. ISBN :

Selle, P.H. & Ravindran, V. (2007). Microbial phytase in poultry nutrition : Review Article. *Animal Feed Science and Technology*, Vol. 135, No. 1-2, pp. 1-41, ISSN: 0377-8401

Shepard, T. H. ; Pyne, G.E. ; Kirschvink, J.F. & McLean, M. (1960). Soybean Goiter — Report of Three Cases. *The New England Jounal of Medicine*, vol. 262, pp.1099-1103, ISSN : 1533-4406

Smits, C.H.M. & Annison, G. (1996). Non-starch plant polysaccharides in broiler nutrition - towards a physiologically valid approach to their determination. World's Poultry Science Journal, vol. 52, pp. 203-221, ISSN 0043-9339

Taylor, M. ; Hartnell, G. ; Lucas, D. ; Davis, S. & Nemeth, M. (2007). Comparison of Broiler Performance and Carcass Parameters When Fed Diets Containing Soybean Meal Produced from Glyphosate-Tolerant (MON 89788), Control, or Conventional Reference Soybeans. *Poultry Science*, vol.86, pp.2608–2614, ISSN 0032-5791

Wright, K. N. (1968). Soybean meal processing and quality control. *Journal of the American Oil Chemists' Society*, Vol. 58, No. 3, pp. 294-300, ISSN: 0003-021X

Yoshiki, Y.; Kudou, S. & Okubo, K. (1998). Relationship between chemical structures and biological activities of triterpenoid saponins from soybean (Review). *Bioscience Biotechnology and Biochemistry,* vol. 62, pp. 2291–2299, ISSN : 1347-6947

Zanella, I.; Sakomura, N. K.; Silverside, F. G. & (1999). Effect of enzyme supplementation of broiler diets based on corn and soybeans. *Poultry Science.*, v. 78, pp.561–568, ISSN 0032-5791

Zhenyu, G. ; Jianrong, L. ; Ping, Y. ; Xinle, L. & Peilin, C. (2000). Removal of Goitrogen in Soybean. Journal of the Chinese Cereals and Oils Association., vol. 01, ISSN : 1003-0174

9

Protein Sources in Ruminant Nutrition

Monica I. Cutrignelli, Serena Calabrò,
Raffaella Tudisco, Federico Infascelli and Vincenzo Piccolo
Department of Animal Science and Food Control,
University of Naples Federico II, Naples,
Italy

1. Introduction

Since 2001 the European Commission banned the use of meat and bone meal and its by-products in diets for livestock animals (EC directive 999/2001) in order to assure consumer safety on animal products. Consequently, soybean meal became the most utilised protein source in the intensive livestock systems.

Moreover, the proteins of this source are low degradable in the rumen and well proportioned to the non structural carbohydrates (NSC).

Soybean meal solvent extract (s.e.) is a by-product of oil industry, where soybean seeds are treated with organic solvents (e.g. hexane) and subsequently with high temperature. For this reason soybean meal has been banned in the organic livestock (EC directive 2092/1991; EC directive 834/2007).

Even if in Europe the high part of soybean is imported, soybean solvent extract represents the less expensive protein source for its high crude protein content (44-50 % as fed). However soybean meal costs and availability are strongly related with the price development of agricultural commodities on the world market (Jezierny et al., 2010). Factors which may influence world market prices include variations in population and economic growth, changes in consumer's product preferences, but world market prices are also dependent on weather conditions (Gill, 1997; Trostle, 2008).

Finally, another factor has to be evaluated that is the genetically modification (GM) of soybean. Indeed, public concerns are increasing in GM food consumption due to the fact, even if for several years no direct evidence that it may represent a possible danger for health has been reported, recently, a number of papers have been published with controversial results.

Thus, the search for alternative protein sources has led to an increasing interest in the use of grain legumes, as they supply the important source of plant protein.

The botanical family of grain legumes is known as *Fabaceae*, also referred to as *Leguminosae*. Grain legumes are cultivated primarily for their seeds which are harvested at maturity, and which are rich in protein and energy. The mature dry seeds of grain legumes are used either as animal feed ingredient or for human consumption (Singh et al., 2007). Beans, lentils and chickpeas are utilised exclusively for human nutrition, while the other grains are used in animal feeding too.

In Italy grain legumes cultivation is progressively increased due the presence of new cultivars more hardy and productive. These new cultivars were selected principally in

France and are characterised by lower water requirements, higher production and higher resistance to the parasitic infestations and to the adverse environmental conditions.

Generally, legumes are characterised by their ability to use atmospheric nitrogen as a nutrient due to the symbiosis with nitrogen-fixing bacteria from the *Rhizobium* species (Sprent and Thomas, 1984; Zahran, 1999). Therefore, unlike other cultivated plants, legume crops need less nitrogen fertiliser for optimal growth, and the use of legumes in crop rotation systems reduces the need of nitrogen fertiliser in subsequent crops (López-Bellido et al., 2005). Nitrogen benefits in legume-cereal rotation systems have been attributed not only to the transfer of biologically fixed nitrogen (Díaz-Ambrona and Mínguez, 2001; Evans et al., 2001), but also to lower immobilisation of nitrate in the soil during the decomposition of legumes compared to cereal residues (Green and Blackmer, 1995), also termed as the nitrogen-sparing effect. Thus, nitrogen benefits may result from a combination of legume nitrogen sparing effects and the bacterial nitrogen fixation (Chalk et al., 1993; Herridge et al., 1995). In addition, crop rotation and intercropping with legumes may provide successful strategies for weed suppression (Liebman and Dyck, 1993; Bulson et al., 1997). Weed growth and development may be disrupted due to varying cultivation conditions prevailing for the different crops used (e.g. fertiliser requirements, planting or maturation dates), thereby preventing domination of only a few weed species (Froud-Williams, 1988; Liebman and Janke, 1990). Due to these crop effects, cultivation of grain legumes is an important part of crop rotation, particularly in organic farming (Badgley et al., 2007).

In animal nutrition, grain legumes are mainly used as protein supplements, but also as a valuable energy source, due to their partly high contents of starch (faba bean, peas) and lipids (lupins) (Gatel, 1994; Bach Knudsen, 1997; Salgado et al., 2002a).

However, the use of grain legumes in animal nutrition has been hampered due to partially high concentrations of secondary plant metabolites, also referred as antinutritional factors (ANFs), including condensed tannins, protease inhibitors, alkaloids, lectins, pyrimidine glycosides and saponins. Possible negative effects of these secondary plant metabolites include, for example, feed refusals (tannins, alkaloids), reduced nutrient digestibility (tannins, protease inhibitors, lectins) or even toxic effects (alkaloids) (Rubio and Brenes, 1995; Lallès and Jansman, 1998; Huisman and Tolman, 2001).

The objectives of the following chapters are the comparison of chemical composition as well as the nutritive value of soybean, soybean meal solvent extract and several legume grain (e.g.: peas, lupine, faba bean). In addition, in order to evaluate the opportunity of soybean replacement with grain, the results of *in vitro* studies are described and the influence of protein sources on meat quality are discussed.

2. Nutritional characteristics of soybean meal solvent extract and legume grains

The chemical composition of some grain legumes in comparison to soybean and soybean meals s.e are pictured in Table 1.

Soybeans (*Soja hispida*) are characterized by high protein (380 g/kg dry matter) and lipid (200 g/kg dry matter) concentrations, which provide high energy density (1.1-1.2 UFL/kg dry matter). In ruminant feeding, soybean integral seeds could represent the only source of protein supplementation for cattle fattening. However, in dairy cows it is preferable not to exceed the dry matter administration of soybean seeds as it may modify fatty acid profile of milk fat and worse butter consistency and conservation.

Soybean meal s.e. proteins, due to the heat treatment, are medium to low degradable in the rumen (Chaubility et al., 1991; Infascelli et al., 1995). In addition, soybean meal shows an elevated ratio of protein/non structural carbohydrates.

Legume grains are characterised by high energy density allowed to the high protein, starch and/or fat concentrations, as more than sufficient is their calcium concentration.

The proteins of legume grain are highly degradable in the rumen and digestible in the intestine. Notwithstanding, large part of legume grain shows anti-nutritional factors (i.e. lecithin, trypsin inhibitors, tannins, saponin, phytase), that are inactivated by the enzymes produced by the bacteria present in the rumen. Within the grain legumes, lupins have higher amounts of crude protein (324–381 g/kg dry matter), compared to faba beans (301 g/kg dry matter) and peas (246 g/kg dry matter) (Degussa, 2006). Jezierny et al. (2007) reported similar contents of crude protein in different batches of lupins, faba beans and peas averaging 387, 308 and 249 g/kg dry matter, respectively. In comparison to soybean meal, faba beans and peas contain between 45 to 55% and lupins (L. albus) even up to 70% of its crude protein content (Degussa, 2006).

The ether extract content in peas and faba beans is generally rather low compared to lupins; crude fat contents of faba beans and peas range from 15 to 20 g/kg dry matter, thus being in a similar range as values for soybean meal (15–28 g/kg dry matter) (DLG, 1999; Jezierny et al., 2007). In lupins, the crude fat content varies between cultivars, with values of about 57 g/kg dry matter (*L. luteus, L. angustifolius*) to 88 g/kg dry matter (*L. albus*) (DLG, 1999).

The carbohydrate fraction includes the low molecular-weight sugars, starch and various non-starch-polysaccharides (NSP) (Bach Knudsen, 1997). The NSP and lignin are the principal components of cell walls and are commonly referred to as dietary fibre (Theander et al., 1989; Canibe and Bach Knudsen, 2002). Generally, faba beans and peas are rich in starch (422–451 and 478–534 g/kg dry matter, respectively) (DLG, 1999; Jezierny, 2009), whereas lupins have comparatively low levels of starch (42–101 g/kg dry matter) (DLG, 1999; Jezierny, 2010). However, it needs to be emphasized that the determination of starch in grain legumes may be confounded by the analytical method used (Hall et al., 2000).

Faba beans and peas contain rather low amounts of fibre fractions in comparison to lupins (Bach Knudsen, 1997; Jezierny, 2009), and, with regard to lignin content, faba beans and *L. angustifolius* have similar amounts of lignin (1 to 7 and 6 to 9 g/kg dry matter, respectively), whereas the lignin content in peas is of minor importance (0.4–3 g/kg dry matter) (Salgado et al., 2002a; Jezierny, 2010).

The NSP fraction of faba beans consists mainly of cellulose (89–115 g/kg dry matter), with lower levels of hemicellulose (21–57 g/kg dry matter) (Salgado et al., 2002a,b; Jezierny, 2009).

Hemicellulose contents in peas range from 23 to 95 g/kg dry matter and cellulose contents range from 52 to 77 g/kg dry matter (Salgado et al., 2002a,b; Jezierny, 2009).

Lupins contain high levels of NSP, with contents of cellulose generally being higher than hemicellulose (131 to 199 vs. 40 to 66 g/kg dry matter) (Bach Knudsen, 1997; Salgado et al., 2002a,b; Jezierny, 2009), and they also have considerable amounts of oligosaccharides (Bach Knudsen, 1997; Salgado et al., 2002a).

Comparing the different batches in each species, lupin show lower variability than faba bean and peas, probably because the genetic selection in this species was addressed principally on the reduction of secondary plant metabolites (Colombini, 2004) than to the improving of chemical characteristics.

The protein of faba beans and peas contains similar or even higher proportions of lysine (70 and 80 g/kg crude protein, respectively), when compared to protein from soybean meal s.e. (69 g/kg crude protein) or lupins (51 to 54 g/kg crude protein) (Degussa, 2006).
Comparing the different batches in each species, lupin show lower variability than faba bean and peas, probably because the genetic selection in this species was addressed principally on the reduction of secondary plant metabolites (Colombini, 2004) than to the improving of chemical characteristics.
Comparing the different batches in each species, lupin show lower variability than faba bean and peas, probably because the genetic selection in this species was addressed principally on the reduction of secondary plant metabolites (Colombini, 2004) than to the improving of chemical characteristics.

Protein sources	DM	CP	EE	Ash	NDF
Soybean	93.30	33.2	25.0	5.57	14.5
Soybean meal s.e.	89.80	41.3	1.36	6.00	22.2
Lupin Lublanc	93.26	36.92	6.78	4.33	4.33
Lupin Lutteur	94.31	35.30	5.73	4.19	4.19
Lupin Multitalia	93.97	36.67	9.54	3.92	3.92
Faba bean Irena	89.22	25.62	1.05	3.99	20.94
Faba bean Lady	88.48	25.17	1.02	4.21	25.82
Faba bean Scuro di Torre Lama	90.55	26.91	0.90	4.32	21.75
Faba bean Chiaro di Torre Lama	90.80	24.69	1.01	4.19	21.23
Faba bean ProtHABAT69	90.60	28.69	1.10	4.51	18.21
Faba bean Sicania	90.29	26.52	0.95	3.89	21.43
Peas Alembo	88.49	31.27	0.73	3.94	21.60
Peas Alliance	89.20	28.47	0.56	3.82	20.40
Peas Attika	89.68	25.04	0.81	4.01	18.09
Peas Corallo	88.80	28.50	0.80	3.69	21.82
Peas Iceberg	90.08	27.30	0.78	4.12	22.75
Peas Ideal	89.94	28.28	0.88	4.17	18.37
Peas Spirale	93.00	28.74	0.55	4.30	19.05

Table 1. Mean values of chemical composition (%dry matter) of different protein sources (Calabrò et al., 2001; Calabrò et al., 2009; Calabrò et al., 2010).

Comparing the different batches in each species, lupin show lower variability than faba bean and peas, probably because the genetic selection in this species was addressed principally on the reduction of secondary plant metabolites (Colombini, 2004) than to the improving of chemical characteristics.
The protein of faba beans and peas contains similar or even higher proportions of lysine (70 and 80 g/kg crude protein, respectively), when compared to protein from soybean meal (69 g/kg crude protein) or lupins (51 to 54 g/kg crude protein) (Degussa, 2006).
The proportion of threonine in grain legume protein (38 to 42 g/kg crude protein) is similar to that in soybean meal (45 g/kg soybean meal) (Degussa, 2006), however, there is a severe deficiency in the sulphur containing AA methionine + cystine, while tryptophan is marginally deficient to fulfill nutrient requirements for pigs (20 to 50 kg body weight) (NRC, 1998; Degussa, 2006). In fact, apart from *L. albus*, the seeds of faba beans, peas and lupins

contain less than 50% of these AA in comparison to soybean meal (Table 2), thus constraining the use of grain legumes as sole protein source in pig diets.

	Vicia faba	*Pisum sativum*	*Lupinus albus*	*Lupinus angustifolius*	*Lupinus luteus*	SBM
CP	301	246	381	324	361	541
Indispensable AA						
Arginine	26.4	21.0	39.3	33.5	38.0	39.7
Histidine	7.8	6.1	9.3	8.8	9.7	14.4
Isoleucine	11.8	10	15.3	12.7	14.2	24.3
Leucine	21.4	17.4	27.5	21.5	24.1	40.9
Lysine	18.4	17.3	18.2	15	16.3	33.1
Methionine	2.2	2.2	2.5	2.0	2.0	7.3
Phenyl-alanine	12.6	11.7	14.9	12.5	13.6	27.2
Threonine	10.5	9.1	13.3	10.9	11.9	21.3
Tryptophan	2.6	2.2	3.0	2.6	3.0	7.4
Valine	13.3	11.4	14.5	12.5	13.6	25.5
Dispensable AA						
Alanine	11.9	10.5	12.5	10.9	11.8	23.3
Aspartic acid	31.6	28.2	38.5	31.5	35.1	62.0
Cystine	3.5	3.5	6.7	4.3	4.8	8.0
Glutamic acid	46.9	40.0	79.3	65.6	72.5	97.6
Glycine	12.2	10.6	15.0	13.4	14.3	23.0
Proline	11.8	10.2	15.3	13.5	14.3	27.5
Serine	14.1	11.5	19.0	15.3	17.0	27.3

Table 2. Amino acid contents of grain legumes compared to soybean meal (g/kg dry matter) (Jezierny et al., 2010) SBM= soybean meal; CP= crude protein; AA= amino acids.

As concerns the fatty acid profile, the rather high proportion of essential unsaturated fatty acids of some grain legumes, e.g. some *Vicia* species (Akpinar et al., 2001) or *L. albus* (Erbas¸ et al., 2005) may be attractive both from the human and animal nutrition perspective (Bézard et al., 1994), while adverse effects of unsaturated fatty acids on meat quality should be taken into account (Wood et al., 2003). For example, in faba beans a ratio of saturated to unsaturated fatty acids of 40-60 has been reported (Akpinar et al., 2001), whereas in *L. albus*, a ratio of saturated, monounsaturated and polyunsaturated fatty acids of 13.5 to 55.4 to 31.1 has been established (Erbas¸ et al., 2005).

As concerns the mineral composition of soybean and grain legumes, only few data are reported in literature.

The calcium concentration ranges between 1.0 g/kg (faba beans) and 1.9 g/kg (*L. angustifolius*). Phosphorus concentration varies between 4.2 g/kg (*L. angustifolius*) and 7.6 g/kg (*L. luteus*). No extreme differences in trace mineral concentrations occurred except for the manganese concentration of *L. albus*, which containes approximately 10 times more manganese than the other legume grains (Brand et al., 2004).

Concerning the trace elements, Cabrera et al (2003) reported that in legumes their levels ranged from 1.5-5.0 µg Cu/g, 0.05-0.60 µg Cr/g, 18.8-82.4 µg Fe/g, 32.6-70.2 µg Zn/g, 2.7-45.8 µg Al/g, 0.02-0.35 µg Ni/g, 0.32-0.70 µg Pb/g and not detectable—0.018 µg Cd/g. In nuts, the levels ranged from 4.0-25.6 µg Cu/g, 0.25-1.05 µg Cr/g, 7.3-75.6 µg Fe/g, 25.6-69.0 µg Zn/g, 1.2-20.1 µg Al/g, 0.10-0.64 µg Ni/g, 0.14-0.39 µg Pb/g, and not detectable—0.018 µg Cd/g. The authors found a direct statistical correlation between Cu-Cr, Zn-Al and Cr-Ni ($P<0.05$), and Al-Pb ($P<0.001$).

Sankara Rao and Deosthale (2006) comparing for their total ash, calcium, phosphorus, iron, magnesium, zinc, manganese, copper, and chromium contents five grain legumes (indian legumes, chick pea, pigeon pea, green gram, and black gram), found significant varietal differences only for chromium content in black gram. The cotyledons of these legumes were significantly lower in calcium content as compared to the whole grains. The authors concluded that for human nutrition, differences in mineral composition of whole grain and cotyledons were marginal except for calcium. These legumes as whole grain and cotyledons, appeared to be significant contributors to the daily requirements of magnesium, manganese, and copper in the diet.

As concerns the mineral composition of soybean and grain legumes, only few data are reported in literature.

The calcium concentration ranges between 1.0 g/kg (faba beans) and 1.9 g/kg (*L. angustifolius*). Phosphorus concentration varies between 4.2 g/kg (*L. angustifolius*) and 7.6 g/kg (*L. luteus*). No extreme differences in trace mineral concentrations occurred except for the manganese concentration of *L. albus*, which containes approximately 10 times more manganese than the other legume grains (Brand et al., 2004).

Concerning the trace elements, Cabrera et al (2003) reported that in legumes their levels ranged from 1.5-5.0 µg Cu/g, 0.05-0.60 µg Cr/g, 18.8-82.4 µg Fe/g, 32.6-70.2 µg Zn/g, 2.7-45.8 µg Al/g, 0.02-0.35 µg Ni/g, 0.32-0.70 µg Pb/g and not detectable—0.018 µg Cd/g. In nuts, the levels ranged from 4.0-25.6 µg Cu/g, 0.25-1.05 µg Cr/g, 7.3-75.6 µg Fe/g, 25.6-69.0 µg Zn/g, 1.2-20.1 µg Al/g, 0.10-0.64 µg Ni/g, 0.14-0.39 µg Pb/g, and not detectable—0.018 µg Cd/g. The authors found a direct statistical correlation between Cu-Cr, Zn-Al and Cr-Ni ($P<0.05$), and Al-Pb ($P<0.001$).

Sankara Rao and Deosthale (2006) comparing for their total ash, calcium, phosphorus, iron, magnesium, zinc, manganese, copper, and chromium contents five grain legumes (indian legumes, chick pea, pigeon pea, green gram, and black gram), found significant varietal differences only for chromium content in black gram. The cotyledons of these legumes were significantly lower in calcium content as compared to the whole grains. The authors concluded that for human nutrition, differences in mineral composition of whole grain and cotyledons were marginal except for calcium. These legumes as whole grain and cotyledons, appeared to be significant contributors to the daily requirements of magnesium, manganese, and copper in the diet.

3. *In vitro* evaluation of soybean and legume grains

The nutritive value of a feed is assessed by its chemical composition, digestibility and level of voluntary intake. Feed evaluation methods are use to express nutritive value of feed. It is basically description of feeds interns that allow for a prediction of the performance of animals offered the feeds (Medsen et al., 1997).

Several methods are used in feed evaluation such as chemical analysis, rumen degradability measurement using the nylon bag technique, digestibility measurement and feed intake prediction. However, the tables supply mean values, which cannot be used for individual lots, and all the *in vivo* techniques are very expensive and time consuming while accuracy may be low.

The *in vitro* gas production technique (IVGPT, Theodorou et al., 1994) has proved to be a potentially useful technique for ruminant feed evaluation (Herrero et al., 1996; Getachew et al., 2004), as it is capable of measuring rate and extent of nutrient degradation (Groot et al., 1996; Cone et al., 1996).

To evaluate a feedstuff by IVGPT, it is incubated at 39°C and under anaerobiosis condition with buffered rumen fluid and gas produced is measured as an indirect indicator of fermentation kinetics. During the incubation the feedstuff is first degraded and the degraded fraction may either be fermented to produce gas (CO^2 and methane) and fermentation acids, or incorporated into microbial biomass.

The IVGPT is considered the most complete *in vitro* technique, because it allows to estimate the fermentation kinetics and contemporary gives information on the fermentation products (degradability of dry matter, volatile fatty acids).

The IVGPT has been used by our research group for many years in order to investigate feed fermentation kinetics.

In particular, in this chapter we report some results obtained incubating different protein sources (i.e. soybean and legumes grains) in order to compare their fermentation kinetics (Calabrò et al., 2001a), to assess the effect of some technological treatments (i.e. crushing and flaking) on the carbohydrates fermentation kinetics (Calabrò et al., 2001b) and to test different legumes grain cultivars (Calabrò et al., 2009).

Calabrò et al. (2001a), in order to study the *in vitro* fermentation characteristics and kinetics proposed the following protocol: the samples (about 1.00 g), ground to pass a 1 mm screen, were incubated in triplicate at 39°C in 120 ml serum bottles under anaerobic conditions. Rumen liquor for the *inoculum* was collected from four buffaloes fed a standard diet, and immediately transported to the laboratory where it was homogenised and filtered.

The gas measurements was made at 2-24 time intervals using a manual a pressure transducer (figure 1).

The cumulative gas produced at each time was fitted to the Groot et al. (1996) model which estimates the asymptotic value (A, ml/g), the time after incubation at which A/2 is formed (B, h), the time to reach the maximum rate (tmax, h) and the maximum rate (Rmax, ml/h).

At the end of incubation, the degraded organic matter (dOM, %) was calculated as a difference between incubated and residual OM (filtering the bottle content through pre-weighed glass crucibles and burning at 550°C for 3 hours), and pH and volatile fatty acid concentration (VFA, mM/g) were determined, using a pH-meter and a gas chromatography, respectively.

Several concentrate ingredients such as cereals and grain legumes, used in ruminant diets in order to increase production levels, were evaluated. In particular barley, maize, hard wheat, soft wheat, oats, faba bean and pea were used as test substrates.

Carbohydrates fractionation was carried out according to Cornell Net Carbohydrate and Protein System (CNCPS, Sniffen *et al.,* 1992).

The Cornell Net Carbohydrate and Protein System was developed to predict requirements, feed utilization, animal performance and nutrient excretion for dairy and beef cattle and sheep, using accumulated knowledge about feed composition, digestion, and metabolism in supplying nutrients to meet requirements.

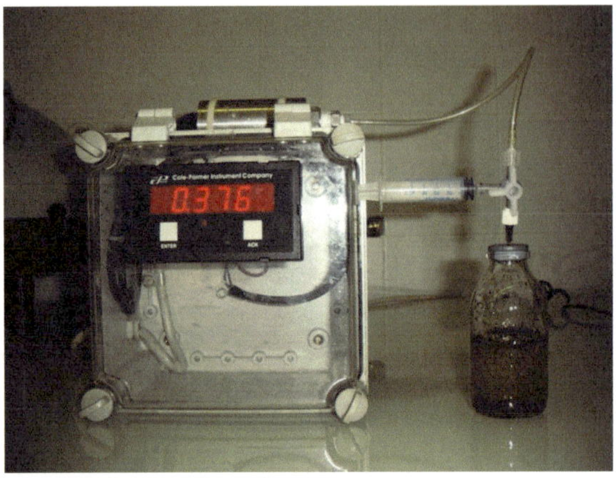

Fig. 1. Pressure transducer.

The CNCPS partitions crude protein into fractions A, B, and C, depending on their rate and extent of degradability in the rumen (NRC, 2001). Fraction A represents the non-protein N (NPN) (ammonia, peptides, amino acids) and is considered to be completely soluble; fraction B, subdivided into B1, B2, and B3, consists of true protein with progressively declining ruminal degradability. Fraction C is unavailable true protein. Broadly, these crude protein fractions are categorized into rumen degradable protein (RDP) and rumen undegradable protein (RUP). The rumen degradable protein meets protein requirements for ruminal microbial growth and protein synthesis.

Once reaching the rumen, feed and protein degradation is a function of microbial activity. Rumen microbial activity, growth and protein synthesis is primarily limited by the rate and extent of carbohydrate fermentation in the rumen. Consequently, dietary fiber fractions in the forage determine the animal response to feed.

Microbial protein and rumen undegradable protein reaching the small intestine are absorbed to meet the ruminant's protein requirement. When rumen degradable protein exceeds the capacity of the rumen microbes to assimilate it, ammonia builds up in the rumen. This is followed by absorption of ammonia into the blood, conversion into urea by the liver, and excretion in the urine. The conversion of ammonia to urea costs the dairy cow energy that could otherwise be used for milk production. This loss of dietary crude protein and energy reduces the utilization efficiency of rumen degradable protein and therefore, reduced ruminant production (NRC, 2001). It also causes a negative energy balance that leads to a reduced fertility.

In the study of Calabrò et al. (2001a), the CNCPS carbohydrate fractions (table 3) was consistent with the values reported by Sniffen *et al.* (1992).

Oats (table 4) had the lowest potential gas production (A: 251 *ml/g*, P<0.0) and a very fast fermentation process (evidenced by low B and high RM). This result was probably due to its high soluble sugars contents (A fraction). Interestingly, oat values were very high compared to *in situ* observations. As reported by Van Soest *et al.*, (1992), maize proved to be the slowest because of its high starch content (B_1 fraction), which degraded slowly. Hard and soft wheat showed very similar fermentation characteristics according to their chemical composition. Barley has fermentation kinetics between those of wheat and maize. Results related to these last three grains were similar to the results obtained *in vitro* using the IGPT (Mould *et al.*, 2005).

	Barley	Maize	Oats	Hard wheat	Soft wheat	Faba bean	Pea
CP (% DM)	11.0	8.79	12.2	14.4	13.5	29.8	23.9
A	3.80	0.49	12.6	2.40	5.00	6.00	18.0
B1	66.4	75.4	45.6	67.6	65.7	43.6	40.8
B2	11.4	6.14	16.0	9.86	9.64	9.99	7.92
C	3.60	2.57	6.77	2.64	2.16	6.41	4.18

Table 3. Chemical composition and CNCPS fraction of the tested grains.

Legumes were characterised by high gas production alongside a rather slow fermentative process. However, interpretation of faba bean and pea data was complicated by their high protein content, whose degradation can interact with gas production (Schofield., 2000).
Organic matter degradability for all tested grains was always higher than 95%.
At 72 h the pH always remains good to guarantee microbial activity.
Oats reached the maximum rate before all the other grains (figure 2) and remained the highest until 12 incubation hours, subsequently decreasing below the soft wheat level. Until 6 hours of incubation barley was faster than the two wheat, later becoming slower until the end; up to this time the three sharp curves were practically overlapping, suggesting that the fermentation trend was similar for the three grains. Maize, faba bean and pea always showed the slowest fermentation rate with overlapping trends: fermentation slowly reaches its maximum rate and slowly decreases. The results of this investigation evidenced the validity of the IVGPT in describing the kinetics fermentation of the examined grains.
The range of rates obtained with the GPT (oats>wheat>barley>maize>legumes) approximately reflects that reported by other authors cited by Sniffen *et al.* (1992): wheat>oats>barley>maize.

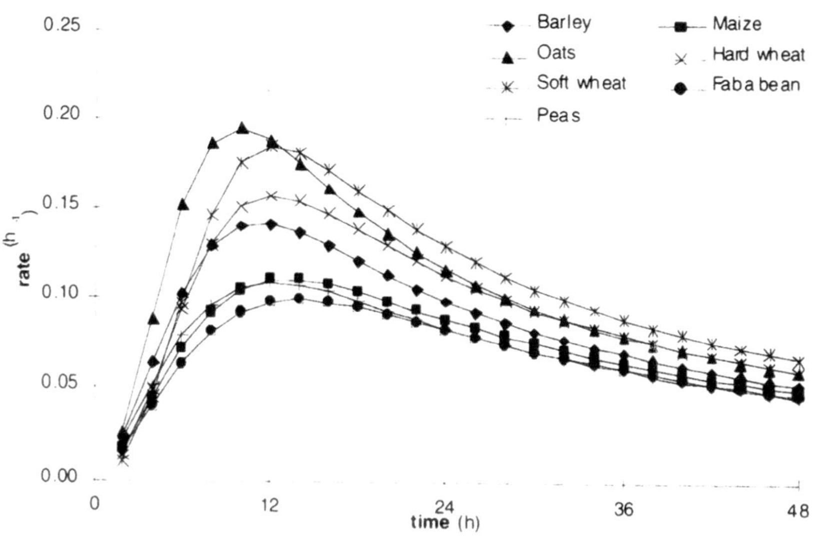

Fig. 2. Gas production fractional rate plots of different grains.

In the second investigation Calabrò et al. (2001b) evaluated the effects of flaking and crushing on some cereal and legume grains, using the *in vitro* cumulative gas production technique (Theodorou *et al.*, 1994).

The grains used in animal nutrition are commonly subjected to different technological treatments (i.e.: grounding, crushing, flaking, extrusion, cooking, micronization). Flaking is a hydrothermal-mechanical process which causes starch gelatinization and increases the enzymatic hydrolysis rate of the polysaccharide, which in turn favours the rumen microbial attack. Crushing differs from flaking in the shorter exposure to vapour. Also in this case, the obtained product presents a larger surface area compared to the primary grain, and a modified starch, which promotes faster fermentation kinetics.

Several investigations to evaluate the effect of flaking on organic matter degradability in the rumen, indicated the considerable influence of the primary grain (Arieli *et al.*, 1995; Bittante *et al.*, 1989); by contrast, little was known about the effects of crushing.

		VC_{SOI}	A	B	RM	dOM	Y	VFA
		ml/g	*ml/g*	*h*	*h⁻¹*	%	*ml/g*	*mM/g*
Maize	wholegrain	340	362	14.9	0.078	92.9	363	99.1
	flaked grain	338	347	13.2	0.096	96.6	343	83.3
	whole + flaked	339^A	355^A	14.1^B	0.087^B	94.8	353^A	91.2^A
Barley	wholegrain	304	323	10.2	0.125	90.6	341	92.4
	flaked grain	346	369	11.7	0.104	97.9	359	94.8
	whole + flaked	325^B	346^A	11.0^C	0.115^A	92.6	350^A	93.6^A
Soybe an	wholegrain	204	236	15.5	0.051	96.7	209	71.6
	flaked grain	225	280	19.8	0.039	86.8	257	76.0
	whole + flaked	215^C	258^B	17.6^A	0.045^C	91.8	233^B	73.8^B
Main effect								
	grain	***	***	***	***	n.s.	***	**
	treatment	***	**	**	n.s.	n.s.	**	n.s.
Interaction								
	grain. x treat.	***	***	***	***	***	***	n.s.
Var.er.		87.9	135	1.21	0.00009	10.5	206	103

OMCV, cumulative gas production related to incubated OM at 120 h; A, potential gas production; B, time at which A/2 is produced; Rmax, maximum fermentation rate; Y, yield related to the degraded OM; VFA, volatile fatty acids related to degraded OM.
*a,b,c: P<0.05; A,B: P<0.001ns, not significant;***: p < 0.001.*

Table 4. *In vitro* fermentation characteristics of whole and flaked grains.

In this experiment three whole and flaked grains (barley, maize, soybean) and four whole and crushed grains (barley, oats, maize, faba bean) were used as test substrates.

Fermentation characteristics (Table 4 and 5) were affected by the grain type and treatment; in addition, the two factors interacted with each other.

		OMCV	A	B	Rmax	dOM	Y	VFA
		ml/g		*H*	h^{-1}	%	*ml/g*	*mM/g*
Oats	whole grain	321	334	10.6	0.107	87.8	362	90.6
	crushed grain	272	311	10.3	0.106	86.9	332	90.5
	whole +crushed	296^{B}	322^{B}	10.4^{C}	0.0106^{Aa}	87.3^{aCD}	347^{B}	90.5^{AB}
Faba bean	whole grain	335	360	14.4	0.071	96.7	344	94.3
	crushed grain	325	367	14.8	0.062	89.9	359	79.3
	whole +crushed	330^{A}	364^{A}	14.6^{A}	0.066^{B}	93.3^{B}	352^{B}	86.8^{AbB}
Maize	whole grain	324	348	12.2	0.099	97.6	331	90.6
	crushed grain	327	350	13.4	0.092	96.7	338	100
	whole +crushed	326^{Ab}	349^{A}	12.8^{B}	0.096^{AbC}	97.2^{A}	335^{B}	95.5^{aA}
Barley	whole grain	310	327	8.68	0.138	81.5	398	77.3
	crushed grain	370	386	9.59	0.128	90.1	395	91.7
	whole +crushed	340^{Aa}	357^{A}	9.13^{D}	0.133^{D}	85.8^{bD}	397^{A}	84.5^{B}
Main effect								
	Grain	***	***	***	***	***	***	***
	Treatment	n.s.	n.s.	*	n.s.	n.s.	n.s.	n.s.
Interaction								
	gran x treat	***	***	n.s.	n.s.	***	***	n.s.
	Var. er	204	398	0.73	$1 \cdot 10^{-1}$	1.85	437	48.19

OMCV, cumulative gas production related to incubated OM at 120 h; A, potential gas production; B, time at which A/2 is produced; Rmax, maximum fermentation rate; Y, yield related to the degraded OM; VFA, volatile fatty acids related to degraded OM.
*a,b,c: P<0.05; A,B: P<0.001ns, not significant;***: p < 0.001.*

Table 5. *In vitro* fermentation characteristics of whole and crushed.

Flaking increased the potential gas production (A) for soybean and barley (P< 0.01), and slowed their fermentation kinetics (increased t/1/2 and decreased RM values). By contrast, maize flakes showed lower gas production (P<0.05) and the fermentation process was faster (P<0.01), compared to whole grain.

Organic matter degradability decreased after the treatment for soybean (P<0.01), while it increased for barley (P<0.01) and, not significantly, for maize. The slower fermentative process for barley flakes (increased t/1/2and decreased RM values) agrees with the results of other authors on treated cereals for both *in vitro* (Bittante et al., 1989) and *in situ* experiments (Arieli et al., 1995).

Comparing each grain with the respective flakes, OM degradability did not strictly follow the trend of potential gas production, confirming that not all the degraded OM is fermented. This result holds mainly for soybean flakes, which had a higher A (P<0.01) and a lower dOM (P<0.01) compared to the whole grain.

Such decreased OM degradability was probably due to the thermal treatment causing a decrease in the rapidly degradable fraction (Sarubbi, 1999).

Overall, the two cereals showed higher gas and VFA production, as well as a faster fermentative process.

Crushing showed slighter effects compared to flaking, and also the results of the grain x treatment interaction were less important.
The different technological condition of the two treatments (less drastic for crushing) may well have contributed to this trend. In particular, crushed barley presents, with respect to the whole grain, the same behaviour as the flakes (higher A and dOM, P<0.001; higher *t/1/2* and lower RM, not significant).
Interestingly, there was a decrease in faba bean degradability due to the treatment (P<0.01), which agrees with the results of a contemporaneous *in situ* trial (Sarubbi, 1999). Besides, faba bean also showed a lower VFA concentration, which was probably also caused by the considerable decrease in their protein degradability (P<0.001) as found *in situ* (Sarubbi, 1999). As usually observed *in vivo* in young bulls fed a cereal-rich diet, pH values were always quite low.
In the third study Calabrò *et al.* (2009) evaluated the fermentation characteristics of different cultivars of grain legumes using IVGPT.
Three grain legumes were tested: lupine (*Lupinus* spp.) (Lublanc, Luteur, Multitalia), faba bean (*Vicia faba* L.) (Chiaro di Torre Lama, Irena, Lady, ProtHABAT69, Scuro di Torre Lama, Sicania) and peas (*Pisum sativum*) (Alembo, Alliance, Attika, Corallo, Iceberg, Ideal, Spirale).
The fermentation characteristics are reported in table 6.
The values of pH ranged between 6.35 and 6.72, indicating a normal pattern of fermentation, and were consistent to the crude protein content.
As regards faba bean, the cultivar "Scuro di Torre Lama" showed significantly (P<0.01) lower values of dOM and OMCV than the other 5 cultivars.
In the case of lupine the cultivar "Lublanc" had lower (P<0.01) OMCV than the other 2 cultivars and for peas the cultivar "Spirale" produced less gas and showed a faster kinetics than the other 6 cultivars.
As expected, the OM degradability resulted very high in any case. However, comparing the pools of the grain legumes, dOM was in each case lower than that of soybean meal. OMCV was significantly (P<0.01) higher for pea than faba bean (330 vs. 316 ml/g, P <0.05) and lupine (330 vs. 258 ml/g, P <0.01). Gas production of peas was always higher than that of lupine, faba bean and also soybean meal according to the results of Buccioni et al. (2007) who studied the *in vitro* fermentation of soybean meal, faba bean and pea, and found in the latter the best balance between energy and nitrogen inputs.
The slower fermentation kinetics of faba bean may be due to the content in polyphenols while that of lupine may be caused by the very low starch content (INRA, 1988).
From the data obtained, the authors concluded that the tested grain legumes show only few differences compared to soybean meal (higher dOM and lower OMCV), consequently they may be considered in replacing, totally or partially, soybean.
The reported results are of particular importance as well as highlight the differences in dietary and nutritional characteristics of soybean and legume grains, provide data on the effects of the treatments of these feedstuffs and the differences among the cultivars present on the European market.
With this information, the nutritionist may from time to time choose the most suitable protein source in order to satisfy animal requirements ensuring the simultaneous availability of nitrogen and energy for the bacteria present in the rumen with beneficial effects on livestock production and environmental impact.

Cultivar	pH	dOM	OMCV	Yield	A	B	tmax	Rmax
		%	*ml/g*	*ml/g*	*ml/g*	*h*	*h*	*ml/h*
Faba bean								
Irene	6.46	92.9	370	397	328	22.9	12.42	9.14
Lady	6.35	93.3	354	363	333	24.5	15.67	9.26
Scuro di Torre Lama	6.49	87.8	308	351	269	22.0	15.39	8.84
Chiaro di Torre Lama	6.41	91.8	348	379	310	23.1	13.20	8.69
ProtHABAT69	6.47	93.8	359	383	303	20.2	12.49	10.02
Sicania	6.40	92.9	324	349	299	21.0	12.83	9.71
MSD	0.135	3.19	60.1	49.3	60.1	3.90	3.65	1.89
Lupine								
Lublanc	6.63	93.4	256	279	283	26.1	10.25	6.73
Luteur	6.69	92.4	275	298	309	25.6	5.58	7.83
Multitalia	6.72	91.2	273	297	303	27.0	8.48	7.08
MSD	0.219	5.40	26.6	91.5	13.9	45.8	15.6	4.74
Peas								
Alembo	6.57	99.0	406	410	361	20.6	12.52	11.73
Alliance	6.49	99.3	397	396	358	20.1	11.99	11.72
Attika	6.57	98.4	397	404	360	20.5	11.82	11.46
Corallo	6.53	98.9	393	394	365	22.3	11.42	10.38
Iceberg	6.55	98.8	381	385	347	21.0	12.45	10.86
Ideal	6.58	97.0	371	383	336	20.7	13.24	11.06
Spirale	6.58	98.8	344	343	310	17.1	10.66	12.14
MSD	0.188	2.81	52.5	53.5	3.68	74.1	5.22	2.47
Faba bean[1]	6.52Ab	90.9b	368B	405A	321a	21.1ab	12.4	10.0Ab
Lupine[1]	6.64B	91.8ab	284C	309B	293b	24.4a	9.03	7.42B
Peas[1]	6.60a	95.1a	394A	413A	336a	18.2b	11.4	12.6Aa
Soybean meal[2]	6.73	96.5	295	306	323	18.7	6.01	10.67
MSE	0.001	2.31	56.2	30.5	101	4.40	2.99	0.56

MSD: Minimum Significant Differences for P<0.01. MSE: Mean Square Error.,In the column A,B,C: P<0.01; a,b,c: P<0.05. [1]Data obtained from the grain legumes incubated in vitro as a pool. [2]Data not statistically assessed.

Table 6. Fermentation characteristics of the different grain legume cultivars and soybean meal.

4. Effect of different protein sources on animal performance

Several researches have been carried out in order to compare the nutritional characteristics of soybean solvent extract and legume grains, as faba bean, peas and lupine, for ruminant feeding.

Di Francia et al. (2007) evaluated the effect of partial replacement of soybean cake with extruded peas in the diet of lactating buffalo cows on milk yield and quality over the first 100 days of lactation. Their results showed that peas could represent an attractive GMO free

protein feed when approaching the problem of the choice of a protein source alternative to soybean in diet formulation for buffalo cows raised in organic farms.
On the other hand Morbidini et al (2005) investigating on the effect of two different fattening diets with different protein sources - soybean meal and flaked faba bean - on slaughtering performance and carcass quality in light Apennine and Italian Merino lambs, observed that the use of faba beans lightly depressed growth performance (Morbidini et al., 2004) and slaughtering weight, even if did not affect the carcass quality.
In order to make a contribution on this issue, our research group conducted a trial on the effect of protein source on growth performance and meat quality of Marchigiana young bulls (Cutrignelli et al 2008 a, b).
The trial was carried out on a farm situated at 700 m a.s.l. in Campania Region (Southern Italy), where 12 weaned young bulls (129 d of age) were equally divided into two groups. Each animal was placed in individual box up to the slaughtering weight (620 kg).
The groups were fed diets with the same protein and energy concentrations and the same forage/concentrate ratios (F/C), but differing in protein source: faba bean (*Vicia faba minor* L.) vs soybean meal (*Soja hispida*).
All the animals were regularly weighed until the body weight (BW) of 620 kg fixed in advance as slaughter weight, was reached. All animals were slaughtered in an authorized slaughterhouse according to EU legislation (EU Regulation EC No 882/2004).
Live animals and carcasses were weighed and measured according to ASPA (1991). After 9 days of refrigeration at 4 ± 1 °C, dissection of the carcasses was carried out.
Samples of *Longissimus thoracis* (LT), *Semitendinosus* (ST), *Iliopsoas* plus *Psoas minor* (IP) muscles, perirenal (PF) and subcutaneous (SF) adipose tissues were collected and rapidly transported, upon refrigeration temperature, to the laboratories for the chemical analysis in order to evaluate the rheological (water holding capacity) and nutritional (chemical composition fatty acid profile, cholesterol and hydrossiproline contents) characteristics of meat.
The protein source did not affect any *infra vitam* parameters except the body weight (BW) at 180 d of age (173 vs. 186 kg, for group FB and SB, respectively; $P<0.01$).
The difference was probably due to the higher non-protein nitrogen (NPN) concentration of the faba bean than the soybean meal (about 12 vs 1.3% of crude protein, respectively). Indeed, in the months immediately following the weaning it should be preferable to administer diets with higher rumen undegradable protein content because at this age the rumen is not yet perfectly functional and microbial protein synthesis is less efficient. Furthermore, for the same reason, the animals in this period are probably unable to neutralize possible anti-nutritional factors of faba beans.
As respects the influence of the replacement of soybean meal solvent extract with legumes seeds, the literature results are contrasting. Moss et al. (1997) found no significant effects on weight gain and feed intake when soybean meal was replaced on an iso-protein basis by lupin seeds in diets for growing bulls (BW from 182 to 243 kg); similar results are reported by Kwak and Kim (2001) on Korean native bulls (BW from 247 to 427 kg) utilising two different concentrations (15 and 30%) of flaked lupin. Instead, according to our results, Murphy and McNiven (1994) found significantly higher weight gain in growing steers (BW from 235.2 to 343.7 kg) fed soybean meal vs raw or roasted lupin although the differences were not significant in the finishing phase (final BW 503.4 kg).

No significant differences between the groups were found for carcass measurements. In each case the carcass measurements of this trial ranged into the interval indicated by Keane (2003) for European/North American breeds.

No differences were found in dressing out, organs and tare incidence on net weight.

Protein source influenced neither body and carcasses conformations nor dressing out with the exception of the incidence of long bones showed a significant difference (6.2 vs 6.7 for faba and soybean, respectively; $P<0.05$).

The first quality meat cuts were acceptable in both groups (58.1 and 57.8%, for faba and soybean, respectively). Concerning the comparison between the two protein sources, only the incidence of long bones showed a significant difference (6.2 vs 6.7 for faba and soybean, respectively; $P<0.05$). It is important to underline that the sample cut measurements were contradictory and conflicted with the data obtained from total carcass dissection. The results of the sample cut dissection indicated a significant ($P<0.05$) difference between groups faba and soybean in meat incidence (69.4 vs 66.9, for faba and soybean, respectively); while no differences were found between groups at carcass dissection. Moreover, sample cut of soybean group showed in the meantime the smallest meat incidence and the highest LT area (88.4 vs. 84 for soybean and faba group, respectively).

The animals fed faba bean showed significantly higher water losses, measured with the compression method (WHC 7.6 vs. 5.7% for faba and soybean group, respectively; $P<0.01$).

Our grilling loss data (31.7 vs. 28.3 % for faba and soybean group, respectively; $P>0.01$) were higher than those reported by Sami *et al.* (2004) in Simmental young bulls but they were in agreement with those reported by Pen *et al.* (2005) cooking ST samples of Holstein steers in an oven, on the contrary our drip loss data resulted higher than that reported by these authors.

The chemical composition of LT was not statistically different between groups. Meat from both groups showed a very low fat content (<3%) and higher protein concentrations than the Holstein steers (Pen *et al.*, 2005), confirming the high quality of the Marchigiana meat.

Hydroxyproline (60.0 vs. 62.6 mg $100g^{-1}$ of meat, in group faba and soybean, respectively) contents were not influenced by protein sources. Regarding the differences registered for this parameter among the texted muscles (LT: *Longissimus thoracis*; ST: *Semitendinosus*; IP: *Iliopsoas* plus *Psoas minor*) the IP samples, which correspond to the tenderloin, showed in all the groups significantly ($P<0.01$) lower hydroxyproline concentrations than the other two muscles; also between the LT and ST muscles the differences were statistically significant ($P<0.05$) being lower for the former.

Cholesterol content (56.3 vs.55.1 mg $100g^{-1}$ of meat, in group faba and soybean, respectively) was not influenced by protein sources according to the observations of Cutrignelli (2000) on Podolian young bulls and by Poli *et al.* (1996) on Chianina young bulls.

Considering the differences among the muscles, cholesterol values were significantly ($P<0.05$) lower for IP. Cifuni *et al.* (2004) found no differences in cholesterol contents among muscles, while Rusman *et al.* (2003) found significant differences. This contradiction is probably due to the different muscles analysed in each experiment. As theorised by Wheeler *et al.* (1987) the cholesterol content may be affected by the different physiological function of the muscles. In both groups, and especially for LT and ST muscles, the cholesterol contents were slightly higher than the value (less than 50 $mg100g^{-1}$ of muscle) indicated by the Protected Geographical Indication (PGI) of the "Vitellone Bianco dell'Appennino Centrale" (Council Regulation EEC No 2081/92; Floroni, 2002). Nevertheless, our results are very close

to those reported for Italian meat breeds (Poli *et al.* 1996; Cifuni *et al.*, 2004) and lower than those from other breeds (Migdal *et al.*, 2004).
Regarding the fatty acids profile in both groups and in each analysed tissues (intramuscular, perirenal and subcutaneous adipose tissues) palmitic (C16:0), stearic (C18:0) and oleic (C18:1) acids were the most widely represented fatty acids. In particular, in intramuscular fat, the sum of oleic, stearic and palmitic acids represents over 50% of total fatty acids, according to the observations of Cifuni *et al.* (2004) and Migdal *et al.* (2004).
Comparing the data of groups faba and soybean, the only significant ($P<0.01$) difference was for stearic acid being higher for bulls fed faba bean than for those fed soybean meal solvent extract. However this result did not significantly affect the SFA concentration, or AI and TI indexes (table 7).
Protein source did not affected the fatty acids composition of the analysed adipose tissues. This is probably due to the very low concentration of phospholipids in subcutaneous and perirenal fat tissue.
Comparing the fatty acid composition of the three adipose tissues the following differences were noted:

- the erucic (C22:1), docosahexaenoic (DHA, C22:6, ω-3) eicosapentaenoic (C 20:5, ω-3) and docosapentaenoic (C 22:5, ω-3) acids, were present only in intramuscular fat, due to the lower level of intramuscular fat compared to the two adipose tissues. Lower fat content is associated with fewer and smaller adipocytes containing fewer triglycerides, accompanied by a relative increase in the proportion of phospholipids in total lipids and an increased PUFA content (Scollan *et al.*, 2006);
- the proportion of total SFA was minimal in LT and maximal in SF (mean values 46.56 *vs* 54.50 and 59.35% of total fatty acids; in LT, PF and SF, respectively, Figure 1); the single saturated fatty acids amounts also showed a similar trend. These data are partly in contrast with those of Kim *et al.* (2004) who in Hanwoo cattle found a higher SFA proportion in perirenal than in subcutaneous fat. Even if the differences in the fatty acids profile between internal and external fat deposits have not been fully elucidated, Eguinoa *et al.* (2003) suggested that the lipogenic enzyme activities per cell could be influenced by several factors such as adipocite size, nutrient supply, etc.;
- the proportion of total polyunsaturated fatty acids (PUFA) was higher in LT and lower in the perirenal adipose tissue (mean values 22.23 *vs* 11.36 *vs* 11.02% of total fatty acids, respectively in LT, SF and PF, Figure 1). This difference may be primarily ascribed to the lower concentration of ω-6 PUFA and to the absence DHA in SF and PF adipose tissues;
- the atherogenic and thrombogenic indexes were considerably higher in SF and PF tissues than in LT. This is due to the high SFA concentration of SF and PF adipose tissues as well as to the high PUFA concentration of LT.

The LT fatty acid profile was similar to those reported by Raes *et al.* (2003) for *Longissimus lumborum* of Belgian Blue and Limousin beef (SFA: 338 and 506 mg $100g^{-1}$ edible portion; MUFA: 323 and 554 mg $100g^{-1}$ edible portion; PUFA: 195 and 195 mg $100g^{-1}$ edible portion; in Belgian Blue and Limousin bulls, respectively) and different from values found in Irish and Argentine beef. The latter showed significantly higher total intramuscular fatty acid content compared to the former, probably due to genetic selection.
This observation confirms that of Carnovale and Nicoli (2000) who concluded that Italian meat showed a favourable intramuscular fatty acid composition with high PUFA content. Muscle with a high percentage of unsaturated fatty acids (UFA) generally scored higher in

taste panel evaluation (Westerling and Hedrick, 1979) and food with high UFA, especially PUFA, is good for human health (Rusman *et al.*, 2003). The ω-6/ω-3 ratio was higher than the value (less than 3) reported by Scollan *et al.* (2006) but lower than that registered by Warren *et al.* (2003) for steers fed corn silage and concentrates (8.9).

It has to be underlined that the fatty acid profile o the food arouse high interest in human medicine due to their influence on the functionality of the cardio-circulatory apparatus. A number of epidemiological researches put in evidence that diets with high content of saturated fatty acids (SFA) were associated with high levels of serum cholesterol (especially of low density lipoprotein, LDL) which appear important in atheroma. Successively, Ulbricht and Southgate (1991) reported that:

1. diets high in C18:0, stearic acid, do not raise serum cholesterol;
2. short-chain SFA (C 10 and below) likewise do not raise blood cholesterol, so the putative atherogenic SFA are C12:0 (lauric), C14:0 (myristic) and C16:0 (palmitic). Myristic acid is the most atherogenic, with about four times the cholestrol-raising potential of palmitic acid.

PUFA are considered protective factors: ω-6 fatty acids show mainly anti-atherogenic activity while ω-3 fatty acids have anti-thrombogenic activity. More recently, high prominence is attributed to the role developed by the MUFA, and particularly by the oleic acid that, reducing the oxidation of the cholesterol LDL, may slow the progression of atherosclerosis.

As was pointed out above, the P/S (poliunsaturated/saturated) ratio is not suitable measure of the atherogenicity or thrombogenicity of a diet or foods. Currently they are expressed as follows:

Index of atherogenicity: C12:0 + (4 x C14:0) + C16:0/ ω-3 + ω-6 + MUFA

Index of thrombogenicity: C14:0 + C16:0 +C18:0/(0,5 x C18:1) + (0,5 other MUFA) + 0,5(ω-6) + 3(ω-3) + (ω-3/ ω-6).

The AI of the meat in this trial was particularly interesting, rather lower than the data reported by Ulbricht and Southgate (1991) for raw minced beef and than those reported by Badiani *et al.* (2002) for cooked beef (0.72 and 0.77, respectively); our data were similar to those of Poli *et al.* (1996) on Chianina young bulls (AI: 0.58).

	FB	SB	Significance
SFA	365.1 ± 39.1	330.6 ± 26.9	Ns
MUFA	244.3 ± 39.5	271.7 ± 29.4	Ns
PUFA	186.6 ± 20.5	201.9 ± 15.7	Ns
ω-6	174.2 ± 19.9	188.3 ± 12.9	Ns
ω-3	12.39 ± 1.5	13.57 ± 0.8	Ns
AI	0.54 ± 0.04	0.49 ± 0.07	Ns
TI	1.46 ± 0.12	1.20 ± 0.09	Ns

FB: faba bean; SB: soybean meal solvent extract;AI: atherogenic index; TI thrombogenic index.
ns: not significant.

Table 7. Fatty acid profile of Longissimus thoracis muscle (mg $100g^{-1}$ of edible part).

However, the index of thrombogenicity in this trial was higher than the findings of the above-cited authors (1.27 and 1.30 for Ulbricht and Southgate, 1991 and Poli *et al.*, 1996, respectively); only the TI (1.77) reported by Badiani *et al.* (2002) was similar to our data.
These results show that the faba bean could be used as an alternative protein source to soybean meal solvent extract as it did not affect the growth rate (body weight, daily weight gain and biological efficiency of growth) or the feed conversion indexes during the whole experimental period, and offers decided agronomical, economical and healthy advantages.
Nevertheless, in the first period after weaning the faba bean reduced the growth rate, probably due to the higher concentrations of NPN and anti-nutritional factors. It might be useful in this period to use this protein source associated with other richer in rumen undegradable.
Our results contribute to show that both protein sources (soybean meal and faba bean) could be utilised in the diet for young Marchigiana bulls ensuring the high quality of the meat obtained by this breed.
Although the meat of group fed faba bean had significantly higher concentrations of stearic acid compared to the level found in soybean group, neither the atherogenic and thrombogenic indexes, nor the cholesterol content were influenced.
As regards the effect of protein source on organolepitc characteristics, our results on hydroxyproline content and water holding capacity were conflicting. While the meat of the group fed soybean meal solvent extract showed a potential low tenderness (higher level of hydroxiproline) the water holding capacity measured by compression was lower for the group receiving faba bean.
From our results it is also possible to formulate a favourable assessment of the nutritional characteristics of the of meat Marchigiana young bulls. Indeed, the cholesterol values were very close to those indicated by the PGI of the "Vitellone Bianco dell'Appennino Centrale" and lower than those found in other breeds. Moreover, the fatty acids profile of LT confirms that the meat of the Italian breed specialised in meat production has higher unsaturated fatty acids concentration and lower saturated fatty acids levels, which in turn ensures medium-low atherogenic and thrombogenic indexes.

5. Conclusions

The protein source largely used in ruminant nutrition (soybean, soybean meal solvent extract and grain legumes) are characterised by several differences in chemical composition, in particular as concerns crude protein, ether extract and carbohydrates concentrations. Also the technological treatments as flaking and crushing could affect these parameters. In addition, mainly concerning the grain legumes, the contents of different nutrients is affected by the cultivar.
In order to better estimate the nutritive value of protein source, the *in vitro* gas production technique seems to be the most useful methods, as it allow to study also the kinetics of degradation in the rumen. The knowledge of the latter phenomenon is particularly useful, since the nutritionist may choose to treat a particular protein source in order to synchronize the availability of nitrogen and energy for rumen bacteria.
From our *in vitro* data, faba bean, lupin and peas, even if from different cultivars, may be considered in replacing, totally or partially, soybean as only few differences were found among these feeds.

In addition, our *in vivo* studies concerning the *infra vitam* and *post mortem* performances of young bulls fed either faba bean or soybean showed:

- no influence of protein source on growth rate (body weight, daily weight gain and biological efficiency of growth) and feed conversion indexes;
- no influence of protein source on the nutritional characteristics of meat. Indeed, even if the meat of group fed faba bean had significantly higher concentrations of stearic acid compared to the level found in soybean group, neither the atherogenic and thrombogenic indexes, nor the cholesterol content were influenced;
- conflicting results on hydroxyproline content and on water holding capacity. Indeed, while the meat of the group fed soybean showed a potential low tenderness (higher level of hydroxiproline), the water holding capacity measured by compression was lower for the group receiving faba bean.

From these results, it can be concluded that the use of grain legumes as a protein source in ruminant diets could be used as alternative to soybean meal. However, mainly in the intensive livestock system the soybean meal solvent extract represents the protein source par excellence for the high protein content (in particular undegradable fraction), the absence of anti-nutritional factors and the extremely favourable biological value of its proteins. Not insignificant, then is the economic assessment, soybean is, especially in the areas where the ruminants breeding has been particularly developed, the protein source with the quality/price ratio more favourable.

By contrast, it has to be underlined that biotech herbicide tolerant soybean continued to be the principal biotech crop in 2010, occupying 73.3 million hectares or 50% of global biotech area. Farm animals are currently fed soybean and soybean meal developed from genetic transformation. Europe is strongly dependent upon the American continent for its protein requirements amounting up to 90 to 95%.

Thus, the possible risk connected to genetically modified organisms use in animal breeding has led to the reconsideration of animal production processes with special reference to the use of alternative protein sources (e.g. faba beans, dried peas, lupine seeds, chickpeas) able to replace soybean. These legumes have agronomic importance because they improve soil fertility and reduce nitrogenous dressing, with positive effects on environmental pollution. Moreover, they need a limited initial investment for their modest requirements of chemical and energetic inputs and their short culture cycle.

6. References

A.S.P.A. (1991). Metodologie relative alla macellazione degli animali di interesse zootecnico ed alla valutazione e dissezione della loro carcasse. Ed. ISMEA, Roma.

Akpinar, N., Akpinar, M.A. & Türkoglu, S. (2001). Total lipid content and fatty acid composition of the seeds of some Vicia L. species. *Food Chem.*, 74, pp. 449–453.

Arieli, A., Bruckental, I., Kedar. & I., Sklan, D. (1995). *In sacco* disappearance Of starch nitrogen and fat in processed grains. *Anim. Feed sci. Techn.*, 51, pp. 287-295.

Bach Knudsen, K.E. (1997). Carbohydrate and lignin contents of plant materials used in animal feeding. *Anim. Feed Sci. Technol.*, 67, pp. 319–338.

Badgley, C., Moghtader, J., Quintero, E., Zakem, E., Chappell, M.J., Avilés-Vázquez, K., Samulon, A. & Perfecto, I., (2007). Organic agriculture and the global food supply. *Renew. Agric. Food Sys.*, 22, pp. 86–108.

Badiani, A., Stipa, S., Bitossi, F., Gatta, P.P., Vignola, G. & Chizzolini, R., (2002). Lipid composition, retention and oxidation in fresh and completely trimmed beef muscles as affected by common culinary practices. *Meat Sci.*, 60, pp. 169-186.

Bézard, J., Blond, J.P., Bernard, A. & Clouet, P., (1994). The metabolism and availability of essential fatty acids in animal and human tissues. *Reprod. Nutr. Dev.*, 34, pp. 539–568

Bittante, G., Chies, L., Fasone, V., Sinatra, M.C. & Gallo, L. (1989). Degradabilità e digeribilita *in vitro* della sostanza organica di cereali fioccati e/o macinati. *Zoot. Nutr. Anim.*, 15, pp. 613-626.

Brand, T.S., Brandt, D.A. & Cruywagen, C.W. (2004). Chemical composition, true metabolisable energy content and amino acid availability of grain legumes for poultry. *South African Journal of Animal Science,* 34, pp. 116-122.

Boccioni, A., Minieri, S., Petacchi, F. & Antongiovanni, M. (2007). Il favino (*Vicia faba minor*) e il pisello proteico (*Pisum sativum*) in sostituzione della farina di estrazione di soia nell'alimentazione zootecnica. Valutazione comparativa delle caratteristiche nutrizionali con metodi di laboratorio. *Arsia*, Toscana.

Bulson, H.A.J., Snaydon, R.W. & Stopes, C.E., (1997). Effects of plant density on intercropped wheat and field beans in an organic farming system. *J. Agric. Sci.*, 128, pp. 59–71.

Cabrera, C., Lloris, F., Giménez, R., Olalla, M. & López, M.C. (2003). Mineral content in legumes and nuts: contribution to the Spanish dietary intake. *The Science of total environment*, 308, pp. 1-14.

Calabrò et al. (2010) www.progettobufale.it/pubblicazionenew.html

Calabrò, S., Marchiello, M. & Piccolo, V. (2001a). *In vitro* fermentation of some cereals and legumes grains. *Proc. XIII Congr. Naz. ASPA* , Firenze (Italy), pp. 153-155.

Calabró, S., Piccolo, V., Bovera, F. & Sarubbi, F. (2001b). *In vitro* fermentation kinetics of crushed and flaked grains. *Proc. VI World Buffalo Congr.*, Maracaibo (Venezuela), pp. 478-484.

Calabrò, S., Tudisco, R., Balestrieri, A., Piccolo, G., Infascelli, F. & Cutrignelli, M.I. (2009). Fermentation characteristics of different grain cultivars with the *in vitro* gas production technique. *Ita, J. of Anim. Sci.*, 8, pp. 280-282.

Canibe, N., Bach Knudsen, K.E., (2002). Degradation and physicochemical changes of barley and pea fibre along the gastrointestinal tract of pigs. *J. Sci. Food Agric.*, 82, pp. 27–39.

Carnovale, E. & Nicoli, S., (2000). Changes in fatty acid composition in beef in Italy. *J. Food Compos. Anal.*, 13, pp. 505-510.

Chalk, P.M., Smith, C.J., Hamilton, S.D. & Hopmans, P., (1993). Characterization of the N benefit of a grain legume (Lupinus angustifolius L.) to a cereal (Hordeumvulgare L.) by an in situ 15N isotope dilution technique. *Biol. Fertil. Soils,* 15, pp. 39–44.

Chaubility, U.B., Gupta, R., Bird, S.H. & Mudgal, V.D., (1991). In sacco dry matter disappearance of different protein cakes in the rumen of male buffalo and their solubility in water. *Proceedings of the 3rd World Buffalo Congress*, Paestum (Italy), p. 165, Abstract.

Chumpawadee, S., Chantiratikul, A. & Chantiratikul, P. (2007). Chemical composition and nutritional evaluation of protein feeds for ruminants using an *in vitro* gas production technique. *Journal of Agricultural Technology* pp. 191-202

Cifuni, G.F., Napolitano F., Riviezzi, A.M., Braghieri, A. & Girolami, A., (2004). Fatty acid profile, cholesterol content and tenderness of meat from Podolian young bulls. *Meat Sci.,* 67, pp. 289-297.

Colombini, S.(2004). Resa proteica di alcune varietà di pisello, fava e lupino nell'Italia settentrionale. *Informatore agrario*, 60, pp. 73-75
Cone, J.W., van Gelder, A.H., Visscher, G.J.W. & Oudshoorn, L. (1996). Influence of rumen fluid and substrate concentration on fermentation kinetics measured with a fully automated time related gas production apparatus. *Animal Feed Science Technology*, 61, pp. 113-128.
Council Regulation (EC) No 834/2007 on organic production and labelling of organic products and repealing Regulation (EEC) No 2092/91 L 189/1- L 189/23.
Council Regulation (EEC) No 2092/91 of 24 June 1991 on organic production of agricultural products and indications referring thereto on agricultural products and foodstuffs L 198 , 22/07/1991 P. 0001 - 0015
Cutrignelli, M.I., (2000). Caratteristiche del taglio campione di mezzene di vitelloni Podolici sottoposti a differenti sistemi di allevamento. *Proc. Nat. Congr: Parliamo di... on Breeding in the third Millennium*, Fossano (CN), Italy, pp. 127-134
Cutrignelli, M.I., Piccolo, G., Bovera, F., Calabrò, S., D'Urso, S., Tudisco, R. & Infascelli, F. (2008a) Effects of two protein sources and energy level of diet on the performance of young Marchigiana bulls. 1. Infra vitam performance and carcass quality. *Ital.J.Anim.Sci.*, 7, pp. 259-270
Cutrignelli, M.I., Calabrò, S., Bovera, F., Tudisco, R., D'Urso, S., Marchiello, M., Piccolo, V. & Infascelli, F. (2008b) Effects of two protein sources and energy level of diet on the performance of young Marchigiana bulls. 2. Meat qualità. *Ital.J.Anim.Sci.*, 7, pp. 271-285.
Degussa, (2006). The amino acid composition of feedstuffs. 5th rev. ed., Degussa AG, Feed Additives, Hanau, Germany.
Di Francia, A., De Rosa, G., Masucci, F., Romano, R., Borriello, I. & Grassi, C. (2007). Effect of Pisum sativum as protein supplement on buffalo milk production. *Ital.J.Anim.Sci.*, 6, pp. 472-475
Díaz -Ambrona, C.H. & Mínguez, M.I. (2001). Cereal-legume rotations in a Mediterranean environment: biomass and yield production. *Field Crops Research*, 70, pp. 139- 51
DLG, (1999). Kleiner Helfer für die Berechnung von Futterrationen. Wiederkäuer und Schweine. DLG-Verlag, 10. Auflage, Frankfurt am Main, Germany.
Erbas, M., Certel, M. & Uslu, M.K., (2005). Some chemical properties of white lupin seeds (Lupinus albus L.). *Food Chem.*, 89, pp. 341–345.
European Commission directive, 999/2001. Regulation (EC) No. 999/2001 of the European Parliament and of the Council of 22 May 2001 laying down rules for the prevention, control and eradication of certain transmissible spongiform encephalopathies. *Official Journal of the European Communities* L 147/1-L 147/40.
European Commission, 1992. Council Regulation of 14 July 1992 concerning the protection of geographical indications and designations of origin for agricultural products and foodstuffs, 92/2081/EC. *Official Journal*, L 208, 24/07/1992, pp. 1-8.
European Commission, (2004). European Parliament and Council Regulation of the of 29 April 2004 concerning the official controls performed to ensure the verification of compliance with feed and food law, animal health and animal welfare rules, 882/2004/EC. *Official Journal*, L 165, 30/04/04, pp 1-141.
Evans, J., McNeil', A.M., Unkovich, M.J., Fettell, N.A. & Heenan, D.P., (2001). Net nitrogen balances for cool-season grain legume crops and contributions to wheat nitrogen uptake: a review. *Aust. J. Exp. Agric.*, 41, pp. 347–359.
Floroni, A. (2002). I.G.P. "Vitellone Bianco dell'Appennino Centrale". *L'Allevatore*, 58, pp. 12-13.

Froud-Williams, R.J., (1988). Changes in weed flora with different tillage and agronomic management systems. In: Altieri, M.A. & Liebman, M. (Eds.), *Weed Management in Agroecosystems: Ecological Approaches*. CRC, Boca Raton, Florida, pp. 213–236.

Gatel, F., (1994). Protein quality of legume seeds for non-ruminant animals: a literature review. *Anim. Feed Sci. Technol.*, 45, pp. 317–348.

Getachew, G., Robinson, P.H., DePeters, E.J. & Taylor, S.J. (2004). Relationships between chemical composition, dry matter degradation and *in vitro* gas production of several ruminant feeds. *Animal Feed Science Technology*, 111, pp. 57-71.

Gill, C., (1997). World feed panorama. High cost of feedstuffs: global impact, response. *Feed International*, 18, pp. 6–16.

Green, C.J. & Blackmer, A.M., (1995). Residue decomposition effects on nitrogen availability to corn following corn or soybean. *Soil Sci. Soc. Am. J.*, 59, pp. 1065–1070.

Groot, J.C.J., Cone, J.W., Williams, B.A. & Debersaques, F.M.A., (1996). Multiphasic analysis of gas production kinetics for *in vitro* fermentation of ruminant feedstuff. *Anim. Feed Sci. Tech.*, 64, pp. 77-89.

Herrero, M., I. Murray, R.H. Fawcett & Dent, J.B. (1996). Prediction of *in vitro* gas production and chemical composition of kikuyu grass by near-infrared reflectance spectroscopy. *Animal Feed Science Technology*, 60, pp. 51-67.

Herridge, D.F., Macellos, H., Felton, W.L., Turner, G.L. & Peoples, M.B., (1995). Chickpea increases soil-N fertility in cereal systems through nitrate sparing and N_2 fixation. *Soil Biol. Biochem.*, 27, pp. 545–551.

Huisman, J. & Tolman, G.H., (2001). Antinutritional factors in the plant proteins of diets for non-ruminants. In: Garnsworthy, P.C. & Wiseman, J. (Eds.), Recent Developments in Pig Nutrition, vol. 3. Nottingham University Press, Nottingham, pp. 261–322.

Infascelli, F., Di Lella, T. & Piccolo, V., (1995). Dry matter, organic matter and crude protein degradability of high protein feeds in buffaloes and sheep. *Zoot. Nutr. Anim.*, 21, pp. 89–94.

INRA (1988). Alimentation des bovins, ovins & caprins. Paris.

Jezierny, D., Mosenthin, R. & Bauer, E. (2010). The use of grain legumes as a protein source in pig nutrition: A review. *Animal Feed Science and Technology*, 157, pp. 111–128.

Jezierny, D., (2009). *In vivo* and *in vitro* studies with growing pigs on standardised ileal amino acid digestibilities in grain legumes. Ph.D. Thesis. University of Hohenheim, Stuttgart, Cuvillier Verlag Göttingen, Germany.

Jezierny, D., Mosenthin, R., Eklund, M. & Rademacher, M., (2007). Determination of standardized ileal digestibilities of crude protein and amino acids in legume seeds for growing pigs. *Proceedings of the 16th International Science Symposium on Nutrition of Domestic Animals*, Radenci, pp. 198–203.

Keane, M.G., (2003). Beef production from Holstein-Friesian bulls and steers of New Zeland and European/American descent, and Belgian Blue x Holstein-Fresians slaughtered at two weights. *Livest. Prod. Sci.*, 84, pp. 207-218.

Kim, C.M., Kim, J.H., Chung, T.Y. & Park, K.K., (2004). Effect of flaxseed diets an fattening response of Hanwoo cattle: 2 fatty acid composition of serum and adipose tissues. *Asian-Aust. J. Anim. Sci.*, 9, pp. 1246-1254.

Lallès, J.P. & Jansman, A.J.M., (1998). Recent progress in the understanding of the mode of action and effects of antinutritional factors from legume seeds in non-ruminant farm animals. In: Jansman, A.J.M., Hill, G.D., Liebman, M. & Dyck, E.,. Crop rotation and intercropping strategies for weed management. *Ecol. Appl.* 3, pp. 92–122.

Liebman, M. & Janke, R., (1990). Sustainable weed management practices. In: Francis, C.A., Flora, C.B. & King, L.D. (Eds.), Sustainable Agriculture in Temperate Zones. John Wiley&Sons, New York, pp. 111-143.

López-Bellido, F.J., López-Bellido, L. & López-Bellido, R.J., (2005). Competition, growth and yield of faba bean (Vicia faba L.). *Eur. J. Agron.*, 23, pp. 359-378.

Medsen, J., T. Hvelplund & Weisbjerg, M.R. (1997). Appropriate method for the evaluation of tropical feeds for ruminants. *Animal Feed Science Technology,* 69, pp. 53-66.

Morbidini, L., Rossetti, E., Cozza, F. & Pauselli, M. (2005). Different protein source (soybean or faba bean) in postweaning diets for Apennine and Sopravissana (Italian Merino) light lamb: slaughtering performances. *Ital.J.Anim.Sci.*, 4, pp. 360-362.

Morbidini, L., Pauselli, M., Rossetti, E. & Cozza, F., (2004). Diete post-svezzamento con differente fonte proteica (favino o soja) per la produzione di agnelli leggeri di razza Appenninica e Sopravissana: rilievi *in vivo. Proc. Congr. "Parliamo di ... nuove normative in campo zootecnico"*, Cuneo, Italy, pp. 113-122.

Moss, A.R., Givens, D.I., Grundy, H.F. & Wheeler, K.P.A., (1997). The nutritive value for ruminants of lupin seeds from determinate plants and their replacement of soybean meal in diets for young growing cattle. *Anim. Feed Sci. Tech.*, 68, pp. 11-23.

Mould, F.L., Kliem, K.E. Morgan, R. & Mauricio, R.M, (2005). *In vitro* microbial inoculum: A review of its function and properties *Anim. Feed Sci. Technol.,* 123, pp. 31-50

Murphy, S.R. & McNiven, M.A., (1994). Raw or roasted lupin supplementation of grass silage diets for beef steers *Anim. Feed Sci. Tech.*, 46, pp. 23-35.

NRC (2001). Nutrient Requirements of Dairy Cattle: Seventh Revised Edition, 2001. National Academy Press, Washington, D.C.

NRC, (1998). Nutrient requirements of swine. 10th ed., National Academy Press, Washington, D.C.

Pen, B., Oyabu, T., Hidaka, S. & Hidari, H., (2005). Effect of potato by-products based silage on growth performance, carcass characteristics and fatty acid composition of carcass fats in Holstein steers. *Asian-Aust. J. Anim. Sci.*, 4, pp. 490-496.

Poli, B.M, Giorgetti, A., Bozzi, R., Funghi, R., Balò, F.& Lucifero, M., (1996). Quantity and quality of lipid fractions for human nutrition in Chianina muscles as influenced by age and nutritive level. In: G. Enne and G.F. Greppi (eds.) Food & Health: role of animal products, *Proc. 31st Int. Symp. of Società Italiana per il Progresso della Zootecnia* held in Milano, Italy, Elsevier publ., Paris, France, pp. 199-204.

Raes, K., Balcaen, A, Dirinck, P., De Winne, A., Claeys, E., Demeyer, D. & De Smet, S., (2003). Meat quality, fatty acid composition and flavour analysis in Belgian retail beef. *Meat Sci.*, 65, pp. 1237-1246.

Rubio, L.A. & Brenes, A., (1995). Utilization de leguminosas-grano en nutricion animal: problemas y perspectives. Proc. XI Curso de Especializacion FEDNA, Barcelona, Spain. www.cirval.asso.fr.

Rusman, S. & Setiyono, A.S., (2003). Characteristics of Biceps femoris and Longissimus thoracis muscles of five cattle breeds grown in a feedlot system. *J. Anim. Sci.*, 74, pp. 59-65.

Salgado, P., Freire, J.P.B., Mourato, M., Cabral, F., Toullec, R. & Lallès, J.P., (2002a). Comparative effects of different legume protein sources in weaned piglets: nutrient digestibility, intestinal morphology and digestive enzymes. *Livest. Prod. Sci.*, 74, pp. 191-202.

Salgado, P., Martins, J.M., Carvalho, F., Abreu, M., Freire, J.P.B., Toullec, R., Lallès, J.P. & Bento, O., (2002b). Component digestibility of lupin (Lupinus angustifolius) and pea (Pisum sativum) seeds and effects on the small intestine and body organs in anastomosed and intact growing pigs. *Anim. Feed Sci. Technol.*, 98, pp. 187–201

Sami, A.S., Augustini, C. & Schwarz, F.J., (2004). Effect of feeding intensity and time on feed on performance, carcass characteristics and meat quality of Simmental bulls. *Meat Sci.*, 67, pp. 195-201.

Sankara Rao, D.S. &, Deosthale & Y.G. (2006). Mineral Composition of Four Indian Food Legumes. *Journal of Food Science*, 46, pp. 1962–1963

Sarubbi, F. (1999). Studi sull'impiego di proteasi tioliche per la previsione della degradabilita ruminale delle proteine e sule modificazioni dell' attivita ruminale indotte nei bufali da razioni arricchite con *Aspergillus oryzae*. PhD Thesis, Univ. Napoli, 1-23. 1999.

Schofield, P. (2000). Gas production methods. In "Farm Animal Metabolism and Nutriion", J.P.F. D'Mello, ed. CAB International, Wallingford, Oxon, U.K, pp. 209-232.

Scollan, N., Hocquette, JF., Neuerbetg, K., Dannenberger, D, Richardson, I. & Moloney, A., (2006). Innovations in beef production system that enhance the nutritional and health value of beef lipids and their relationship with meat quality. *Meat Sci.*, 74, pp. 179-33.

Singh, R.J., Chung, G.H. & Nelson, R.L., (2007). Landmark research in legumes. *Genome*, 50, pp. 525–537.

Sniffen, C. J., J. D. O'Connor, P. J. Van Soest, D. G. Fox & J. B. Russell. (1992). A net carbohydrate and protein system for evaluating cattle diets: 11. Carbohydrate and protein availability. *J. Anim. Sci.*, 70, pp. 3562.

Sprent, J.I. & Thomas, R.J., (1984). Nitrogen nutrition of seedling grain legumes: some taxonomic, morphological and physiological constraints. *Plant Cell Environ.*, 7, pp. 637–645.

Theander, O., Westerlund, E., Åman, P. & Graham, H., (1989). Plant cell walls and monogastric diets. *Anim. Feed Sci. Technol.*, 23, pp. 205–225

Theodorou, M.K. (1994.) A new laboratory procedure for determining the fermentation Kinetics of ruminants feeds. *Ciencia e Investigcion Agraria*, 20, pp. 332-334

Trostle, R., (2008). Global agricultural supply and demand: Factors contributing to the recent increase in food commodity prices. A Report from the Economic Research Service. United States Department of Agriculture Economic Research Service, Washington, DC

Ulbricht, T.L.V. & Southgate, D.A.T., (1991). Coronary heart disease: seven dietary factors. *Lancet*, 338, pp. 985-992.

Warren, HE., Enser, M., Richardson, I., Wood, JD.& Scollan, N.D., (2003). Effect of breed and diet on total lipid and selected shelf-life parameters in beef muscle. *Proc Nat. Congr. BSAS*, York, UK. pp. 43.

Westerling, D.B. & Hedrick, H.B., (1979). Fatty acid composition of bovine lipids as influenced by diet, sex and anatomical location and relationship to sensory characteristics. *J. Anim. Sci.*, 48, pp. 1343-1348.

Wheeler, T.L., Davis, G.W., Stoeker, B.J.& Hatmon, C.J., (1987). Cholesterol concentration of Longissimus muscle, subcutaneous fat and serum of two beef cattle breed type. *J. Anim. Sci.*, 65, pp.1531-1537.

Wood, J.D., Richardson, R.I., Nute, G.R., Fisher, A.V., Campo, M.M., Kasapidou, E., Sheard, P.R. & Enser, M., (2003). Effects of fatty acids on meat quality: a review. *Meat Sci.*, 66, pp. 21-32.

Zahran, H.H., (1999). Rhizobium-legume symbiosis and nitrogen fixation under severe conditions and in an arid climate. *Microbiol. Mol. Biol. Rev.*, 63, pp. 968–989.

10

Soybean as a Feed Ingredient for Livestock and Poultry

H.K. Dei
Department of Animal Science, Faculty of Agriculture, University for Development Studies, Tamale, Ghana

1. Introduction

The need to meet animal protein demand of ever growing world population, currently at approximately 6.8 billion (US Census Bureau, 2010), is set to increase at an even greater rate as the economies of developing countries improve and their growing affluent populace alter their dietary habits. This means production of soybean, which is used extensively as animal feed, must increase beyond current production level of about 246 million metric tonnes (FAS/USDA, 2009).

Soybean (*Glycine max,* L) is not only a source of high quality edible oil for humans, but also a high quality vegetable protein in animal feed worldwide. Its universal acceptability in animal feed has been due to favourable attributes such as relatively high protein content and suitable amino acid profile except methionine, minimal variation in nutrient content, ready availability year-round, and relative freedom from intractable anti-nutritive factors if properly processed. Also, attention has been focused on soybean utilisation as an alternate protein source in animal diets due to the changing availability or allowed uses of animal proteins coupled with relatively low cost.

Despite soybean's pivotal role in animal production, it cannot be fed raw because there are a number of anti-nutritive factors (ANFs) present that exert a negative impact on the nutritional quality of the protein. The main ANFs are protease inhibitors (trypsin inhibitors) and lectins (Liener, 1994), which fortunately can be destroyed by heat treatment. The trypsin inhibitors cause pancreatic hypertrophy/hyperplasia with consequent inhibition of growth, while lectins inhibit growth by interfering with nutrient absorption (Liener, 1994). The elimination of these ANFs and those of less significance can be achieved through various processing methods. These methods have different impact on the nutritional quality of the products derived such as full-fat soybeans, soybean meal and soybean protein concentrates. Of these, soybean meal has been the major ingredient in both poultry and livestock diets.

This chapter discusses soybean production and consumption, primary soybean products and their nutritional value for feeding animals, anti-nutritive factors present and ways of eliminating them, and utilisation in animal feeds as well as future challenges of using soybeans as a major source of animal feed.

2. Soybean production and consumption

Soybean (*Glycine max,* L) is an annual crop that belongs to the Fabaceae or Leguminosae family. It originated from East Asia, but now grown over a wide geographical area worldwide with United States of America, Brazil and Argentina being the leading producers (Table 1). It is used primarily for production of vegetable oil and oilseed meal for animal feeding. The surge in the use of soybean meal in feeding animal as replacement protein source for animal protein feeds has been the main driving force in soybean production.
Table 1 shows the major soybean producing countries and their relative supplies. Generally, there has been an increase of supply with a slight depression in most producing countries between 2006 and 2008 cropping seasons. The US and China tend to consume virtually what they produce, while Argentina and Brazil are major exporters with exports largely to the EU (Table 2).

Major producing countries	2005/06	2006/07	2007/08	2008/09	2009/10 October
United States	83,507	87,001	72,859	80,749	88,454
Brazil	57,000	59,000	61,000	57,000	62,000
Argentina	40,500	48,800	46,200	32,000	52,500
China	16,350	15,967	14,000	15,500	14,500
India	7,000	7,690	9,470	9,100	9,000
Paraguay	3,640	5,856	6,900	3,900	6,700
Canada	3,161	3,460	2,700	3,300	3,500
Other	9,512	9,337	8,004	9,090	9,413
World Total	220,670	237,111	221,133	210,639	246,067

Source: FAS/USDA (2009)

Table 1. World soybean supply (million tonnes) and distribution.

Countries	2009/10		2010/11	
	Production	Consumption	Production	Consumption
China	37.42	35.82	41.71	40.36
US	37.31	27.22	35.41	27.58
Argentina	27.13	0.70	29.95	0.60
Brazil	24.41	12.80	25.42	13.38
EU	9.85	31.49	9.77	32.30
India	4.85	2.85	6.08	3.08
Others	20.58	49.60	21.30	51.20
World Total	161.63	159.77	169.64	167.89

Source: FAS/USDA (2009)

Table 2. World soybean meal production and consumption outlooks for 2010/11 in million tonnes.

3. Primary soybean products for animal feeding

Figure 1 shows a schematic processing of soybeans into various high quality protein products. The processes involved either reduce or eliminate the ANFs in the beans and improve the nutritional value substantially for all classes of animals. Several steps involved in processing these products can have either positive or negative effect on the quality of the protein depending on the conditions used in processing. The heat applied in processing is identified as the single most important factor that affects soybean meal protein quality. Proper processing conditions such as moisture content, heating time and temperature inactivate ANFs such as trypsin inhibitors and lectins, which results in improved performance when fed to monogastric animals (Araba, 1990). High processing temperatures of oilseeds has deleterious effects on proteins and amino acids due to formation of Maillard reaction products (Hurell, 1990) or denaturation (Parsons *et al.*, 1992).

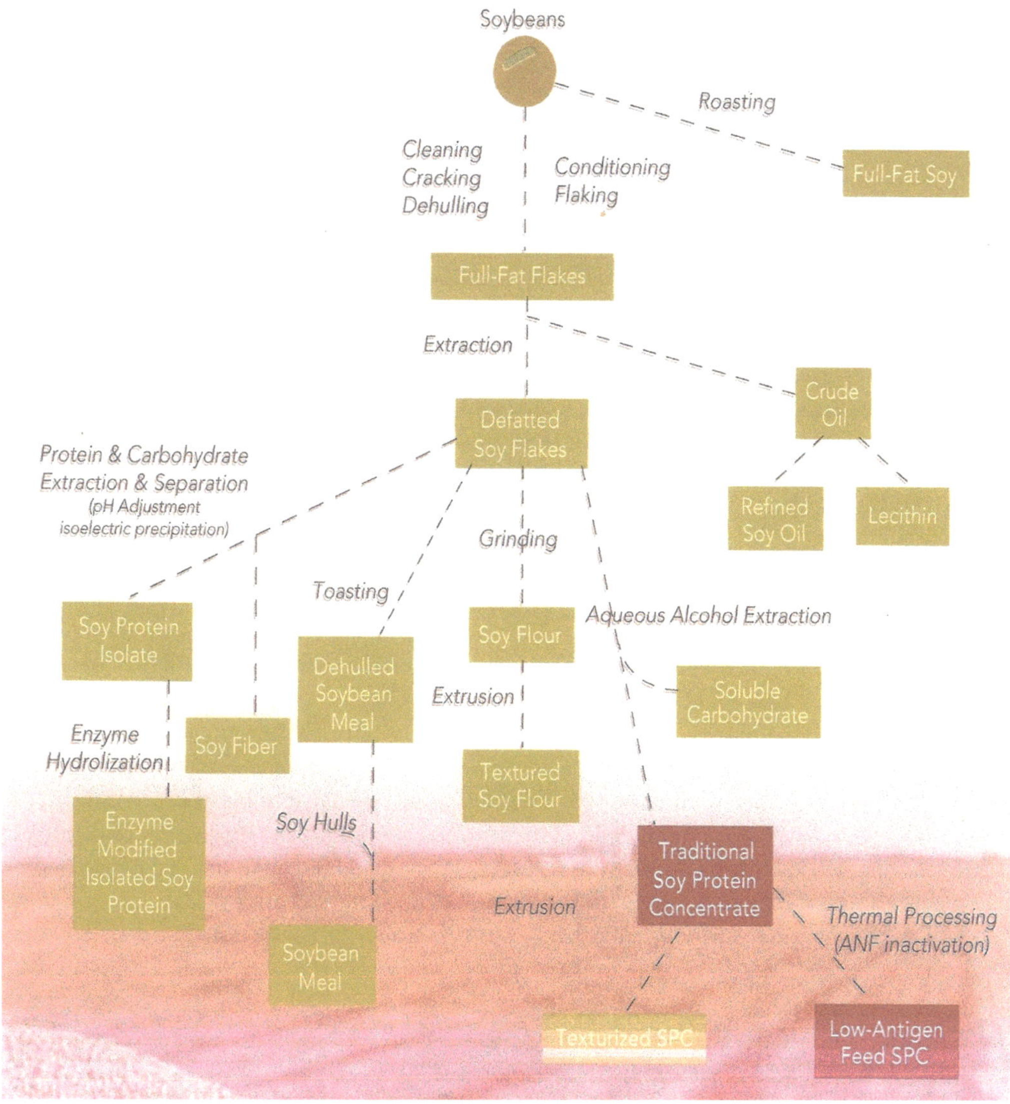

Fig. 1. Processing of soybeans into soybean products (USSEC, 2008).

3.1 Full-fat soybeans

These are whole soybeans in which the oil is not extracted. These products are produced by a variety of processes such as extruding (dry or wet), cooking/autoclaving, roasting/toasting, micronizing and jet-sploding to inactivate the ANFs. All of these processes have a different impact on the nutritive value of the products depending on heat damage or degree of inactivation of ANFs. Normally, soybeans are processed into defatted meals for feed formulation, particularly for poultry and pigs. However, the amount of full-fat soybeans used has been increasing in the livestock industry due to development of new varieties with limited number or levels of ANFs (Gu *et al.*, 2010). Also, properly processed full-fat soybeans are a valuable feed ingredient for animal feeding because of their high energy content.

3.2 Soybean meal

Soybeans yield 18.6% of oil and 78.7% of soybean meal with the rest being waste (FEFAC, 2007). The oil can be extracted either mechanically or by solvent means. There are two main types of soybean meal. The dehulled soybean meal and soybean meal, depending on whether the testa (seed coat) is removed or not. Both products vary in their nutrient composition, but are quite high in protein content with a good amino acid balance except methionine, low in fibre, high in energy, and have little or no anti-nutritive factors when properly processed.

The amino acid profile of soybean meal is close to that of fishmeal, except methionine (INRA, 2004). This deficiency can easily be corrected in monogastric diets using synthetic source of methionine. Also, soybean meal is superior to other vegetable protein sources in terms of crude protein content and matches or exceeds them in both total and digestible amino acid content (Table 3a). Soybean meal protein digestibility in poultry is approximately 85% (Woodworth *et al.*, 2001), ranging between 82% and 94% for individual amino acid digestibility. Among the vegetable protein sources, soybean meal is used to meet the animal's requirement for limiting amino acids in cereal-based (e.g. maize) diets (Table 3b), because it is usually the most cost-effective source of amino acids (Kerley and Allee, 2003).

The carbohydrates in soybean meal are incompletely digested by colonic microbiota in monogastrics (Kerley and Allee, 2003). Thus removal of raffinose and stachyose improved metabolisable energy content by 12% (Graham *et al.*, 2002).

3.3 Soybean protein concentrate (SPC)

SPC is produced from the defatted flakes by the removal of the soluble carbohydrates. This can be achieved by two methods, either by ethanol extraction or enzymatic degradation (Figure 1). SPC is valuable as milk replacer feed for calves and as piglet pre-starter feed. This is because it contains only traces of the heat-stable oligosaccharides and the antigenic substances (Table 5). In milk replacer feed, it has been largely substituted for dried skim milk; whilst in pig starter feeds it can replace dried skim milk, whey powder and fishmeal.

3.4 Soybean oil

Soybean oil is produced primarily for human consumption. However, it has become a useful source of feed-grade fat for animals due to a need to formulate high-energy diets for modern breeds. Feed-grade soybean oil is popularly used in high energy diets, particularly for

	Soybean	Canola	Cottonseed	Palm kernel	Peanut	Sunflower
Crude protein	43.0	36.2	39.6	13.2	45.2	32.8
Amino acid						
Lysine	90.7	78.6	62.8	58.9	78.1	80.4
Methionine	90.6	88.6	71.9	83.7	85.6	91.2
Cystine	82.1	73.1	70.9	66.6	78.5	79.2
Threonine	84.1	77.6	67.2	69.2	83.8	83.7
Tryptophan	87.9	80.0	80.3	-	75.6	-
Arginine	91.1	90.6	85.3	88.6	89.6	93.1
Isoleucine	91.2	89.0	72.8	81.0	89.3	88.9
Leucine	90.7	94.1	74.8	85.0	89.7	88.7
Valine	88.9	87.8	76.3	80.1	88.9	85.8
Histidine	88.5	88.5	64.1	80.3	85.4	86.1
Phenylalanine	91.6	91.6	84.0	85.3	92.3	90.8

Source: Ajinomoto Heartland Lysine LLC Revision 7 (2006)

Table 3a. True digestibility (%) of essential amino acids in common oilseed meal proteins for poultry.

Amino Acid[1]	Soybean meal (475.0 g/kg CP)	Palm kernel meal (200.0 g/kg CP)	Maize (85.0 g/kg CP)
Arginine	73.3	135.0	44.7
Histidine	26.9	23.0	27.1
Isoleucine	44.6	32.0	34.1
Leucine	78.7	60.0	117.6
Lysine	62.3	36.0	30.6
Methionine	14.1	20.0	21.2
Cystine	15.2	15.0	21.2
Phenylalanine	49.3	39.0	44.7
Threonine	39.4	35.0	34.1
Tryptophan	15.6	10.0	7.1
Valine	46.7	57.0	47.0

[1]Data are adapted from Elkin (2002).

Table 3b. Comparative amino acid composition (g/kg protein basis) of soybean meal with palm kernel meal and maize.

poultry, because of its high digestibility and metabolisable energy content compared with other vegetable fats/oils (Table 4a). It is used widely in rations for broiler chickens and growing turkeys as a feed-grade fat to increase energy density of feeds and improve efficiency of feed utilisation (Sell *et al.*, 1978). The high energy value of soybean oil is attributed to its high percentage of (poly) unsaturated fatty acids (Table 4b), which are well absorbed and utilised as a source of energy by the animal (Huyghebaert *et al.*, 1988). Also, the high polyunsaturated fatty acids (PUFA) in soybean oil appears to have an energy independent effect on improving reproduction in dairy cattle (Lucy *et al.*, 1990; Kerley and Allee, 2003), and this has been attributed to the role of linoleic acid in reproduction (Staples *et al.*, 1998).

Source	Digestibility (%)		Metabolisable energy (MJ/kg)	
	3-4 weeks	>4 weeks	1-3 weeks	7.5 weeks
Soybean oil	96+	96+	38.5*	38.5*
Corn (maize) oil	84+	95+	-	41.3#
Lard	92+	93+	30.8***	-
Beef tallow	70+	76+	30.9*	32.9*
Menhaden oil	88+	97+	35.9#	37.6#
Palm oil	74**	-	27.7*	32.3*
Sunflower oil	85****	88****	-	40.4#

+Leeson and Summers (2001), *Wiseman and Salvador (1991), ** Zumbado *et al.* (1999), ***Huyghebaert *et al.* (1988), ****Ortiz *et al.* (1998), #NRC (1994)

Table 4a. Comparison of digestibility and metabolisable energy values of triglycerides in broiler chickens fed soybean oil and selected dietary fats/oils.

Fatty acid	Soybean oil[2]	Palm oil[2]	Sunflower Oil[3]	Corn Oil[4]	Tallow[2]	Lard[5]	Poultry oil[4]
Lauric acid (C12:0)	1	3	-	-	2	-	-
Myristic acid (C14:0)	2	14	-	-	23	16	6
Palmitic acid (C16:0)	161	488	60	112	249	224	232
Palmitoleic acid (C16:1)	6	1	-	-	39	21	71
Stearic acid (C18:0)	61	55	64	21	206	177	64
Oleic acid (C18:1)	251	364	284	269	405	461	430
Linoleic acid (C18:2)	452	73	581	579	66	80	179
Linolenic acid (C18:3)	66	2	1	8	10	21	6
Arachidic acid (C20:0)	-	-	6	5	-	-	2

[1]Values may not total 1000 g due to trace amounts of other fatty acids not reported or rounding of figures

[2]Wiseman and Salvador (1991), [3]Ortiz et al. (1998), [4]Waldroup et al. (1995), [5]Huyghebaert et al. (1988)

Table 4b. Comparison of fatty acid composition of soybean oil with selected dietary fats/oils (g/kg total fatty acids)[1].

4. Chemical composition of commonly used soybean products in animal diets

There are variations in the reported chemical composition of soybean products that can be attributed to differences in processing methods (Table 5). Also, genetic variations have been observed in the soybean biotypes of *Glycine* (Yen *et al.*, 1971; Gu *et al.*, 2010), which may vary in their chemical compositions. The use of soybean products in non-ruminant diets can give reasonable performance only if diets are formulated correctly or their anti-nutritive factors removed. In this regard, nutrient levels, bioavailability, and anti-nutritive factors and their effects on animal performance must all be considered in determining the usefulness of any of the soybean products as a feed ingredient. Table 5 shows composition of some soybean products commonly used in animal feed. It is clear that soybean is a source of high protein content and quality as well as energy with little or no ANFs. It appears the quality of soybean proteins improves when subjected to multiple processing procedures. This is shown by increases in concentrations of limiting essential amino acids such as lysine and methionine for monogastric animals (Table 5). However, the cost of such improved products may limit their use in animal feeds.

	Full-fat soybean	Soybean Meal	Soy protein concentrate	Soy protein isolate
Dry matter	89.4	87.6 - 89.8	91.8	93.4
Crude protein	37.1	43.9 - 48.8	68.6	85.9
Crude fibre	5.1	3.4 - 6.3	1.7	1.3
Ether extract	18.4	1.3 - 5.7	2.0	0.6
Ash	4.9	5.7 - 6.3	5.2	3.4
NDF	13.0	10.0 - 21.4	13.5	-
ADF	7.2	5.0 - 10.2	5.4	-
ADL	4.3	0.4 - 1.2	0.4	-
Starch	4.7	3.3 - 7.0	-	-
Total sugars	-	9.1 - 9.3	-	-
Gross energy (MJ/kg)	20.95	17.22 – 17.41	17.89	22.45
Lysine	2.34	2.85 - 3.50	4.59	5.26
Methionine	0.52	0.62 - 0.80	0.87	1.01
Cystine	0.55	0.68 - 0.77	0.89	1.19
Tryptophan	0.49	0.56 - 0.74	0.81	1.08
Calcium	0.26	0.27 - 0.31	0.24	0.15
Phosphorus	0.57	0.64 - 0.66	0.76	0.65
Linoleic acid	9.7	0.6 - 2.9	-	-
Urease activity (pH-rise)	2.0	0.05 - 0.5	<0.05	<0.05
Trypsin inhibitor (mg/g)	45-50	1 – 8	2	<1
Glycinin (ppm)	180.000	66.000	<100	-
β-conglycinin (ppm)	>60.000	16.000	<10	-
Lectins (ppm)	3.5000	10 – 200	<1	0
Oligosaccharides (%)	14	15	3	0
Saponins (%)	0.5	0.6	0	0

Data are adapted from NRC (1994), INRA (2004), Peisker (2001)

Table 5. Per cent composition of some soybean products used in animal feed.

4.1 Anti-nutritive factors

Anti-nutritive factors are natural compounds in feedstuffs that impair utilisation of nutrients with consequent undesirable effects on animal performance. The ANFs in soybeans exert a negative impact on the nutritional quality for animals (Table 6). Fortunately, those ANFs with significant impact such as trypsin inhibitors and lectins are easily destroyed by heat. Of lesser significance are the anti-nutritional effects produced by relatively heat stable factors, such as goitrogens, tannins, phytoestrogens, oligosaccharides, phytate, and saponins (Liener, 994). Heat stable ANFs with the exception of oligosaccharides and the antigenic factors are low in soybeans and not quite likely to cause problems under practical feeding conditions. The removal of the oligosaccharides and antigens in the manufacture of soybean protein concentrates further improves the nutritional value.

Anti-nutritional factor	Mode of action	Method of detoxification
Protease inhibitors	Combines with trypsin or chymotrypsin to form an inactive complex and lower protein digestibility Causes hypertrophy of the pancreas Counteracts feedback inhibition of pancreatic enzyme secretion by trypsin	Heat treatment Germination Fermentation
Lectins (Phytohaemagglutinins)	Agglutinates red blood cells	Heat treatments
Anti-vitamin factors (rachitogenic factor and anti-vitamin B12 factor)	These factors render certain vitamins (e.g. vitamins A, B_{12}, D, and E) physiologically inactive	Cooking Supplementation of vitamins
Goitrogens	Enlargement of the thyroid	Heat treatment in some cases Administration of iodide
Metal-binding factors (phytate)	These factors decrease availability of certain minerals (e.g. P, Cu, Fe, Mn, Zn)	Heat treatment Addition of chelating agents Use of enzymes
Saponins	Bitter taste, hemolyze red blood cells	Fermentation
Estrogens	Cause an enlargement of the reproductive tract	
Cyanogens	Cause toxicity through the poisonous hydrogen cyanide	Cooking
Oligosaccharides	Impair digestion (e.g. intestinal cramps, diarrhoea, and flatulence)	Ethanol/water extraction
Antigens (glycinin and β-conglycinin)	Cause the formation of antibodies in the serum of calves and piglets. Prevent proliferation of beneficial bacteria in the GIT	Ethanol/water extraction

Sources: Liener (1977), Ensminger and Olentine Jr (1978), Peisker (2001)

Table 6. Anti-nutritive factors in soybeans.

Soybean meal contains high levels of phosphorus, but much of it is present in a complex form due to the presence of phytic acid. However, the use of phytase can increase phosphorus retention by 50% and reduction in excretion by 42% (Lei *et al.*, 1993).

5. Utilisation of soybean in animal production

The major farmed animal species diets containing soybean include poultry, pigs, cattle and aquatic. The global animal feed production by species a decade ago included pigs (31%), broiler (27%), dairy cattle (17%), beef cattle (9%), layer (8%), aquatic (5%) and 3% of others (Hoffman, 1999). Thus soybean meal is used relatively more in some types of animal feed than in others. The major aim is to provide high quality protein to poultry and pigs.

Of all plant protein sources, soybean cultivation alone occupies most land needed for production of animal products. For example, soybean meal is used extensively in animal compound feed in the United States (Table 7a) and European Union (Table 7b). The annual EU livestock consumption alone demands soybean acreage of 5.0 million hectares in Brazil and 4.2 million hectares in Argentina (Table 7c).

Species	Million metric tons	Percent of total
Poultry - broilers	12.36	44
Poultry - layers	1.88	7
Swine	6.69	24
Cattle - beef	3.45	13
Cattle - dairy	1.61	6
Pet animals	0.74	3
Aquaculture	0.18	1
Other	0.65	2
Total	27.56	100

Source: United Soybean Board (1999/2000)

Table 7a. Utilisation of soybean meal by livestock in the United States

Type of animal compound feed	Production volume (1,000 tonnes)	Estimated soy bean meal content (%)	Volume of soybean meal in compound feed (1,000 tonnes)
Cattle - meat	12,148	13.9	1,683
Cattle - dairy	27,852	10.4	2,893
Pigs	51,440	28.8	14,815
Poultry - broilers	30,929	36.8	11,389
Poultry - layers	15,532	22.4	3,477
Other animals (e.g. sheep, goats, ducks, etc)	9,522	16.6	1,577
Total	147,423	24.3	35,834

Source: PROFUNDO (2008)

Table 7b. Soybean meal used in types of animal compound feed in the European Union-27.

Livestock product	Soybean Equivalent[1] (1,000 tonnes)	Acreage (ha)	Country of origin	Soybean Equivalent (1,000 tonnes)[1]	Acreage (ha)
Beef and veal	1,557	595,519	United States	2,102	781,256
Milk	621	237,642	Canada	463	182,290
Pork	10,341	3,956,061	Argentina	11,450	4,240,559
Poultry meat	7,934	3,035,314	Brazil	12,789	4,995,608
Eggs	3,247	1,242,109	Paraguay	585	263,553
Cheese	1,156	442,402	Uruguay	53	26,319
Other products	2,764	1,057,330	Other countries	180	76,791
Total	27,620	10,566,377	Total	27,621	10,566,377

Source: PROFUNDO (2008) 1,000 tonnes of soybean meal = 771 tonnes of soybeans.

Table 7c. Soybean acreage needed for livestock consumption in the European Union-27 and by country of origin.

6. Future challenges of soybean utilisation in animal diets

Future challenges confronting soybean utilisation in animal diets have been discussed by Kerley and Allee (2003). The major challenges include the following:

- Increased demand for vegetable oil for biodiesel production may in turn reduce overall production of soybean in favour of other oilseed crops that produce more oil per acre. For instance, soybean produces about 36 litres of oil per acre compared to 72 litres of safflower, 84 litres of sunflower and 108 litres of canola (United Soybean Board, 2011). Even though the nutritional values of meals from these oilseeds are lower than that of soybean, the increased value of the oil may shift production to these crops at the expense of soybean.
- Competition between the bio-fuel industry and animal agriculture has increased the prices of feed ingredients with consequent increase in feeding cost. Also, by-products from ethanol and biodiesel production (e.g. distillers dried grains with soluble) are now competing with maize and soybean meal for their place in animal diets.
- Demands on animal production exerted by environmental regulations as a result of nitrogen waste, malodour and excretion of phosphorus into the environment by the use of soybean in diets.
- Pressures to improve nutritional value of soybean through breeding to modify aspects such as anti-nutritive factors, fatty acid profile, and oligosaccharide or protein synthesis in order to allow greater levels of soybean meal in animal diets.

7. Conclusion

Soybean is the major vegetable protein source in the animal feed industry. Its universal acceptability in animal feed is as a result of important attributes such as relatively high protein content and suitable amino acid profile except methionine, minimal variation in nutrient content, ready availability year-round, and relative freedom from intractable anti-nutritive factors if properly processed, limited allowable uses of animal proteins in feed and its relatively low cost. Therefore, its production and consumption will continue to grow as a preferred source of alternate high quality protein in animal diets.

Commonly used soybean products as protein source in animal feed are soybean meal, full-fat soybean and soybean protein concentrates, which are obtained through various heat processing methods that reduce anti-nutritive factors present such as trypsin inhibitors and lectins. Of these products, soybean meal is most preferred due to its relatively low cost. It is used extensively in feeds for poultry, pigs and cattle.

Soybean is also a major source of vegetable fat in animal feed. Feed-grade soybean oil is popularly used in high energy diets, particularly for poultry, because of its high digestibility and metabolisable energy content compared with other vegetable fats/oils.

Soybean production and utilization for animal feed is bound to face future challenges as a result of increased demand of vegetable oil for biofuel production; of which soybean is less competitive. There is also increased research to use co-products from biofuel production as substitutes for soybean meal in animal diets. Thus, there is a need to overcome these and other challenges in order not to jeopardise cheap meat production for ever increasing world population.

8. References

Ajinomoto Heartland Lysine LLC Revision 7. True digestibility of essential amino acids in poultry. http://www.lysine.com/new/Technical%20Reports/Poultry/PoultryDigTableV7.pdf.

Araba, M. and N.M. Dale. 1990. Evaluation of protein solubility as an indicator of under processing soybean meal. *Poultry Science,* 69:1749-1752

Elkin, R.G. 2002. Nutritional components of feedstuffs: a qualitative chemical appraisal of protein. In: J.M. McNab and K.N. Boorman, eds. *Poultry Feedstuffs: Supply, Composition and Nutritive Value, pp. 57-86.* CAB International, UK

Ensminger, M.E. and C. G. Olentine Jr. 1978. *Feeds and Nutrition Complete, 1st ed.* Ensminger Publishing Co., California, USA.

FAS/USDA (Foreign Agricultural Service/United States Department of Agriculture). 2009. World soybean supply and distribution. FAS/USDA, Washington, D.C.

FEFAC (European Feed Manufacturers Federation). 2007. Industrial compound feed production, FEFAC Secretariat General, Brussels, May 2007.

Graham, K.K., Kerley, M.S., Firman, J.D., and G.L. Allee. 2002. The effect of enzyme treatment of soybean meal on oligosaccharide disappearance and chick growth performance. *Poultry Science, 81: 1014-1019.*

Gu, C., Pan, H., Sun, Z. and G. Qin. 2010. Effect of soybean variety on antinutritional factors content, and growth performance and nutrients metabolism in rat. *International Journal and Molecular Science, 11: 1048-1056.*

Hoffman, P. 1999. Prospects for feed demand recovery. International Feed Markets '99, October 1999, Agra Europe.

Hurrell, R.F. 1990. Influence of the Maillard Reaction on nutritional value of foods. In: *The Maillard Reaction in foods processing, human nutrition and physiology, pp 245-358.* Birkhauser Verlag, Basel, Switzerland.

Huyghebaert, G., G.D. Munter and G.D. Groote. 1988. The metabolisable energy (AMEn) of fats for broilers in relation to their chemical composition. *Animal Feed Science and Technology, 20: 45-58.*

INRA (Institut Scientifique de Recherche Agronomique). 2004. *Tables of composition and nutritional value of feed materials, 2 ed.* (Sauvant, D., Perez, J.M. and Tran, G., eds.), p.186. Wageningen Academic Publishers, Netherlands.

Kerley, M.S. and G.L. Allee. 2003. Modifications in soybean seed composition to enhance Animal feed use and value: Moving from dietary ingredient to a functional dietary component. *AgBioForum, 6 (1&2)14-17.*

Lei, X.G., Ku, P.K., Miller, E.R. and M.T. Yokoyama. 1993. Supplementing corn-soybean meal diets with microbial phytase linearly improves phytate phosphorus utilisation by weanling pigs. *Journal of Animal Science, 71:3359-3367.*

Leeson, S. and J.D. Summers. 2001. *Nutrition of the chicken, 4th ed.* University Books, Ontario, Canada.

Liener, I.E. 1977. Removal of naturally occurring toxicants through enzymatic processing. In: R.E. Feeney and J.R. Whitaker (Eds.), *Food proteins: Improvements through chemical and enzymatic modifications (p.7-57).* Advances in Chemistry Series 160. Washington, D.C.

Lucy, M.C., T.S. Gross and W.W. Thatcher. 1990. Effect of intravenous infusion of a soybean oil emulsion on plasma concentration of 15-keto-13, 14-dihydro-prostaglandin F2a and ovarian function in cycling Holstein heifers. *In*: Livestock reproduction in Latin America. Internaqtional Atomic Energy Agency, Vienna, Austria.

National Research Council. 1994. *Nutrient Requirements of Poultry, 9th .rev. ed.* National Academy Press, Washington, DC.

Ortiz, L.T., Rebole, A., Rodriquez, M.L., Trevino, J., Alzueta, C and B. Isabela. 1988. Effect of chicken age on the nutritive value of diets with graded additions of full-fat sunflower seed. *British Poultry Science, 39: 530-535.*

Parsons, C.M., Hatsimoto, K., Wedeking, K.J., Han, Y. and D.H. Baker. 1992. Effect of over-processing on availability of amino acids and energy in soya bean meal. *Poultry Science, 71,* 133-140.

Peisker, M. 2001. Manufacturing of soy protein concentrate for animal nutrition. *Cahiers Options Mediterraneennes, 54: 103-107.*

PROFUNDO. 2008. Soy consumption for feed and fuel in the European Union (J.W. van Gelder, K. Kammeraat, and H. Kroes, eds.). The Netherlands.

US Census Bureau. 2010, World Population, International database, Washington, DC United Soybean Board 2011. Soybean meal-Demand, Soybean meal information centre, 2010, United Soybean Board, Washington, D.C.

USSEC (United States Soybean Export Council). 2008. Processing of soybeans into soybean products. Annual Report, 2008. Washington, D.C.

Sell, J.L., L.G. Tenesaca, and G.L. Bales. 1978. Influence of dietary fat on energy utilisation by laying hens. *Poultry Science, 58: 900-905.*

Staples, C.R., J.M. Burke, and W.W. Thatcher. 1998. Influence of supplemental fats on reproductive tissues and performance of lactating cows. *Journal of Dairy Science,* 81: 856-871

Waldroup, P.W., S.E Walkins, and EA. Saleh, 1995. Comparison of two blended animal-vegetable fats having low or high free fatty acid content. *Journal of Applied Poultry Research, 4:4143.*

Woodworth, J.C., Tokach, M.D., Goodband, R.D., Nelssen, J.L., O'Quinn, P.R., Knabe, D.A. and N.W. Said. 2001. Apparent ileal digestibility of amino acids and digestible and metabolisable energy content of dry extruded-expelled soybean meal and its effect on growth performance of pigs. *Journal of Animal Science, 79:1280-1287.*

Wiseman, J. and F. Salvador. 1991. The influence of free fatty acid content and degree of saturation on the apparent metabolisable energy value of fats fed to broilers. *Poultry Science,* 70, 573-582.

Yen, J.T., T. Hymowitz and A.H. Jensen. 1971. Utilization by rat of proteins from a trypsin-inhibitor variant soybean. *Journal of Animal Science, 33: 1012-1017.*

Zumbado ME, Scheele CW, C. Kwakernaak. 1999. Chemical composition, digestibility, and metabolizable energy content of different fat and oil by-products. *Journal Applied Poultry Research, 8:263–271.*

11

Productivity and Nutritional Composition of *Lentinus strigosus* (Schwinitz) Fries Mushroom from the Amazon Region Cultivated in Sawdust Supplemented with Soy Bran

Ceci Sales-Campos, Bazilio Frasco Vianez
and Raimunda Liége Souza de Abreu
National Institute for Amazonian Research –INPA/ Department of Forest Products – CPPF, Aleixo CEP-Manaus, AM Brazil

1. Introduction

The cultivation of edible mushrooms is a biotechnological process that uses various residues to produce food of high nutritional value. It is an activity of economical importance, in particular, the production of *Agaricus*, *Pleurotus* and *Lentinus* species (Guzmán et al., 1993). Nutraceutical properties of mushrooms are increasing its economic value. The cultivation of mushrooms can be a solution to problems of global importance, such as the lack of protein in developing countries and the possibility of environmental management. The use of organic materials for growing mushrooms is an indication of its extraordinary metabolic activity.

In the Amazon region there are large amounts of wood and agricultural residues whose potential has been underestimated. In the timber industry, raw material waste can be as high as 60%. In the agricultural industry there is no data on how much waste their activities produce. Vianez and Barbosa (2003), suggest several alternatives for the use of wood residues, including the use for the cultivation of edible mushrooms. This activity could contribute to a sustainable regional development. In this way, the objective of this study is to study the feasibility of using sawdust supplemented with soy bran for growing *L. strigosus*, a native mushroom of the Amazon region.

2. Literature review

The fungus proposed in the present study is a wild edible and a wood decomposer (white-rot fungus), whose domestication was sought for the production of mushroom. As there is no cultivation of this fungus with the proposed wild strain, for comparison in the literature review, a parallel association was made with the cultivation of species of edible fungi of related genera that have similar physiology and cultivation conditions, being considered mainly the genera *Lentinus* and *Pleurotus*.

2.1 History

Fossil finds have revealed that fungi exist since the Cretaceous period (approximately 130 million years ago), long before humanity (Chang, 1993). Fungi (mushrooms), also called macromycets, belong to the Fungi Kingdom, being known by man since the most remote period of human history. Edible mushrooms were first collected by man in China and dates from 5000-4000 BC. (Zhanxi and Zhanhua, 2001). It is estimated that the first cultivation of edible mushrooms in China started in the early 7th century, with the species *Auricularia auricula* (Chang and Miles, 1987). China is a country with a long tradition in cultivation and consumption of mushrooms and according to Zhanxi and Zhanhua (2001), it has more than ten species of fungi which are currently cultivated in several countries of the world. That country is a pioneer in the cultivation and consumption of edible and medicinal mushrooms, followed by Japan, Europe and The United States (Urben et al., 2001).

Edible mushrooms were described as the "food of the Gods" and as such, confirmed by Roman gourmets who appreciated them as a kind of spice. The Chinese considered them as the "elixir of life". The Greeks believed that the mushrooms were able to give strength to warriors in battles and the Egyptian pharaohs also nourished themselves on these spices (Chang and Miles, 1984). Mushrooms had a wide acceptance, and some species are considered as "Kings of the dining table" or "kitchen diamonds" (Zhanxi and Zhanhua, 2001).

The Greeks Euripides, Theophrastus and Plinio have described the consumption of edible mushrooms in their time (Guzmán et al., 1993). In some societies, the mushroom was a royal food, probably by its pleasant flavor and texture (Miles and Chang, 1997).

The Romans knew several edible and poisonous fungi. There is a story about the Emperor Julius Caesar who was very fond of *Amanita caesarea* mushroom, whose scientific name was a homage to him and for that reason, it became known as "Mushroom of the Caesars" (Guzmán et al., 1993).

According to Molena (1986), the species *Polyporus tuberoster* (*fungaie stone*) and *Polyporus corilinus* are among the first cultivated mushrooms, collected from the wood of hazels and eucalyptus. These fungi were consumed in 4-5 cm slices, and their production demanded about six months, yielding sometimes one or two mushrooms at a time. There was neither any knowledge about their nutritional requirements, nor about their growth cycle. The only thing that was known was that rubbing a mature mushroom on those woods, and leaving them in a wet environment during a particular period of the year could produce appreciable mushrooms (Molena, 1986).

During the Roman Empire the fungaie stone (stone that produces the mushroom) appeared in Italy, which was composed of a cluster of humus, leaves, twigs and limestone rocks, forming a compact mass, which was cut in blocks in the form of bricks and transported to the royal palaces. They were kept in a damp place and irrigated daily until harvest time to serve the senators and other members of the Roman aristocracy (Molena, 1986). In France, the mushroom cultivation began during the reign of Luis XIV, according to Molena (1986). However, the cultivation of *Agaricus bisporus*, the "Champignon de Paris", the most widely cultivated and commercialized species, has been produced since about 1650 (Delmas, 1978; Chang and Miles, 1984).

With the advances of knowledge and technology of mushroom cultivation, commercial production of dozens of species became viable in several countries in recent decades (Guzmán et al., 1993; Stamets, 1993; 2000; Vedder, 1996; Eira, 2000), reaching a production of approximately 4.3 million tons of edible mushrooms in 1991 (Miles and Chang, 1997). The world current production is around 6.2 million tons (Chang, 2003).

2.2 An overview of the commercial cultivation of edible fungi in the world and in Brazil

After World War II, the edible mushroom industry grew from 350,000 tons in 1965 to 4.3 million tons in 1991, from which 3.4 million tons belong to the six most worldwide important genera: *Agaricus, Pleurotus, Lentinula, Auricularia, Volvariella* and *Flammulina*. The major producers are China, Japan, USA and France (Miles and Chang, 1997). The most cultivated genera are *Agaricus, Pleurotus* and *Lentinula*. This increase was due to several factors, among them: a) the increase in the number of species on a commercial scale; b) the development of cultivation techniques using plastic bags, which allowed many wood decomposers edible fungi to be grown on lignocellulosic residues, preferably the cultivation on logs, reducing considerably the cultivation time; c) due to the marketing techniques highlighting the nutritional merits of mushrooms as an important part of the diet, so they wouldn't be marketed as simple accompaniments or delicacies, but as a food of high nutritional value (Miles and Chang, 1997).

The literature cites approximately 200,000 species of fungi existing in the world, from which, about 2,000 are potentially edible species. However, only 25 of them are commonly used as food, and fewer still are commercially cultivated (Chang and Miles, 1984; Chang, 1980; Bononi, 1999).

In the early 1980s, only *Agaricus bisporus* (Champignon de Paris) and other species of this genus and "shiitake" (*Lentinus edodes*, currently named *Lentinula edodes*) had a modern technology for commercial production, where 70% of the world production was represented by *Agaricus* and 14% by *Lentinula* (Chang and Miles, 1984). However, according to the same authors, the world's attention is turning to the development of new technologies for different species of worldwide known edible mushrooms, especially considering the difficulties of production in tropical and subtropical climates. Special technologies are being developed in several countries allowing the cultivation of: *Volvariella volvaceae* in China, Taiwan, Japan, Philippines and Indonesia; *Kuehneromyces mutabilis, Flamulina velutipes, Hypholoma capnoides* and *Coprinus comatus* in some countries of Europe and Asia; *Pleurotus ostreatus* in Italy, Hungary, West Germany, Mexico and Brazil (Chang and Miles, 1984; Guzman et al., 1993; Eira and Minhoni, 1997; Bononi et al., 1999; Zhanxi and Zhanhua, 2001; Urben et al., 2001). This way, the overview of the world production has changed suddenly, showing a considerable increase in cultivation and consumption of *Pleurotus* as reported by Eira (1997) adaptated by Fermor (1993).

An adaptation based on Fermor (1993), made by Eira et al. (1997), the world production of cultivated mushrooms in the early 1990s was 1,424,000 tons for *Agaricus bisporus*, 900,000 tons for *Pleurotus* spp, 393,000 tons for *L. edodes* and 887,000 tons for other mushrooms, representing, respectively, 39.51%, 24.98%, 10.91% and 24.61%. The current trend is to increase production.

Concerning the production of mushrooms in Brazil, there is not a precise documentation that could allow us to determine when the cultivation of mushrooms started in the country (Fidalgo and Guimarães, 1985). Its popularization in the Center-South region of Brazil dates back to 50 years ago. Bononi (1999) reports that the cultivation of champignon (*Agaricus*) began in 1953, when the Chinese immigrants settled in Mogi das Cruzes and the Italian Oscar Molena in Atibaia, brought technology and imported strains of their countries. For Molena (1985), mushroom cultivation began in 1953 and developed after the poultry crisis in the period of 1955-1959, when breeders began to use chicken sheds for the cultivation of mushrooms, without proper technical conditions.

The commercial cultivation of edible fungi in Brazil is limited to *Agaricus bisporus* (champignon), *Lentinula edodes* (shiitake) and *Pleurotus* spp, known as oyster mushroom, giant mushroom or caetetuba (Bononi et al., 1999; Eira, 2000). Varieties or strains of mushrooms of the *Pleurotus* genus gave origin to the "hiratake" (mushrooms with very large basidiocarp, harvested in mature stage with opened basidiocarps, before they turn their edges upwards and with more than 5 cm in diameter) and the "shimeje" (with long stipes, harvested with their basidiocarps very young and dark, smaller than 5 cm, and can be harvested in bunches) (Eira and Minhoni, 1997).

There are few Brazilian research reports about the subject, and the Botanical Institute of São Paulo was one of the pioneers, creating a research center of edible mushrooms in Mogi das Cruzes in 1985 and a teaching, research and extension nucleus was created in the Faculty of Agronomic Sciences/UNESP in Botucatu, in 1986, named Module of Mushrooms (Eira, 2000). Other centers are springing up in many universities and research institutions.

The production of edible mushrooms in Brazil is difficult to be evaluated. Producers give preference (90%) to the cultivation of *A. bisporus*, (Bononi et al., 1999). Among producers, the majority, almost 90%, are from the East of Taiwan, China, Korea, Japan, working in small properties, in a family system, with all family members operating in all stages of cultivation, in a collective way. The region of the city of Mogi das Cruzes, São Paulo, is responsible for approximately 70% of the edible mushrooms commercialized in Brazil. The remainder are produced by other municipalities, most of them also in São Paulo, Ribeirão Pires, Suzano, Cabreúva, Atibaia, Mariporã, Sorocaba (Bononi et al., 1999; Souza, 2011). There are some important producers in Porto Alegre and some producing installations in southern Minas Gerais and Paraná States.

In 1990, the production of Mushrooms was only 3,000 tons according to Eira et al. (1997), being estimated at 10,000 tons per year until 1997. According to the APAN (Natural Agriculture Producers Association), the Brazilian production of shiitake in late 1995 among its associates, was approximately six tons per month. The official data are underestimated, because they include only the mushrooms marketed by CEAGESP (General Supply Center of São Paulo State) and those intended for export, which are recorded by the CACEX-Department of Foreign Trade (Eira et al., 1997; 2004; Bononi et al., 1999). It is known, however, that significant amounts are marketed directly by the producers with restaurants, pizzerias, snack bars and other establishments, as well as street markets.

Brazilian productivity of *Agaricus bisporus* "Champignon de Paris" in Mogi das Cruzes (in São Paulo State) until 2000 was of the order of 5 to 7 kg of fresh mushrooms/100 kg of moist substrate (4 to 6 kg of fresh mushroom/m^2) (Eira, 2000). In Europe, however, in countries such as Belgium, Holland, Germany and France, the average productivity of mushroom at that time was 30 kg/100 kg of substrate.

Currently in Brazil, farms with more technology get on average a productivity (substrate conversion in mushrooms) for "Champignon de Paris" ranging from 18 to 24% in 20 to 30 days of the crop cycle. In more rustic crops this conversion varies from 12 to 15% in 70 to 90 days. While the numbers seem to have greatly increased, yet it is little when compared to Asian and European productions, where they manage 30 to 40% of conversion (Souza, 2011).

Even today there are no official data concerning the production of mushrooms in Brazil, but some unofficial sources report that 12,050 tons a year of mushrooms "in natura" (table 1). Since 1995, there is an annual import of 12,000 tons per year, on average, most of it cooked *Agaricus bisporus*, to meet market demand. Therefore, it can be concluded that, Brazilian consumption is much higher than its production, reaching 24,050 tons per year. In this context Brazilian people consume around 130 g per capita (Souza, 2011). The world production is around 6.2 million tons (Chang, 2003).

Agaricus bisporus	8,000 ton
Pleurotus ostreatus	2,000 ton
Lentinula edodes	1,500 ton
Agaricus blazei	500 ton
Other species	50 ton

Table 1. Annual production of mushrooms in Brazil. Source (Souza, 2011).

2.3 The importance of fungi

The importance of fungi is unlimited in the terrestrial ecosystem and consequently in man's life. However, these organisms can be beneficial or not, according to the results of their actions. If we consider the decomposing action of fungi on food and the associated production of toxic substances (mycotoxins), the decomposition of other materials such as wood, pathogenicity caused to plants, animals and man, this is the negative aspect of it. On the other hand, if we consider the important role in the decomposition, which along with other microorganisms, participate in the mineralization of organic matter, as well as the symbiosis with plants in the process of mycorrhizae formation, bioremediation, biological control, food, and medicinal properties, one can see the positive side of these organisms.

In nature, the fungi do not participate only in the role of providing a food source for humans and other animals; they also play an important role in the cycling of carbon and other elements, by breaking the lignocellulosic residues and animal excrements which serve as a substrate for saprophytic fungi. This way, these decomposing agents play a very important environmental role along with other organisms, complementing the cycling of plants and animals. Simultaneously, they produce multiple enzymes that degrade complex substances that allow the absorption of soluble substances used for their own nutrition (Chang, 1993).

Trufem (1999) and Matheus and Okino (1999) highlight the importance of fungi in the context of biotechnology, where they are widely used in the food industry, pharmaceutical industry, bioremediation, in biosorption (removal of heavy and radioactive metals), in agriculture as arbuscular mycorrhizal fungi (AMF), where they are used in techniques that help the development of plants of economic interest, biological control, xenobiotics biodegradation, bioremediation of the soil, treatment of industrial effluents and bioconversion of lignocellulosic residues.

One of the most important processes from an economic point of view is the use of fungi in the conversion of lignocellulosic residues in edible mushrooms by fungus X substrate interaction, enabling the solid fermentation process, through enzymatic system of these microorganisms (Matheus and Okino, 1999).

The cultivation of edible mushrooms has become an increasingly important practice in modern society due to the biotechnological process of bioconversion of various residues in edible mushrooms or in dietary supplements of high nutritional value, enabling a more efficient utilization of materials, besides, it can reduce the volume of waste or accelerate the decomposition process. This way, the residual substrate obtained from the cultivation of edible mushrooms can also be used as soil conditioner, natural fertilizer, or food for animals, closing the exploitation cycle of raw materials (Miles and Chang, 1997), which today is called "zeri" technology, trying to get the maximum use of such material, eliminating the residue of the residue (Chang, 2003).

2.3.1 Nutritional importance

Man has constantly realized the nutritional value of mushrooms, as well as their healthy properties compared to other foods, such as red meat, where mushrooms are more advantageous and important as they are great sources of carbohydrates, proteins, mineral salts, vitamins and essential amino acids, which can help to maintain a good nutritional balance (Crisan and Sands, 1978; Garcia et al., 1993; Miles and Chang, 1997).

Nutritional Analyses of mushrooms have shown their importance. They contain more protein than vegetables. Sources of protein such as meat, chicken, have a high level of cholesterol and fat, which are known to cause increase in weight and cardiovascular diseases. For this reason, the proteins from other sources became more popular in recent years, such as proteins from fungi, algae, bacteria and yeast (Lajolo, 1970; Chang and Haynes, 1978; Urben et al., 2003).

Studies carried out by Lintzel (1941; 1943), according to Crisan and Sands (1978), indicated that approximately 200 g of mushrooms (dry weight) are sufficient to feed a normal human being weighing approximately 70 Kg, providing a good nutritional balance. Nutritionally, these macrofungi are a good food source. The composition of fats, carbohydrates, vitamins, etc., varies according to species, the cultivation method and also with the substrate used in cultivation (Crisan and Sands, 1978; Przybylowicz and Donoguue, 1990; Bononi et al. 1999; Miles and Chang, 1997; Andrade, 2007).

Mushrooms are excellent foods for the diets, because they nourish and do not accumulate fat in the organism. They are sources of all essential and some nonessential amino acids. They contain minerals like calcium, potassium, iodine, phosphorus and vitamins including thiamine, riboflavin, niacin, and ascorbic acid, and others related to the B complex (Molena, 1986; Miles and Chang, 1997 Bononi et al., 1999). They also have a high unsaturated fat content (Miles and Chang, 1997).

Mushrooms with larger nutrition index (based on essential amino acid index) have nutritional value similar to meat and milk, while those with a smaller nutrition index compare to some vegetables such as carrots and tomatoes. The nutritional index of these fungi outperforms those of plants and vegetables, except soy (Crisan and Sands, 1978). In general, the protein content of fresh mushrooms is twice higher than cabbage, four times greater than the content of protein of the orange and twelve times that of the Apple (Chang, 1980).

Research carried out in India by Garcia, et al. (1993), where the authors compared the nutritional levels of *Agaricus* and *Pleutotus*, revealed the importance of the amino acids of these mushrooms for people that are lacking animal protein, for religious reasons, and whose main food source comes from vegetables and grains usually poor in essential amino acids. Food supplementation with mushrooms is of fundamental importance in the diet of this kind of people.

In addition, there is also a great interest in the cultivation of the mycelium in a submerged condition to obtain flavoring and fragrant compounds of great value to the food industry. For this purpose, the mycelium is grown submerged, using a variety of substrates, according to the type of the desired compound. This flavoring property is characteristic of some lignolitic mushrooms, such as the *Pleurotus* genus (Gurtiérrez et al., 1994)

2.4 Factors inherent to the nutritional needs of the mushroom

2.4.1 Carbon

The main source of carbon and energy of a plant tissue, used by fungi for their development, are the polysaccharides and lignin in the cell wall, although other polymeric

compounds such as lipids and proteins can also be used. Approximately 50-60% of the dry weight of wood is made of cellulose; 10-30% of hemicellulose and 20-30% of lignin. Cellulose, which is attacked by both brown-rot fungi as well as white-rot fungi, is made up of glucose molecules. On the other hand, the hemicellulose consists of molecules of arabinnose, galactose, mannose, xylose and uronic acids. The lignin has a more complex structure and has not yet been fully described, being basically units of phenyl-propane with a benzene ring bonded to a hydroxyl group and one or two metoxilic groups. The links in this molecule are highly resistant to chemical degradation. Therefore, there are few microorganisms that can use this substance for their nutrition (Mason, 1980).
In relation to the degradation of wood and other lignocellulosic materials, it is generally known that the most efficient natural decomposers of lignin are the white rot-fungi, which are mostly the basidiomycets. This name comes from the white color that wood acquires in advanced stages of degradation (Capelari, 1996). Such organisms degrade cellulose, hemicellulose and lignin, but the lignin is preferentially attacked and these are the only organisms able to metabolize the molecule of lignin in CO_2 and water (Zadrazil, 1978). The degradation is derived from the excretion of enzymes metabolized through the hyphae of fungi (Miles and Chang, 1997).
As a typical white-rot fungi, with decomposing activities of wood, the fungus studied in this work: *Lentinus strigosus*, grows in nature, in favorable conditions, and produces mushrooms through the degradation of the wood substrate or any substrate containing cellulose. From this degradation, the fungi can absorb the nutrients needed for their development and reproduction. The success of mushroom production depends on the understanding of the biology of the fungus and how the environment can influence its growth and development. The domestication of a strain is not a very easy task, when trying to reproduce in the laboratory the ideal conditions for its development, which requires preliminary tests to try to understand its physiology.

2.4.2 Nitrogen

Although wood is the natural substrate for fungi, this substrate does not have a high nitrogen content, and this is necessary for the synthesis of all nitrogen compounds (proteins, purines, pyrimidines and the cell wall chitin of the fungus). The main sources are: salts of ammonia, nitrate, urea nitrogen, and organic compounds like amino acids (Miles and Chang, 1997). However, the need for nitrogen by wood-rot fungi is not very great.
It should be taken into account that when using a salt as a source of nitrogen, there is the release of the ion that integrates the substrate molecules, and this can change the pH of the medium if it is not metabolized at the same rate as nitrogen, since an accumulation of this ion will take place. The same phenomenon occurs when other salts are used as a supplement. Therefore, the various species and strains may respond differently to the addition of these supplements. Urea, ammonia phosphate, tartarate of ammonia and potassium nitrate, apparently are those with best results according to a research carried out by Maziero (1990). Peptone provides better growth of the fungus when compared with other sources of organic nitrogen.
Some authors (Rangaswami et al., 1975; Ginterová and Lazarová, 1987) cited by Maziero (1990), argued that *Pleurotus* has the ability to fix atmospheric nitrogen into organic compounds, because some experiments conducted with pasteurized substrates showed that the total nitrogen content has increased. Kurtzman (1979), cited by Maziero (1990) however, discussed the improbable ability of an eukaryote organism to fix nitrogen. The author

suggested the hypothesis that the spores of nitrogen fixing bacteria are stimulated to develop during the process of pasteurization of the substrate, generating bacteria responsible for the nitrogen fixation.
Care should be taken to avoid excessive nitrogen supplements, which can inhibit the development of the fungus. Montini (2001) reports that tested substrates with high concentrations of cereal bran inhibited the formation of the mushroom and consequently, the number of cultivated mushrooms *Lentinula edodes* in axenic conditions (cultivation with substrate sterilized and under controlled environmental conditions). In Taiwan, the substrate for cultivation of *Pleurotus* mushroom is prepared with 84% of sawdust, 5% of rice bran, 5% wheat straw, 3% soya bran and 3% calcium oxide (Przybylowicz and Donoghue, 1990).

2.4.3 Mineral salts

In general, the mineral elements necessary for the fructification of the mushroom are the same as those required by any cultivated plant, which are major elements and microelements (Molena, 1986). Phosphorus, potassium, magnesium and sulphur are major nutrients needed for the growth of various fungi (Miles and Chang, 1997). Molena, (1986), cites the calcium as one of these elements. In addition to increased growth of mycelium, some minerals such as sodium chloride, magnesium, and calcium also stimulate the early formation of fruiting bodies (Kurtman and Zadrazil, 1989).
Among the more studied microelements (trace elements) and essential for the growth of many species of fungus are: iron, zinc, aluminium, manganese, copper, chrome and molybdenum (Molena, 1986; Miles and Chang, 1997). Experimentally, it is not easy to determine the required quantity of these elements because the element under test may be present in sufficient quantities in an impure form in any ingredient of the cultivation medium or may have been introduced through the inoculum. These elements are constituents or enzyme activators (Miles and Chang, 1997).

2.4.4 Vitamins

Vitamins play an important role in the metabolism of fungi, acting as coenzymes. Fungi are capable of producing sufficient quantities of most of the vitamins they need (Miles and Chang, 1997).
Maziero (1990), in some studies testing several vitamins (Vitamin C, folic acid, calcium pantothenate, niacin, pyridoxine, riboflavin and thiamine) in relation to the mycelial growth of *Pleurotus*, observed a better growth of the mycelium on all vitamins tested, but the best result was to thiamine. Kurtzman and Zadrazil (1989) say that there is no need for the addition of thiamine or other vitamins in "not sterile" substrates, because the other present organisms will normally synthesize them. Molena, (1986) experimented various combinations of vitamins, but their high cost did not compensate for the increased production of mushrooms. (Eira and Minhoni, 1991), report that the vitamins and other growth factors are normally excreted by many microorganisms that live in synthrofy during composting, pasteurization and incubation of the substrate, therefore there was no need of vitamin supplements.

2.5 Physical factors

The growth and development of the fungus are not affected only by nutritional factors, but also by physical factors such as temperature, humidity, light, aeration and gravity. There is a range that varies from minimum, maximum, and optimum growth in relation to these physical factors. Certainly these factors are influenced by other factors such as nutrition, medium conditions, genetic characteristics of the strain and mycelial growth stage (Miles and Chang, 1997).

2.5.1 Temperature

The influence of temperature on mycelial growth and production of fruiting bodies is dependent on the species and strains in question, i.e. there is an ideal temperature for the proper development of the metabolism of the fungus, which is a characteristic of each strain. Nonetheless, there is an interval that varies between 10-40° C, which must be respected, because exceeding these limits, it is going to cause the death of the mycelium (Maziero, 1990). The optimum temperature for growth also varies with the purpose of cultivation. So, the ideal temperature to produce the fruiting body (Miles and Chang, 1997) is different from that intended for the production of metabolic products such as those intended for medicinal compounds as polysaccharides/polypeptides immune-regulatory compounds (PSPC). Temperature extremes are important in determining the survival and dispersion of species in nature (Miles and Chang, 1997).

Kaufer (1935), cited by Maziero (1990) cultivated *Pleurotus corticatus* in laboratory and according to their results, the ideal temperature for the growth of mycelium was 27°C. Clock et al., (1959) according to Maziero (1990), obtained a good growth of mycelium of *P. ostreatus* in the range of 22-31° C. At 37° C the mycelium still was able to grow, but abnormally, while at 17° C no growth was observed. The lethal temperature for *P.* "florida", *P. ostreatus* and *P. eringii* is 40° C when exposed to more than 24 hours (Zadrazil, 1978). Maziero (1990) studying different strains of *Pleurorus* observed that the better mycelial growth happened between 25 and 30° C.

In relation to the emergence of primordia, Block et al. (1959) cited by Maziero (1990) report that the strain of *P. ostreatus* fructified at a 26° C, however at 31° C, although the fruiting body continues to develop, there was no emergence of primordia. For Kurtzman and Zadrazil (1989), the authors must have used in their work, a strain of *P. "florida"* since the fruiting temperature at 26° C is very high, being more appropriate for *P. "florida"*.

Eira and Minhoni (1997) report that the control of temperature in a cultivation chamber is decisive for a good harvest. For a good growth of the mycelial mass on substrate cultivation, the ideal temperature for *Pleurotus* spp should be between 24 and 26° C. After that the primordia initiation and growth phases start, when the temperature inside the cultivation chamber must be between 15 and 24°C, considering that, the lowest are ideal for cultivation of shimeji or *Pleurotus* spp strains that are more demanding and also minimizes the incidence of pests and diseases. According to the same authors, some strains of hiratake usually fructify in hot weather (up to 30° C). For the most demanding strains, temperature and relative humidity control in the chamber of cultivation can be achieved with an automated central air-conditioning associated with a ventilation system, to ensure the ideal climatic conditions for the development of the mushroom.

2.5.2 Moisture

Most fungi require high moisture content. Guzmán et al. (1993) report that fungi have an optimum growth on substrates with 70 to 80% humidity. Urben et al. (2003) cite a good humidity range for *Lentinula edodes* cultivated with Jun-Cao technique between 55-70%. It has to be taken into consideration, not only the moisture content of the substrate, but also the relative humidity of the air. It should also be taken into account that the mushrooms are composed of approximately 90% water, therefore, water is very important to its development, besides the fact that they do not have special structures to protect themselves against water loss, since they lose water easily to the environment, mainly the vegetative mycelium (Maziero, 1990).

There is an optimum water content, both in the compost and in the air. Low relative air humidity causes the mushroom to lose water to the environment, which can even prevent it from growing properly. The outer layers of the mushroom begin to dry and yellow. This way, there is a loss of quality or a loss of production. Low air humidity also causes the compost to lose moisture to the environment, reducing the availability of water for the formation of the mushroom. In the case of *Pleurotus,* if the superficial mycelium of the compost suffers a very intense dryness, it dies and the primordia are aborted (Eira and Minhoni, 1997; Bononi et al., 1999). The relative humidity of the production room is around 80-90% and can be maintained that way by waterproof walls and by sprinkling water (Eira and Minhoni, 1997). There are highly sophisticated systems of cultivation on a commercial scale in Europe, Canada, United States and Japan, where patterns of moisture, temperature, O_2 and CO_2 are monitored by computers. Currently there are automated systems in South and Southeast of Brazil, but not as much as in those countries. In rustic cultivations in Brazil, it is customary to keep the floor and sides of the cultivation shed damp, so that normal evaporation maintains the relative humidity the air. We consider that, in addition to other factors already mentioned, the humidity is the key factor in the cultivation process of edible mushrooms.

2.5.3 Lighting

Even though it is not a photosynthesizer organism, luminosity is essential to many species of fungi. It can retard the primordia formation in some species while in others, it is essential for fruiting. For *Pleurotus* and *Lentinus* cultivation, as well as for many other edible fungi, there must be some light to induce the formation of primordia and also for the normal development of fruiting bodies. The recommended luminosity for *Pleurotus,* after the incubation period and the opening of the cultivation bags is 2000 lux/hour, 12 hours a day (Bononi et al., 1999). Nevertheless, it can vary according to the mushroom species.

Miles and Chang (1997) mention that ultraviolet light in the range from 200 to 300nm affects the growth of the fungus, it can be lethal or induce mutation, since this wavelength is absorbed by the DNA. The authors report that the effects of ultraviolet light can be reversible by the photo reactivation process, provided that these mycelia are exposed to visible light at a wavelength between 360 and 420nm.

For Przybylowicz and Donoghue (1990), shiitake mushroom needs light in both stages: vegetative growth and fruiting. Light exposure during vegetative growth, according to Ishikawa (1967) cited by Przybylowicz and Donoghue (1990), is a prerequisite for the fruiting stage. The duration is not well defined. However, Przybylowicz and Donoghue (1990) suggest that a brief exposure of 20 minutes per day can be enough. For these authors, the growth of shiitake responds well to a range between 180-940 lux, with an optimum value of 500 lux. Rajarathnan and Bano (1987), cited by Eira and Minhoni (1997), stated that the presence of light is required for the formation of fruiting bodies. However, there may be changes in the color of the pileus, where *Pleurotus* species can change from white to opaque and dark color in the presence of light, due to the release of fenoloxidases that oxidize phenol and form melanoidins.

Urben et al. (2003) report that the light affects the growth of mycelium and spores of *Lentinula edodes*, and therefore, it needs a dark environment for its development. Under a light intensity of 50 to 270 lux and a suitable temperature, the mycelium forms a membranous brown layer for substrates made with Jun-Cao and sawdust. According to the same authors, for the formation of the fruiting body (mushroom), little diffused light is necessary. On the contrary, in a very bright environment the fruiting body becomes pale with a long stipe and a deformed pileus. In very bright environments the authors advise to use plastic bags in the green house to cover the mushrooms during the day.

2.6 Chemical factors

2.6.1 Gaseous exchanges

Requirements during the growth phases of the vegetative mycelium of a fungus are different from those during the fruiting stage. The rate of CO_2 that occurs naturally inside a trunk colonized by *Pleurotus* in the forest will surely be higher than the rate of fruiting. However, it does not cause damage to growth. It is a self-regulated system.
Zadrazil (1975), studied various *pleurotus* species, relating to the effect CO_2, and noted that all studied species grew faster in higher concentrations of CO_2, limited to approximately 22%. The good performance of these strains in high rates of CO_2 demonstrates their significant competitive advantage against other microorganisms which do not grow or do not survive in such conditions, especially if the substrate is colonized in not axenic conditions. On the other hand, high concentrations cause a deformation of the fruiting body, being similar to that which occurs when there is light deficiency in the development of the fruiting body. The stipe grows sharply and the pileus stays reduced, similar to the process of etiolation in plants (Zadrazil, 1978). Oxygen also influences the growth of mycelium. Despite the fact that *Pleurotus* mycelium develops in semi-anaerobic conditions, a certain rate of O_2 is required, otherwise, the growth will be nil (Zadralil, 1978). For the development of fruiting body oxygen is essential.
Adequate ventilation is essential to reduce the carbon dioxide content (generated during the development stages of the fungus) to a desirable level in the mushroom production phase. Concentrations above 2% may cause delays in the mycelial growth and, consequently, decrease productivity (Eira and minhoni, 1997). Concentrations of CO_2 below 0.2% are considered optimum for development. During a peak of growth, the ventilation must be intense and constant, since large quantities of mushrooms in rapid growth give off large amounts of CO_2 (Eira and minhoni, 1997). The same authors reported that in cultivations carried out in The Mushrooms Module of the Faculty of Agricultural Sciences of the "Universidade Estadual Paulista" (FCA/UNESP) it is possible to cultivate strains that usually demand cold weather, provided that climatic chambers for thermal shock are used.

2.6.2 pH

Its importance is primarily related to the metabolism of nutrients. Most mushrooms have a good development with pH levels between 6.5 and 7, but there are variations according to the species and strains (Miles and Chang, 1997). The microbiota present in the substrate, according to Zadrazil and Grabbe, (1983), is distinctly influenced by the initial pH level: values below 7.0 usually are good for the development of the mushroom mycelium, but most fungi can develop at pH levels above 7.0. Urben et al. (2003), reporting about the cultivation of *Lentinula edodes* by Jun-Cao technique, stated that the mycelium can grow in pH levels between 3.0 to 6.5, while the ideal range is between 4.0 and 5.5. However, the pH value between 3.5 and 5.0 is the best for the formation of primordia and development of the fruiting body. For this reason, the pH value should always be monitored when choosing the materials that compose the substrate, the cultivation and the source of water supply (Urben et al., 2003).

The pH is directly linked to the enzymatic reaction of fungus and wood. Each enzyme has its optimum pH value. The pH affects the solubility of the compost which in turn determines its availability to the fungus (Przybylowicz and Donoghue, 1990). The optimum pH value for the wood-rotting fungi *Lentinus* and *Pleurotus* is between 4.5 and 5.5. The pH of the wood is usually 4.5 to 5.0, increasing the acidity with its decomposition. The optimum pH value for fruiting lies between 3.5 to 4.5 (for laboratory culture or artificial medium) and 5.0 for compost with sawdust for *Lentinula edodes* (Przybylowicz and Donoghue, 1990)

2.7 Steps to be followed in the cultivation of mushrooms

For the cultivation of edible fungi the following steps are generally adopted: obtaining primary matrix, the production of seed or Spawn (matrix that will serve as inoculum for the substrate), preparation of substrate or compost, sterilization or pasteurization (when cultivation is done in natural conditions), inoculation and colonization of substrate, inducing primordia (with thermal or water shock when necessary), fruiting and harvest. The production aspects of primary decomposition fungi like *Pleurotus*, *Lentinula edodes*) will be covered here.

2.7.1 Obtaining primary matrix and spawn production

For most mushrooms, the production matrix or mycelium follows the same techniques and recommendations for the cultivation of champignon (*Agaricus*), oyster mushroom (*Pleurotus*), shiitake (*Lentinula edodes*) and jewish ear (Auricularia), with some exceptions (Urben et al., 2003). Two distinct steps are fundamental for the preparation of the matrix: obtaining pure inoculum of the fungus and the preparation of the "spawn" or matrix itself.

Obtaining the primary matrix of mushroom can be performed both by sexual or by asexual process. In this work it will be related as an asexual process. It is relative to the mycelial or vegetative phase of the fungus colonizing a previously sterilized nutritional substrate (growth medium). Its production starts by the isolation of a fungus using tiny fragments of a mushroom, placed in sterile culture medium under aseptic conditions. After mycelial growth in the dark, with a temperature of 24 ± 1^{0}C (depending on the strain), fragments of this culture (primary matrix) are transferred to the cereal grain or bran or sawdust enriched with bran and incubated for 30 days in the dark at 24 ± 1 ^{0}C. This step corresponds to the production of the "seed" or spawn (Molena, 1986; Eira and Minhone, 1997; Eira and Montini, 1997). The main function of the grain is to serve as means of dispersion of mycelium, since it is impossible to handle the mycelium without damaging the fragile structure of the hyphae walls (Maziero, 1990).

Although the most used media to obtain the primary matrix are potato-dextrose-agar and malt extract (Bononi et al., 1999), the sawdust-dextrose-agar (SDA) medium is the most indicated by avoiding the physiological adaptation that can occur when the used culture medium has very different characteristics of production substrate (Eira and Minhoni, 1997, Eira and Montini, 1997).

The current trend is to produce inoculum from the cultivation substrate. When working with sawdust it is possible to produce the inoculum ("seed") with grain mixed with sawdust or with sawdust only. In this case "spawn" or "seed" is the substrate colonized by mushroom mycelium, with the goal of facilitating the distribution of the inoculum in different points of cultivation, thereby contributing to a more uniform and rapid colonization of the substrate, reducing the possibility of contamination.

2.7.2 Substrate cultivation

Currently there is a growing tendency to use agro-industrial residues for the cultivation of edible and medicinal fungi. However, traditional methods are still being used like the cultivation of *Lentinula edodes* (shiitake), by some Japanese and Chinese farmers, using oak and hazel logs, although the cultivation in cylindrical tubes (in polypropylene or high density polyethylene-HDPE bags) with enriched sawdust is the most widely used technique.

The technique for the production in sawdust was developed mainly in Japan. Other countries like the Netherlands and the United States are also using this method for the production of *Lentinula edodes* (shiitake), on a large-scale (Bononi at al., 1999). In Brazil the traditional cultivation is done normally on eucalyptus logs and it may also be grown on logs of avocado, mango, walnut, hazel and oak (the last two being widely used for cultivation in Japan) (Eira and Minhoni, 1997). Eucalyptus sawdust is already used in Brazil for the production of this mushroom.

The material used for production of mushroom has to be preferably a residue, easily available, and produced not far away from the cultivation place to lower the production costs. Care should be taken to observe that the waste should be free of chemicals that could affect the growth of the mycelium and not offering toxicity. If a low productivity residue is used, supplementation has to be made with cereal grains or cereal bran (Eira and Minhoni, 1997; Bononi et al., 1999; Przybylowicz and Donoghue, 1990; Stames and Chilton, 1983; Stames, 2000).

The supplements contain a mixture of protein, carbohydrate and fat, where the protein is the main source of nitrogen. They contain minerals and vitamins that also influence the growth of the fungus. The addition of these supplements aims mainly to increase the levels of nitrogen and carbohydrates available. Sugars and starch which are readily available carbohydrates, speed up colonization and the consequent degradation of the substrate, reducing the time of fruiting since the mycelium easily converts these carbohydrates in reserve for the fructification, increasing productivity (Przybylowicz and Donoghue, 1990). Other supplements like limestone ($CaCO_3$) must be added to the cultivation medium, to get the correct pH favorable for the growth of the fungus during the last stages of decomposition since there is an increase in acidity caused by the fungus metabolism. Gypsum is widely used in the mushroom industry to improve the physical structure of the compost and to change the pH value, also acting as a source of calcium (Przybylowicz and Donoghue, 1990). The concentration of 5% (in relation to the dry weight of the substrate) is ideal for the cultivation of shiitake in sawdust, improving structure and porosity of the substrate (Stames and Chilton, 1993)

When working with primary-decomposer fungi as *Pleurotus* and *Lentinus*, i.e. fungi that degrade the structural elements of the residue, it is important to ensure that the material to be used in cultivation has not undergone decomposition by microorganisms during storage. If it is already degraded, colonization by these fungi will be hampered and the attack of other organisms will be facilitated, causing a reduction in productivity (Maziero, 1990).

The sawdust used to prepare the substrate is usually from hardwoods. Sawdusts of conifers are used for *Lentinula* cultivation (shiitake) in areas where there is shortage of hardwood sawdust and it is therefore necessary to make a mixture of the two kinds of sawdust (Przybylowicz and Donoghue, 1990). Many conifers contain resin and phenolic compounds which inhibit the growth of the fungus. These compounds must be degraded or removed before using this kind of sawdust, or it can be changed with the addition of sodium carbonate to remove these compounds (Przybylowicz and Donoghue, 1990).

Various types of substrates have been used for the production of edible fungi (Guzmán and Martinez; 1986; Guzmán et al., 1993; Maziero, 1990; Bononi et al., 1999; Eira and minhoni, 1997; Miles and Chang, 1997; Stames and Chilton, 1983; Stames, 1993; Urben, 2001; Urben et al., 2003; Zhanhua and Zhanxi, 2001). The most used are: sawdust, wheat straw, corn, rice; corn cobs, sugar cane bagasse, various grasses, supplemented with cereal grain or bran. The choice of one or more residues as supplement to sawdust will depend, among other factors, on cost and availability of these materials (Maziero, 1990; Eira and Minhoni, 1997; Eira and Montini, 1997; Guzmán et al., 1993; Urben et al., 2003; Stames and Chilton, 1983; Stames, 1993).

When using the bagasse of sugar cane it is important to make sure that this residue is not very old, which can reduce productivity. However, the fresh ground bagasse is rich in carbohydrates, allowing other competitors or pathogens to colonize the substrate more quickly. To avoid this problem, the residue should be pre-treated through a process of fermentation or washing (Kurtzman and Zadrazil (1989)

Japanese producers of *Flammulina Velutipes, Auricularia* and *Pleurotus ostreatus* use a standard formula with a ratio 4:1 of sawdust and bran respectively, where the sawdust is aged for one year, with the purpose to improve the water retention capacity. An immersion of sawdust in water before mixing with the bran is an effective way used by these producers to achieve an optimum of 60% humidity (Samets and Chilton, 1983). This method, according to Lizuka and Takeuchi (1978) cited by Przybylowicz and Donoghue (1990) is widely used in Asia. In the United States, 80% of sawdust, 10% bran and 10% grain (usually wheat or millet). In Taiwan, the substrate for the cultivation of shiitake is done with 84% of sawdust, 5% of rice bran, 5% wheat straw, 3% soy bran and 3% calcium oxide (Przybylowicz and Donoghue, 1990). The substrate formulations have become unlimited in terms of raw material and agro-industrial residues (Stames, 1993).

Currently, studies are trying to develop a technology that allows the cultivation of edible mushrooms in substrates of low cost and easily available. Perhaps this explains why in Brazil the largest edible fungi producing region is located in São Paulo, where there is a big sugar and alcohol production. This bagasse comes from sugar cane mills, it is homogeneous, it has a fibrous characteristic, and when it is pressed allows aeration for mycelial growth (Rossi, 1999). The sawdust is also a material in abundance in the Amazon region, because of its timber industry.

2.7.3 Pasteurization/sterilization

The pasteurization process is a heat treatment given to the compost for the removal of possible organisms that could compete with the fungus to be cultivated (Maziero, 1990). It can be done in a natural way, in a pasteurizing tunnel or room without heated steam, using only the thermogenesis, with the control of the air that gets in and out in order to control the temperature inside the room, or it can be made with heated steam produced by boilers heated with firewood, diesel or gas.

Pasteurization for the cultivation of lignicol fungi as *Pleurotus*, occurs when after the revolving process of the compost, the temperature of thermogensis, produced by the action of microorganisms, falls below 45 to 50° C (Eira and Minhoni, 1997). The compost is then introduced in the pasteurization chamber. The pasteurization temperature for *Pleurotus* is more severe than for *Agaricus*, being raised to 75°C during the first 6 hours. After cutting the steam, the temperature falls to 40 to 45° C, maintaining a constant ventilation to cool the compost and then proceeding to the inoculation process (Eira and Minhoni, 1997). Depending on the type of cultivation, the substrate to be inoculated can be packed in plastic bags, put in wooden or plastic boxes, shelves or "bed", pressed blocks covered with plastic sheets or in special containers (Maziero, 1990).

2.7.4 Inoculation of the substrate

The sterilization process (used in axenic cultivation) or pasteurization (when working with composted natural substrate) is followed by the inoculation of the substrate, which is made immediately after the cooling of the substrate in aseptic conditions (laminar flow chamber) in case of cultivation in totally axenic conditions. Under these conditions, the substrate to be inoculated is autoclaved at 121° C for 2 to 4 hours (Eira and Minhoni, 1997). For cultivation under natural conditions (not axenic), the substrate is inoculated after pasteurization and cooling of substrate that is around 30° C, in aseptic place.

There are several types of inoculum ("seed"): with grains, grains with sawdust, and less used liquid inoculum. There are still those that are made of small wooden dowels (wooden rods inoculated with fungus, used for the cultivation of shiitake in logs). Inoculum of sawdust/grain is used for *Pleurotus* and *Lentinus* cultivation, when using sawdust for cultivation. It is important that the inoculum is the same sawdust from the cultivation substrate. The quantity used for the cultivation substrate is also variable. It is usually 0.5 to 5% (v/v) (Chang and Miles, 1997). (Zadrazil and Grabbe, 1983), recommend 0.5 to 5% of wet substrate. Urben et al. (2003), recommend 0.5 to 5% of the wet weight of the substrate for the *Pleurotus* cultivation when using the Jun-Cao technique. Gonçalves (2002) studying the effect of mycelial fragmentation in order to obtain inoculants in suspension (liquid fermentation) for cultivation of shiitake in axenic cultivation, found that inoculants fragmented up to 10 seconds provided greater biological productivity and efficiency in comparison with usual solid inoculants. The author reports that cultivation using liquid inoculants has the advantage of reducing the time for fruiting. However, it has the disadvantage of having predisposition for degeneration and mutation after successive crops (Itaavara, 1993), cited by Gonçalves (2002).

2.7.5 Substrate incubation

Incubation period, also known as the "mycelial race", is the development of the vegetative mycelium on the substrate (Przybylowicz and Donoghue, 1990). It is the mechanism in which the mycelium of the fungus, through an enzymatic process, digests the substrate and stores reserves for fruiting. During this process the mycelium develops and colonizes the whole compost, forming a compact white mass. It is a complex process, characterized by intense biological activity in which molecules of cellulose, hemicellulose and lignin of the compost are attacked by fungal enzymes such as cellulase and lacase that reduce these molecules to phenols and simple sugars which are more easily assimilated. This enzymatic activity lasts from the beginning of colonization until the production of mushrooms, but during the period of growth of the mycelium production it is greater (Bononi et al., 1999). Incubation usually occurs in a room that can be dark or not, depending on the light requirement of the fungus, at a temperature between 22 to 25° C for *Pleurotus* (Maziero, 1990).

Guzmán et al. (1993) uses the range 25 to 30⁰ C for the cultivation of several *Pleurotus* species in Mexico. Bononi et al. (1999) report that the ideal temperature for the incubation of these fungi varies between species, but in general it should be kept between 25 and 28⁰ C (the temperature of the compost). The range 25-30⁰ C is also used for the *Pleurotus* cultivation, and 22-25 ⁰ C for *L. edodes* cultivation in Jun-Cao (Urben et al., 2003). Przybylowicz and Donoghue (1990), reported temperatures of 25⁰ C that are ideal for *L. edodes*.

It is important to monitor the temperature during the mycelial race to maintain the optimum temperature for the growth of the mycelium. If there is an excessive rise in temperature (a phenomenon that occurs during metabolic activity of *Pleurotus* and micro-organisms present in the substrate) mycelial growth retardation or even its death may occur. Containers with large amounts of substrate mass are avoided since of heat loss is hampered, and generates an increase in temperature. (Maziero, 1990).

The incubation period is approximately three weeks for *Pleurotus* (Maziero, 1990; Bononi et al., 1999). At low temperatures, such as 4-5⁰ C, the mycelium of most species ceases its activity, entering "latency", and at temperatures over 35-40⁰ C can be lethal to certain species (Bononi et al., 1999). To avoid excessive internal temperature of the substrate during the incubation period, Bononi et al. (1999) recommend keeping the room temperature between 20-22⁰ C, and avoid to clutter the bags of the substrates.

The incubation period is variable, because the development of mycelium occurs within variable time, according to the type of the inoculum, the quality of the compost and conditions of the cultivation chamber, but it generally oscillates between 20 and 30 days for *Pleurotus* (Eira and Minhoni, 1997). Urben et al. (2003) report 20-45 days for the total development of the mycelium with Jun-Cao technique. For the cultivation of shiitake in logs, Eira and Minhoni (1997) report that, after two to three months from log incubation, there is already a significant mycelial growth, which can be indicated by a yellow color in the region of the inoculated holes and the region around those holes become soft. In natural conditions of cultivation on logs, this period of maturity of the mycelium for mycelia production, ranges from six months to a year (Przybylowicz and Donoghue, 1990).

During the colonization of the substrate in the cultivation of *Lentinula edodes* using sawdust enriched with rice bran, packed in plastic bags, Bononi et al. (1999) recommend cycles of alternating light and dark, with at least 8 hours of light per day during a period of four to six weeks. The wavelengths between 370 to 420nm and light intensity between 180 to 500 lux are more efficient during the process of colonization, and it can be achieved with cold fluorescent lamps (Przybylowicz and Donoghue, 1990). According to Bononi et al. (1999), after the total colonization of the substrate, the plastic bags are cut and the surface of mycelium begins to turn into a brown skin. The air humidity must be maintained around 80 to 90% and between 40 to 50 days after the opening of the bags the production starts, after the induction of primordia through thermal shock for 24 to 48 hours at 10⁰ C.

At the end of the mycelial race for shitake there is a period of mycelial stability or mycelium maturation, which lasts until the hardening and darkening of the mycelial skin that becomes brownish grey (Chang and Miles, 1989). The formation of mycelial cover is very important because it acts as a barrier to moisture loss, being also a defense against contaminants, resulting from the oxidation of poliphenol oxidase, a reaction to light and oxygen (Przybylowicz and Donoghue, 1990).

2.7.6 Induction of primordia, fruiting and harvesting

The induction of the primordia occurs naturally in nature. The sudden change of external physical conditions stimulates primordia formation, which will develop, forming the fruiting body (Bononi et al., 1999). On the cultivation of mushrooms, it is used to stimulate or speed their formation. During the induction phase and the production of mushrooms, physical factors such as temperature, lighting, gas exchange, water availability in the compost, relative humidity and the methods of induction are aspects that influence the production and the quality of mushrooms (Zadrazil and Grabbe, 1983).

Sudden changes in temperature usually cause induction of primordia. However, there are differences according to the strains (Przybylowicz and Donoghue, 1990). Low temperatures may indirectly induce fruiting in strains of shiitake, because of the reduction of metabolic activity, reducing therefore the available nutrients, leading to a condition of "stress". On the other hand, in other fungi, temperature can have a direct effect, favoring specific metabolic processes that trigger the induction (Przybylowicz and Donoghue, 1990).

Hawker (1966), cited by (Przybylowicz and Donoghue, 1990) reports that studies with various fungi showed that reducing sugars readily available on the substrate (end of vegetative growth) favors the fruiting. During the entire cycle of fruiting, the primordia phase is the most sensitive to environmental changes. The moisture content of the substrate, temperature and relative humidity are important in this process. On the cultivation of shiitake in logs, moisture for primordia induction should be around 55-65% and the temperature depends on the strain (Przybylowicz and Donoghue, 1990).

There are several artificial induction mechanisms of primordia. It can be done by changing the temperature of incubation (± 25^0 C) to lower temperatures (±16^0 C) in *Pleurotus* cultivation "shimeji" (Eira and Minhoni, 1997). Some strains respond well to this temperature variation, others produce more when subjected to thermal shock.

Marino (2002) in a study about genetic improvement with *Pleurotus ostreatus* aiming the axenic cultivation of strains resistant to heat obtained strains that stood out by their early fruiting and productivity, with two production cycles and without the need for thermal shock, using water immersion only.

For the cultivation of shiitake in sawdust, according to Leatham (1985), cited by Przybylowicz and Donoghue (1990), the thermal shock can be done by cooling the cultivation blocks (packed in bags of polypropylene) at a temperature of 5^0-8^0 C for five to twelve days or by putting them into cold water (5-16^0 C) for 12 to 24 hours, packing them later in the fruiting room (16^0C). After some time primordia will appear at the top of the bags. After the development (3-4 days), according to Eira and Minhoni (1997), the mushrooms are ready to be collected.

Additional flushes of fruiting will emerge without the need for new inductions, provided they are kept in conditions of fruiting. Producers can control the flush making synchronized induction by heating the blocks, followed by reduction of temperature or thermal shock. Sprinkling or immersion can also induce the flush (Przybylowicz and Donoghue, 1990).

Treatment for production of mushrooms (Eira and Minhoni, 1997) is done by reducing temperature and/or water logging (covering with clean cold water for 2 to 4 hours) and removing the bag after water drainage.

The thermal shock for *Lentinula edodes* in modified Jun-Cao technology is made by dipping the miceliated substrates in cold or icy water during 7-8 hours. Then, the bags are packed in a shed or green house. When the buttons (primordia) begin to emerge, the plastic bags (high-density polyethylene) are removed, and the substrates are watered twice a day. After spraying, the bags are covered with a plastic for two hours or until the environment is agreeable (Urben et al., 2003). Regional climatic variations need to be considered. The relative humidity varies with the location of cultivation.

In Brazil, the necessary time for the complete development of the shiitake mushroom is not well defined due to climatic variations. The fruiting occurs over a period between three and twelve months after the inoculation, depending on the temperature of the region and the maintenance of moisture in the log (Eira and Minhoni, 1997; Eira and Montini, 1997). To accelerate this process in the cultivation on eucalyptus logs, the authors recommend soaking the miceliated logs for induction, after the incubation period when the first signs of primordia emission (callus or popcorns) which usually appear after 2 to 3 months. Mineral supplementation in water immersion increased the productivity of this mushroom. However, the increase of productivity and the efficiency of energy conversion were only possible in logs well colonized by the fungus (Queiroz, 2002; Eira and Minhoni, 1997; Eira and Montini, 1997).

In relation to water temperature for immersion, there is a controversy, probably because of environmental differences, and observations often without experimental parameters (Eira and Minhoni, 1997; Eira and Montini, 1997). Some Brazilian producers who own cooling bath system report positive results since this system causes a steady temperature differential of 5-10^{0} C. However, experiments performed in the Module of Mushrooms of the Faculty of Agricultural Sciences "Universidade Estadual Paulista" (FCA/UNESP), in Botucatu, São Paulo State, these same authors report that, in regions with mild climate and thermal amplitude greater than 10^{0} C, the use of ice for cooling did not show significant difference in relation to normal bath immersion.

Induction time depends on environmental conditions and age of the logs and the fruiting temperature varies from 5 to 30^{0} C depending on the strain and the spawn used for cultivation. The relative humidity of the location of the logs should be between 80 and 90%. The emergence of primordia will be within two to three days and harvesting can be made after seven to ten days, and in cool seasons the metabolism of the fungus is reduced, increasing the time before harvest (Eira and Minhoni, 1997). The induction bath can be done in stages, depending on the needs of the producer, thus inducing bath of logs can be programmed as a function of demand (Eira and Minhoni, 1997; Eira and Montini, 1997).

3. Material and methods

The study was carried out at the Edible Mushroom Cultivation Laboratory from the Department of Forest Products in the "Instituto Nacional de Pesquisas da Amazônia" - INPA, in the following steps:

3.1 Collection, drying and preparation of material

Wood residue (sawdust) was chosen based on the generation of the wood waste produced by local lumber industry. Collection, drying and preparation of materials were done at CPPF/INPA, using sawdust of *Anacardium giganteum Hanck* ex Engl (cajuí). After the collection of the residues, they were dried (12% of humididy) in a solar dryer at CPPF/INPA, and packaged into plastic bags of 100 L until the preparation of the substrates.

3.2 Production of a primary and secondary matrix and the "spawn"

The strain of *Lentinus strigosus (Schwinitz) Fries* was taken from the collection of fungi at INPA Institute. Mycelial fragments of fungus (stored in test tubes) were transferred to a Petri dish containing malt medium and incubated at 27 °C until colonization by the fungus (primary matrix) that was used as a source of inoculum for the secondary matrix . Mycelial disks, 9 mm in diameter, were removed from the primary matrix and transferred to Petri dishes containing SDA medium (sawdust-dextrose-agar), prepared according to Sales-Campos (2008), named secondary matrix. "Spawn" is the source for inoculation of the cultivation substrate, considered here as a tertiary matrix. This matrix was produced from cajuí sawdust, with humidification of 75%. The pH was corrected to approximately 6.5, by adding $CaCO_3$. Then that substrate was deposited on glass bottles of 500 mL, in 200 g portions, which were autoclaved at 121 °C for 45 minutes. After cooling, the substrate was inoculated with the secondary matrix. The bottles were partially closed, and kept in special chamber with biochemical oxygen demand (BOD) at 25 ± 2 °C until the complete colonization of the substrate by fungus. This matrix served as a source of inoculation for the cultivation substrates for the production of *L. strigosus* mushrooms

3.3 Preparation of cultivation substrate and processing

The cultivation substrate was prepared from the same residue (cajuí sawdust) as the spawn inoculums. It consisted of 88% of sawdust + 10% of the soy bran as a protein source + 2% of $CaCO_3$, for pH adjustment (6.5). The material was homogenised and humidified to 75%, and packed into bags of high density polyethylene-HDPE (1 kg capacity). Only 500 g of the substrate (wet basis) were put into each bag, with ten repetitions. The substrates were autoclaved at 121 °C for one hour. After that, they were cooled and inoculated with a tertiary matrix under axenical conditions. Each experimental unit (the bag containing the substrate) received 3% of the inoculum in relation to the wet weight of the substrate. They were taken to an incubating chamber until the colonization of the substrate by the fungus. Afterwards, they were transferred to a production chamber. The control samples were also prepared as above, but without inoculation by the fungus. The bags were taken to an oven with air circulation at 55 ± 5 °C and dried to a constant weight, in order to obtain the dry mass of the initial substrate (DMIS) so that they were used to calculate the productivity, based on the biological efficiency index of substrate (BE) and the loss of organic matter (LOM).

3.4 Experimental conditions

The experiment was conducted indoors. The bags contained the substrates were incubated in a climatic chamber at the temperature of 25 ± 3 °C, in the absence of light and at around 80-85% humidity, in order to allow substrate colonization until the production of primordia. Then, they were transferred to the production chamber. The temperature was reduced from 25 °C to 22 °C to induce primordial emission and to allow the production of basidioma (fruit body of the mushrooms) in a way that it would be as uniform as possible. Light intensity was maintained at 2000 Lux, with a photoperiod of 12 hours per day. The relative humidity was scheduled to 95% during the "fructification. The total period of cultivation was 100 days. After "fructification", the mature mushrooms were collected and weighed, and then oven dried for the determination of moisture, dry mass and chemical analyzes. During cultivation, the variables analyzed were: biological efficiency (BE), and loss of organic matter (LOM). Biological efficiency (used to express the productivity of fungus), was calculated according to Tisdale et al. (2006) and Das and Mukherjee (2007):

$$BE = \frac{FMM}{DMS} * 100$$

Where:
BE= Biological efficiency, %
FMM= Fresh mass of mushrooms, g
DMIS= Dry mass of the initial substrate, g

The loss of organic matter (LOM) is the index that evaluates the substrate decomposition by the fungus. It was evaluated according to Sturion (1994), expressed by the following formula:

$$LOM = \frac{DMIS - DMSS}{DMSS} * 100$$

Where:
LOM = loss of organic matter,%
DMSS = Dry mass of the spent substrate, g
DMIS = Dry mass of the initial substrate, g

4. Results and discussion

Table 2 presents the development of the *L. strigosus*. The fructification happened three to five days after the primordia initiation (Table 2), with the development of vigorous mushrooms.

Period (days)						
Mycelial growth	Primordium emission	Fruiting	Cultivation time	Number of flushes	Pileus cm	Stipe cm
11 to 15	33 to 34	35 to 38	100	6	2 to 7	1

Table 2. Profile of the *Lentinus strigosus* mushroom cultivated in cajuí sawdust with soy bran as protein source during 100 days.

The high biological efficiency (BE) of the substrate formulated with cajuí sawdust, supplemented with soy bran demonstrates good productivity of the substrate (Table 3). The result (80%) is superior to other substrates used in cultivation of *P. ostreatus* mushroom formulated with sawdust of *Fagus orientalis* (Yildz et al., 2002), and with *Eucalyptus* sp. according to Marino et al. (2002) which presented BE equal to 8.6 to 64.3% and 11.4 to 43% in their respective studies. The good productivity of this substrate is the result of the quantity of material readily available and absorbed by fungus during the mycelial development process and the soy bran was a good source of protein.

Philippoussis *et al.* (2003) showed that the mycelial growth rate is related to the bio-availability of nitrogen and that the formulation of the substrate influences nutritional levels and porosity (availability of O_2) and Gbolagade *et al.* (2006) stated that each fungus utilizes a specific C/N ratio. The soy bran provided a good source of protein for the fungus as we can see at the table 3 (20%). The results are very important for this edible mushroom, since they present low lipid content (2.5%) and a high fiber level (18%).

The use of alternative substrates, easily obtained at low cost for the cultivation of edible mushrooms have been investigated in many publications (Özçelik and Pekşen, 2007; Philippoussis *et al.*, 2007; Royse and Sanchez, 2007). The supplementation of the substrates with a nitrogen source, mainly with cereal bran, has been adopted to achieve a C/N ratio good for the production of mushrooms.

Özçelik and Pekşen (2007), analyzing the application of hazelnut shells in the formulation of substrate for mushroom cultivation *Lentinula edodes*, reported that the biological efficiency of the substrate made with hazelnut shells only, was considered to be low (43.73%). However, when the proportion of hazelnut shells was reduced and combined with wheat straw (25:75) the biological efficiency was considered good (62.24%). The result however, is less than that reported in this study (Table 3).

Philippoussis et al. (2007) tested the productivity of agricultural residues (sawdust of oak, wheat straw and corncobs) in the cultivation of *Lentinula edodes* and found that corncobs and wheat straw presented higher rates of biological efficiency: 80.64% and 75.23% respectively, which were similar to those presented in this research. Alberto and Lechner (2007) however, obtained lower BE (61.93%), cultivating *Lentinus tigrinus* with Salix sawdust.

Royse and Sanchez (2007) tested three formulations for the cultivation of *L. edodes*, combining wheat straw and oak residues. They found that the substrate with higher proportions of wheat straw (in relation to oak residue), provided the best biological efficiency (98.9%) at the end of 4 harvests. These results are superior to the ones obtained in this work. However, 80% of Biological Efficiency presented by *L. strigosus* cultivated in cajuí sawdust supplemented with soy bran in the present study is considered high.

	Productivity Results		Nutritional Composition			
Substrate Cajui Sawdust	Biological Efficiency (BE) (%)	Loss of Organic Matter (LOM) (%)	Total Protein (%)	Total Fiber (%)	Lipid (%)	Ash (%)
Standard deviation	6.94	4.85	1.00	2,00	0.30	1.00
Average	**80.0**	**47.51**	**20**	**18**	**2.5**	**5**

Table 3. Results of Productivity and Nutritional Composition of the edible mushroom *Lentinus strigosus* cultivated on cajuí wood waste, supplemented with soy bran.

5. Conclusions

The high biological efficiency of the mushroom in this substrate, formulated with the cajuí sawdust supplemented with soy bran, makes its use feasible for the cultivation of *Lentinus strigosus* mushroom from the Amazon Region. The soy bran provided a good source of protein for the fungus.

The findings presented herein point out the utilization of the Amazon wood waste as substrate for the mushroom cultivation, which will certainly promote the improvement of the social and economical conditions of its people and the sustainability of the biodiversity resources, enabling the establishment of a new economical niche in the region.

L. strigosus can be considered an important food in terms of their characteristics: rich in protein and low in fat, important for nutrition and human health.

6. References

Bononi, V. L.; Capelari, M.; Maziero, R. & Trufem, S. F. B. S. (1999). *Cultivo de cogumelos comestíveis*. 2nd Ed., Ícone , ISBN: 85-274-0339-0, S. Paulo.

Crisan, E. V.; Sands, A. (1978). A nutritional value, In: *The biology and cultivation of edible mushroom*, Chang, S. T.; Hayes, W.A. (Eds), pp.137-168, Academic Press, ISBN: 0-12-168050-9, New York.

Chang, S. T.; Hayes, W.A. (1978). *The biology and cultivation of edible mushroom,* Academic Press, ISBN: 0-12-168050-9, New York.

Chang, S. T. (1980). Mushroom as human food. *Bioscience,* vol. 30, No. 6, jun, pp. 399-401, ISSN: 0006-3568

Chang, S. T.; Miles, P. G. (1984). A new look at cultivated mushroom *Bioscience,* vol. 34, No. 6, pp. 358-362, 1984, ISSN: 0006-3568

Chang, S. T. O (2003). Mushroom cultivation using the "Zeri" principle: potential for application in Brazil, Proceeding of First International Symposium on Mushroom in food, health, technology and the environment in Brazil, ISBN: 0102-0110 Brasília-DF, dec, 2002.

Chang, S. T.; M*iles,* P. G. (1987). Historical record of the early cultivation of *Lentinus* in China. *Mushroom Journal for the Tropics,* vol. 7, pp. 31-37.

Delmas, J. (1978). Cultivation in Westerns Countries: Growing in Caves. In: *The biology and cultivation of edible mushroom,* Chang, S. T.; Hayes, W.A. (Eds), pp. 252-299, Academic Press, ISBN: 0-12-168050-9, New York.

Chang, S. T. (1993). Mushroom biology: The impact on mushroom production and mushroom products, In: *Mushroom biology and mushroom products,* Proceeding of the first international conference on mushroom biology and mushroom products, Chang, S. T.; Buswell, J. A.; Chiu, S. W. (Eds), pp. 3-20, The Chinese University Press, ISBN: 962-201-610-3, Hong Kong.

Eira, A. F.; Minhoni, M. T. A. (1991). *Manual teórico/ prático de biotecnologia e microbiologia agrícola: cultivo de cogumelos comestíveis.* Fundação de Estudos e Pesquisas Agrícolas e Florestais, Universidade Estadual Paulista FEPAF-UNESP, Botucatu, São Paulo.

Eira, A. F.; Minhoni, M. T A. (1997). *Manual do cultivo do "Hiratake" e "Shimeji" (Pleurotus spp).* Fundação de Estudos e Pesquisas Agrícolas e Florestais, Universidade Estadual Paulista FEPAF-UNESP, Botucatu, São Paulo.

Eira, A. F.; Montini, R. M. C. (1997). *Manual do cultivo do shiitake (Lentinula edodes (Berk) Pegler).* Fundação de Estudos e Pesquisas Agrícolas e Florestais, Universidade Estadual Paulista FEPAF-UNESP, Botucatu, São Paulo.

Eira, A. F.; Minhoni, M. T. A.; Braga, G. C.; MontinI, R. M. C.; Ichida, M. S.;Marino, R. H.; Colauto, N. B.; Silva, J.; & Neto, F. J.(1997). *Manual teórico/ prático do cultivo de cogumelos comestíveis,* Fundação de Estudos e Pesquisas Agrícolas e Florestais, Universidade Estadual Paulista FEPAF-UNESP, Botucatu, São Paulo.

Eira, A. F. (2000). Cultivo de cogumelos (compostagem condução e ambiente). In: *Anais da III Reunião Itinerante de Fitossanidade do Instituto Biológico,* pp. p. 83-95. Mogi das Cruzes, São Paulo.

Das, N. and Mukherjee, N. (2007), Cultivation of *Pleurotus ostreatus* on weed plant. *Bioresource Technology,* vol. 98, No. 14, pp. 2723-2726, ISBN: 0960-8524

Fidalgo, O., Guimarães, S. M. P. B. (1985). A situação do cogumelo comestível no Brasil e no exterior, *Anais do I Encontro Nacional sobre Cogumelos Comestíveis,* pp. 7-23, Secretaria de agricultura e abastecimento- Instituto de Botânica, Mogi das Cruzes, São Paulo, 15-17 out, 1980.

Garcia, H. S., Khanna, P. K. & Soni, G. L.(1993). Nutritional importance of the mushrooms, In: *Mushroom biology and mushroom products,* Proceeding of the first international conference on mushroom biology and mushroom products, Chang, S. T.; Buswell, J.

A. & Chiu, S. W. (Eds), pp. 227-236, The Chinese University Press, ISBN: 962-201-610-3, Hong Kong.

Gbolagade, J., Ajayi, A., Oku, I., Wankasi, D. (2006), Nutritive value of common wild edible mushrooms from northern Nigeria. *Global Journal of Biotechnology & Biochemistry*, vol. 1, No. 1, pp. 16-21, ISSN: 2078-466X.

Gonçalves, C. R. (2002). *Efeito da fragmentação do micélio visando a obtenção de inoculantes em suspensão para cultivo de shiitake.* Thesis (Doctorate in Biotchnology) Instituto de Química. Universidade Estadual Paulista, UNESP, Araraquara, São Paulo.

Guzmán, G., Mata, G., Salmones, D., Soto-Velasco,C. & Guzmán-Dávalos, L (1993). *El cultivo de los hongos comestibles.* Instituto Politécnico Nacional, ISBN: 968-29-4492-9. México, D. F.

Gutiérrez, A. Caramelo, L., Prieto, A., Martinez, M. J., Martinez, A. T. (1994). Anisaldehyde production and aryl-alcohol oxidase and dehydrogenase activities in lignolytic fungi of the genus Pleurotus. *Applied and Environmental Microbiology*, Vol. 60, n. 6, pp. 1783-1788, 1994, ISSN: 0099-2240.

Kurtzman JR. R. H., ZadraziL, F. (1989). Physiological and taxonomic consideration for cultivation of *Pleurotus* mushrooms, In: *Tropical Mushrooms*, Chang, S. T., Quimio, T. H., (Eds.), pp. 299-348, The Chinese University Press, ISBN: 962-201-264-7, Hong Kong.

Lajolo, R. M (1970). Fungos como alimentos, In: *O grande mundo dos cogumelos*, Lacaz, C. S., Minami, P. S., Purchio, A.(Eds.), p. 113-124, Nobel, São Palo.

Lechner, B. E. and Albertó, E. (2007). Optimal conditions for the fruit body production of natural occurring strains of *Lentinus tigrinus*. *Bioresource Technology*, vol. 98, No. 9, pp. 1866-1869, ISSN: 0960-8524.

Maziero, R. (1990). *Substratos alternativos para o cultivo de Pleurotus* spp. 1990. 136p., Dissertação (Mestrado em Ciências Biológicas/Botânica), Instituto de Biociências USP, Universidade de São Paulo, São Paulo.

Matheus, D. R., Okino, L. K. (1999). Utilização de Basidiomicetos em processos biotecnológicos, In: *Zigomiceto, Basidiomicetos e Deuteromicetos: noções básicas de taxonomia e aplicações biotecnológicas*, Bononi, V. L. R.; Grande, R. A. P. (Eds.), pp.107-139, Instituto de Botânica/Secretaria do Meio Ambiente, ISBN: 85-85662-08-5, São Paulo.

Mason, C. F. (1980). *Decomposição.* 1rst Ed, EPU/EDUSP, ISBN: 10 8512921803, São Paulo.

Montini, R. M. C. (2001). *Efeito de linhagens e substratos no crescimento micelial e na produtividade em cultivo axênico de shiitake* (*Lentinula edodes* (Berk.) Pegler). Thesis (Doctorate) Faculdade de Ciências Agronômicas. Universidade Estadual Paulista, UNESP, Botucatu, São Paulo.

Marino, R. H (2002). *Melhoramento Genético de Pleurotus ostreatus visando o cultivo axênico de linhagens resistentes ao calor.* Thesis (Doctorate) - Instituto de Química, Universidade Estadual Paulista, UNESP, Araraquara - São Paulo.

Molena, O. (1986). *O moderno cultivo de cogumelos.* Nobel, ISBN: 85-213-0377-7, S. Paulo,

Özçelik, E. and Pekşen, A. (2007), Hazelnut husk as a substrate for the cultivation of shiitake mushroom (*Lentinula edodes*), *Bioresource Technology*, vol. 98, No.14, pp. 2652–2658, ISSN: 0960-8524.

Philippoussis ,A., Diamantopoulou, P. & Zervakis, G. (2003). Correlation of the properties of several lignocellulosic substrates to the crop performance of the shiitake mushroom

Lentinus edodes", *World Journal of Microbiology and Biotechnology*, vol. 19, No. 6, pp. 551-557, ISSN: 0959-3993.

Philippoussis, A., Diamantopoulou, P. & Israilides, C. (2007). Productivity of agricultural residues used for the cultivation of the medicinal fungus *Lentinula edodes*, *International Biodeterioration & Biodegradation*, vol. 59, No.3, pp. 216–219, ISSN: 0964-8305

Przybylowicz, P., Donoghue, J. (1990). *Shiitake growers handbook: The art and science of mushroom cultivation*, Kendall/Hunt Publishing Company, ISBN: 0-8403-4962-9 Dubuque, Iowa.

Royse, D. J., Sanchez, J. E. (2007). Ground wheat straw as a substitute for portions of oak wood chips used in shiitake (*Lentinula edodes*) substrate formulae. *Bioresource Technology*, vol. 98, No. 11, pp. 2137–2141, ISSN: 0960-8524

Souza, E. (2011). Cultivo de cogumelos no Brasil: situação atual e perspectivas futuras, In: *Bioconversão de resíduos lignocelulolíticos da Amazônia para cultivo de cogumelos comestíveis*, Sales-Campos, C. (Ed.), PP. 77-94, Editora INPA, ISBN: 978-85-211-0068-3, Manaus-AM, Brasil.

Stamets, P. Chiiton, J. S. (1983) *The Mushroom cultivator: a practical guide to growing mushroom at home*, Agarikon Press, ISBN: 0-9610798-0-0, Olympia, Washington.

Stamets, P. (2000). *Growing gourmet and medicinal mushrooms*, 3 ed, Ten Speed Press, ISBN: 1-58008-175-4, Barkeley, Ca.

Sturion, G. L. (1994). *Utilização da folha da Bananeira como substrato para o cultivo cogumelo (Pleurotus spp.)*, Dissertação (MSc.), ESALQ/ USP, Piracicaba, São Paulo. 147pp.

Tisdale, T. E.; Miyasaka, S. C. & Hemmes, D.E. (2006), Cultivation of oyster mushroom (*Pleurotus ostreatus*) on wood substrates in Havai, *Word Journal of Microbiology & Biotechnology*, vol. 22, No. 3, pp. 201-206, ISSN: 0959-3993.

Trufem, S. F. B. (1999). Utilização de zigomicetos em processos biotecnológicos. In: *Zigomiceto, Basidiomicetos e Deuteromicetos: noções básicas de taxonomia e aplicações biotecnológicas*, Bononi, V. L. R.; Grande, R. A. P. (Eds.), pp.51-67, Instituto de Botânica/Secretaria do Meio Ambiente, ISBN: 85-85662-08-5, São Paulo.

Urben, A. F.; Oliveira, H. C. B.; Vieira, W., Correa, M. J. & Uriartt, A. H (2001). *Produção de cogumelos por meio de tecnologia chinesa modificada*, Urben, A. F. (Ed.), Editora EMBRAPA - Recursos Genéticos e Biotecnologia, ISBN: 85-87697-08-0, Brasília-DF.

Urben A. F.; URIARTT, A. H.; AMAZONAS, M. A. L.; OLIVEIRA, H. C. B.; CORREA,M. J.; VIEIRA, V. (2003). *Cultivo de cogumelos comestíveis e medicinais* (apostila do curso) EMBRAPA. 169p Brasília, DF.

Vianez, B. F., Barbosa, A. P. (2003). *Estudo de alternativas de uso dos resíduos gerados pela indústria madeireira em Manaus e Itacoatiara, Estado do Amazonas.* Report from Department of Forest Products /INPA, Manaus, 50 p.

Vedder, P.J.C. (1996). *Cultivo moderno del champiñon.* Tradução: Martinez, J. M. G. Ediciones Mundi-Prensa, ISBN: 84-7114-074-8, Madrid.

Zhanxi, L., Zhanhua, L. (2001) *Juncao Technology*, Xiangzhou, L. (Ed.), translated by Dongmei, L; Rui, T, China Agricultural Scientech Press, ISBN: 7-80167-210-0, the People's Republic of China.

Zadrazil, F. (1978). Cultivation of *Pleurotus* , In: *The biology and cultivation of edible mushrooms*, Ghang, S. T.; Hayes W. A. (Eds.), pp. 521-557, Academic Press, ISBN: 0- 12- 168050-9, New York.

ZadraziL, F.; Grabbe, K. (1983) Edible mushrooms. *Biotechnology*, Weinheim, Vol.3, N. 1, pp.145-187, ISSN: 0733-222X.

12

Farming System and Management

Morio Matsuzaki
National Agricultural Reseach Center (NARC),
National Agriculture and Food Research Organization (NARO)
Japan

1. Introduction

In the cultivation of soybean, it is necessary to pay attention to nitrogen absorption and soil organism. Nitrogen of 11 - 31kg (an average of 16 kg) is necessary to produce soybean grains of 200 kg (Salvagiotti et al., 2008) because soybean has high grain protein content. Soybean and rhizobia (*Bradyrhizobium japonicum*) form symbiosis for N2 fixation (Gray & Smith, 2005). Soybean N2 fixation is approximately half of the soybean nitrogen uptake (Salvagiotti et al., 2008), and soybean absorbs the other half from fertilizer or soil. If there is much inorganic nitrogen in soil, the N2 fixation is suppressed (Ray et al., 2006), and the amount of fertilizer application for soybean is a little. Therefore, soybean yield probably depends on the quantity of soil organic nitrogen which is mineralized during crops growing period. This means nitrogen which a soil microbe holds, and it is called "biomass nitrogen" (Jenkinson & Parry, 1989).

Soybean is influenced by a biologic factor. Soybean forms symbiosis not only rhizobia but also arbuscular mycorrhizal (AM) fungi (Antunes et al., 2006; Troeh & Loynachan, 2003). The biologic factors such as nematodes, soil-borne diseases become the problem in soybean. In continuous cropping of soybean, soybean cyst nematode (SCN : *Heterodera glycines*) reduce soybean yield approximately 30% (Donald et al., 2006). Sudden death syndrome (SDS) due to the coinfection of SCN and *Fusarium solani* becomes the problem in U.S.A. (Rupe et al., 1997; Xing & Westphal, 2009).

Soybean secretes flavonoids such as daidzein or genistein, and they are key signal compounds for control of symbiosis with rhizobia and AM fungi (Antunes et al., 2006). Glycinoeclepin which kidney beans (*Phaseolus vulgaris*) secrete promotes the hatching of the SCN (Kushida et al., 2002). Crops influence soil organism by various compounds to secrete from root (Faure et al., 2009). Therefore, the growth and yield of soybean are probably influenced by the preceding crop. In soybean, yield decrease is remarkable by continuous cropping (Matsuda et al., 1980; Matsuguchi & Nitta, 1988).

The continuous cropping experiment with five crops including soybean was conducted in northern Japanese Hokkaido for 16 years (Memuro continuous cropping experiment). Organic matter application and soil fumigation were conducted in the experiment. Soybean continuous cropping will influence soil microbe (Kageyama et al., 1982; Matsuguchi & Nitta, 1988). Soil biomass nitrogen increases by organic matter application (Sakamoto & Oba, 1993). Soil fumigation promotes mineralization of soil nitrogen and suppresses nitrification

(Neve et al., 2004). Soil fumigation sterilizes Fungus (Asano et al., 1983) and nematodes. This experiment is a good example to study the influence of the soil microbe on soybean. About a subject picked up in this experiment, the knowledge of past were surveyed.

	C0	W15	W30	W50	B15	B30	B50	F30	F15	F0	R
Continuous cropping	○	○	○	○	○	○	○	○	○	○	
Rotation											○
Wheat straw manure		15	30	50				30	15		
Burk compost					15	30	50				
Soil fumigation								○	○	○	

* The application rate of organic matter is expressed in t/ha.
Soil fumigation (D-D) was applied from 1990 to 1995.

Table 1. Treatments in Memuro continuous cropping experiments.

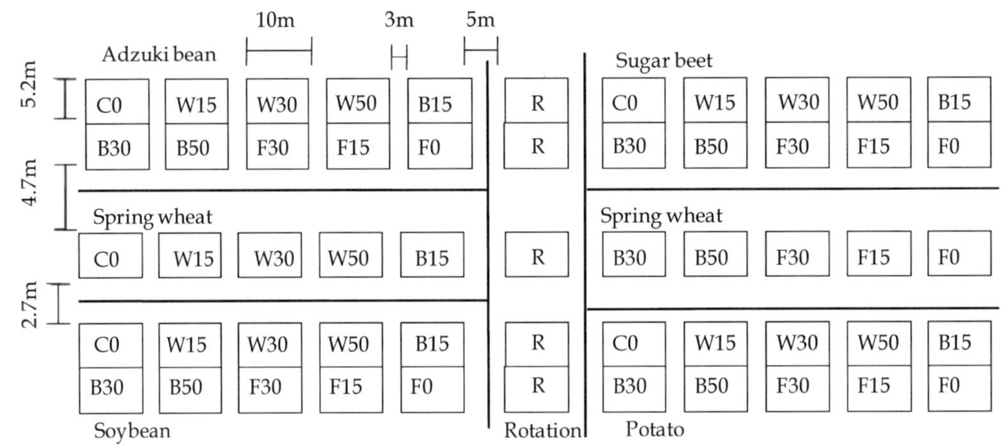

The cropping sequence of rotation plot is sugar beet - Potato - Adzuki bean - Spring wheat - Soybean ·

Fig. 1. The treatment plots in Memuro continuous cropping experiment.

2. Memuro continuous cropping experiment

2.1 Eexperimental design

Memuro continuous cropping experiment (42°53' N, 143°04' E) was conducted from 1980 to 1995. Soybean (*Glycine max*), adzuki bean (*Vigna angralis*), sugar beet (*Beta valgaris*), potato (*Solanum tuberosum*) and spring wheat (*Triticum aestivum*) were cultivated in the experiments. The eleven plots were established in each crop (Fig. 1). One plot was rotation plot, and other plots were continuous cropping plots (Table 1). Soybean was cultivated only chemical fertilizer in rotation plot (R) and the control plot in continuous cropping plots (C0). Wheat straw manure was applied in every year from 1980 at 15, 30, 50 t/ha, respectively (W15, W30, W50). Burk compost was applied in every year from 1981 at 15, 30, 50 t/ha, respectively (B15, B30, B50). D-D (1,3-dichloropropene) was applied from 1990 as a soil fumigation. In the soil fumigation plots, wheat straw manure was applied in every year from 1980 at 0, 15,30t/ha (F0, F15, F30).

2.2 The effect of continuous cropping to soybean cyst nematode (SCN)

The time course changes of egg density of SCN were showed in Fig. 2A. Closed symbols showed the value of rotation plot, and open symbols showed that of continuous cropping plots. The nematode susceptibility cultivar "Kitami-shiro" was used from 1980 to 1991, and nematode-resistant cultivar "Toyo-musume" was used from 1992. D-D was applied from 1990. In the continuous cropping plots, the egg density of SCN tended to decrease from 1985. Thereafter, the egg density in the rotation plot was higher than that in the continuous cropping plots.

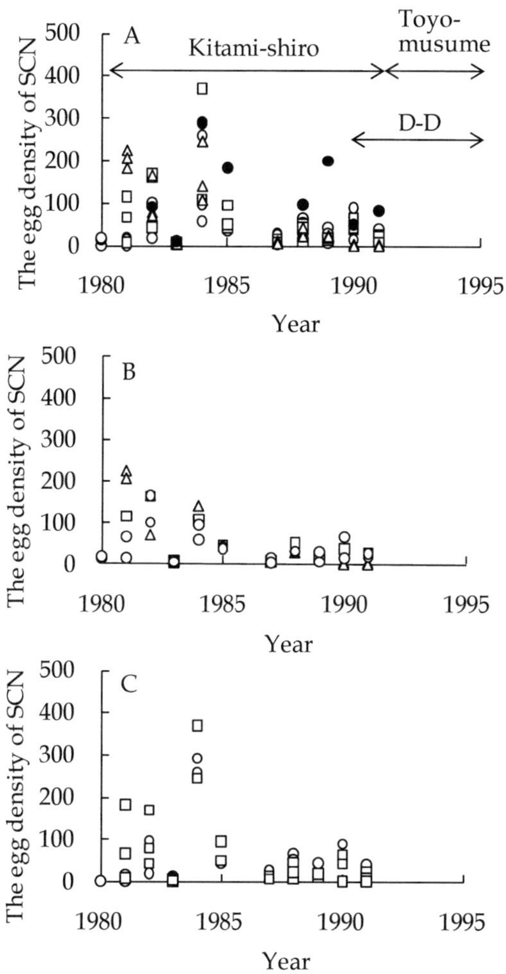

○ : C0, M15, M30, M50, □ : B15, B30, B50, △ : F0, F15, F30, ● : R.
Graph A: The changes of all plots.
Graph B: The changes of plots in which the egg density peaked at 1981.
Graph C: The changes of plots in which the egg density peaked at 1984.
The unit of the egg density of SCN is number / g dry soil.

Fig. 2. The time course changes of egg density of SCN.

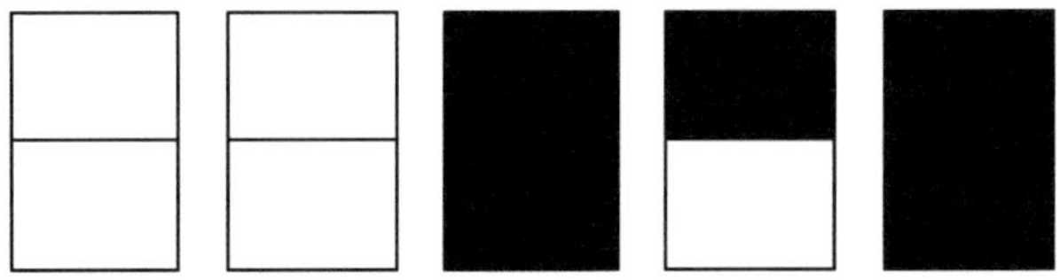

■ : plots in graph B, □ : plots in graph C.

Fig. 3. The placement of soybean continuous cropping plots in Graph B and C.

Fig. 2B and 2C show the time course changes of egg density of SCN in the continuous cropping plots. Fig. 2B shows the plots which the egg density of SCN was highest in 1981-1982. Fig. 2C shows the plots which the egg density of SCN was highest in 1984. The placement of each plots are showed in Fig. 3. The closed squares show the plots in Fig. 2B, and the open squares show the plots in Fig. 2C. The plots which belonged in Fig. 2B were located in the south side of experiment field, and the plots which belonged in Fig. 2C were located in the north side. The difference of time course changes of egg density probably depended on the position of plots. In both plots, the egg density of SCN decreased by 2 - 5 years continuous cropping.

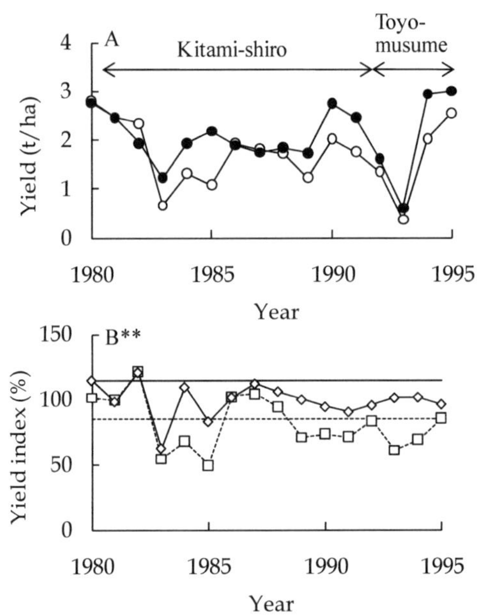

Graph A: The changes in yield of C0 and R.
○ : C0, ● : R.
Graph B: The time course changes in yield index.
□ : C0 / R × 100.
◇ : The means of (W15, W30 W50, B15, B30 or B50) / R × 100,
Solid line: the value of ◇ in 1980.
Dotted line: solid line - least significant difference from Stutentized range.
** is significantly in 1%.

Fig. 4. The time course changes of yield and yield index of soybean.

2.3 The effects of treatments to soybean yield

2.3.1 Time course change

Soybean yield was measured from 1980 to 1995. The time course changes of yield of R and C0 plots were showed in Fig. 4A. In Hokkaido, soybean yield decrease by cool summer damage. It was a cool summer in 1983 and 1993, and soybean yield decreased. To examine the effects of continuous cropping and organic matter application, analysis of variance (ANOVA) was conducted for the soybean yield data. First, the yield of continuous cropping plots except soil fumigation plots (C0, W15, W30, W50, B15, B30, B50) were converted into the index by the yield of rotation plot (R). For the index of the continuous cropping + organic matter application plots (W15, W30, W50, B15, B30, B50), ANOVA was conducted as treatment replication (Fig. 4B). The solid line is a value of 1980, and the dotted line is solid line - least significant difference. In the year when index significantly decreased than that of 1980, soybean yield probably decreased by continuous cropping. For reference, the indexes of C0 were shown.

The mean of indexes of organic matter application plots did not decrease significantly except 1983. However, the indexes of C0 of 1983~1985 were lower than a dotted line. This time was almost coincided with the time when the egg density of SCN increased. The indexes of C0 increased again afterwards. The egg density of SCN decreased, too. Therefore, it is suggested that the yield decrease of continuous cropping was influenced by SCN. The organic matter application probably increased soybean yield.

2.3.2 Treatment effects

For the soybean yield data, ANOVA was conducted as year replication. Using a yield of all plots except soil fumigation plots (C0, W15, W30, W50, B15, B30, B50, R) of 1981~1995, the effects of continuous cropping and organic matter application were investigated. Using a yield of all plots in 1990~1991, the effects of continuous cropping, organic matter application and D-D on nematode susceptibility cultivar "Kitami-shiro" were investigated. Using a yield of all plots in 1992~1995, the effects of continuous cropping, organic matter application and D-D on nematode resistant cultivar "Toyo-musume" were investigated.

By ANOVA for the data of 1981-1995, yield decreased significantly by continuous cropping, and the decrease was approximately 20% (Fig. 5A). Organic matter application increased yield. By ANOVA for soil fumigation period (Fig. 5B, 5C), yield did not decrease significantly by continuous cropping. However, the trend in Fig. 5A was similar in those figures. Yield decreased by continuous cropping, and increased by organic matter application. By D-D, yield increased at the same level as the rotation plot. D-D might remove the effect of continuous cropping as a nematocide. However, the egg density of SCN declined before D-D application period. The effect of D-D was found to "Toyo-musume" that was nematode resistant variety. Therefore, it was suggested that D-D influenced the factor except SCN.

3. Factors to influence continuous cropping soybean

3.1 Soybean cyst nematode (SCN)

SCN forms the cyst containing a large number of eggs (Ichinohe, 1955a). Two or three generations of SCN can grow up in the soybean growing period of Hokkaido (Ichinohe, 1955a). SCN inhibits rhizobial adherence, too (Ichinohe 1955a). The damage of SCN is most remarkable if SCN invaded to soybean at 2-3 weeks after sowing (Okada, 1968). The damage of SCN is reduced by fertilization (Okada, 1966). SCN reduces the growth of soybean, but

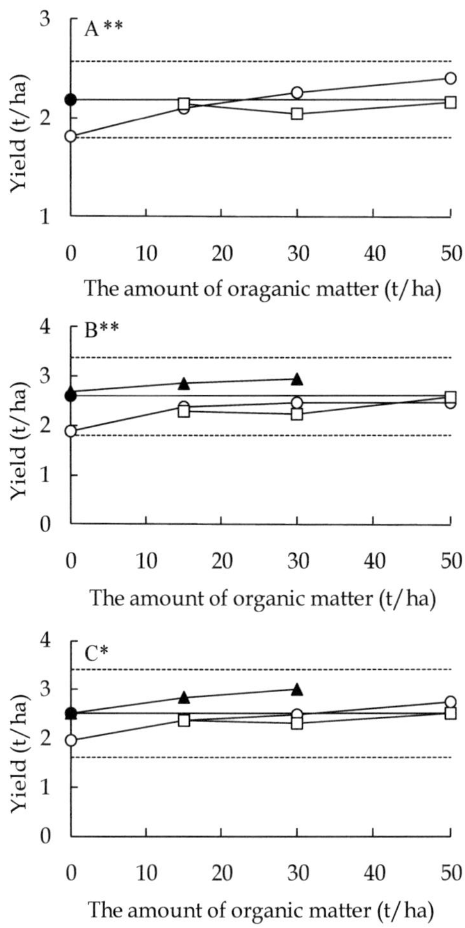

○ : C0, M15, M30, M50, □ : B15, B30, B50, ▲ : F0, F15, F30, ● : R.
Graph A : ANOVA for 1981 – 1995 (All period of continuous cropping),
Graph B : ANOVA for 1990 – 1991 (D-D for "Kitami-shiro"),
Graph C : ANOVA for 1992 – 1995 (D-D for "Toyo-musume").
Solid line : the value of R,
Dotted lines : solid line ± l.s.d. from Studentized range.
** is significantly in 1%, * is 5%.

Fig. 5. The treatment effects to soybean yield.

SCN cannot increase with poor growth soybean (Ichinohe, 1955b). Therefore, if the soybean growth is less, the density of SCN may be less. In Memuro continuous cropping experiment, the egg density of SCN was low in the cool summer damage year (1983).

Soil fumigation, organic matter application and cultivation of non-host crops or the resistant variety is effective for control of SCN. D-D etc are used for soil fumigation. Organic acid or ammonia released from organic matter suppresses a nematode (Oka, 2010). By organic matter including the chitinous substance, chitinase activity in the soil rises, and a nematode is suppressed (Akttar & Malik, 2000; Oka, 2010). Brassicaceae crops including glucosinolates release isothiocyanates, and suppress a nematode (Oka, 2010). Probably Marigold suppresses a nematode by α- terthienyl (Oka, 2010).

Host crops of SCN is soybean, adzuki bean and kidney bean (Ichinohe, 1953). SCN is not parasitic on a non-leguminous crop. SCN is parasitic on other leguminous crops, but cannot become the adult (Ichinohe, 1953). After five years cultivation of corn (*Zea mays*) which is non-host crop, SCN increases by soybean cultivation (Porter et al., 2001). By the planting of the single resistant variety, the races adapted to the resistant variety increased (Shimizu & Mitsui, 1985). In contrast, the leguminous crop red clover (*Trifolium pratense*) may be used as trap crop (Kushida et al., 2002). Red clover hatch the egg of SCN, but the hatched larva cannot become the adult (Kushida et al., 2002). Therefore, after red clover cultivation, the density of SCN decreased (Kushida et al., 2002).

In Memuro continuous cropping experiment, SCN density decreased by 5 years continuous cropping of soybean. In other experiments, the density of SCN decreased by continuous cropping, too (Hashimoto et al., 1988). This phenomenon is called "SCN decline". SCN decreases in wheat-soybean double cropping (Bernard et al., 1996). It is suggested that these phenomena are caused because fungus or bacteria are parasitic on SCN. *Hirsutella* is parasitic on the second larva of SCN (Liu & Chen, 2000). *Fusarium* and *Verticillium* are parasitic on a female, cyst and egg of SCN (Bernerd et al., 1996; Sayre, 1986, Siddiqui & Mahmood, 1996). The nematode control using these microorganisms is possible. However, in Memuro continuous cropping experiment, soybean might be cropped continuously before the experiment station establishment. Long term continuous cropping may be needed to "SCN decline".

3.2 Nitrogen supply and soybean yield

In American Corn Belt, the potential yield of soybean is estimated at 6 - 8t/ha (Salvagiotti et al., 2008). Because soybean is crop which is high grain protein content, soybean need a large quantity of nitrogen. Nitrogen of 106 - 310 kg N / ha is necessary to get a yield of 2 t / ha (Salvagiotti et al., 2008). In the Tokachi district, the fixed nitrogen of soybean is 4 - 127 kg N /ha (Nishimune et al., 1983). There is negative correlation between amount of applied fertilizer and N2 fixation (Salvagiotti et al., 2008). Fertilizer nitrogen for soybean is less than 40kgN/ha. Therefore, soybean needs to absorb nitrogen from soil. Soybean yield will increase with soil nitrogen absorption.

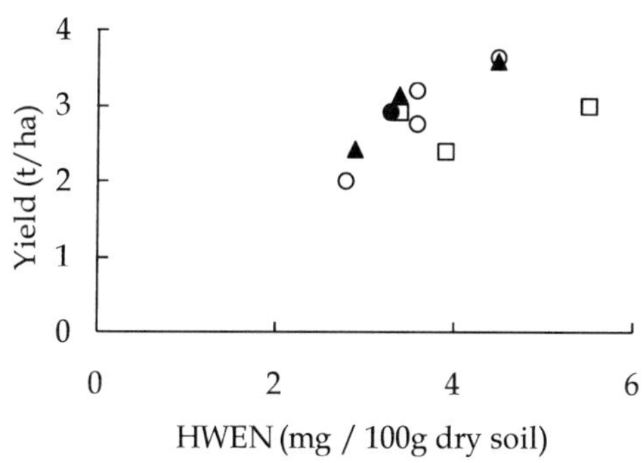

○ : C0, M15, M30, M50, □ : B15, B30, B50, ▲ : F0, F15, F30, ● : R.

Fig. 6. The relationship of the hot water extractable nitrogen (HWEN) of postharvest soil and soybean yield (1994).

Soil Nitrogen can be divided into inorganic, humus and biomass nitrogen (Jenkinson & Parry, 1989). Crops absorb inorganic nitrogen. The available nitrogen is organic nitrogen at sowing, but it is mineralized during a growing period. It is suggested that most of available nitrogen come from biomass nitrogen (Sakamoto & Oba, 1993). Organic matter application increases biomass nitrogen (Sakamoto & Oba, 1993). Available nitrogen and the heated water extraction nitrogen (HWEN) have corelation (Akatsuka & Sakayanagi, 1964). In Memuro continuous cropping experiment, the relationship of the HWEN of postharvest soil and soybean yield was investigated in 1994 (Fig. 6). With increase of the HEWN, soybean yield tended to increase. It is suggested that organic matter application increases biomass nitrogen, and contributes to yield increase.

D-D promotes mineralization from biomass nitrogen (Neve et al., 2004). D-D suppresses nitrification from ammonia nitrogen (Neve et al., 2004). Therefore, much ammonia nitrogen in soil will be kept by D-D. In the Memuro continuous cropping experiment, D-D was applied before one month of sowing. Therefore, it is thought that these effects influenced a soybean.

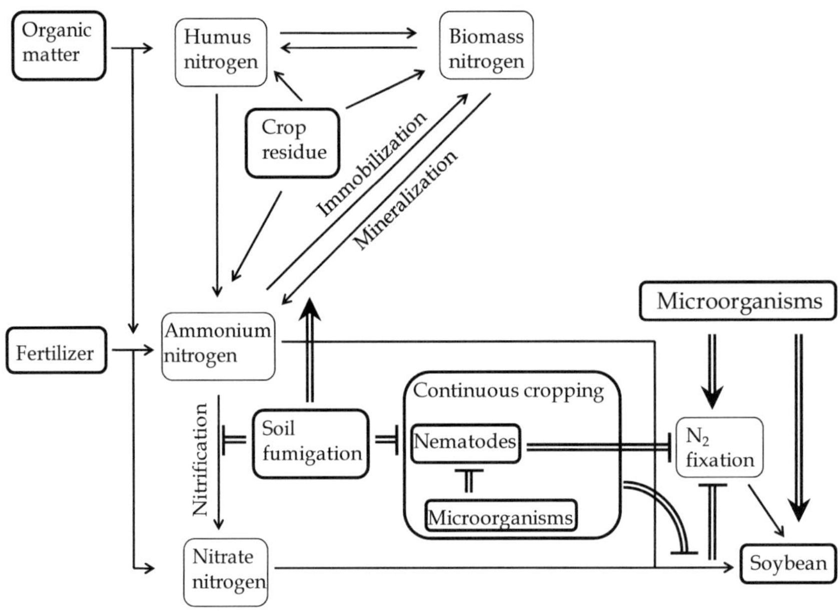

Fig. 7. The nitrogen flow and the factors influence to it.

4. Farming system and management to influence soybean yield

Farming system and management influencing soybean are described. Fig. 7 illustrates the results provided in Memuro continuous cropping experiment and the knowledge of the past.

4.1 Soil organic matter (SOM)

Soybean needs nitrogen absorption to get high yield. The Brazilian soybean absorbs approximately 100% of N need by N2 fixation, but in soybean of U.S.A. or Japan, fixed nitrogen is approximately 50% of N need (Graham & Vance, 2000). This may be connected with that soybean N2 fixation decrease at low temperature (F. Zhang et al., 1995). In other words, the nitrogen supply from soil probably become important so as to be a cold area.

4.1.1 Farming system

The soil organic matter (SOM) is broken down by the soybean planting (Cheng et al., 2003). This effect is called "priming effect". Soil carbon and nitrogen decrease in soybean continuous cropping in comparison with Gramineae - soybean rotation (Kelley et al., 2003; Wright & Hons, 2004). However, soybean-corn rotation can reduce the fertilizer nitrogen of 60kgN/ha/year in comparison with corn continuous cropping (Varvel & Wilhelm, 2003). The soil nitrogen mineralization quantity increases in soybean- corn rotation in comparison with soybean continuous cropping (Carpenter-Boggs et al., 2000). It is suggested that soybean breaks down SOM and will increase inorganic nitrogen, but soybean continuous cropping will cause a decrease of SOM.

In soybean rotation which incorporated alfalfa (*Medicago sativa*), the quantity of soil nitrogen mineralization increases greatly (Carpenter-Boggs et al., 2000). In soybean introducing to the permanent grass pasture, soybean yielded 3 t / ha at no chemical fertilizer (Diaz et al., 2009). One of the causes of these phenomena will be that pasture plant leaves much organic matter in soil.

4.1.2 Management

SOM increases by organic matter application. SOM is maintained by no-tillage (Wright & Hons, 2004). Because mineralization of soil nitrogen decreases by no-tillage, N2 fixation probably increases (van Kessel, 2000). These treatments are suggested to increase soybean yield.

4.2 Soybean cyst nematode (SCN)

4.2.1 Farming system

SCN inhibits the production of soybean. SCN does not increase by the cropping of non-host crop, but SCN increases by soybean cropping again (Asai & Ozaki, 1965). Gramineous crops such as corn and wheat are non-host crop of SCN. Soybean- corn rotation carried out in the northern part of U.S.A. (Varvel & Wilhelm, 2003; Xing & Westphal, 2009), but SDS by SCN and *Fusarium solani* occurs in this rotation (Rupe et al., 2003; Xing & Westphal, 2009). SDS can lead to defoliation of the leaflets, leaving the petioles attached to the plant after flowering (Rupe et al., 2003; Xing & Westphal, 2009). Pythium have a pathogenicity in soybean and corn and cause damping-off (B.Q. Zhang et al., 1998).

In the Southern U.S.A., soybean is cultivated by no-tillage in soybean - wheat double cropping (Bernard et al., 1996). No-tillage is used to corn, soybean, wheat and etc in U.S.A., and the cultivated area occupies 23% in U.S.A. (Triplett. Jr. & Dick, 2008). No-tillage reduces nematode density in soybean - wheat double cropping (Bernard et al., 1996). However, take-all (*Gaeumannomyces graminis*) of wheat cannot be reduced in soybean- wheat double cropping (Cook, 2003).

Some plants are able to control nematodes. Probably Marigold controls nematodes with chemical substances such as α -terthienyl (Oka, 2010). The Brassicaceae plants control nematodes with isothiocyanates which is broken down from glucosinolates (Oka, 2010). The leguminous crops such as red clovers are probably available as trap crop reducing the egg density of SCN (Kushida et al., 2003).

4.2.2 Management

Nematodes may be controlled by organic matter application. The organic matter including inorganic nitrogen or chitinous substance is effective for nematodes control (Akhtar & Malik, 2000; Oka, 2010). The application of organic matter including antagonism microorganism will be effective (Oka, 2010). However, these effects will vary according to materials or adjustment methods.

4.3 Arbuscular Mycorrhizal (AM) fungi

With soybean, plant growth promoting rhizobacteria (PGPR) such as rhizobia and *Bacillus* form symbiosis (Bai et al., 2003; Cattelan et al., 1999; Gray & Smith 2005), and arbuscular mycorrhizal (AM) fungi form symbiosis, too (Troeh & Loynachan, 2003). The AM fungi increase N2 fixation (Antunes et al., 2006). The symbiosis with AM fungi helps phosphorus acid absorption of the crop (Harrison, 1998). The crops doing N2 fixation need a lot of phosphoric acid (Graham & Vance, 2000), and the symbiosis with AM fungi will be effective for soybean. Soybean and corn form symbiosis with AM fungi (Troeh & Loynachan, 2003), and AM fungi probably increases in soybean - corn rotation. Sugar beet was non-host crop of AM fungi, and soybean growth after non-host crop such as sugar beet was suppressed (Karasawa, 2004).

5. Conclusion

Soybean forms symbiosis with rhizobia. However, in low temperature area, the nitrogen absorption of soybean may not be served only in N2 fixation, and soil nitrogen probably becomes important. Soybean stimulates the decomposition of SOM, and can absorb nitrogen from soil. The crops such as gramineous crop or pasture plant supply organic matter to soil. No-tillage maintains SOM. Organic matter application increases SOM. Soybean yield probably increases by the combination of these treatments.

SCN inhibits the nitrogen absorption of soybean. By cropping of non-host crops or resistant varieties, SCN does not increase, but the cyst of SCN does not decrease. Soybean is affected by not only SCN but also pathogenic fungi such as *Fusarium*. On the other hand, soybean growth is promoted by PGPR (rhizobia, *Bacillus* and etc) and AM fungi. In soybean, it is necessary to decide farming system and management while considering these organisms.

6. References

Akatsuka K. and Sakayanagi, M. (1964) Consideration on the Soil Nitrogen Test. Research Bulletin of Hokkaido National Experiment Station, Vol. 83, (March 1964) pp. 64-70, ISSN 0018-3415 (In Japanese)

Akhtar, M. and Malik, A. (2000) Roles of Organic Soil Amendments and Soil Organisms in the Biological Control of Plant-Parasitic Nematodes: A Review. Bioresource Technology, Vol. 74, No. 1, (August 2000) pp. 35-47, ISSN 0960-8524

Antunes, P.M., de Varennes, A., Rajcan, I. and Goss, M.J. (2006) Accumulation of Specific Flavonoids in Soybean (*Glycine max* (L.) Merr.) as a Function of the Early Tripartite Symbiosis with Arbuscular Mycorrhizal Fungi and *Bradyrhizobium japonicum* (Kirchner) Jordan. Soil Biology & Biochemistry, Vol. 38, No. 6 (June 2006) pp.1234–1242, ISSN 038-0717

Asai, K. and Ozaki, K. (1965) Studies on the Rotation Systems. Part 6. Behavior of Population Density of Soybean Cyst Nematode by Different Cropping Systems. Research Bulletin of Hokkaido National Experiment Station, Vol. 87 (December 1965) pp. 66-73, ISSN 0018-3415 (In Japanese)

Asano, M., Kato, T., Kinoshita, T. and Arisawa, M. (1983) Studies on the Behavior of Nutrients in Soil Resulting from Soil Disinfection (II) Effects of Soil Disinfection on Soil Microbes and the Growth of Fruit Vegetables in Greenhouse. Research Bulletin of the Aichi Agricultural Research Center, Vol. 15, (October 1983) pp. 216-222, ISSN 0388-7995 (In Japanese)

Bai, Y., Zhou, X. and Smith, D.L. (2003) Enhanced Soybean Plant Growth Resulting from Coinoculation of *Bacillus* Strains with *Bradyrhizobium japonicum*. Crop Science, Vol. 43, No. 5 (September 2003) pp. 1774–1781, ISSN 0011-183X

Bernard, E.C., Self, L.H. and Tyler, D.D. (1996) Fungal Parasitism of Soybean Cyst Nematode, *Heterodera glycines* (Nemata: Heteroderidae), in Differing Cropping-Tillage Regimes. Applied Soil Ecology, Vol. 5, No. 1, (January 1997) pp. 57–70, ISSN 0929-1393

Carpenter-Boggs, L., Pikul, J.L., Vigil, M.F. and Riedell, W.E. (2000) Soil Nitrogen Mineralization Influenced by Crop Rotation and Nitrogen Fertilization. Soil Science Society of America Journal, Vol. 64, No, 6 (November 2000) pp. 2038–2045, ISSN 0361-5995

Cattelan, A.J., Hartel, P.G. and Fuhrmann, J.J. (1999) Screening for Plant Growth–Promoting Rhizobacteria to Promote Early Soybean Growth. Soil Science Society of America Journal, Vol. 63, No. 6 (November 1999) pp. 1670-1680, ISSN 0361-5995

Cheng, W., Johnson, D.W. and Fu, S. (2003) Rhizosphere Effects on Decomposition Controls of Plant Species, Phenology, and Fertilization. Soil Science Society of America Journal Vol. 67, No. 5, (September 2003) pp. 1418-1427, ISSN 0361-5995

Cook, R. J. (2003) Take-All of Wheat. Physiological and Molecular Plant Pathology, Vol. 62, No. 3, (March 2003) pp. 73–86, ISSN 0885-5765

Diaz, D.A.R., Pedersen, P. and Sawyer, J.E. (2009) Soybean Response to Inoculation and Nitrogen Application Following Long-Term Grass Pasture. Crop Science, Vol. 49, No. 3, (May 2009) pp. 1058–1062, ISSN 0011-183X

Donald, P.A., Pierson, P.E., ST. Martin, S.K., Sellers, P.R., Noel, G.R., MacGuidwin, A.E., Faghihi, J., Ferris, V.R., Grau, C.R., Jardine, D.J., Melakeberhan, H., Niblack, T.L., Stienstra, W.C., Tylka, G.L., Wheeler, T.A. and Wysong, D.S. (2006) Assessing *Heterodera glycines* -Resistant and Susceptible Cultivar Yield Response. Journal of Nematolgy, Vol. 38, No. 1 (March 2006) pp. 76–82, ISSN 0022-300X

Faure, D., Vereecke, D. and Leveau, J.H.J. (2009) Molecular Communication in the Rhizosphere. Plant and Soil Vol. 321, No. 1-2, (August 2009) pp. 279–303, ISSN 0032-079X

Graham, P.H. and Vance, C.P. (2000) Nitrogen Fixation in Perspective: An Overview of Research and Extension Needs. Field Crops Research, Vol. 65, No.2-3, (March 2000) pp. 93–106, ISSN 1378-7290

Gray, E.J., and Smith, D.L. (2005) Intracellular and Extracellular PGPR: Commonalities and Distinctions in the Plant–Bacterium Signaling Processes. Soil Biology & Biochemistry, Vol. 37, No. 3, (March 2005) pp. 395–412, ISSN 0038-0717.

Harrison, M.J. (1998) Development of the Arbuscular Mycorrhizal Symbiosis. Current Opinion in Plant Biology, Vol. 1, No. 4, (August 1998) pp. 360–365, ISSN 1369-5266

Hashimoto, K., Inagaki, H., Momota, Y., Sakai, S., Nagasawa, T. and Kokubun, K. (1988) Variation in Populations of *Heterodera glycines*, and the Growth and Yield of Soybean Cultivars with Various Levels of Resistance Grown under Continuous Cropping. Bulletin of the Tohoku National Agricultural Experiment Station, Vol. 78, (August 1988) pp. 1–14, ISSN 0495-7318 (In Japanese)

Ichinohe, M. (1953) On the Parasitism of the Soy Bean Nematode, *Heterodera glycines*. Research Bulletin of Hokkaido National Experiment Station, Vol. 64, (February 1953) pp. 113-124, ISSN 0018-3415 (In Japanese)

Ichinohe, M. (1955a) Studies on the Morphology and Ecology of the Soy Bean Nematode, *Heterodera glycines*, in Japan. Report of Hokkaido Agricultural Experiment Station, Vol. 48, (March 1955) pp. 1–72, ISSN 0018-3407 (In Japanese)

Ichinohe, M. (1955b) A Study on the Population of the Soy Bean Nematodes (*Heterodera glycines*) 1. An Observation on the Relation between the Crop Damage and the Female Infestation. Research Bulletin of Hokkaido National Experiment Station, Vol. 68, (January 1955) pp. 67-70, ISSN 0018-3415 (In Japanese)

Jenkinson, D.S. and Parry, L.C. (1989) The Nitrogen Cycle in the Broadbalk Wheat Experiment: A Model for the Turnover of Nitrogen through the Soil Microbial Biomass. Soil Biology & Biochemistry, Vol. 21, No. 4 (1989) pp. 535-541, ISSN 0038-0717

Kageyama, K., Ui, T., Narita, Y. and Yamaguchi, H. (1982) Relation of *Pythium* spp. to Monoculture Injury of Soybean. Annals of the Phytopathological Society in Japan, Vol. 48, No. 3, (July 1982) pp. 333–335, ISSN 0031-9473 (In Japanese)

Karasawa T. (2004) Arbuscular Mycorrhizal Associations and Interactions in Temperate Cropping Systems. Research Bulletin of National Agricultural Research Center for Hokkaido Region, Vol. 179, (March 2004) pp. 1–71, ISSN 1347-8117. (In Japanese)

Kelley, K.W., Long Jr., J.H. and Tod, T.C. (2003) Long-Term Crop Rotations Affect Soybean Yield, Seed Weight, and Soil Chemical Properties. Field Crops Research, Vol. 83, No. 1 (June 2003) pp. 41–50, ISSN 1378-7290

Kushida, A., Uehara, T. and Momota, Y. (2002) Effect of Red Clover Hatching and Population Density of *Heterodera glycines* (Tylenchida: Heteroderidae). Japanese Journal of Nematology, Vol. 32, No. 2 (December 2002) pp. 69–76, ISSN 0919-6765

Liu, X.Z. and Chen, S.Y. (2000) Parasitism of *Heterodera glycines* by *Hirsutella* spp. in Minnesota Soybean Fields. Biological Control, Vol. 19, No. 2, (October 2000) pp. 161–166, ISSN 1049-9644

Matsuda, M., Horie, M., Honda, K. and Shimura, E. (1980) Studies on the Rotation of Upland Crops XI. Yield Trends of Crops in Continuous Cropping and Several Rotation Systems over 42-Year Experiments. Japanese Journal of Crop Science, Vol. 49, No. 4, (December 1980) pp. 548–558, ISSN 0011-1848 (In Japanese)

Matsuguchi, T. and Nitta, T. (1988) Organic Amendment to Upland Soil as a Conditioner of the Rhizosphere Ecosystem (Part 3) Effects of Organic Amendments on Root Development and the Rhizosphere Microflora of Monocultured Upland Crops. Japanese Journal of Soil Science and Plant Nutrition, Vol. 59, No.1. (February 1988) pp.1–11, ISSN 0029-0610 (In Japanese)

Neve, S. D., Csitári, G., Salomez, J. and Hofman, G. (2004) Quantification of the Effect of Fumigation on Short- and Long-Term Nitrogen Mineralization and Nitrification in Different Soils. Journal of Environmental Quality, Vol. 33, No. 5, (September/ October 2004) pp. 1647–1652, ISSN 0047-2425

Nishimune, A., Konno, T., Saito, G. and Fujita, I. (1983) The Relationship between Nitrogen Fixation and Grain Yields of Legumes on the Main Upland Soils in Tokachi, Hokkaodo. Research Bulletin of Hokkaido National Experiment Station, Vol. 137, (March 1983) pp. 81-106. ISSN 0367-5955 (In Japanese)

Oka, Y. (2010) Mechanisms of Nematode Suppression by Organic Soil Amendments—A Review. Applied Soil Ecology, Vol. 44, No. 2 (February 2010) pp. 101–115, ISSN 0929-1393.

Okada, T. (1966) Studies on the Parasitic Distribution of Nematodes. Part 3. Effects of Amounts of Fertilizer on the Growth of Soybeans and the Parasitic Distribution of Soybean Cyst Nematodes. Research Bulletin of Hokkaido National Experiment Station, Vol. 89, (March 1966) pp. 30-36, ISSN 0018-3415 (In Japanese)

Okada, T. (1968) Effects of the Inoculation Time of Soybean Cyst Nematodes, *Heterodera lycines* ICHINOHE, on the Growth of Soybean. Research Bulletin of Hokkaido National Experiment Station, Vol. 93, (October 1968) pp. 32-38, ISSN 0018-3415 (In Japanese)

Porter, P.M., Chen, S.Y., Reese, C.D. and Klossner, L.D. (2001) Population Response of Soybean Cyst Nematode to Long Term Corn–Soybean Cropping Sequences in Minnesota. Agronomy Journal, Vol. 93, No. 3, (May 2001) pp. 619–626, ISSN 0002-1962

Ray, J.D., Heatherly, L.G. and Fritschi, F.B. (2006) Influence of Large Amounts of Nitrogen on Nonirrigated and Irrigated Soybean. Crop Science, Vol. 46, No. 1, (January 2006) pp. 52-60, ISSN 0011-183X

Rupe, J.C., Robbins, R.T. and Gbur Jr., E.E. (1997) Effect of Crop Rotation on Soil Population Densities of *Fusarium solani* and *Heterodera gl'ycines* and on the Development of Sudden Death Syndrome of Soybean. Crop Protection, Vol. 16, No. 6 (September 1997) pp. 575–580, ISSN 0261-2194

Sakamoto, K. and Oba, Y. (1993) Relationship between Available N and Soil Biomass in Upland Field Soils. Japanese Journal of Soil Science and Plant Nutrition, Vol. 64, No. 1, (February 1993) pp. 42–48, ISSN 0029-0610 (In Japanese)

Salvagiotti, F., Cassman, K.G., Specht, J.E., Walters, D.T., Weiss, A. and Dobermann, A. (2008) Nitrogen Uptake, Fixation and Response to Fertilizer N in Soybeans: A Review. Field Crops Research, Vol. 108, No. 1, (July 2008) pp. 1-13, ISSN 1378-7290

Sayre, R.M. (1986) Pathogens for Biological Control of Nematodes. Crop Protection, Vol. 5, No. 4, (August 1986) pp. 268-276, ISSN 0261-2194

Shimizu, K. and Mitui, Y. (1985) Races of the Soybean Cyst Nematode in Tokachi Province, Hokkaido. Research Bulletin of Hokkaido National Experiment Station, Vol. 141, (January 1985) pp. 65–72, ISSN 0367-5995 (In Japanese)

Siddiqui, Z.A. and Mahmood, I. (1996) Biological Control of Plant Parasitic Nematodes by Fungi: A Review. Bioresource Technology, Vol. 58, No. 3, (December 1996) pp. 229–239, ISSN 0960-8524.

Triplett Jr., G.B. and Dick, W.A. (2008) No-Tillage Crop Production: A Revolution in Agriculture! Agronomy Journal, Vol. 100, Supplement, (May 2008) pp. 153–165, ISSN 0002-1962

Troeh, Z.I. and Loynachan, T.E. (2003) Endomycorrhizal Fungal Survival in Continuous Corn, Soybean, and Fallow. Agronomy Journal, Vol. 95, No. 1, (January 2003) pp. 224-230, ISSN 0002-1962

Varvel, G. E. and Wilhelm, W.W. (2003) Soybean Nitrogen Contribution to Corn and Sorghum in Western Corn Belt Rotations. Agronomy Journal, Vol. 95, No. 5, (September 2003) pp. 1220-1225, ISSN 0002-1962

van Kessel C. and Hartley, C. (2000) Agricultural Management of Grain Legumes: Has It Led to an Increase in Nitrogen Fixation? Field Crops Research, Vol. 65, No. 2-3, (March 2000) pp. 165–181, ISSN 1378-7290

Wright, A.L. and Hons, F.M. (2004) Soil Aggregation and Carbon and Nitrogen Storage under Soybean Cropping Sequences. Soil Science Society of America Journal, Vol. 68, No. 2 (March 2004) pp. 507 - 513, ISSN 0361-5995

Xing, L. and Westphal, A. (2009) Effects of Crop Rotation of Soybean with Corn on Severity of Sudden Death Syndrome and Population Densities of *Heterodera glycines* in Naturally Infested Soil. Field Crops Research, Vol. 112, No. 1, (April 2009) pp. 107-117, ISSN 1378-7290

Zhang, B.Q., Chen, W.D. and Yang, X.B. (1998) Occurrence of *Pythium* Species in Long-Term Maize and Soybean Monoculture and Maize/Soybean Rotation. Mycological Research, Vol. 102, No. 12, (December 1998) pp. 1450–1452, ISSN 0953-7562.

Zhang, F., Lynch, D. and Smith D. L. (1995) Impact of Low Root Temperatures in Soybean (*Glycine max* (L.) Merr.) on Nodulation and Nitrogen Fixation. Environmental and Experimental Botany, Vol. 35, No. 3 (July 1995) pp. 279–285, ISSN 0098-8472

Part 2

Modern Processing Technologies

13

Rationality in the Use of Non Renewable Natural Resources in Agriculture

Rogério de Paula Lana
Universidade Federal de Viçosa
Brazil

1. Introduction

The increase in human population and the demand for life quality have induced the growing production of food and alternative vegetal energy sources in replacement to petrol. Soybean responds to more than 80% of biodiesel production, and will reach 5% inclusion in the fossil diesel in the next years in Brazil. This trend will increase pressure to new areas for soybean production on actually human food production areas, as well on pasture and untouched forests areas.

The progress of agriculture has been based on increase in animals and plants productivity per unit of area, which only has application when land availability is the sole limiting factor. However, the efficiency of use of limiting resources (including water, fertilizers and petrol) has to be considered. This mistaken vision is leading to excessive use of non renewable natural resources and environmental pollution. The reserves of phosphate in the world that can be explored at low cost are enough for 40 to 100 years and the world reserves of potassium are enough for 50 to 200 years. The situation is worse for micronutrients, in which the reserves of copper and zinc are enough for 60 years, manganese for 35 years and selenium for 55 years (Herring & Fantel, 1993; Roberts & Stewart, 2002; Aaron, 2005).

In addition to the depletion of natural reserves, the excessive use of fertilizers can contribute to soil and water courses contamination with nitrate (Angus, 1995; Bumb, 1995), soil acidification (Helyar & Poter, 1989), and emissions of carbon dioxide (CO_2), nitrous oxide (N_2O) and ammonia to the atmosphere. The pollution with nitrate has being an actual preoccupation in Europe and North America. The fertilization with phosphorus and nitrogen cause decrease in water oxygenation by excessive increase in the population of toxic algae in the oceans (Kebreab et al., 2002).

The agriculture participates in 20% of annual increase in the anthropogenic emission of greenhouse gases, mainly CH_4 and N_2O. Approximately 70% of all anthropogenic emission of N_2O is attributed to agriculture. The current methodology used in Canada to estimate the flow of N_2O is based in the direct relation between the emission of N_2O and the application of nitrogen fertilizers (Lemke et al., 1998).

The possible deleterious effects of emissions of N_2O are global warming and catalytic destruction of the ozone chain in the stratosphere, in which the N_2O retains 13 times more heat than methane (CH_4) and 270 times more than CO_2 (Granli & Bockman, 1994). The atmospheric level of N_2O has increased in growing fashion since 1960, associated with increase in utilization of nitrogen fertilizers (Bumb, 1995; Strong, 1995).

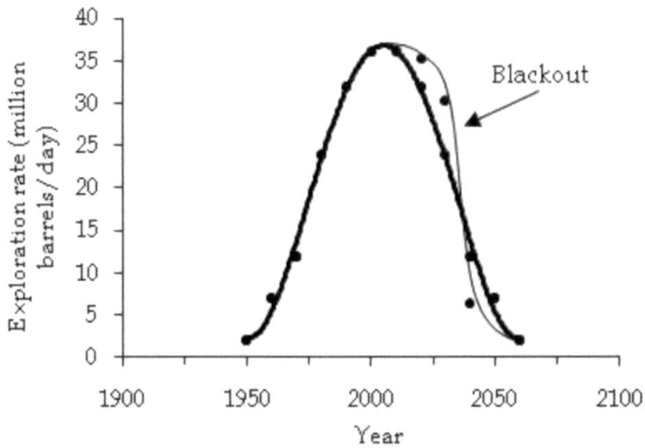

Fig. 1. Hubbert curve of exploration of petrol (non renewable natural resource), and altered curve by artificial maintenance of peak of production.

An worried phenomena about the use of non renewable natural resources can be visualized in the Hubbert curve (by Dr. Marion King Hubbert), which was a Shell Geologist, who predicted in 1956 that the global production of petrol would present a peak in the beginning of the XXI century (Hubbert Peak theory, accessed in March 02, 2009), and the curve of exploration follows the bell shape (Figure 1).

The phenomena observed by Hubbert related to petrol exploration is applicable to any other limiting natural resource, such as fertilizers, soil, and water and, consequently, food production. As more persistent is the maintenance of the maximum exploration of a resource, more drastic is the fall in the exploration of the final reserves in a short space of time, occurring the called blackout or sharply decay in the rate of production (Figure 1). Therefore, after the peak of exploration, if there is no new reserves to be discovered, no alternatives to produce more food without dependence on the available resources, or control of excessive exploration based on efficiency of use of these resources, catastrophic consequences can occur with mankind in some time of this century, as predicted by the Club of Rome in 1972, in the known publication "The limits of growth" (Meadows et al., 1972). The alert of the Club of Rome was based in the model associating accelerated industrialization, rapid population growth, depletion of non renewable natural resources, widespread malnutrition, and environmental pollution.

The objective of this work is to demonstrate the application of saturation kinetic models to improve efficiency of use of non renewable natural resources in agriculture, avoid the complete depletion as predicted by the Hubbert curve, and minimize the problems related to environmental pollution.

2. Population growth curve

The growth curve of populations of life beings in the absence of factors that affects the physical integrity, such as sickness and predation, has sigmoid curve (Gompertz or hyperbolic curve), including latency, exponential growth, plateau and senescence or death. The plateau occurs due to the saturation phenomena associated with depletion of nutrients or in some cases by environmental pollution, which acts in feedback against the uncontrolled growth (Figure 2).

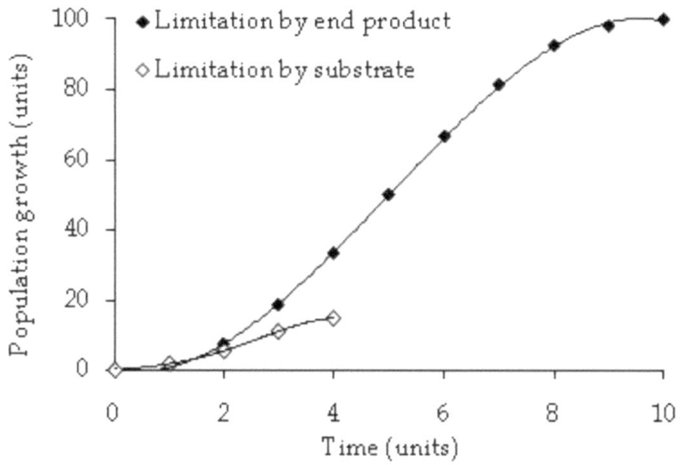

Fig. 2. Theoretical growth curve of life beings as a function of saturation by limitation of nutrients (under nutrition) or by products of metabolism (environmental pollution).

As an example, a bacterium with volume of 1 μm^3 and duplication time of 20 minutes has potential to reach a biomass much greater than the earth in only 48 hours or 144 generations (Russell, 2002; p.57-58). Therefore, the archaea and bacteria are the first life beings of the planet, are hungry most of time, and can be the survivors of a biological collapse, such as those that occurred 65,000,000 years ago with the dinosaurs and 250,000,000 years ago, when more than 90% of life beings were extinct, leading to the formation of petrol reservoirs that are being explored actually.

Another example of cessation of growth and death of population by environmental pollution is in the silage production, where the bacteria die and the nutrients are conserved to be used by ruminants, as a consequence of the acidity caused by accumulation of fermentation end products - the volatile fatty acids.

Speaking of food production crisis lead we back to the Malthus theory, which although there are some conceptual errors, it will threaten the humanity and all life beings forever. According to Malthus, the population growth curve follow geometric progression and the food production arithmetic progression, leading to the crisis of food supply in some situations or in some periods of our existence (Thomas Malthus, accessed in March 02, 2009).

However, both population growth and food production follow a sigmoid curve up to the plateau or in form of a bell or double sigmoid over time (the second goes down hill, similar to the first in a mirror). The population growth curve is cumulative, as a consequence of the sum of the annual growth rates (Figure 3), which depend on the annual rate of food production (productivity), that by its time is consequence of the annual rate of soil utilization and exploration of non renewable natural resources (fertilizers and petrol), that follow the Hubbert curve. Changes in these curves can be caused by men or naturally, with discoveries of new food production technologies, population death (caused by diseases, wars, predations, among others), proliferation of plagues and diseases in plants, climatic changes, among others.

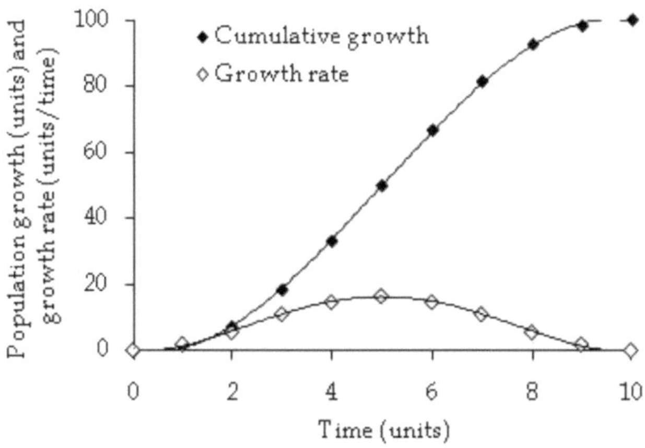

Fig. 3. Theoretical population growth (cumulative and growth rate) as function of time.

Studies of these phenomena lead us to understand the need for rational use of non renewable natural resources. The kinetic saturation models are important tools generated by science in order to evaluate the efficiency and allow the rational use of non renewable natural resources (Lana, 2005; Lana et al., 2005; Lana, 2007a,b; Lana et al., 2007a,b; Lana, 2008). As a result, they can avoid the complete depletion of the resources and the collapse in food and energy supply, with dramatic consequences for our civilization, as predicted by Malthus, Hubbert and Club of Roma.

3. Models of biological responses to nutrients

The first studies on the limiting factors in the plants growth were developed by Carl Sprengel in 1826 and 1828, and by Liebig in 1840, leading to the rejection of humus theory and formulation of Law of minimum (van der Ploeg et al., 1999). The Law of minimum or Law of response is associated with the absence of nutrient replacement, linear response in production by increase in the quantity of the limiting factor and a maximum plateau of response, in which the plants do not respond satisfactorily anymore to the limiting nutrient.
A posterior mark was the Law of diminishing return of Mitscherlich (1909). The convex exponential equation of Mitscherlich, with a model that includes the maximum asymptotic yield, allows calculating the optimum economic level of fertilization, based in the benefit-cost ratio.
The Michaelis-Menten model (Michaelis & Menten, 1913) was developed to describe the enzymatic kinetic in the beginning of the 20 century. The Lineweaver-Burk model (Lineweaver & Burk, 1934), an equation of the linear regression of the reciprocal of Y (enzymatic activity) as a function of the reciprocal of X (concentration of substrate), was used to obtain the kinetic constants of the Michaelis-Menten model: k_s (the amount of substrate needed to reach half of maximum enzymatic activity) and k_{max} (maximum enzymatic activity).
Later, researchers verified that the microbial growth rate was dependent of substrate concentration and both were related to the saturation kinetic typical of enzymatic systems (Monod, 1949; Russell, 1984).

Although the use of saturation kinetic model to explain the nutrients responses by the superior forms of life is not being adopted (Morgan et al., 1975), the Michaelis-Menten model allows to explain the curvilinear relationship of plants and animals to the nutrients and the model of Lineweaver-Burk allows to obtain the kinetic constants, k_s (the amount of substrate needed to reach half theoretical maximum response in rate of growth or production of milk, wool, eggs, among others) and k_{max} (theoretical maximum response in rate of growth or production), according to Lana et al. (2005).

The responses of plants and animals to nutrients as saturation phenomena have important implications in addition to calculation of the rate of decreasing economical return and estimates of nutrients recommendations, such as the consciousness about the excessive use of non renewable natural resources; soil, water, and air pollution; and global warming.

The knowledge about the efficiency of utilization of fertilizers in agriculture will play an important role in the political decisions about the rational use of non renewable natural resources in the future. The natural fertilizer sources have to be used with maximum efficiency and with minimum negative effects in the environment.

4. Marginal response or Law of diminishing return in plants

Recommendations of fertilization are mostly based in the method of calculation of nutrients requirements of a culture and the mineral contribution of the soil. The fertilizers are then calculated to supply the deficiencies. This method allows recommendation of the lower level that maximize the production. However, the method does not indicate changes in the recommendation based on changes in the costs of nutrients and grains. Also, it does not give direct information of the effect of application of other level than the recommended one (Makowski et al., 1999).

It has being utilized a variety of empirical models to predict the responses to nutrients and to calculate the optimum levels of nutrients. Among then, it is included the model of Mitscherlich, square root (Mombiela et al., 1981; Sain & Jauregui, 1993), exponential, linear-plus-plateau, linear-plus-hyperbola, quadratic and quadratic-plus-plateau (Cerrato & Blackmer, 1990; Bullock & Bullock, 1994; Makowski et al., 1999, 2001).

The use of saturation kinetics to explain the nutritional responses to nutrients by superior life beings are rarely employed (Morgan et al., 1975). The model of Michaelis-Menten has not being evaluated to make recommendations of fertilization. This model has a great potential in recommendation of use of nutrients in agriculture, by considering the efficiency of use of nutrients and the Law of diminishing return, as observed by Mitscherlich (1909). This model can aggregate important concepts such as responses to different levels of nutrients, benefit-cost ratio, efficiency of use of nutrients, rationality of use of non renewable natural resources and consciousness about environmental pollution.

Linear regressions of reciprocal of plants responses as a function of reciprocal of nutrients supply, methodology known as data transformation of Lineweaver-Burk (Lineweaver & Burk, 1934; Champe & Harvey, 1994), were proposed by Lana et al. (2005) as follow:

$$1/Y = a + b * (1/X)$$

where:

Y = responses of plants (grain yield, x 1,000 kg/ha),

a = intercept,

b = coefficient of linear regression,
X = amount of nutrient (kg/ha/year).
The theoretical maximum grain production (k_{max}) is obtained by the reciprocal of intercept (1/a). The amount of nutrient (X) needed to reach half of theoretical maximum response (k_s) is obtained by the model presented above, replacing Y by 1/a x 50(%) x 0.01, or dividing the coefficient of the linear regression by the intercept (b/a).
The efficiency of use of fertilizers is calculated dividing the accretion in grain production (Y_2 - Y_1) by the accretion in fertilization (X_2 - X_1), from a specific level of fertilizer in relation to the previous level.
Simulations of biological responses to nutrients in the absence or presence of a second limiting nutrient are presented in Table 1 and Figure 4, in which are expected changes in the maximum yield (k_{max}) and k_s of the first limiting nutrient (increase, no effect or decrease). The Figure 4A illustrates four kind of responses in production and models of double-reciprocal are presented in Table 1 and Figure 4B, demonstrating the combination of two values of k_{max} by two of k_s.
The best effects that a second limiting nutrient can cause are by increasing k_{max}, decreasing k_s, or both changes that is even better. However, the most common kind of response is by increasing both k_{max} and k_s. Increase in k_{max} by increase in productivity with a second nutrient lead to increase the efficiency of use of the first limiting nutrient (Figure 4C), but this benefit decreases sharply by increase in the amount of the first limiting nutrient, especially when k_s is low.

Equation	Symbol	Intercept (a)	Coefficient (b)	r^2	k_s	k_{max}
1	O	0.8163	79.789	1.00	98	1.2
2	△	0.9195	39.591	1.00	43	1.1
3	□	0.4082	39.894	1.00	98	2.4
4	◇	0.4768	19.483	1.00	41	2.1

Table 1. Constants of linear regression of reciprocal of grain production (x1,000 kg/ha) as a function of reciprocal of amount of fertilizers (kg/ha/year) in hypothetic situations of high or low values of the saturation constants k_s (kg of fertilizer/ha) and k_{max} (x1,000 kg/ha) – see Figure 4B

The plants responses to fertilization depend on soil fertilization, in which high responses occur when soil fertility is low (Figure 5A) and in low level of fertilization, that is the main factor that affects the efficiency of use of fertilizers (Figure 5B).
Equations of data transformation of Lineweaver-Burk were used to explain the effect of fertilization and the effect of a second factor in the yield, k_s, k_{max} and efficiency of use of fertilizers in soybean, bean, wheat and cotton production (Tables 2, 3 and 4).
When limestone was the second factor, there was change in k_s and k_{max} in 34 and 85%; -75 and -10%; and 33 and 22% for soybean fertilized with P_2O_5 (Table 2). Limestone as a second factor changed k_s and k_{max}, respectively, in -55 and -12% for wheat fertilized with P_2O_5, and in 9 to 87% in cotton fertilized with K_2O.
As seen above, increase or decrease in k_{max} is associated with the same effect in k_s, but increase in k_{max} associated with exaggerated increase in k_s is not desirable because it requires more fertilizer to reach the plateau. In other words, the greater values of k_s present greater response to the use of fertilizers in high level of fertilization, but it cannot be advantageous due to the increase in the cost of fertilization.

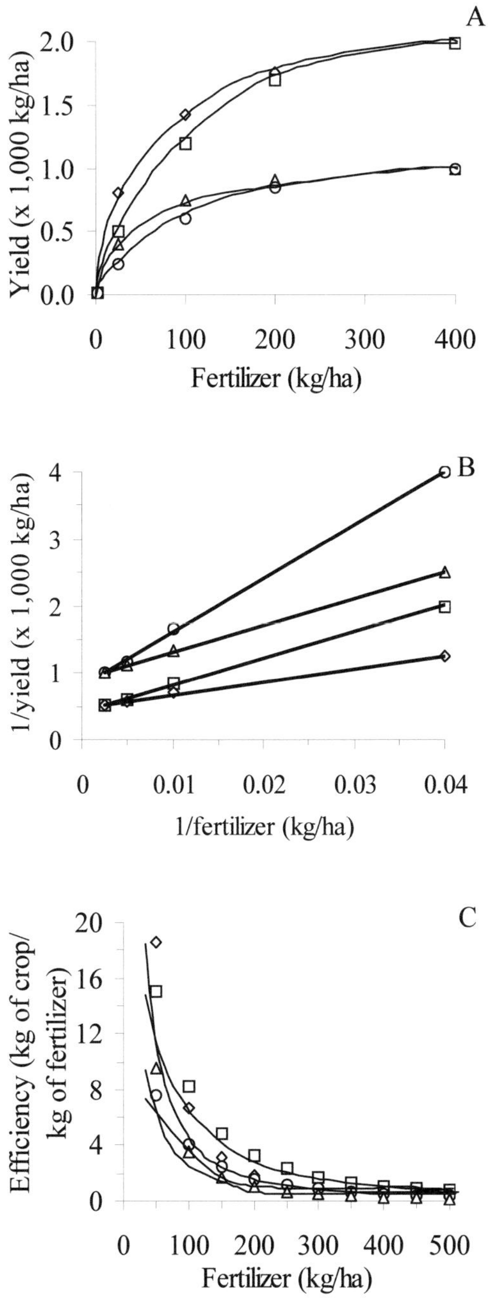

Fig. 4. Biological responses to nutrients as a function of a second limiting nutrient (A) - control (O), decrease in k_s (△), increase in k_{max} (□) and decrease in k_s and increase in k_{max} (◇); reciprocal of production as a function of reciprocal of fertilizer level – plot of Lineweaver-Burk (B); and effect of a second limiting nutrient in the efficiency of use of the first one (C)

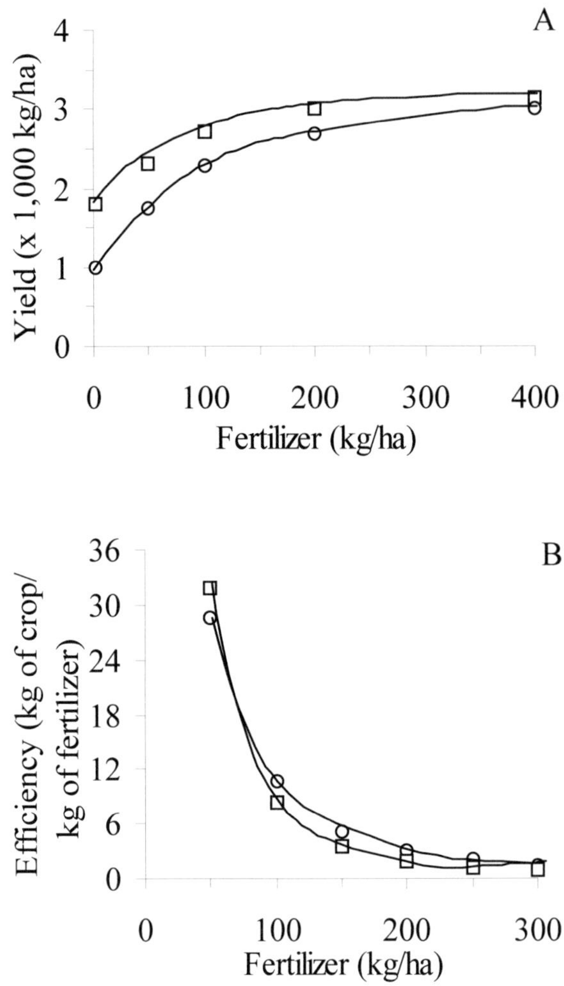

Fig. 5. Plants responses to fertilizers in low (O) and high (□) soil fertility (A); and the effect of soil fertility in the efficiency of use of fertilizer (B)

In the case of soybean (Table 2), considering US$1.208/kg of P_2O_5 and US$0.178/kg of soybean, it is necessary 6.8 kg of soybean to pay 1 kg of fertilizer. Therefore, it is viable to use 50 to 100 kg of P_2O_5 in the absence of limestone and 100 kg of P_2O_5 in the presence of limestone. Above 150 kg of P_2O_5, although in some cases there was still response to fertilizer, especially in high values of k_s, the response is not viable economically.

In bean production (Table 3), the second factor (P_2O_5) increased the k_{max} of nitrogen from 0.1-0.5 to 1.0-1.6 x 1,000 kg/ha of bean, but also increased the k_s (1 to 13 and 17 to 29 kg/ha of nitrogen). When the second factor was nitrogen, this increased the k_{max} of P_2O_5 from 0.7-0.8 to 1.5 x 1,000 kg/ha of bean, but also increased the k_s (5 to 15 and 136 to 199 kg/ha of P_2O_5). In the second case, the high values of k_s for P_2O_5 caused low improvement in the efficiency of use of this fertilizer in low level of fertilization (Table 4). The extra production in this case cannot be enough to pay the extra amount of fertilizers.

Product	Fertilizer (kg/ha/year)	Second factor [1]	Intercept (a)	Coefficient (b)	r^2	k_s [2]	k_{max} [3]	Source of data [4]
Soybean	P_2O_5	-	0.7536	57.766	1.00	77	1.3	1
		+	0.4096	42.198	1.00	103	2.4	1
Soybean	P_2O_5	-	0.3502	30.524	0.98	87	2.9	2
		+	0.3801	8.2987	0.99	22	2.6	2
Soybean	P_2O_5	-	0.3103	3.6726	0.68	12	3.2	2
		+	0.2535	3.9962	0.53	16	3.9	2
Wheat	P_2O_5	-	0.4169	174.48	1.00	419	2.4	1
		+	0.4781	91.00	1.00	190	2.1	1
Cotton	K_2O	-	0.622	4.6865	0.91	7.5	1.6	1
		+	0.3284	2.7052	0.97	8.2	3.0	1

	Fertilizer	Efficiency of use of fertilizers (kg of grains/kg of fertilizer)[5]					
	(kg/ha)	50 [2]	100	150	200	250	300
Soybean	P_2O_5 - Lim [1]	10.0	4.5	2.5	1.6	1.1	0.8
	+ Lim	16.0	8.1	4.9	3.3	2.4	1.8
Soybean	P_2O_5 - Lim	15.5	9.7	5.6	3.7	2.6	1.9
	+ Lim	15.4	6.6	2.7	1.5	1.0	0.7
Soybean	P_2O_5 - Lim	28.0	5.5	2.1	1.1	0.7	0.5
	+ Lim	37.0	8.2	3.2	1.7	1.1	0.7
Wheat	P_2O_5 - Lim	5.0	4.1	3.4	2.9	2.4	2.1
	+ Lim	8.3	5.7	4.0	3.0	2.3	1.8
Cotton	K_2O - Lim	10.5	2.0	0.7	0.4	0.2	0.2
	+ Lim	20.8	4.0	1.5	0.8	0.5	0.3

[1] Limestone: without (-) or with (+) 4,000 to 7,000 kg/ha; [2] Kg of fertilizer/ha - P_2O_5 or K_2O; [3] x1,000 kg/ha of grain; [4] 1 = Malavolta (1989), p.61, 275 and 283; 2 = Oliveira et al. (1982), p.36; [5] Considering US$1.208/kg of P_2O_5 and US$0.178/kg of soybean, is necessary 6.8 kg soybean to pay one kg of fertilizer. Efficiency lower than 6.8 kg of soybean/kg of P_2O_5 is not viable. These calculations can be used to choose the level of fertilization.

Table 2. Changes in the constants of linear regression of the reciprocal of grain production (x1,000 kg/ha) as a function of the reciprocal of amount of fertilizer (kg/ha/year), by the second factor, and the respective efficiency of use of fertilizers (kg of grains/kg of fertilizer)

5. Marginal response in bovines

The weight gain in growing bovines in pasture in the dry season is curvilinear as a function of supplement supply, based on corn and soybean meal, in which the supplement conversion (kg of supplement/kg of accretion in weight gain) becomes worse with increase in the supplementation (Lana et al., 2005; Keane et al., 2006; Lana, 2007b) (Figure 6).

The milk production by supplemented cows in pasture or in feedlot is also curvilinear as a function of increase in the concentrate supply, based on corn and soybean meal (Figure 7A), in which the marginal increase in milk production per kg of concentrate decreases with increase in the amount of concentrate (Bargo et al., 2003; Pimentel et al., 2006a; Sairanen et al., 2006; Lana et al., 2007a,b), as shown in Figure 7B, and in some studies the milk response to concentrate was satisfactory only up to 2-4 kg of concentrate/animal/day (Fulkerson et al., 2006).

The curvilinear response can also be verified with specific nutrients, such as the observed positive curvilinear response in milk production and negative curvilinear response in the efficiency of use of nitrogen by increasing the dietary crude protein content from 11 to 19% in cows with mean production of 38 kg of milk/day (Baik et al., 2006). In the third experiment of Figure 7, in addition to decreasing response in milk production, there was decreasing response in body weight variation with increase in the concentrate level (0.20, 0.12, and 0.095 kg of body weight gain per additional kilogram of concentrate intake; Teixeira et al., 2006).

Fertilizer (kg/ha/year)	Second factor	Intercept (a)	Coefficient (b)	r^2	k_s [1]	k_{max} [2]	Sorce of data [3]
	P_2O_5(kg/ha)						
	0	2.044	2.794	0.82	1	0.5	
N	40	0.782	8.516	1.00	11	1.3	1
	80	0.710	6.718	0.99	9	1.4	
	100	0.630	8.205	1.00	13	1.6	
	P_2O_5(kg/ha)						
	0	10.764	183.36	0.23	17	0.1	
N	50	2.5229	46.235	0.78	18	0.4	2
	150	1.3364	15.623	0.99	12	0.7	
	250	0.9539	28.056	0.98	29	1.0	
	N (kg/ha)						
P_2O_5	0	1.3812	6.7411	0.98	5	0.7	1
	30	0.8181	8.8241	1.00	11	1.2	
	60	0.6842	10.186	1.00	15	1.5	
	N (kg/ha)						
P_2O_5	0	1.3257	180.38	0.95	136	0.8	2
	50	1.0402	90.586	1.00	87	1.0	
	120	0.6684	132.74	1.00	199	1.5	

[1] Kg of fertilizer/ha; [2] Ton of grain/ha; [3] 1 = Bolsanello et al. (1975) and Oliveira et al. (1982), p.155; 2 = Malavolta (1989), p.273.

Table 3. Changes in constants of linear regression of reciprocal of bean production (x1,000 kg/ha) as a function of reciprocal of amount of fertilizer (kg/ha/year), by a second factor

According to the Biotechnology and Biological Sciences Research Council (1998), formerly known as AFRC (Agricultural and Food Research Council), all currently feed systems calculate the dietary requirements of energy and protein to meet the animals needs for maintenance and production. However, in practice, the situation is different, because there is no need for the farmer to meet the cow's nutritional requirements if it is against the economical interest. So, it is evident that studies in animal response to increasing levels of concentrate or specific nutrients are needed, as suggested by Lana (2003; p.87).

Although the animal's responses to nutrients are curvilinear, the daily weight gains estimated by the level 1 of NRC (1996) of beef cattle are linear as a function of intakes of metabolizable energy and protein (Figure 8A). In the same way, the milk production estimated by the model CNCPS 5.0 as a function of intakes of metabolizable energy and protein, and model NRC (2001) of dairy cattle as a function of intakes of net energy for lactation and metabolizable protein, were linear by using increasing levels of concentrate

Fertilizer (kg/ha)	Second factor	Efficiency of use of fertilizers (kg of grains/kg of fertilizer)[1]					
		50[2]	100	150	200	250	300
	P_2O_5 (kg/ha)						
N	0	1.1	0.1	0.0	0.0	0.0	0.0
	40	10.9	2.1	0.8	0.4	0.3	0.2
	80	11.5	2.0	0.8	0.4	0.2	0.2
	100	14.2	2.9	1.1	0.6	0.4	0.3
	P_2O_5 (kg/ha)						
N	0	0.6	0.2	0.1	0.0	0.0	0.0
	50	2.8	0.9	0.4	0.2	0.1	0.1
	150	4.5	1.3	0.5	0.3	0.2	0.1
	250	7.9	3.0	1.3	0.7	0.5	0.3
	N (kg/ha)						
P_2O_5	0	5.4	0.6	0.2	0.1	0.1	0.0
	30	12.2	2.0	0.7	0.4	0.2	0.2
	60	15.3	2.9	1.1	0.6	0.4	0.3
	N (kg/ha)						
P_2O_5	0	4.0	2.3	1.5	1.1	0.8	0.6
	50	6.9	3.3	1.9	1.2	0.9	0.6
	120	6.0	4.0	2.9	2.1	1.7	1.3

[1] Considering US$0.966/kg of N, US$1.208/kg of P_2O_5 and US$0.36/kg of bean, it is necessary 2.68 and 3.36 kg of bean to pay 1 kg of N or P_2O_5. Efficiency worse than 2.68 or 3.36:1 for N or P_2O_5 is not economically desirable. These calculations can be used to choose the level of fertilization. [2] Level of fertilizer (kg/ha) - N in the first two cases or P_2O_5 in the last two cases.

Table 4. Efficiency of use of fertilizers (kg of bean/kg of fertilizer) calculated with base in the equations of Table 3

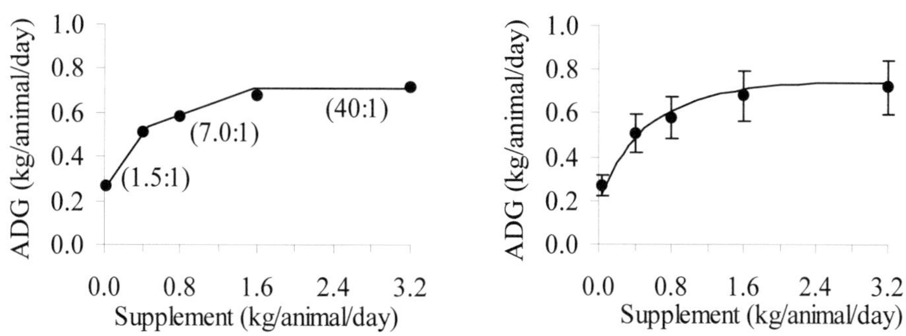

Fig. 6. Body weight gain (BWG) of growing bovines in pasture during the dry season, as a function of daily intake of supplement with 24% CP, in which the values among parenthesis represent the differential in kilograms of supplement given daily divided by the differential in weight gain, in relation to the previous treatment (Lana, 2005; Lana et al., 2005)

(Figure 8B), as suggested by Lana (2005; p.290-291) and Lana (2007b; p.39 a 43). Therefore, in order to these systems be compatible with the tropical conditions, in which it is more evident the curvilinear responses to nutrients, it is necessary modifications in future versions, by adopting models of saturation kinetics.

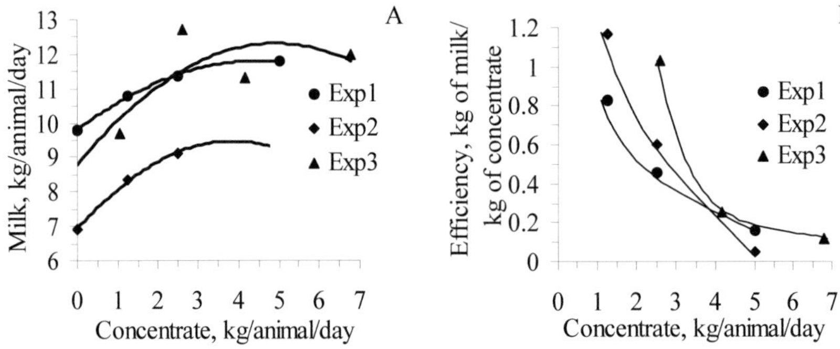

Fig. 7. Production of milk (A) and efficiency of use of concentrate (B) as a function of intake of increasing level of concentrate in three experiments (Pimentel et al., 2006b, 2006c; Teixeira et al., 2006)

6. Production versus productivity

Two studies were conducted to evaluate the factors that affect milk production in Brazil, on farmer level or by state of federation (Guimarães et al., 2008; Lana et al., 2009). In the first case, data were collected from fifty producers that sell milk for a dairy plant in the south region of Rio de Janeiro state, including data of daily milk production by producer, with the respective data of production per cow and per hectare, farmer size and size area designated to the herd, total of milking cows and herd size, and breed (Table 5). In the second case, data were collected from EMBRAPA and IBGE in the years of 2004-2006, in which the emphasis was in milk production per state instead of production per producer.

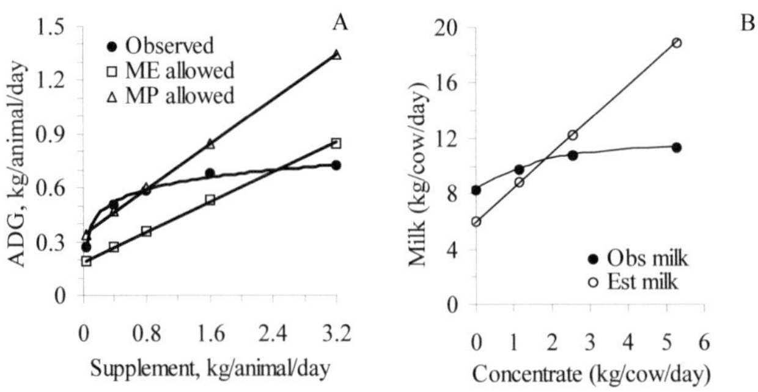

Fig. 8. Mean daily weight gain of steers in pastures, observed and estimated by level 1 of NRC (1996) as a function of intake of metabolizable energy and protein in the supplement (A); and observed milk production (mean of data of Figure 7A) and estimated by CNCPS 5.0 and NRC (2001) as a function of intakes of metabolizable energy or net energy of lactation, respectively, and metabolizable protein (B)

Production (Kg of milk/ producer/day)	Number of producers	Production per producer (Kg of milk/day)	Area for the herd (ha)	Herd (number of animals)	Milking cows (number)
Up to 150	6	117	37	56	14
151-300	12	238	70	73	28
301-600	14	451	106	94	43
601-1200	14	821	111	159	66
1201-2400	3	1667	316	633	132
2401-4800	1	4000	300	1800	300
Kg of milk/ producer/day	**Milking cows/ha**	**Milk (Kg/ha/day)**	**Milk (Kg/milking cow/day)**	**Milk (Kg/total herd/day)**	**Milking cows/total herd**
Up to 150	0.38	3.17	8.33	2.08	0.25
151-300	0.40	3.40	8.57	3.27	0.38
301-600	0.41	4.27	10.53	4.77	0.45
601-1200	0.59	7.41	12.51	5.15	0.41
1201-2400	0.42	5.27	12.66	2.63	0.21
2401-4800	1.00	13.33	13.33	2.22	0.17

Table 5. Number of producers, daily mean milk production by producer, area for the herd, number of animals and milking cows in the herd, and productivity indexes per area and per cow, as a function of production levels

The milk production in farmer level (first case) ranged from 60 to 4000 kg/producer/day. The increase in milk production was highly correlated with the number of milking cows (r = 0.94), followed by moderate correlation with the size of pasture (r = 0.67) and, surprisingly, the productivity per cow and per unit of area did not correlate with the milk production per producer (r = 0.11 and 0.06, respectively; Table 6). In the nation level (second case), the result repeated, in which there was high correlation of milk production/state/year with the total of milking cows in relation to productivity of milk/km^2/year and milk/cow/year (r = 0.95, 0.55 and 0.51, respectively; Table 7).

Parameter	Correlation (r)	Parameter	Correlation (r)
Total animals in the herd	0.94	Milk (kg/ha)	0.13
Total milking cows	0.93	Milk (kg/cow/day)	0.11
Area for the herd (ha)	0.67	Milking cows/ha	0.06
Total area in the farm	0.20	Milk (kg/total of animals/day)	-0.11

Table 6. Linear correlation of daily milk production by producer with: total of animals in the herd, total of milking cows, area for the herd, farmer size and some productivity indexes (daily milk production per hectare and per cow, milking cows per hectare and daily milk production per total animals in the herd)

Therefore, the milk production by Brazilian farmers is more dependent on the farmer size and pasture extension, than on productivity indexes, and similar effect occur with crop production (Lana, 2009). Agricultural showed the same results verified with dairy cattle, in which cultivated area generally presents more than 90% correlation with crop production (Table 8 and Figure 9). Then, the concepts about agricultural production need to be revised, facing the actual problems related with the inadequate use and depletion of the non renewable natural resources, and environmental pollution.

Parameters	Liters of milk/state/year	Total of milked cows	Liters of milk/km^2/year	Liters of milk/cow/year
Total of milking cows	0.95			
Liters of milk/km^2/year	0.55	0.39		
Liters of milk/cow/year	0.51	0.31	0.88	
Surface of state (in km^2)	0.11	0.21	-0.37	-0.26

Table 7. Linear correlation (r) of annual milk production by Brazilian states with total of milked cows, liters of milk/km^2/year, liters of milk/cow/year and surface of the state (in km^2).

Item	n	Production (ton/year) X planted area (ha)	Production (ton/year) X productivity (ton/ha)	productivity (ton/ha) X planted area (ha)
Corn	110	0.95	0.35	0.14
Bean[1a]	110	0.87	0.18	0.02
Bean[1b]	110	0.96	0.42	0.31
Sugarcane	101	0.99	0.34	0.30
Coffee	95	0.98	0.07	0.00
Banana[2]	90	0.96	-0.02	-0.14
Cassava	86	0.98	0.21	0.09
Orange[3]	84	0.91	0.32	0.03
Rice[4a]	76	0.88	0.22	-0.05
Rice[4b]	71	0.91	0.53	0.38
Tomato	51	0.98	0.46	0.37
Coconut	18	0.95	0.31	0.12
Potato[5a]	12	0.99	0.47	0.42
Potato[5b]	11	0.88	0.22	-0.19
Potato[5c]	10	0.99	-0.20	-0.29
Mean		0.94	0.26	0.10

Source: Lana & Guimarães (2010); n = number of municipalities; 1a = first harvest; 1b = mean of second and third harvest; 2 = racemes instead of ton (1,000 kg); 3 = number (x1000) of oranges instead of ton (1,000 kg); 4a = rice planted in wet land; 4b = rice planted in dry land, without or with irrigation; 5a,b,c = harvest 1, 2 and 3, respectively; Source of data: www.cidadesnet.com.br (year of 2003).

Table 8. Linear correlation (r) among some variables related to agricultural production, in municipalities of Zona da Mata and Central of Minas Gerais state, Brazil.

7. Conclusions

The agriculture progress is based in improvements of animals and plants productivity per unit of area, which is only applicable when land is the limiting factor, but other factors are emerging as limiting, such as water, fertilizer and petrol.

Models of saturation kinetics are important tools to improve the efficiency and decrease costs of utilization of non renewable natural resources in agriculture, allowing the conservation of these resources for the future generations, and decreasing the negative impacts in the environment.

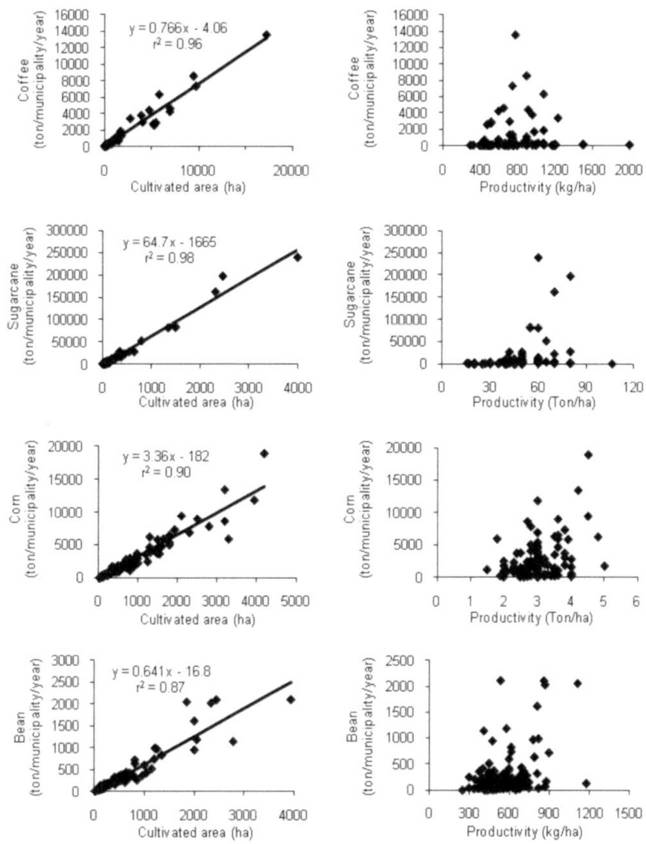

Fig. 9. Effect of cultivated area and productivity on production of some main cultures (coffee, sugarcane, corn and bean), in municipalities of Zona da Mata and Central of Minas Gerais state, Brazil.

8. References

Aaron, S. (2005). *Some Statistics on Limited Natural Resources*, 31.07.2006, available from Http://scotaaron.com/resources2.html

Angus, J.F. (1995). Modeling N Fertilization Requirements for Crops and Pasture, In: *Nitrogen Fertilization in the Environment*, P.E. Bacon, (Ed.), 109-127, Marcel Dekker inc., ISBN 0-8247-8994-6, New York, NY, United States

Baik, M.; Aschenbach, J.R.; Vandehaar, M.J. & Liesman, J.S. (2006). Effect of Dietary Protein Levels on Milk Production and Nitrogen Efficiency in Dairy Cattle. *Journal of Dairy Science*, Vol.89, Suppl. 1, pp.81, ISSN 0022-0302

Bargo, F.; Muller, L.D.; Kolver, E.S. & Delahoy, J.E. (2003) Invited Review: Production and Digestion of Supplemented Dairy Cows on Pasture. *Journal of Dairy Science*, Vol.86, No.1, pp.1-42, ISSN 0022-0302

Biotechnology and Biological Sciences Research Council (1998). *Responses in the Yield of Milk Constituents to the Intake of Nutrients by Dairy Cows*, CAB International, ISBN 0-8519-9284-6, Wallingford, United Kingdon

Bolsanello, J.; Vieira, C.; Sediyama, C.S. & Vieira, H.A. (1975). Ensaios de Adubação Nitrogenada e Fosfatada no Feijão (*Phaseolus vulgaris* L.) na Zona Metalúrgica de Minas Gerais. *Revista Ceres*, Vol.22, pp.423-430, ISSN 0034-737X

Bullock, D.G. & Bullock, D.S. (1994). Quadratique and Quadratic-Plus-Plateau Models for Predicting Optimal N Rate for Corn. A comparison. *Agronomy Journal, Vol.86, pp.*191-195, ISSN 1435-0645

Bumb, B.L. (1995). World Nitrogen Supply and Demand: an Overview, In: *Nitrogen Fertilization in the Environment*, P.E. Bacon, (Ed.), 1-40, Marcel Dekker inc., ISBN 0-8247-8994-6, New York, NY, United States

Cerrato, M.E. & Blackmer, A.M. (1990). Comparison of Models for Describing Corn Yield Response to Nitrogen Fertilizer. *Agronomy Journal, Vol.*82, pp.138-143, ISSN 1435-0645

Champe, P.C. & Harvey, R.A. (1994). *Biochemistry*, J.B. Lippincott Company, ISBN 0-3975-1091-8, Philadelphia, PA, United States

Fulkerson, W.J.; Nandra, K.S.; Clark, C.F. & Barchia, I. (2006). Effect of Cereal-Based Concentrates on Productivity of Holstein-Friesian Cows Grazing Short-Rotation Ryegrass (*Lolium multiflorum*) or Kikuyu (*Pennesitum clandestinum*) Pastures. *Livestock Science*, Vol.103, pp.85-94, ISSN 1871-1413

Granli, T. & Bockman, O.C. (1994). Nitrogen Oxide from Agriculture. *Norwegian Journal of Agricultural Sciences*, Vol.12, pp.7-127, ISSN 0801-5341

Guimarães, G.; Lana, R.P.; Guimarães, A.V.; Santos, M.A.; Andrade, F.L.; Fialho, C. & Castro, T.R. (2008). Sustentabilidade da Agricultura Familiar na Produção de Leite, *Proceedings of 10º Minas Leite*, EMBRAPA, CDD 637.1, Juiz de Fora, MG, Brazil

Helyar, K.R.; Poter, W.M. (1989). Soil Acidification, its Measurement and the Processes Involved, In: *Soil acidity and plant growth*, A.D. Robson, (Ed.), 61-100, Academic Press, ISBN 0-1259-0655-2, Sydney, Australia

Herring, J.R. & Fantel, R.J. (1993). Phosphate Rock Demand into the Next Century: Impact on World Food Supply. *Nonrenewable Resources*, Vol.2, No.3, pp.226-246, ISSN 0961-1944

Keane, M.G.; Drennan, M.J. & Moloney, A.P. Comparison of Supplementary Concentrate Levels with Grass Silage, Separate or Total Mixed Ration Feeding, and Duration of Finishing in Beef Steers. *Livestock Science*, Vol.103, pp.169-180, ISSN 1871-1413

Kebreab, E.; France, J.; Mills, J.A.N.; Allison, R. & Dijkstra, J. (2002). A Dynamic Model of N Metabolism in the Lactating Cow and an Assessment of Impact on N Excretion on the Environment. *Journal of Animal Science*, Vol.80, pp.248-259, ISSN 0021-8812

Lana, R.P. (2003). *Sistema Viçosa de Formulação de Rações*, Editora UFV, ISBN 978-85-7269-314-1, Viçosa, MG, Brazil

Lana, R.P. (2005). *Nutrição e Alimentação Animal (Mitos e Realidades)*, Suprema Gráfica, ISBN 978-85-9050-672-0, Viçosa, MG, Brazil

Lana, R.P. (2007a). Plants Responses to Nutrients Follow a Michaelis-Menten Relationship, *Proceedings of ASA, CSSA, SSSA International Annual Meetings*, ASA-CSSA-SSSA, ISSN 1529-9163, New Orleans, United States

Lana, R.P. (2007b). *Respostas Biológicas aos Nutrientes*, Editora CPD, ISBN 978-85-9050-673-7, Viçosa, MG, Brazil

Lana, R.P. (2008). Plants Responses to Nutrients Follow the Saturation Kinetic Typical of Enzyme Systems: Biological, Economical and Environmental Implications. *Online Journal of Biological Sciences*, Vol.8, No.1, pp.19-24, ISSN 1608-4217

Lana, R.P. (2009). Uso Racional de Recursos Naturais não Renováveis: Aspectos Biológicos, Econômicos e Ambientais. *Revista Brasileira de Zootecnia*, Vol.38, pp.330-340, Supplement Especial, ISSN 1806-9290

Lana, R.P.; Abreu, D.C.; Castro, P.F.C. & Zamperline, B. (2007a). Milk Production as a Function of Energy and Protein Sources Supplementation Follows the Saturation Kinetics Typical of Enzyme Systems, *Proceedings of 2nd International Symposium on Energy and Protein Metabolism and Nutrition*, European Association for Animal Production, ISBN 978-90-8686-041-8, Vichy, France

Lana, R.P.; Abreu, D.C.; Castro, P.F.C.; Zamperlini, B. & Souza, B.S.B.C. (2007b). Kinetics of Milk Production as a Function of Energy and Protein Supplementation. *Journal of Animal Science*, Vol.85, Suppl. 1, pp.566, ISSN 0021-8812

Lana, R.P.; Goes, R.H.T.B.; Moreira, L.M.; Mancio, A.B. & Fonseca, D.M. (2005). Application of Lineweaver-Burk Data Transformation to Explain Animal and Plant Performance as a Function of Nutrient Supply. *Livestock Production Science*, Vol.98, pp.219-224, ISSN 0301-6226

Lana, R.P.; Guimarães, G.; Guimarães, A.V. & Santos, M.A. (2009). Factors Affecting Milk Production in Brazil. *Journal of Dairy Science*, Vol.92, Suppl. 1, pp.427, ISSN 0022-0302

Lana, R.P. & Guimarães, G. (2010). Produção Média e Correlações entre Algumas Variáveis que Afetam a Produção de Diversas Culturas em Municípios da Zona da Mata e Central de Minas Gerais, In: *2º Simpósio Brasileiro de Agropecuária Sustentável – Anais de Resumos Expandidos*, R.P. Lana & G. Guimarães, (Eds.), ISSN 2176-0772, Viçosa, MG, Brazil

Lemke, R.L.; Izaurralde, R.C.; Malhi, S.S.; Arshad, M.A. & Nyborg, M. (1998). Nitrous Oxide Emissions from Agricultural Soils of the Boreal and Parkland Regions of Alberta. *Soil Science Society of America Journal, Vol.*62, pp.1096-1102, ISSN 0361-5995

Lineweaver, H. & Burk, D. (1934). The Determination of Enzyme Dissociation Constants. *Journal of the American Chemical Society*, Vol.56, pp.658-666, ISSN 0002-7863

Makowski, D.; Wallach, D. & Meynard, J.-M. (1999). Model of Yield, Grain Protein, and Residual Mineral Nitrogen Responses to Applied Nitrogen for Winter Wheat. *Agronomy Journal, Vol.*91, pp.377-385, ISSN 1435-0645

Makowski, D.; Wallach, D. & Meynard, J.-M. (2001). Statistical Methods for Predicting Responses to Applied Nitrogen and Calculating Optimal Nitrogen Rates. *Agronomy Journal, Vol.*93, pp.531-539, ISSN 1435-0645

Meadows, D.H.; Meadows, D.L.; Randers, J. & Behrens III, W.W. (1972*). The Limits of Growth. A Report for the Club of Rome's Project on the Predicament of Mankind*, Universe Books, ISBN 0-87663-165-0, New York, NY, United States

Michaelis, L. & Menten, M.L. (1913). Kinetics of Invertase Action. *Biochemistry Zournal*, Vol.49, pp.333-369, ISSN 1470-8728

Mitscherlich, E.A. (1909). Das Gesetz des Minimuns und das Gesetz des Abnehmenden Bodenertrages. *Landwirtschaftliches Jahrbuch*, Vol.38, pp.537-552, ISSN 0005-7150

Malavolta, E. (1989). *ABC da Adubação*, Editora Ceres, ISBN 85-318-0002-1, São Paulo, SP, Brazil

Mombiela, F.; Nicholaides III, J.J. & Nelson L.A. (1981). A Method to Determine the Appropriate Mathematical form for Incorporating Soil Test Levels in Fertilizer Response Models for Recommendation Purposes. *Agronomy Journal, Vol.*73, pp.937-941, ISSN 1435-0645

Monod, J. (1949). The Growth of Bacterial Cultures. *Annual Review of Microbiology,* Vol.3, pp.371-394, ISSN 0066-4227

Morgan, H.P.; Mercer, L.P. & Flodin, N.W. (1975). General Model for Nutritional Responses of Higher Organisms. *Proceedings of the National Academy of Sciences,* Vol.72, No.11, pp.4327-4331, ISSN 0027-8424

NRC. *Nutrient Requirements of Beef Cattle.* (1996). National Academy Press, ISBN 0-309-06934-3, Washington, DC, United States

NRC. *Nutrient Requirements of Dairy Cattle.* (2001). National Academy Press, ISBN 0-309-06997-1, Washington, DC, United States

Oliveira, A.J.; Lourenço, S. & Goedert, W.J. (1982). *Adubação Fosfatada no Brasil,* EMBRAPA-DID, CDD 631.850981, Brasília, DF, Brazil

Pimentel, J.J.O.; Lana, R.P.; Zamperlini, B.; Paulino, M.F.; Valadares Filho, S.C.; Teixeira, R.M.A. & Abreu, D.C. (2006a). Milk Production as a Function of Nutrient Supply Follows a Michaelis-Menten Relationship. *Journal of Dairy Science,* Vol.89, Suppl. 1, pp.74-75, ISSN 0022-0302

Pimentel, J.J.O.; Lana, R.P.; Zamperlini, B.; Valadares Filho, S.C.; Abreu, D.C.; Silva, J.C.P.M. & Souza, B.S.B.C. (2006b). Efeito do Teor de Proteína e Níveis de Suplementação com Concentrado na Produção e Composição do Leite em Vacas Leiteiras Confinadas, *Proceedings of 43ª Reunião Anual da Sociedade Brasileira de Zootecnia,* Sociedade Brasileira de Zootecnia, ISSN 1983-4357, João Pessoa, PB, Brazil

Pimentel, J.J.O.; Lana, R.P.; Teixeira, R.M.A.; Zamperlini, B.; Sobreira, H.F.; Paulino, M.F.; Leão, M.I. & Oliveira, A.S. (2006c). Produção de Leite em Função de Níveis de Suplementação com Concentrado para Vacas Leiteiras sob Pastejo, *Proceedings of 43ª Reunião Anual da Sociedade Brasileira de Zootecnia,* Sociedade Brasileira de Zootecnia, ISSN 1983-4357, João Pessoa, PB, Brazil

Roberts, T.L. & Stewart, W.M. (2002). Inorganic Phosphorus and Potassium Production and Reserves. *Better Crops,* Vol.86, No.2, pp.6-7, ISSN 0006-0089

Russell, J.B. (1984). Factors Influencing Competition and Composition of the Ruminal Bacterial Flora, In: *The Herbivore Nutrition in the Subtropics and Tropics,* F.M.C. Gilchrist & R.I. Mackie, (Eds.), 313-345, Science Press, ISBN 10 0907997031, Craighall, South Africa

Russell, J.B. (2002). *Rumen Microbiology and its Role in Ruminant Nutrition,* James B. Russell, Ithaca, NY, United States

Sain, G.E. & Jauregui, M.A. (1993). Deriving Fertilizer Recommendations with a Flexible Functional Form. *Agronomy Journal, Vol.*85, pp.934-937, ISSN 1435-0645

Sairanen, A.; Khalili, H. & Virkajarvi, P. (2006). Concentrate Supplementation Responses Of The Pasture-Fed Dairy Cow. *Livestock Science,* Vol.104, No.3, pp.292-302, ISSN 1871-1413

Strong, W.M. (1995). Nitrogen fertilization of upland crops, In: *Nitrogen Fertilization in the Environment,* P.E. Bacon, (Ed.), 129-169, Marcel Dekker inc., ISBN 0-8247-8994-6, New York, NY, United States

Teixeira, R.M.A; Lana, R.P.; Fernandes, L.O.; Veloso, R.G.; Ferreira, M.B.D. & Paiva, V.R. (2006). Efeito da Adição de Concentrado em Dietas de Vacas Gir Leiteiro Confinadas sob a Produção de Leite, *Proceedings of 43ª Reunião Anual da Sociedade Brasileira de Zootecnia,* Sociedade Brasileira de Zootecnia, ISSN 1983-4357, João Pessoa, PB, Brazil.

Van Der Ploeg, R.R.; Böhm, W. & Kirkham, M.B. (1999). On the Origin of the Theory of Mineral Nutrition of Plants and the Law of the Minimum. *Soil Science Society of America Journal, Vol.*63, pp.1055-1062, ISSN 0361-5995

14

Chemical Conversion of Glycerol from Biodiesel into Products for Environmental and Technological Applications

Miguel Araujo Medeiros[1], Carla M. Macedo Leite[2] and Rochel Montero Lago[2]
[1]Universidade Federal do Tocantins,
[2]Universidade Federal de Minas Gerais,
Brasil

1. Introduction

Currently, fossil fuels represent over 80% of energy consumption in the world. However, due to environmental and geopolitical issues the development of new energy sources is mandatory. For example, only the Middle East holds 63% of global reserves, which directly influences in the final price of fuel.

In developed nations there is a growing trend towards employing modern technologies and efficient bioenergy conversion using a range of biofuels, which are becoming cost competitive with fossil fuels (Puhan et al., 2005). In Brazil, this work is focused on the production of bioethanol and biodiesel.

There are discussions around the world on the feasibility of using renewable fuels, which may cause a much smaller impact to global warming, because the balance of CO_2 emissions decreases when using these fuels. (Demirbas, 2008)

In 1997 at a meeting in Kyoto, Japan, many of the developed nations agreed to limit their greenhouse gas emissions, relative to the levels emitted in 1990. In this occasion Brazil established social and environmental policies to collaborate with those global goals (Puhan et al., 2005). An example is the biodiesel program which in 2008 implemented the use of B2 (2% biodiesel into conventional diesel). In other countries, like Germany, it is possible to supply only with B100 biodiesel (100% biodiesel).

1.1 Biodiesel

Biodiesel, a renewable biofuel produced from biomass, is biodegradable and does not cause significant contamination with emissions containing sulfur or aromatics. Biodiesel, is an viable alternative for compression-ignition engines (Puhan et al., 2005), in total or partial substitution of fossil diesel (Chiang, 2007).

The use of biodiesel as fuel should occupy a prominent place in the world, with a market that is booming because of its enormous contribution to the environment, such as qualitative and quantitative reduction of environmental pollution (Ferrari et al., 2005). Furthermore, this fuel is a strategic source of renewable energy to replace petroleum products.

Biodiesel is fuel produced mainly by transesterification of vegetable oils, but can also be obtained by the reaction of animal fat (Pinto et al. 2005; Puhan et al., 2005; Chiang, 2007) soybean (Costa Neto & Rossi, 2000), Cotton (Pinto et al., 2005; Puhan et al., 2005), castor bean (Pinto et al., 2005), canola (Pinto et al., 2005; Catharino et al., 2007; Kocak et al., 2007; Puhan et al., 2005), palm (Pinto et al., 2005; Catharino et al., 2007; Puhan et al., 2005), sunflower (Pinto et al., 2005; Catharino et al., 2007; Puhan et al., 2005; Costa Neto & Rossi, 2000), peanut and babassu.

Synthesis of biodiesel can be accomplished by using acid, basic (Costa Neto & Rossi, 2000; Puhan et al., 2005; Chiang, 2007; Pinto et al., 2005) or enzymes (Talukder et al., 2007; Schuchardt, 1990) catalysts or even in supercritical methanol (Puhan et al., 2005).

1.2 Biodiesel production

Transesterification (Figure 1) is the reaction of triglycerides with an alcohol to form esters and glycerol (Chiang et al., 2007; Georgogianni et al., 2007; Krishna et al., 2007; Wu et al., 2007; Talukder et al., 2007; Aparício et al., 2007; Zuhair, 2005; Vicente et al., 2005; Medeiros et al., 2008; Stern & Hillion, 1990; Freedman et al., 1984; Encinar et al, 2002; Vicente et al., 2006; Bunyakiat et al., 2006; Karinen & Krause, 2006). This process decreases the viscosity of the oil and transforms the large, branched molecular structure of bio-oils into smaller molecules, of type required in regular diesel engines.

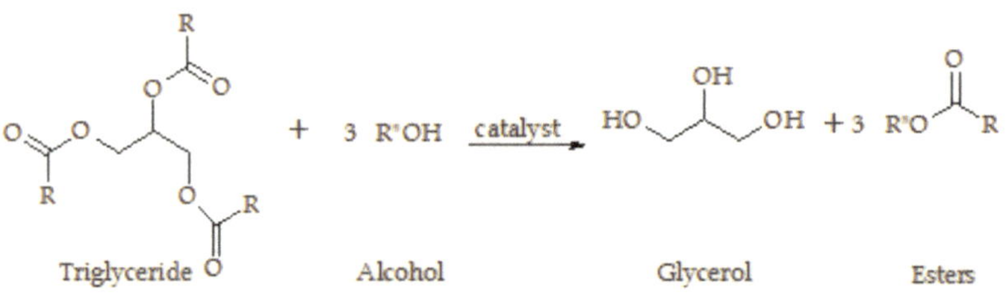

Fig. 1. Synthesis of biodiesel by transesterification of triglyceride.

In the transesterification for biodiesel production, a large amount of glycerol as a byproduct (about 10% compared to the mass of ester produced) (Puhan et al., 2005; Medeiros et al., 2010) is produced. The separation step of glycerol can be accomplished by decanting, in which the lower phase has the glycerol, the catalyst of the process (usually homogeneous and high polar character), alcohol and oil residue without reacting (crude glycerol, Figure 2, a). The biodiesel separates from the upper stage, almost pure.

The transesterification using methanol is the most used process around the world (Chiang, 2007) offering several advantages, such as: (i) small volume of alcohol recovery, (ii) lower cost of alcohol compared to ethanol (not in Brazil) and (iii) shorter reaction times (Pinto et al., 2005). The use of ethanol proves more advantageous, when considering its lower toxicity.

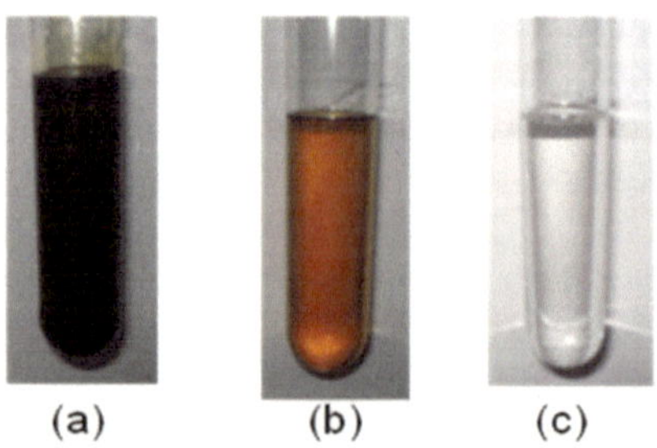

Fig. 2. (a) crude glycerol, (b) pre-purified glycerol, (c) glycerol purified.

1.3 Use of the glycerol

The investigation of new uses for glycerol is critical for the success of the biodiesel program, especially in relation to the crude glycerol, which has few direct uses and market value marginalized. Currently, the demand for purified glycerol PA for the pharmaceuticals, food additives, personal care (Puhan et al., 2005), industry is supplied by the petrochemical industry.

The biodiesel production will produce a large increase in, the amount of glycerol in the market, causing a decrease in the prices significantly, in the world. In the European Union, for example, the price of glycerol, in 1995 was € 1500 t^{-1} and reduced to 330 € t^{-1} in 2006 (Puhan et al., 2005). In Brazil, in 2005 the price of glycerol reached € 1270 t^{-1}, but already in 2007 the price dropped to 720 € t^{-1}. And in regions close to the price of biodiesel plants did not exceed € 300 t^{-1}, in 2010.

Different routes have been investigated to transform this glycerol to new products and new applications. Some of these processes are listed in Table 1.

Process	Conditions	Products	Ref.
Polymerization	T = 210-230°C; reduced pressure (~0,3atm); 0,5-1,5% NaOH.	Cyclic polymers.	(Blytas & Frank, 1993)
Pyrolise	T = 650°C.	CO; acetaldehyde; acrolein.	(Chiang, 2007)
steam reforming	T = 200-250°C; 1% cat. Níquel-Raney (Ni-Sn).	50-70% H_2; 30-40% CO; 2-11% of alkanes.	(Stein et al., 1983)
Esterification	T = 200-240°C; 0,1-0,3% NaOH; t =0,5 h; 100% methanol.	Carbohydrates and esters.	(Noureddini & Medikonduru, 1997)
Oxidation	T = 50°C; Pd/C (5-8% of Pd), t = 8h; pH = 5-11.	Dihydroxyacetone.	(Garcia et al., 1995)
Etherization	T = 90°C; 1-7,5% amberlist 15; t = 2-3h.	70% of 3-tert-butoxi-1,2-propanodiol (*mono ether*). 87% of mono ether.	(Klepácová et al., 2003, 2006)
Oligomerization	T = 260°C; 2% $Mg_{25}Al_{20}$ (cat.); t = 8h.	65% of diglycerol; 20% of triglycerol and 15% of tetraglycerol.	(Barrault et al., 2004)

Table 1. Conversion of glycerol to different products.

Table 1 can be summarized in Scheme 1, which shows some reactions that originate from glycerol.

Oxidation products of glycerol, for example, can be used in cosmetics and pharmaceuticals intermediates (Davis et al., 2000; Pachauri & He, 2006; Krishna et al., 2007) and even suntan lotion (Kimura, 1993).

The products of oligomerization of glycerol can be used as additives for cosmetics and foods, the raw material for resins and foams (Shenoy , 2006; Lemke, 2003; Werpy, 2004; Pagliaro & Rossi, 2008), lubricants (Pagliaro & Rossi, 2008), cement additives (retains moisture) and are synthetic intermediates and possible substitutes of polyols, e.g. polyvinyl alcohol, in some applications (Werpy, 2004; Pagliaro & Rossi, 2008; Medeiros et al., 2008).

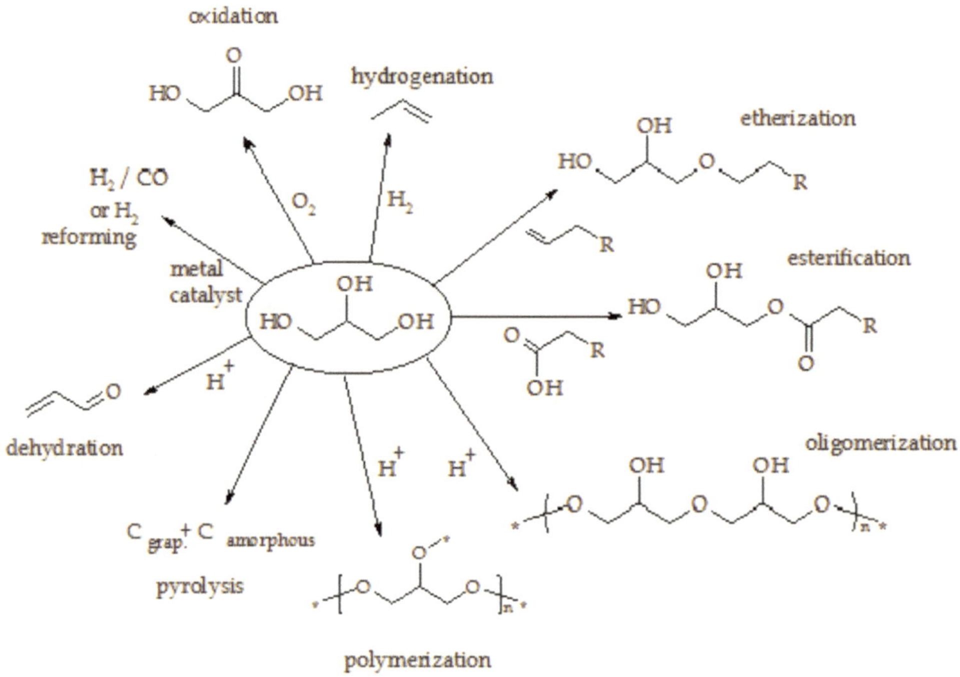

Scheme 1.

It is noteworthy that many of the applications mentioned for the glycerol require high degree of purity, which for glycerol derived from biodiesel requires several stages of treatment, increasing its cost. The main impurities in the glycerol from biodiesel is methanol or ethanol, water, inorganic salts and catalyst residues, free fatty acids, unreacted mono, di and triglycerides and various other matter organic non-glycerol (MONG) (Pagliaro & Rossi, 2008). Thus, it is necessary to develop new routes for the consumption of glycerol from biodiesel.

In this chapter, will be treated the transformation of glycerol based on production of ethers from condensation of two (or more) glycerol's molecules. One of the mechanisms Favorable for the formation of ethers is by alcohol protonation (ROH_2^+), followed by condensation of other alcohol and water loss. The condensation reaction of glycerol (Scheme 2), is usually catalyzed by acids or bases producing small polymers called oligomers and water. Along the text will be described the oligomers, polymers and carbons obtained from polyglycerol and its applications.

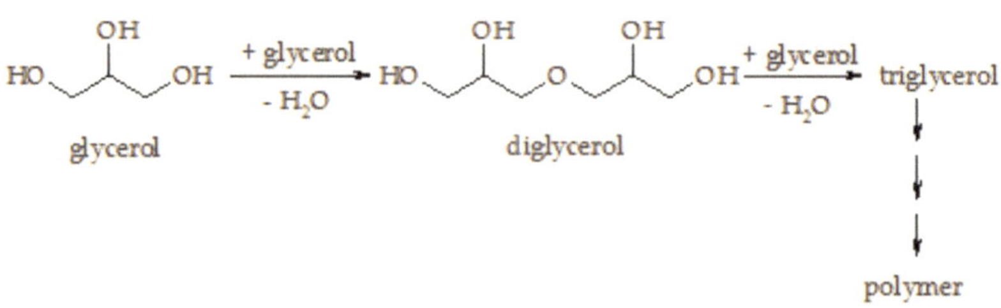

Scheme 2.

2. Oligomers

Oligomerization of glycerol (Scheme 2) is an alternative to the use of byproduct of biodiesel, because their products have wide application. For a better understanding of oligomerization (and polymerization), was accompanied through ESI-MS (Electrospray Ionization Mass Spectrometry in the positive ion mode) a typical reaction of oligomerization - glycerol PA catalyzed by 1% H_2SO_4 at 280°C/2h, in reflux.
Analysis of the sample (2h) is shown in Figure 3. The presence of an intense ion of *m/z* 93 (protonated glycerol = [glycerol + H]$^+$) is clearly noticeable indicating the subsistence of glycerol in the reaction medium even after 2 h reaction.

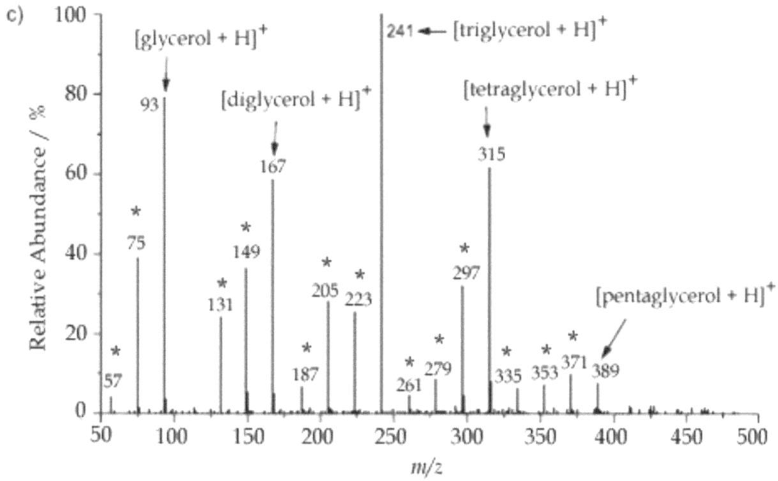

Fig. 3. ESI(+)-MS of the acid-catalyzed oligomerization of glycerol conducted in aqueous medium at 280 °C, 2 h. The ions marked with an asterisk (*) refer to dehydration products.

A remarkable presence of an ion of *m/z* 167 is also noticed in Figure 3. This corresponds to the protonated form of diglycerol, i.e. [(glycerol)$_2$ – H_2O], formed under these reaction conditions via the condensation of two molecules of glycerol and loss of water. This condensation can occur via the primary or secondary hydroxyl groups at the glycerol molecule to yield linear (α, α -diglycerol) and branched (α, β -diglycerol; β, β -diglycerol) isomers, as displayed in Scheme 3.

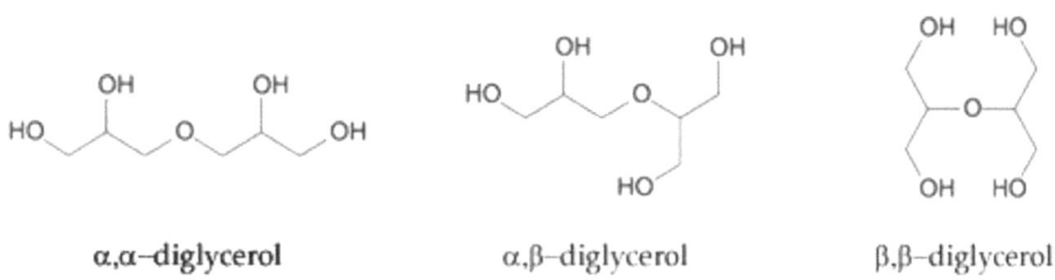

Scheme 3.

Across of the fragmentation of the ion of *m/z* 167 are yield mainly product ions from losses of one or two molecules of water (*m/z* 149 and 131, respectively) besides to other product

ions, such as [glycerol + H]$^+$ (*m/z* 93), [glycerol – H_2O + H]$^+$ (*m/z* 75), and [glycerol – 2 H_2O + H]$^+$ (*m/z* 57). To illustrate the formation of such fragments, the dissociation pathways for protonated α, α-diglycerol are shown in Scheme 4.
In the Table 2, are showed ions ascribed to be the protonated forms of products formed by successive dehydrations of di, tri, tetra and pentaglycerol.

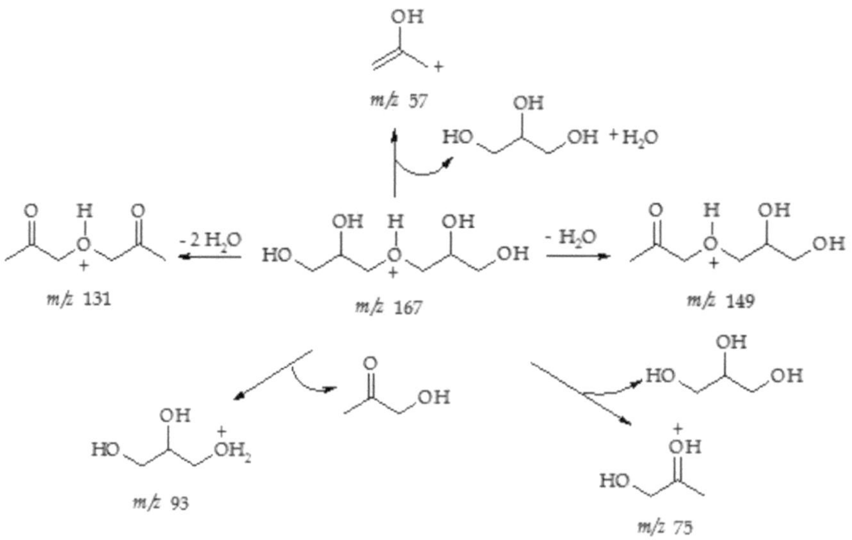

Scheme 4.

Primary Oligomers (m/z of the protonated forms)	Dehydration Products (m/z of the protonated forms)
diglycerol (167)	[diglycerol – H_2O] (149) [diglycerol – 2 H_2O] (131)
triglycerol (241)	[triglycerol – H_2O] (223) [triglycerol – 2 H_2O] (205) [triglycerol – 3 H_2O] (187)
tetraglycerol (315)	[tetraglycerol – H_2O] (297) [tetraglycerol – 2 H_2O] (279) [tetraglycerol – 3 H_2O] (261)
pentaglycerol (389)	[pentaglycerol –H_2O] (371) [pentaglycerol – 2 H_2O] (353) [pentaglycerol – 3 H_2O] (335)

Table 2. Primary products (diglycerol, triglycerol, tetraglycerol and pentaglycerol) and their dehydration products formed upon acid-catalyzed oligomerization of glycerol at 280°C. All these products were observed as their protonated forms in the ESI(+)-MS (Fig. 3).

These findings thus indicate that under acidic medium and heating, olygomers can easily lose one or two molecules of water to form a myriad of isomeric products. Scheme 5 shows, for instance, products possibly formed as a result of the mono-dehydration of diglycerol, such as the cyclic species **1a-c** (their formation have been reported by Barrault and coworkers (Barrault et al., 2004, 2005) that submitted glycerol to similar reaction conditions than those employed herein) besides the acyclic carbonyl compounds **2a-b** and the alkene **2c**.

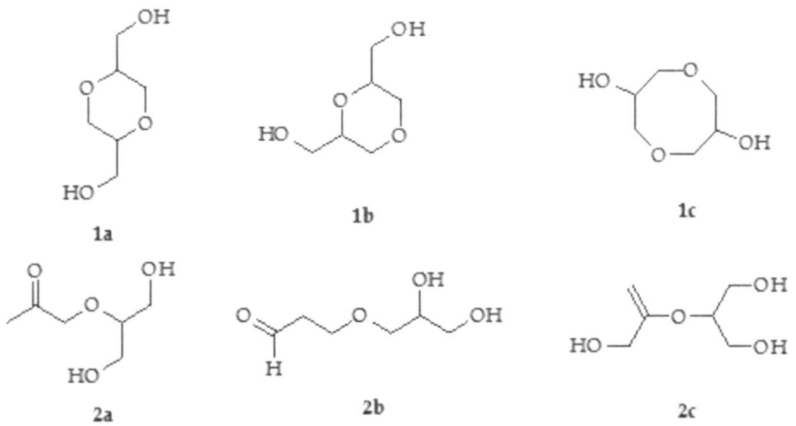

Scheme 5.

All the products resulting from the mono-dehydration of diglycerol, including the ones shown in Scheme 3 (**1a-c** and **2a-c**), possess the same chemical formula ($C_6H_{12}O_4$) and bear similar functional groups (especially hydroxyl substituents). Hence, these protonated molecules lose mainly water and other small molecules, being unfeasible the unambiguous characterization of a particular product based exclusively on your fragmentation profile.

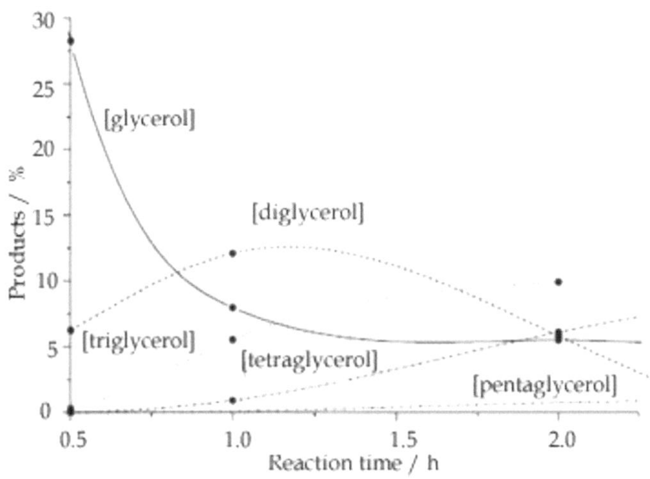

Fig. 4. Fractions of the ions of *m/z* 167, 241, 315, and 389 as a function of reaction time. Each fraction was calculated as the quotient ratio between the absolute intensity of one of such ions and the sum of the absolute intensities of the whole set of ions.

In Figure 4, the fraction of the ions of *m/z* 93, 167, 241, 315 and 389 (protonated glycerol, di, tri, tetra and pentaglycerol, respsctively), given as a quotient ratio between the absolute intensity of one of such ions and the sum of the absolute intensities of the whole set of ions, are plotted against the reaction time. These results show that after 2 h reaction more than 90% of glycerol is consumed. Furthermore, during the first 30 min a relatively high concentration of diglycerol is formed. At longer reaction times, however, its concentration decreases whereas the amount of the heavier oligomers (tri, tetra and pentaglycerol) concomitantly increases. The result shows that glycerol is continuously converted into the heavier oligomeric compounds.

3. Polymerization of glycerol

In open system, the polycondensation (condensation of many molecules to create larger molecules - polymers) that the glycerol suffers in the presence of H_2SO_4 at 150 ° C, is a type of polymerization in which mingle the three stages: initiation, propagation and termination, which are characteristic of polymerization reactions (Mano & Mendes, 1999). The condensation polymerization, when employ monomers (molecules susceptible to undergo polymerization) with more than two functional groups (glycerol has three OH groups), tends to form crosslinked or branched polymers (structures with crosslinks between chains). In this case, the polymerization is complex because it is formed gel (polymer molecular weight too large), in the same setting of the sol (the fraction that remains soluble and can be extracted from the middle). As the sol will turn into gel, the mixture becomes increasingly viscous until elastic consistency, and finally rigid. In this transformation of glycerol in hard polymer, the catalyst concentration has an important role. An example of the participation of the catalyst in the polymerization of glycerol is shown in Figure 5, in which the viscosities of solution reaction is monitored by 60 minutes, with different concentrations of catalyst (0.5, 1, 3 and 5 mol%) .

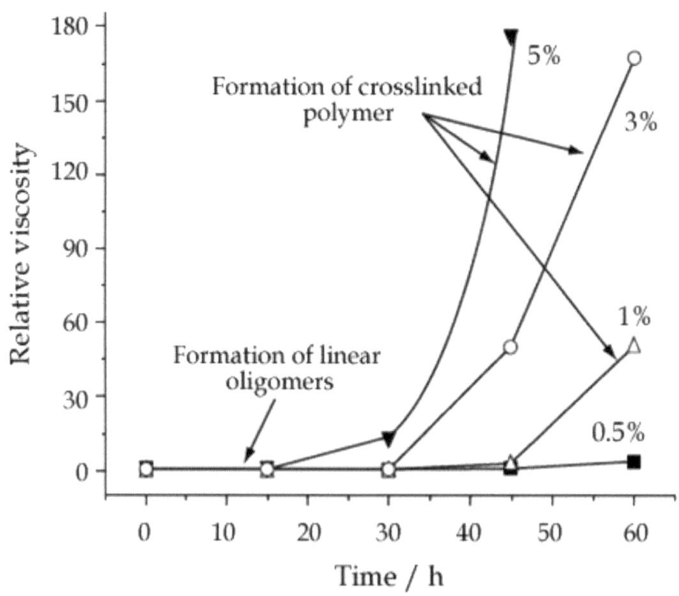

Fig. 5. Variation of relative viscosity of the solution for the polymerization of glycerol, with 0.5,1, 3 and 5 mol% H_2SO_4 (viscosity values are relative to the glycerol).

The curves shown in Figure 4 indicate a significant increase in viscosity of the solution, by varying the mole percentage of catalyst of 0.5-5%. However, this increase is gradual, as it rises the concentration of H_2SO_4. It is interesting to note that the system promoted by 5 mol% of catalyst is very active, because it took only 45 minutes to produce a solid polymer (unable to measure the viscosity, since the material solidified, Figure 5), whereas in other systems it took at least 120 minutes.

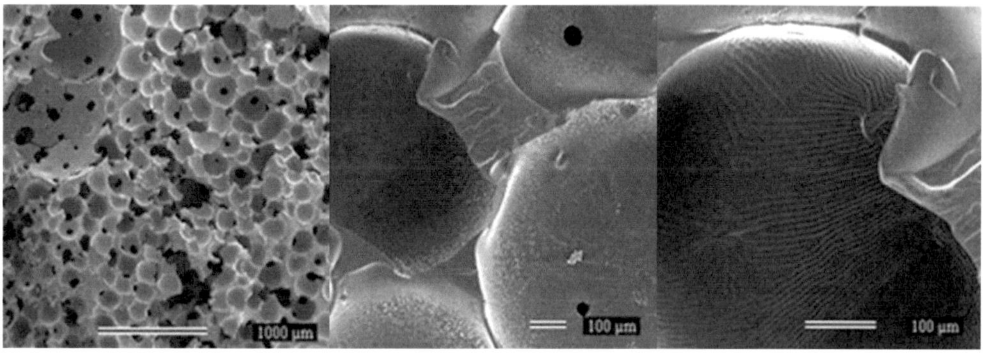

Fig. 6. Images of scanning electron microscopy (SEM) of polyglicerol with 5 mol% of H_2SO_4, after 45 min of reaction.

The system promoted by 0.5, 1 and 3 mol% H_2SO_4 show similar viscosity curves (Figure 4) but with varying slopes (the higher the catalyst concentration, greater the slope of the viscosity). With the increase of H_2SO_4 concentration (0.5-3%) is expected larger number of simultaneous condensation, therefore, the selectivity of the catalyst to reactive hydroxyl groups of glycerol decreases, leading to complex structures, which offer increased viscosity of the solution (in 60 minutes: system promoted by 0.5 mol% → relative viscosity of 4,1%; 1 mol% → relative viscosity of 51 e 3 mol% → relative viscosity of 169 times that of glycerol.
In the first 15 minutes of reaction, the viscosity of the medium practically does not change. It is believed that during this period, is occurring the formation of linear oligomers and products of dehydration. However, as the polymerization reaction progresses, the ethers formed formed become larger and more complex, mainly due to the formation of branches and some bonds between parallel chains of oligomers and/or polymers. And this is the increase size and complexity of structures of the ethers formed which increased the solution viscosity, reaching 169 times the viscosity of glycerol in just 60 minutes (system promoted by 3 mol% H_2SO_4), because the move of the structures is becoming increasingly difficult.
To confirm the nature (thermoplastic or thermosetting) polymeric material formed by polymerization of glycerol, are carried out two separate tests: heating in the direct flame of a Bunsen burner, to ensure that it is malleable (suffers fusion) or undergo thermal decomposition, and washing the polymer in solvents with different polarities (hexane, THF and ethanol). The results of these tests showed that all the polymers (0.5, 1, 3 and 5 mol% H_2SO_4) are thermosets, because not suffers fusion, but rather, thermal decomposition and did not dissolve in any solvent tested.
The polyglycerol may be used as a substitute for thermosetting phenolic resins, used in home utensils as well as controlled release fertilizers.

4. High surface area carbons

The thermosetting polymers have the property that thermally decomposes, producing carbon in quantities that can vary with the degree of crosslinking of the polymer. As previously discussed, the degree of crosslinking (polymerization in all directions, linking parallel chains of the polymer) is influenced by the concentration of the catalyst. Hence, a polymer obtained with 5 mol% H_2SO_4 is more reticulated and produces more carbon than a polymer with only 1 mol% catalyst.

The thermal decomposition of polyglycerol (obtained with 5 mol% H_2SO_4) yields 16% of carbon (relative to initial mass of polymer), which has extremely low surface area (2 m^2g^{-1}). For environmental applications, are typically used carbonaceous materials with high surface area. Thus, it was necessary to increase the surface area of carbons derived from polyglycerol, performing physical activation (850°C) with a flow of CO_2 by 3, 5, 10, 15 and 18 h (Figure 7).

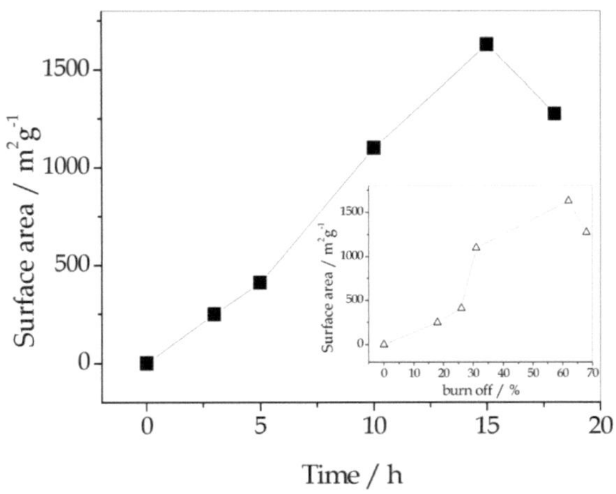

Fig. 7. Surface area of carbons derived from polyglycerol. Detail: surface area as a function of burn off.

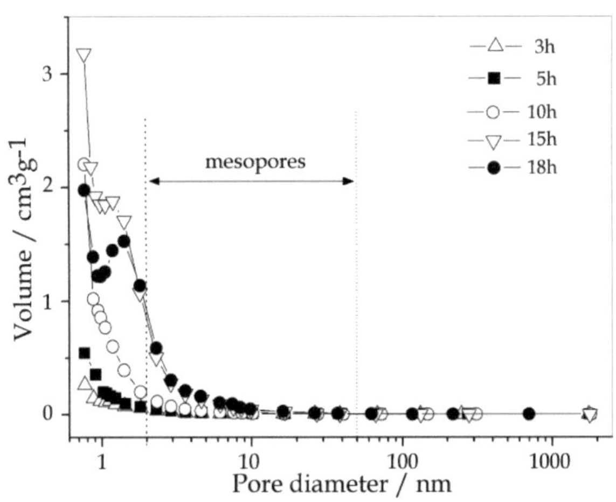

Fig. 8. Distribution of pores in activated carbonaceous material for 3, 5, 10, 15 and 18 h.

The Figure 7 shows a gradual increase in surface area of carbonaceous material up to 1830 m2 g-1, 15 h of activation. After that time, the surface area begins to decrease, reaching a value of 1275 m2 g-1 at 18 h. A similar behavior, but not linear is observed for the surface area as a function of burn off (mass loss of carbon during activation) (detail of Figure 7). The analysis of distribution of pores indicates that the materials are essentially microporous (internal diameter of less than 2 nm) (Figure 8).
As the surface area of carbonaceous material derived from polyglycerol increased with activation time by 15 hours, tests were made to adsorb organic contaminants (methylene blue) during the activation process (0, 3, 5, 10 and 15 h). The results of these tests are shown in Figure 9.
It is evident the relationship between the activation time of the carbonaceous material (surface area) and the adsorption of organic contaminant. During the activation process of the carbonaceous material, there was a gradual increase of its surface area, which is intimately related to the growth of its adsorption capacity. The sample of the material that was activated for 15 hours showed better results in the removal of organic contaminant (in 20 minutes, the removal of contaminant was 90%, while others samples took at least 60 minutes to obtain the same results).

5. Vermiculite composites/activated carbon

In order to facilitate the application of carbonaceous material in environmental problems, was produced a composite based in vermiculite clay and activated carbon, derived from the polyglycerol. This composite was designed, considering (i) some properties of the expanded vermiculite clay (Figure 10), which has low cost and ability to float in water, (ii) large adsorption capacity of Activated carbon derived from polyglycerol and (iii) facility removal of the composite in case of water application, requiring only one net.

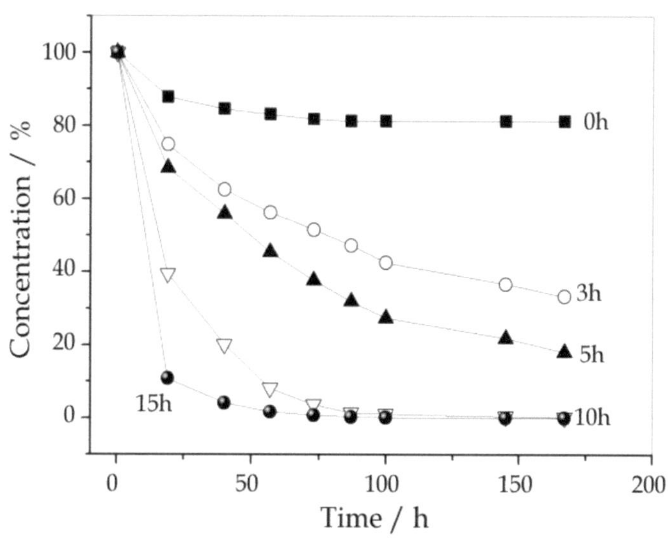

Fig. 9. Adsorption of methylene blue by carbonaceous material derived from the polyglycerol at different activation times (0, 3, 5, 10 and 15 h).

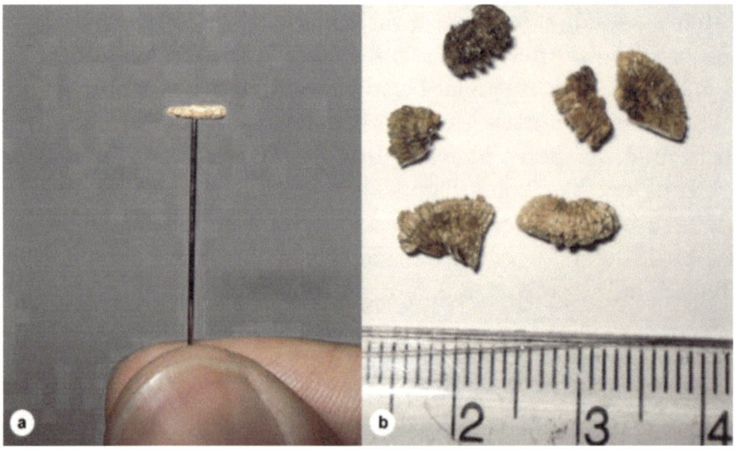

Fig. 10. (a) Sample vermiculite *in natura*; (b) Sample expanded vermiculite.

The composite vermiculite/carbon is prepared the same way that the pure carbon, except that the clay expanded vermiculite (EV) is added before the initial stage of polymerization of glycerol, which will occur on the surface of clay.

The best condition for prepare of the composite (GVE4), which has carbon content of 25% (compared to the mass of the composite) is 3 mol% H_2SO_4 and 580°C/3h and ratio (by mass) glycerol/VE = 4. This condition was obtained after tests with different reaction conditions. The images of scanning electron microscopy (SEM) for pure EV and composite (GVE4) showed significant differences in their surfaces (Figure 11).

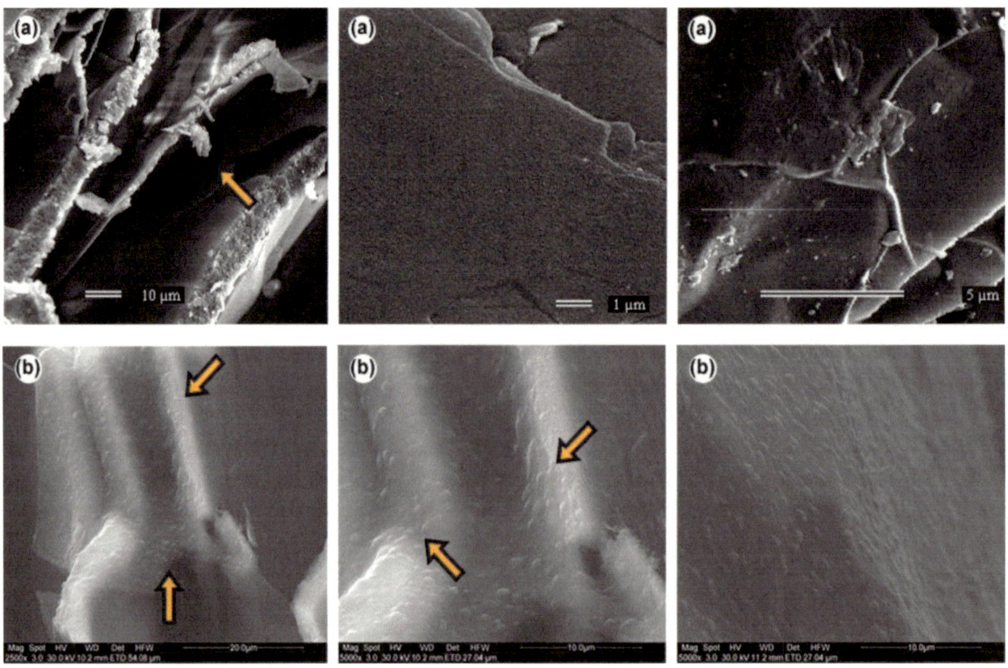

Fig. 11. SEM images of: (a) EV; (b) GVE4.

The SEM images for pure EV (Figure 11 (a)) show regular flat surfaces with an interlamellar space between 10-100 μm with some fragments attached to the edges (arrow). Figure 11 (b) shows large amounts of regular deposits on the EV layers with materials connecting some layers.

To obtain a good adsorbent material, the composite GVE4 was submitted to physical activation with CO_2 for periods of 0.5 (GVE4CA0.5); 1 (GVE4CA1), 2 (GVE4CA2) and 4 hours (GVE4CA4). Table 3 shows the surface area and burn off the composite GVE4, activated at different times.

To observe the data presented in Table 3 and Figure 12, perceives a linear increase in surface area, depending on the activation time, until the limit value of 835 m^2g^{-1} (2h of activation), when the value of surface area begins to decrease to 143 m^2g^{-1}, 4h of activation. A similar performance is observed for the surface area as a function of burn off (detail of Figure 12).

Sample	Burn off/ %	Surface Area / m^2g^{-1}
GVE4CA0	0	9
GVE4CA0.5	23,4	387
GVE4CA1	44,2	648
GVE4CA2	58,0	835
GVE4CA4	73,0	146

Table 3. Data for surface area GVE4CA0, GVE4CA0.5, GVE4CA1, GVE4CA2 and GVE4CA4, obtained by BET method.

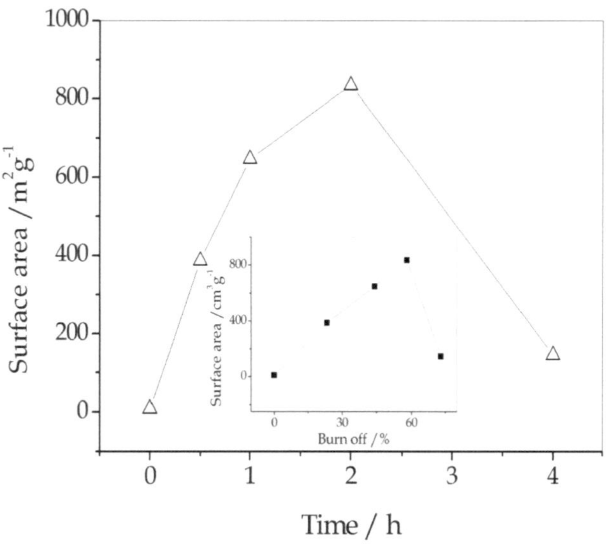

Fig. 12. Surface area of the activated composites. Detail: surface area as a function of burn off.

SEM images presented in Figure 13 show how the carbon deposits on the surface of the composites was changed during activation. After the first hour of activation, the amount of material deposits on the surface of the composite is significantly lower when compared to the composite without activation (Figure 11 (b)), because of the oxidizing action of CO2,

850°C/1h. SEM images show that 2 h of activation are sufficient to make large part of the surface of the EV is exposed, reducing the carbonaceous deposits, although the surface area is the largest obtained (835 m^2g^{-1}). But it's after 4 h of activation that the composite loses most part of the carbon deposits and therefore reduces the surface area to only 146 m^2g^{-1}.

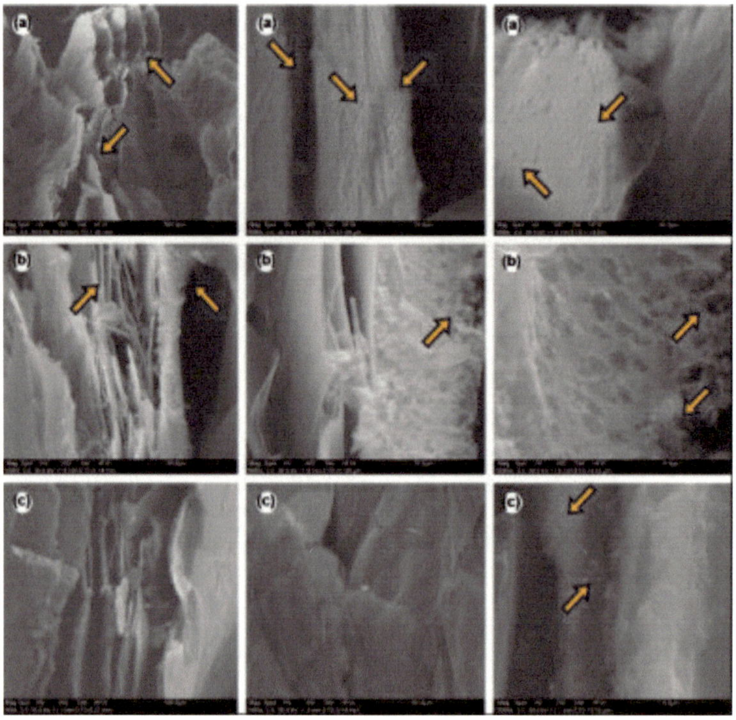

Fig. 13. SEM images of GVE4, actived by: (a) 1h; (b) 2h and (c) 4h.

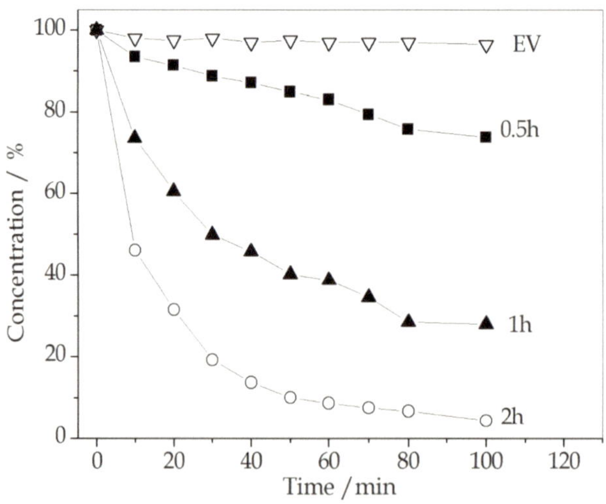

Fig. 14. Adsoption of methilene blue by EV and composites actived by 0.5, 1 and 2h.

After the activation process of the composite GVE4, samples of 0.5, 1 and 2h of activation were tested as adsorbents for organic contaminants (methylene blue) (figure 14).
It is possible notice that the EV practically no adsorbs or reacts with the contaminant during all the test period. The composite GVE4CA2 adsorbs 90% of the dye in only 60 minutes, tending to 100% until the end of the test (100 min). It is also notable, the action of the composite GVE4CA1, which absorbs more than 50% of the dye in the first 30 minutes of testing, tending to an equilibrium around 70% of contaminants adsorbed to the end of the test. Already the composite GVE4CA0.5, has unsatisfactory result, with only 20% of adsorbing dye after 100 minutes of testing.

6. Conclusions

Glycerol is a very versatile chemical species which can produce different materials to distinct applications. In this chapter, we discussed some possibilities for the glycerol that boost its use in the production of polymers and adsorbents for organic contaminants.
Study of oligomerization of glycerol, by ESI-MS, is an important step in understanding how the molecules of glycerol were initially organized to enable the formation of thermosetting polymers and, later, special carbonaceous materials.
The preparation of carbonaceous materials from glycerol, for environmental applications is a way to consume an important portion of glycerol introduced in the market from the production of biodiesel.

7. Acknowledgements

UFMG, UFT, Capes, CNPq and Fapemig.

8. References

Aparício, C., Guignon, B., Rodriguez-Antón, L.M., Sanz, P.D. *J. Agric. Food Chem.*, 55, 7394-7398, 2007
Barrault, J., Clacens, Y., Pouilloux, Y. *Topics in Catalysis*, 27, 137-142, 2004.
Barrault, J., Jerome, F., Pouilloux, Y. *Lipid Technol.*, 17, 131-135, 2005.
Blytas, G.C., Frank, H. US Pat. US5401860, 1993 - atribuída à Shell Oil Company.
Bunyakiat, K., Makmee, S., Sawangkeaw, R., Ngamprasertsith, S. *Energy Fuels*, 20, 812-817, 2006.
Catharino, R.R., Milagre, H.M.S., Saraiva, S.A., Garcia, C.M., Schuchardt, U., Eberlin, M.N., Augusti, R., Pereira, R.C.L., Guimarães, M.J.R., de Sá, G.F., Caxeiro, J.M.R., Souza, V. *Energy Fuels*, 21, 3698-3701, 2007.
Chiang, W. H. *Biodiesel - Sebrae*, 2007
Costa Neto, P.R., Rossi, L.F.S. *Quím. Nova*, 23 (4), 531-537, 2000.
Davis, W. R., Tomsho, J., Nikam, S., Cook, E. M., Somand, D., Peliska, J. A. *Biochemistry*, 39, 14279–14291, 2000.
Dorado, M.P., Ballesteros, E., Mittelbach, M., López, F.J. *Energy Fuels*, 18, 1457-1462, 2004.
Encinar, J.M., Gozález, J.F., Rodrigues, J.J., Tejedor, A. *Energy Fuels*, 18, 443-450, 2002.
Ferrari, R.A., Oliveira. V.S., Scabio, A. *Quim. Nova*, 28, 19-23, 2005.
Freedman, B., Pryde, E.H., Mounts, T.L. *J. Am. Oil Chem. Soc.*, 61, 1638, 1984.
Garcia, R., Besson, M., Gallezot, P. *Applied Catalysis A: Gen.*, 127, 165-176, 1995.

Georgogianni, K.G., Kontominas, M.G., Tegou, E., Avlonitis, D, Gergis, V. *Energy Fuels*, 21, 3023-3027, 2007.
Huber, G.W., Shabaker, J. W., Dumesic, J. A. *Science* 300(5628), 2075-2077, 2003.
Karinen, R.S., Krause, A.O.I. *Appl. Cat. A: Gen.*, 306, 128-133, 2006.
Kimura, H., *Appl.Catal. A: Gen.*, 105 (2), 147-158, 1993.
Klepácová, K., Miravec, D., Hájeková, E., Bajus, M. *Petroleum and Coal*, 45, 54-57, 2003.
Klepácová, K., Miravec, E., Bajus, M. *Chem. Pap.*, 60(3), 224,230, 2006.
Kocak, M.S., Ileri, E., Utlu, Z. *Energy Fuels*, 21, 3622-3626, 2007.
Krishna, C.R., Thomassen, K., Brown, C., Butcher, T.A., Anjom, M., Mahajant, D. *Ind. Eng. Chem. Res.*, 46, 8846-8851, 2007.
Lemke, D.W. US Patent US 6620904 (2003).
Mano, E. B.; Mendes, L. C.; *Introdução a Polímeros*, Blücher: São Paulo, 1999.
Medeiros, M.A., Oliveira, L.C.A., Gonçalves, M., Araújo, M.H., Lago, R.M. *Estudo por ESI-MS das reações de oligomerização catalítica do glicerol*, abstract XXI SICAT, Málaga, 2008.
Medeiros, M.A.; Oliveira, D.L.; Sansiviero, M.T.C.; Araujo, M.H. ; Lago, R.M. ; *J. Chem. Technol. Biotechnol.* 2010, 85, 447.
Noureddini, H., Medikonduru, V. *J. Am. Oil Chem. Soc.*, 74(4), 419-425, 1997.
Ott, L., Bicker, M., Vogel, H. *Green Chem.*, 8, 214-221, 2006.
Pachauri, N.; He, B. *Encontro Anual da American Society Biological and Agricultural Engineering*, 2006, *Paper Number: 066223*.
Pagliaro, M., Rossi, M. *The Future of Glycerol* - New uses of a versatile raw material, RSC Publishing: Cambridge, 2008.
Pinto, A.C., Guarieiro, L.L.N., Rezende, M.J.C., Ribeiro, N.M., Torres, E.A., Lopes, W.A., Pereira, P.A.P., Andrade, J.B. *J. Braz. Chem. Soc.*, 16, 1313-1330, 2005.
Puhan, S., Vedaraman, N., Rambrahaman, B.V., Nagarajan, G. 2005. Mahua (*Madhuca indica*) seed oil: a source of renewable energy in India. J Sci Ind Res 64:890−896.
Schuchardt, U., Sercheli, R., Vargas, R.M. *J.Braz. Chem. Soc.*, 9, 199-210, 1998.
Shenoy, M. A., Sabnis, A., D'Melo, D. J., *Pigment & Resin Technology*, 35, 326–333, 2006.
Stein, Y.S., Antal, M. J. J., Jones, M. *J. Anal. and Appl. Pyrolysis*, 4(4), 283-296, 1983.
Stern, R., Hillion, G. Eur. Pat. EP356317, 1990.
Talukder, M.M.R., Beatrice, K.L.M., Sond, O.P., Wu, S.P.J.C., Won, C.J., Chow, Y. *Energy Fuels*, 21, 2007.
Vicente, G., Martinez, M., Aracil, J. *Energy Fuels*, 20, 394-398, 2006.
Vicente, G., Martinez, M., Aracil, J., Estaban, A. *Ind. Eng. Chem. Res.*, 44, 5447-5454, 2005.
Werpy, T., Petersen, G. *Top Value Added Chemicals From biomass*, U.S. Department of Energy. Oak Ridge, Richland, 2004.
Wu, H., Fu, Q., Giles, R., Bartle, J. *Energy Fuels*, 21, 2007.
Zuhair, S.A., *Biotechnol. Prog.*, 21, 1442-1448, 2005.

15

Use of Soybean Oil in Energy Generation

Roberto Guimarães Pereira[1], Oscar Edwin Piamba Tulcan[2],
Valdir de Jesus Lameira[3], Dalni Malta do Espirito Santo Filho[4]
and Ednilton Tavares de Andrade[5]
[1]*Fluminense Federal University/TEM/PGMEC/MSG, Niterói, RJ*
[2]*National University of Colombia, Bogota*
[3]*INESC, Coimbra*
[4]*LAFLU/DIMEC/INMETRO, RJ*
[5]*Fluminense Federal University/TER/PGMEC*
[1,4,5]*Brazil*
[2]*Colombia*
[3]*Portugal*

1. Introduction

This chapter deals with the possibility of using soybean oil in energy generation. The environmental, energetic and social-economic aspects are discussed. The steps for obtaining biodiesel from soybean oil are presented as well as the characterization of soybean oil and soybean biodiesel. Results for performance and emissions of using soybean oil and soybean biodiesel in a stationary engine are also presented.

Vegetable oils are obtained predominantly from grains of different plant species. The oil extraction can be made by physical process (pressing) or chemical (solvent). The solvent extraction produces better results, but the more traditional way is physical extraction, which uses mechanical and hydraulic presses to crush the grains. A mixed extraction (mechanical/solvent) can also be done. Selecting the type of extraction depends on two factors: the productive capacity and oil content.

Soybean (*Glycine max* (L.) Merrill) is a very versatile grain that gives rise to products widely used by agro-chemical industry and food industry. Besides is a raw material for extraction of oil for biofuel production. Soybean has about 25% of oil content in grain.

In the agribusiness world, soybean production is, among the economic activities in recent decades, the most prominent. This can be attributed to several factors, such as structuring of a large international market related to trade in products of soybean, oilseed consolidation as an important source of vegetable protein and increased development and delivery of technologies that made possible the expansion of soy exploration for various regions of the world. The largest producers of soybeans are: United States, Brazil, Argentina, China and India.

One possible use of vegetable oil is in the power generation engines. The vegetable oil can be used directly in diesel engines, preferably mixed with diesel. It may also undergo a chemical reaction (transesterification), yielding biodiesel and glycerol. In literature, several works are related to the use of vegetable oil and biodiesel for power generation, as evidenced below.

A review of the use of vegetable oils as fuel in compression ignition (CI) engines is presented (Hossain & Davies, 2010). The review shows that a number of plant oils can be used satisfactorily in CI engines, without transesterification, by preheating the oil and/or modifying the engine parameters. As regards life-cycle energy and greenhouse gas emission analyses, these reveal considerable advantages of raw plant oils over fossil diesel and biodiesel.

It is pointed out (Grau et al., 2010) that straight vegetable oil can be used directly in diesel engines with minor modifications. It is proposed a small-scale production system for self-supply in agricultural machinery.

It is emphasized (Misra & Murthy, 2010) that the ever increasing fossil fuel usage and cost, environmental concern has forced the world to look for alternatives. Straight vegetable oils in compression ignition engine are a ready solution available, however, with certain limitations and with some advantages.

It is presented (Sidibé et al., 2010) a literature review on the use of crude filtered vegetable oil as a fuel in diesel engines. It is emphasized the potential and merits of this renewable fuel. Typically, straight vegetable oils produced locally on a small scale, have proven to be easy to produce with very little environmental impact. However, as their physico-chemical characteristics differ from those of diesel oil, their use in diesel engines can lead to a certain number of technical problems over time.

A review on the utilization of used cooking oil biodiesel is presented (Enweremadu & Rutto, 2010). There were no noticeable differences between used cooking oil biodiesel and fresh oil biodiesel as their engine performances, combustion and emissions characteristics bear a close resemblance.

A review on biodiesel production, combustion, emissions and performance is shown (Basha et al., 2009). A vast majority of the scientists reported that short-term engine tests using vegetable oils as fuels were very promising, but the long-term test results showed higher carbon built up, and lubricating oil contamination, resulting in engine failure. It was reported that the combustion characteristics of biodiesel are similar as diesel. The engine power output was found to be equivalent to that of diesel fuel.

An overview of political, economic and environmental impacts of biofuels is presented (Demirbas, 2009a). Biofuels provide the prospect of new economic opportunities for people in rural areas in oil importer and developing countries. Renewable energy sources that use indigenous resources have the potential to provide energy services with zero or almost zero emissions of both air pollutants and greenhouse gases. Biofuels are expected to reduce dependence on imported petroleum with associated political and economic vulnerability, reduce greenhouse gas emissions and other pollutants, and revitalize the economy by increasing demand and prices for agricultural products.

It is emphasized (Sharma & Singh, 2009) that Biodiesel, a renewable source of energy seems to be an ideal solution for global energy demands.

The soy, which cultivation is widespread in the world, can be used to produce vegetable oil in order to be used as fuel, according some following examples.

It is pointed out (Liu et al., 2008) that Soybean (*Glycine max* (L.) Merrill), one of the most important crops in China, has been known to man for over 5000 years. The largest production areas in China are in the Northeast China's three provinces, where soybean is spring seeded and grown as a full-season crop.

A study about large-scale bioenergy production from soybeans in Argentina is presented (van Dam et al., 2009), showing the potential and economic feasibility for national and international markets.

Inedible vegetable oils and their derivatives can, also, be used as alternative diesel fuels in compression ignition engines (No, 2011).

The advantages of using vegetable oil as fuel are evidenced in the literature (Alonso et al. 2008; Misra & Murthy, 2010) as described below.

One of the main advantages of vegetable oils is its life cycle, as it is a closed cycle. Crops take CO_2 via photosynthesis from the atmosphere. Oil is extracted from these crops which can be used directly as fuel or, after the pertinent transformations, a fuel can be obtained which, when burned, generates CO_2 that can be absorbed by the plants.

Environmental advantages of vegetable oils are: minor influence on the greenhouse effect when used instead of fossil fuels; biodegradability; lower sulfur and aromatic content; at lower percentages of vegetable oil blends with diesel have shown better results than the fossil diesel in terms of engine performance and exhaust emissions.

Energetic advantages of vegetable oils are: renewable energy source; reduction of the dependence on fossil fuels; positive energy balance; the fuel production technology is simple and proven; heating values of various vegetable oils are nearly 90% to those of diesel fuel; higher flash point allows it to be stored at high temperatures without any fire hazard; additional oxygen molecule in its chemical structure helps in combustion process.

Social-economic advantages of vegetable oils are: use of marginal land for energy purposes; maintains employment and income levels in rural areas; avoids population migrations; encourages job creation in various agro-industries; contributes to the creation of new jobs; straight vegetable oils are available normally in rural area where its usage is advantageous especially in smaller engines in agricultural sector; improve the living conditions of the rural people and offer greater income opportunities through enhanced rural employment; plant leafs and cake can be used as organic manure which can be source of additional income farmers; selected crops can be grown on arid and semi-arid lands which are presently not cultivable; having carbon credit value (Kyoto protocol).

The main disadvantages of vegetable oils, compared to diesel fuel, are higher viscosity, lower volatility, and the reactivity of unsaturated hydrocarbon chains. The problems meet in long term engine use (Misra & Murthy, 2010). The land required for commercial production is vast.

Many countries have potential to produce biodiesel from vegetable oil. The top 10 countries in terms of absolute potential are: Malaysia, Indonesia, Argentina, USA, Brazil, Netherlands, Germany, Philippines, Belgium and Span (Sharma & Singh, 2009).

Some reasons for sufficiently strengthen the program for the use of biofuels are: the variation in prices for oil, which in recent years has fluctuated between $ 40 and 150 per barrel, the political and social pressure in order to reduce the emission of gases that cause global warming, the development of the market for carbon credits, the need to strengthen the agriculture industry and the possibility of strengthening the energy matrix, reducing dependence on foreign energy sources and giving relief to the trade balance with clear effects on the macroeconomics of the country.

The possibility to produce hydrocarbons from different raw material to replace the diesel makes the definition of biodiesel a complex and legal nature. Many documents of an academic nature define biodiesel as a monoalkyl ester of vegetable oil or animal fat, but official documents and international standards are more specific, defining the process by which one can obtain the ester and the characteristics that it must have to be considered biodiesel.

The regulations for biodiesel have been developed in different countries where its use is permitted. In the U.S. the standard for biodiesel is set by the technical standard ASTM D 6751, the European Union is related with the standard EN 14214 and in Brazil is set in the ANP (National Petroleum Agency) No. 07 from 19.03.2008.

Biodiesel can be produced from different oilseeds, according to the design of the plant, market conditions and availability of raw material in the region. Each oilseed production has a different culture method and a different destination. The disposal of oils for biodiesel production must take into account beyond the capacity of oil production, market competitiveness in relation to the cost of oil and the price of a barrel of fuel. Biodiesel can be obtained, too, from used cooking oil, animal fats and algae.

Just as vegetable oil, the use of biodiesel as fuel in partial or total replacement to diesel has many advantages that have been highlighted in the literature (Demirbas, 2009b; Pereira et al., 2007).

Environmental advantages of biodiesel are: greenhouse gas reductions; biodegradability; higher combustion efficiency; improved land and water use; carbon sequestration; lower sulfur content; lower aromatic content; less toxicity.

Energetic advantages of biodiesel are: supply reliability; higher flash point; reducing use of fossil fuels; ready availability; renewability.

Social-economic advantages of biodiesel are: sustainability; fuel diversity; increased number of rural manufacturing jobs; increased income taxes; increased investments in plant and equipment; agricultural development; international competitiveness; reducing the dependency on imported petroleum.

2. Theoretical foundations

All vegetable oils consist primarily of triglycerides. The triglycerides have a three-carbon backbone with a long hydrocarbon chain attached to each of the carbons. These chains are attached through an oxygen atom and a carbonyl carbon, which is a carbon atom that is double-bonded to second oxygen. The differences between oils from different sources relate to the length of the fatty acid chains attached to the backbone and the number of carbon-carbon double bonds on the chain. Most fatty acid chains from plant based oils are 18 carbons long with between zero and three double bonds. Fatty acid chains without double bonds are said to be saturated and those with double bonds are unsaturated (Misra & Murthy, 2010).

In general, vegetable oils are made especially of fatty acids with chains between 12 and 24 carbons: Lauric (C12:0); Myristic (C14:0); Palmitic (C16:0); Palmitoleic (C16:1); Stearic (C18:0); Oleic (C18:1); Linoleic (C18:2); Linolenic (C18:3); Arachidic (C20:0); Gadoleic (C20:1); Behenic (C22:0); Erucic (C22:1); Lignoceric (C24:0). The proportions of the fatty acid composition can be determined by gas chromatography method.

Triglycerides are hydrocarbons with physical and chemical characteristics that can be classified as liquid fuels in the majority.

Vegetable oils have a high heating value, near the heating value of conventional diesel fuel, which makes them an important energy resource.

To improve the properties of vegetable oils and in order to use them as substitutes for diesel fuel, triglycerides are converted into esters (biodiesel) by transesterification process, modifying the composition of molecules and changing the characteristics of the fluids. The composition of the resulting esters has approximately the same proportion of fatty acids present in oils before the transformation process.

Biodiesel is known as monoalkyl, such as methyl and ethyl, esters of fatty acids. Biodiesel can be produced from a number of sources, including recycled waste vegetable oil, oil crops and algae oil. Biodiesels play an important role in meeting future fuel requirements in view of their nature (less toxic), and have an edge over conventional diesel as they are obtained from renewable sources (Demirbas, 2009b).

The transesterification process is used to transform triglycerides into esters, or biodiesel. In the process of transesterification, the triglycerides found in different kinds of oils and fats react with alcohol, usually methanol or ethanol to produce esters and glycerin. For the reaction to occur it is necessary to use a catalyst. Processes performed the supercritical conditions of methanol transesterification can be conducted without the catalyst. (Demirbas, 2003; Saka & Kusdiana, 2001)

In the transesterification process, a triglyceride molecule reacts with an alcohol molecule causing the separation of one of the fatty acids of the triglyceride, producing a diglyceride and an ester. This diglyceride reacts with a second molecule of alcohol that takes another fatty acid, forming a second ester and a monoglyceride. Finally a third molecule of alcohol reacts with the monoglyceride, forming the third ester and a molecule of glycerin. The reactions occurring are reversible, and the stoichiometric ratio is three moles of alcohol for each mole of oil being processed. The reaction can be carried out with concentrations of alcohol in excess, as this reduces time and increases the conversion efficiency of the process. (Lang et al., 2001)

The transesterification process using methanol and base catalyst is the most commonly used to produce biodiesel. The catalysts commonly used are sodium hydroxide (NaOH) or potassium hydroxide (KOH). The catalyst is diluted in alcohol and then added the oil. The product is the ester (biodiesel) and crude glycerin. The glycerin is separated from the ester by decanting or centrifuging.

Repeated washing processes are performed by adding acidified water in the reaction products. This mixture is stirred lightly and serves to remove residual glycerine soap, catalyst and serves as a neutralizing agent of the fuel. The washing process is repeated until the biodiesel becomes clear. The main drawbacks of the process are the presence of water on some of the reagents and the high level of free fatty acids in the raw material. In both cases, the transesterification reaction is replaced by a saponification reaction.

The transesterification process can also be developed using acid catalysts and no homogeneous catalysts. In the case of acid catalysts, the process times are longer, but do not have drawbacks with the water content and free fatty acids. In the case of no homogeneous catalysts, these bring benefits in: reducing the washing process; product separation and reuse of catalysts.

Methanol and ethanol are produced on an industrial scale and their use in transesterification reactions has been reported.

The biodiesel used in several countries of Europe and in the United States is a mixture of methyl esters. Methanol is usually obtained from non-renewable fossil fuels, but can also be obtained by distillation of wood; this route, however, produces smaller quantities. The technology of biodiesel production using methanol is fully understood, however, this route has the disadvantage that methanol is extremely toxic.

The transesterification using ethanol is more difficult because the use of alcohol, even if anhydrous, involves problems in the separation of glycerin from the reaction medium. However, the use of ethanol is advantageous, since it is produced on a large scale in countries like Brazil and the United States, being from a renewable source of energy, resulting in environmental gains could generate carbon credits. As for the difficulties in the separation of phases in reactions employing ethanol in biodiesel synthesis, they can be bypassed by adjustments in reaction conditions.

3. Soybean oil and soybean biodiesel obtaining and characterization procedures

The fuel blends used were prepared at the Thermo-sciences Laboratory of Mechanical Engineering Department at Fluminense Federal University (UFF).

It was used Oxx as the nomenclature to designate mixtures of soybean oil with diesel, being xx the percentage by volume of vegetable oil added to diesel. The following mixtures were used: O5; O10; O15 and O20, besides pure diesel (O0).

It was used Bxx as the nomenclature to designate mixtures of biodiesel with diesel, being xx the percentage by volume of biodiesel added to diesel. The following mixtures were used: B5; B10; B15; B20; B50 and B75, besides pure diesel (B0) and pure biodiesel (B100).

Diesel fuel used is from the Laboratory of Distributed Power Generation at Fluminense Federal University. This fuel is used as reference for the tests.

The soybean oil used in the tests was obtained in the food market and the biodiesel used was produced at the Thermo-sciences Laboratory of Mechanical Engineering Department at Fluminense Federal University.

Refined oils are free of substances inhibiting the transesterification process. They have low amount of free fatty acids, less than 0.5%. The beta-carotene and phosphatides of the raw material are eliminated in the process of bleaching and degumming the oil. Thus the processing of these oils for conversion to biodiesel is a process that does not present major technical difficulties. The process is performed at atmospheric pressure.

3.1 Basic procedures for biodiesel obtaining

A guide for biodiesel obtaining from refined oils is given as follows.

Raw material:

- 100 mL of refined vegetable oil (commercial oil, whose acidity is less than 0.5%, degummed, microfiltered, deodorized and bleached);
- 25 mL of anhydrous methyl alcohol;
- 1 g of potassium hydroxide;

Steps:

1. Put 100 mL of vegetable oil in a glass container and lead to heating to a temperature of 45°C;
2. Place the container on the balance. Tare, and put 1g of KOH in it and close the container to avoid hydration of the reagent;
3. Measure a volume of 25 mL of anhydrous methyl alcohol using a test tube (methanol is toxic and must be handled with care in appropriate place);
4. Mixing the methanol with KOH to achieve a uniform solution of methoxide;
5. Add the methoxide to the solution of vegetable oil and stir for a period of two hours;
6. Turn off the mixer and check whether it produces a phase separation, it can be seen that a fluid dark (glycerin) deposits at the bottom of the container;

7. After checking the reaction, put the fluid in the decanting funnel, allow the glycerin settle for a period exceeding 30 minutes. Separate the glycerin and ester (biodiesel) produced in clean containers;
8. Wash the decanting funnel to remove the glycerin stuck on the walls and then fill with the ester obtained;
9. Perform the washing process of ester, adding 30 mL of water, preferably hot (50-60°C), inside the funnel, stirring to ensure the contact of two fluids;
10. Decant the water and remove it from the funnel;
11. Perform the washing process three more times to ensure complete removal of the glycerin;
12. Perform the drying process putting the ester in an oven, heating up to 110 ° C for 10 minutes
13. Cool and bottle the product (biodiesel)

3.2 Characterization of soybean oil and soybean biodiesel and stationary engine tests

The properties of soybean oil, soybean biodiesel and diesel were determined at the Thermo-sciences Laboratory and Rheology Laboratory of Fluminense Federal University. The heating values were determined at the Laboratory of fuels at the National University of Colombia. The characterization tests followed the standards, as detailed in Table 1.

The tests conducted in the stationary engine were made at constant speed of 3600 rpm and variable power in the Fluminense Federal University

For each fuel tested, a test was performed and repeated. Performance data and emissions were measured continuously during the test and the series of data were analyzed to obtain values representative of engine performance

PROPERTY	STANDARD
Viscosity	ASTM D 445 Standard Test Method for Kinematic Viscosity of Transparent and Opaque Liquids
Density	ASTM D 4052 Density and Relative Density of Liquids by Digital Density Meter.
Flash point	ASTM D 93 Standard Test Methods for Flash Point by Pensky-Martens Closed Cup Tester
Cloud point	ASTM D2500 Test Method for Cloud Point of Petroleum Products
Pour point	ASTM D97 Test Method for Pour Point of Petroleum Products
Copper strip corrosion	ASTM D130 Test Method for Copper strip corrosion of Petroleum Products
Heat of combustion	ASTM 240 Test Method for heat of combustion of Petroleum Products.

Table 1. Technical standards associated with the characterization tests

The stationary engine used (Figure 1) is formed by an engine, a generator and a control panel, with the possibility of producing electricity at 115V and 230V. The generator has a control system to regulate the motor rotation. The characteristics of the diesel engine are: 3600rpm; four-stroke; direct injection; one cylinder; air cooling system; 0.211L displacement

volume; 2.0kW maximum output; 1.8kW nominal power; 2.5L fuel capacity and 47 kg weight.
The engine was modified in order to have a fuel consumption control by gravity, changing the original fuel tank by a remote tank, being possible to be placed on a balance
The electrical load was simulated on a load bank, where 150W power lamps were activated to modify the load (Figure 1). The measurements of instantaneous power, current frequency, voltage and electrical current were made using a measuring device (CCK 4300) manufactured by CCK Automation Ltda (São Paulo, Brazil).
Emissions were measured using the gas analyzer Greenline 8000 built by Eurotron Instrument S.A. The equipment has measurement system of gas concentration by non-dispersive infrared (NDIR) and electrochemical method, in addition to measuring temperature, pressure and temperature of gases. The equipment has RS232 communication system for data acquisition and algorithms for calculating the efficiency indicators for different fuels. The resolution and the error limits of the equipment for measured gases are: electrochemical CO, 1 ppm and ± 10 ppm; NDIR CO_2, 0.01% and ± 0.3%; electrochemical NO, 1 ppm and ± 5 ppm; electrochemical NO_2, 1 ppm and ± 5 ppm and calculated NOx, 1 ppm. The SO_2 measurements were made by the method of molecular absorption spectroscopy (Tulcan, 2009)

Fig. 1. Stationary engine and control panel (operating).

4. Energy generation using soybean oil and soybean biodiesel

The yield of the process of producing biodiesel from soybean was 0.91L of biodiesel per 1L of used oil.

Table 2 shows the properties of diesel, soybean oil and soybean biodiesel, obtained in accordance with the standards shown in Table 1.

The expanded uncertainty of measurements are: density = ± 0.00008 kg/L; viscosity = ± 0.006 mm^2/s; flash point = ± 2.1 °C; cloud point = ± 1.5°C and pour point = ± 1.8°C (Santo Filho, 2010a; Abreu, 2010).

The weight composition of fatty acids found in soybean oil was: C16:0=11.6%, C16:1=0.1%, C18:0=3.2%, C18:1=20.4%; C18:2=59.7% and C18:3=5%.

Based on the physicochemical characterization performed in the soybean oil and soybean biodiesel, some correlations were determined.

PROPERTY	DIESEL	SOYBEAN OIL	SOYBEAN BIODIESEL
Density kg/L (20 °C)	0.85519	0.92037	0.88230
Viscosity mm^2/s (40 °C)	4.689	30.787	4.161
Flash point °C	82	332	150
Cloud point °C	2	-2	0
Pour point °C	-12	-14	-6
Copper strip corrosion	1a	1b	1a
Higher heating value kJ/kg	42800	-	41685

Table 2. Properties of diesel, soybean oil and soybean biodiesel.

The equation that best describes the behavior of flash point for blending diesel-soybean biodiesel is (Tulcan, 2009):

$$FP(mixture) = \left[(FP(a) - FP(d)).(\%mixture)^{\left(\frac{FP(a)}{FP(d)}\right)} \right] + FP(d) \tag{1}$$

where: FP (a) is the flash point of the additive (biodiesel) and FP (d) is the flash point of diesel.

The best fitting function representing the behavior of soybean oil density is, as follows (Santo Filho et al., 2010b):

$$\rho = -0.00069T + 0.93420 \tag{2}$$

where: ρ is the density of soyben oil (kg/L) and T is the temperature (${}^0C$).

The best fitting function representing the behavior of soybean biodiesel density is, as follows (Santo Filho et al., 2010c):

$$\rho = -0.00073T + 0.89691 \tag{3}$$

where: ρ is the density of soyben biodiesel (kg/L) and T is the temperature (${}^0C$).

The behavior of soybean oil viscosity is, as follows (Santo Filho, 2010a):

$$\nu = (0.1115)e^{(953.1562/(T+129.5732))} \tag{4}$$

where: the viscosity of soyben oil is given in mm^2/s and the temperature (T) is in (^0C).
The behavior of soybean biodiesel viscosity is, as follows (Santo Filho, 2010a):

$$\nu = (0.1246)e^{(609.1440/(T+133.6409))} \tag{5}$$

where: the viscosity of soyben biodiesel is given in mm^2/s and the temperature (T) is in ^0C.
Figures 2-7 show the values of specific fuel consumption (SFC) and emissions of NO, NO_x, CO, CO_2 and SO_2 for diesel, soybean biodiesel and mixtures of diesel-soybean oil and diesel-soybean biodiesel. The reported values represent the average for four values of load (400W, 700W, 1000W and 1300W).
Figure 2 shows the behavior of SFC for diesel-soybean oil mixtures and diesel-soybean biodiesel blends. The specific fuel consumption is lower for mixtures of 5% soybean oil than that for diesel. For this percentage, the oil has an oxygenating effect which improves engine performance, with an average of 1.9% decrease from the SFC. For larger percentages of mixture, the SFC increases, indicating a drop in engine performance. This is a consequence of lower heating value of soybean oil and of the increasing of the difficulties to burn fuel in the combustion chamber, requiring more fuel. For mixtures of 20% soybean oil, the increase in SFC is 4.5%. In the case of diesel-soybean biodiesel blends, a slight decrease in the SFC can be observed for smaller proportions of the mixture (5% to 10% soybean biodiesel). This decrease is due to the oxygenating capacity of biodiesel. For larger values of the mixture (15% to 100% soybean biodiesel), the SFC increases. This increase in SFC is due to the lower heating value of biodiesel, requiring more fuel. As the mixing ratio increases, the specific fuel consumption increases. The SFC hits an increase of 14% for the use of pure biodiesel compared to diesel.
Figures 3 and 4 present the average values of NO and NO_x emissions for different mixing ratios of soybean oil with diesel and soybean biodiesel with diesel. The NO and NO_x emissions increase with the use of blends up to 10% of soybean oil in diesel. From this amount of mixture, the NO and NO_x emissions decrease reflecting a decrease in the temperature of combustion chamber. However, the values are still higher than the emissions of diesel. For diesel-soybean biodiesel blends, the production of NO and NO_x is higher for blends superior to 5% than for diesel. Emissions of NO and NO_x grow rapidly for blends above 15% of soybean biodiesel. For mixtures between 20% and 75% soybean biodiesel, the emission levels of NO are an average of 200 ppm. For pure biodiesel the emission levels fall, this is a consequence of the low temperature in the combustion chamber due to the lower heating value of fuel.
Figure 5 shows the mean values of CO for soybean oil and soybean biodiesel blended in diesel. It may be noted that in proportions of up to 5% of soybean oil, the mixtures have an advantage in relation to diesel. For higher proportions, the CO emission increases to a level of 405 ppm for the mixture with 20% of soybean oil, 18% higher than the levels achieved by diesel emissions. The increase in the amount of CO shows a less efficient combustion. In the case of diesel-soybean biodiesel blends, it can be observed that the emission of CO for mixtures was lower than for diesel. The CO emission decreases by increasing the proportion of soybean biodiesel in the blend. In the case of pure biodiesel, the reduction in CO emission was 21% compared with diesel.

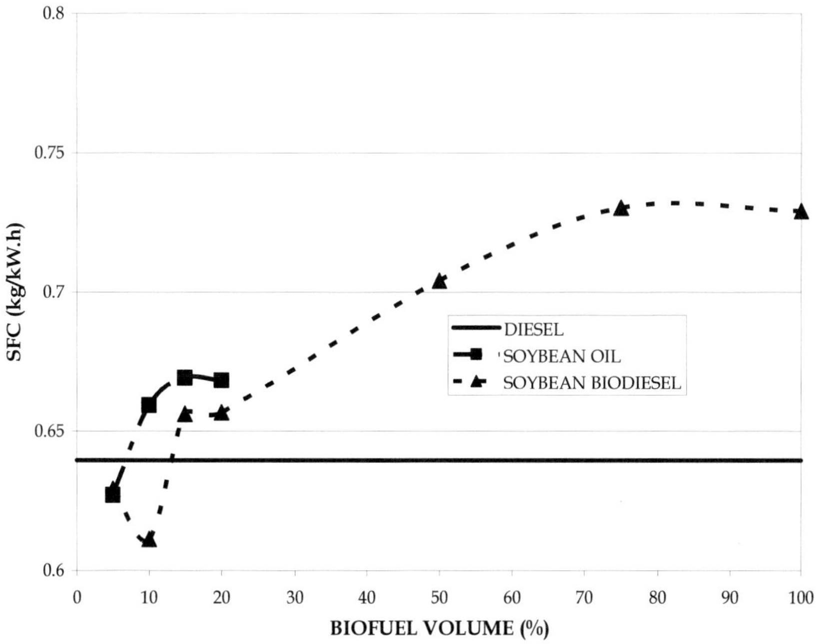

Fig. 2. Mean values of SFC for soybean biodiesel and soybean oil blended in diesel.

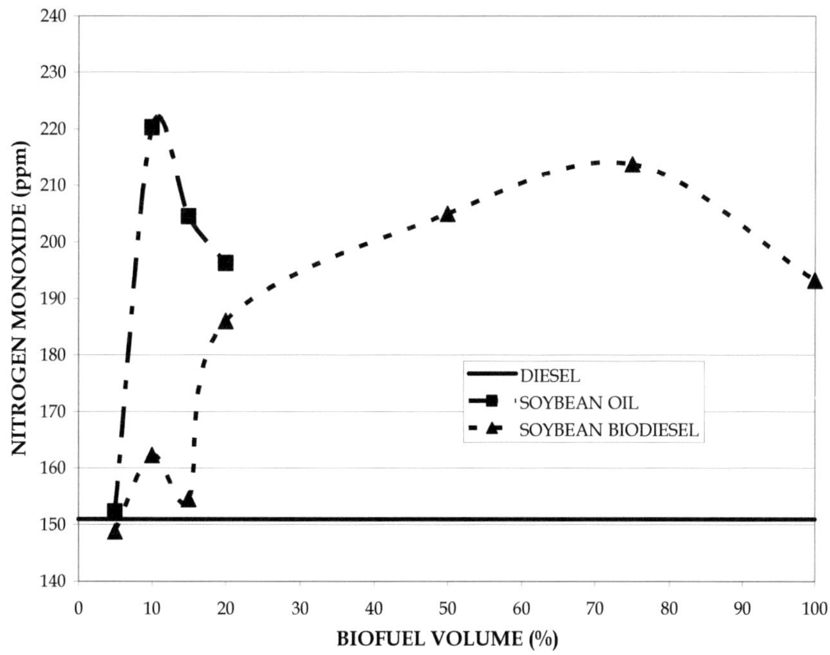

Fig. 3. Mean values of NO for soybean biodiesel and soybean oil blended in diesel.

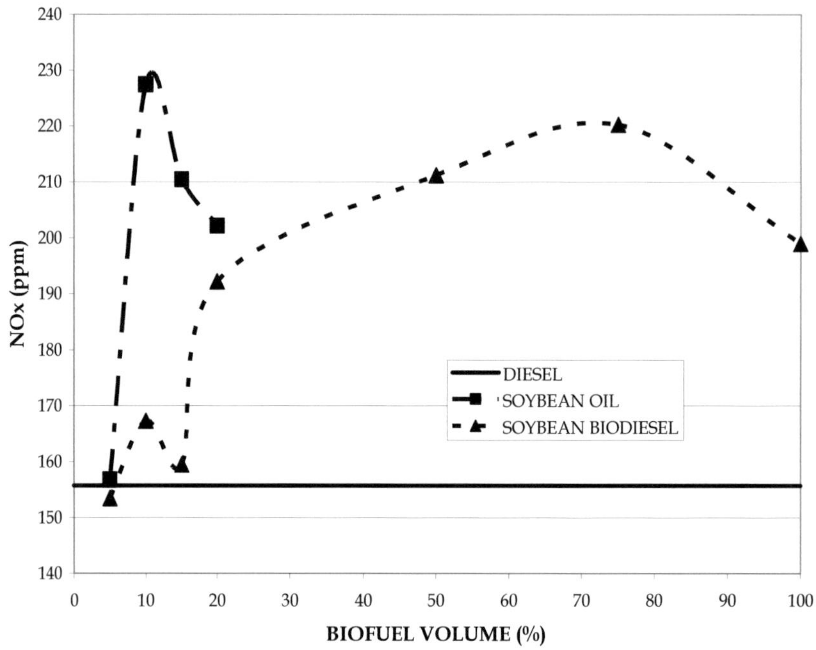

Fig. 4. Mean values of NO_x for soybean biodiesel and soybean oil blended in diesel.

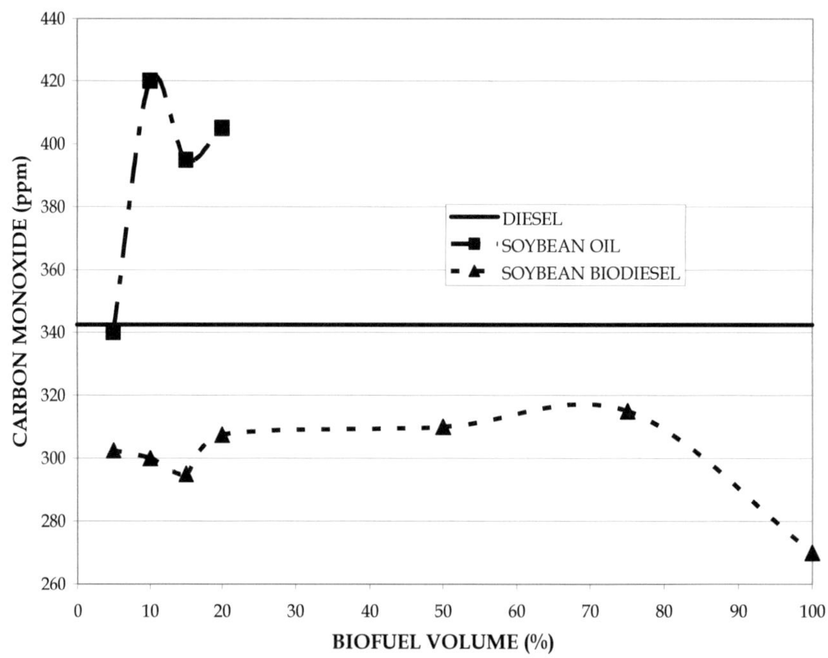

Fig. 5. Mean values of CO for soybean biodiesel and soybean oil blended in diesel.

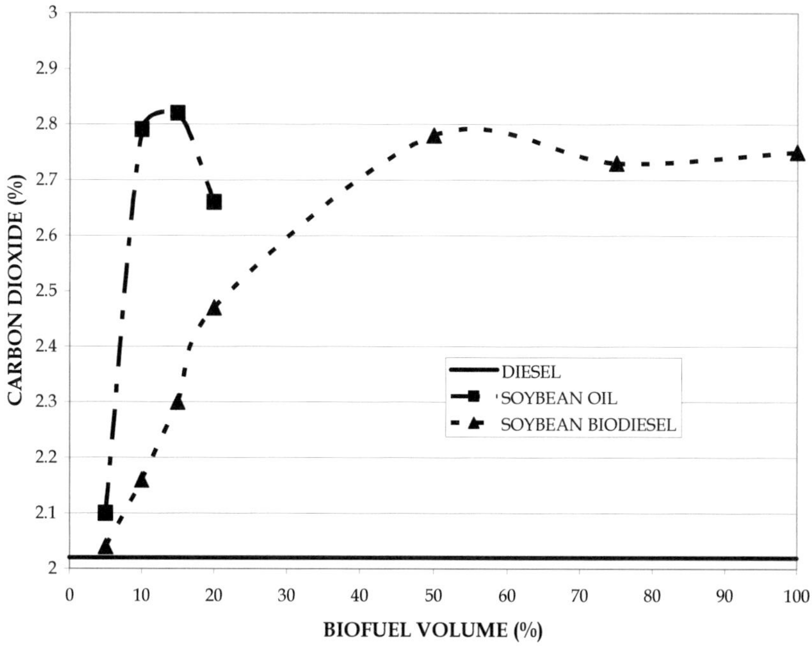

Fig. 6. Mean values of CO_2 for soybean biodiesel and soybean oil blended in diesel.

Figure 6 shows the CO_2 average emissions for soybean oil and soybean biodiesel blended in diesel. It can be observed that the percentages of CO_2 emissions for the mixtures are always higher than for diesel. Compared with diesel, a mixture of 15% of soybean oil increases the production of CO_2 by 40% and the mixture of 20% of soybean oil by 31%. The production of CO_2 increases with the addition of soybean biodiesel reaching a value of 2.77% to 50% soybean biodiesel, representing an increase of 37% compared to diesel. From this value a slight reduction in CO_2 occurs.

Figure 7 shows average results of the SO_2 emission obtained by the method of molecular absorption spectroscopy. In the figure it can be observed that, in some cases, the addition of soybean oil increases the production of SO_2. The production of SO_2 is caused by oxidation of sulfur in the fuel. Although the addition of soybean oil reduces the presence of sulfur in fuel, the cause of production of sulfur oxides may be due to rising temperatures in the combustion chamber and the consequent degradation of lubricating oils in the engine and volatilization of no burning fuel inside the combustion chamber. For fuel mixtures in proportions greater than 10% of soybean oil, the levels of SO_2 emissions begin to decrease. For mixtures of 15% to 20% of soybean oil, the levels of SO_2 emissions are lower than for diesel. In the case of 20% soybean oil, the reduction of SO_2 is 69%. For diesel-soybean biodiesel blends, the emission of sulfur dioxide, in some cases, is higher than for diesel, as also happened in the case of mixtures with soybean oil. Moreover, one can observe that the emission of SO_2 decreases for fuel mixtures in proportions greater than 50% of soybean biodiesel. In the case of 100% soybean biodiesel, the reduction of SO_2 is 71%. The presence of SO_2 in the burning of pure biodiesel can be due to burning of lubricating oil and residual diesel into the combustion chamber.

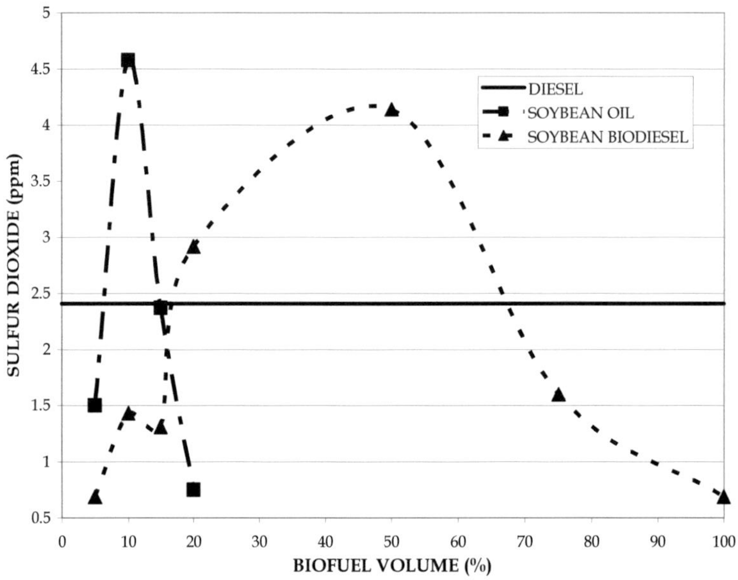

Fig. 7. Mean values of SO_2 for soybean biodiesel and soybean oil blended in diesel.

Figures 8 to 13 show the emission behavior as a function of load in the stationary engine for mixtures of 20% soybean oil and 20% soybean biodiesel with diesel.

The behavior of the SFC is similar in the cases of diesel and the mixtures 20% soybean oil-diesel and 20% soybean biodiesel-diesel. As shown in Figure 8 the value of SCF for the fuels studied decreases with increasing load.

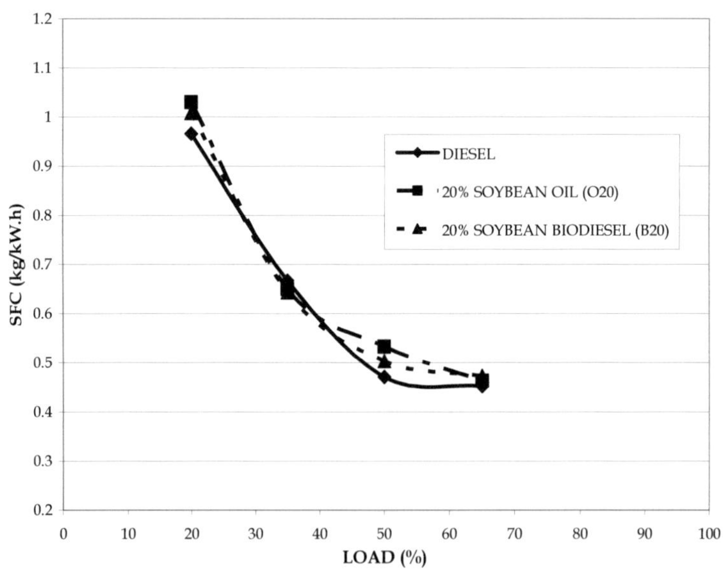

Fig. 8. Values of SFC for 20% soybean biodiesel and 20% soybean oil blended in diesel.

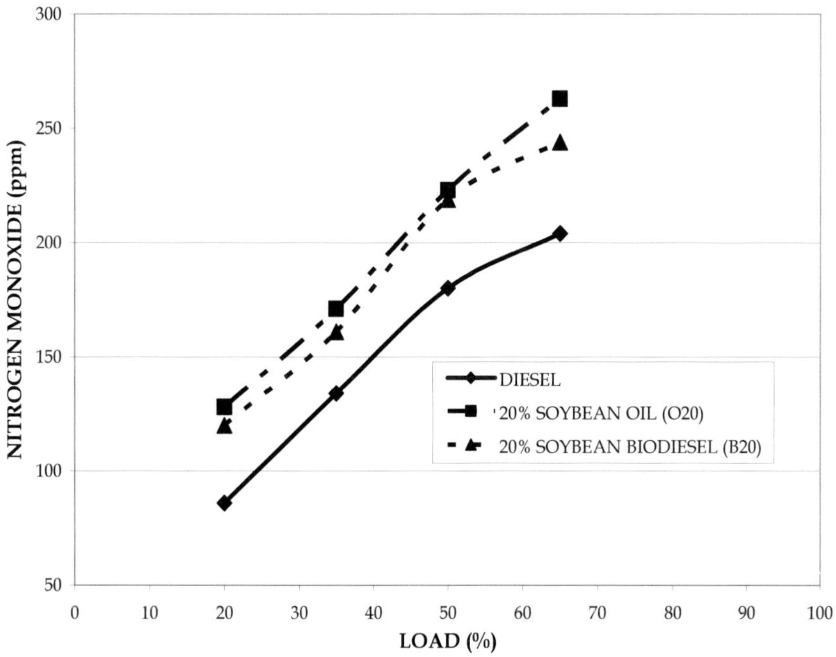

Fig. 9. Values of NO for 20% soybean biodiesel and 20% soybean oil blended in diesel.

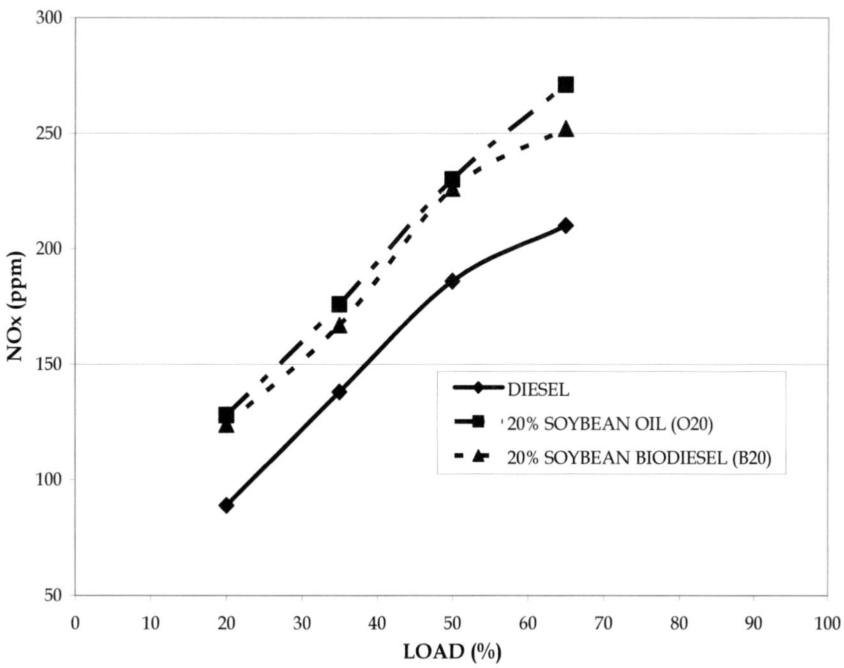

Fig. 10. Values of NO_x for 20% soybean biodiesel and 20% soybean oil blended in diesel.

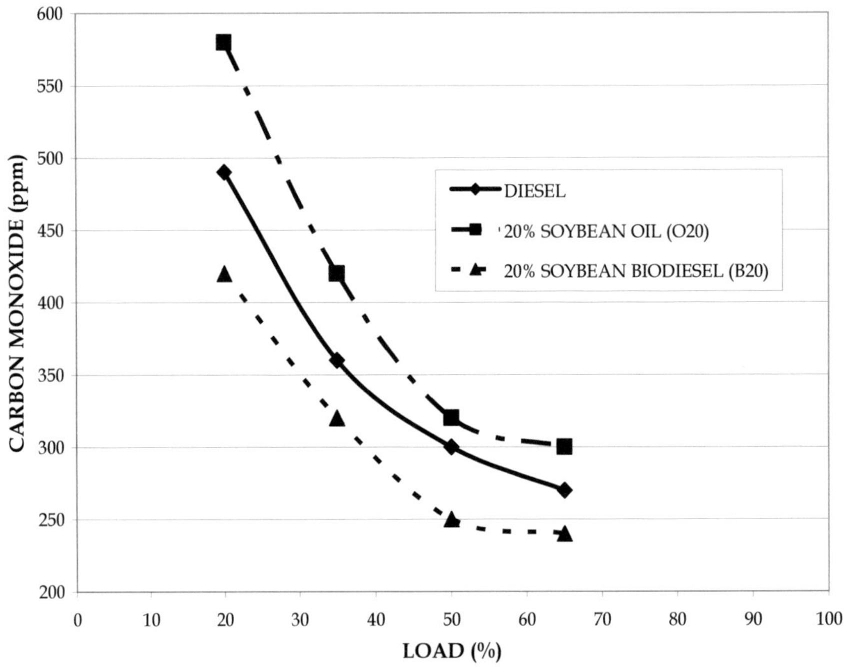

Fig. 11. Values of CO for 20% soybean biodiesel and 20% soybean oil blended in diesel.

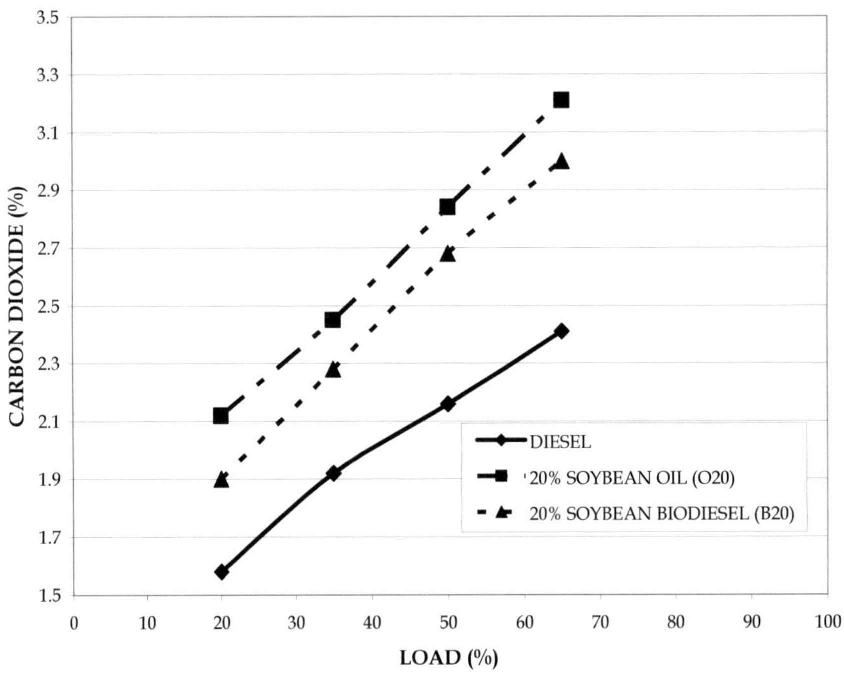

Fig. 12. Values of CO_2 for 20% soybean biodiesel and 20% soybean oil blended in diesel.

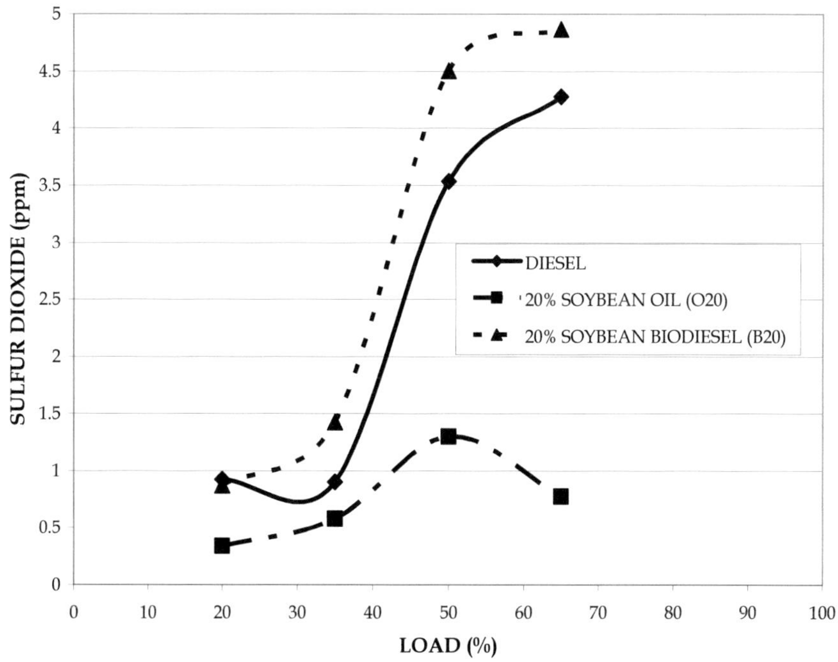

Fig. 13. Values of SO_2 for 20% soybean biodiesel and 20% soybean oil blended in diesel.

The Figures 9 and 10 show that the emission of NO and NO_x, for mixtures of 20% soybean oil-diesel and 20% soybean biodiesel-diesel, are higher than those of diesel at all loads studied, probably due to increased temperature in the combustion chamber.

In Figure 11 it can be observed that the CO emissions decrease with increasing load for all fuels tested, showing a better combustion in these cases. The lower values of CO occur with the 20% blend of soybean biodiesel (B20).

As shown in Figure 12, as the load increases the amount of CO_2 emissions increases for all fuels studied.

The emissions of SO_2 with the load are shown in Figure 13. The lowest emissions occur at lower loads.

5. Conclusion

Soybean oil and soybean biodiesel can be added to diesel fuel to be burned in combustion engines. These compounds have an oxygenate capacity that is useful to improve engine performance, but this ability only gives you an edge when the mix ratio is 5% for vegetable oil and 10% for biodiesel. The gains made in reducing the SFC using the oxygenating additives affect about 2% in the case of 5% soybean oil blended with diesel and about 4.5% for 10% soybean biodiesel blended with diesel. Using a larger proportion of mixture generates increases in SFC by 9% on average when pure biodiesel is used, and 3% when mixture of 20% soybean oil is used.

The emission of NO and NO_x increases with the addition of oxygenated components (vegetable oil and biodiesel). The use of 20% soybean oil blended with diesel (O20) increases

the NO emission by 30%. But the use of pure soybean biodiesel (B100) promotes an increase in NO emission by 28%.
The addition of soybean oil in diesel reduces emissions of CO only for mixtures of up to 5% soybean oil. In the case of biodiesel addition in diesel, CO emissions decrease with the mixture reaching a 21% reduction, when pure soybean biodiesel (B100) is used.
The addition of soybean oil and soybean biodiesel in diesel increases the emission of CO_2 which however is compensated by the absorption of CO_2 by the plants (raw material for production of vegetable oil and biodiesel).
The addition of soybean oil and soybean biodiesel in diesel reduces the sulfur content in fuel and consequently reduces the emission of sulfur dioxide.
Soybean oil can be successfully applied in CI engine blending with diesel up to 20% of soybean oil. Soybean oil can also be converted in biodiesel and applied in CI engines neat or blended with diesel in any proportion. Concerning the exhaust emissions it is better use soybean biodiesel or blends of soybean biodesel with diesel instead of blends of soybeans oil with diesel.
The possibility of using soybean oil in power generation, leads to the concept of energy farms soybeans that can be an opportunity for farmers who can not meet the quality standards required for selling the soybeans to food industry.

6. Acknowledgment

The authors are grateful to the National Research Council of Brazil, CNPq, for the financial support received.

7. References

Abreu, F.L.B. (2010). Power generation and analysis of emissions in stationary engine using biodiesel, blends of biodiesel and blends of biodiesel containing additive, produced via ethylic route and its characterization, (in portugues), *Ph.D. Thesis*, Fluminense Federal University, Niterói, RJ, Brazil, December 2010

Alonso, J.; Sastre, J.; Ávila, C. & López, E. (2008). A Note on the Combustion of Blends of Diesel and Soya, Sunflower and Rapeseed Vegetable Oils in a Light Boiler. *Biomass and Bioenergy*, Vol. 32, Issue 9, (September 2008), pp. 880-886,ISSN 0961-9534

Basha, S.; Gopal, K. & Jebaraj, S. (2009). A Review on Biodiesel Production, Combustion, Emissions and Performance. *Renewable and Sustainable Energy Reviews*, Vol. 13, Issues 6-7, (August-September 2009), pp. 1628-1634, ISSN 1364-0321

Demirbas, A. (2003). Biodiesel Fuels from Vegetable Oils via Catalytic and non-Catalytic Supercritical Alcohol Transesterifications and other Methods and Survey. *Energy Conversion and Management*, Vol 44, Issue 13, (August 2003), pp. 2093-2109, ISSN 0196-8904

Demirbas, A. (2009a). Political, Economic and Environmental Impacts of Biofuels: A Review. *Applied Energy*, Vol. 86, Supplement 1, (November 2009), pp. S108-S117, Bio-fuels in Asia, ISSN 0306-2619

Demirbas, A. (2009b). Biofuels Securing the Planet's Future Energy Needs. *Energy Conversion and Management*, Vol. 50, Issue 9, (September 2009), pp. 2239-2249, ISSN 0196-8904

Enweremadu, C. & Rutto, H. (2010). Combustion, Emission and Engine Performance Characteristics of Used Cooking Oil Biodiesel -A Review. *Renewable and Sustainable Energy Reviews*, Vol. 14, Issue 9, (December 2010), pp. 2863-2873, ISSN 1364-0321

Grau, B.; Bernat, E.; Antoni, R.; Jordi-Roger, R. & Rita, P. (2010). Small-Scale Production of Straight Vegetable Oil from Rapeseed and its use as Biofuel in the Spanish Territory. *Energy Policy*, Vol. 38, Issue 1, (January 2010), pp. 189-196, ISSN 0301-4215

Hossain, A. & Davies, P. (2010). Plant Oils as Fuels for Compression Ignition Engines: A Technical Review and Life-Cycle Analysis. *Renewable Energy*, Vol. 35, Issue 1, (January 2010), pp. 1-13, ISSN 0960-1481

Lang, X.; Dalai, A.; Bakhshi, N.; Reaney, M. & Hertz, P. (2001) Preparation and Characterization of Bio-Diesels from Various Bio-Oils. *Bioresource Technology*, Vol. 80, Issue 1, (October 2001), pp. 53-62, ISSN.0960-8524

Liu, X.; Jin, J.; Wang, G. & Herbert, S.(2008). Soybean Yield Physiology and Development of High-Yielding Practices in Northeast China. *Field Crops Research*, Vol. 105, Issue 3, (February 2008), pp. 157-171, ISSN 0378-4290

Misra, R. & Murthy, M. (2010). Straight Vegetable Oils usage in a Compression Ignition Engine - A Review. *Renewable and Sustainable Energy Reviews*, Vol. 14, Issue 9, (December 2010), pp. 3005-3013, ISSN 1364-0321

No, S. (2011), Inedible Vegetable Oils and their Derivatives for Alternative Diesel Fuels in CI Engines: A Review. *Renewable and Sustainable Energy Reviews*, Vol. 15, Issue 1, (January 2011), pp. 131-149, ISSN 1364-0321

Pereira, R.; Oliveira, C.; Oliveira, J; Oliveira, P., Fellows, C. & Piamba, O. (2007). Exhaust Emissions and Electric Energy Generation in a Stationary Engine using Blends of Diesel and Soybean Biodiesel. *Renewable Energy*, Vol. 32, Issue 14, (November 2007), pp. 2453-2460, ISSN 0960-1481

Saka, S. & Kusdiana,D.(2001) Biodiesel Fuel from Rapeseed Oil as prepared in Supercritical Methanol. *Fuel*, Vol. 80, Issue 2, (January 2001), pp. 225-231, ISSN.0016-2361

Santo Filho, D. M. E. (2010a). Metrology applied to analysis of biodiesel (in portugues), *Ph.D. Thesis*, Fluminense Federal University, Niterói, RJ, Brazil, July 2010

Santo Filho, D; Abreu, F; Pereira, R.; Santos Junior, J; Siqueira, J.; Ferreira, P.; Barbosa, T.; Lima, L. & Baldner, F. (2010b) The Influence of the Addition of Oils in the Diesel Fuel Density. *Journal of ASTM International (Online)*, Vol. 7, Issue 8, (September 2010), pp. 1-9, ISSN 1546-962X

Santo Filho, D. ; Abreu, F.; Pereira, R..; Siqueira, J.; Santos Junior, J. & Daroda, R. (2010c). Characterization of Density of Biodiesel from Soybean, Sunflower, Canola and Beef Tallow in Relation to Temperature, Using a Digital Density Meter, with a Metrological Point of View. *Journal of ASTM International (Online)*, Vol. 7, Issue 2, (February 2010), pp. 1-6, ISSN 1542-962X

Sharma, Y. & Singh, B.(2009). Development of Biodiesel: Current Scenario. *Renewable and Sustainable Energy Reviews*, Vol. 13, Issues 6-7, (August-September 2009), pp. 1646-1651, ISSN 1364-0321

Sidibé, S.; Blin, J; Vaitilingom, G. & Azoumah. Y. (2010). Use of Crude Filtered Vegetable Oil as a Fuel in Diesel Engines State of the Art: Literature Review. *Renewable and*

Sustainable Energy Reviews, Vol. 14, Issue 9, (December 2010), pp. 2748-2759, ISSN 1364-0321

Tulcan, O. E. P., Performance of a stationary engine using different biofuels and emissions evaluation (in portugues), *Ph.D. Thesis*, Federal Fluminense University, Niterói, RJ, Brazil, November 2009

van Dam, J.; Faaij, A.; Hilbert, J.; Petruzzi, H. & Turkenburg, W. (2009). Large-Scale Bioenergy Production from Soybeans and Switchgrass in Argentina: Part A: Potential and Economic Feasibility for National and International Markets. *Renewable and Sustainable Energy Reviews*, Vol. 13, Issue 8, (October 2009), pp. 1710-1733, ISSN 1364-0321

16

Soybean Oil De-Acidification as a First Step Towards Biodiesel Production

C. Pirola, D.C. Boffito, G. Carvoli,
A. Di Fronzo, V. Ragaini and C.L. Bianchi
Università degli Studi di Milano – Dipartimento di Chimica Fisica ed Elettrochimica, Milano Italy

1. Introduction

According to the predictive studies of the World Energy Outlook 2009, the global demand of energy is expected to increase till 2030 of about 1.5 percentage points per year. Fossil fuels are expected to remain the main energy source in the world, but in the meantime renewable energy sources (wind, solar, geothermal, bioenergy) will be characterized by a rapid growing rate. Their use and development is strongly encouraged by most of the recent regulations. For instance, as reported by the European Environment Agency Transport (EEA), 2009, the European Union required in the same year to achieve by 2020 at least 10% of mixture of hydrocarbons from renewable and conventional sources for what concerns the energy employed for the transports. In addition, the increase in oil price and the growing interest in environmental issues have recently given a considerable impetus to the research for cleaner and renewable energy sources, in order to ensure a sustainable future.

Biodiesel (BD) is a renewable energy source in liquid form that has many advantages over normal diesel, including lower emissions of gases harmful to humans and environment. The UE directive 2003/30/EC, defines the Biodiesel as "a methyl ester produced from vegetable or animal oil, of diesel quality, to be used as biofuel". Moreover, the National Biodiesel Board (NBB), 1996, responsable for biodiesel ASTM standards, define biodiesel as "the mono alkyl esters of long chain fatty acids derived from renewable lipid feedstock's, such as vegetable oils or animal fats, for use in compression ignition (diesel) engines."

The processes for BD production are well known. According to the NBB, 2007, there are three main routes to BD production from oils and fats:

- Base- catalyzed transesterification;
- Direct acid- catalyzed transesterification;
- Conversion of the oil into fatty acids and then into biodiesel.

At the present BD is mainly produced through the base-catalyzed transesterification for many different reasons:

- Mild reaction conditions, i.e. low temperature and pressure may be adopted;
- High conversions (up to 98.5%) are usually achieved in short times with minimization of side reactions;
- The conversion into BD is direct and no intermediate steps are required;

- No expensive construction materials are required.

According to the base-catalyzed process, BD is produced through the transesterification of triglycerides contained in oils or fats, with methanol and in the presence of an alkaline catalyst, also yielding glycerin as a by-product (Fig. 1).

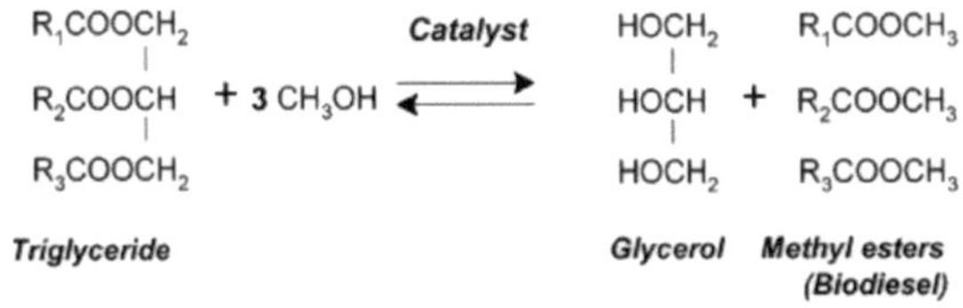

Fig. 1. Transesterification of a triglyceride for biodiesel production.

Although food-grade oils with low acidity can be employed with few practical problems, their use is strongly discouraged to avoid interference with the human food requirements, besides being not cost-wise competitive with the petroleum-based diesel.

To overcome this problem, waste materials, such as waste cooking oils or animal fat, can be employed.

The use of not refined or waste oils as a feedstock represents a very convenient way in order to lower biodiesel production costs. Crude vegetable oil, waste cooking oils and animal fat are examples of alternative, cheaper, raw materials. The main problem associated with the use of this type of low-cost feedstock lies in its high content of FFA, leading to the formation of soaps during the final transesterification step.

The presence of soaps during the transesterification complicates the reaction resulting in hindering the contact between the reagents and causing difficulties in products separation

Not refined or waste fats require therefore to be standardized by the reduction of the acidity prior to be processed through the transesterification reaction (Bianchi et al., 2010).

However, while BD (pure or mixed) as an alternative fuel to diesel for use in diesel engines is a reality in many states (in France it is usually used in a 5% blend with diesel fuel, in Germany pure, in the USA in the "fleet"), the same cannot be stated for what concerns the use of biofuels as boilers in small, medium or large size plant.

The EU has also published some very restrictive parameters in collaboration with the CEN (*European Committee for Standardization*) to ensure an adequate performance and consequently a higher quality of the BD as biofuel. The required limits for biodiesel properties are listed in the paragraph 4 (European Standard EN 14214).

2. Different technologies for FFA removal from triglycerides (TG)

The starting materials for the biodiesel production are usually vegetable oils and animal fats, indeed constituted mainly by triglycerides (TG) and Free Fatty Acids (FFA), linear carboxylic acids in the range C_{14}-C_{22} with different unsaturation levels. FFA are contained in the oils in their free form as a result of the spontaneous hydrolysis of the starting TG molecules.

Fats have more saturated fatty acids as compositional building blocks than oils. This gives rise to higher melting points and higher viscosities for fats in comparison to oils. The FFA content varies among different lipid sources and also depends on the treatments and storage conditions. In Tab.1 a comparison among soybean oil and other kinds of feedstock is reported, from Lotero et al. (2005).

Fatty acid	Fatty acid composition, wt%							
	Myristic 14:00	Palmitic 16:00	Palmitoleic 16:01	Stearic 18:00	Oleic 18:01	Linoleic 18:02	Linolenic 18:03	Sat. (%)
Soybean oil	0.1	10.6	-	4.8	22.5	52.3	8.2	15.5
Palm oil	1.2	47.9	-	4.2	37	9.1	0.3	53.3
Sunflower oil	-	6.0	-	4.2	18.7	69.3	-	10.2
Lard	1.7	17.3	1.9	15.6	42.5	9.2	0.4	34.6
Yellow grease	2.4	23.2	3.8	13	44.3	7.0	0.7	38.6
Brown grease	1.7	22.8	3.1	12.5	42.4	12.1	0.8	37.0

Table 1.Typical FFA composition of soybean oil and others raw materials for BD production.

As already discussed in the introduction, the chemical transformation of these lipids into biodiesel involves the transesterification of glycerides with alcohols to alkylesters.

Nowadays BD is industrially obtained using alkaline homogeneous catalysts, such as sodium and potassium methoxides and hydroxides. Other possible routes to obtain biodiesel through transesterification and exploiting different catalytic systems are reported in a recent review (Vyas et al., 2010). These include: 1) homogeneous acid catalysis, 2) heterogeneous alkali or acid catalysis; 3) enzymatic catalysis, 4) supercritical conditions without catalyst, 5) microwave or ultrasound assisted reactions. All these methods will be presented in the paragraph 2.2.

Any TG or FFA source (vegetable oil, animal fat or waste grease) may be potentially used as source for biodiesel production through alkali or acid-catalyzed transesterification reaction. In spite of this, a feedstock characterized by a low impurities level and low water and FFA content is required to obtain a valuable, marketable product. In particular, the base-catalyzed transesterification requires high purity reactants (FFA < 0.5wt%, water< 0.1-0.3 wt%), having demonstrated to be very sensitive to the impurities contained in the feedstock (Strayer et al., 1983).

As a matter of fact, the raw material contributes 60-70% to the final manufacture cost of BD obtained from soybean oil. As a consequence, the utilization of expensive raw materials is responsible for the lack of economic competitioness of BD with fossil fuel.

In paragraph 2.1 different methods of performing the transesterification reaction are described, while in paragraph 2.2 various processes to lower the acidity content of the oil are reported.

2.1 Non alkali-catalyzed transesterification for BD production from feedstock with high FFA content

Synthesis of biodiesel via homogeneous acid catalysis: the homogeneous acid-catalyzed reaction rate is reported to be about 4000 time slower than the homogeneous one (Srivastava and Prasad, 2000). Nevertheless, adopting this technology it is possible to perform TG transesterification of not refined oils. Sulphuric acid is reported to be the best performing catalyst. Other homogeneous catalytic systems, such as HCl, BF_3, H_3PO_4 and organic sulphonic acids have also been studied (Liu, 1994). Homogeneous systems require a large molar ratio alcohol to oil (30:1 at least) to reach acceptable reaction rates. On the other hand, by increasing the alcohol amount, the separation costs increase as well.

Reaction rates may also be increased using higher amounts of catalyst. Common catalysts loadings are in the range 1- 5 wt%, while higher catalyst's loadings result in promoting ether formation by alcohol dehydration (Lotero et al., 2005).

The amount of water content in the oil is more critical in the case of the acid-catalyzed transesterification than in the base-catalyzed one (Canakcy and Van Gerpen, 1999). Canakcy reports that esters production can be affected by a water concentration as little as 0.1 wt% and can be almost totally inhibited by water concentrations higher than 5 wt%. This can be explained supposing that water molecules form a sort of shield around the catalyst, preventing its coming in contact with the hydrophobic TG molecules, so inhibiting the reaction. Water can in fact bind acid species in solution more effectively than alcohol. For this reason, in acid catalyzed processes, the water removal step has to be taken into account. The most economical method for water removal from oils is the one acting under gravity separation.

Synthesis of biodiesel via enzymatic transesterification: the enzymatic methods require expensive enzymes such as lipase. On the other hand these methods are affected by water to a less extent than acid-catalyzed process and can tolerate FFA concentration till 30 wt% (Vyas et al, 2010). Besides some advantages such as the mild reaction conditions (50°C for 12-24 h), easy products separation, minimal wastewater treatment and absence of side reactions, there are also some drawbacks such as the contamination of the final product with the residual enzymatic activity and the high costs of this technology (A. Sulaiman, 2007).

Low water contents in the production of BD from soybean oils using lipase as catalyst are reported to lower the enzyme activity (A. Sulaiman, 2007). Nevertheless, an excess of water is not convenient using immobilized lipase (Yuji et al., 1999). Indeed, the water content is a crucial factor, which requires to be optimized basing on the used reaction system.

In any case, the cost of lipase is still the major concern for the industrialization of this technology.

Synthesis of biodiesel via supercritical transesterification: when a fluid or gas is subjected to temperatures and pressures exceeding its critical point, a single fluid phase is present. Solvents containing hydroxyl (OH) groups, as methanol or water, when subjected to supercritical conditions, gain super-acids properties which can be exploited for some kinds of catalysis.

Transesterification reaction of soybean oil in supercritical methanol conditions (350°C; 200 bar) is reported to have been completed in about 25 minutes (Huayang et al., 2007). In supercritical conditions, the use of an excess of alcohol (ratio soybean oil/methanol =1: 40) is also possible, as a single homogeneous phase is present. This results in accelerating the reaction rate as no limitations to the mass transfer due to the presence of interphases occur.

The use of such high temperatures and pressures undoubtedly leads to very huge capital and operating costs and high energy consumption. The scale-up of this process may be therefore very difficult.

Synthesis of biodiesel via Microwave or Ultrasound assisted transesterification: microwave (MW) irradiation activates the smallest degree of variance of polar molecules such as alcohol through the continuous changing of the magnetic field. The production of BD using MW leads to some advantages as short reaction times, low oil/methanol ratio and general reduction of energy consumption (Vyas et al., 2010). MW assisted processes have been studied both in homogeneous and heterogeneous alkali- and acid-catalyzed BD syntheses (Leonelli and Mason, 2010). For this reason MW might be a suitable solution to process feedstock characterized by high initial FFA content. The main problem of the use of MW is the scale-up from the laboratory scale to the industrial plant. The crucial issue is represented by the penetration depth of MW radiation into the absorbing material. Another critical point is the safety aspect concerning the use of this technology, in particular on an industrial scale. Ultrasound (US) is well known as a powerful tool to enhance the reaction rate in a variety of chemical reactions. At high ultrasonic intensities and frequencies between 20 kHz and 100 MHz, a small gas cavity present in the liquid may grow rapidly generating oscillating bubbles; when these bubbles collapse they produce local hot spots of high temperature and pressure able to promote chemical and mechanical effects (Leonelli and Mason, 2010; Colucci et al., 2005). In the biodiesel reacting media, the collapse of these bubbles may be moreover able to disrupt the phase boundary causing emulsification, so impinging one liquid to another as a consequence of the formation of ultrasonic jets (Stavarache et al., 2005). US also introduce turbulence in the system resulting in an improved mechanical mixing: the activation energy required for initiating the reaction can be so easily achieved. Both the advantages and the drawbacks of the US-assisted transesterification are the same described for MW-assisted reaction.

2.2 FFA removal to make oil feedstock suitable to the alkali-catalyzed transesterification

Alkali refining method: in this technology, the removal of FFA is performed adding caustic soda and water to the oil before carrying out the transesterification reaction. In this way the FFA are transformed in fatty acid soaps and then removed by washing. This is a well-established practice in the soybean processing industry (Erikson, 1995). The soybean oil is heated to 70°C and mixed with a caustic solution to form soap and free fatty acids. The amount of FFA measured in the oil determines the flow rate of caustic soda to be added. The washing step is also carried out at 70°C, at a rate of 15% of the crude oil soybean mass flow rate. A certain yield loss occurs as result from the saponification of tryglicerides. The resulting mixture (oil, soap and wash water) is sent to a centrifuge to separate soap and water from the oil. A total quantity of about 1% of oil is lost in the soap and water mixture. The loss of product represents the main drawback of this method. Moreover, this technology gives often rise to problems during the separation phase.

Solvent extraction method: the FFA can be transferred into another phase from the oil one exploiting the difference of solubility in a solvent (e.g. methanol) between the fatty acids and the triglycerides (Ganquli et al., 1998). The oil and the solvent are fed counter-current with a high ratio solvent/oil. After the extraction process, the esterification using H_2SO_4 as acid homogeneous catalyst is performed. The solvent is then separated from the final product, purified and re-used. The main drawbacks of this technology are represented by the high costs lying in the separation and re-use cycle, also due to the presence of emulsions in the extraction reactor.

Hydrolization method: this technology is based on the hydrolyzation of the starting TG into pure FFA and glycerine. This process is typically performed in a counter-current reactor using sulphuric/sulfonic acids and steam. Then, pure FFA undergo to the acid-catalyzed esterification in another counter-current reactor and are converted into methylesters. In this case yields can be higher than 99%. The equipment to be adopted requires being highly acid-resistant.

Glycerolysis: this technique involves the addition of glycerol to the starting TG and the consequent heating to high temperatures (200°C). Zinc chloride is often used as catalyst. This reaction produces mono and dyglicerides, i.e. low FFA oil suitable for the based-

catalysed transesterification. A recent patent by Parodi and Marini (WO 2008/007231 A1) deals with this technology and its improvement with a new optimized process design and new kinds of catalysts.
Pre-esterification method: this method will be deeply described in the next paragraphs. The involved technology is based on the esterification of FFA with an alcohol in presence of a homogeneous or heterogeneous acid catalyst. Transesterification is then performed in a second step by using an alkaline homogeneous catalyst.
The not alkali-catalyzed systems for BD production are today used only on the laboratory scale. Moreover, the possibility to maintain and improve the alkali-catalyzed transesterification process as main route to BD production is an important requirement by all the currently working BD plants.
The pre-esterification process is the only method not resulting in a loss of final product, differently from all the other technologies previously described in this paragraph. These last give moreover rise to problems during the separation phase and require therefore high energy exploitation.
The main drawback of the pre-esterification method, if performed with the use of a homogeneous acid catalyst, consists in the necessity of the catalyst's removal from the oil before the transesterification step. This problem, as will be discussed in the next paragraphs, can be solved using a heterogeneous catalyst.

3. Pre-esterification methods by heterogeneous acid catalysis

Nowadays BD synthesis using homogeneous catalysis is considered not advisable. In fact, all processes involving homogeneous catalysis give raise to problems such as product purification and catalyst recovery. In addition, homogeneous acid catalysts are strongly corrosive.
Even a very small amount of residual acid catalyst in the final BD could cause engine problems; hence, an extensive washing with water is required to remove the catalyst residuals from the systems and obtain marketable products.
The use of heterogeneous catalysts prevents neutralization and separation costs, besides being not corrosive, so avoiding the use of expensive construction materials. Another important advantage is that the recovered catalysts can be potentially used for a long time and/or multiple reaction cycles. For all these reasons, the FFA pre-esterification method using heterogeneous acid catalysts is usually preferred to the homogeneously-catalyzed process.
Different solid acid catalysts have been studied in the recent years (Goodwin Jr. et al., 2005). They can be classified into two main categories: inorganic materials and ion-exchange resins functionalized with $-SO_3H$ groups. The main advantage of inorganic materials is represented by their higher thermal stability compared to the resin-based ones. For these last the maximum operating temperatures are in fact around 140° C. Nevertheless, the effectiveness of the $-SO_3H$ active groups for the catalysis of the FFA esterification reaction has been proved even at low temperatures (< 100°C).
Among the different types of inorganic solid materials used for the production of esters, the most popular are the zeolitic compounds. The acid strength of these materials can be modulated changing the Si/Al ratio. In addition, by adopting zeolites as catalysts, it is possible to choose among different pore structures and surface hydrophobicity. Only large-pore zeolites have been used in FFA esterification to avoid limitation to the mass transfer of

both reactants and products inside the catalyst's pores. Their use along with high operating temperatures may lead to the formation of undesired by-products (Corma and Garcia, 1997). Silica molecular sieves with amorphous pore walls, as MCM-41, are not sufficiently acid to catalyze the esterification process. The introduction of aluminum, zirconium or titanium into the silica matrix to improve the acid strength is not advisable, due to the easy deactivation to which these materials are usually subjected when water is present in the reaction system (Goodwin Jr. et al., 2005). Another possibility is represented by sulfated zirconia (SO_4^{2-}/ZrO_2), which has already been experimented in other kinds of esterification reactions (Bianchi et al., 2003), both in monophasic and biphasic systems. The main drawback of this type of catalyst lies in its fast deactivation due to the sulphate groups leaching, which may be favored by the presence of water in the system. Others similar materials that can be employed in the FFA esterification reaction are: sulfated tin oxide (SO_4^{2-}/SnO_2), prepared from meta-stanic acid, which is characterized by higher acidity compared to sulfated zirconia, and tungstaned zirconia, characterized by lower acidity but higher resistance to deactivation (Di Serio et al, 2008).

Recently, sulfonated carbons (the so called "sugar catalysts", derived by incomplete carbonization of simple cheap sugar), were reported to have a good performance in the FFA esterification (Takagaki et al., 2006). These carbon-based acids are thermally stable up to 230°C, and are characterized by very low surface area (1-2 $m^2\ g^{-1}$) and amorphous structure. Their high acid strength, due to the electron-withdrawing capacity of the polycyclic aromatic rings, besides to the surface hydrophobicity, makes these catalysts highly suitable for FFA esterification in oils (Goodwin Jr. et al., 2005).

Mixed zinc and aluminium oxide (Bournay et al., 2005) is an inorganic material industrially adopted in the Hepsterfip-H technology, developed by the Institute Français du Petrol and used in a plant producing 160000 t/y started up in 2006 (Santacesaria et al., 2008). In this case the range of the operating temperature is 200-250 °C.

The ion-exchange resins are characterized by a gel structure of microsphere that forms a macroporous polymer (generally copolymers of divinylbenzene and styrene) with sulfonic Brønsted acid groups as active sites. Due to their polymeric matrix, such materials have limited thermal stability (< 140°C) and low structural integrity at high pressure. Their swelling capacity controls substrate accessibility to the acid sites and for some kinds of reactor the effective operating volume of the catalytic bed. Once swelled in a polar medium, such as methanol, the resins pores are able to become macropores, so contributing to reduce the diffusive limitations in the working conditions. Recent studies dealing with the use of acid ion exchange resins demonstrated the possibility to obtain excellent results in FFA esterification in mild temperature and pressure conditions, as reported in the following papers: (Santacesaria et. al., 2005; Pirola et al., 2010) (T= 85°C) and (Bianchi et al., 2010) (T = 65°C). The total pressure inside the system is given by the methanol vapour pressure at the reaction temperature.

Several kinds of ion-exchange resins are commercially available from various producers and differ to each other for what concerns acidity strength, surface area, porosity, swelling, characteristics and disposal of acid groups. In Table 2, some features of a series of Amberlysts by Dow Chemical ® and D5081 resin by Purolite are reported.

A distinguishing feature of A46 and D5081 is represented by the location of the active acid sites: these catalysts are in fact sulphonated only on their surface and not inside the pores. Consequently, A46 and D5081 are characterized by a smaller number of acid sites per gram if compared to other Amberlysts®, which are also internally sulphonated.

Catalyst	A15d	A36d	A39w	A40w	A46w	A70w	D5081
Surface area (m^2/g)	53	33	32	33	75	36	n.d.
Average pore diameter (Å)	300	240	230	170	235	220	n.d.
Total pore volume (cc/g)	0.40	0.20	0.20	0.15	0.15	0.20	n.d.
Acidity (meq H+/g)	4.7	5.4	5.0	2.2	0.43	2.55	1.00
Max. operating temperature (K)	393	423	403	413	393	463	403

[a]: Nitrogen BET; [b]: Dry weight

Table 2. Characteristics of some ion-exchange resins (Amberlyst® - Dow Chemical).

The main advantage represented by the use of these catalysts lies in the possibility of adopting very mild reaction conditions. In particular, working at temperatures lower than the methanol boiling point (64,7°C), FFA esterification can be performed without overpressure. In this way no expensive and complex plants are required, making this technology adaptable also for little biodiesel manufacturers. Another interesting aspect of these catalysts is their small deactivation even after long operating periods. In fact, if no particular critical conditions are present in the system during the process (e.g. mechanical fragmentation of the catalyst (Pirola et al., 2010) or presence of metallic ions as Fe^{3+} in the starting TG (Tesser et al., 2010)), no remarkable diminution of the catalytic performance is observed for several operating hours.

Different types of reactors exploiting these ion-exchange resins have been proposed for FFA esterification (Santacesaria et al., 2007; Pirola et al., 2010). The most studied system is a slurry configuration reactor.

The main drawback of the slurry system lies in the fragmentation of the catalyst's particles due to their collision one against the other and against the inner reactor's walls (Pirola et al., 2010).

Alternatives to the slurry reactor are the PFR (Plug Flow) reactor, the Carberry-type reactor, the chromatographic reactor or spray tower loop reactor.

4. Experimental part

4.1 Oil characterization

Oil characterization before proceeding with the standardization of the raw material is a very important issue. Some properties remain in fact unchanged from the starting material to the finished biodiesel, or they are anyway predetermined. It is so important to check that the values of such chemical and physical oil properties are in range with those required by the standard regulations (see Table 3).

The experimental procedures to get the values of such properties are also standardized and are indicated in the regulations. The following are parameters for starting oil that can affect the quality of the final biodiesel.

- Sulfur and Phosphorous content:

High sulphur fuels cause greater engine wear and in particular shorten the life of the catalyst. Biodiesel derived from soybean oil, as well as from pure rapeseed oil, is known to contain virtually no sulphur (Radich, 2004; Zhiyuan et al., 2008).

The phosphorus content of the vegetable oil depends mainly on the grade of refined oil and arises mainly from phospholipids within the starting material. Measurement of the SO_2 from sulphur is accomplished by ultraviolet fluorescence (ASTM D5453, 2002), whereas the analytical method to determine phosphorous requires an Inductively Coupled Plasma Atomic Emission Spectrometry (ASTM International, 2002).

- Linoleic acid methyl ester and polyunsaturated methyl esters

Soy, sunflower, cottonseed and maize oils contain a high proportion of linoleic fatty acids, so affecting the properties of the derived ester with a low melting point and cetane number. Quantitative determination of linoleic acid methyl ester is accomplished by gas chromatography with the use of an internal standard after the substrate has been transesterificated and allows also the quantification of the other acid methyl esters (Environment Australia, 2003).
A typical fatty acid methyl esters composition of soybean oil and other feedstock oils is given in Tab. 1, paragraph 2.

- Iodine Value

The iodine value (IV) is an index of the number of double bonds in biodiesel, and therefore is a parameter that quantifies the degree of unsaturation of biodiesel. Both EN and ASTM standard methods measure the IV by addition of an iodine/chlorine reagent.
Soybean oil is reported to have an IV ranging from about 117 to 143 (Knothe, 1997), having quite the same unsaturation level of sunflower oil.

- Cold Filter Plugging Point

The cold filter plugging point (CFPP) is the temperature at which wax crystals precipitate out of the fuel and plug equipment filters. At temperatures above this point, the fuel should give trouble free flow. These limits are to be decided by each EU member state according to its climate conditions, whereas the US ASTM D 6751 does not set any limit.
The test requires that the sample is cooled and, at intervals of fixed temperature, is drawn through a standard filter so determining the temperature at which the fuel is no longer filterable within a specified time limit.
The CFPP of soybean oil is reported to be around -5°C (Georgianni et al., 2007; Ramos et al., 2009), i.e. accomplishing only a part of the EU members countries (Meher et al., 2006).

- Cetane Number

The cetane number (CN) measures the readiness of a fuel to auto-ignite when injected into the engine. It is also an indication of the smoothness of combustion. The CN of biodiesel depends on the distribution of fatty acids in the original oil. The CN determination is accomplished with the use a diesel engine called *Cooperative Fuel Research* (CFR) engine, under standard test conditions. The CFPP of soybean oil is reported to be higher than 50 (Ramos et al., 2009), so matching in the most cases the limit required by both EN and ASTM biodiesel standards.

4.2 Oil standardization: the esterification reaction

As already remarked in paragraph 3, pre-esterification of FFA in oils assumes great importance to obtain a feedstock suitable to be processed in the transesterification reaction.
In the recent years the authors have deepened the study of the pre-esterification process investigating the effect of the use of different kinds of oils, different types of reactors and catalysts and different operating conditions.
In the following paragraphs, the most relevant aspects of the experimental work and the results obtained by the authors for what concerns the pre-esterification process are reported.

Specification	Units	limits		Method
		Min	Max	
Ester content	% (m/m)	96.5		EN 14103
Density 15°C	kg/m^3	860	900	EN ISO 3675 EN ISO 12185
Viscosity 40°C	mm^2/s	3.50	5.00	EN ISO 3104
Sulphur	mg/kg	-	10.0	preEN ISO 20846 preEN ISO 20884
Carbon residue (10% dist.residue)	% (m/m)	-	0.30	EN ISO 10370
Cetane number		51.0		EN ISO 5165
Sulphated ash	% (m/m)	-	0.02	ISO 3987
Water	mg/kg	-	500	EN ISO 12937
Total contamination	mg/kg	-	24	EN 12662
Cu corrosion max		-		EN ISO 2160
Oxidation stability, 110°C	h (hours)	6.0		EN 14112
Acid value	mg KOH/g	-	0.5	EN 14104
Iodine value	gr I_2/100 gr	-	120	EN 14111
Linoleic acid ME	% (m/m)	-	12.0	EN 14103
Methanol	% (m/m)	-	0.20	EN 14110
Monoglyceride	% (m/m)	-	0.80	EN 14105
Diglyceride	% (m/m)	-	0.20	EN 14105
Triglyceride	% (m/m)	-	0.20	EN 14105
Free glycerol	% (m/m)	-	0.02	EN 14105
Total glycerol	% (m/m)	-	0.25	EN 14105
GpI metals (Na+K)	mg/kg	-	5.0	EN 14108 EN14109
Gp II metals (Ca+Mg)	mg/kg	-	5.0	EN14538
Phosphorous	mg/kg	-	5.0	EN 14538

Table 3. Standard specifications for biodiesel (automotive fuels).

In the following table (Table 4) the IV obtained by the authors using the standard procedure are listed for different kinds of not refined feedstock.

Oilseed	Iodine Value (g I_2/100 g fat)
Brassica juncea (Indian mustard)	111
Brassica napus (Rapeseed)	115
Cartamus tinctorius (Safflower)	109
Heliantus annus (Sanflower)	143
Nicotiana tabacum (Tobacco)	137
Waste Cooking Oil	54.0

Table 4. Iodine values of some potential feedstock for biodiesel production.

4.2.1 General reaction conditions

A remarkable aspect of the proposed process is represented by the mild operative conditions, i.e. low temperature (between 303 and 338 K) and atmospheric pressure.

Each single reaction has been carried out for six hours withdrawing samples from the reactor at pre-established times and analysing them through titration with KOH 0.1 M. The percentage of FFA content per weight was calculated as otherwise reported (Marchetti & Errazu, 2007, Pirola et al. 2010).

Unless otherwise specified, all the esterification experiments have been conducted using a slurry reactor as the one represented in Fig. 2a. A slurry reactor is the simplest type of catalytic reactor, in which the catalyst is suspended in the mass of the regents thanks to the agitation. In Fig. 2b a typical kinetic curve for the esterification reaction performed with soybean oil is displayed.

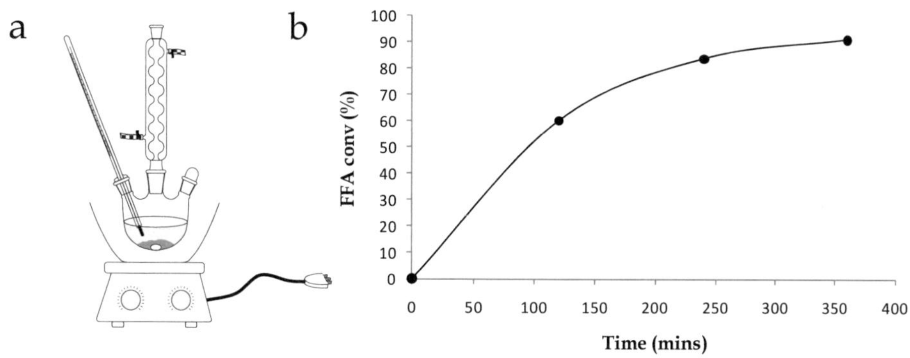

Fig. 2. a) Scheme of the slurry reactor; b) Example of kinetic curve of crude soybean oil with initial acidity=5% (wt.): FFA conversion (%) vs. time, slurry reactor, T=338K, catalyst: Amberlyst® 46, weight ratio alcohol/oil= 16:100, weight ratio catalyst/oil=1:10.

Much attention has been paid by the authors to the use of acid ion exchange resins. Amberlysts ®, i.e. a commercial product by Dow Advanced Materials, and D5081, a catalyst at the laboratory development stage by Purolite® have been successfully applied in this reaction. The main features of the employed catalysts are reported in Tab. 2 and described in paragraph 3.

4.2.2 Effect of the use of different kinds of oil

In Fig. 3 the results from the esterification reaction of different starting oils are shown.

From the graph it can be noted that the lowest final acidity values are obtained with the refined materials, in spite of their initial acidities are the highest due to the addition of pure oleic acid. Refined oils are undoubtedly more easily processable with the esterification in comparison to crude oils, probably due to their lower viscosity which does not result in limitations to the mass transfer of the reagents towards catalysts.

This result has been confirmed by the addition of rapeseed oil, less viscous, to the waste cooking oil in different ratios: increasing the ratio of rapeseed oil to waste oil, the FFA conversion after 6 hours increases.

The differences in the acidic composition seem not to affect the yield of the reaction; in fact, similar values of FFA conversions are obtained for both the soybean oil and the animal fat, in spite of their different acidic compositions. Indicative compositions of some oils used in the experimentation are given in Tab.1

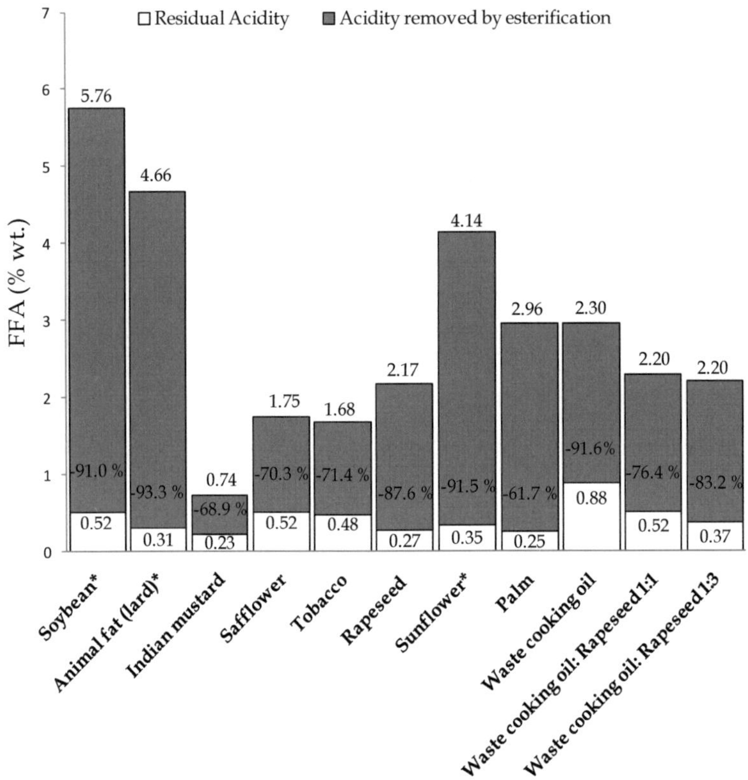

Fig. 3. Acidity removed by esterification (6 h) and residual acidity of different oils used as raw material: slurry reactor, T=338K, catalyst: Amberlyst® 46 weight ratio alcohol/oil= 16:100, weight ratio catalyst/oil=1:10; *commercial, refined fats with the addition of pure oleic acid.

4.2.3 Comparison among different catalysts at different loadings and temperatures

A comparison among the different kinds of Amberlyst at different temperatures has been performed by the authors in a recently published paper dealing with the de-acidification of animal fat (Bianchi et al., 2010). In Fig. 4 the results of this study are summarized.

Under the applied conditions, all the catalysts perform quite well in the esterification reaction, with the exception of A40. Its unsatisfactory performance can be explained taking into account its lower specific surface area and a lower acid site concentration if compared to other Amberlysts. Being these two parameters directly connected to catalytic activity, their simultaneous deficiency is clearly the cause of the unsatisfactory performance.

The catalytic performances of the sample A46 appear to be remarkable, in spite of the low concentration of active acid sites. This result can be explained considering the particular configuration of the catalytic particles, where acid sites are located only on its surface, thus being immediately and easily available for the reaction.

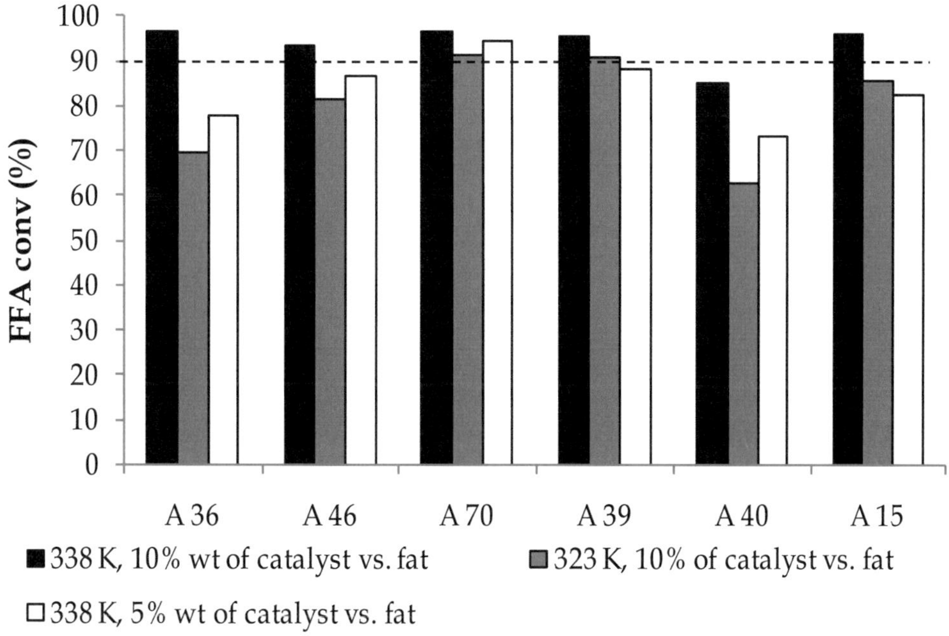

Fig. 4. FFA conversion (%) after 6 h. Comparison of Amberlysts in different operative conditions, initial acidity=5%, slurry reactor, T=338K, weight ratio alcohol/oil= 16:100, weight ratio catalyst/oil=1:10. The dotted line represents the value of FFA conversion necessary to obtain a feedstock with FFA content 0.5% per weight, i.e. suitable for industrial applications.

A70 shows the best performance in all the operative conditions. For this reason, it was further tested to evaluate its catalytic activity in milder operating conditions, i.e. lower temperatures and lower catalyst/fat ratio. The results thus obtained are displayed in Figs. 5a and b, showing that the activity of the catalyst decreases as the reaction temperature or its concentration decrease. However it is worth remarking that even at room temperature.

Catalysts A46 and D5081, have been compared at different temperatures and catalyst's loadings. The results of this study are summarized in the Figs. 6 a and b.

As can be seen from the graphs, catalyst D5081 shows better results than A46 in milder operative conditions. This can be easily explained by the higher number of acid sites located on its surface (compared with Tab. 2). In particular, the use of a ratio of 10% of catalyst D5081 vs. oil allows reaching the maximum conversion in 2 hours. The outcome of this study suggested that a fixed amount of acid active sites per gram of FFA was required to reach the maximum of conversion in 4 hours. Based on the experimental data, this amount was found to be equal to 1,2 meq of H^+.

To verify this hypothesis, different batches of sunflower oil with different initial acidity were prepared and then de-acidified by loading a quantity of A46 corresponding to 1,2 meq of H^+ per gram of FFA. The obtained results are shown in Fig. 7 and confirm the hypothesis set out above. In fact, a complete conversion, corresponding to a FFA concentration lower than 0,5%, is reached after 4 hours regardless of the initial FFA amount.

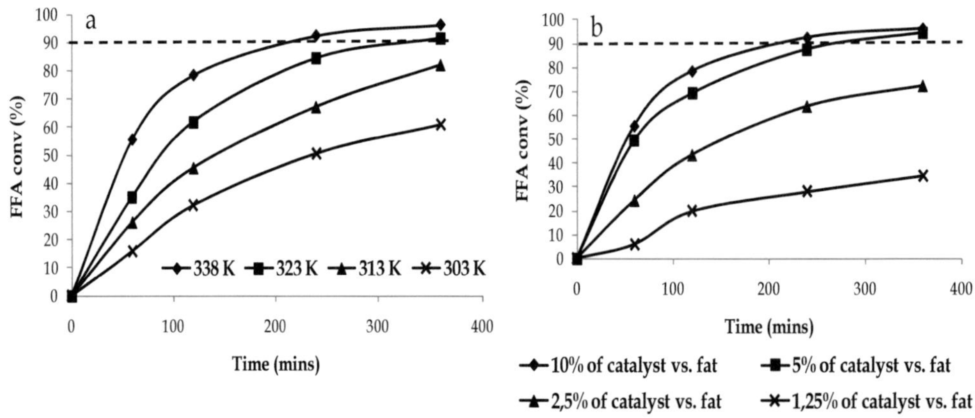

Fig. 5. a) FFA conversion (%) vs reaction time in presence of A70 for different reaction temperatures, initial acidity=5%, slurry reactor, 10 % per weight (wt) of catalyst vs. fat, weight ratio alcohol/oil= 16:100; b) FFA conversion (%) vs. time in presence of A70, slurry reactor, weight ratio alcohol/oil= 16:100, T = 338K with different amounts of the catalysts. The dotted line represents the value of FFA conversion necessary to obtain a feedstock with a FFA content 0.5% per weight, i.e. suitable for industrial applications.

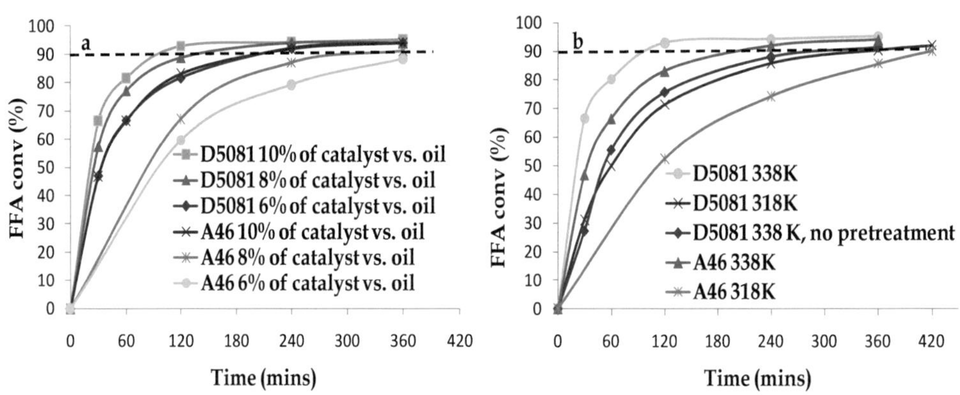

Fig. 6. a) FFA conversion (%) vs reaction time for different amounts of catalysts A46 and D5081, rapeseed oil with initial acidity=5%, slurry reactor, weight ratio alcohol/oil= 16:100, T=338K; b) FFA conversion (%) vs. time at different temperatures, slurry reactor, 10 % per weight (wt) of catalyst vs. fat, weight ratio alcohol/oil= 16:100, T = 338K. The dotted line represents the value of FFA conversion necessary to obtain a feedstock with a FFA content 0.5% per weight, i.e. suitable for industrial applications.

4.2.4 Study of catalysts' lifetime

A crucial parameter for the industrial application is the catalyst lifetime; this parameter has been evaluated by the authors in a recent work (Pirola et al., 2010) by performing ninety consecutive batch de-acidification runs, each lasting 6 hours, were conducted using crude palm oil or soybean oil as a feedstock and Amberlyst® 46 as a catalyst. The final FFA

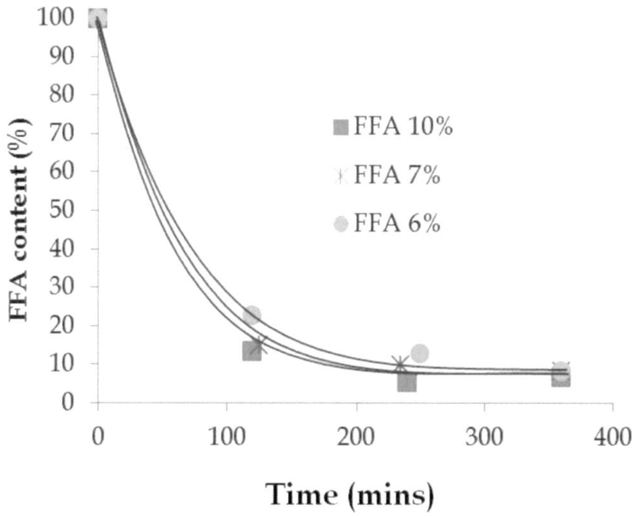

Fig. 7. FFA content (%) vs. reaction time: kinetic curves of the FFA esterification in sunflower oil with different initial acidities using a fixed catalyst/FFA ratio, slurry reactor, weight ratio alcohol/oil= 16:100, T = 338K.

conversions, measured at the end of each of the 6-hour reactions, are reported in Fig. 8 as a function of the run number. At the end of the recycles, a decrease of activity of about 25% was observed, to be probably ascribed to some fragmentation of catalyst's particles. For further details, (Pirola et. al., 2010).

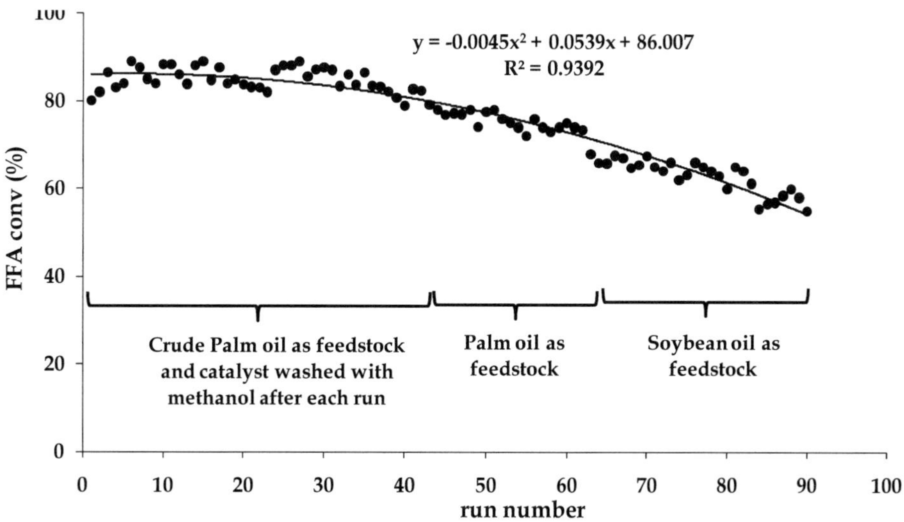

Fig. 8. FFA conversion % after 6 h during 90 successive runs with the same Amberlyst 46® sample. Slurry reactor, T= 338 K, weight ratio alcohol/oil= 16:100, weight ratio catalyst/oil=1:10.

The recycle of the use of catalyst was also performed for D5081 in the FFA esterification of rapeseed oil. The obtained results are outlined in the following graphs (Figs. 9a and b).

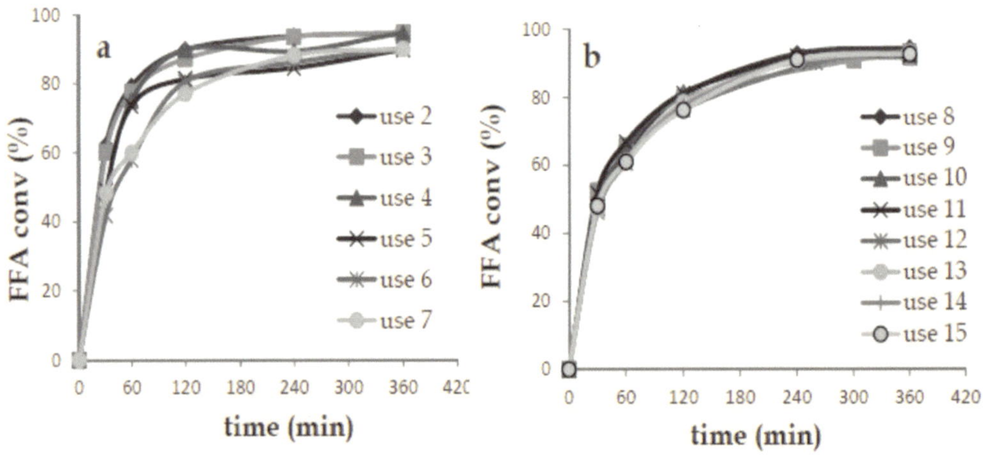

Fig. 9. FFA conversion % vs. reaction time of a) recycles 2 to 7 and b) 8 to 15, rapeseed oil, initial acidity=5% (oleic acid), slurry reactor, T = 338 K, weight ratio alcohol/oil= 16:100, weight ratio catalyst/oil=1:10.

In the initial stage (use 2÷7) the resin does not results in well defined kinetic curves. The reason of slight diminution of FFA conversion with time is probably ascribable to catalyst's settling in the system as it has to adapt to the ambient of reaction before giving a stable performance. From the 8th recycle of catalyst's use on, the curves of FFA conversion overlap: the conversion reached after pre-established times is the same for different runs with the use of the same batch of catalyst.

4.2.5 Reactors

The experimental results discussed in the previous paragraph suggested that a packed-bed reactor, where the catalyst particles are immobilized inside it, could eliminate the mechanical stress of the catalyst particles typical of a slurry reactor. On the other hand, a packed-bed reactor makes the contact between the organic phase (oil/FFA mixture) and methanol less effective. For this reason, in the employed experimental setup a mixing chamber was located just before the catalytic reactor.

The reaction in both continuous and semicontinuous modes was conducted using the experimental setup shown in Fig. 10.

The methanol/oil mixture is taken from the vessel (a), where it is continuously mechanically stirred (b), and then it is admitted into the mixing chamber (d) by a pump (c). This chamber (0.2 L) is located just before the catalytic reactor in order to obtain the maximum contact between oil and methanol (not fully soluble) inside the catalytic bed (e). The catalytic reactor (0.5 L) contains a packed bed of Amberlyst 46® (7 g). The pump flow is maintained at 10 mL min^{-1}, so obtaining a contact time in the catalytic bed equal to 1 min.

In the semi-continuous experiments the reaction stream, leaving the catalytic reactor (e) was returned to the vessel (a); in continuous experiments the reaction stream from the catalytic reactor (e) was continuously discharged from the system.

In Fig. 11, the results obtained for both crude palm oil and soybean oil are reported using the experimental setup shown in Fig. 11 as a semicontinuous reactor.

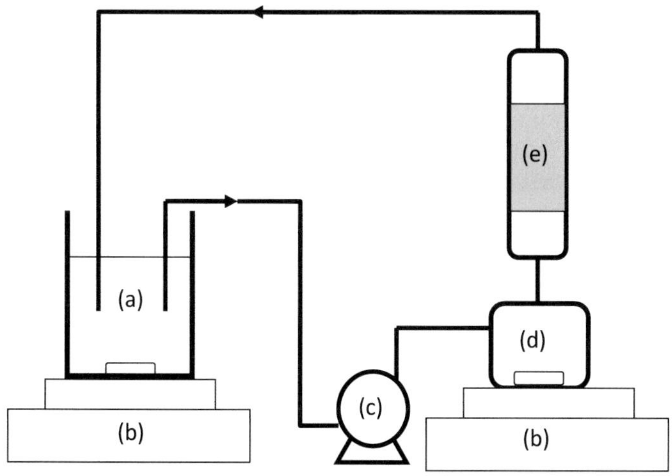

Fig. 10. Continuous or semi-continuous experimental set-up: (a) feeding vessel, (b) mechanic stirrer, (c) pump, (d) mixing chamber, (e) catalytic packed bed reactor.

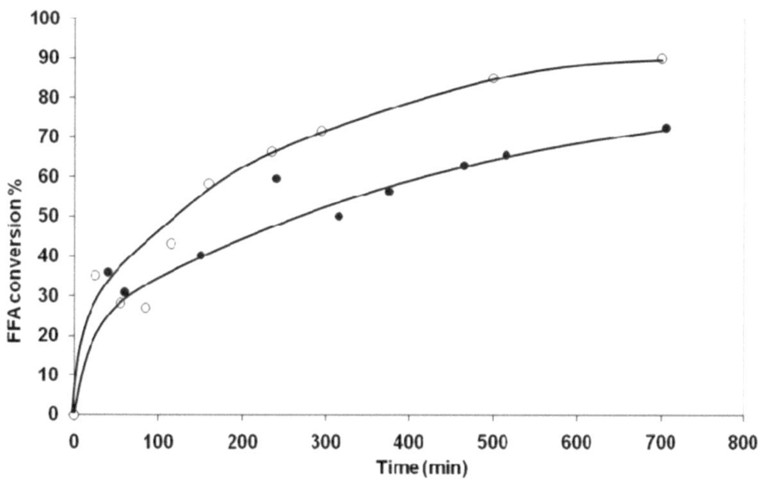

Fig. 11. Semi-continuous experiments with 7 g of catalyst; FFA conversion % vs. time using crude palm oil (o) or soybean oil (•). Pump flow = 10 mL min^{-1}; T = 338 K, molar ratio FFA/alcohol = 1: 6, weight ratio catalyst/ oil = 1: 10.

The FFA conversion increases with time using both oils, but it is higher using the crude palm oil (FFA conversion after 700 min: about 90% and 70% for palm and soybean oil, respectively). This result can be explained considering the difference between the two different raw materials, which concerns both the composition of the substrate (different unsaturation levels between the two oils) and the composition of the FFA (only oleic acid for soybean oil and a mixture of C_{12}-C_{18} acids for crude palm oil). These differences obviously affect both the lifetime of the oil/methanol emulsion and the diffusional aspects along the catalytic bed.

To improve the stability of the methanol/oil emulsion, the authors substituted the classical mixing chamber with an emulsificator based on five co-axial rotating ring gears which are able to break the biphasic mixture into very tiny drops. Using this device and starting from two entirely separated liquid phases (oil and methanol) it was possible to obtain a much more stable emulsion. In Fig. 12 a comparison between FFA conversion obtained with the classical mixing chamber and the emulsificator is reported: the better results reached using the emulsificator are evident.

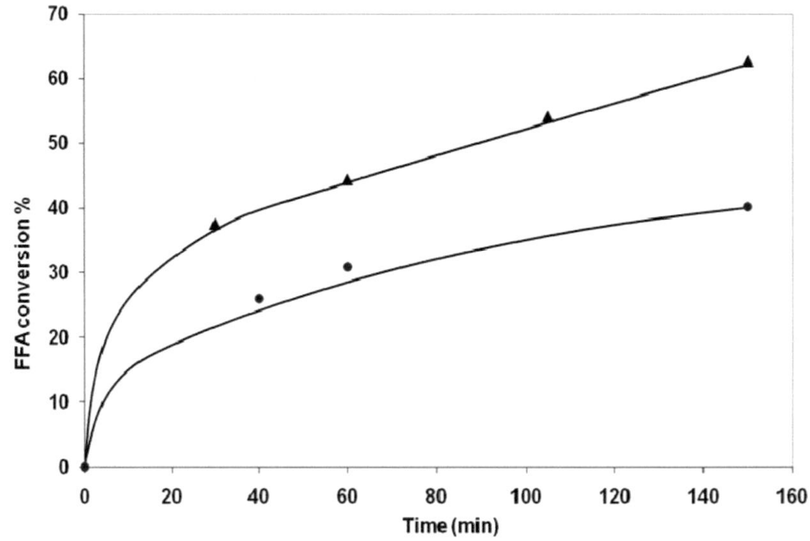

Fig. 12. Semi-continuous experiment: FFA conversion % vs. time using crude palm oil as feedstock and having before the catalytic bed: classical mechanic stirring reactor (•) or emulsificator at 500 rpm (▲). Pump flow = 10 mL min^{-1}; T = 338 K, molar ratio FFA: alcohol = 1: 6, weight ratio catalyst: oil = 1: 10.

The methanol/oil emulsion lifetime without the emulsificator device, measured in the same way at the exit of the mixing chamber, is of about 15 seconds. Referring to our system, the contact time between the catalyst and the reactant is 1 min for 7 g of catalyst (pump flow = 10 mL min^{-1}): for this reason without the emulsificator device a significant part of the catalytic bed (about 75%) does not work.

4.3 Oil transformation: the transesterification reaction and biodiesel characterization

The transesterification reaction has been performed by the authors on the raw materials de-acidified with the esterification process described in the previous paragraph.
Sodium methoxide (MeONa) was employed as catalyst. MeONa is known to be the most active catalyst for triglycerides transesterification reaction, but it requires the total absence of water (Schuchardt, 1996). For this reason, the unreacted methanol and the reaction water were evaporated from the de-acidified oils before processing them with the transesterification reaction.
The employed experimental setup was the same as displayed in Fig. 2.

Being the transesterification an equilibrium reaction, it was performed in two steps, removing the formed glycerine after the first step. The adopted conditions were the following:

- 1st step: weight ratio methanol/oil=20:100, weight ratio MeONa/oil=1:100, 233 K, 1,5 h
- 2nd step: weight ratio methanol/oil=5:100, weight ratio MeONa/oil=0.5:100, 233 K, 1 h

The total ester content is a measure of the completeness of the transesterification reaction. Many are the factors that affect ester yield in the transesterification reaction: molar ratios of glycerides to alcohol, type of catalyst(s) used, reaction conditions, water content, FFA concentration, etc. (Environment Australia, 2003).

The European preEN14214 biodiesel standard sets a minimum limit for ester content of >96.5% mass, whereas the US ASTM D 6751 biodiesel standard does not set a specification for ester content.

Mono- and di-glycerides as well as tri-glycerides can remain in the final product in small quantities. Most are generally reacted or concentrated in the glycerine phase and separated from the ester.

The analyses of methyl esters and unreacted mono-, di- and triglycerides are accomplished through gas chromatography.

The detailed requirements for biodiesel according to both EN 14214 and US ASTM D 6751 are listed in paragraph 1.

5. Simulation

In order to develop a process simulation of the FFA esterification, able to predict the reaction progress, a thermodynamic and kinetic analysis was performed.

5.1 Thermodynamic aspects

The considered reaction system turns out to be a highly non-ideal system, being formed by a mixture of oil, methylester, methanol, FFA and water. The interactions among these molecules are absolutely not ideal, in fact they are only partially soluble and a two phase system is formed if the quantity of methanol is greater than 6-8 wt%.

Indeed, the activity coefficients are used not only for the phase and chemical equilibria calculations, but also for the kinetic expressions. Modified UNIFAC model was used adopting the parameters available in literature and published by Gmehling et al. (2002).

5.2 Reaction kinetics

The considered reaction is the following (Fig.13).

$$\text{FFA} + \text{methanol} \rightleftharpoons \text{methylester} + \text{water}$$

Fig. 13. FFA esterification reaction and hydrolysis (indirect reaction) considered in the process simulation.

The oil is considered as non-reacting solvent, being present in large quantity.

A pseudohomogeneous model was used for describing the kinetic behaviour of the reaction (Pöpken et al., 2000). The adopted model is displayed in the following equation:

$$r = \frac{1}{m_{cat}} \frac{1}{\upsilon_i} \frac{dn_i}{dt} = k_1 a_{FFA} a_{methanol} - k_{-1} a_{methylester} a_{water}$$

where:
r= reaction rate
m_{cat}= dry mass of catalyst, gr
υ_i= stoichiometric coefficients of component i
n_1= moles of component i
t = reaction time
k_1= kinetic constant of direct reaction
k_{-1}= kinetic constant of indirect reaction
a_i= activity of component i
The temperature dependence of the rate constant is expressed by the Arrhenius law:

$$k_i = k_i^0 \exp\left(\frac{-E_{a,i}}{RT}\right)$$

where k_i^0 and $E_{a,i}$ are the pre-exponential factor and the activation energy of the reaction i, respectively (i=1 for the direct reaction, i=-1 for the indirect reaction), T is the absolute temperature and R the Universal Gas Constant.
The adopted parameters set is the same reported by Steinigeweg (Steinigeweg & Gmehling, 2003). The absolute values of pre-exponential factors were corrected as reported in Table 5, so to take into account the presence of both a second liquid phase and a different type of catalyst.

Reaction (i)	E_A (kJ mol^{-1})
Esterification (1)	68.71
Hydrolisis (-1)	64.66

Table 5. Kinetic Parameters for the adopted pseudohomogeneous Kinetic Model.

The ratio of pre-exponential factors is: $K_{eq}^T = k_1^0/k_{-1}^0 = 60.7841$.
All the simulations were carried using Batch Reactor of PRO II by Simsci – Esscor.
In the next picture (Fig. 14) the comparison between the experimental and calculated FFA conversion is shown for the esterification reaction of an acid rapeseed oil.
The initial composition (% by weight) entered in the simulation program is: oil (83.2%), FFA (3.47 %), methanol (13.3%).
The model turned out to be able to reproduce qualitatively the behaviour of different systems, characterized by different starting acidities values, at different temperatures and with an high impurities content.

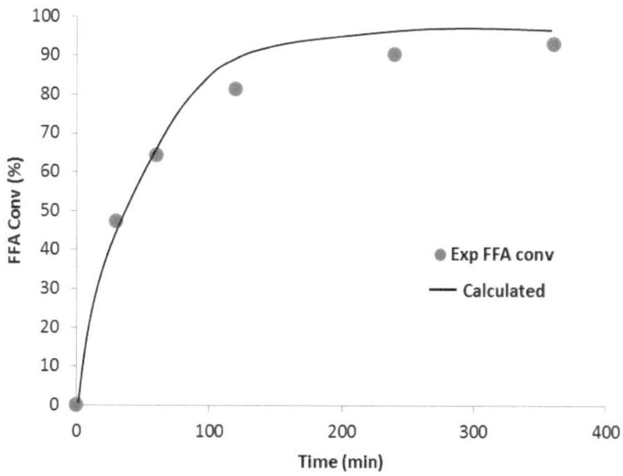

Fig. 14. Experimental and calculated (pseudohomogeneous model) FFA conversion % vs. reaction time, rapeseed oil, initial acidity= 3.47% (oleic acid), slurry reactor, T= 338 K, weight ratio alcohol/oil=16:100, weight ratio catalyst/oil=1:10.

6. Conclusions

The use of not refined or waste oils as a feedstock represents a very convenient way in order to lower biodiesel production costs. The main problem associated with the use of this type of low-cost feedstock lies in its high content of FFA, leading to the formation of soaps during the final transesterification step. These materials require therefore to be standardized by the reduction of their acidity and different de-acidification methods have been described in this chapter. Among them, a new technology based on an esterification reaction heterogeneously catalyzed and performed at mild operative conditions, i.e. low temperature (between 303 and 338 K) and atmospheric pressure has been proposed and described. Several kinds of ion-exchange resins, commercially available, have been used as heterogeneous catalysts, different one form the other for what concerns acidity strength, surface area, porosity, swelling, characteristics and disposal of acid groups.

The experimental tests were performed using different reactors (CSTR or PFR), starting oils (in comparison with the results obtained for soybean oil), catalyst/oil ratio and working temperature. All these experimental parameters have been optimized in order to obtain, at the end of the reaction, a concentration of FFA suitable for the transesterification reaction for the biodiesel production (FFA < 0.5 wt%). A crucial parameter for the industrial application is the catalyst lifetime and this parameter has been evaluated by performing ninety consecutive batch de-acidification runs, each lasting 6 hours, with the same catalyst sample (Amberlyst 46®) in a slurry reactor. At the end of the recycles, a decrease of activity of about 25% was observed, to be ascribed to some fragmentation of catalyst's particles, that collide against one another and against the reactor walls. To overcome this problem a packed bed configuration have been adopted and optimized. At last, a process simulation of the FFA esterification, able to predict the reaction progress, through a thermodynamic and kinetic analysis, was successfully performed. A pseudohomogeneous model was used for describing the kinetic behaviour of the reaction, using a modified UNIFAC model for the calculation of the activity coefficients (used not only for the phase and chemical equilibria calculations, but also for the kinetic expressions).

7. Acknowledgements

The authors gratefully acknowledge the financial support by the Italian Ministero delle Politiche Agricole, Alimentari e Forestali (project SUSBIOFUEL - D.M. 27800/7303/09).

8. References

Allen, C.A.W. Watts, K.C., Ackman, R.G. & Pegg M.J. (1999). Predicting the Viscosity of Biodiesel Fuels from their Fatty Aid Ester Composition. *Fuel*, Vol. 78, (September 1999), pp. 1319-1326

ASTM International (2002). Standard Specification for Biodiesel Fuel - Blend Stock (B100) for Distillate Fuels - ASTM D 6751-02, in *Annual Book of ASTM Standards*, Vol. 05.04

ASTM International (2002). Determination of Total Sulfur in Light Hydrocarbons, Motor Fuels and Oils by Ultraviolet Fluorescence Method D5453 - 00 (2002), in *Annual Book of ASTM Standards*, Vol. 05.03

Bianchi, C.L., Boffito, D.C., Pirola, C., Ragaini, V. (2010). Low temperature de-acidification process of animal fat as a pre-step to biodiesel production. *Catalysis Letters*, Vol. 134, (November, 2009), pp. 179-183.

Bianchi, C.L., Ragaini, V., Pirola, C., Carvoli, G. (2003). A new method to clean industrial water from acetic acid via esterification, *Applied Catalysis B Environmental*, Vol. 40, (June 2003), pp. 93-99.

Bournay, L., Casanave, D., Delfort, B., Hillion, G., Chodorge, J.A., (2005). New heterogeneous process for biodiesel production: a way to improve the quality and the value of the crude glycerin produced by biodiesel plants. *Catalysis Today*, Vol. 106, (2005). pp. 190-192.

Canakci, M., Van Gerpen, J. (1999). Biodiesel production via acid catalysis. *Trans ASAE*, Vol. 42, (1999), pp. 1203-1210.

Colucci, J.A., Borrero, E., Alape, F. (2005). Biodiesel from an alkaline transesterification reaction of soybean oil using uyltrasonic mixing. *Journal of the American Oil Chemists' Society,*, Vol. 82, (2005), pp. 525-530.

Corma, A., Garcia, H. (1997). Organic reactions catalyzed over solid acids. *Catalysis Today*, Vol. 38, (April 1997), pp. 257-308. ISNN 0920-5861

Di Serio, M., Tesser, R., Pengmei, L., Santacesaria, E. (2008). Heterogeneous catalysis for biodiesel production. *Energy & Fuel*, Vol. 22. (September 2008), pp. 207-217.

Environment Australia (2003). National Standards for Biodiesel - Discussion Paper, In *Setting National Fuels Quality Standards*, 0 642 54908 7 pp. 1-196, Retrieved from <www.ea.gov.au/atmosphere/transport/biodiesel/index.html>

Erickson, D.R., Alkaline Refining. (1995). *Practical Handbook of Soybean Processing and Utilization*, Erickson, D.R., ed, AOCS Press, Champlain, Illinois, 1995.

Ganquli, K.L. ,van Immersel, A.R., Michaelides, G.C., van Putte, K.P., Turksma, H., (1998), U. S Pat. Nos. 1,371,342

Georgianni, K.G. Kontominas, M.G. Tegou, E. Avlonitis, D. & Gergis V. (2007). Biodiesel Production: Reaction and Process Parameters of Alkali-Catalyzed Transesterification of Waste Frying Oils. *Energy Fuels*, Vol 21, No. 5, pp. 3023-3027

Gmehling, J. Wittig, R. Lohmann, J. Joh, R. (2002). A Modified UNIFAC (Dortmund) Model. 4. Revision and Extension. *Industrial & Engineering Chemistry Research*, Vol. 41, pp. 1678-1688.

Huayang, H., Tao, W., Shenlin, Z. (2007). Continuous production of biodiesel fuel from vegetable oil using supercritical methanol process. *Fuel,* Vol. 86, (2007), pp. 442-447.

Knothe, G. Dunn, R.O. & Bagby M.O. (1997). Biodiesel: the Use of Vegetable Oils and their Derivates as Aternative Fuels, In *Fuels and Chemicals from Biomass*, Badal, C. Saha, J., pp. 172-208, American Chemical Society, , ACS Symposium Series, Retrieved from <http://pubs.acs.org/doi/abs/10.1021/bk-1997-0666.ch010>

Leonelli, C., Mason, T.J. (2010). Microwave and ultrasonic processing : Now a realistic option for industry. *Chemical Engineering and Processing*, Vol. 49, (June 2010), pp. 885-900.

Liu, K.S. (1994). Preparation of Fatty-Acid Methyl Esters for Gas-Chromatographic Analysis of Lipids in biological materials. *J. Am. Oil Chem Soc.*, Vol. 71 (1994), pp. 1179-1187.

Lotero, E., Liu, Y., Lopez, D.E., Suwannakarn, K., Bruce, D.A., Goodwin Jr., J.G. (2005). Synthesis of Biodiesel via acid Catalysis. *Industrial & Engineering Chemistry Research,* Vol. 44, (January 2005), pp. 5353-5363.

Marchetti, J.M. & Errazu, A.F. (2008) Comparison of Different Heterogeneous Catalysts and Different Alcohols for the Esterification Reaction of Oleic Acid. *Fuel*, Vol. 87, (June 2008), pp. 3477-3480, ISSN 0016-2361

Meher, L.C., Vidaya Sagar, D. & Naik S.N. (2006). Technical Aspects of Biodiesel Production by transesterification - a Review. *Renewable and Sustainable Energy Reviews*, Vol. 10, (2006), pp. 248-268

Moser, B.R. Haas, M.J. Jackson M.A. Erhan S.Z. & List G.R. (2007). Evaluation of Partially Hydrogenated Methyl Esters of Soybean Oil as Biodiesel. *European Journal of Lipid Science and Technology*, Vol. 109, (2007), pp. 17-24

Öezbay, N. Oktar, N. &. Tapan, N. (2008). Esterification of Free Fatty Acids in Waste Cooking Oils (WCO): Role of Ion-exchange Resins. *Fuel*, Vol. 87, No. 10-11, pp. 1789-1798

Pirola, C., Bianchi, C.L., Boffito, D.C., Carvoli, G., Ragaini, V. (2010). Vegetable oil deacidification by Amberlyst : study of catalyst lifetime and a suitable reactor configuration. *Industrial & Engineering Chemistry Research,* Vol. 49 (2010), pp. 4601-4606.

Pöpken, T. Götze, L. Gmehling, J. (2000). Reaction Kinetics and Chemical Equilibrium of Homogeneously and Heterogeneously Catalyzed Acetic Acid Esterification with Methanol and Methyl Acetate Hydrolysis. *Industrial & Engineering Chemistry Research*, Vol. 39, pp. 2601-2611.

Prankl H & Worgetter M. (1996). Influence of the Iodine Number of Biodiesel to the Engine Performance. *Proceedings of the Alternative Energy Conference*, American Society of Agricultural Engineers, 1996

Radich, A. (1998). Biodiesel Performance, Costs, and Use. Energy Information Administration Retrieved from <http://www.eia.doe.gov/oiaf/analysispaper/biodiesel/pdf/biodiesel.pdf>

Ramos, M.J. Fernández, C.M. Casas, A. Rodríguez, L. & Pérez, Á. (2009). Influence of Fatty Acid Composition of Raw Materials on Biodiesel Properties. *Bioresource Technology,* Vol. 100, pp. (261-268), ISSN 0960-8524

Santacesaria, E., Tesser, R., Di Serio, M., Guida, M., Gaetano, D., Garcia Agrda, A., Cammarota, F. (2007). Comparison of different reactor configurations for the reduction of free acidity in raw materials for biodiesel productio. *Industrial & Engineering Chemistry Research*, Vol. 46 (March, 2007), pp. 8355-8362.

Schäfer A. (1991). Pflanzenölfettsäure-Methyl-Ester als Dieselkraftstoffe / Mercedes-Benz, *Proceedings of Symposium Kraftstoffe aus Pflanzenöl für Dieselmotoren*, Technische Akademie, Esslingen, Germany, 1991

Schuchardt, U. Vargas, R.M. & Gelbard, G. (1996). Transesterification of soybean oil catalyzed by alkylguanidines heterogenized on different substitude polystyrenes. *Journal of Molecular Catalysis A: Chemical A*, Vol. 109, pp. 1381-1169

Srivastava, A., Prasad, R. (2000). Tryglicerides-based diesel fuel. *Renewable Sustainable Energy Reviews*, Vol. 4, (2000), pp. 111-133.

Stavarache, C., Vinatoru, M., Nishimura, M., Maeda, Y. (2005). Fatty Acids methyl esters from vegetable oil by means of ultrasonic energy. *Ultrasonics Sonochemistry*, Vol. 12, (2005), pp. 525-530.

Steinigeweg, S. & Gmehling, J. (2003). Esterification of a Fatty Acid by Reactive Distillation. *Industrial & Engineering Chemistry Research*, Vol. 42, pp. 3612-3619

Steinigeweg, S. & Gmehling, J. (2003). Esterification of a Fatty Acid by Reactive Distillation. *Industrial & Engineering Chemistry Research*, Vol. 42, pp. 3612-3619.

Strayer, R. C., Blake, J. A., Craig, W. K. (1983). Canola and high erucic rapeseed oil as substitutes for diesel fuel: Preliminary tests. *Journal of the American Oil Chemists Society*, Vol. 60, (1983),pp. 1587-1592.

Sulaiman, A. (2007). Production of biodiesel: possibilities and challenges. *Biofuel Bioprod. Bioref.* Vol. 1, (2007), pp. 57-66.

Takagaki, A., Toda, M., Okamura, M., Kondo, J. N., Hayashi, S., Domen, K., Hara, M. (2006). Esterification of higher fatty acids by a novel strong solid acid. *Catalysis Today*, Vol. 116 (2006), pp. 157-161.

Tate, R.e. Watts, K.C. Allen, C.A.W. & Wilkie K.I. (2006). The Viscosities of Three Biodiesel Fuels at Temperatures up to 300°C. *Fuel*, Vol. 85, pp. 1010-1015, ISSN 0016-2361

Tesser, R., Di Serio, M, Casale, L., Sannino, L., Ledda, M., Santacesaria, E. (2010). Acid exchange resins deactivation in the esterification of free fatty acids. *Chemical Engineering Journal*, Vol. 161, (April, 2010), pp. 212-222.

Tesser, R., Di Serio, M., Guida, M., Nastasi, M., Santacesaria, E. (2005). Kinetics of oleic acid esterification with methanol in the presence of triglycerides. *Industrial & Engineering Chemistry Research*, Vol. 44 (2005), pp. 7978-7982.

Vyas, A.P., Verma, J.L., Subrahmanyam, N. (2010). A review on FAME production processes. *Fuel*, Vol. 89, (August 2009), pp. 1-9

Yuji, S., Yomi, W., Taichi, S., Akio, S., Hideo, N., Hideki, F., Yoshio, T. (1999). Conversion of vegetable oil to biodiesel using immobilized Candida Antarctica Lipase. *Journal of the American Oil Chemists' Society*, Vol. 76, (1999), pp. 789-793.

Zheng, S. Kates, M. Dubé M.A. & McLean D.D. (2006) Acid-catalyzed Production of Biodiesel from Wastre Frying Oil. *Biomass and Bioenergy*, Vol. 30, No 3, pp. 267-272

Zhiyuan, H. Piqiang, T. Xiaoyu & Y. Diming, L. (2008). Life cycle energy, environment and economic assessment of soybean-based biodiesel as an alternative automotive fuel in China. *Energy*, Vol. 33, pp. 1654-1658, ISSN 0360-5442

17

Processing of Soybean Oil into Fuels

Joanna McFarlane
Oak Ridge National Laboratory[1]
USA

1. Introduction

1.1 Rationale for processing soybean oil into a fuel

Abundant and easily refined, petroleum has provided high energy density liquid fuels for a century. However, recent price fluctuations, shortages, and concerns over the long term supply and greenhouse gas emissions have encouraged the development of alternatives to petroleum for liquid transportation fuels (Van Gerpen, Shanks et al. 2004). Plant-based fuels include short chain alcohols, now blended with gasoline, and biodiesels, commonly derived from seed oils. Of plant-derived diesel feedstocks, soybeans yield the most of oil by weight, up to 20% (Mushrush, Willauer et al. 2009), and so have become the primary source of biomass-derived diesel in the United States and Brazil (Lin, Cunshan et al. 2011). Worldwide ester biodiesel production reached over 11,000,000 tons per year in 2008 (Emerging Markets 2008). However, soybean oil cannot be burned directly in modern compression ignition vehicle engines as a direct replacement for diesel fuel because of its physical properties that can lead to clogging of the engine fuel line and problems in the fuel injectors, such as: high viscosity, high flash point, high pour point, high cloud point (where the fuel begins to gel), and high density (Peterson, Cook et al. 2001).

Industrial production of biodiesel from oil of low fatty-acid content often follows homogeneous base-catalyzed transesterification, a sequential reaction of the parent triglyceride with an alcohol, usually methanol, into methyl ester and glycerol products. The conversion of the triglyceride to esterified fatty acids improves the characteristics of the fuel, allowing its introduction into a standard compression engine without giving rise to serious issues with flow or combustion. Commercially available biodiesel, a product of the transesterification of fats and oils, can also be blended with standard diesel fuel up to a maximum of 20 vol.%. In the laboratory, the fuel characteristics of unreacted soybean oil have also been improved by dilution with petroleum based fuels, or by aerating and formation of microemulsions. However, it is the chemical conversion of the oil to fuel that has been the area of most interest. The topic has been reviewed extensively (Van Gerpen, Shanks et al. 2004), so this aspect will be the focus in this chapter. Important aspects of the chemistry of conversion of oil into diesel fuel remain the same no matter the composition of

[1] This manuscript has been authored by UT-Battelle, LLC, under Contract No. DE-AC05-00OR22725 with the U.S. Department of Energy. The United States Government retains and the publisher, by accepting the article for publication, acknowledges that the United States Government retains a non-exclusive, paid-up, irrevocable, world-wide license to publish or reproduce the published form of this manuscript, or allow others to do so, for United States Government purposes.

the triglyceride. Hence, although the focus in this book is on soybean oil, studies on other plant based oils and simulated oils have occasional mention in this chapter. Valuable data can be taken on systems that are simpler than soybean based oils, with fewer or shorter chain components. Sometimes the triglycerides will behave differently under reaction conditions, and when relevant, these have been noted in the text.

1.2 Transesterification and homogeneous base catalysis

Processing of soybean oil into a diesel compatible fuel through transesterification has received much recent attention as the most likely route to large-scale adoption of bio-based diesel. To improve flow characteristics, the triglyceride that constitutes the soybean oil has to be broken apart into smaller molecules. Fragmentation of the triglyceride takes place through a transesterification mechanism, a three step process that yields a molecule of esterified fatty acid at each step, shown below in Reaction (1) (Freedman, Pryde et al. 1984). Initially, the soybean oil reacts with a molecule of methanol, in the form of a reactive methylate in the case of base catalysis, to cleave a long-chain fatty acid fragment from the glycerine backbone that becomes a methyl ester, depicted as R^1 in the reaction below. The residual chains (R^2,R^3) attached to the backbone comprise a diglyceride after the first step, a monoglyceride after the second step, before the final decomposition to glycerine, or 1,2,3-propanetriol, and an ester at the last step. Commercially, a base such as sodium hydroxide or methylate is used to catalyze the transesterification process, promoting the reaction between the alcohol and the oil.

$$\begin{aligned} &C_3H_5(CO_2R^1)(CO_2R^2)(CO_2R^3) + CH_3OH \rightarrow CH_3O_2R^1 + C_3H_5(OH)(CO_2R^2)(CO_2R^3) \\ &C_3H_5(OH)(CO_2R^2)(CO_2R^3) + CH_3OH \rightarrow CH_3O_2R^2 + C_3H_5(OH)_2(CO_2R^3) \\ &C_3H_5(OH)_2(CO_2R^3) + CH_3OH \rightarrow CH_3O_2R^3 + C_3H_8O_3 \end{aligned} \tag{1}$$

Triglyceride + 3 Methanol → 3 Methyl Esters + Glycerine

In commercial parlance, the glycerine that is produced by transesterification is termed free glycerine, and the unreacted tri-, di-, and monoglycerides are called bound glycerine, usually expressed as wt.%. The base catalyst is usually introduced as anhydrous sodium methylate, to minimize the amount of water in the system as this leads to saponification, Reaction (2). The amount of base catalyst typically used is only slightly over 1 vol.% of the methanol, again to reduce formation of soapy emulsions. While the stoichiometry of the process demands a mole ratio of methanol to oil of 3, commercially the ratio is doubled to push the reaction to completion. In the US, biodiesel must have a bound glycerine content of less than 0.24 wt.% and a free glycerine content of less than 0.3 wt.% to be sold commercially. Standards for biodiesel purity are based either on the removal of contaminants before the oil feedstock is esterified or on the separation of unwanted by-products (ASTM 2007; ASTM 2008).

$$CH_3O_2R + NaOH \xrightarrow{H_2O} NaO_2R + CH_3OH \tag{2}$$

Methanol and base catalyst (in the form of NaOH or sodium methylate) are the reagents of choice in industrial production because of their being less expensive than other reagents. Potassium hydroxide has the advantage of a lower rate of saponification. Other alcohols can be used, primarily ethanol. Longer chain alcohols have better miscibility with the oil and hence higher yields, however, the product esters also become more difficult to separate from glycerine.

1.3 Esterification and homogeneous acid catalysis

The conversion of fatty acids to esters can also be catalyzed by acids in the esterification reaction scheme shown in Equation (3). The mineral acids commonly used as catalysts include sulfuric or hydrochloric acid. This chemical route is less popular when working with good quality soybean oil as a feedstock, because of the low free fatty acid content of the oil. For degraded or lower quality feedstocks, however, the advantages of avoiding large amounts of bound and free glycerine production as happens during transesterification can be desirable. Before triglycerides can be subjected to esterification, they must be saponified using base, such as NaOH, to strip apart the acylglyceride chains. Treatment with acid follows to protonate and form fatty acids. In the expression (3) below, the acid allows a complex to form between the triglyceride and the alcohol, which then falls apart to give a methyl ester and the diglyceride. Similarly to transesterification, the reaction progresses sequentially through a number of steps, not all shown in (3).

$$\begin{aligned} & C_3H_5(CO_2R)_2(CO_2R) + CH_3OH \underset{}{\overset{H^+}{\rightleftharpoons}} C_3H_5(CO_2R)_2(COROH(CH_3OH)) \\ & C_3H_5(CO_2R)_2(COROH(CH_3OH)) \rightarrow C_3H_5(CO_2R)_2OH + CH_3O_2R \end{aligned} \tag{3}$$

1.4 Conventional processing of soybean oil into methyl esters

Vegetable oils, including soybean oil, have complex compositions, which include a variety of fatty acid chain lengths. Soybean oil consists primarily of palmitic, oleic, linoleic, and linolenic acid chains, with a typical mixture given in Table 1 (Holčapek, Jandera et al. 2003). The actual composition depends on the source of the oil and can vary from one variety to another (Mello, Pousa et al. 2011). In addition to the variation in the fatty acid chains linked to the glyceryl backbone, processing of soybean oil will induce some degradation in a fraction of the triglyceride molecules to yield free fatty acid fragments. These compounds will not undergo base-catalyzed transesterification and must be esterified under acidic catalytic conditions.

Separation of the free fatty acids from the intact triglyceride molecules prior to conversion to esters is one of the challenges of processing soybean oil to biodiesel. Food grade soy oil can have very low free fatty acid content, less than 4%, in comparison with other oils, such as olive oil with up to 20%. However, lower quality feedstocks being considered for fuel production have higher free fatty acid content, with the highest concentration being present in waste oil that has usually been subjected to repeated heating cycles before being salvaged for biodiesel production. To process waste soybean oil, a combination of transesterification and esterification can be used, shown schematically in Figure 1. The waste oil passes through a centrifugal separator to remove water and suspended solids. The oil then moves to a tank of acid catalyst, H_2SO_4, and methanol. Upon esterification, three phases will form and separate: a rag layer containing acid, water, and methanol, a layer of unreacted oil, and the esterified products on the bottom of the tank. The lower two layers go through to the transesterification reactor, a reaction environment that does not degrade the already-formed methyl esters.

Triglyceride Chains	Normalized Mole Fraction	Ln (Linolenic) C18:3	L (Linoleic) C18:2	O (Oleic) C18:1	P (Palmitic) C16:0
LLLn	0.18	1	2	0	0
LLL	0.34	0	3	0	0
OLL	0.27	0	2	1	0
LLP	0.21	0	2	0	1
Mole fraction		0.083	0.751	0.083	0.083
Molecular weight, g	278	278	280	282	256

Table 1. Triglyceride Composition of Soybean Oil.

Recent expansion of biodiesel manufacture has resulted in increased interest among commercial enterprises to minimize the cost of feedstock materials and waste production and to maximize the efficiency of production. Hence, the technical issues limiting the feasibility of biodiesel production have received a lot of attention in the last decade. The next section discusses new approaches to converting soy oil to biodiesel highlighting the advantages that new technologies give over standard homogeneous base or acid-catalysis. Some ideas for improvement focus on gains in chemical kinetics or mass transfer, and others seek to reduce the amount of reagent methanol or simplify separations in pretreatment or posttreatment (preparation for sale). The next sections also present some of the drivers for advances in conversion technologies, along with recently published discoveries in making fuel from soybeans.

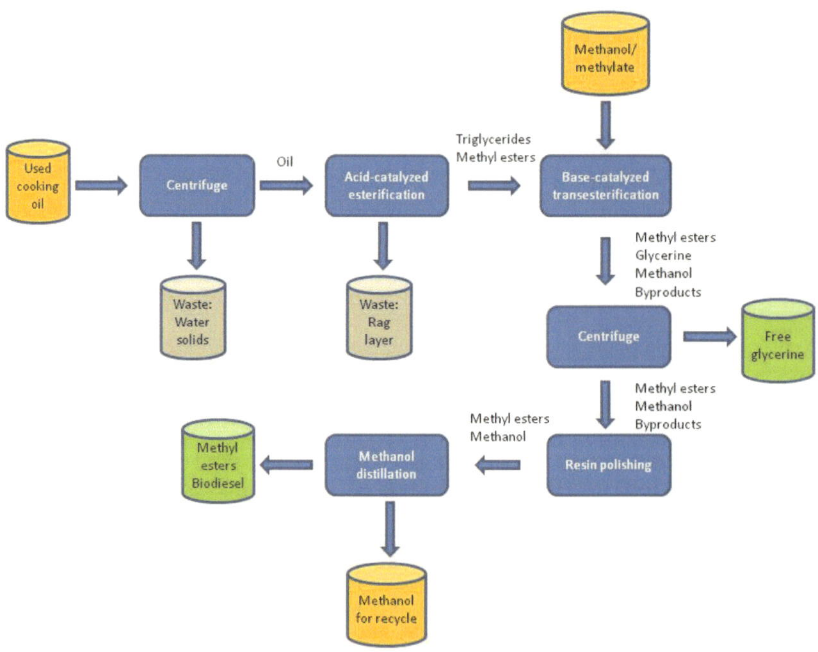

Fig. 1. A simplified flowsheet for the conversion of waste oil into methyl esters.

2. Advances in conversion technologies

Although homogeneous catalysts are used to promote the kinetics of the conversion of soybean oil to biodiesel, speeding up the process has inherent limitations. The reagent oil and alcohol, usually methanol, have limited miscibility, and so the reaction occurs primarily at the interface between the two liquid phases. Mass transport to the interface can be increased by rapidly mixing or forcing methanol into solution using higher pressures. Commercial processes are carried out at about 80°C, a temperature above the normal boiling point of methanol, to ensure conversion to the required ASTM specification. Even under pressurized conditions, the conversion takes at least half an hour in a batch reactor. At lower temperatures conversion typically take several hours. Several studies have investigated the use of supercritical methanol in the transesterification of soybean oil. Process intensification methods have been applied such as the use of rapid mixing and separation of products in a centrifugal contactor.

In addition to the issues with mass transfer, the kinetics of the three step transesterification can limit the overall conversion of the triglycerides to the esters. As the concentrations of intermediates increase, the rate of the back reactions can become significant with respect to the rates of the forward reactions. The online removal of glycerine to drive the process to completion has been attempted with some success; however, conversion to ASTM specification still takes minutes to complete. Addition of catalyst gives rise to saponification of the esters, resulting in phase separation and foaming, causing difficulties in processing. The reaction kinetics of transesterification has been modeled successfully with a three step forward and backward mechanism (Freedman, Butterfield et al. 1986; Noureddini & Zhu 1997). Although rate constants do vary with the type of oil and the processing conditions, the success with the model suggests that the constraints on reaction rate can be predicted and mitigated in a developing a optimized process flowsheet for soybean oil conversion to biodiesel.

Also problematic are multiple separation steps required during the conversion of soy oil to biodiesel. These include: pretreatment and removal of contaminants, separation of free-fatty acids from triglycerides, removal of free glycerine, washing to remove base catalyst, polishing of product in a resin bed, capture and recycle of unreacted methanol. In conventional processing, these steps can take several hours. However, the goal of some of the newer technologies being investigated in the laboratory is to minimize separation requirements.

A variety of new conversion technologies are being investigated to facilitate the conversion of soybean oil into biodiesel. Many of these ideas have been captured in recent reviews (i.e., (Lin, Cunshan et al. 2011), and a summary is given in the next section. Table 2 gives a synopsis of some of the new technologies, listing advantages and disadvantages for development on a commercial scale.

2.1 Advances in catalysis

2.1.1 High temperature cracking and heterogeneous catalysis

Thermal cracking of triglycerides, as opposed to transesterification discussed earlier, has been carried out for over 100 years, with a recent focus on converting fats and oils to liquid fuels (Maher & Bressler 2007). The cracking process takes place at high temperatures, 300-500°C, and atmospheric pressure producing alkanes, alkenes, aromatics and carboxylic acids, that can be separated by distillation (Lima, Soares et al. 2004). The resulting mixture has a lower viscosity than the parent oil. Yields tend to be low in comparison with transesterification, although up to 77% conversion of soybean oil has been observed with the use of a high quality, edible oil as the starting material. Although pyrolysis has been tested successfully on used cooking oil, fatty acid salts, and soaps, the low yields and the wide variety of chemicals produced in pyrolysis have made this process uneconomical. Difficulties with the pyrolytic method include the formation of char and cokes, as well as oxygenated compounds that need to be removed if the products are to be used as diesel substitutes. However, if the complex chemistry can be understood, and the decomposition pathways leading to aromatics and olefins as well as to more desirable alkanes identified, better control of reactor conditions to give desired products should be possible.

Processing Technology	Conditions & Reagents	Advantages	Disadvantages
Homogeneous catalysis	80°C. Mole ratio[2] ~6 (more needed for acid catalysis).	Used on commercial scale. Base catalysis has high yields	Batch process. Many separations needed. Corrosion.
Heterogeneous catalysis	160°C. Mole ratio ~12.	Flash separation of H_2O & excess alcohol. Catalyst recyclable. H_2O tolerant.	Small batch process. Lower yields <1 h. 80-90% yield at 3 h.
Supercritical alcohol	250°C. 200 bar. Mole ratio 30-80.	Complete conversion. Processing time < 1 h.	High mole ratio. Small scale only. High temperature. High pressure.
Enzymatic catalysis	50°C. 2 h reaction time. Mole ratio is 3.	Low temperature. Low mole ratio. Catalyst regeneration. Tolerant to H_2O & free fatty acids.	Long reaction time. Rate depends on enzyme loading. Scaling difficult. Low yields. Expensive catalyst.
Hydrodeoxygenation	300-350°C. 45-70 bar H_2. Mix with petroleum oil.	Diesel fraction linear alkanes produced. 100% conversion.	Long reaction time (hundreds of hours). High temperature. Mid pressure H_2.
Ultrasonic	45°C. 1 h reaction time. Mole ratio is 6.	Low temperature. Continuous process. Scalable.	Longer reaction time/multiple passes (bound glycerine after one pass is 80-90 wt%).
Centrifugal reactor/separator	80°C. 2.6 bar. Mole ratio is 4.8.	90% conversion after 2 min. Staged approach to get ASTM spec.	Pressurized system.
Metal foam	100°C. 5 bar. Mole ratio is 10. 1 wt.% catalyst.	95% after 3.3 min. Low power (<1% conventional). 0.9 L/h throughput.	Pressurized system.

Table 2. Summary of soybean oil conversion technologies to biodiesel.

[2] Methanol-to-oil or alcohol-to-oil mole ratio unless otherwise specified.

Better selectivity may be achieved through the use of catalysts in the pyrolysis process (Maher & Bressler 2007). For instance, molecular sieve materials, being porous with high surface area, exhibit high catalytic reactivity. The tetrahedral structures of zeolites, or crystalline aluminosilicate AlO_4-SiO_4 materials, show localized areas of high reactivity associated with the cations in the structure. Heterogeneous catalysis for pyrolysis of oils carried out at temperatures of 300-500°C over zeolites produces paraffins, olefins, carboxylic acids and aldehydes (Lima, Soares et al. 2004). Other studies using a protonated zeolite H-ZSM5 (85kPa He) have shown relatively more olefins and aromatics being produced from a variety of lipid starting materials and very little formation of oxygenated species. The reaction only generates a small amount of alkanes, and what is produced comes in the form of gases such as propane, and so is not appropriate for diesel fuel.

Metal catalysts have also been used for deoxygenation, Pt and Pd on activated carbon, at 300°C under nitrogen. Detailed analysis of the chemistry show that fragmentation of the triglyceride occurs more quickly than decarboxylation of the fatty acid chains. The fatty acid chains eventually form alkanes, with lighter hydrocarbons coming from β-fission at the double bonds. Lighter alkanes, CO_2, and CO come from the glycerol backbone. These studies demonstrate alkane production in the gasoline and diesel fraction range, with yields as high as 54% at 92% conversion for soybean oil (Morgan, Grubb et al. 2010).

A drawback to using catalyzed heterogeneous pyrolysis has been coking of the catalyst, requiring frequent cycles of oxidative regeneration (Milne, Evans et al. 1990). Fractionating batch reactors may allow the selective removal of alkanes, increasing their relative abundance, but yields are still relatively low (62 wt%) with high coke production (38 wt%) (Dandik & Aksoy 1999). Using mesoporous MCM-41 (1.93 nm pore size) as a catalyst showed lower gas production than H-ZSM5, with the best results being observed for palm oil, with 97.72% being converted overall and a yield of linear hydrocarbons C13-C17 in the diesel range of 42.52 wt.% (Twaiq, Zabidi et al. 2003). Palm oil differs from soybean oil with a higher fraction of shorter chain triglycerides, 50% C12 and 16% C14 and so these results may not relate directly to the conversion of soybean oil. In the same study, the authors show that experimentation with an oil of higher average molecular weight, in this case palm olein oil, showed a lower conversion and higher coke formation.

2.1.2 Low-temperature heterogeneous catalysis

Heterogeneous catalysis at low temperatures promises advantages over conventional processing in phase separation and avoidance of the use of strong caustic or acidic reagents. Some catalytic systems have proven to be more robust to fatty acids and to water than heterogeneous base catalysis (Zeng, Deng et al. 2009). The catalysts commonly used include transition metals and inorganic oxide systems that promote esterification and transesterification, besides the molecular sieves that are used in pyrolysis (discussed in Section 2.1.1).

A recent review gives details on supported solid metal oxides that have been used both for the transesterification and the esterification of oils to biodiesel (Zabeti, Daud et al. 2009). The transition metal oxides (alumina, tin, and zinc) form Lewis acids with the metal atoms acting as electron accepters. Alkaline earth oxides (magnesium, calcium, and strontium) form Brønsted bases through the oxygen atoms in the structure. Because of the colocation of acidic and basic sites, the activity of the catalyst is often described in both of these terms (Yan, DiMaggio et al. 2010). In a series of steps, Figure 2, the metal atom coordinates with both the oxygen of the carbonyl group in the acylglyceride or fatty acid and the alcohol, liberating a water molecule. The basic site can stabilize transfer of a proton from a fatty acid to water. The product ester forms within the supported complex or transition state, which decomposes regenerating the active metal oxide. Oxides such as alumina or silica can exhibit catalytic activity at acidic sites, dehydrating and decarboxylating fatty acids and triglycerides (Boz, Degirmenbasi et al. 2009). Acid-base catalysts can also be used in high temperature pyrolysis as well as for transesterification reactions.

Yields tend to be lower with heterogeneous catalysis in comparison with homogeneous catalysis (Section 1) because of reduced interfacial contact, not only between the oil and alcohol phases but also with the catalytic surfaces. To mitigate this limitation, methanol-to-oil ratios are usually high, 12 or greater; several wt% catalyst is often used; and reactions continue for a number of hours to drive the conversion to completion. Hence, studies are carried out in batch microreactors or autoclaves where extreme conditions can be controlled. Co-solvents have been used to improve the miscibility of the reagents (Yang & Xie 2007). Another way of improving reaction rate to get higher yields is to use high surface area catalysts and catalyst supports. For instance, nanoscale MgO has been used to achieve a 99% yield of methyl ester at 523°C and 24 MPa (Wang & Yang 2007). In autoclave studies of esterification at 160°C, a mass ratio of methanol: fatty acid: catalyst of 4: 10: 0.1 generated yields of up to 74% after only 1 h of residence time. The lowest yield, 32%, occurred in systems without the hetereogeneous catalyst, with methanol: fatty acid: catalyst 4: 10: 0.0, showing that the catalyst had a significant effect on reaction rate (Mello, Pousa et al. 2011). The same group showed that higher yields could be achieved after 3 h of reaction time. They also demonstrated that the catalyst could be regenerated at least ten times using centrifugation and cleaning in solvent without an observable loss in performance.

Heterogeneous catalysis continues to generate much interest in the research community. Surface area and morphology appear to have a greater influence over catalyst activity than the chemistry of the catalysts. Although some of the conversions show promise, the extreme temperatures or pressures currently required for effective heterogeneous catalysis, as well as the relatively low yields in comparison with homogeneous catalysis, preclude them from being used on a large or commercial scale. However, many of the catalysts being considered appear quite robust, and although subject to coking and other deactivation processes, can be regenerated many times.

2.1.3 Enzymatic catalysis

Lipases, naturally occurring enzymes, have been used to catalyze the transesterification of triglycerides. The mechanism is thought to be a two step process, where the lipase reacts with one substrate to form a product and an intermediate enzyme, followed by reaction with another substrate to give a final product and the regenerated enzyme (Varma, Deshpande et al. 2010). The advantages of enzymatic processing are high yields of methyl esters, milder reaction conditions, high tolerance of water contamination, and easy separation of free glycerine. The lipase process can be done in a number of different solvents, including supercritical CO_2. In the case of enzymatic catalysis, the loading of the enzyme has a profound effect on the initial rate of the reaction, and loadings of 5-10% w/w were found to be optimal. Enzymatic catalysis can be used for both esterification and transesterification, and a variety of oils and alcohols as feedstocks; however, processing conditions can be different depending on the starting material and desired product. Processing can take hours to reach equilibrium, typically achieved at when the reaction reaches about 50-70% conversion. Yields have been limited by inhibition of the catalyst by the alcohol, although the enzyme can be regenerated by driving off the alcohol to regain its activity.

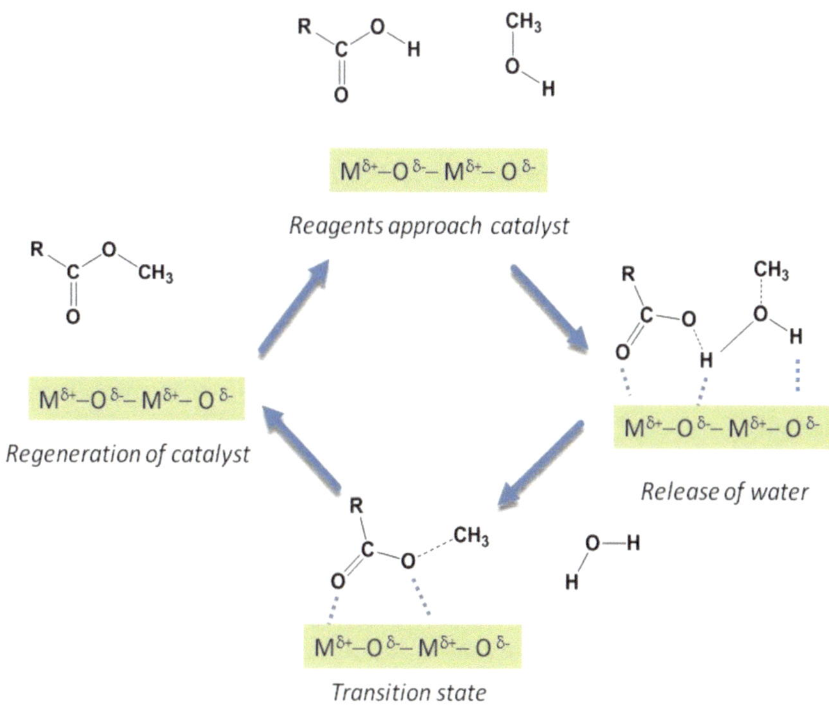

Fig. 2. Heterogeneous catalytic formation of a methyl ester from a fatty acid precursor.

2.2 Supercritical alcohols

A way of driving the transesterification reaction to completion without requiring catalyst is to perform the reaction under supercritical conditions. Many types of oils have been esterified in this way, including soy oil (Zhou, Wang et al. 2010). Both methanol and ethanol have been used as reagents (Rathore and Madras 2007). Pressures and temperatures are high for these processes, so that the conditions in the reactor exceed the critical point of the alcohols involved in the reaction. Pressures greater than 200 bar and temperatures exceeding 300°C are typical, although conversions of soybean oil have been successful at temperatures as low as 250°C. Because of the extreme conditions, these processes have only been demonstrated in the laboratory at bench scale. With a large excess of alcohol, the transesterification process can be described as a pseudo-first order reaction, and rate constants have been measured for a number of different alcohols reacting with a variety of oils (Varma, Deshpande et al. 2010). Rates of conversion in ethanol are greater than in methanol because of the greater miscibility of ethanol and the oil reagent. The rates also depend on the fatty acid content of the oil, being inversely proportional to the saturated fatty acid content.

2.3 Continuous and intensified processing

Conventional batch processing of soybean oil to biodiesel can take several hours, especially when post-conversion separation and polishing steps are included, see Figure 1. Each batch has to be tested against ASTM specifications before sale and a batch that has been compromised must be recycled back into the feed loop, adding cost. Properties of the fuel product can change because variations in the feedstock or changes in process condition. If implemented, a continuous process has the advantage of allowing online control of reagent flows, temperature and pressure conditions, to achieve good conversion, reducing the need for recycling of impaired product.

To achieve high conversion in a continuous process, however, issues such as the non-miscibility of reagents and mass transfer limitations in the transesterification process have to be overcome. Process intensification, an engineering concept that gained attention through investigations in the 1970s at the University of Newcastle (Stankiewicz & Moulijin 2002), is a way of enhancing mass transfer, thus reducing the capital cost of a chemical plant through a smaller plant size and reagent inventory and reducing operating costs through decreased energy consumption and feedstock required per unit mass of the product. Centrifugal phase contact and separation is an example of an intensified technique that enhances mass transfer at high throughput and minimizes the inventory of solvents (Tsouris & Porcelli 2003). Another example is to use bubble formation to increase the interfacial area of immiscible fluids, which can be induced by introducing energy to the system through acoustic coupling (Cintas, Mantegna et al. 2010).

Process intensification methodology has been adapted to enhance the pretreatment of biodiesel feedstocks, the conversion reactions, or the posttreatment separation of reaction products. A cavitation reactor was used in the process intensification of the homogeneous acid (H_2SO_4) catalyzed esterification of simulant fatty acids (Kelkar, Gogate et al. 2008). High throughput ultrasonic irradiation at 21.5 kHz coupled with a stirred tank was used to make a fine emulsion of oil and methanol, thereby increasing the interfacial area. The reactor achieved a yield of >80% methyl esters from soybean oil (Cintas, Mantegna et al. 2010). In this apparatus, temperatures were kept low, ~45°C, to prevent boiling of methanol in the microwave reactor. A sonochemical reactor has also been used to enhance the base-catalyzed transesterification of lightly used cooking oil as well as food grade vegetable oil (Hingu, Gogate et al. 2010).

Centrifugal mixing has been applied to biodiesel production (Peterson, Cook et al. 2001), because of its ease of operation, rapid attainment of steady state, high mass transfer, phase separation efficiencies, and compact size (Leonard, Bernstein et al. 1980). The high shear force and turbulent mixing achieved in a contactor minimize the effect of diffusion on the reaction rate of transesterification, pushing it to be limited only by the reaction kinetics. The contactor has been used as a low-throughput homogenizer, employing very low flow rates to increase residence times to tens of minutes (Kraai, van Zwol et al. 2008; Kraai, Schuur et al. 2009).

At ORNL, we have combined the reaction of oil and methoxide with the online separation of biodiesel and glycerol into one processing step, using a modified centrifugal contactor. Two distinct phases enter the reactor (reagents: methanol and base catalyst; and vegetable oil), and two distinct phases leave the reactor/separator (products: glycerol and methyl ester), thus demonstrating process intensification in high-throughput biofuel production. The ORNL reactor separator was modified from a commercial unit, Figure 3a, to increase the residence time from a few seconds to a few minutes by achieving hold-up in the mixing zone, Figure 3b. (Birdwell, Jennings et al. 2009). In the ORNL tests, base-catalyzed transesterification of soybean oil was carried out at continuous flow conditions at 60°C and in static pressurized tests at 80°C (McFarlane, Tsouris et al. 2010).

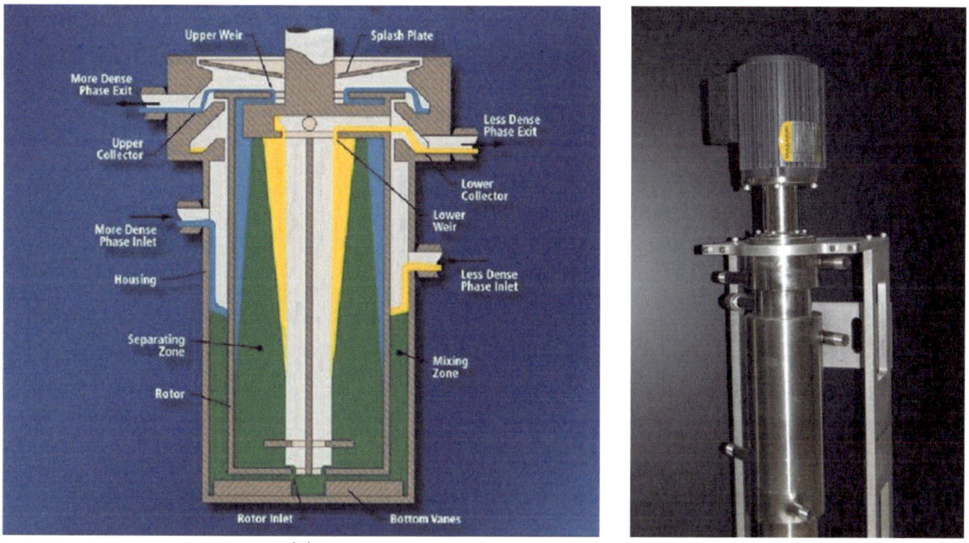

(a) (b)

Fig. 3. Reactor-separator housing: a) commercial unit schematic, b) modified contactor housing.

Besides bubble formation and stirring, another way of achieving high turbulence and good mass transfer for the production of biodiesel is through the use of reactors involving tortuous flow pathways. These concepts were first tested on microreactors, involving zigzag channels (Wen, Yu et al. 2009). Although high conversions were achieved, 99.5% at 28s residence time, scaling the reactor up from microliter $\cdot s^{-1}$ flow rates has not been possible. More recently, turbulence has been achieved by passing the reagents through porous metal foam, which can be made to have a high pore density (50 pores per inch) and a relatively low pressure drop (0.6 MPa). At 100°C and with a methanol-to-oil mole ratio of 6, a conversion of 90.5% was observed (Yu, Wen et al. 2010). With the foam, the arithmetic mean drop size of the disperse phase was about 3 μm. By balancing the effect of smaller, high surface area bubbles at high flow rates, with the lower residence time, conversions were pushed to 95 mol% with a flow rate of 0.9 L $\cdot h^{-1}$. While high for a microreactor, this flow rate is much lower than for competing continuous technologies.

In all continuous processes, the conversion of soybean oil to esters is limited by residence time in the reactor. Producers and investigators have focused on the kinetics of transesterification to determine if conversions to methyl ester are limited by mass transfer effects or by slow kinetics (Darnoko & Cheryan 2000; Karmee, Mahesh et al. 2004). In the transesterification reaction, mass transfer limitations early in the process become superseded by kinetic limitations when trying to achieve high yields of methyl esters. In the case of the Oak Ridge experiments, although 90% conversion was achieved in 2 min, a 22 min residence time at 80°C was needed to achieve ASTM specification grade fuel, ~98% conversion, Figure 4. Hence, in both the centrifugal processing and the ultrasonic reaction, multiple stages were found to shorten reaction time and reduce energy consumption. The online

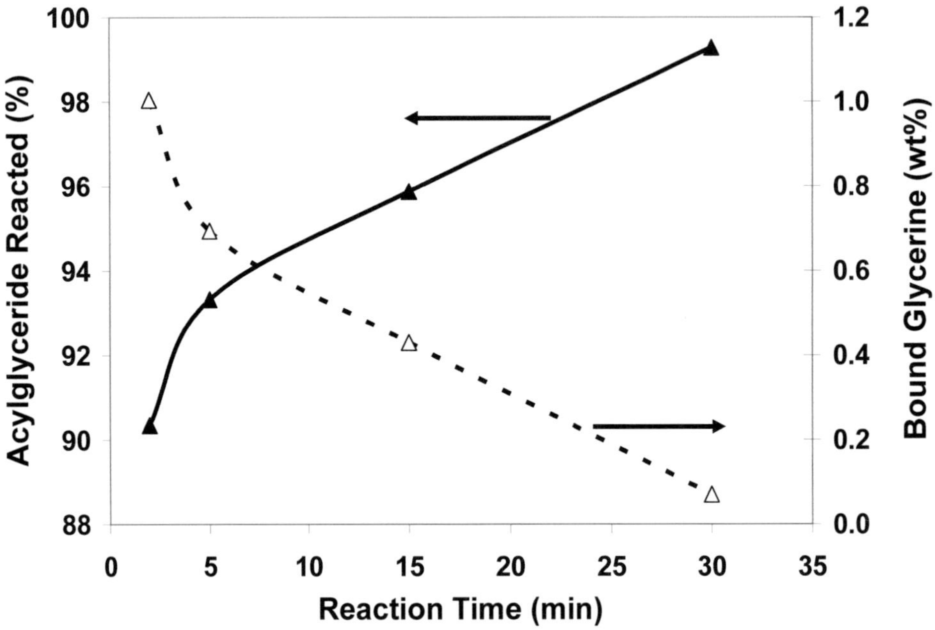

Fig. 4. Yield of batch transesterification reaction in continuous contactor in terms of the weight percent of triglyceride reacted (▲) and remaining total bound glycerine (Δ) as a function of reaction time (80°C, above ambient pressure to 2.6 bar, 3600 rpm rotor speed). The arrows indicate the conversion goal of <0.24 wt% bound glycerine, or 97.8% conversion of acylglyceride.

separation of free glycerine removes a sink for the base catalyst (Cintas, Mantegna et al. 2010), as well as reduces back reactions to form bound glycerine species (McFarlane, Tsouris et al. 2010). The accelerated reaction achieved with online separation also prevents thermal degradation of the methyl esters, arising from beta scission adjacent to the carbonyl group and cleavage of the unsaturated bonds in the fatty acid chains (Nawar & Dubravcic 1968; Osmont, Catoire et al. 2010).

3. Generation of fungible fuels from plant oils and new technologies for de-oxygenation

Even after esterification, the product biodiesel can be substituted directly for standard diesel fuel only to a limited percentage and is normally restricted from portions of the United States common carrier distribution system[3]. Although biodiesel has a similar cetane number to hexadecane, the higher oxygen content causes changes in the combustion profile and can enhance corrosion of engine seals (Haseeb, Fazal et al. 2011). The higher oxygen content also means that the heating value of methyl esters is slightly lower than standard diesel, although the reduction is not nearly as large as is when comparing ethanol to gasoline. The

[3] ASTM specifications allow 5 vol.% fatty acid methyl esters (FAME) in commercial diesel fuel.

lower volatility and higher oxygen content of biodiesel change the injection profile in a compression engine, and hence the ignition timing and production of pollutants, for instance decreasing soot and increasing the NO_x in the exhaust (Ra, Rietz et al. 2008; Toulson, Allen et al. 2011). This active area of study has an impact on high efficiency clean combustion engines, the vanguard of advanced diesel engine design. In most standard vehicles, biodiesel concentrations are limited to a blend of 20% to mitigate the effects of its physical properties being different from those of standard diesel fuel (Mushrush, Willauer et al. 2009), such as poor cold flow. In addition, biodiesel has a limited shelf life and can form precipitates and go rancid in storage, causing problems in distribution.

One method of producing deoxygenated products from soybean oil is to use a high temperature (350-450°C) hydrogenation process rather than transesterification to make fuels. This hydroprocessing, carried out over supported catalysts, is different than the pyrolytic schemes described in some detail in Section 2.1.1 because hydrogen is introduced directly into the reactor. Heavier paraffinic fragments are produced rather than the small gaseous alkanes made in pyrolysis. The process, as applied to triglycerides, has been reviewed by Donnis and colleagues (Donnis, Egeberg et al. 2009). Hydrotreating experiments on triglycerides have used the same conventional catalysts used in hydroprocessing oil, such as sulfided NiMo or CoMo on alumina under relatively low pressures of H_2S/H_2 mixtures (Huber, O'Connor et al. 2007). The process includes several chemical steps to give alkanes as a final product, including: hydrogenation of C=C bonds; decarboxylation (removal of CO_2); decarbonylation (removal of CO); and dehydration (hydrodeoxygenation (HDO) to convert COOH to H_2O). The glycerin backbone may react to form methane or propane (Donnis, Egeberg et al. 2009). By carefully controlling temperature and reaction time the yield of the paraffinic diesel-fraction, or straight chain C15-C18, can be maximized. Although some studies show that catalyzed hydroprocessing over nickel generates too many aromatics and cyclic compounds, tailoring of HDO products by additional isomerization steps has been suggested to produce branched alkanes (Jakkula, Niemi et al. 2004). This would give a biorefinery the ability to produce the desired fuel properties for vehicular use without the need for blending, giving a product similar to Fischer-Tropsch diesel fuel from natural gas.

Huber and colleagues have also shown that the bio-derived oils can be hydrotreated along with petroleum oils, suggesting that a processing can take place within an existing refinery to lower the capital cost. Issues with hydroprocessing vegetable oils rather than petroleum include: the high oxygen content of biomass can increase heat load in the reactor and cause leaching of sulfur from the catalyst; water and CO_2 generated during the hydrotreatment can reduce catalyst lifetime and must be removed from the product; and also the large triglyceride molecules can clog catalysts with pore sizes of less than 2 nm (Tiwari, Rana et al. 2011). Mesoporous molecular sieves, such as MCM-41, or alumina can have the advantage of a high surface area and activity, but also have much larger pore diameters than zeolites (Kubicka, Simacek et al. 2009), and so may be useful in a combined bio-petro refinery.

Another route to achieving a hydrocarbon rich fuel from soybean oil is through deoxygenation of the esters after the transesterification process has taken place. In this case the biodiesel produced from soybean oil is further reacted to form a hydrocarbon fuel. The processing involves deoxygenation to remove the ester moiety from the hydrocarbon chain. With this step, the product becomes completely miscible with standard diesel fuel and can be introduced at any step in the supply chain, either at the refinery or at the filling station. Note that if blending is done at the terminal or filling station, the product has to meet completely ASTM specifications. Some of these processes involve hydrogen and some do not.

The hydrogenation of methyl octanoate, as a simulant for methyl esters from biodiesel, has been carried out over an N-ZSM5 zeolite catalyst under atmospheric pressure H_2 (Danuthai, Jongpatiwut et al. 2009). The experiments were run over a few hours at temperatures up to 500°C, and showed 99.7% conversion of the ester to C1-C7 alkanes - a third comprising ethane, and small aromatics (C6-C9). Residual oxygenated species comprised only 2.8%. The group also found that the aromatic fraction increased with the time in the reactor, and that H_2O promoted the catalytic activity of the zeolite by enhancing production of an acid byproduct, obviously undesirable as a fuel component. Tests with methyl octanoate, a smaller molecule than methyl esters derived from soybean oil, showed conversion to alkanes and aromatics through formation of a high molecular weight ketone intermediate. The patent literature suggests that similar results have been achieved with longer fatty acid chain methyl esters from soybean and other oils (Craig 1991). Reaction 4 shows the overall conversion process of a methylester to a linear alkane by hydrodeoxygenation: step A) removing the oxygen as CO_2 or methanol followed by formation of the enol, and step B) involving hydrogenation and dehydration of the enol to the linear alkane (Donnis, Egeberg et al. 2009).

$$
\begin{aligned}
&\text{A)}\\
&CH_3OCOC_nH_{2n+1} \rightarrow CH_2{=}C_{n-1}H_{2(n-1)} + CO_2 + CH_4 \xrightarrow{H_2} C_nH_{2n+2}\\
&CH_3OCOC_nH_{2n+1} \xrightarrow{H_2} CHOCH_2C_{n-1}H_{2(n-1)+1} + CH_3OH\\
&CHOCH_2C_{n-1}H_{2(n-1)+1} \xrightleftharpoons{\text{enol rearrangment}} CH(OH){=}CHC_{n-1}H_{2(n-1)+1}\\
&\text{B)}\\
&CH(OH){=}CHC_{n-1}H_{2(n-1)+1} \xrightarrow{H_2} CH_2(OH)C_nH_{2n+1} \xrightarrow[-H_2O]{+H_2} C_{n+1}H_{2(n+1)+2}\\
&CH(OH){=}CHC_{n-1}H_{2(n-1)+1} \xrightarrow[-H_2O]{+H_2} CH_2{=}CHC_{n-1}H_{2(n-1)+1} \xrightarrow{H_2} C_{n+1}H_{2(n+1)+2}
\end{aligned}
\qquad (4)
$$

As discussed in Section 2.1.1, non-hydrogenated direct catalytic cracking of triglycerides can lead to products with greater oxygen content than desirable for fuels. Better control of the cracking process can be engineered when starting with an esterified feedstock. A recent example is the use of supported platinum and bimetallic platinum-tin catalysts in the deoxygenation of methyl octanoate, methyl dodecanate, and soybean oil by reactive distillation at 320 to 350°C. By manipulating the residence time and the catalyst properties, selectivity for paraffins of 80% was achieved. Overall yields were low, suggesting this process requires more investigation before commercialization (Do, Chiappero et al. 2009; Chiappero, Do et al. 2011).

4. Feasibility of using plant oils for fuels in comparison with petroleum, ethanol, and lignocellulosic feedstocks

The use of soybean oil in production of biodiesel has been primarily limited by economic factors, in particular the cost of the feedstock. Less expensive fuel can be made from degraded starting material such as waste oil. Energy crop alternatives to seed oils have also been proposed (Vinokurov, Barkov et al. 2010). However, the processing of feedstock with higher free fatty acid content adds complexity to the manufacturing process, particularly because of the variability in composition and treatment prior to conversion. The solution to tightening of petroleum supply will likely involve liquid fuel generation from a variety of sources. As should have been apparent from the previous discussion, the processing of biomass-derived oils into burnable esters depends on the chemical composition of the feedstock: the relative concentration of free fatty acids, the saturated versus unsaturated fatty acid chains, impurities and water content.

An additional cost is associated with the alcohol used to convert the seed oil to biodiesel, typically used in amounts well above stoichiometric to push the reaction to completion. An analysis was recently done at ORNL where the cost of a three stage biodiesel manufacturing process was assessed based on the reactor-separator reactor discussed in Section 2.3 (Ashby & McFarlane 2010). In order to optimize the process, environmental conditions such as temperature, pressure, and the starting proportion of methanol–to–oil were all varied individually. Each of these aspects of the production affected the residence time and the fraction of soybean oil converted during the reaction, hence the economics of the process. The analysis gave the projected capital cost for a new plant and its projected profit in the first five years. These analyses revealed that reactions run at higher temperatures needed less time to convert a larger fraction of triglyceride. Reactions with a greater proportion of methanol–to–oil had a higher yield at a residence time of 600s than those with a lower ratio. Figure 5 shows the effects of adding additional methanol at various stages of a five stage process.

An economic analysis shows that production of biodiesel should be more profitable in a three contactor series than a single reactor given similar process conditions, i.e., temperature and ratio of methanol–to–oil, in spite of the costs associated with the reactor and pumps for each additional stage. In the long term, the feedstock soybean oil comprised the highest fraction of the operating expenses, ranging from 70-80% of the total. The cost of the alcohol was also found to be significant, but could be minimized through recycling, thereby also reducing the carbon footprint of the process. In this analysis, the production of biodiesel from soybean oil could only become profitable if the product could be sold at about 1.5 times the cost of the soybean oil feedstock (assuming a 300,000 gal/year operation amortized over 5 years). Based on simulation of the chemical kinetics of soybean oil transesterification, the highest yield of methyl esters in the shortest time arose from using three reactor-separators in series, each with a 200s residence time, recycling of all excess methanol, a 4.5–to–1 initial proportion of methanol–to–oil, and an operating temperature of 100°C. Similar analyses have been done for other reactor configurations, feedstocks, alcohols, and catalysts, to assess the viability of these process designs for commercial production of biodiesel (Peterson, Cook et al. 2001).

In the case of soybean utilization, the feedstock costs appear to dominate the potential use of biomass conversion to supplant petroleum-derived diesel in any of the reactor configurations being considered (Lin, Cunshan et al. 2011). However, the economics of biodiesel production can be improved if value added products can be developed from the byproduct glycerine. Janaun and Ellis give many of these in their review: catalytic conversion to oxidized products such as propylene glycol; biological conversion to lipids and citric acid; fuel oxygenates; gasification to H_2 and syngas; remediation of acid mine drainage; and in agriculture as animal feed (Janaun & Ellis 2010).

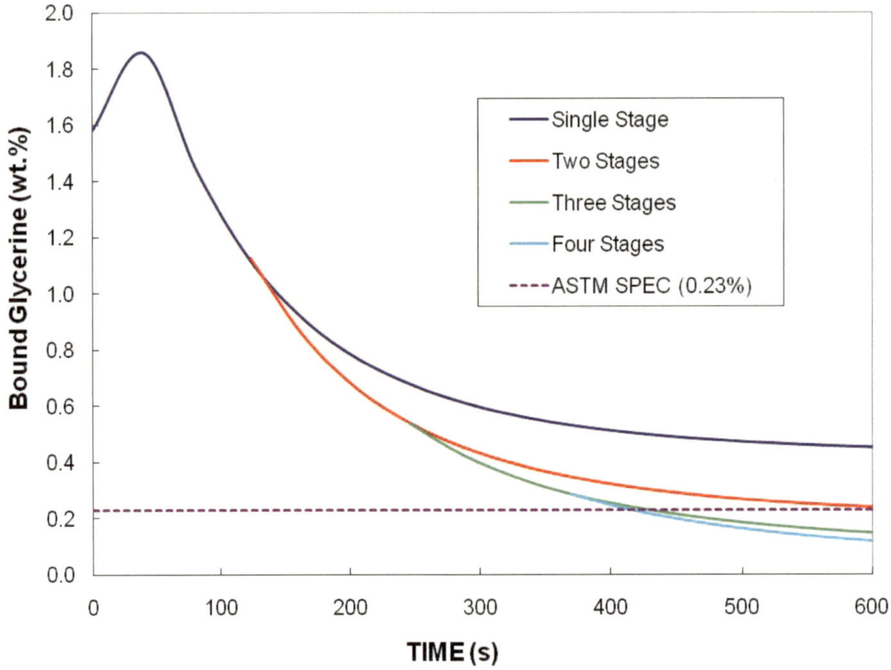

Fig. 5. Bound glycerine content as a function of time for various options of methanol addition, at 80°C: Single stage involves no additional input of methanol over 600s reaction time; Two stages involves adding 10% original methanol charge at 2nd stage; Three stages involves adding an extra 5% original methanol charge at 3rd stage; Four stages has no additional methanol as the effect is minimal by this point.

Another aspect worth consideration is that unless specifically designed to do so, compression engines are not constructed to handle the higher oxygen content of biofuels such as biodiesel or ethanol. Hence, many alternative fuels under consideration are blended to give the properties needed for engine performance and fuel stability, 10% ethanol in gasoline being a common example. However, fuel from different sources may not be compatible. Biodiesel, with its high oxygen content, mixes well with standard diesel, but not with purely paraffinic Fischer-Tropsch fuel. The aromatics in standard diesel solubilize the olefinic chains and electron-rich esters, where as tertiary carbons in the Fischer-Tropsch paraffins appear to form stable hydroperoxides with degradation products in the biodiesel (Mushrush, Willauer et al. 2009). If the biodiesel contains unreacted free-fatty acids, phase separation and precipitates are likely to form. One possibility is to hydrogenate the biodiesel to create a fully hydrocarbon fuel, as discussed earlier in Section 3. Another is to exploit the properties of other biomass-derived fuels to produce a blend with properties that meet the requirements for compression ignition engines. For instance, lignin has the potential to become a biofuel feedstock can be broken down into appropriately sized aromatic fragments, which can be used as additives to diesel fuel or to biodiesel methyl esters (Gluckstein, Hu et al. 2010). The properties of the blend will have the high cetane number and the high lubricity of the biodiesel methyl esters, but with the reduced viscosity and low cloud point of the aromatics. Hence, while a pure biofuel may have some undesirable characteristics, mixtures of alternative fuels may be compatible with standard diesel engines. An assessment of mixtures of diesel compatible formulations has been performed by the Fuels for Advanced Combustion Engines (FACE) Project and target properties are presented in Table 3. The average properties of marketed diesel fuel are shown in brackets (Gallant, Franz et al. 2009).

Property	Range for FACE project	Standard Diesel Fuel	Soybean Methyl Esters
Cetane number	30–55	43–51	51
Aromatics (%)	20–45	32	0
T90 Distillation (°C)	270–340	320	Not applicable
Specific gravity ($g \cdot cm^{-3}$)	0.803–0.869	0.82–0.86	0.884
Heat of combustion (kJ/kg)	7790-7980	7850	Reduced by 9-13%
Kinematic viscosity ($mm^2 \cdot s^{-1}$)	1.319-3.218	1.90-4.1	4.08
Cloud Point (°C)	-19.5--55.5	-18--30	-0.5
Flash Point (°C)	53–74	55	131
Pour point (°C)		-25	-4
Constraints	< 15 ppm sulfur, <4% olefins Smooth distillation curve		

Table 3. Fuel Formulation Property Targets for Compression Ignition Engines.

Biofuel production in the US and Brazil is dominated by ethanol, where as biodiesel has greater importance in Europe (Rusco & Walls 2009). In some respects the issues with ethanol and biodiesel are similar, competition for agricultural resources with food, oxygen content and lower heating value, and distributed production (Kalnes, Marker et al. 2007). Varying fuel standards can further complicate distribution, leading to lower pipeline capacity and increased storage requirements. For instance, ethanol, even blended with gasoline, currently is not transported through pipelines because of its high affinity for water resulting in corrosiveness and phase separation. However, ethanol is a simple molecule that has the same composition no matter the source, and its impact on petroleum refining can be assessed on a large scale. This is not the case for plant-based biodiesels, from which a variety of fuels can be produced depending on the plant variety and processing conditions. Depending on the regulatory environment and governing standards, this may further break up the markets for biodiesel production and distribution. For example, southern regions will better be able to tolerate higher cloud points than northern, both for pipeline, truck and rail transport, as well as for combustion in passenger vehicles. The cost of the adoption of biofuels needs to be assessed along with benefits, such as reduction in greenhouse gas emissions, energy security, or support of US agriculture (Rusco & Walls 2008).

5. Conclusions

Although the price of diesel fuel has increased, economical production of biodiesel is a challenge because of (1) the increasing price of soybean oil feedstocks and reagent methanol, (2) a distributed supply of feedstocks that reduces the potential for economies of scale, (3) processing conditions that include pressures and temperatures above ambient, and (4) multiple processing steps needed to reduce contaminant levels to ASTM specification D6751 limits (Vasudevan & Briggs 2008). Much of the cost of biodiesel production is related to the conversion of the oil to the methyl ester and so there has been an emphasis to research improved methods of converting soybean oil to biodiesel. However, most of these studies have taken place at the bench scale, and have not demonstrated a marked improvement in yield or reduced oil-to-methanol ratio in comparison with standard base-catalyzed transesterification.

One aspect that has a short term chance of implementation is the improvement of the conversion process by the use of a continuous rather than batch process, with energy savings generated by combined reaction and separation, online analysis, and reagent methanol added by titration as needed to produce ASTM specification grade fuel. By adapting process intensification methods, recycled sources of soybean oil may also be used for diesel production, taking advantage of a lower priced feedstock material.

Even if the economics of production are feasible, biodiesel distribution is complicated by thermal stability and degradation over time, and the physical properties of methyl esters make them undesirable for standard compression ignition engines in concentrations greater than 20% in a blend with diesel fuel. Generation of truly fungible fuel from biomass is now being investigated through a variety of routes. However, it is too early to judge which will become the most viable.

The promise of soybean-generated biodiesel is that of a truly fungible, thermodynamically and economically viable technology providing a biomass replacement for a petroleum product. The use of biodiesel has the potential to reduce the amount of CO_2 released to the atmosphere by the transportation sector; to provide an additional source of liquid fuel that can be produced in small distributed operations; and to allow the processing of waste oil-to-energy that can result in enhanced lifecycle efficiencies as well as reduced environmental footprint.

6. Acknowledgments

Research on biodiesel manufacture at Oak Ridge National Laboratory was sponsored by the Laboratory Directed Research and Development Program; the United States Department of Energy Office of Efficiency and Renewable Energy Technology Commercialization and Deployment Program's Technology Commercialization Fund; and by Nu-Energie, LLC, under CRADA #01377. Oak Ridge National Laboratory is managed by UT-Battelle, LLC, for the U.S. Department of Energy. The author would also like to thank Dr. Bruce Bunting of the National Transportation Research Center at ORNL for his helpful comments.

7. References

Ashby, E. & McFarlane, J. (2010). *Optimization of Biodiesel Production in a Centrifugal Contactor,* Poster Presentation, May 5, 2010. Oak Ridge National Laboratory, Oak Ridge.

ASTM. (2007). *Standard Test Method for the Determination of Free and Total Glycerin in B-100 Biodiesel Methyl Esters by Gas Chromatography,* ASTM International. ASTM Method D6854.

ASTM. (2008). *Standard Specification for Biodiesel Fuel Blend Stock (B100) for Middle Distillate Fuels*, ASTM International. D6751-07b.

Birdwell, J. F. J., Jennings, H. L., McFarlane, J. & Tsouris, C. (2009). *Integrated reactor and centrifugal separator and uses thereof*, UT-Battelle. US Patent Application WO/145954.

Boz, N., Degirmenbasi, N. & Kalyon, D. M. (2009). Conversion of biomass to fuel: Transesterification of vegetable oil to biodiesel using KF loaded nano-gamma-Al_2O_3 as catalyst. *Applied Catalysis B-Environmental*, Vol. 89, No. 3-4, pp. (590-596), 0926-3373.

Chiappero, M., Do, P. T. M., Crossley, S., Lobban, L. L. & Resasco, D. E. (2011). Direct conversion of triglycerides to olefins and paraffins over noble metal supported catalysts. *Fuel*, Vol. 90, No. 3, pp. (1155-1165), 0016-2361.

Cintas, P., Mantegna, S., Gaudino, E. C. & Cravotto, G. (2010). A new pilot flow reactor for high-intensity ultrasound irradiation. Application to the synthesis of biodiesel. *Ultrasonics Sonochemistry*, Vol. 17, No. 6, pp. (985-989), 1350-4177.

Craig, W. (1991). *Production of hydrocarbons with a relatively high cetane raing*, US Patent 4,992,605, USA.

Dandik, L. & Aksoy, H. A. (1999). Effect of catalyst on the pyrolysis of used oil carried out in a fractionating pyrolysis reactor. *Renewable Energy*, Vol. 16, No. 1-4, pp. (1007-1010), 0960-1481.

Danuthai, T., Jongpatiwut, S., Rirksomboon, T., Osuwan, S. & Resasco, D. E. (2009). Conversion of methylesters to hydrocarbons over an H-ZSM5 zeolite catalyst. *Applied Catalysis a-General*, Vol. 361, No. 1-2, pp. (99-105), 0926-860X.

Darnoko, D. & Cheryan, M. (2000). Kinetics of palm oil transesterification in a batch reactor. *Journal of the American Oil Chemists Society*, Vol. 77, No. 12, pp. (1263-1267), 0003-021X.

Do, P. T., Chiappero, M., Lobban, L. L. & Resasco, D. E. (2009). Catalytic deoxygenation of methyl-octanoate and methyl-stearate on Pt/Al_2O_3. *Catalysis Letters*, Vol. 130, No. 1-2, pp. (9-18), 1011-372X.

Donnis, B., Egeberg, R. G., Blom, P. & Knudsen, K. G. (2009). Hydroprocessing of bio-oils and oxygenates to hydrocarbons. Understanding the reaction routes. *Topics in Catalysis*, Vol. 52, No. 3, pp. (229-240), 1022-5528.

Emerging Markets (2008). Biodiesel 2020, In: *Emerging Markets Online* 2nd Ed. Date of access March 24, 2011, Available from: <www.emerging-markets.com>

Freedman, B., Butterfield, R. O. & Pryde, E. H. (1986). Transesterification kinetics of soybean oil. *Journal of the American Oil Chemists Society*, Vol. 63, No. (1375-1380), 0003-021X.

Freedman, B., Pryde, E. H. & Mounts, T. L. (1984). Variables affecting the yields of fatty esters from transesterified vegetable oils. *Journal of the American Oil Chemists Society*, Vol. 61, No. 10, pp. (1638-1643), 0003-021X.

Gallant, T., Franz, J. A., Alnajjar, M. S., Storey, J. M. E., Lewis, S. A., Sluder, C. S., Cannella, W. C., Fairbridge, C., Hager, D., Dettman, H., Luecke, J., Ratcliff, M. A. & Zigler, B. T. (2009). Fuels for advanced combustion engines research diesel fuels: Analysis of chemical and physical properties. *SAE International Journal of Fuels and Lubricants*, Vol. 2, No. 2, pp. (262-272), 978-0-7680-1800-4 (proceedings), 1946-3952 (journal).

Gluckstein, J. A., Hu, M. Z., Kidder, M., McFarlane, J., Narula, C. K. & Sturgeon, M. R. (2010). *Final report: Investigation of catalytic pathways for lignin breakdown into monomers and fuels*, (ORNL/TM-2010/281), Oak Ridge National Laboratory.

Haseeb, A. S. M. A., Fazal, M. A., Jahirul, M. I. & Masjuki, H. H. (2011). Compatibility of automotive materials in biodiesel: A review. *Fuel*, Vol. 90, No. 3, pp. (922-931), 0016-2361

Hingu, S. M., Gogate, P. R. & Rathod, V. K. (2010). Synthesis of biodiesel from waste cooking oil using sonochemical reactors. *Ultrasonics Sonochemistry*, Vol. 17, No. 5, pp. (827-832), 1350-4177.

Holčapek, M., Jandera, P., Zderadička, P. & Hrubá, L. (2003). Characterization of triacylglycerol and diacylglycerol composition of plant oils using high-performance liquid chromatography - Atmospheric pressure chemical ionization mass spectrometry. *Journal of Chromatography A*, Vol. 1010, No. (195-215), 0021-9673.

Huber, G. W., O'Connor, P. & Corma, A. (2007). Processing biomass in conventional oil refineries: Production of high quality diesel by hydrotreating vegetable oils in heavy vacuum oil mixtures. *Applied Catalysis A-General*, Vol. 329, No. (120-129), 0926-860X.

Janaun, J. & Ellis, N. (2010). Perspectives on biodiesel as a sustainable fuel. *Renewable & Sustainable Energy Reviews*, Vol. 14, No. 4, pp. (1312-1320), 1364-0321.

Kalnes, T., Marker, T. & Shonnard, D. R. (2007). Green diesel: A second generation biofuel. *International Journal of Chemical Reactor Engineering*, Vol. 5, No. (A48), 1542-6580.

Karmee, S. K., Mahesh, P., Ravi, R. & Chadha, A. (2004). Kinetic study of the base-catalyzed transesterification of monoglycerides from pongamia oil. *Journal of the American Oil Chemists Society*, Vol. 81, No. 5, pp. (425-430), 0003-021X.

Kelkar, M. A., Gogate, P. R. & Pandit, A. B. (2008). Intensification of esterification of acids for synthesis of biodiesel using acoustic and hydrodynamic cavitation. *Ultrasonics Sonochemistry*, Vol. 15, No. 3, pp. (188-194), 1350-4177.

Kraai, G. N., Schuur, B., van Zwol, F., Haak, R. M., Minnaard, A. J., Feringa, B. L., Heeres, H. J. & de Vries, J. G. (2009). Process intensification. Continuous two-phase catalytic reactions in a table-top centrifugal contact separator. In: *Catalysis of Organic Reactions*, 123, M. L. Prunier, pp. (39-49), 978-1-4200-7076-7, CRC Press-Taylor & Francis Group, Boca Raton.

Kraai, G. N., van Zwol, F., Schuur, B., Heeres, H. J. & de Vries, J. G. (2008). Two-phase (bio)catalytic reactions in a table-top centrifugal contact separator. *Angewandte Chemie-International Edition*, Vol. 47, No. 21, pp. (3905-3908), 1433-7851.

Kubicka, D., Simacek, P. & Zilkova, N. (2009). Transformation of vegetable oils into hydrocarbons over mesoporous-alumina-supported CoMo catalysts. *Topics in Catalysis*, Vol. 52, No. 1-2, pp. (161-168), 1022-5528.

Leonard, R. A., Bernstein, G. J., Ziegler, A. A. & Pelto, R. H. (1980). Annular centrifugal contactors for solvent-extraction. *Separation Science and Technology*, Vol. 15, No. 4, pp. (925-943), 1520-5754.

Lima, D. G., Soares, V. C. D., Ribeiro, E. B., Carvalho, D. A., Cardoso, E. C. V., Rassi, F. C., K.C., M. & Suarez, P. A. Z. (2004). Diesel-like fuel obtained by pyrolysis of vegetable oils. *Journal of Analytical and Applied Pyrolysis*, Vol. 71, No. 2, pp. (987–996), 0165-2370.

Lin, L., Cunshan, Z., Vittayapadung, S., Xiangquian, S. & Mingdong, D. (2011). Opportunities and challenges for biodiesel fuel. *Applied Energy*, Vol. 88, No. 4, pp. (1020-1031), 0306-2619.

Maher, K. D. & Bressler, D. C. (2007). Pyrolysis of triglyceride materials for the production of renewable fuels and chemicals. *Bioresource Technology*, Vol. 98, No. (2351-2368), 0960-8524.

McFarlane, J., Tsouris, C., Birdwell, J. F. J., Schuh, D. L., Jennings, H. L., Pahmer Boitrago, A. M. & Terpstra, S. M. (2010). Production of biodiesel at kinetic limit achieved in a centrifugal reactor/separator. *Industrial and Engineering Chemistry Research*, Vol. 49, No. 7, pp. (3160–3169), 0888-5885.

Mello, V. M., Pousa, G., Pereira, M. S. C., Dias, I. M. & Suarez, P. A. Z. (2011). Metal oxides as heterogeneous catalysts for esterification of fatty acids obtained from soybean oil. *Fuel Processing Technology*, Vol. 92, No. 1, pp. (53-57), 0378-3820.

Milne, T. A., Evans, R. J. & Nagle, N. (1990). Catalytic conversion of microalgae and vegetable oils to premium gasoline, with shape-selective zeolites. *Biomass*, Vol. 21, No. 3, pp. (219-232), 0144-4565.

Morgan, T., Grubb, D., Santillan-Jimenez, E. & Crocker, M. (2010). Conversion of triglycerides to hydrocarbons over supported metal catalysts. *Topics in Catalysis*, Vol. 53, No. 11-12, pp. (820-829), 1022-5528.

Mushrush, G. W., Willauer, H. D., Bausermann, J. W. & Williams, F. W. (2009). Incompatibility of fischer-tropsch diesel with petroleum and soybean biodiesel blends. *Industrial and Engineering Chemistry Research*, Vol. 48, No. 15, pp. (7364-7367), 0888-5885.

Nawar, W. W. & Dubravcic, M. F. (1968). Thermal decomposition of methyl oleate. *Journal of the American Oil Chemists Society*, Vol. 45, No. 2, pp. (100-102), 0003-021X.

Noureddini, H. & Zhu, D. (1997). Kinetics of transesterification of soybean oil. *Journal of the American Oil Chemists Society*, Vol. 74, No. 11, pp. (1457-1463), 0003-021X.

Osmont, A., Catoire, L. & Dagaut, P. (2010). Thermodynamic data for the modeling of the thermal decomposition of biodiesel. 1. Saturated and monounsaturated FAMES. *Journal of Physical Chemistry A*, Vol. 114, No. 11, pp. (3788-3795), 1089-5639.

Peterson, C. L., Cook, J. L., Thompson, J. C. & Taberski, J. S. (2001). Continuous flow biodiesel production. *Applied Engineering in Agriculture*, Vol. 18, No. 1, pp. (5-11), 0883-8542.

Ra, Y., Rietz, R. D., McFarlane, J. & Daw, S. (2008). Effects of fuel physical properties on diesel engine combustion using diesel and biodiesel fuels. *SAE International Journal of Fuels and Lubricants*, Vol. 1, No. 1, pp. (703-718), 1946-3952.

Rusco, F. W. & Walls, W. D. (2008). *Biofuels, Petroleum Refining, and the Shipping of Motor Fuels*, University of Calgary, TP-08005, Calgary AB.

Rusco, F. W. & Walls, W. D. (2009). *Biofuels and the Fungibility of Motor Fuels*, International Association for Energy Economics. Third Quarter: 49-52.

Stankiewicz, A. & Moulijin, J. A. (2002). Process intensification. *Industrial & Engineering Chemistry Research*, Vol. 41, No. 8, pp. (1920-1924), 0888-5885.

Tiwari, R., Rana, B. S., Kumar, R., Verma, D., Kumar, R., Joshi, R. K., Garg, M. O. & Sinha, A. K. (2011). Hydrotreating and hydrocracking catalysts for processing of waste soya-oil and refinery-oil mixtures. *Catalysis Communications*, Vol. 12, No. 6, pp. (559-562), 1566-7367.

Toulson, E., Allen, C. M., Miller, D. J., McFarlane, J., Schock, H. J. & Lee, T. (2011). Modeling the autoignition of fuel blends with a multistep model. *Energy & Fuels*, Vol. 25, No. 2, pp. (632-639), 0887-0624.

Tsouris, C. & Porcelli, J. V. (2003). Process intensification—Has its time finally come? *Chemical Engineering Progress*, Vol. 99, No. 10, pp. (50-55), 0360-7275.

Twaiq, F. A., Zabidi, N. A. M., Mohamed, A. R. & Bhatnagar, A. K. (2003). Catalytic conversion of palm oil over mesoporous aluminosilicate MCM-41 for the production of liquid hydrocarbon fuels. *Fuel Processing Technology*, Vol. 84, No. 1-3, pp. (105-120), 0378-3820.

Van Gerpen, J., Shanks, B., Pruszko, R., Clements, D. & Knothe, G. (2004). *Biodiesel Production Technology*, National Renewable Energy Laboratory, NREL/SR-510-36244, Golden CO.

Varma, M. N., Deshpande, P. A. & Madras, G. (2010). Synthesis of biodiesel in supercritical alcohols and supercritical carbon dioxide. *Fuel*, Vol. 89, No. 7, pp. (1641-1646), 0016-2361.

Vasudevan, P. T. & Briggs, M. (2008). Biodiesel production - current state of the art and challenges. *Journal of Industrial Microbiology and Biotechnology*, Vol. 35, No. 5, pp. (421-430), 1367-5435.

Vinokurov, V. A., Barkov, A. V., Krasnopol'skaya, L. M. & Mortikov, E. S. (2010). New methods of manufacturing alternative fuels from renewable feedstock sources. *Chemistry and Technology of Fuels and Oils*, Vol. 46, No. 2, pp. (75-78), 0009-3092.

Wang, L. & Yang, J. (2007). Transesterification of soybean oil with nano-MgO or not in supercritical and subcritical methanol. *Fuel*, Vol. 86, No. 3, pp. (328–333), 0016-2361.

Wen, Z., Yu, S.-T., Yan, J. & Dahlquist, E. (2009). Intensification of biodiesel synthesis using zigzag micro-channel reactors. *Bioresource Technology*, Vol. 100, No. 12, pp. (3504-3560), 0960-8524.

Yan, S. L., DiMaggio, C., Mohan, S., Kim, M., Salley, S. O. & Ng, K. Y. S. (2010). Advancements in heterogeneous catalysis for biodiesel synthesis. *Topics in Catalysis*, Vol. 53, No. 11-12, pp. (721-736), 1022-5528.

Yang, Z. & Xie, W. (2007). Soybean oil transesterification over zinc oxide modified with alkali earth metals. *Fuel Processing Technology*, Vol. 88, No. 6, pp. (631–638), 0378-3820.

Yu, X. H., Wen, Z. Z., Lin, Y., Tu, S. T., Wang, Z. D. & Yan, J. Y. (2010). Intensification of biodiesel synthesis using metal foam reactors. *Fuel*, Vol. 89, No. 11, pp. (3450-3456), 0016-2361.

Zabeti, M., Daud, W. M. A. W. & Aroua, M. K. (2009). Activity of solid acid catalysts for biodiesel production: A review. *Fuel Processing Technology*, Vol. 90, No. 6, pp. (770-777), 0378-3820.

Zeng, H. Y., Deng, X., Wang, Y. J. & Liao, K. B. (2009). Preparation of Mg-Al hydrotalcite by urea method and its catalytic activity for transesterification. *AIChE Journal*, Vol. 55, No. 5, pp. (1229-1235), 0001-1541.

Zhou, C., Wang, C. W., Wang, W. G., Wu, Y. X., Yu, F. Q., Chi, R. A. & Zhang, J. F. (2010). Continuous production of biodiesel from soybean oil using supercritical methanol in a vertical tubular reactor: I. Phase holdup and distribution of intermediate product along the axial direction. *Chinese Journal of Chemical Engineering*, Vol. 18, No. 4, pp. (626-629), 1004-9541.

18

Soybean Biodiesel and Metrology

Vanderléa de Souza, Marcos Paulo Vicentim, Lenise V. Gonçalves, Maurício Guimarães da Fonseca and Viviane Fernandes da Silva
INMETRO- National Institute of Metrology, Standardization and Industrial Quality; Directorate of Industrial and Scientific Metrology; Division of Chemical Metrology
Brazil

1. Introduction

Biodiesel is a renewable fuel defined as a monoalkyl ester derived from vegetable oils, animal fats or microbial oils (algae, bacteria and fungi). The conversion of the fats or oils from these raw materials into biodiesel is possible through enzymatic or chemical reactions, which the most widely employed and studied is the transesterification reaction, involving alcohol and a catalyst. Such process converts triacylglycerols into esters of fatty acids molecules, which present physical-chemical properties and cetane number similar to diesel (Krawczyk, 1996; Ma & Hanna, 1999; Li *et al.*, 2008; ASTM D6751, 2008; Moser, 2009; Knothe *et al.*, 2005; Knothe & Steidley, 2005).

Vegetable oils were first tried for combustion in engines since the early creation of Diesel engines, in the end of 19th century. At that age, the higher cost and lower availability of these oils compared to the just developed petroleum derivates, associated to the higher homogeneity and efficiency gain up to 35% utilizing diesel, led to the complete abandonment of vegetable oils for combustion in engines. However, in the last century, the supply stability of petroleum by some countries has changed, causing drastic petroleum price raise. Thus, worldwide discussions concerning petroleum dependence were retaken, and since the second half of 90's utilization of fuels derived from renewable sources, including biodiesel, has increased in Brazil, Europe, USA and Asia (Costa *et al.*, 2003). In Brazil, social factors, such as new job opportunities, also stimulated biodiesel production.

The direct use of vegetable oils as fuel in compression ignition engines could be considered, but they are problematic due to their high viscosity (about 11-17 times greater than diesel fuel) and low volatility. These oil types do not burn completely and form carbon deposits in the fuel injectors of diesel engines. The viscosity of vegetable oils can be better improved with transesterification reaction, a process which seems to insure very good outcomes in terms of lowering viscosity and enhancing other physicochemical properties. Transesterification is a chemical reaction which proceeds under heat and involves triacylglycerols and an alcohol of lower molecular weights (typically methanol, ethanol, isopropanol or butanol) using homogeneous or heterogeneous substances as catalyst, which typically is an acid or a base, to yield biodiesel and glycerol (Ferella *et al.*, 2010), as presented in Figure 1.

Almost all biodiesel is produced from virgin vegetable oils using the base-catalyzed technique as it is the most economical process for treating virgin vegetable oils, requiring only low temperatures and pressures and producing over 98% conversion yield (provided the starting oil is low in moisture and free fatty acids). However, biodiesel produced from other sources or by other methods may require acid catalysis which is much slower (Ataya *et al.*, 2007).

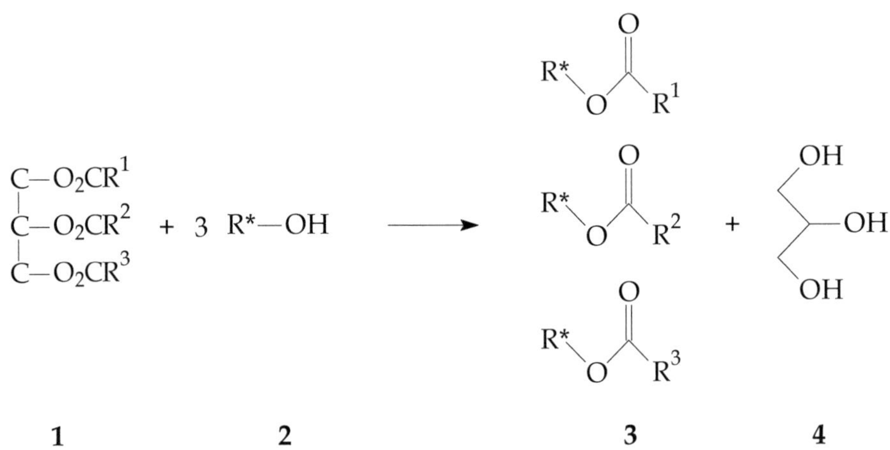

Fig. 1. A general representation of transesterification reaction between a triacylglycerols (1) and an alcohol (2) to give alkyl esters of fatty acids (3) and glycerol (4).

The purification of biodiesel is a crucial step for production of a high-quality product, and choosing of the appropriate techniques is important for this biofuel to become economically viable. Biodiesel and glycerol are typically mutually soluble, but a notable difference in density between biodiesel (880 kg/m3) and glycerol (1050 kg/m3, or more) phases is a property that allows the employment of simple separation techniques such as gravitational settling or centrifugation. Washing can be also applied to remove free glycerol, soap, excess alcohol, and residual catalyst. But in this case, drying of alkyl ester is needed to achieve the stringent limits of biodiesel specification on the amount of water (Atadashi *et al.*, 2011).
Biodiesel presents physical-chemical properties and cetane number similar to diesel, but this biofuel has several advantages over the fossil fuel (petrodiesel). Biodiesel is biodegradable, its sources are renewable, it respects the carbon cycle, and it presents lower toxicity, essentially no sulfurous and no aromatic compounds. The substitution of conventional diesel by biodiesel would reduce sulfur emissions by 20%, carbonic anhydride by 9.8%, non-burned hydrocarbonates by 14.2%, particulate material by 26.8%, and nitrogen oxide by 4.6%, thus reducing most regulated exhaust emissions. Biodiesel presents superior lubricity, higher flash point and positive energy balance (Albuquerque, 2006).
Biodiesel has been utilized blended to petrodiesel for internal combustion engines. Most countries utilize a system known as "factor B" to indicate the volumetric concentration of biodiesel in the blends. So, B100 indicates that a sample is pure biodiesel, while B20 or B5, for example, indicates that the blend has 20% (v/v) or 5% (v/v), respectively, of biodiesel. Some authors report that mixtures containing up to 20% biodiesel can generally be employed in diesel engines without modifications, but most of the authors and the vehicle produces do not recommend employing mixtures containing more than 5% biodiesel (Biodieselbr.com, 2011; Fueleconomy.gov, 2011; Biodiesel.org, 2009). Nowadays, biodiesel sold in Brazil and Europe is a B5 fuel (Biodieselbr.com, 2011; Biopowerlondon.co.uk, 2011).

However, a primary disadvantage of biodiesel is inferior oxidative and storage stability versus petrodiesel, lower volumetric energy content, reduced low temperature operability, susceptibility to hydrolysis and microbial degradation, as well as higher nitrogen oxide emissions (Albuquerque, 2006; Moser, 2009; Knothe *et al.*, 2005; Knothe & Steidley, 2005). Also, the esters which biodiesel is composed can attach to water attributing to this biofuel the hygroscopic property. Some water content also comes from the extraction and transesterification processes. The presence of water in biodiesel reduces the calorific value and enhances engine corrosion. Moreover, water promotes the growth of microorganisms and increases the probability that oxidation products are formed during long-term storage. These oxidation products can cause disturbances in the injection system and in the engine itself (Schlink & Faas, 2009).

The most significant vegetable oils produced worldwide during 2009 were palm (45.13 MMT), soybean (37.69 MMT), rapeseed/canola (21.93 MMT), and sunflower (11.45 MMT) oils (United States Department of Agriculture, 2010). Generally, the most abundant oils or fats in a region are most commonly used as feedstocks for biodiesel production. Thus, for production of biodiesel, rapeseed/canola and sunflower oils are principally used in Europe, palm oil predominates in tropical countries, and soybean oil and animal fats are most commonly used in the USA and Brazil (Moser, 2009; Knothe *et al.*, 2005).

As globalization increases, there is a need for harmonization of technical parameters regarding several products and services (especially commodities) provided around the world. In order to achieve greater transparency, reliability and suitability among these products and services, International Standards have been developed and must be followed by those countries and companies that take part in this worldwide trade. For biodiesel it is not different. Biodiesel must be certified as compliant with accepted fuel standards before combustion in diesel engines. In Europe, specifications for this biofuel are described by the European Committee for Standardization (CEN) through standard EN 14214:2008 (EN 14214:2008, 2009); in the United States the specifications must be according to ASTM D6751-08a (ASTM D6751, 2008); and in Brazil, fuels and biofuels are regulated by National Agency of Petroleum, Gas and Biofuels (ANP), through Resolution ANP no. 7 from March 19, 2008 (ANP Resolution n°7, 2008).

However, international standards are not conclusive, and many times they also are not suitable or accurate for many parameters of many products and services. Several parameters have to be exhaustively studied for the convergence of some rules. In the case of biodiesel, standards that have been applied for this biofuel were originally developed for diesel analysis and adapted for biodiesel. Just a few standards have been developed specifically for biodiesel up this time. Standards have to be constantly attending to the modernization of products, markets and methodologies. In some cases, technical methods employed to analyze some kind of product is simply adapted to analyze another similar one. An example for Biodiesel concerns the water determination method. The water content in biodiesel is ruled by EN 14214:2008 (EN 14214:2008, 2009), ASTM D6751-08a (ASTM D6751, 2008) and in Brazil by Resolution ANP no. 7 (ANP Resolution N°7, 2008). All of them settle the maximum water content as 0.05% (w/w). These standards require Karl Fischer titration for water determination, as described in ISO 12937:2000 (ISO 12937:2000, 2000). Otherwise, this ISO standard was created considering petrodiesel analyses and specifications, but now it has been adapted for water content determination in biodiesel. These adaptations of methods cause errors or at least low accuracy in analyses. Furthermore, many standards, like this one, do not describe a method setting exactly the parameters to be employed by the apparatus.

Determining such low water content in non-aqueous substances with high accuracy is not an easy task and just a few works have been published regarding moisture in biodiesel. Assessments for high accuracy determination of water in biodiesel have been performed by the Laboratory of Organic Analyses from INMETRO (Brazilian Institute of Metrology), which optimized some parameters for commercial soybean biodiesel utilizing Coulometric Karl Fischer Titration coupled to Auto-sampler Oven (Vicentim *et al.*, 2010). Experiments ongoing by this group are still verifying the necessity of further optimizations for biodiesel samples produced form another sources (data unpublished yet).

Complete discussions regarding the need for International Standards, their applications, the mechanisms for biodiesel obtainment, its chemical and physicochemical properties, and considerations about the importance of metrology and its influence on biodiesel quality will be presented in the next sessions of this chapter.

2. Biodiesel synthesis

All over the world there are many research groups searching for fuels from renewable sources due to the imminent serious depletion of fossil resources and also due to an increasing societal ecological environmental awareness. Many types of alternative energy sources have been studied, as solar, wind, water, nuclear (through the cleavage of radioisotopes) and plant biomass. However, nowadays, the only ready-to-use technologies for automotive renewable energy supply, and that has presented excellent results, are the production and utilization of the so-called biofuels, like the bioethanol from sugar cane, corn starch, sugar beet and the biodiesel, especially that one produced from oily crops.

Biodiesel can be defined as mono alkyl esters of fatty acids derived from animal fat and vegetable oils (researches are ongoing for utilization of microbial oils), and obtained mainly through the transesterification reaction. In a general way, this reaction involves triacylglycerols (which are esters) reacting with a small chain aliphatic alcohol, generally methanol, ethanol, isopropanol or butanol, producing a new ester and an alcohol, as shown in Figure 1 (Pinto et al, 2005).

Biodiesel can be derived from the following processes: pyrolysis, cracking, alcoholysis, esterification and transesterification of fats and oils which is the most commonly used process.

Processes like pyrolysis and cracking produce many side products, the reactions are not very selective and the processes require many steps, like removing ash and solid products for example. Pyrolysis, strictly defined, is the conversion of one substance into another by means of heat or by heat with the aid of a catalyst. It involves heating in the absence of air or oxygen and cleavage of chemical bonds to yield small molecules. Pyrolytic chemistry is difficult to characterize because of the variety of reaction paths and the variety of reaction products. The pyrolyzed material can be vegetable oils, animal fats, natural fatty acids and methyl esters of fatty acids. The first pyrolysis of vegetable oil was conducted in an attempt to synthesize petroleum from vegetable oil. Since World War I, many investigators have studied the pyrolysis of vegetable oils to obtain products suitable for fuel. Catalysts have been used in many studies, largely metallic salts, to obtain paraffines and olefins similar to those present in petroleum sources. Soybean oil was thermally decomposed and distilled in air and nitrogen sparged with a standard ASTM distillation apparatus. The total identified hydrocarbons obtained from the distillation of soybean and high oleic safflower oils were 73-77 and 80-88% respectively. The main components were alkanes and alkenes, which

accounted for approximately 60% of the total weight. Carboxylic acids accounted for another 9.6-16.1% (Fangrui & Milford, 1999).
Esterification is a process that consists in two main steps. In the first one the oil is saponified with sodium hydroxide followed by acidification, washing and drying, obtaining a mix of fatty acids. In the final steps the fatty acids are esterified with a small chain alcohol, like methanol, ethanol or isopropyl alcohol.

2.1 Transesterification using catalysts

In a general way transesterification reaction occur catalyzed by an acid (Gerpen, 2005), alkali (Rinaldi et al, 2007), enzyme (Mendes et al, 2011 & Watanabe et al, 2002) or employing heterogeneous catalysis (Mell et al, 2011). The main heterogeneous catalysts are zeolites (Suppes et al, 2004), clays (Jaimasith et al, 2007), ion-exchange resins (Honda et al, 2007) and oxides.
The most used way of catalysis is employing an alkali. The reaction mechanism under alkaline condition occurs in two steps: In the first step sodium hydroxide reacts with methanol, in an acid-base reaction producing a strong base, sodium methoxide and water. In the second step sodium methoxide reacts as a nucleophile and attacks the three carbonyl carbons from the triacylglycerol. A very unstable tetrahedral intermediate is obtained. As a result, the cracking of the triacylglycerol occurs, obtaining three methyl esters (biodiesel) and glycerol.
The most employed transesterifying agent is methanol. Other alcohols may also be used in the preparation of biodiesel, such as ethanol, propanol, isopropanol, and butanol. Ethanol is of particular interest primarily because it is less expensive than methanol in some regions of the world, and biodiesel prepared from bioethanol is completely bio-based. Butanol may also be obtained from biological materials, thus yielding completely bio-based biodiesel as well. Methanol, propanol, and isopropanol are normally produced from petrochemical materials such as methane obtained from natural gas in the case of methanol. Some conditions utilized in these reactions are described below:
Methanolysis: The classic reaction conditions for the methanolysis of vegetable oils or animal fats are 6:1 molar ratio of methanol to oil, 0.5 wt.% alkali catalyst (with respect to TAG), 600 rpm, 60°C reaction temperature, and 1 h reaction time to produce FAME and glycerol.
Ethanolysis: The classic conditions for ethanolysis of vegetable oils or animal fats are 6:1 molar ratio of ethanol to oil, 0.5 wt.% catalyst (with respect to TAG), 600 rpm, 75°C reaction temperature, and 1 h reaction time to produce fatty acid ethyl esters (FAEE) and glycerol.
Butanolysis: The classic conditions for butanolysis of vegetable oils or animal fats are 6:1 molar ratio of butanol to oil, 0.5 wt.% catalyst (with respect to TAG), 600 rpm, 114°C reaction temperature, and 1 h reaction time to produce fatty acid butyl esters and glycerol. Butanol is completely miscible with vegetable oils and animal fats because it is significantly less polar than methanol and ethanol. Consequently, transesterification reactions employing butanol are monophasic throughout. The monophasic nature of butanolysis reactions also complicates purification of the resultant butyl esters (Moser, 2009).

2.1.1 Homogeneous catalysts

Conventional processes include the use of homogeneous alkaline catalysts—NaOH, KOH, NaOMe and KOMe—under mild temperatures (60–80 ◦C) and atmospheric pressure. There are two main factors that affect the cost of traditional biodiesel production: the cost of raw materials and the cost of processing (multiple steps), though the commercialization of resultant glycerol can share the production costs with biodiesel, improving the overall process profitability. In order to reduce the costs associated with feedstock, waste cooking oils, animal fats or non-edible oils could be used. However, the use of homogeneous alkaline catalysts in the transesterification of such fats and oils involves several troubles due to the presence of large amounts of free fatty acids (FFAs). Of course, alkaline catalysts can be used to process these raw materials, but a large consumption of catalyst as well as methanol is compulsory to achieve biodiesel of standard specifications. Thus, FFA concentration in the oil inlet stream is usually controlled below 0.5% (w/w), avoiding the formation of high soap concentrations as a consequence of the reaction of FFAs with the basic catalyst. The soap causes processing problems downstream in the product separation because of emulsion formation. Usually, this problem is overcome through a previous esterification step where FFAs are firstly esterified to FAMEs using a homogeneous acid catalyst, and then, once the acid homogeneous catalyst has been removed, transesterification of triacylglycerols is performed as usual by means of an alkaline catalyst. Likewise, homogeneous acid catalysts (H_2SO_4, HCl, BF_3, H_3PO_4) have been proposed to promote simultaneous esterification of FFAs and transesterification of triacylglycerols in a single catalytic step, thus avoiding the pre-conditioning step when using low cost feedstock with high FFA content. However, these catalysts are less active for transesterification than alkaline catalysts and therefore higher pressure and temperature, methanol to oil molar ratio and catalyst concentration are required to yield adequate transesterification reaction rates. Hence, despite its insensitivity to free fatty acids in the feedstock, acid-catalyzed transesterification has been largely ignored mainly due to its relatively slower reaction rate. (Melero et. al, 2009).

2.1.2 Heterogeneous catalysts

The use of heterogeneous catalysts (Wang et al, 2007 and Leclercq et al, 2001) has as main advantage the reaction work-up, the post reaction treatment, the purification steps and the separation steps. These catalysts can be easily removed from the reaction medium and even can be reused. Another interesting factor is based in the fact that these catalysts avoid the formation of undesirable side products, like the saponification products (Botts et al, 2001; Thomasevic & Marincovic, 2003). The biggest difficulty at this type of reaction is the diffusion between the systems oil/catalyst/solvent (Gryglewics, 1999). For the soybean biodiesel production, the catalysts that have been commonly employed are tin, zinc and aluminum, as Al_2O_3, ZnO and $(Al_2O_3)_8(SnO_2)$, for example (Mello et al, 2011). Other processes have used an heterogeneous catalyst of a spinel mixed oxide of two (non noble) metals, which eliminate several neutralization and washing steps needed for process using heterogeneous catalysts (Helwani et. al, 2009).

2.1.3 Alkaline catalysts

Alkaline catalysis (Zhou et al, 2003) is a procedure that generally uses sodium and potassium alkoxides, and some times sodium and potassium hydroxides or carbonates. Among these three groups, alkoxides have the advantage of performing reactions at mild temperatures, they provide high yields of esters derived from fatty acids and they are not corrosives like the acid catalysts. On the other hand, these catalysts are hygroscopic, more expensive and usually result in side products, such as the saponification ones.

2.1.4 Acid catalysts

Sulfuric and hydrochloric acid compounds are the main catalysts. This catalysis (Mohamad and Ali, 2002) has the advantage of avoiding the formation of side products and obtaining high yield formation of alkyl esters. However, reactions in acid media are highly corrosive and the work up is more difficult, seen it needs a special treatment to neutralize the reaction medium

2.1.5 Enzymes catalysts

A fourth class of catalysts employed for biodiesel production is enzyme (Fukuda et al, 2001). The enzymes allow the use of mild temperature reactions, between 20 and 60 °C, excess of alcohol is dispensed, the reactions can be performed with or without a solvent and the catalyst can be reused several times (Shimada et al, 2002). Other great advantages are the easy work-up, dispensing neutralization and deodorization of the reaction medium (Bielecki et al, 2009). The disadvantages are related to the fact that enzymes are very specifics, expensive and very sensible to alcohols, causing their deactivation (Manduzzi et al, 2008). The literature (Modi et al, 2007) shows that this problem can be solved by the use of small amounts of water (Kaieda et al, 2001 and Ban at al, 1999). Another research group showed that the use of organic solvents (Narasisham et al, 2008) can activate the enzymes, in special the use of dioxane and petroleum ether (Dennis et al, 2008), for example. Watanabe and his research group has developed a methodology to produce biodiesel from soybean degummed oil by the use of the lipase (an enzyme specific for hydrolysis of lipids, like triacyglycerides) from *Candida antarctica* in a free solvent system (Watanabe et al, 2002). Another procedure was performed by Liu et al. (2005) studying the acyl group migration with immobilized lipozyme TL catalyzing the production of biodiesel from soybean oil (Noureddini et al, 2005).

2.2 New process for biodiesel obtainment

2.2.1 Microwave and Ultrasound

Many researches seek for the improvement of catalysts in biodiesel production. Reactions employing ultrasound (Santos et al, 2009) and microwave (Leadbeater & Stencel, 2006) techniques represent a great advance. Ultrasound (Chand et al, 2010) and microwaves (Barnard et al, 2007) as auxiliary techniques facilitate the interaction between methoxide ions and reagents, increasing the process efficiency, obtaining higher yields in a shorter reaction time. Reaction employing these techniques can be performed at mild temperatures due to a higher kinetic energy in the reaction medium, facilitating also the miscibility among the reactants (Fukuda, 2001).

2.2.2 Transesterification using supercritical fluids

This is a non catalytic method to produce biodiesel, which has the several advantages. One of them concerns the shorter reaction time than the traditional catalyzed transesterification. This is possible because the initial reaction lag time is overcome due to the reaction is proceeded in a single homogeneous phase since the supercritical methanol is fully miscible with the vegetable oils. Moreover, the reaction rate is very high and the subsequent purification is much simpler than that of the conventional process. The supercritical route is also characterized by high yield because of simultaneous transesterification of triacylglycerols and esterification of fatty acids. This process is environmentally friendly seen that waste water containing alkali or acid catalysts is not produced. The disadvantages of this process regard the high costs, the necessity of a high pressure system (200-400 bar), high temperatures (350-400°C) and high methanol/oil rates (Balat, M.H., 2008; Melero et. al, 2009).

2.2.3 Hydrotreating

Hydrotreating is a process that produces biodiesel through a hydrotreatment of triacylglycerols. The hydrocarbons are produced by two reaction pathways: hydrodeoxygenation (HDO) and hydrodecarboxylation (HDC). n-Alkanes originating from HDO have the same carbon number as the original fatty acid chain, i.e., even carbon number, typically 16 or 18. Water and propane are the main reaction by-products of this route (Snare et al, 2007).

2.3 Biodiesel purification process

Separation and purification of biodiesel is a critical task. Normally, the crude biodiesel produced by homogeneous catalysis can be separated from glycerol by simple gravitational settling or centrifugation, due to their notable difference in phase density (biodiesel 880 kg/m3 and glycerol 1050 kg/m3 or more). Washing ester phase with water or an acid mineral or base solution to remove base/acid catalyst residues, for example, can be also applied to remove free glycerol, soap, excess alcohol, and residual catalyst. Finally the biodiesel is dried after neutralization. For methanol recycle vacuum distillation can be used prior to glycerin purification. When the biodiesel is obtained by a heterogeneous catalysis, this one is removed by a filtration process.

However, these conventional technologies and other ones like decantation, washing with ether and the use of absorbents have proven to be inefficient, time and energy consumptive, and less cost effective. On the other hand, the involvement of membrane reactor and separative membrane shows great promise for the separation and purification of biodiesel. Membrane technology needs to be explored and exploited to overcome the difficulties usually encountered in the separation and purification of biodiesel (Zhang et al, 2003; Atadashi et al., 2011).

3. Biodiesel properties and their influence on engine performance

3.1 The fatty acid composition of feedstocks and the influence on the properties of biodiesel fuel

The most common feedstocks for biodiesel production are commodities such as vegetable oils derived from soybean, palm and sunflower seed. These materials possess fatty acid profiles consisting primarily of five fatty acids with carbon chains containing 16 to 18 carbon atoms (C16 to C18) namely palmitic acid (hexadecanoic-C16: 0), stearic acid (octadecanoic, C18: 0), oleic acid (9 (Z)-octadecenoic - C18: 1), linoleic (9 (Z), 12 (Z)-octadecadienoic acid - C18: 2), linolenic (9 (Z), 12 (Z), 15 (Z) , octadecatrienoic acid- C18: 3). The proportions of different fatty acids in feedstocks influence the properties of biodiesel. Some of the most relevant properties to be considered for a biodiesel candidate to be used as a substitute for diesel fuel (or blended with the same) are cetane number, viscosity, cold flow properties and oxidative stability. Lubricity is another important parameter for a fuel but it is independent on the fatty acid composition.

Two major problems to be overcome in biodiesel are the poor properties at low temperatures and low oxidative stability. In most cases these two problems occur with the same sample. They result from physical and chemical properties of fatty esters, the major components of biodiesel and minor constituents that arise during the transesterification reaction or are from raw materials.

The profile of methyl esters found in greater proportion in soybean is about 11% C16: 0, 4% C18: 0, 21-24% C18: 1, 49-53% C18:2, 7-8% C18: 3 which provides cetane number in the range of 48-52, kinematic viscosity at 40 °C equal to 4.10 to 4.15 mm^2s^{-1} and cloud point approximately equal to 0 °C (Knothe et al., 2005, Mittelbach and Remschmidt, 2004). Rapessed (canola) methyl esters have a fatty acid profile approximately 4% C16:0, 2% C18:0, 58-62% C18:1, 21-24% C18:2, 10-11% C18:3 and present cetane number in the range of 51-55, kinematic viscosity at 40 °C around 4,5 mm^2s^{-1} and cloud point of approximately -3 °C (Knothe et al., 2005, Mittelbach and Remschmidt, 2004). Thus the difference in fatty acid profile, more specifically concerning C18:1 and C18:2 contents, which had their values almost reversed in the case presented, causes a noticeable change in fuel properties.

Many researches have focused on resolving or at least reducing problems related to low oxidative stability and cold flow properties of biodiesel. Some trials in this way involves the addition of additives and changes in the composition of fatty esters, that can be reached varying either the reactive alcohol or the oil fatty acid profile. Changing the fatty acid profile can be achieved by physical methods, genetic modification of feedstock or use of alternative feedstocks with different fatty acid profiles.

Important features regarding the use of neat biodiesel or its blends with diesel fuel include reduced emissions, with the exception of nitrogen oxides, compared to petrodiesel (petroleum-derived diesel fuel), biodegradability, absence of sulfur, inherent lubricity, positive energy balance, higher flash point, compatibility with existing infrastructure for distribution of fuel, to be renewable and a domestic source. The American ASTM D6751-08a, the European EN 14214:2008 and the Brazilian ANP nº 7 standards deal with the technical specifications for biodiesel to be used in internal combustion cycle diesel engine taking into account the advantage of utilizing the existing infrastructure for distribution of diesel ensuring fuel quality for the final consumer. Table 1 shows the specifications recommended by American, European and Brazilian standards aiming biodiesel utilization as fuel.

3.2 The influence of cetane number on combustion and atmospheric emissions

The cetane number (CN) is a dimensionless parameter related to the ignition delay time after fuel injection into the combustion chamber of a diesel engine. A higher cetane number results in a shorter ignition delay time and vice versa. A cetane scale was established, being hexadecane commonly used as reference compound, with CN = 100, and 2,2,4,4,6,8,8-heptamethylnonane, a highly branched compound with poor ignition quality in a diesel engine, with CN =15.

The cetane scale explains why the triacylglycerols, such as those found in vegetable oils, animal fats and their derivatives, are suitable alternatives to diesel fuel. The reason is the long chain, linear and unbranched fatty acids, chemically similar to those in n-alkanes of conventional diesel fuels with good quality.

The cetane number of fatty esters increases with the increase of saturation and carbon chain. Thus, the CN of methyl palmitate and methyl stearate (C16: 0 and C18: 0) is greater than 80 (Knothe et al., 2003), the CN of methyl oleate (C18: 1) is in the range of 55-58, the methyl linolenate is (C18: 2) around 40 and the methyl linolenato is (C18: 3) around 25. Esters derived from branched alcohols such as isopropanol have CN values comparable to methyl esters or other ester with alkylic chain (Knothe et al., 2003, Zhang & Gerpen, 1996) linear, although the cost of production once isopropyl alcohol is more expensive than methanol and ethanol costs.

In general biodiesel does not require additives to improve cetane number, because its cetane number generally reaches the minimum values established in the international technical specifications. An exception may be the methyl esters of soybean that did not reach the minimum of 51 set by EN 14214:2008 (EN 14214:2008, 2009) but usually reach the minimum set of 47 recommended in ASTM D6751 (ASTM D6751-08a, 2008), as shown in Table 1.

Cetane number may influence both the quality of combustion and vehicle emissions. Several international agencies like the EPA (Environmental Protection Agency - USA) and the CONAMA (National Environment Council - Brazil) set limits and goals for reducing pollutants automotive emissions. In diesel cycle engines, the main pollutants are hydrocarbons, carbon monoxide, nitrogen oxides (NOx) and particulate matter. Reducing these emissions requires improving the combustion process, the treatment of exhaust gases from existing engines and technical fuels specifications. A low cetane number leads to difficulties in cold starting, increases emissions and noise level of combustion. If the cetane number is high may occur an increase in particulate emissions but NOx emissions decrease. Samples of biodiesel with low level of triacylglicerols, especially those with polyunsaturated fatty acids of C18:3, should show low levels of NOx emissions. Linear correlation was obtained between the level of unsaturation of biodiesel indicated by iodine number, the density of biodiesel and NOx emissions (McCormick et al., 2001). Thus little amounts of unsaturated fatty acids may reduce the density and the NOx emissions. An important property of biodiesel is its ability to reduce total particulate emissions of the engine and also carbon monoxide and hydrocarbons contents of exhaust gases. However biodiesel causes an increase in NOx emissions. Increasing CN to a certain level (around 60) implies in the reduction of NOx emissions (Landommatos et al., 1996).

An experiment was conducted with the OM 611 diesel engine light load of Damler Benz with ultra low sulfur content diesel (ULSD), conventional diesel and B20 blend of pure methylic soybean biodiesel and ULSD. The results obtained with the B20 blend showed no differences in NOx content compared to the two reference diesel fuels. Reductions of particulate matter by 32% and 14%, respectively, compared to conventional diesel fuel and USLD were observed with B20 blend (Sirman, et al., 2000).

The causes for the increase of NOx associated with biodiesel for fuel injection systems are related to a small displacement in the range of fuel injection which is caused by differences in mechanical properties of biodiesel compared to conventional diesel (Tat & van Gerpen, 2003; Monyem et al., 2001). Due to the higher modulus of compressibility (or sound speed) of biodiesel, there is a faster transfer of the pressure wave of the injection pump to the injector needle resulting in anticipation of lifting the needle and the production of a small advance in the injection interval. It was observed that samples of B100 derived from soybeans produces an increase of one degree in the injection interval, which was accompanied by a four degree at the start of combustion (Sybist & Boehman, 2003). Strategies that can be used to reduce NOx emissions to a level equivalent to that of conventional diesel involve increase of cetane number by use of additives.

Specification	Standards					
	ASTM D6751-08a		EN 14214:2008		ANP n º7	
	Test method	Limit	Test method	Limit	Test method	Limit
Cetane number	ASTM D613; D6890	47 minimum	EN ISO 5165	51 minimum	ASTM D613; D6890 EN ISO 5165	Report
Kinematic viscosity	ASTM D445	1.9-6.0 mm^2s^{-1}	EN ISO 3104	3.5-5.0 mm^2s^{-1}	ASTM D445; EN ISO 3104; ABNT NBR10441	3.0-6.0 mm^2s^{-1}
Oxidative stability	EN 14112	3h minimum	EN 14112	6h minimum	EN 14112	6h minimum
Cloud Point	ASTM D2500	Report	-	-	-	-
Cold filter plugging point	-	-	EN 116	Depending on time of year and location	ASTM 6371; EN 116; ABNT NBR14747	19 ºC
Cold soak/Filterability	Annex to D6751	Report	-	-	EN 12662	24 mg/Kg (max.)

ASTM - American Society for Testing and Materials; ISO - International Standards Organization; ANP - National Agency of Oil, Gas and Biofuels ; NBR - Brazilian Standard; ABNT - Brazilian Association of Technical Standards

Table 1. Specifications of biodiesel standards that affect the properties of alkyl esters as fuel in diesel cycle engines [(a)].

It was observed that B20 blends of soy diesel respond well to conventional peroxide di-t-butyl, a cetane improver, when tested on DDC Series 60 engines of 1991 (McCormick , et al., 2001). The biodiesel NOx was reduced by 6.2% without the contribution of 9.1% in reducing emissions of particulate matter to be compromised and B20 blend produced no noticeable increase in NOx of this engine. The peroxide, di-t-butyl nitrate and 2-ethylhexyl were tested in a similar engine (Sharp, 1994) and the reduced levels of NOx in exhaust emissions were confirmed. Notice the economy of this procedure if necessary high levels of additives.

3.3 The importance of viscosity in the use of biodiesel as fuel

Viscosity is one of the properties that most affect the use of biodiesel as a fuel since the atomization process, the initial stage of combustion in a diesel engine, is significantly affected by the viscosity of the fuel. The viscosity of the transesterified oils, ie, biodiesel is less than their vegetable oil sources, which explains the failure to use pure vegetable oils as alternative fuels to diesel. The high viscosity of untransesterified oils leads to operational problems in diesel engine for example increased engine deposits. Viscosity in the form of Kinematic viscosity is specified in quality standards of biodiesel, which exhibit a range with minimum and maximum values for this parameter. Although there are standardized methods for determining the kinematic viscosity as shown in Table 1, several studies have been conducted to predict the viscosity of biodiesel from its composition of fatty ester.

Assuming a soy biodiesel made from soybean oil containing 0.1% C14: 0, 10.3% C16: 0, 4.7% C18: 0, 22.5% C18: 1, 54 1% of C18: 2 and 8.3% of C18: 3, Allen et al., 1999 using the equation of Grunberg & Nissan, 1949, modified, predicted the value of 3.79 mm^2s^{-1} for the viscosity of methylic soybean biodiesel. The viscosity of fatty esters increases with the chain length and with increasing degree of saturation (Kern and Van Nostrand, 1949).

This rule also applies to alcohol used in the reaction, since the viscosity of ethyl esters is slightly higher than that of methyl esters. The configuration of double bonds also influences the viscosity. If there are only double bonds in cis configuration is observed remarkable reduction in viscosity, as well as esters with double bonds in the trans configuration have viscosities similar to the corresponding saturated esters (Kern & Van Nostrand, 1949).

3.4 Cold flow properties of biodiesel and its blends with diesel

3.4.1 Biodiesel and cold flow properties

The mixtures such as biodiesel do not possess defined melting points, but melting ranges. This fact reflects in the specifications used in biodiesel standards.

The cloud point (CP) is the temperature at which the first solids appear, but the fuel can still flow, although these solids can lead to fuel filter plugging (Dunn and Bagby, 1995).

The pour point, usually a few degrees below the cloud point, is the temperature at which the fuel can no longer be freely poured.

Several other methods exist for determining the low-temperature properties of biodiesel. These are the cold filter plugging point (CFPP) and low-temperature flow test (LTFT) (Dunn and Bagby, 1995). The CP and CFPP are included in biodiesel standards without severity since in ASTM D6751 for the value of CP only a report is required and the CFPP value in EN 14214 can vary with time of year and geographic location. The low-temperature properties of biodiesel are also influenced by the properties of individual components. The melting point of fatty esters generally increase with chain length (although chains with odd numbers of carbon have slightly lower melting points that the preceding even-number chain) and increasing saturation (Knothe, 2009).

Intending to provide the industry with an independently generated set of cold flow information on a variety of fuels in the market of the United States in 2009 with the new Ultra Low Sulfur Diesel Fuel (ULSD), cloud point, cold filter plugging point (CFPP) and low temperature flow test (LTFT) methods were used to assess the cold-flow properties for seven different biodiesel fuels blended with four different ULSD fuels representing the span of the market in 2009 (Heck, Thaeler, Howell and Hayes, 2009). The neat fuels were tested in addition to biodiesel blends with ratios of 2% biodiesel (B2), 5% biodiesel (B5), 11% biodiesel (B11), 20% biodiesel (B20) and 50% biodiesel (B50) for cloud point, CFPP, and LTFT. The pour point of the neat biodiesel and B50 blends were also analyzed. Three petrodiesel fuels with cloud points of -47,5 ^{o}C, -16 ^{o}C and -11 ^{o}C were used to produce petrodiesel having target cloud points of -40 ^{o}C, -34,4 ^{o}C, -26,1 ^{o}C, and -12,2 ^{o}C. Seven biodiesel (B100) samples were selected. Three of them with low cloud point, vegetable oil base, in this case soybean oil from various manufacturing processes (distilled biodiesel, non distilled biodiesel from hexane extracted oil, non distilled biodiesel from extruder-expeller oil) were collected from commercial biodiesel producers. Four additional biodiesels of mid to mid-high to high cloud points from commercial biodiesel producers representing mixed saturation levels, blends of soybean, animal, and recycled oil based biodiesel. The cloud point values of the various B100 samples were -2,5 ºC, -2,0 ºC, -1,5 ºC, 1,0 ºC, 7,0 ºC , 8,0 ºC , and +12,0 ºC. In some cases the impact on cold flow properties of blending biodiesel with petrodiesel appeared to be mostly linear, while in others the impact was curvilinear. In all cases except a few with low blends of biodiesel where the CFPP of the blend was slightly below that of the petrodiesel, the blended fuel values fell between the pure petrodiesel and pure biodiesel values for all three cold flow measurements.

Biodiesel blends primarily B20 of soybean biodiesel (SME) and diesel n º2 have also been used in a variety of climates including some of the coldest weather on record without cold flow problems. A study to determine the CFPP of blends containing up 20% (SME) and Number 2 (nº 2) diesel was conducted. The University of Missouri prepared the samples that were analyzed in the Cleveland Technical Center in Kansas City, United States. The characteristics of the nº 2 Diesel and the SME are shown in Table 2. The results suggest that the blends with the highest content of biodiesel begin to gel first. Higher concentrations of biodiesel, eg, above 20% may not be suitable for use in cold climates without mixing large quantities of kerosene in combination with cold flow proven enhancers specific to the conventional diesel (National Biodiesel Board- 2007/2008).

Blends's components	Cloud Point (ºC)	Pour Point (ºC)	CFPP (ºC)
n º2 Diesel	-15,6 ºC	-34,4 ºC	-17,2 ºC
Soy Methyl Esters (SME)	0 ºC	-3,9 ºC	-5,6 ºC

Table 2. Characteristics of the components of the blends.

Literature data show that ethyl esters and specially isopropyl esters improve low-temperature properties of biodiesel compared to methyl esters. Isopropyl and isobutyl esters of common soybean oil exhibited crystallization temperatures 7-11 and 12-14 ºC lower than the corresponding methyl esters (Lee at all., 1995). The data suggest that the fuel blend that begins to gel first contains the highest concentration of biodiesel. Higher concentrations of biodiesel, eg, above 20% may not be suitable for use in cold climates without mixing large quantities of kerosene in combination with proven enhancers of cold-flow properties specific to the conventional diesel.

3.4.2 Diesel fuel background information relevant to biodiesel

The cold weather operability of diesel fuel is defined as the lowest temperature which a vehicle will operate without loss of power due to waxing of the fuel delivery system. Diesel fuels composition and cold flow properties vary greatly across the United States. Cold flow characteristics of diesel fuels are influenced by the source of the crude oil they are made from, how they are refined and if they are blended to improve performance during cold weather. The cold temperature properties of diesel fuel vary across the country depending on the time of year the fuel is produced and the climate. Generally, diesel fuels used in cold climates have better cold flow characteristics than diesel fuels used in warmer regions. Both of these statements have a direct impact on the operability of biodiesel blends in cold weather.

The refining process separates the crude oil into mixtures of its constituents, based primarily on their volatility. Diesel fuels are on the heavy end of a barrel of crude oil. This gives diesel fuel its high BTU content and power, but also causes problems with diesel vehicle operation in cold weather when this conventional diesel fuel can gel. This is not an issue for gasoline vehicles. A tremendous amount of effort has been spent over the years to understand how to deal with the cold flow properties—or the low temperature operability--of existing petroleum based diesel fuel. The low temperature operability of diesel fuel is commonly characterized by the cloud point, and the cold filter plugging point (CFPP) or the low temperature filterability test (LTFT). In general, Number 2 diesel fuel will develop low temperature problems sooner than will Number 1 diesel fuel. Number 1 diesel fuel is sometimes referred as kerosene. The gelling of diesel fuel in cold climates is a commonly known phenomenon and diesel fuel suppliers, as well as customers and diesel engine designers, have learned over time to manage the cold flow problems associated with Number 2 diesel fuel in the winter time. The leading options to handle cold weather with diesel fuel are: -Blending with kerosene;-Utilization of an additive that enhances cold flow properties;-Utilization of fuel tank, fuel filter or fuel line heaters;-Storage of the vehicles in or near a building when not in use. In most diesel engine systems today, excess diesel fuel is brought to the engine and warm fuel that has come close to the engine is recycled back to the fuel tank. This assists in keeping the fuel from gelling in cold weather. This is, in part, why diesel engines are kept running overnight at truck stops in cold climates (Bickell, 2008; Krishna & Butcher, 2008; Joshi & Pegg, 2007).

3.4.3 The impact of minor components of the vegetable oils in the cold flow properties of biodiesel

The presence of sterol glucosides (SGs), wax, monoglycerides, saturated fatty acids and polymers in both B100 and blends with petrodiesel can limit the application of these fuels due to problems with precipitation. In the last years researches have shown that biodiesel precipitations can arise even if specifications of this biofuel are met. Special attention has been dedicated to the SGs, components commonly found in vegetables and in oils derived from soybean, rapeseed and palm that will be processed to produce biodiesel. Usually the concentration of SGs in vegetable oils is not significant since they are mostly found as sterol glucosides acylated (ASGs). The ASGs have average solubility in vegetable oils, but after transesterification they are broken down chemically by removing the side chain containing the fatty acid and they are converted partially to SGs. This class of compounds is not soluble in biodiesel and its crystallization is extremely slow and depends on temperature, other impurities (as crystallization nuclei) and surface effects. However, even a brand new biodiesel, meeting all the specifications, presents precipitation of SGs after a few days of storage/transport.

The spontaneous clogging of the filters in the production unit or in the supply chain has been observed in Minnesota, United States with a B2 blend of soybean methyl esters and diesel. Several other places in the world have also observed this occurrence with B5 blend. The precipitates do not contain only SGs but also ASGs and other substances. In some cases sources containing higher concentrations of ASGs could be responsible for the deposits. The concentration of SGs and ASGs in vegetable oils depends on the feedstock and on the process used by industry to obtain them. The literature suggests that the highest concentrations of SGs and ASGs will be found in soybean and palm oil. Rapeseed oil usually has low concentrations of these compounds. In order to minimize the effect of SGs and ASGs on the FAME-Biodiesel (Fatty Acid Methyl Esters -biodiesel) and their blends with petrodiesel, special treatments of the esters in an oil refinery, or the use of adsorbents could be possible solutions for a post-processing of FAME (Haupt et al., 2009).

3.5 The effect of the impurities in the quality of biodiesel

Biodiesel is composed by alkyl esters (generally methyl esters) that can be analyzed as tool for controlling the transesterification yield. Low concentration of triacylglycerols is an indicative that transesterification is almost complete. Seen that this kind of reaction is reversible, excess of alcohol must be added to ensure that transesterification will prone to esters production. If great amount of alcohol remains in the biodiesel its flash point decreases and problems with storage and transport can occur. Glycerine is a by-product of transesterification that must be recovered in order to avoid solid in diesel engines. Standard methods that must be employed to determine triacylglycerols, alcohol (methanol), total ester and glycerine are presented in Figure 2.

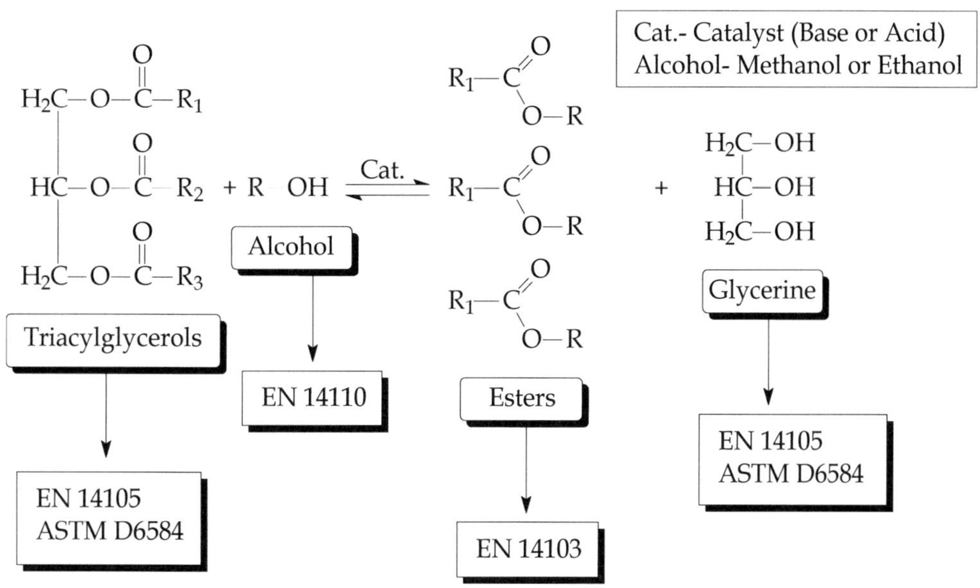

Fig. 2. Methods to quantify some organic impurities present in biodiesel.

Transesterification reaction proceeds in three main steps, shown in Figure 3. Firstly the triacylglycerols are transformed into diacylglycerols and then, these ones are converted into monoacylglycerols, which in turn reacts with alcohol yielding glycerine and an ester. Glycerine can be present in biodiesel in a free form or combined with glycerides. The total glycerine is the sum of these 2 glycerine forms. Maximum limits of methanol, glycerides, free and total glycerine contents in biodiesel, as standard methods for determination of these parameters are shown in Table 3.

All the methods described in Table 3, for determination of the concentration of the organic compounds in biodiesel, employ gas chromatography. So, a typical soybean biodiesel chromatogram, acquired in accordance with EN 14105 standard, is presented in Figure 4. This chromatogram shows the peak of free glycerin (1) and of the internal standards (butanetriol (2) and tricaprine (5)) utilized to quantify free glycerine and mono, di and triglycerides, respectively. It is also observed the regions where the methyl esters (3), monoglycerides (4), diglycerides (6) and triacylglycerols (7) are eluted. This standard method was developed for rapeseed methyl esters determination, but they have been applied successfully for the same determination in soybean and sunflower derivate. In Brazil, Resolution ANP nº7 demands the method validation when EN 14105 is employed to analyze biodiesel samples derived from feedstocks other than rapessed, or when biodiesel was produced from by ethylic route.

$$\text{Triacylglycerol} + \text{R* OH} \xrightleftharpoons{\text{Cat.}} \text{Diacylglycerol} + \text{RCOOR*}$$

$$\text{Diacylglycerol} + \text{R* OH} \xrightleftharpoons{\text{Cat.}} \text{Monoacylglycerol} + \text{RCOOR*}$$

$$\text{Monoacylglycerol} + \text{R* OH} \xrightleftharpoons{\text{Cat.}} \text{Glycerine} + \text{RCOOR*}$$

Fig. 3. Steps of the transesterification reaction.

Specification	Standards					
	ASTM D6751-08a		EN 14214:2008		ANP nº7	
	Test method	Limit (g/100g)	Test method	Limit (g/100g)	Test method	Limit (g/100g)
Methanol	EN 14110	0,20 max.	EN 14110	0,20 max.	EN 14110	0,20 max.
Free glycerine	ASTM D6584	0,02 max.	EN 14105	0,02 max.	ASTM D6584 EN 14105	0,02 max.
Monoglycerides	-	-	EN 14105	0,80 max.	ASTM D6584 EN 14105	0,80 max.
Diglycerides	-	-	EN 14105	0,20 max.	ASTM D6584 EN 14105	0,20 max.
Triglycerides	-	-	EN 14105	0,20 max.	ASTM D6584 EN 14105	0,20 max.
Total glycerine	ASTM D6584	0,24	EN 14105	0,25 max.	ASTM D6584 EN 14105	0,25 max.

ASTM - American Society for Testing and Materials; ISO - International Standards Organization; ANP - National Agency of Oil, Gas and Biofuels

Table 3. Methods and limits of the impurities present in biodiesel.

3.6 Oxidative stability

The oxidation of fatty acid chain is a complex process proceeded by a variety of mechanisms. Oxidation of biodiesel is due to the unsaturation in fatty acid chain and presence of double bonds in the molecule which offers high level of reactivity with O_2, especially, when it is placed in contact with air/water. The primary oxidation products of double bonds are unstable allylic hydroperoxides which are unstable and easily form a variety of secondary oxidation products. This includes the rearrangement of product of similar molecular weights to give short chain aldehydes, acids compounds and high molecular weight materials.

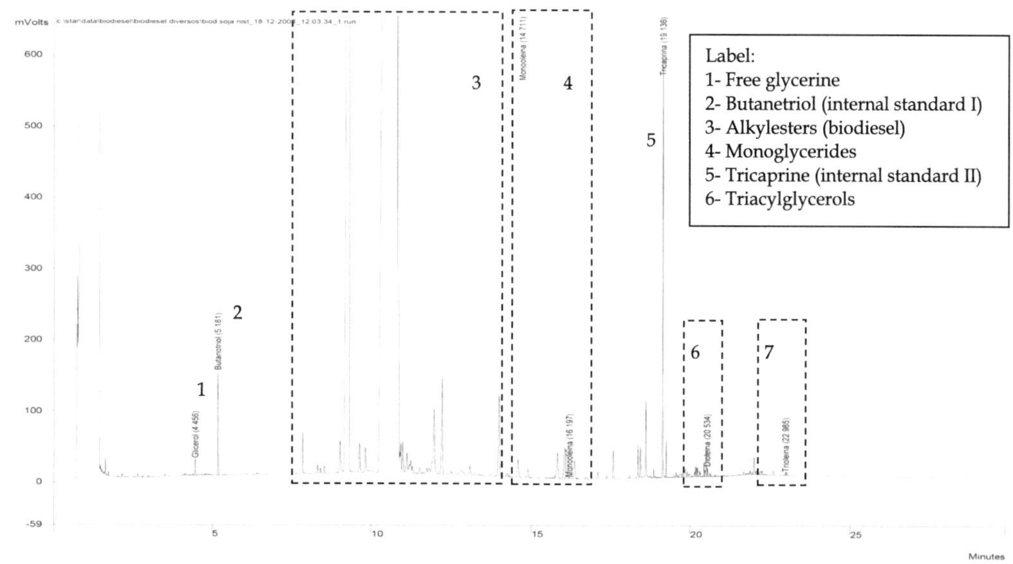

Fig. 4. Soybean biodiesel chromatogram obtained in accordance with EN 14105 standard (Source: Organic Analysis Laboratory- INMETRO - 2008).

Peroxidation occurs by a set of reactions categorized as initiation, propagation, and termination, as shows Figure 5. The reaction mechanism involved in the first step is the removal of hydrogen from a carbon atom to produce a carbon free radical. If diatomic oxygen is present, the subsequent reaction to form a peroxyl radical becomes extremely fast, not allowing significant alternatives for the carbon-based free radical. The peroxyl free radical is not reactive compared to carbon free radical, but is sufficiently reactive to quickly abstract hydrogen from a carbon to form another carbon radical and a hydroperoxide (ROOH). The new carbon free radical can then react with diatomic oxygen to continue the propagation cycle. This chain reaction terminates when two free radicals react with each other to yield stable products.
Fatty oils that contain more poly-unsaturation are more prone to oxidation. Literature reveals the relative rate of oxidation for the methyl esters of oleic (18:1), linoleic (18:2), and linolenic (18:3) acids to be 1:12:25 (Siddharth & Sharma, 2010).

Initiation: $RH + I \rightarrow R\cdot + IH$

Propagation: $R\cdot + O_2 \rightarrow ROO\cdot$

$ROO\cdot + RH \rightarrow ROOH + R\cdot$

Termination: $R\cdot + R\cdot \rightarrow R\text{–}R$

$ROO\cdot + ROO\cdot \rightarrow$ Stable Products

Fig. 5. Mechanism of peroxidation of fatty acids

The biodiesel oxidative stability study is a very important parameter to measure the product quality, mainly about its feedstock. This parameter is a measure of time required to reach the point where the oxidation increases sharply. This methodology is useful to determinate the final biodiesel stability under several oxidative conditions. Useful appropriate oxidative automatic techniques are Petrooxy, differential scanning calorimetry (DSC), Pressure

Differential Scanning Calorimetry (PDSC) (Dufaure et al, 1999) and mainly Rancimat technique. At the Rancimat technique, oxidative stability is based on the electrolytic conductivity increase (Hadorn & Zurcher, 1974.). The biodiesel is prematurely aged by the thermal decomposition. The formed products by the decomposition are blown by an air flow (10L/ 110 °C) into a measuring cell that contains bi-distilled, ionized water. The induction time is determined by the conductivity measure and this is totally automatic. Rancimat is the most used technique to determine finalized biodiesel stability, under oxidative accelerated conditions, according to standard EN14112.
At the PetroOxy Technique, the sample is inducted to oxidation through an intense oxygen flow, manipulating by this way the stability conditions through a specific apparatus. The analysis time is recorded as the required time to the sample absorbs 10% of oxygen pressure.
The differential scanning calorimetry (DSC) monitors the difference in energy provided/released between the sample (reagent system) and the reference system (inert) as a function of temperature when both the system are subjected to a controlled temperature program. Changes in temperature sample are caused by rearrangements of induced phase changes, dehydration reaction, dissociation or decomposition reactions, oxidation or reduction reaction, gelatinization and other chemical reactions.
The Pressure Differential Scanning Calorimetry (PDSC) is a thermo analytical technique that measures the oxidative stability using a differential heat flow between sample and reference thermocouple under variations of temperatures and pressure. This technique differs from the Rancimat for being a fast method and presents a more variable - the pressure, allowing to work at low temperatures and using a small amount of sample (Candeia, 2009).

3.6.1 Antioxidants used in biodiesel

Most of biodiesel has a lower value of oxidative stability than recommended by current legislation (Ji-Yeon, 2008 & Ferrari, 2009) (Table 4), the soybean derivative has also the same inconvenient. This characteristic is due to the rich composition in mono and polyunsaturated fatty acids from the soybean oil.

Source of Biodiesel	Oxidative Stability (h)
Sunflower	1,17
Jatropha	3,23
Soybean	3,87
Palm	11,00

Table 4. Oxidative stability of biodiesel samples produced from different sources.

Compounds containing allylic and bis-allylic have greater reaction fragility with oxygen due to the formation of stable resonance structure, as shows Figure 6.

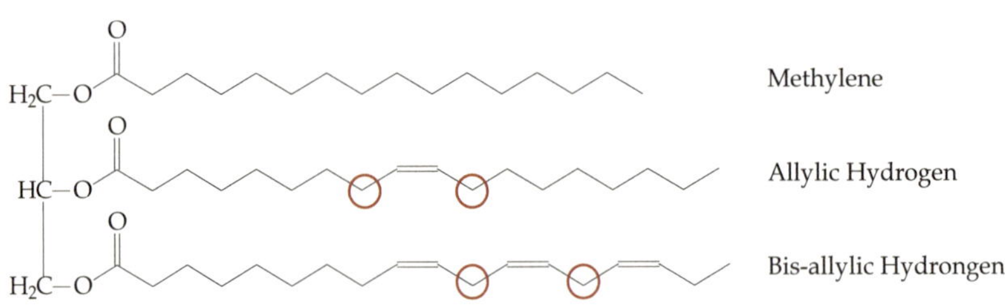

Fig. 6. Methylenic ,Allylic and bis allylic hidrogens at triacylglycerol (Asadukas et al, 2007).

The allylic hydrogen reactivity is 40 times greater than the methylene hydrogen and the bis-allylic is 100 times more reactive than the methylene hydrogen (Knothe, 2007).
The main characteristic for a substance be considered a good antioxidant is its capacity to react with oxygen faster than the biodiesel components, mainly the unsaturated compounds. Moreover, the radicals generated in this reaction have to be stable enough and less reactive with the initial biodiesel components or even with the generated products from the biodiesel reaction.
Denisov & Khudyakov (1987) divide the antioxidants class in four groups:
Group 1 - Inhibitors that terminate chains through reactions with peroxyl radicals, including phenols, aromatic amines, diamines, and aminophenols;
Group 2 - Inhibitors that terminate chains through reactions with alkyl radicals, including stable radicals, quinones, quinone imines, methylenequinones, nitro compounds, and condensed aromatic hydrocarbons (these inhibitors are effective when dissolved oxygen concentration is low);
Group 3 - Agents that decompose peroxides without generating free radicals, including sulfides, disulfides, phosphites, metal thiophosphates, and carbamates;
Group 4 - Complexing agents that deactivate heavy metals are capable of catalyzing hydroperoxide decomposition to free radicals, thereby promoting oxidation, including diamines, amino acids, hydroxy acids and other bifunctional compounds.
Compounds from the group 3 contain sulfur that turn difficult its use as biodiesel for environmental reasons. Compounds from the group two are effective only for oxygen low concentrations.
Actually there are two substances classes are very useful for this purpose: phenols and aromatics amines. These compounds are cheap and very useful at oil and polymers industry and are the most useful at biodiesel industry.
Figure 7 shows the oxidative stability of biodiesel containing different types of phenol antioxidants. Its stability can be up to 5 times higher when *tert*-Butylhydroquinone (TBHQ) is added, Karavalis, 2011.

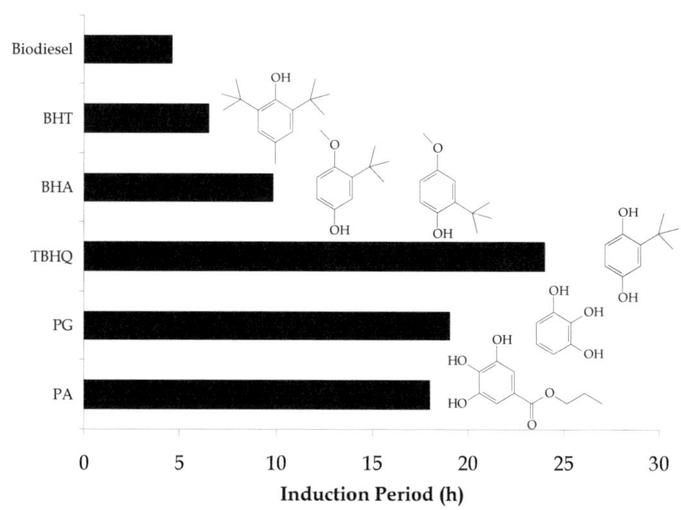

Fig. 7. PA= Propylgallate (3,4,5-trihydroxybenzoate) ; PG= Pyrogallol (benzene-1,2,3-triol) ; BHA= mixture of the isomers 2 and 3-*tert*-butyl-4-hydroxyanisol; BHT= di-*tert*-butyl-metil-phenol (Butylated hydroxytoluene) ; TBHQ= *tert*-Butylhydroquinone.

4. The importance of metrology for biodiesel quality

4.1 Efforts for harmonization of biodiesel standards

In 2006, the Government of Brazil, the European Commission (representing the European Union) and the Government of the United States of America, during trilateral discussions, affirmed their belief that the current market for biofuels is viable. The market will continue to grow within these regions and the international trade in biofuels would increase significantly by the end of this decade (Tripartite Task Force, 2007). However, a potential barrier to global trade in biofuels concerns the differences among the standards describing and ruling their composition and properties.

To overcome these potential barriers, a conference was organized by the European Commission and the European Committee for Standardization (CEN), with the active participation of the U.S. National Institute of Standards and Technology (NIST) and the Brazil's National Institute of Metrology, Standardization, and Industrial Quality (INMETRO). This meeting, held in Brussels in February, 2007, convened a broad range of private-sector biofuels experts and government representatives from the EU, US and Brazil. The participants confirmed that differing standards for biofuels were a potential handicap to the free circulation of biofuels among the three regions.

To support the global trade of biofuels, representatives of Brazil, the EU and the U.S. agreed to promote, whenever possible, the compatibility of biofuels-related standards in their respective regions. Such compatibility would not only facilitate the increasing use of biofuels in each of the regional markets, but also would support both exporters and importers of biofuels by helping to avoid adverse trade implications in a global market. Subsequently, the International Biofuels Forum (IBF) - a governmental initiative among Brazil, China, the European Commission, India, South Africa, and the United States - was launched in March, 2007 to promote the sustained use and production of biofuels around the globe. The IBF also concluded that trade will play an increasing role in providing adequate supplies of biofuels to the markets where the energy demand for transport fuel is rising at an accelerated rate.

In June, 2007, a NIST and INMETRO-sponsored Biofuels Symposium in Washington, DC, convened representatives from Brazil, the EU and the U.S. to build on the work begun in Brussels. These representatives agreed to review existing documentary standards for biofuels and identify areas where greater compatibility could be achieved in the short, medium and long term. According to the tripartite agreement, the standards to be considered were those produced by Brazilian Association for Technical Standards (ABNT), Brazilian Petroleum, Gas and Biofuels Agency (ANP), European Committee for Standardization (CEN) and American Society for Testing and Materials (ASTM International) and in effect before the end of 2007. It was further agreed that only standards pertaining to the biofuels being currently traded - biodiesel and bioethanol - would be addressed; this was further limited to pure biofuels and not to ready-made blends.

Comprised of representatives from the private and public sectors, the Biodiesel Tripartite Task Force and the Bioethanol Tripartite Task Force each started their technical work in July. The immediate task was to classify the various specifications into three categories:

- Category A: specifications that are already similar;
- Category B: specifications with significant differences between parameters and methods, but which might be aligned by work on documentary standards and measurement standards; and

- Category C: specifications with fundamental differences, perhaps due to emissions or environmental regulations within one or more regions, which are not deemed bridgeable in the foreseeable future.

There were commonalities with the approach and methodology used by both of the Task Forces. Each of the two groups assembled and translated existing standards from ABNT, ASTM International and CEN, and the units for specifications were converted to a common basis. Each Task Force first compared the standards as they presently exist. Since it was noted that many parameters were different, the Task force members entered into discussions and negotiations and were able to make specific recommendations to address these differences. They further agreed that these recommendations should be forwarded to standards bodies for consideration and possible implementation. Here, we will only present all biodiesel discussions to compatibility biodiesel standards. Summary results from each group are listed below in Table 5.

4.2 General considerations for biodiesel standards

The current standards established to govern the quality of biodiesel on the market are based on a variety of factors which vary from region to region, including characteristics of the existing diesel fuel standards, the predominance of the types of diesel engines most common in the region, and the emissions regulations governing those engines. Europe, for example, has a much larger diesel passenger car fleet, while United States and Brazilian markets are mainly comprised of heavier duty diesel engines. It is therefore not surprising that there are some significant differences among the three sets of standards.

Category A Similar	Category B Significant Differences	Category C Fundamental Differences
Sulfated Ash	Total glycerol content	Sulfur content
Alkali and alkaline earth metal content	Phosphorus content	Cold climate operability
Free glycerol content	Carbon residue	Cetane number
Copper strip corrosion	Ester content	Oxidation stability
Methanol and ethanol content	Distillation temperature	Mono, di, and tri-acylglycerides
Acid number	Flash point	Density
	Total Contamination	Kinematic viscosity
	Water content and sediment	Iodine number
		Linolenic acid content
		Polyunsaturated methyl ester

Table 5. Classification of the Various Biodiesel Specifications.

Other sources of regional differences in biodiesel standards arise from the following factors. The biodiesel standards in Brazil and the U.S. are applicable for both fatty acid methyl esters (FAME) and fatty acid ethyl esters (FAEE), whereas the current European biodiesel standard is only applicable for fatty acid methyl esters (FAME). Also, the standards for biodiesel in Brazil and the U.S. are used to describe a product that represents a blending component in conventional hydrocarbon based diesel fuel, while the European biodiesel standard describes a product that can be used either as a stand-alone diesel fuel or as a blending component in conventional hydrocarbon based diesel fuel.

It should also be noted that some specifications for biodiesel are feedstock neutral and some have been formulated around the locally available feedstocks. The diversity in these technical specifications is primarily related to the origin of the feedstock and the characteristics of the local markets. Though this currently translates into some significant divergence in specifications and properties of the derived fuels - which could be perceived as an impediment to trade - in most cases it is possible to meet the various regional specifications by blending the various types of biodiesel to the desired quality and specifications.

The Task Force members classify the various specifications according with the limits for each parameter. They have collaboratively assembled a definitive and widely vetted list of Brazilian, EU and US standard specifications that are similar. In addition, they have identified a list of specifications that have significant, but alignable differences. Perhaps even more importantly, some indirect benefits have been derived. There is widespread agreement amongst the participating experts that the discussions and commitment to cross-border cooperation have been a major accomplishment that will support the increase in global trade of biofuels. The experts now have a better understanding of reasons why regional differences exist, and a new atmosphere of collegiality has been created - not only between countries but also between the private and public sector representatives. These positive outcomes foster a working environment that will support ongoing movement towards enhanced compatibility among the biofuels standards.

After the discussions, it was concluded that:

- standardization bodies of the tripartite agreement (i.e., ABNT, ANP, CEN and ASTM International) as a basis for ongoing discussions and cooperation that will promote alignment and mitigate divergence among evolving standards and specifications.
- other members of the International Biofuels Forum as a basis for ongoing discussions on more closely aligning their respective specifications and prioritizing future efforts for maximum impact.
- request the standardization bodies of the Tripartite Agreement to consider adapting existing national standards wherever appropriate. Furthermore the standardization bodies should attempt where possible, when developing and updating their standards on biodiesel from now on to consider the opportunity to align with the other standards in question;
- support efforts to initiate an analysis of the categorized specifications to study trade implications and appropriate next steps for harmonization;
- support the development of internationally-accepted reference methods and certified reference materials for improving the accuracy of measurement results that underpin assessment of product quality, and help facilitate trade.

4.3 Development of Internationally-accepted reference methods and certified reference materials

Beyond the difference among the standards, an unacceptable barrier to trade, are the measurement disagreements between countries. To overcome such problems it is necessary to have an international infrastructure within which it is possible to make comparable measurements (Wielgoz & Kaarls, 2009).

This is true for all areas of measurements including chemical ones. Such a system requires measurement standards that have long-term stability and are internationally recognized. The International System of Units (SI) represents such a system, and by the use of traceable measurements provides an international infrastructure for comparable measurements. This system is demonstrated in Fig. 8.

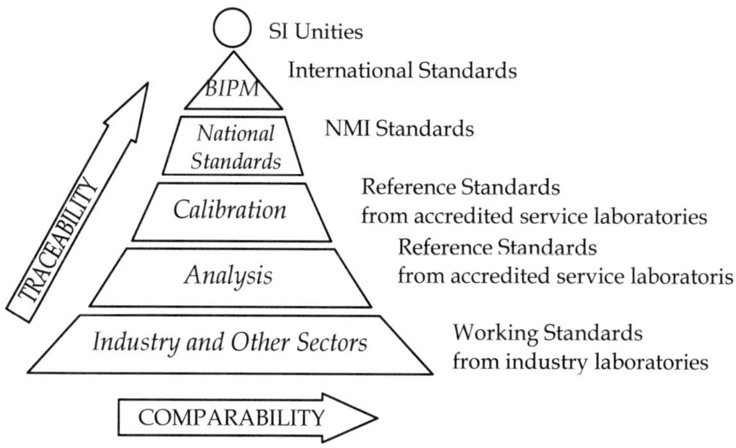

Fig. 8. Traceability scheme.

The International System of Units (SI) is at the top of the system (Dube, 2001). Its units are realized by standards. A measurement is a process, in the course of which the measurand is compared to a standard. For practical measurements, usually a working standard not a primary standard is used. To state the uncertainty of the measurement result, the uncertainty of the value assigned to the working standard must be known. It results from the uncertainty of the comparison measurement of the working standard with the reference standard. The uncertainty of the value assigned to the reference standard results from the uncertainty of the comparison measurement of the reference standard with the primary standard. This chain of comparison measurements is exactly what the definition of the term "traceability" means. If the traceability of a measurement result is guaranteed, its uncertainty can be stated. From this considerations it follows that metrology can provide the tools, necessary to get reliable measurement results.

Metrology is the science of measurement, embracing both experimental and theoretical determinations at any level of uncertainty in any field of science and technology. Within a robust metrological system the values of measurement standards and measurement results are linked *via* comparisons or calibrations which take into account the measurement uncertainty of the linking processes. Measurement uncertainty is the parameter associated with the results of a measurement, that characterizes the dispersion of the values that could reasonably be attributed to the measurand.

The ISO Guide to the Expression of Uncertainty in Measurement (GUM) (BIPM et al., 2008) and the Eurachem (CITAC, 2000) guide on measurement uncertainty provide guidance on the evaluation of measurement uncertainty. The property of the result of a measurement or the value of a measurement standard whereby it can be related to stated references, usually national or international measurement standards, through an unbroken chain of comparisons all having stated uncertainties, is termed (metrological) traceability. Where these stated references are realizations of the SI units the term SI-traceable is used. Traceability is the basis of the comparability of a measurement: whether the result of a measurement can be compared to the previous one, a measurement result a year ago, or to the result of a measurement performed anywhere else in the world. Traceability is most often obtained by calibration, establishing the relation between the indication of a measuring instrument and the value of a measurement standard.

In the field of analytical chemistry the term Certified Reference Material (CRM) is more often used than measurement standard. A CRM is a reference material, accompanied by a certificate, one or more of whose property values are certified by a procedure which establishes traceability to an accurate realization of the unit in which the property values are expressed, and for which each certified value is accompanied by an uncertainty at a stated level of confidence.

Metrological traceability (Eurachem/CITAC, 2003) may also be established to a reference method, defining the measurand and fixing a number of influence parameters, the results of which are expressed in SI units, and an approach that has been documented for the field of laboratory medicine. In the field of metrology in chemistry the role of a National Metrology Institute (NMI) involves: realization, maintenance and dissemination of the units; development and application of primary measurement methods; establishment of traceability structure, guaranteeing the equivalence of measurement standards using programs which facilitate traceable measurements to be achieved, including the provision of certified reference materials, both as pure materials and calibration solutions as well as matrix reference materials for method validation or calibration.

For producers of CRMs, there are three ISO Guides that assist the set-up of a facility to produce and certify RMs and to ensure that the quality of thus-produced CRMs meet the requirements of the end-users (ISO, 2006). ISO Guide 34 (ISO, 2009) outlines the requirements to be met by a CRM producer to demonstrate competence, whereas the Guide 35 provides assistance on how to meet these requirements. At a fairly generic level, this Guide provides models for homogeneity testing, stability testing, and the characterization of the candidate CRM.

ISO Guide 31(ISO, 2000) describes the format and contents of certificates for CRMs. In some ways, this Guide can be seen as an application of the *Guide to the Expression of Uncertainty in Measurement* (GUM) with respect to the peculiarities of the production of CRMs. Where possible, the Guide 35 makes reference to the GUM, as the latter describes in detail how to evaluate measurement uncertainty of a value obtained from measurement. This Guide complements the GUM in a sense that it provides additional guidance with respect to the inclusion of the uncertainties due to the (remaining) batch inhomogeneity and instability of the CRM in the uncertainty of the property values, and the determination of these uncertainty contributions.

Thorough knowledge of the material and its properties, and of the measurement methods used during homogeneity testing, stability testing and characterization of the material, along with a thorough knowledge of the statistical methods (Eurachem/CITAC, 2000), are needed for correct processing and interpretation of experimental data in a typical certification project. It is the combination of these required skills that makes the production and certification of RMs so complex. The greatest challenge in these projects is to combine these skills to allow a smooth implementation of RM certification.

Three categories of values can be assigned for certified reference materials producers: Certified values fulfill the highest standards for reliability. They are traceable to stated references and are accompanied by a GUM (BIPM, JCGM 100:2008) compatible expanded uncertainty statement valid for the entire shelf life of the CRM. Indicative values are not certified due to either a larger uncertainty than required for the intended use or insufficient variety of methods used in the characterization. The information is therefore unsuitable for certification at the accuracy required for certified values. Additional material information are values created during the certification exercise, which are usually the result of one method only and indicate the order of magnitude rather than an accurate value.

In summary, certified values are those values the certifying body is confident in assigning with the highest accuracy, while indicative values display higher uncertainties and/or lack a full traceability statement. This hierarchy in reliability is shown by the fact that only certified values are on the first page of the certificate. It follows that certified values are more assured than indicative values which in turn are more assured than additional material information.

The measurement method used for the homogeneity study should have very good repeatability and selectivity. The main purpose of the homogeneity assessment is, however, to detect unexpected problems, for example due to contamination during packaging (Linsinger et al., 2000, Van der Veen et al., 2000, 2001a, 2001b). To establish the homogeneity, a statistically defined number of bottles is randomly selected and analyzed for all relevant property values. For evaluation of homogeneity results, unifactorial analysis of variance ("one-way ANOVA") (Van der Veen & Pauwels, 2000) is applied. The stability of the reference materials have to be assessed for all parameters, by measuring the property values periodically during the course of the project. In this case, where samples are measured on different days, the selectivity and the reproducibility of the measurement method are of great importance. Therefore, methods for homogeneity and stability studies are not necessarily the same. This is not a problem so long as traceability of the results of the homogeneity and stability studies and characterization to a common reference are established. Such a reference may be a material that is suitable for assessing the various calibrations or results from different measurement methods. Ensuring the traceability of all measurements in a certification project is an important requirement.

For the characterization of the candidate reference material, the producer shall use and document technically valid procedures to characterize its reference materials. It shall comply with the requirements of ISO Guide 35 and ISO/IEC 17025 for testing, calibration and related activities. There are several technically valid approaches for characterizing a reference material. These include carrying out measurements using: a) a single (primary) method in a single laboratory; b) two or more independent reference methods in one or several laboratories; c) one or more methods of demonstrable accuracy, performed by a network of competent laboratories; d) an approach providing method-specific, operationally defined property values, using a network of competent laboratories.

Depending on the type of reference material, its intended use, the competence of the laboratories involved and the quality of methods employed, one approach may be chosen as appropriate. Results obtained from proficiency testing can be used only if the competence of the laboratories involved has been checked and it has been ensured that the measurements done comply with ISO/IEC 17025. The single (primary) method approach shall be carried out only when the procedure and expertise enable it to ensure metrological traceability. More usually, a property value can be reliably assessed when its value is confirmed by several laboratories working independently and using more than one method, for each of which the accuracy has been well established.

Primary methods play an essential role in the practical realization of the base units of the SI and hence in establishing traceability to the SI throughout metrology (Milton, 2001). There are seven SI units (meter, kilogram, second, ampere, Kelvin, mole, candela). They are the essential first link in such a chain of traceability because they do not require prior knowledge of any measurement of the same quantity. Explained briefly, a primary method of measurement allows a quantity to be measured in terms of a particular SI unit without reference to a standard or measurement already expressed in that unit. It is thus, in principle, completely independent of measurements of the same quantity, but calls upon measurements expressed in other units of the SI. By their nature, primary methods are unbiased (their results are accurate) but they may not necessarily be precise. Those primary methods that are at the same time precise are the ones that are useful in practice. Put in a different way, a primary method provides the means to transform the abstract definition of an SI unit into practical measurements made in terms of that unit. This is sometimes referred to as a "realization" of that unit, but this statement can be confusing because it gives the impression that the result is in some way a concrete materialization of the unit. In some cases such a concrete materialization can be made (although it is not possible for the mol unit) but, more importantly, a primary method allows measurements to be made in terms of that unit and this is the main characteristic of a primary method. The potential primary methods available for the chemical area are very few, they are: Gravimetry, Titrimetry, Coulometry, Calorimetry (Differential Scanning Calorimetry), Isotope Dilution Mass Spectrometry (IDMS), Instrumental Neutral Activation Analysis (INAA).

The biggest challenge involving the characterization of a biodiesel CRM is the lack of primary methods for all parameters that involves the biodiesel quality assessment. Therefore, some efforts are being made in that sense. Inmetro and NIST have teamed up to develop a CRM for several parameters for biodiesel derived from soybean and animal fat. The composition of the materials had to be close to specification levels, for those parameters where international agreement on these levels exists. Characterization is carried out for those parameters where SI traceability of the measurement results is possible. NIST and Inmetro reported the results accompanied by a complete uncertainty statement calculated according to the "Guide to the expression of uncertainty in measurement"(GUM) (BIPM, JCGM 100:2008). These CRMs are available at NIST homepage.

Additional effort concerns BIOREMA project (Inmetro et al., 2008) which involved not only Inmetro and NIST, but also several European National Metrology Institutes as Laboratory of Government for Chemistry - LGC and National Physical Laboratory- NPL (England), Van Swinden Laboratory - VSL (Netherlands), Institute for Reference Material and Measurements - IRMM (Belgium). Initially, this group intended to develop a (Certified) Reference Material for biodiesel from rapeseed/canola and conducted homogeneity, stability and characterization studies in test samples to obtain certified values.

However NMI's results for several parameters were not harmonized as expected, so these test samples were not possible to be used as CRM. On the other hand, these results were not dismissed, and they can be used as reference value for intercomparisons. Thus, the

BIOREMA group decided to promote a proficiency testing within Brazilian, American and European testing laboratories. In this proficiency testing, soybean and animal fat biodiesel CRM's, developed by Inmetro and NIST, were included. The agreement of results provided by participating laboratories, and the consensus related to reference values, were perceived as satisfactory. A clear demand was expressed for RMs of biodiesel derived not only from one feedstock but from the various ones. This fact happens due to the considerable difference in chemical and physical properties among biodiesels deriving form the diverse sources, which make harder the accommodation of all aspects relevant in quality control and the validation of analytical methods.

This project concluded that this was a positive example of collaboration among metrology institutes and useful exercise for establishing a common approach to the production of biofuels reference materials. However, for several specifications, the assignment of an SI traceable value still need further research due to the complexity of the material.

5. Conclusion

The use of fuels derived form renewable sources, like crops, microorganisms or animal derivates, bring several advantages related to the environment, economy and the fewer dependence on the main fossil energy source, petroleum. Concerning environmental aspects the main fact that has been constantly emphasized, and actually is one the greatest advantage of using biofuels, is related to the reduction of greenhouse gases exhausts. Since biofuels are derived form renewable sources, the burning of biofuels in internal combustion engines linked to the renovation of those sources which biofuels are derived from, allows the establishment of CO_2 recycling in environment. Moreover, biodiesel production, especially in Brazil, has presented a great social advantage, seen that small farmers are encouraged to produce and furnish the raw material, which in turn stimulate the local economy of small cities and create new job positions.

From all the possible and already studied biodiesel sources, soybean presents great prominence. Soybean is a widespread culture, adapted to cultivation in several climates and countries, there is cultivation technology available and the logistics chain is well established. Otherwise, some aspects related to biodiesel production, economy, supplying and technical-economic availability need more profound studies. Literature points that a change in fatty acids profile of some feedstocks, using biologically modified species, in order to attend both human consumption and its utilization as fuel, would be one the most appropriate alternative for biodiesel replaces efficiently petrodiesel. Other aspect to be considered regards the byproduct glycerine recycling, as transforming it into products with commercial interest. Other point to be addressed is the gradual substitution of feedstocks that today are commodities (like soybean) for alternative raw materials that have less commercial importance, as macauba (*Acrocomia aculeata*) and jatropha (*Jatropha curcas*). This substitution would undervaluate biodiesel production costs. Biodiesel quality control has also crucial importance. Concerning this subject there is a great need for the development of robust and specific analytical methods, since the existing ones are adapted from the petrodiesel quality control. Moreover, in order to ensure biodiesel quality, although some efforts have been done, only two certified reference materials (soybean and animal fat + soybean based ones) are available. Another challenge concerns the development or enhancement for getting more efficient biodiesel production processes, seen that the actual ones, like those employing homogeneous or heterogeneous catalysis, still present several drawbacks.

6. References

Albuquerque, G. A. (2006). *Obtainment and Physical-Chemical Characterization of Canola (Brassica napus).* M.S. thesis, Centro C. Exatas e da Natureza, Univ. Fed. Paraíba.

Allen, C.A.W., Watts, K.C., Ackman, R.G. & Pegg, M.J. (1999). Predicting the viscosity of biodiesel fuels from their fatty acid ester composition, *Fuel*, Vol. 78, pp. 1319-1326

Antczak, M. S.; Kubiak, A.; Antczak, T. & Bielecki, S. (2009). Enzymatic biodiesel synthesis – Key factors affecting efficiency of the process, *Renewable Energy*, Vol.34, pp.1185–1194

Asadauskas, S.J.; Griguceviciene, A. & Stoncius, A. (2007). Review of late stages of oxidation in vegetable oil lubricant basestocks. *Proceedings of the international conference.*

ASTM D6751-08a. (2008). Standard specification for biodiesel (B100) fuel blend Stock (B100) for middle distillate fuels. In: *ASTM (American Society for Testing and Materials) annual book of standards.* West Conshohocken: ASTM International.

Atadashi, I.M.; Aroua, M.K.; Aziz, A.A. (2011). Biodiesel separation and purification: A Review. *Renewable Energy*, 36, 437-443.

Ataya, F; Dubé, M.A.; Ternan, M. (2007). Acid-Catalyzed Transesterification of Canola Oil to Biodiesel under Single- and Two-Phase Reaction Conditions. *Energy and Fuels*, 21, 2450-2459.

Balat, M & Balat, H. (2008). A critical review of bio-diesel as a vehicular fuel, *Energy conv. Mgmt.*, Vol.49, No.10, pp.2727–2741.

Barnard, M. T.; Leadbeater, N. E.; Boucher, M. T.; Stencel, L. M. & Wilhite, B. A. (2007). Continous flow preparation of biodiesel using microwave heating, *Energy & Fuels*, Vol.21, No.3, pp.1777-1781

Bickell, K. (2008). Cold Flow Properties of Biodiesel and Biodiesel Blends - A Review of Data - University of Minnesota Center for Diesel Research, Available form http: //www.biodiesel.org/pdf_files/fuelfactsheets/Cold%20flow.PDF

Biodiesel.ogr. (2009). Biodiesel Basics. *Biodiesel.ogr*, http://www.biodiesel.org/resources/biodiesel_basics/.

Biodieselbr.com. (2011). O que é o Biodiesel. *Biodieselbr.com*, http://www.biodieselbr.com/biodiesel/definicao/o-que-e-biodiesel.htm.

Biopowerlondon.co.uk. (2011). What is Biodiesel. *Biopowerlondon.co.uk. Energy, Power and Fuels*, http://www.biopowerlondon.co.uk/biodiesel.htm.

BIPM, IEC, IFCC, ISO, IUPAC, IUPAP, OIML. (2008). *Guide to the expression of uncertainty in measurement*, first edition, ISO Geneva.

Candeia, R. A.; Silva, M. C. D.; Carvalho F. J. R.; Brasilino, M. G. A.; Bicudo, T. C., Santos, I. M. G. & Souza, A. G. (2009). Influence of soybean biodiesel content on basic properties of biodiesel diesel blends. *Fuel (Guildford)*. Vol. 88, pp. 738-743

Chand, P.; Chintareddy V. R.; Verkade J. G. & Grewell D. (2010). Enhacing biodiesel production from soybean oil using ultrasonics, *Energy & Fuels*, Vol.24, pp.2010-2015

Du, W.; Xu, Y.; Liu, D. & Li, Z. (2005). Study on acyl migration in immobilized lipozyme TL-catalyzed transesterification of soybean oil for biodiesel production, *J Mol Catal B: Enzym.*, Vol.37, pp.68–71

Dube, G. (2001). Metrology in Chemistry - a public task, *Accreditation and Quality Assurance*, Vol. 6, pp. 26-30

Dufare, C.; Thamrin, U. & Mouloungui, Z. (1999). *Thermochim. Acta.* Vol. 388.

Dunn, R.O. & Bagby, M. O. (1995). Low-Temperature Properties of Triglyceride_Based Diesel Fuels: Transesterified Methyl Esters and Petroleum Middle Distillate/Ester Blends, *J. Am. Oil Chem. Soc.*, Vol. 72, pp. 895-904.

EN 14214:2008, (2009). Automotive fuels - fatty acid methyl esters (FAME) for diesel engines - requirement methods. Brussels, Belgium: *European Committee for Standardization (CEN).*

ERM (2008). Application Note 6, *European Reference Material (ERM)*. http://www.erm.org.

Eurachem/CITAC. (2000). *Quantifying uncertainty in analytical measurement*, 2nd edition, http://www.eurachem.ul.pt/

Eurachem/CITAC. (2003). Traceability in chemical measurement - a guide to achieving comparable results in chemical measurement, http://www.eurachem.ul.pt/

Evgeniy T. & Denisov, I. V. (1987). Mechanisms of Action and Reactivities of the Free Radicals of Inhibitors. *Chem. Rev.* Vol. 87; pp. 1313-1357

Fangrui, M.A.; Milford, A.H. (1999). Biodiesel production- A Review. *Bioresource Technology*, Vol. 70, pp. 1-15.

Ferella, F.; Mazziotti, G.; De Michelis, I.; Stanisci, V.; Veglio, F. (2010). Optimization of the transesterification reaction in biodiesel production. *Fuel*, 89, 36-42.

Ferrari, R. A. & Souza, W. L. (2009) Avaliação da estabilidade oxidativa de biodiesel de óleo de girassol com antioxidantes. *Quim. Nova.* Vol. 32, No. 1, pp. 106-111
for midle distillate fuels, ASTM, West Conshohocken, PA

Fueleconomy.gov. (2011). Biodiesel. *Fueleconomy.gov, US Departament of Energy, US Environmental Protection Agency*,
http://www.fueleconomy.gov/feg/biodiesel.shtml.

Fukuda, H.; Konda, A. & Noda, H. (2001). Bioidesel fuel production by transesterification of oils, *J. Biosc. Bioeng.*, Vol.92, pp.405-416

Gerpen, J.V. (2005). Biodiesel processing and production;*Fuel Process Technol*, Vol.86, pp.1097–107

Grunberg, L. & Nissan, A.H. (1949), Mixture law for viscosity, *Nature*, Vol. 164, pp. 799-800

Gryglewicz, S. (1999). Rapeseed oil methyl esters preparation using heterogeneous catalysts, *Bioresour Technol*, Vol. 70, pp.249–253

Hadorn, H. & Zurcher, K. (1974). Zurbestimmung der oxydationsstabilitat von olen und fetten. *Deustsche Ledensmittel Rundschau.* Vol. 70, No. 2, pp. 57-65

Harding, K. G.; Dennis, J. S.; von Blottnitz, H. & Harrison, S. T. L. (2008). A life-cycle comparision between inorganic and biological catalyse for the production of biodiesel, *J. Clean. Produc.*, Vol.16, No.13, pp.1368-1378

Haupt, J., Brankatschk, G. & Wilharm, T. (2009). Sterol Glucoside Content in Vegetable Oils as a Risk for the Production of Biodiesel - Study of The technological Chain Impact - Arbeitsgemeinschaft Qualitätsmanagement Biodiesel e.V. (AGQM) Final Report to American Soybean Association (ASA).

Heck, D.A., Thaeler, J., Howell, S. & Hayes, J.A. (2009). Quantification of the Cold Flow Properties of Biodiesels Blended with ULSD, http://www.biodiesel.org/pdf_files/Cold_Flow_Database_Report.pdf.

Helwani, Z.; Othman, M. R.; Aziz, N.; Fernando, W. J. N. & Kim, J. (2009). Technologies for production of biodiesel focusing on green catalytic techniques: A review, *Fuel Processing Technology*, Vol.90, pp.1502, 1514,

Honda, H.; Kuribayashi, H.; Toda, T.; Fukumura, T.; Yonemoto, T. & Kitakawa, N. S. (2007). Biodiesel production using anionic ion-exchange resin as heterogeneous catalyst. *Bioresource Technology*, Vol. 98, pp. 416–421

ISO 12937:2000. (2000). Determination of Content. Coulometric Karl Fischer Titration Method. *Methods of Test for Petroleum and Its Products: BS 2000-438.*

ISO Guide 31. (2000). Reference Materials - Contents of certificates and labels, 3rd edition, ISO Geneva

ISO Guide 34. (2009). General requirements for the competence of reference material producers, 3rd edition, ISO Geneva

ISO Guide 35. (2006). Reference Materials - General and statistical principles for certification, 3rd edition, ISO Geneva

Jaimasith, M. & Phiyanalinmat, S. (2007). Biodiesel synthesis from transesterification by Clay basic catalyst"; *Chiang Mai J Sci*, Vol.34,No. 2, pp. 201-207

Ji-Yeon P.; Deog-Keun K.; Joon-Pyo L.; Soon-Chul P.; Young-Joo K. & Jin-Suk L. (2008). Blending effects of biodiesels on oxidation stability and low temperature flow properties. *Bioresource Technology.* Vol. 99, pp. 1196–1203

Joshi, R.M. & Pegg, M.J. (2007). Flow Properties of Biodiesel Fuel Blends at Low Temperatures, *Fuels*, Vol. 86, pp. 143-151

Kaieda, M.; Samukawa, T.; Kondo, A. & Fukuda, H. (2001). Effect of methanol and water contents on production of biodiesel fuel from plant oil catalyzed by various lipases in a solvent free system, *J. Biosc. Bioeng.*, Vol.91, No.1, pp. 12-15

Kaieda, M.; Samukawa, T.; Matsumoto, T.; Ban, K.; Kondo, A.; Shimada, Y.; Noda, H.; Nomoto, F.; Obtsuka, K.; Izumoto, E. & Fukuda, H. (1999). Biodiesel fuel production from plant oil catalyzed by rhizopus orizae lipase in a water containing system without an organic solvent, *J. Biosc. Bioeng.*, Vol.88, No.6, pp.627-631

Karavalakis G.; Despina H.; Lida G.; Dimitrios K. & Stamos S. (2011). Storage stability and ageing effect of biodiesel blends treated with different antioxidants. *Energy.* Vol. 36, pp. 369-374.

Kern, D.Q.& Van Nostrand, W. (1948). Heat Transfer Characteristics of Fatty Acids, *Ind, Eng. Chem*, Vol. 41, pp. 2209-2212

Knothe G, Van Gerpen J, Krahl J, editors (2005). *The biodiesel handbook*. Urbana: AOCS Press.

Knothe, G. & Steidley, K. R. (2005). Kinematic Viscosity of Biodiesel Fuel Components and Related Compounds: Influence of Compound Structure and Comparison to Petrodiesel Fuel Components, *Fuel*, 84, 1059–1065.

Knothe, G. (2007). Some aspects of biodiesel oxidative stability. *Fuel Processing Technology.* Vol. 88, pp. 669-667

Knothe, G., (2009). Improving Biodiesel Fuel Properties by Modifying Fatty Ester Composition, *Energy Environ. Sci.*, Vol. 2, pp. 759-766

Knothe, G.; Matheaus, A.C.; Ryan, T.W. (2003), Cetane Numbers of Branched and Straight-Chain Fatty Esters Determined in an Ignition Quality Tester, *Fuel*, Vol. 82, pp. 971-975.

Krawczyk, T., (1996). Biodiesel – alternative fuel makes inroads but hurdles remain. *Inform* 7, 801–829.

Krishna, C.R. & Butcher, T. (2008). Improving Cold Flow Properties of Biodiesel, In: Advanced Energy Conference, 19-20.11.2008, Available from http:www.aertc.org/conference/AEC_Sessions%5CCopy%20of%20Session%202%5CTrack%20B-%20Renewable%5C1

Landommatos, N.; Parsi, M.; Knowles, A. (1996). The effect of fuel cetane improver on diesel pollutant emissions, *Fuel*, Vol. 75, pp. 8-14.

Leadbaeater, N. E. & Stencel, L. M. (2006). Fast, easy, preparation of biodiesel using microwave heating, *Energy Fuels*, Vol.20, No.5, pp.2281-2283

Leclercq, E.; Finiels, A. & Moreau, C. (2001). Transesterification of rapessed oil in the presence of basic zeolites and related solid catalysts, *J. Am. Oil Chem. Soc.*, Vol. 78, pp. 1161-1165

Lee, I., Johnson, L.A. & Hammond, E.G. (1995). Use of branched-chain esters to reduce the crystallization temperature of biodiesel, *J. Am. Oil Chem. Soc.*, Vol. 72, pp. 1155-1160

Li, Q.; Du, W.; Liu, D., (2008). Perspectives of microbial oils for biodiesel production. *Appl. Microbiol. Biotechnol.* 80, 749–756.

Linsinger T.P.J., Pauwels J., Van der Veen A.M.H., Schimmel H., Lamberty A. (2001). Homogeneity and Stability of Reference Materials, *Accreditation and Quality Assurance*, Vol. 6, pp. 20-25

Ma, F. & Hanna, M.A., (1999). Biodiesel production: a review. *Bioresour. Technol.* 70, 1–15.

Mahamuni, N. N. & Adewuyi, Y. G. (2009). Optimization of the Synthesis of biodiesel via Ultrasound-Enhanced Base-Catalyzed Transesterification of Soybean Oil Using a Multifrequency Ultrasonic Reactor, *Energy & Fuels*, Vol.23, No.2757-2766

McCormick, R.L., Graboski, M.S., Alleman T.L. and Herring, A.M. (2001). Impact of biodiesel source material and chemical structure on emissions of criteria pollutants from a heavy-duty engine, *Environ. Sci. Technol.*, Vol. 35, pp. 1742-1747.

Melero, J.A.; Iglesias, J.; Morales, G. (2009). Heterogeneous acid catalysts for biodiesel production: current status and future challenges. *Green Chem.*, Vol. 11, pp. 1285-1308.

Mello, V. M.; Pousa, G. P. A. G.; Pereira, M. S. C.; Dias, I. m. & Suarez, P. A. Z. (2011). Metal oxides as heterogeneous catalysts for esterification of fatty acids obtained from soybean oil. *Fuel Proces. Tech.*, Vol.92, pp.53-57

Mendes, A.A.; Giordano, R.C.; Giordano, R.L.C. & Castro, H.F. (2011). Immobilization and stabilization of microbial lipases by multipoint covalent attachment on aldehyde-resin affinity: Application of the biocatalyst in biodiesel synthesis; *J. Mol. Catlal. B: Enzym*, Vol. 68, pp.109-115

Milton, M.J.T. (2001). Primary Methods for the measurement of Amount of the Substance, *Metrologia* Vol. 38, pp. 289-296.

Mittelbach, M. & Remschmidt, C. (2004), Biodiesel - *The Comprehensive Handbook* M. Mittelbach, Graz, Austria.

Modi, M. K.; Reddy, J. R. C.; Rao, B. V. S. K. & Prasad, R. B. N. (2007). Lipase mediated conversion of vegetables oils in to biodiesel using ethyl acetate as acyl acceptor, *Bioresource. Technology.*, Vol.98, No.6, pp. 1260-1264

Mohamad, I.A.W. & Ali, O.A. (2002). Evaluation of the transesterification of waste palm oil into biodiesel, *Bioresour Technol*, Vol.85, pp.225–256

Monyem, A.; Van Gerpen, J.H.; Canacki, M. (2001). The Effect o Timing and Oxidation on Emissions from Biodiesel. *Fueled Engines, Trans.* , Vol. 44, pp. 35-42

Moser, B.R. (2009). Biodiesel production, properties, and feedstocks. *In Vitro Cell.Dev.Biol. Plant,* Vol. 45, pp. 229–266.

Moser, B.R., (2009). Biodiesel production, properties, and feedstocks. *In Vitro Cell. Dev. Biol.-Plant.* 45, 229-266.

National Biodiesel Board Winter 2007/2008, *Biodiesel Cold Flow Basics* - Information for Petroleum Distributors, Blenders, and End-Users on Issues Affecting Biodiesel in the Winter Months

NIST home page (2011). http://ww.nist.gov/srm

Noureddini, H.; Gao, X. & Philkana, R.S. (2005). Immobilized Pseudomonas cepacia lipase for biodiesel fuel production fromsoybean oil, *Bioresour Technol, Vol.*96, pp.769–77

Raganathan, S. V.; Narasiham, S. L. & Muthukumar, K. (2008). An overview enzymatic production of biodiesel, *Bioresour. Technol.*, Vol.99, No.10, pp.3975-3981

Rakesh S.; Meeta Sharma, S. & Sinharay, R.K. (2007). Jatropha–Palm biodiesel blends: An optimum mix for Asia. *Fuel.* Vol. 86, pp. 1365–1371

Resolution ANP N°7, (2008). ANP, Brazilian National Agency of Petroleum, Natural Gas and Biofuels, "Resolution ANP No 7", from 2008/03/19, http://nxt.anp.gov.br/nxt/gateway.dll/leg/resolucoes_anp/2008/mar%C3%A7o / ranp%207%20-%202008.xml.

Rinaldi R.; Garcia, C.; Marciniuk, L. L.; Rossi, A. V. & Schuchardt U. (2007). Sintese de Biodiesel: Uma proposta contextualizada em experimento para laboratório de química geral. *Quim. Nova*, Vol.30, No.5, pp.1374-1380.

Salis, A.; Pinna, M.; Manduzzi, M. & Solinas, V. (2008). Comparision among immobilised lípases on macroporous polypropilene toward biodiesel synthesis, *J. Molec. Catal. B., Enzim.*, Vol.54, No.1, pp.19-26

Santos, F. F. P.; Rodrigues, S. & Fernandes, F. A. N. (2009). Optimization of production of biodiesel from soybean oil by ultrasound assisted methanolysis, *Fuel Processing Technology*, Vol.90, pp.312-316

Schlink, R. & Faas, B., (2009). Water Content Determination in Biodiesel According to EM ISO 12937. *Metrohm.com,* http://www.metrohm.com/applications/titration/lit/karl-fischer.html.

Sharp, C.A., (1994) Transient Emissions Testing of Biodiesel and Other Additives in a DDC Series 60 Engine. *Southwest Research Report Institute to the National Biodiesel Board.*

Shimada, Y.; Watanabe, Y.; Sugihara, A. & Tominaga, Y. (2002). Enzymatic alcoholysis for biodiesel fuel production and application of the reaction to oil processing, *J Mol Catal B: Enzym* , Vol.17, pp.133–42

Siddharth, J. & Sharma, M.P. (2010). Stability of biodiesel and its blends- A Review. *Renewable and Sustainable Energy Reviews*, Vol. 14, pp. 667-678.

Sirman, M.B.; Owens, E.C.; Whitney, K.A. (2000). Emissions Comparison of Alternative Fuels in an Advanced Automotive Diesel Engine, *SAE Techn. Pap. Ser.* 200-01-2048.

Snåre, M.; Kubic, I.;ková; Arvela, P. M.; Eränen, K.; Warna, J. & Murzin, D. (2007). Production of diesel fuel from renewable feeds: kinetics of ethyl stearate decarboxylation, *Chem. Eng. J.* , Vol.134, No.1, pp.29–34

Suppes, G. J.; Bockwinkel, K.; Lucas, S.; Botts, J. B.; Mason, M. H. & Heppert, J. A. (2001). Calcium carbonate catalyzed alcoholysis of fats and oils, *J. AM. Oil Chem. Soc.*, Vol. 78, No. 2, pp. 139-146

Suppes, G. J.; Dasari, M. A.; Doskocil, E. J.; Mankidy, P. J. & Goff, M. J. (2004). Transesterification of soybean oil with zeolites and metal catalysts; *J. Appl. Catal,.* A., Vol.257, pp. 213-223

Sybist, J.P. & Boehman, A. L., (2003). Behavior of a Diesel Injection System With Biodiesel Fuel, SAE Techn. Pap. Ser. 2003-01-1039

Tat, M.E. & Van Gerpen, J.H. (2003). Measurement of Biodiesel Speed of Sound and Its Impacto on Injection Timing, *National Renewable Energy Laboratory, NREL/SR-510-31462.*

Tomasevic, A. V. & Marinkovic, S. S. (2003). Methanolysis of used frying oils, *Fuel Process Technol.*, Vol.81, No.1–6

Tripartite Task Force. (2007). White Paper on Internationally Compatible Biofuel Standards: Brazil, European Union, and United States of America.

United States Department of Agriculture, (2010). Foreign Agricultural Service, Office of Global Analysis. Oilseeds: World Markets and Trade.

Van der Veen A.M.H., Linsinger T.P.J, Lamberty A., Pauwels J. (2001b). Uncertainty calculations in the certification of reference materials. 3. Stability study, *Accreditation and Quality Assurance*, Vol. 6, pp. 257-263

Van der Veen A.M.H., Linsinger T.P.J, Pauwels J. (2001a). Uncertainty calculations in the certification of reference materials. 2. Homogeneity study, *Accreditation and Quality Assurance*, Vol. 6, pp. 3-7.

Van der Veen A.M.H., Pauwels J. (2000a), Uncertainty calculations in the certification of reference materials. 1. Principles of analysis of variance, *Accreditation and Quality Assurance*, Vol. 5, pp. 464-469

vegetable oils, *J Am Oil Chem Soc*, Vol.80, No.4, pp.367–371

Wang, L. & Yang, J. (2007). Transesterification of soybean oil with nano-MgO or not in supercritical and subcritical methanol, *Fuel*, Vol. 86, No.3, pp. 328-333

Watanabe, Y.; Shimada, Y.; Sugihara, A. & Tominaga Y. (2002). Conversion of degummed soybean oil to biodiesel fuel with immobilized Candida antarctica lipase; *J Mol Catal B: Enzym*, Vol.17, pp.151–155

Wielgoz, R., Kaarls, R. (2009). International Activities in Metrology in Chemistry, *Chimia*,Vol. 63, pp 606-612.

Zhang, Y. & Van Gerpen, J.H. (1996). Performance of Alternative Fuels for SI and CI Engines. *SAE Spec. Publ. SP*- 1160, 1-15.

Zhang, Y.; Dube, M. A.; McLean, D. D. & Kates, M. (2003). Biodiesel production from waste cooking oil: 2. Economic assessment and sensitivity analysis, Bioresour Technol., Vol.90, pp.229–240

Zhou, W.; Konar, S. K. & Boocock, D.G.V. (2003). Ethyl esters from the single-phase base-catalyzed ethanolysis of vegetable oils, *J Am Oil Chem Soc*, Vol.80, No.4, pp.367–371

19

Machine Vision Identification of Plants

George. E. Meyer
University of Nebraska, Department of Biological Systems Engineering, USA

1. Introduction

Weedy and invasive plants cost Americans billions of dollars annually in crop damage and lost earnings. Various Western states have reported annual weed control costs in the hundreds of millions of dollars. Herbicides account for more than 72 per cent of all pesticides used on agricultural crops. $4 billion was spent herbicides in the US in 2006 and 2007 (Grube, et al, 2011). The USDA Economic Research Service reported that adoption of herbicide-tolerant soybeans had grown to 70% from 1996 to 2001, yet significant impacts on farm financial net returns attributable to adoption has yet to be documented. Nebraska is part of regional strategic pest plan published in 2002. During 2001, 97% of the soybean acres in Nebraska were treated with herbicides. One means of improving economic benefit is to develop more efficient management inputs, which may be accomplished with better selection of the kind of pesticide and/or site-specific application of pesticides. Moreover, measuring the impact of various management inputs often depends on manual visual assessment and perhaps this could be automated. One method for estimating impact on crop yield loss includes counting weeds per length of row or determining weed populations by species. In order to improve the weed suppression tactics, accurate mapping and assessment of weed populations within agricultural fields is required. See Figure 1. Weed mapping and taxonomy are major activities and species type found in all regions, which cover much broader ecological areas other than farm fields. These are shown by active websites in Nebraska, Iowa, Pennsylvania, Montana, Nevada, Colorado, and California, as examples. Weed and invasive species mapping also has international implications, (Montserrat, et al, 2003). Efforts of this type support integrated pest management (IPM) programs of both Crops and Risk (CAR) and Risk Avoidance and Mitigation (RAMP) which involve profitability and environmental stewardship and risk management, by providing a tool for timely acquisition of weed information. Research in this area promotes an interdisciplinary, IPM systems approach to weed mapping. There is high labor cost associated with the manual scouting of fields to obtain such maps.

2. Spatial variability of weed populations

Weeds are present in every field and lawn every year. The severity of the weed population is determined by local management practices such as the previous crop in the rotation and the herbicide use. According to a 2002 North Central strategic plan, tillage remained a major tool for controlling perennials, although the dilemma is that tillage contributes to soil erosion. Weed spatial distributions are unique, with monocot infestations more patchy than dicots (Mortensen et al., 1992 and Johnson et al., 1993, 1995). Monocots differ architecturally from dicots. Most weeds are serious competitors for moisture and soil nutrients. By first classifying the weed as either a monocot or dicot, a herbicide could be selected that most effectively controls that type of plant, resulting in better application efficiencies. Most post-emergent herbicides are selective in controlling one plant type or the other. Wiles and Schweizer (1999, 2002) researched the spatial distribution of weed seed banks using soil samples to map locations of weed seed banks in a given field. Seed banks have been found distributed in a patchy manner. Using the maps as a guide, farmers could treat just the weed patches with minimal amounts of the appropriate chemical. Site-specific weed management could mean a significant reduction in herbicide use, which saves the farmer money and benefits the environment. However, a large number of soil and plant samples are needed to get an accurate map—and that can be costly.

Stubbendick, et al (2003) provided a comprehensive compendium of weedy plants found across the Great Plains of the United States. Color plates were provided of canopy architecture and sometimes close-ups of individual leaves, flowers, and fruit. A hand drawing of canopy architecture was also given. In order to recognize a particular species, one needs to understand the concept of inflorescence and various plant taxonomy terms. There are many existing plant image databases around the United States. However, their suitability as reference images has yet to be determined for machine vision applications. An important application using machine vision is site-specific or spot herbicide application systems to reduce the total amount of chemical applied (Lindquist et al., 1998, 2001 a,b; Medlin, et al, 2000). Therefore, a major need for improved weed IPM and ecological assessment of invasive plant species is the development of a low-cost, but high resolution, machine vision system to determine plant incidence, even when imbedded with other plants, and to identify the species type. Machine vision systems should assist in the creation of plant field maps, leading to valid action thresholds (National Roadmap for IPM 2004).

3. Machine vision

Field plants, residue, and soil ecosystems are very complex, but, machine vision technology has the potential to systematically unravel and identify plants using optical properties, shape, and texture of leaves (Meyer et al., 1998). Considerable research has been reported using optical or remote sensing sensors to identify crop health by surface reflectance of green plants in agricultural fields (Gausman et al., 1973; Tucker, et al, 1979; Gausman et al., 1981; Thomas, et al, 1988; Storlie et al., 1989, Tarbell and Reid, 1991.; Franz et al., 1991b; and others). Hagger, et al (1983, 1984) reported the first prototype, reflectance-based plant sensor for spraying weeds. Hummel and Stoller (2002) evaluated a later commercial weed sensing system and noted their problems. Tian, et al (1999) developed a simple weed seeker in Illinois. Unfortunately, subsequent optical, non-image, sensor-based weed seekers and spot sprayers have not gained commercial acceptance for various reasons: first, single-element optical sensors can change the size of their field of view based on lens properties and distance to a target. Secondly, sensed reflectance properties may change according to the spatial contents of target components within the field of view Woebbecke, et al (1994); and finally, these sensors therefore may not always distinguish conclusively between crop, weed, or soil residue background. The voltage signal originating from an optical diode or transistor along with the Gaussian lens system used creating the field of view is a weighted average-problem, where the proportions of contributing reflectance and spatial contents are unknown. That problem can be solved only by spatial image analysis.

Image analysis is a mathematical process to extract, characterize, and interpret tonal information from digital or pixel elements of a photographic image. The amount of detail available depends on the resolution and tonal content of the image. The process is iterative, starting with large features followed by more detail, as needed. However, shape or textural feature extraction first requires identification of targets or Regions of Interest (ROI). These regions are then simply classified as green plants or background (soil, rocks, and residue). ROI's can be also identified with supervised control of the camera or field of view (Woebbecke, et al, 1994, Criner, et al, 1999), using a supervised virtual software window, cropping of selected areas, or unsupervised crisp or fuzzy segmentation procedures. ROI's are then binarized to distinguish target and background. Binarized images are then used for shape analysis or boundary templates for textural feature analysis. The binary image is combined with tonal intensity images of the targets (Gerhards and Christensen, 2003, Meyer et al., 1999; Kincaid and Schneider, 1983; Jain, 1989; Gonzalez and Woods, 1992; and others).
Machine vision offers the best potential to automatically extract, identify, and count target plants, based on color, shape, and textural features (Tillett et al. 2001). However, directing the image analysis process toward the classical botanical taxonomic, plant identification approach has previously required considerable supervised human intervention. A major problem is the presentation of plant features including individual leaves and canopy architecture to a discrimination or classification system. Camargo Neto, et al (2004 a,b; 2005) presented a combination of traditional image processing techniques, fuzzy clustering, pattern recognition, and a fuzzy inference neural network to identify plants, based on leaves. A particular difficult problem was the development of an algorithm to extract individual leaves from complex canopies and soil/residue color images.
If image vegetative/background classification is to be useful for plant species identification, a separated plant region of interest (ROI) must be found to provide important canopy information needed to discriminate at the very least, broadleaf versus grass species (Woebbecke et al., 1995a; Meyer et al., 1998). Four basic steps for a computerized plant species classification system were presented by Camargo Neto (2004). The first step is creating a binary image which accurately separates plant regions from background. The second step is to use the binary template to isolate individual leaves as sub images from the original set of plant pixels (Camargo Neto, et al, 2006a). A third step was to apply a shape feature analysis to each extracted leaf (Camargo Neto, et al, 2006b). The fourth and final step was to classify the plant species botanically using additional leaf venation, textural features acquired during the previous steps (Camargo Neto and Meyer, 2005). Machine vision plant image analysis has been greatly enhanced through the introduction of the automatic color and focusing digital camera (Meyer, et al, 2004). Digital cameras when run in the automatic mode make decisions on "best picture", and thus are extremely popular as consumer products.

4. Vegetation indices

The use of vegetation indices in remote sensing of crop and weed plants is not new. It represents the first step shown in Figure 2. Studies for crop and weed detection have been performed using different spectral bands and combinations for vegetative indices (Woebbecke et al. 1995b, El-Faki, et al., 2000ab, Marchant et al., 2004; Wang et al., 2001, Lamm et al., 2002; Mao et al., 2003; Yang et al., 2003). Color vegetation indices utilize only the red, green and blue spectral bands. The advantage of using color indices is that they

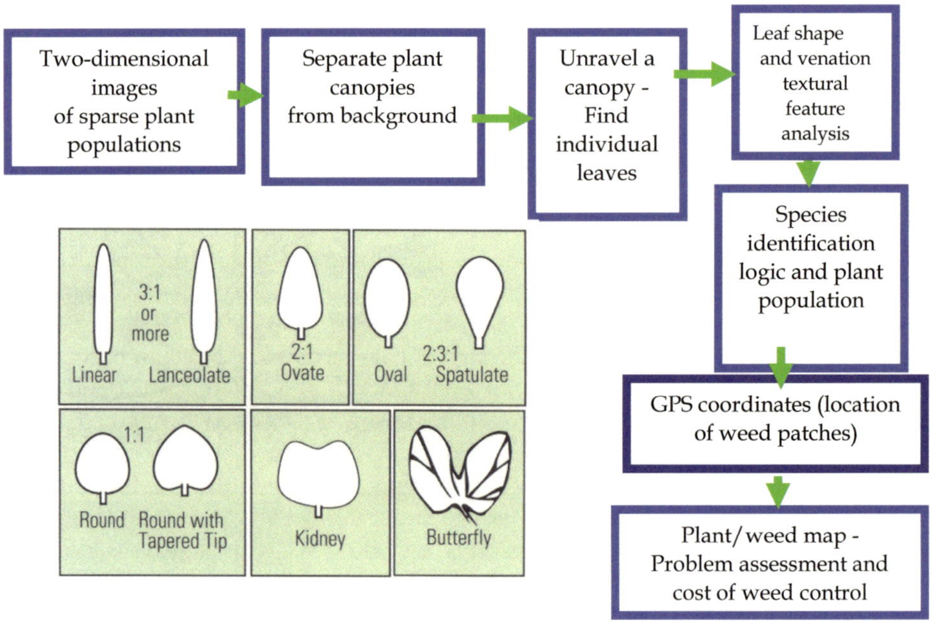

Fig. 1. A strategic approach to weed assessment.

accentuate a particular color such as plant greenness, which should be intuitive by human comparison. Woebbecke et al. (1995a) was one of the first researchers to test vegetation indices that were derived using color chromatic coordinates and modified hue for distinguishing green plant material in images from bare soil, corn residue, and wheat straw residue. Woebbecke's indices (without row and column indices of each pixel) included:

$$\text{Color indices: } \left(r - g,\, g - b,\, \frac{g - b}{r - g},\, \text{and } 2 \cdot g - r - b \right) \tag{1}$$

where: r, g, and b are known as the chromatic coordinates (Wyszecki and Stiles, 1982), given as:

$$r = \frac{R^*}{R^* + G^* + B^*},\; g = \frac{G^*}{R^* + G^* + B^*},\; \text{and } b = \frac{B^*}{R^* + G^* + B^*} \tag{2}$$

and: R^*, G^*, and B^* are normalized RGB values (0 to 1), defined as:

$$R^* = \frac{R}{R_m},\; G^* = \frac{G}{G_m},\; \text{and } B^* = \frac{B}{B_m}$$

R, G, and B are the actual pixel values obtained from color images, based on each RGB channel or band.

R_m, G_m, and B_m = 255, are the maximum tonal value for each primary color.

Woebbecke discovered that the excess green vegetation index ($ExG = 2 \cdot g - r - b$) provided an interesting near-binary, tonal image outlining a plant region of interest. Woebbecke's excess green (ExG) index has been widely cited in the literature and has been tested in recent studies (Giltelson et al., 2002; Lamm et al., 2002; Mao et al., 2003; and others). ExG plant regions of interest could then be completely binarized using a selected contrast threshold value for each image. Thus, an important condition was the selection of the threshold value. Mao et al. (2003) subsequently tested several indices: ExG, normalized difference index (NDI), and the modified hue for separating plant material from different backgrounds (soil and withered plant residue). In his study, the ExG index was found superior to the other indices tested. A critical step was to select a manual threshold value to binarize the tonal image into a black and white image.

Other color vegetation indices have been reported for separating plants from soil and residue background in color images. For example, the normalized difference vegetation index (NDI) by Perez et al. (2000) uses only the green and red channels and is given as:

$$\mathrm{NDI} = \frac{G - R}{G + R} \tag{2}$$

Perez's NDI was improved by adding a one, and then multiplying by a factor of 128. Hunt, et. al (2005) developed a vegetation index, known as the Normalized Green-Red Difference Index (NGRDI) for their model airplane photography for assessing crop biomass. Zhang, et al (1995) and Gebhardt, et al (2003) also used various RGB transforms for their plant image segmentation step.

Color indices have been suggested to be less sensitive to in lighting variations, and may have the potential to work well for different residues backgrounds (Campbell, 1996). However, a disproportionate amount of redness from various lighting sources may overcast a digital image, making it more difficult to identify green plants with simple RGB indices (Meyer et al, 2004b). For example, image redness may be related to digital camera operation and background illumination, but may also be related to redness from the soil and residue itself. An alternate vegetative index called excess red ($\mathrm{ExR} = 1.4 \cdot r - g$) was proposed by Meyer et al.(1998a), but was not tested until later studies.

Meyer and Camargo Neto (2008) reported on the development of an improved color vegetation index: Excess Green minus Excess Red (ExG-ExR). This index does not require a threshold and compared favorably to the commonly used Excess Green (ExG), and the normalized difference (NDI) indices. The latter two indices used an Otsu threshold value to convert the index near-binary to a full-binary image. The indices were tested with digital color images of single plants grown and taken in a greenhouse and field images of young soybean plants. Vegetative index accuracies were compared to a hand extracted plant regions of interest using a separation quality factor algorithm. A quality factor of one represented a near perfect binary match of the computer extracted plant target compared to the hand extracted plant region. The ExG-ExR index had the highest quality factor of 0.88 ± 0.12 for all three weeks, and soil-residue backgrounds for the greenhouse set. The ExG+Otsu and NDI-Otsu indices had similar quality factors of 0.53 ± 0.39. and 0.54 ± 0.33 for the same set, respectively. Field images of young soybeans against bare soil gave quality factors for both ExG-ExR and ExG+Otsu around 0.88 ± 0.07. The quality factor of NDI+Otsu using the same field images was 0.25 ± 0.08. ExG-ExR has a fixed, built-in plant-background zero threshold, so that it does not need Otsu or any user selected threshold value. The ExG-ExR index worked especially well for fresh wheat straw backgrounds, where it was generally 55 per cent more accurate than the ExG+Otsu and NDI+Otsu indices. Once a binary plant region of interest is identified with a vegetation index, other advanced image processing operations may be applied, such as identification of plant species such as would be needed for strategic weed control.

Near-Infrared (NIR) along with color bands have been used in vegetative indices for satellite remote sensing applications. However, NIR is less human intuitive, since the human eye is not particularly sensitive to the NIR spectrum which begins with red light. The human eye is only able to discern color (retinal sensors called cones). The eye also contains rods which are essentially receptive to small amounts of blue light that may exist after sundown. NIR is also not readily available with an RGB color digital camera. NIR usually requires a special monochromatic camera with a silicon-based sensor that can detect light up to one micron in wavelength with an NIR band pass filter. Hunt, at al (2011) has experimented with extracting near infrared out of RGB digital cameras. They developed a low-cost, color and color-infrared (CIR) digital camera that detects bands in the NIR, green, and blue. The issue still remains as to how does one verify the accuracy of infrared-image-based vegetative index without comparison to vegetation observed in a corresponding color visual image? So, the verification process of existence of plant material either returns to color images or some other non-optical method.

Two additional problems tend to exist with previous research regarding vegetative indices (a) the disclosure of the manual or automatic threshold used during the near-binary to binary conversion step, and (b) generally, the lack of reporting of vegetation index accuracy. Gebhardt, et al (2003) suggested that it was not necessary to classify vegetation on a pixel basis with digital imaging. However, if there are too many plant pixels mixed up with background pixels, accuracy may be reduced. Hague, et al (2006) suggested a manual comparison of vegetative areas from high resolution photographs. To date, very few vegetative index studies have reported validation accuracy of detecting plant material in independent images from other sources. This problem becomes particularly apparent, when these indices are applied to the collection of photographic plant databases currently available.

Plant classification might be expanded to hyper spectral imaging (Okamoto, et al. 2007). Wavelet along with discriminant analyses were used to identify spectral patterns of pixel samples for a 75–80 percent classification rate of five young plant species. Typically, hyper spectral cameras are expensive.

In summary, color image classification systems utilize the red (R), green (G), and blue (B) tonal intensity components. Color is a special form of spectral reflectance, which can be derived from spectral measurements (Wyszecki and Stiles, 1982; Murch, 1984; Jain, 1989; Gonzalez and Woods, 1992; Perry and Geisler, 2002). Perceived (human) color is based on the (RGB) primary colors. Woebbecke et al. (1995) discovered that the excess green index (2·G-R-B) could provide excellent near-binary segmentation of weed canopies over bare soil for canopy shape feature analysis. El-Faki et al. (2000b) studied different RGB indices, as potential weed detection classifiers, but none possibly as good as excess green. The best correct segmentation rates (CCR) found were around 62%, while some misclassification rates were less than 3%. Meyer et al. (1999, 2004) proposed an excess red index (1.3·R-G), based on physiological, rod-cone proportions of red and green. This index also provides near-binary silhouettes of plants under natural lighting conditions. Marchant, et al, (2004) proposed additional procedures for dealing with machine vision and natural lighting. The utilization spectral wave bands and color components have been used arithmetically and called vegetation indices. The index Meyer and Camargo Neto (2008) is an advanced color vegetation index.

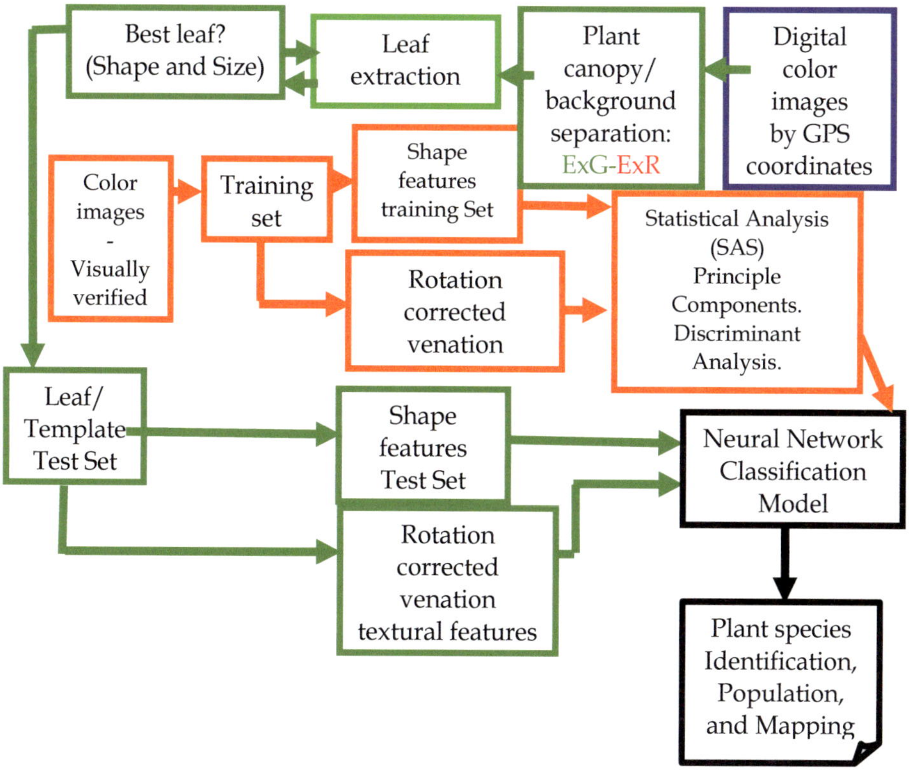

Fig. 2. Prototype Plant Species Identification and Enumeration using Leaf features.

5. Computerized single leaf extraction

Only a few methods of unsupervised leaf extraction from canopy images have been reported in the literature. Franz et al. (1991b) reported the use of curvature functions and the Fourier-Mellin correlation to identify completely visible and partially occluded sets of leaves. Leaf statistical features of mean, variance, skewness, kurtosis were computed, using spectral wavebands of red, green, blue, and near infrared. These features were used to discriminate leaf types of unifoliolate soybean, ivy, morning glory cotyledons, velvetleaf cotyledons, foxtail, first leaf of ivy, morning glory, and the first leaf of velvet leaf. Franz et al. (1995) further developed an algorithm to extract boundaries of occluded leaves using an edge detection technique to link the end points of leaf edge segments. User intervention was required at various steps of the algorithm. The fractions of individual leaves obtained were reported to be 0.91, 0.87, 0.95, and 0.71 for velvetleaf, soybean, ivy leaf morning glory, and foxtail, respectively.

To clarify this issue, occluded or partial fractions of leaves are probably not that useful for species identification. However, all canopies will exhibit whole individual leaves at the canopy apex, which can be seen in overhead photographs. Some leaves may standout by themselves (non-concealed) against the soil-residue background. Others will have vegetation from occluded leaves around them, which we will call concealed leaves. The latter would represent a difficult image processing problem, not easily solved by traditional algorithms such as edge detection, erosion, dilation, and such.

Deformable templates using active contours were used by Manh, et al. (2001) to locate boundaries of green foxtail leaves. Manh's process attempted to combine color separation and shape feature analysis into a single operation. The procedure began with identification of a leaf tip, and followed by shape analysis across the rest of the green material. However, a manually selected energy level or color was needed. Segmentation accuracy for a single species of foxtail leaves was reported to be 84%. No other species were studied.

Individual, whole, and fragments of leaves were isolated using the Gustafson-Kessel fuzzy clustering method over bare soil, corn stalks, and wheat straw color images (Hindman, 2001, Meyer et al., 2004b, Gustafson and Kessel, 1979). Zadeh intensification of the fuzzy cluster membership functions resulted in definitive green canopy areas, but not individual leaves. However, Camargo Neto, et al (2006) used the Gustafsen-Kessel fuzzy leaf cluster fragmentation method on green canopy regions of interest. He also developed a reassembling method of the green cluster fragments resulting in individual leaves using a genetic algorithm (Holland, 1975).

6. Shape feature analysis

If the process of image vegetative/background classification is to be useful, the separated plant region of interest (ROI) must provide important canopy or leaf shape feature or property information to at least discriminate between broadleaf and grass species (Woebbecke et al., 1995b; Meyer et al., 1998a; Meyer et al., 1998b).

Supervised leaf and single plant canopy shape feature analysis has been studied the most. Petry and Kuhbauch (1989) found shape parameters using five canonical indices found distinctly different for several weed species. Guyer et al. (1986, 1993) used image shape feature analysis on individual leaves to distinguish between weed species and corn. Guyer et al. (1993) using only leaf and canopy shapes, reported a 69% correct identification rate for 40 weeds and agricultural crop species. Guyer found that no single shape feature alone could distinguish corn from all other species. Franz et al. (1991 a,b) identified plants based on individual leaf shape at two growth stages using the Fourier-Mellin correlation. Woebbecke et al. (1995a, b) used basic image shape feature analysis to discriminate between broadleaf and grassy plant canopies. Woebbecke found that broadleaf and grass shape features best appeared to a vision system at early stages of growth or within a specific window of time, from 1-4 weeks after emergence. Downey, et al (2004) described a field canopy shape identification system which used a binary canopy erosion technique to discriminate between grasses and broadleaf plants. Yonekawa et al. (1996) presented a set of classical shape features for a leaf taxonomy database. Chi, et al (2002) fitted Bezier curves to different leaf boundary shapes. Mclellan and Endler (1998) compared several morphometric methods for describing complex shapes. They found that approximately 20 harmonics of the elliptic Fourier method accurately depicted shapes of *Acer saccharinum*, *Acer saccharum*, and *Acer palmatum* leaves. A leaf shape image retrieval systems was also reported by Wang, et al (2003).

Du et al (2005, 2006, 2007) proposed the Douglas-Peucker approximation algorithm for leaf shapes and the shape representation was used to form the sequence of invariant attributes.

A modified dynamic programming (MDP) algorithm for shape matching was proposed for the plant leaf recognition. Oide and Ninomiya (2000) used the Elliptic Fourier (EF) method to classify soybean varieties, using a normalized leaf shape. The EF method using a chain-coded, closed contour, invariant to scale, translation, and rotation was first introduced by Kuhl and Giardina (1982). EF has been used in recent studies to describe the shape of objects. Innes and Bates (1999) used an Elliptical Fourier descriptor to demonstrate an association between genotype and morphology of shells. Chen et al. (2000) used Elliptic Fourier descriptors to describing shape changes in the human mandible for male and female at different ages. Most methods previously investigated ignore leaf edge serration. Leaf serration or edgeness is an important morphologic feature used for identifying plant species. For example, the curvature functions developed by Franz et al. (1991b) were found generally inadequate where leaflet serration was quite pronounced. Camargo Neto, et al, 2006b applied the Elliptic Fourier shape feature analysis to extracted leaves of velvet leaf *Abutilon theophrasti*, pig weed *Amaranthus retroflexus L.*, sunflower *Helianthus annus*, and soybean *Glycine max*. A velvet leaf example is shown in Figure 3.
Hearn (2009) used a database of 2,420 leaves from 151 plant species for a plant leaf shape analysis. Using metrics derived during Fourier and Procrustes analyses, it was found that a minimum of ten leaves for each species, 100 margin points, and ten Fourier harmonics were required to develop any accuracy using the leaf shape of a species. His results indicated a success rate of 72% correct species identification for all 151 species used. This may mean that more than leaf shape is needed for classification.

7. Textural feature analysis

Color and/or leaf shape features alone may not be sufficient to consistently distinguish between young weed and crop plant species. Textural features may supply some additional botanical information, such as leaf venation, leaf pubescence, but also leaf disease and insect damage. The color or tonal detail for texture was first described by quantification of co-occurrence of tonal pairs or contrast also known as spatial tonal frequency (Haralick, 1978 and 1979). Wavelet analysis and energy have been recently suggested as a frequency based textural analysis for segmenting weeds imbedded in canopies (Chang and Kuo, 1993, Strickland and Hahn, 1997, Tang, et al, 2003). Shearer and Holmes (1990) used color co-occurrence matrix method to identify the textural features of isolated plants. Shearer and Jones (1991) proposed a texture-alone plant detection system based upon hue-saturation-intensity (HSI) images. Oka and Hinata (1989) used side view images of rice to distinguish between old and new Japanese rice cultivars. Zhang and Chaisattapagon (1995) tested a combination color, shape, and texture approach for detecting weeds in wheat fields and found that leaf –surface- coarseness indices defined by Fourier spectra may be effective in differentiating wheat from broad-leaf weeds. Meyer et al. (1999) showed that combined color, shape, and textural statistical discriminate analysis system could separate grasses from broadleaf canopies against bare soil backgrounds. Major problems for obtaining botanical textural detail involve image resolution, leaf orientation or rotation, shadows, bidirectional reflectance of leaf surfaces, and background lighting. Uneven lighting for example, could obscure venation - mesophyll leaf detail. Diffuse lighting could provide more even illumination than direct-beam lighting. Fu and Chi (2006) presented an algorithm for extracting leaf vein details from detached leaves under artificial light. Park, et al (2008) described a prototype system for classifying plants based on leaf venation features. Their

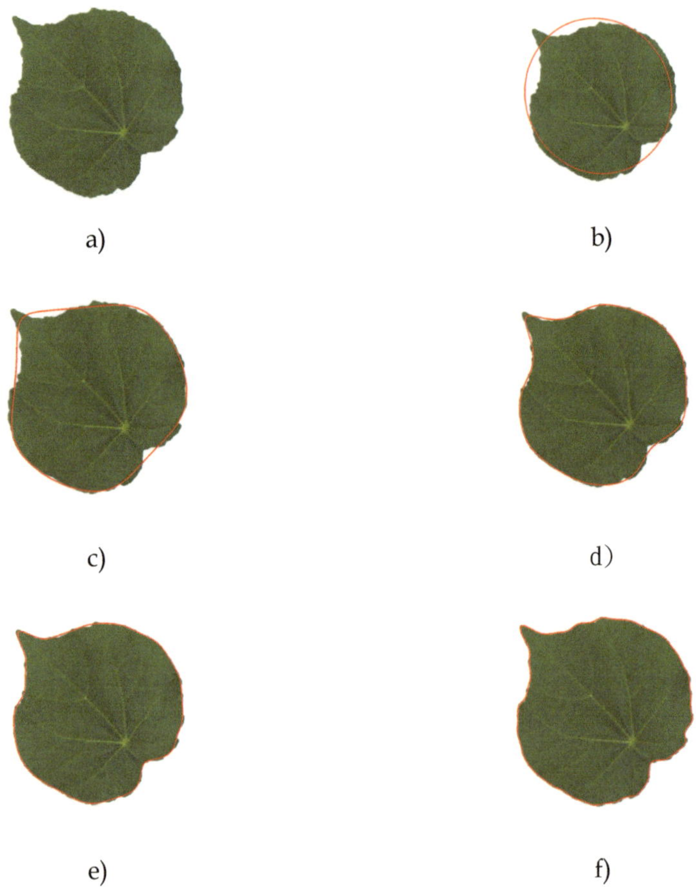

Fig. 3. Elliptic Fourier shape approximations for velvetleaf (*Abutilon theophrasti Medicus*), a) original leaf image, b) 1st EF harmonic, c) 1st + 2nd + 3rd + 4th EF, d) 1st + … + 8th EF harmonics, e) 1st + … + 16th EF harmonics, and f) 1st + … + 30th EF harmonics.

method detected the differences between tree and parallel venations in leaves, and thus could be considered as an enhancement to the classification tool set.

De Oliveira Plotze (2009) combined computer vision techniques and plant taxonomy protocols, these methods are capable of identifying plant species. The biometric measurements are concentrated in leaf internal forms, specifically in the venation system. The methodology was tested with eleven species of passion fruit of the genus *Passiflora*. The features extracted from the leaves were then applied to a neural network system to develop a classification of species. The results were very accurate in correctly differentiating among species with 97% of success. Zheng and Wang (2009, 2010) presented the results of mathematical morphology used on images of single leaf samples. Mathematical morphology provides four fundamental operations of dilation, erosion, opening, and closing in image processing. Their goal was to extract only leaf veins using hue and intensity information. Camargo Neto and Meyer (2005) classified the plant species botanically sing additional leaf venation textural features acquired during the previous steps. One thing is clear, lack of care in the photography of a leaf may affect image textural properties and classification.

8. Plant species classification

Most studies in the last 20-years have addressed the classification of only two crop-weed classes or general cases of broad leaf versus grasses and in other cases, crop row versus between crop row (Tang, et al, 2003). However, to precisely classify a plant species that may be imbedded within other different species of plants in an image is a botanically challenging exercise.

Agarval, et al (2006) described an ongoing project to digitize information about plant specimens that would become available to field botanists and crop managers. They indicated that the first step required acquisition of digital images and possibly plant architectural models, along with an effective retrieval method and mobile computing mechanisms for accessing this information. At that time they had indicated progress in developing a digital archive of the collection of various plant specimens at the Smithsonian Institution.

Analytical tools are improving for classifying plant species. The artificial neural network (ANN) has been proposed for many classification activities. Plotze and Bruno (2009) have also proposed a plant taxonomy system. Yang et al., (2000, 2002, 2003) used RGB pixel intensities as inputs for a fuzzy artificial neural network (ANN) for distinguishing weeds from corn, with success rates as high as 66% for corn and 85% for weeds. To encompass the uncertainty of image classification processes, fuzzy set theory (FST) has been proposed for plant classification (Gottimukkala et al., 1999). FST provides a possibilistic alternative (different, but in many cases complementary) to the probabilistic or statistical approaches. FST embraces virtually all (except one) of the definitions, precepts, and axioms that define classical sets that supports common mathematics, (Ross, 2004). It uses variables in the form of membership functions with degrees of support for fuzziness, incorporating uncertainty (Zadeh, 1965; Mamdani, 1976; Li and Yen, 1995). Pal, et al 1981, 1994, Bezdek, 1973, 1993) summarized the use of a FST neural network for pattern recognition, generating membership functions, performing fuzzy logic (FL) operations, and then deriving inference rule sets. Jang (1993) invented the artificial neural fuzzy inference system (ANFIS) for training membership functions and rule sets that could be used for classification (Figure 4).

Fuzzy logic machine vision classification systems are intended to imitate human perception or vision and to handle uncertainty. In the weed discrimination example, expert human perception or scouting validation is required for ground truthing. Bhutani and Battou (1995) and Tizhoosh (1998, 2000) provide computational overviews and various examples of fuzzy logic applied to image processing. Incorporating unsupervised fuzzy logic clustering and image analysis into site-specific technologies has tremendous potential (Kuhl and Giardina, 1982, Gath and Geva, 1989, De and Chatterji, 1998, Babuska, 1998, Manthalkar, at al, 2003, Meyer, et al. 2004). The very nature of site-specific data collection, image analysis, decision-making, etc., is characterized by uncertainty, ambiguity, and vagueness, which may be over overcome with these techniques.

Hindman and Meyer (2000) demonstrated a prototype fuzzy inference system for plant detection. Jones et al. (2000) used remotely sensed data with FL classification to detect crop status, resulting in a fuzzy description of crop phenology based upon spectral data. Yang et al., (2003) also presented potential herbicide savings in weed control with a fuzzy logic system. Heming and Rath (2001) proposed a fuzzy weed classifier that yielded correct classification accuracies between 51 and 95%. The potential fallacy of any regression, ANN, or fuzzy ANN (ANFIS) model is that they can be designed to mimic signal errors and random noise data too well, especially with an inadequate size of the training data set. Fuzzy inference systems can also incorporate the "I do not know" result.

Fuzzy clustering refers to unsupervised partitioning of data into subclasses for pattern recognition (Ross, 2004). Babuska (1998) presented six different clustering techniques that might be used to organize tonal image data with their limitations. These included the fuzzy c-means, the Gustafson-Kessel, fuzzy maximum likelihood, fuzzy c-varieties, fuzzy c-elliptotypes, and possibilistic clustering that might be used on tonal images. Moghaddamzadeh et al. (1998) described a fuzzy nearest-neighbor, clustering method for segmenting color images. Townsend (2000) discussed methods for making comparisons of fuzzy ecological pattern recognition methods. Beichel, et al. (1999) discussed the use of an unsupervised Gath-Geva clustering method for Landsat thermatic mapper (TM) images. Classification accuracy reached a maximum value of 86 % with five clusters. Individual, whole, and fragments of leaves were isolated using the Gustafson-Kessel fuzzy clustering method over bare soil, corn stalks, and wheat straw color images (Meyer et al., 2004b). Zadeh intensification of the membership functions resulted in definitive green canopy areas.

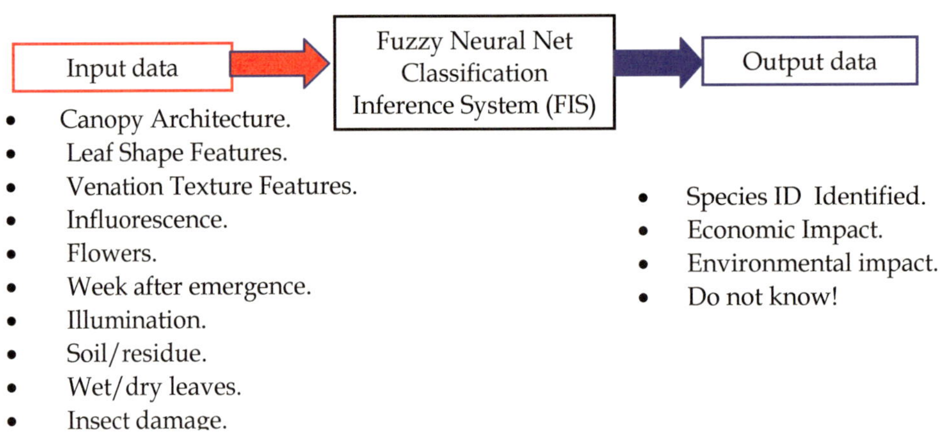

Fig. 4. Advanced Species Classifier Method– Fuzzy Logic- Neural Network using Image metrics and others.

A machine vision system with unsupervised image analysis and mapping of features was presented by Camargo Neto (2006a) and Camargo Neto, et al. (2006b). A classification system was trained using statistical discriminant analysis which was tested using individual test leaves and clusters from several plants. As many as 75 percent of exposed whole leaves were extracted, and can be further species identified at 75% or better. When such a system is improved and validated with scientific-based methods, it could dramatically assist understanding crop-weed relationships, growth, competition, and control. A machine vision system certainly should be able to identify and distinguish weed species that are 7 - 21 days old, a time when post emergence herbicides are most effective.

9. Linking machine vision with weed management systems

Predicting crop yield loss due to weed competition is one critical component of dynamic decision making for integrated weed management. Moreover, spatial variation in weed occurrence must be accounted for to accurately predict crop yield loss (Lindquist et al. 1998, 2001 a,b). The fuzzy logic machine vision classification system will be extremely useful where weeds are distinguished from crop plants and precisely mapped within a farm field. Shape feature analysis also provides a means for determining the relative surface area of weed plants relative to crop plants. Kropff and Spitters (1991) argued that the competitive strength of a species is determined by its share in leaf area at the moment when interspecific competition begins. Kropff et al. (1995) presented an equation that expresses yield loss (YL) as a function of weed and crop LAI. This approach has recently been expanded to relate yield loss to weed and crop relative volume (Conley et al. 2003) and could easily be used to relate yield loss to weed and crop relative surface area obtained from our image analysis. This kind of detail requires close-in imaging within a few meters with current high pixel rate digital cameras.

Holst, et al (2006) reviewed the progress of weed population modeling and of course the use is similar: strategic decision making for weed management. Freckleton and Stephens (2009) discussed the use if dynamic plant models for weed management. They concluded that there exist a discrepancy in the field of weed population modeling; many of the problems faced by weed ecologists require detailed quantitative predictions, but few modelers are attempting to provide such predictions. FST has also been used for modeling biological systems. Ambuel et al. (1994) used FL to develop a crop yield simulator for assessing spatial field variability for accuracy and optimizing pesticide application rates. Weed plant growth and plant population models that also describe the canopy architecture would be very helpful for weed classification.

10. Conclusions

The literature is rich in selected or component ideas for machine vision, plant species identification. Now is the time to put together a complete robust system that essentially mimics the human taxonomic, plant identification keying method. If one returns to Stubbendick, et al (2003), one can verify that the human classification process requires metrics on leaves, stems, flowers, influorescence, and a picture of the plant. Leaf shape and venation images alone may not close the classification process.

Shape analysis for image processing is very well-understood and computer algorithms are readily available. The leaf angle in the plane of the canopy is of interest (the first elliptic Fourier harmonic), and that is a critical angle for rotationally invariant leaf texture or venation analysis. Additional studies regarding leaf orientation relative to the camera lens might help to reduce classification errors. Modern digital cameras are capable of acquiring large amounts of image-pixel data. Future studies need to determine minimal digital image resolutions needed to maintain the highest species discrimination performance.

Fuzzy logic, cluster algorithms and cluster reassembly routines work well for extracting convex leaf shapes from plant canopy images. However, for more botanically diverse leaf shapes, such as species with complex leaves, lobed margins (indented), trifoliolates, etc., new fitness criteria need to be developed to accommodate these leaf shapes. Undoubtedly, integration of specific shape and textural feature analyses as fitness criteria may be a key to improvement of this process. New leaf extraction/species classification algorithm can become especially useful, if acceptance criteria can be designed to accommodate more a extensive leaf taxonomy digital library (shape and texture of single and compound leaves). Work has been extended on utilizing digital canopy architecture metrics in three dimensions which is important plant taxonomy.

Species classification and mapping has been tested using a neural-fuzzy inference model, which can be improved with inclusion of additional training information, including: stage of growth, expected canopy architecture, distance from a designated crop row, crop row spacing and direction.

Studies and discussion should be conducted to determine if older photographic plant image data bases can be used as references for new unknown digital plant images. Considerable field testing and validation are always needed for plant identification studies using machine vision.

11. References

_______, 2004. *National Roadmap For Integrated Pest Management.* http://www.csrees.usda.gov/nea/pest/pdfs/nat_ipm_roadmap.pdf.

Ambuel, J.R., Colven, T.S., Karlen, D.L., 1994. *A fuzzy logic yield simulator for prescription farming.* Trans. ASAE 37 (6), 1999–2009.

Agarwal, G., Belhumeur, P., Feiner, S., Jacobs, D., Kress, W.J., Ramamoorthi, R., Bourg, N.A., Dixit, N., Ling, H., Mahajan, D., Russell, R., Shirdhonkar, S., Sunkavalli, K. and White, S. 2006. *First steps toward an electronic field guide for plants.* Taxon 55: 597–610.

Babuska, R. 1998. Fuzzy Clustering Algorithms. *Fuzzy Modeling for Control.* Kluwer Academic Publishing , Boston 3:49-74.

Beichel, R., R. Bolter, and A. Piz, 1999. *Fuzzy Clustering of a Landsat TM Scene.* Geoscience and Remote Sensing Symposium, IGARSS '99 Proceedings. IEEE 1999 International 1597-1599.

Bezdek, J.C. 1973. *Fuzzy mathematics in pattern classification.* Ph. D. Thesis, Applied Math. Center, Cornell University, Ithaca.

Bezdek, J.C. 1993. *A review of probabilistic, fuzzy, and neural models for pattern recognition.* Journal of Intelligent and Fuzzy Systems. 1(1):1-25.

Bhutani, K.R. and A. Battou, 1995. *An application of fuzzy relations to image enhancement.* Pattern Recognition Letters. 16:901-909.

Camargo Neto, G.E. Meyer , D. D. Jones. 2006a. *Individual Leaf Extractions from Young Canopy Images using Gustafson-Kessel Clustering and a Genetic Algorithm.* Computers and Electronics in Agriculture (Elsevier) 51:65-85.

Camargo Neto, G.E. Meyer, D. D. Jones, A.K. Samal. 2006b. *Plant Species Identification using Elliptic Fourier Analysis.* Computers and Electronics in Agriculture (Elsevier), 50:121-134.

Camargo Neto, J., 2004. *A Combined Statistical -- Soft Computing Approach for Classification and Mapping Weed Species in Minimum Tillage Systems.* Unpublished PhD Dissertation, University of Nebraska, Lincoln, NE.

Camargo Neto, J., G.E. Meyer. 2005. Crop species identification using machine vision of computer extracted individual leaves. In: Chen, Y.R., Meyer, G.E., Tu S. (Eds.), *Optical Sensors and Sensing Systems for Natural Resources and Food Safety and Quality,* Proc. SPIE, Bellingham WA., 5996: 64-74.

Camargo-Neto, J., G.E. Meyer, and D.D. Jones, 2004. *Advances in Color Image Segmentation of Plants for Weed Control.* ASAE Paper Number 043060, The Society for Engineering in Agricultural, Food, and Biological Systems, St Joseph MI.

Campbell, J B, 1996. *Introduction to Remote-Sensing*, The Guilford Press, London).

Chang, T. and C.J. Kuo, 1993. *Texture Analysis and Classification with Tree-Structured Wavelet Transform.* IEEE Transactions on Image processing. 2(4): 429-441.

Chen, S. Y.Y., P.E. Lestrel, W.J.S. Kerr, and J.H. McColl, 2000. *Describing shape changes in the human mandible using elliptic Fourier functions.* European Journal of Orthodonics. 22:205-215.

Chi, Y.T., C.F. Chien, and T.T. Lin, 2002. *Leaf Shape Modeling and Analysis using Geometric Descriptors derived from Bezier Curves.* Trans. ASAE. 46(1):175-185.

Conley, S. P., D. E. Stoltenberg, C. M. Boerboom, and L. K. Binning. 2003. *Predicting soybean yield loss in giant foxtail (Setaria faberi) and common lambsquarters (Chenopodium album) communities.* Weed Science. 51:402-407

Criner, B.R., J.B. Solie, M.L. Stone, R.W.Whitney. 1999. *Field-of-View Determination for a Bindweed Detection Sensor.* Trans. ASAE. 42(5):1485-1491.

De Oliveira Plotze, R., 2009. *Automatic Leaf Structure Biometry: Computer Vision Techniques and Their Applications In Plant Taxonomy.* International Journal of Pattern Recognition and Artificial Intelligence 23(2): 247–262.

De T.K. and B.N. Chatterji, 1998. *An approach to generalized technique for image enhancement using the concept of fuzzy set.* Fuzzy Sets and Systems. 25:145-158.

Downey, D., D.K. Giles, and D.C. Slaughter, 2004. *Weeds are accurately mapped using DGPS and ground-based vision identification.* California Agriculture. 58(4):218-221.

Du, J.X., Huang D.S., Wang, X.F. and Gu, X. 2006. *Computer aided plant species identification (CAPSI) based on leaf shape matching technique.* Trans. Inst. Measurement Control 28: 275–284.

Du, J.X., Wang, X.F. and Gu, X. 2005. *Shape matching and recognition based on genetic algorithm and application to plant species identification.* Lecture Notes Computer Sci. 3644: 282–290.

Du, J.X., Wang, X.F. and Zhang, G.J. 2007. *Leaf shape based plant species recognition.* Appl. Math. Computation 185: 883–893.

El-Faki, M. S., Zhang, N., Peterson, D. E. 2000. *Weed detection using color machine vision.* Trans. ASAE. 43(6):1969-1978.

El-Faki, M. S., Zhang, N., Peterson, D. E. 2000. *Factors affecting color-based weed detection.* Trans. ASAE. 43(4):1001-1009.

Franz, E., L.D. Gaultney, and K.B. Unklesbay. 1995. *Algorithms for extraction leaf boundary information from digital images of plant foliage.* Trans. ASAE 38(2):625 - 633.

Franz, E., M.R. Gebhardt, and K.B. Unklesbay. 1991. *Shape description of completely visible and partially occluded leaves for identifying plants in digital images.* Trans. ASAE. 34(2):673-681.

Franz, E., M.R. Gebhardt, and K.B. Unklesbay. 1991. *The use of local spectral properties of leaves as an aid for identifying weed seedlings in digital images.* Trans. ASAE. 34(2):682-687.

Freckleton, R.P. and P.A Stehens, 2009. *Predictive models of weed population dynamics.* Weed Research. 49:225-232.

Fu, H. and Z. Chi. 2006. *Combined thresholding and neural network approach for vein pattern extraction from leaf images.* IEEE Proceedings Vision, Image and Signal Processing 153: 881–892.

Gath, I., and A.B. Geva, 1989. *Unsupervised optimal fuzzy clustering.* IEEE Transactions on Pattern Analysis and Machine Intelligence. 2(7):773-781.

Gausman, H.W., R.M. Menges, A.J. Richardson, H. Walter, R.R. Rodriguez, and S. Tamez. 1981. *Optical properties of seven weed species.* Weed Science. 29(1):24-26.

Gausman, H.W., W.A. Allen, C.L. Wiegand, D.E. Escobar, R.R. Rodriguez, and A.J. Richardson. 1973. *The leaf mesophylls of twenty crops, their light spectra, and optical and geometrical parameters.* USDA Technical Bulletin No. 1465-59.

Gebhards, R. And S. Christensen, 2003. *Real-time weed detection, decision making, and patch spraying in maize, sugarbeet, winter wheat and winter barley.* Weed Research (European Weed Research Society) 43:385-393.

Giltelson, A.A., Y.J. Kaufman, R. Stark, and D. Rundquist. 2002. *Novel algorithm for remote estimation of vegetation fraction.* Remote Sensing of Environment (Elsevier) 80:76-87.

Gonzalez, R.C. and R.E. Woods. 1992. *Digital Image Processing.* Addison Wesley Publishing Company, Reading, Massachusetts.

Gottimukkala, S., N. Zhang, X. Lin, and N. Wang, 1999. *An application of fuzzy logic in weed detection.* ASAE Paper No. 99-3139 (microfiche). St. Joseph, MI. pp12.

Grube, A., D. Donaldson, T. Kiely, and L. Wu, 2011. *Pesticides Industry Sales and Usage. 2006 and 2007 Market Estimates.* Biological and Economic Analysis Division. Office of Pesticide Programs. Office of Chemical Safety and Pollution Prevention. U.S. Environmental Protection Agency. Washington, DC, 33 pages.

Gustafson, E.E., and W.C. Kessel. 1979. *Fuzzy clustering with a fuzzy covariance matrix.* IEEE CDC, San Diego:761 - 766.

Guyer, D.E., G.E. Miles, L.D. Gaultney, and M.M. Schreiber. 1993. *Application of machine vision to shape analysis in leaf and plant identification.* Trans. ASAE. 36(1):163-171.

Guyer, D.E., G.E. Miles, M.M. Schreiber, O.R. Mitchell, V.C. Vanderbilt. 1986. *Machine vision and image processing for plant identification.* Trans. ASAE. 29(6):1500-1507.

Hagger, R.J., C.J. Stent, and J. Rose, 1985. *Measuring spectral differences in vegetation canopies by a reflectance ratio meter.* Weed Research. 24:59-65.

Hagger, R.J., C.J. Stent, and S. Isaac, 1983. *A prototype hand-held patch sprayer for killing weeds activated by spectral differences in crop/weed canopies.* J. Agricultural Engineering Research. 28:349-358.

Hague, T., N.D. Tillett and H. Wheeler, 2006. *Automated crop and weed monitoring in widely spaced cereals.* Precision Agric. 7:21-32.

Haralick, R.M., 1979. *Statistical and Structural Approaches to Texture.* Proc. IEEE, 67(5): 786-804.

Haralick, R.M., 1978. *Statistical and Structural Approaches to Texture.* Proceedings 4th Int. Joint Conf. Pattern Recognition. 45-69.

Hearn, D.J., 2009. *Shape analysis for the automated identification of plants from images of leaves.* Taxon. 58 (3): 962–981.

Hemming, J. and T. Rath. 2001. *Computer-vision-based weed identification under field conditions using controlled lighting.* J. of Agricultural Engineering Research. 78(3):233-243.

Hindman, T.W. 2001. *A fuzzy logic approach for plant image segmentation and species identification in color images.* PhD Dissertation, University of Nebraska, Lincoln.

Hindman, T.W. and G.E. Meyer, 2000. Fuzzy logic inference systems for discriminating plants from soil and residue with machine vision. In J.A. DeShazer and G.E. Meyer (ed). *Biological Quality and Precision Agriculture II.* SPIE Optical Engineering Press, Bellingham, WA (ISBN 0-8194-3868-5), pp 111-121.

Holland, J.H. 1975. *Adaptation in natural and artificial systems.* Ann Arbor, The University of Michigan, Press.

Holst, N., I.A. Rasmussen, and L. Bastiaans, 2007. *Field weed population dynamics: a review of model approaches and applications.* Weed research. 47:1-14.

Hoppner, F., F. Klawonn, R. Kruse, and T. Runkler. 2000. *Fuzzy cluster analysis methods for classification, data analysis and image recognition.* John Wiley and Sons, LTD, NY.

Hummel, J.W., and E.W. Stoller. 2002. *On-the-go weed sensing and herbicide application for the Northern Cornbelt.* American society of Agricultural Engineers. Paper no. 021021.

Hunt, E.R., M. Cavigelli, C. T. Daughtry, J. McMurtrey III and C. L. Walthall, 2005. *Evaluation of Digital Photography from ModelAircraft for Remote Sensing of Crop Biomassand Nitrogen Status.* Precision Ag (Springer) 6:359-378.

Hunt, E.R., W.D. Hively, G.W. McCarty and C.T. Daughtry, 2011. *NIR-Green-Blue High-Resolution Digital Images for Assessment of Winter Cover Crop Biomass.* GIScience and remote Sensing 48(1)86-98.

Innes D.J. and J.A. Bates, 1999. *Morphological variation of Mytilus edulis and Mytilus trossulus in eastern Newfoundland.* Marinc Biology133,691^699.

Jain, A.K. 1989. *Fundamentals of Digital Image Processing.* Prentice Hall, Inc. Englewood Cliffs, New Jersey.

Jang, J. R. 1993. *ANFIS: Adaptive-network-based fuzzy inference system.* IEEE Trans. 0n Systems, Man, and Cybernerics 23(3):665-684.

Johnson, G. A., D. A. Mortensen, and C. A. Gotway,1993. *Spatial and temporal analysis of weed seeding populations using geostatistics,* Weed Science. 44(3), 704.

Johnson, G.A., D.A. Mortensen, and A.R. Martin, 1995. *A simulation of herbicide use based on weed spatial distribution.* Weed Res. 35:197-205.

Jones, D. and E.M. Barnes, 2000. *Fuzzy composite programming to combineremote sensing and crop models for decision support in precision crop management.* Ag Systems (Elsevier) 65:137-158.

Kincaid, D.T. and R.B. Schneider. 1983. *Quantification of leaf shape with a microcomputer and Fourier transform.* Canadian Journal of Botany. 61:2333-2342.

Kropff, M. J. and C. J. T. Spitters. 1991. *A simple model of crop loss by weed competition from early observations on relative leaf area of the weed,* Weed Res., 31:97-105.

Kropff, M. J., L. A. P. Lotz, S. E. Weaver, H. J. Bos, J. Wallinga and T. Migo. 1995. *A two parameter model for prediction of crop loss by weed competition from early observations of relative leaf area of the weeds,* Ann. Appl. Biol., 126:329-346.

Kuhl, F.P. and C.R. Giardina, 1982. *Elliptic Fourier features of a close contour,* Computer Graphics and Image processing. 18:236-258.

Lamm, R.D., D.C. Slaughter, and D.K. Giles. 2002. *Precision weed control for cotton.* Trans. ASAE 45(1):231-238.

Lee, C.-L. & Chen, S.-Y. 2006. *Classification of leaf images.* Int. J. Imaging Systems and Technology 16: 15–23.

Li, H.X., and V.C. Yen, 1995. *Fuzzy sets and fuzzy decision-making.* CRC.

Lindquist, J. L. and S. Z. Knezevic. 2001. *Quantifying crop yield response to weed population: Applications and limitations. Biotic Stress and Yield Loss.* Pages 205-232, in Peterson R. K. D. and Higley, L. G. (eds.), CRC Press, Boca Raton, FL.

Lindquist, J. L. *Mechanisms of crop loss due to weed competition. 2001. Biotic Stress and Yield Loss. Pages* 233-253, in Peterson R. K. D. and Higley, L. G. (eds.), CRC Press, Boca Raton, FL.

Lindquist, J. L., J. A. Dieleman, D. A. Mortensen, G. A. Johnson, and D. Y. Pester-Wyse. 1998. *Economic importance of managing spatially heterogeneous weed populations.* Weed Technology. 12:7-13.

Mamdani, E.H. 1976. *Advances in the linguistic sysnthesis of fuzzy controllers.* Int. J. of Man-Machine Systems. 8(6):669-678.

Manh, A.G., G. Rabatel, L. Assemat, and M.J. Aldon. 2001. *Weed leaf image segmentation by deformable templates.* Journal of Agricultural Engineering Research 80(2):139 - 146.

Manthalkar, R ; Biswas, P K ; Chatterji, B N, 2003. *Rotation and scale invariant texture features using discrete wavelet packet transform.* Pattern recognition letters. 24,(14): 2455-2463.

Mao, W., Y. Wang, and Y. Wang. 2003. *Real-time detection of between-row weeds using machine vision.* ASAE paper number 031004.

Marchant, J.A., N.D.Tillet, and C.M. Onyango, 2004. *Dealing with color changes caused by natural illumination in outdoor machine vision.* Cybernetics and Systems. 25:19-33.

Mclellan, T. and J.A. Endler, 1998. *The relative success of some methods for measuring and describing the shape of complex objects.* Syst. Biol. 47(2):264-281.

Medlin, C.R., D.R. Shaw, P.D. Gerard, and F.E. LaMastus. 2000. *Using remote sensing to detect weed infestation in Glycine max.* Weed Science 48:393 - 398.

Meyer G.E., T. Mehta, M. F. Kocher, D. A. Mortensen, and A. Samal. 1998. *Textural Imaging and Discriminate Analysis for Distinguishing Weeds for Spot Spraying.* Trans. ASAE., 41(4): 1189-1197.

Meyer, G. E. , J. Camargo Neto, D. D. Jones, T. W. Hindman. 2004. *Intensified fuzzy clusters for determining plant, soil, and residue regions of interest from color images,* Computers and Electronics in Agriculture. (Elsevier), 42(3):161-180.

Meyer, G.E. and J. Camargo Neto, 2008. *Verification of Color Vegetation Indices for Automated Crop Imaging Applications.* Computers and Electronics in Agriculture (Elsevier), 63:282-293.

Meyer, G.E., T.W. Hindman, and K. Lakshmi, 1999. Machine vision detection parameters for plant species identification, in G.E. Meyer and J.A. DeShazer (Ed.) *Precision Agriculture and Biological Quality.* Proceedings of SPIE, Bellingham, WA 3543:327-335.

Meyer, G.E., T.W. Hindman, D.D. Jones, and D.A. Mortensen. 2004. *Digital camera operation and fuzzy logic classification of plant, soil, and residue color images.* Engineering in Agriculture 20(4):519-529.

Moghaddamzadeh, A., D. Goldman, and N. Bourbakis. 1998. *A fuzzy-like approach for smoothing and edge detection in color images.* International Journal of Pattern Recognition and Artificial Intelligence 12(6):801-816.

Montserrat, J., F. Garcia-Torres, A. Garcia-Ferrer, M. Orden, and S. Atenciano, 2003. *Multi-species spatial variability and site-specific management maps in cultivated sunflower.* Weed Sci. 5:319-328.

Mortensen, D.A., G.A. Johnson, and L.J. Young. 1992. Weed distribution in agricultural fields. In *Soil Specific Crop Management*, Agronomy Society of America, 113-124.

Murch, G. M. 1984. *Physiological principles for effective use of color.* IEEE Computer Graphics and Applications. 4(11):49-54.

Oide, M., and S. Ninomiya. 2000. *Discrimination of soybean leaflet shape by neural networks with image input.* Computer and Electronics in Agriculture 29:59 - 72.

Oka, M. and K. Hinata. 1989. *Comparison of plant type between new and old rice cultivars using computer image analysis.* Japanese Journal of Crop Science. 58(2):232-239.

Okamoto, H., T. Mutata, T. Kataoka, and S. Hata, 2007. *Plant classification for weed detection using hyper spectral imaging and wavelet analysis.* Weed Biology and Management. 7:31-37.

Pal N.R. and J.C. Bezdek. 1994. *Measuring Fuzzy Uncertainty.* IEEE Transactions on Fuzzy Systems. 2(2):107-118.

Pal, S.K. and R.A. King, 1981. *Image Enhancement using smoothing with fuzzy sets.* IEEE Transactions on systems, man, and cybernetics. 11(7):494-501.

Park, J.K., E.J. Hwang, and Y.Y. Nam, 2008. Utilizing venation features for efficient leaf image retrieval Systems and Software (Elsevier) 81: 71-82.

Perez, A.J., F. Lopez, J.V. Benlloch, and S. Christiansen, 2000. *Colour and shape analysis techniques for weed detection in cereal fields.* Computers and Electronics in Agriculture. (Elsevier) 25:197-212.

Perry, J.S., and W.S. Geisler. 2002. Gaze-contingent real-time simulation of arbitrary visual fields. In: B. Rogowitz, T. Pappas (Eds.) *Human Visio Electronic Imaging.* SPIE Proceedings, San Jose, CA.

Petry, W. and W. Kuhbauch. 1989. *Automatisierte unterscheidung von unkrauten nach formparametern mit hilfe der quantitativen bild analyse.* Journal of Agronomy and Crop Science. (Berlin) 163:345-351.

Plotze, R.D., and O.M. Bruno, 2009, *Automatic leaf structure biometry: computer vision techniques and their applications in plant taxonomy.* International Journal of Pattern Recognition and Artificial Intelligence 23(2) 247-262.

Ross, T.J., 2004. *Fuzzy logic with engineering applications.* John Wiley and Sons LTD., West Sussex, England. 628 pp.

Shearer, S.A. and P.T. Jones. 1991. *Selective application of post emergence herbicides using photoelectrics.* Trans. ASAE. 34(4):1661-1666.

Shearer, S.A. and R.G. Holmes. 1990. *Plant identification using color co-occurrence matrices.* Trans. ASAE. 33(6):2037-2044.

Storlie, C.A., A. Stepanek, and G.E. Meyer. 1989. *Growth analysis of whole plants using video imagery.* Trans. ASAE. 32(6):2185-2189.

Strickland, R.N. and H.I. Hahn, 1997. *Wavelet Transform Methods for Object Detection and Recovery.* IEEE Transactions on Image Processing. 6(5):724-735.

Stubbendick, J., M.j. Coffin, and L.M. Landholt, 2003.Weeds of the Great Plains. Nebraska department of Agriculture. Lincoln, NE.

Tang, L., L. Tian, B.L.Steward, 2003. *Classification of Broadleaf and Grass Weeds using Gabor Wavelets and an Artificial Neural Network.* Trans. ASAE. 46(4):1247-1254.

Tarbell, K.A. and J.F. Reid. 1991. *A computer vision system for characterizing cotton growth and development.* Trans. ASAE. 34(5):2245-2255.

Tian, F.L., J.F. Reid, and J. Hummel. 1999. *Development of a precision sprayer for site-specific weed management.* Trans. ASAE 42(4):893-900.

Tillet, N.D., T. Hague, and S.J. Miles, 2001. *field assessment of a potential method for weed and crop mapping on the basis of crop planting geometry.* Computers and Electronics in Agriculture (Elsevier), 32(3):229-246.

Tizhoosh, H.R. 1998. *Fuzzy image processing: Potentials and state of the art.* IIZUKA 98, 5th International Conference on Soft Computing, Iizuka, Japan, 1:321-324.

Tizhoosh, H.R. 2000. Fuzzy image enhancement: An overview. In E.E. Kerre and M. Nachtegael (Editors). *Fuzzy techniques in Image processing,* Physica-Verlag (A Springer-Verlag Co.) 5:137-171.

Townsend, P.A. 2000. *A Quantitative Fuzzy Approach to Assess Mapped Vegetation classifications for Ecological Applications.* Remote Sens. Environ (Elsevier). 72:253–267

Tucker, C.J., J.H. Elgin, Jr., and J.E. McMurtrey. 1979. *Temporal spectral measurements of corn and soybean crops.* Photogrammetric Engineering and Remote Sensing. 45(5):643-653.

Wang, Z., W. Chi and D. Feng. 2003. *Shape based leaf image retrieval.* IEEE Proceedings Vision, Image and Signal Processing 150: 34–43.

Wiles, L.J., and E.E. Schweizer. 1999. *The cost of counting and identifying weed seeds and seedlings.* Weed Science 47(6):667-673.

Wiles. L. J, and E.E. Schweizer, 2002. *Spatial dependence of weed seed banks and strategies for sampling.* Weed Sci. 50:595-666.

Woebbeck, D.M., G.E. Meyer, B.K. V., and D.A. Mortensen. 1995. *Shape features for identifying young weeds using image analysis.* Trans. ASAE 38(1):271 - 281.

Woebbeck, D.M., G.E. Meyer, B.K. V., and D.A. Mortensen. 1995. *Color indices for weed identification under various soil, residue and lighting conditions.* Trans. ASAE 38:259-269.

Woebbecke, D.M., A. Al-Faraj, and G.E. Meyer. 1994. *Calibration of Large Field of View Thermal and Optical Sensors for Plant and Soil.* Trans. ASAE 37(2):669-677.

Wyszecki, G. and W.S. Stiles. 1982. *Color Science, Concepts and Methods.* John Wiley and Sons, New York, 117-248.

Yang, C., S.O. Prasher, J.A. Landry, H.S. Ramaswamy, and A. Ditommaso, 2000. *Application of artificial neural networks in image recognition and classification of crop and weeds.* Canadian Ag. Engr. 42(3):147-152.

Yang, C., S.O. Prasher, J.A. Landry, 2002. *Weed recognition in corn fields using back-propagation neural network models* Canadian Biosystems Engr. 22:

Yang, C., S.O. Prasher, J.A. Landry, and H.S. Ramaswamy, 2003. *Development of an Image Processing System and a Fuzzy Algorithm for Site-Specific Herbicide Applications.* Precision Agriculture (Kluwer). 4(1):5-18.

Yonekawa, S., N. Sakai, and O. Kitani. 1996. *Identification of idealized leaf types using simple dimensionless shape factors by image analysis.* Trans. ASAE. 39(4):525-533.

Zadeh, L.A., 1965. *Fuzzy sets, Information and Control.* 8: 338-353.

Zhang, N. and C. Chaisattapagon. 1995. *Effective criteria for weed identification in wheat fields using machine vision.* Trans. ASAE. 38(3):965-974.

Zheng, X. and X. Wang, 2009. *Fast Leaf Vein Extraction using Hue and Intensity Information.* The International Conference on Information Engineering and Computer Science. ICIECS 2009. Wuhan, China. 1-4.

Zheng, X. and X. Wang, 2010. *Leaf vein extraction using a combined operation of mathematical morphology.* The 2nd International Conference on Information Engineering and Computer Science. ICIECS 2010. Wuhan, China. 1-4.

20

Extraction and Analysis of Inositols and Other Carbohydrates from Soybean Plant Tissues

J.A. Campbell[1], S.C. Goheen[1] and P. Donald[2]
[1]*Battelle, Pacific Northwest National Laboratory*
Chemical and Biological Signature Sciences
Richland, WA
[2]*USDA/ARS, Crop Genetics Research Unit*
Jackson, TN
U.S.A.

1. Introduction

An outstanding characteristic of soybean plants is their ability to produce large amounts of the carbohydrate pinitol. Pinitol and the closely related inositols are currently undergoing widespread investigation for their biological and nutritional value. These and all the carbohydrates are typically extracted and analyzed together. Therefore, this review includes a general discussion about the extraction and analysis of carbohydrates in plants as well as a more in depth examination of the biosynthesis and use of compounds related to pinitol. The multiple roles of these substances in plants and animals, and their synergism have not been fully realized. This review discusses not only the extraction and analysis, but also the diverse roles of the inositols with an emphasis on inositols from the soybean plant.

2. Carbohydrate production and nitrogen fixation

Carbohydrates are produced in plants by photosynthesis. Zhu et al. (2010) reviewed photosynthesis in relation to improving crop yield. Agronomically, there has been little benefit in breeding for increased photosynthesis indicating that the relationship of photosynthesis to yield is still not well understood (Farquhar & Sharkey, 1982; Pessarakli, 2005). The relative growth rate of shoots was shown to be correlated to the soluble carbohydrate level in the plant, but shoot growth was also impacted by plant stress (Masle *et al*, 1990). One commonly studied plant stress in relation to carbohydrate production is drought stress. There is confusion regarding the regulation of carbohydrate synthesis when plants are under drought stress. Drought stress in addition to reducing shoot growth, increases root growth (Sharp & Davies, 1979).

Approximately 70 million tons of fixed nitrogen or about 50 % of the total nitrogen that enters the terrestrial ecosystem comes from biological nitrogen fixation (Brockwell *et al.*, 1995; Tate, 1995). The relationship of carbohydrate availability to photosynthesis, phloem sap supply and N_2 fixation in legumes is complex and knowledge is incomplete (Udvardi & Day, 1997).

Carbohydrates are the main energy source for humans. Carbohydrates are classified according to the number of monomers they contain as monosaccharides (simple sugars), oligosaccharides, or polysaccharides. Carbohydrate metabolism in plants has been reviewed (Colowick & Kaplan, 1951; Ochoa & Stern, 1952; and Horecker & Mehler, 1955). Carbohydrate levels in soybean seed are highest at growth stage R 5.5, or when the seed is half-developed (Wilson, 2004). A significant portion of the carbohydrate produced by photosynthesis is respired in the plant roots. (Lambers *et al.*, 1996).

3. Simple sugars

The most common simple sugars are glucose and fructose. Disaccharides consist of two covalently bound sugar molecules. Sucrose, for example, is a disaccharide consisting of glucose and fructose. Sugars have a role in energy, carbon transport molecules, hormone-like signaling factors, and as the source for building proteins, polysaccharides, oils and woody materials (Halford *et al.*, 2010). Plant genotype and environment greatly affect the levels found in plants (Halford *et al.*, 2010).

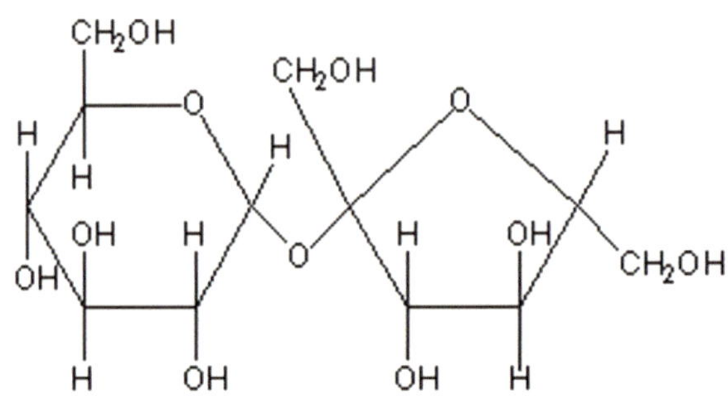

Sucrose

4. Complex carbohydrates

Complex carbohydrates (polysaccharides) are polymers of the simple sugars . Starch is the principal polysaccharide used by plants to store glucose.

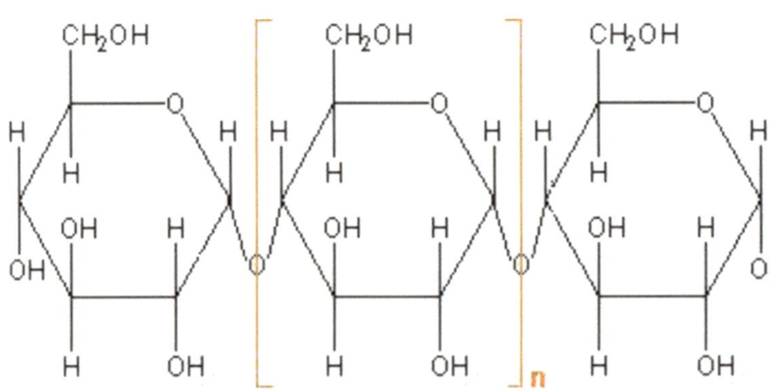

(n is the number of repeating glucose units and ranges in the 1,000's)

Starch

Zeeman *et al.*, 2010 reviewed the role of starches in plants. Starch breakdown commonly occurs when seeds germinate. Starch is also involved in malting (Halford *et al.*, 2010). Glycogen, also a polymer of glucose, is the polysaccharide used by animals to store energy. Another important polysaccharide is cellulose. Cellulose is used as a structural molecule to add support to leaves, stems, and other parts of plants. Although cellulose can't be used as an energy source in most animals, it provides essential fiber in the diet. Cell wall polysaccharides vary with plant groups and can include cellulose, xyloglucan, arabinoxylan, and pectin. In plants they make up the primary biomass and contribute to fiber in the human diet. This area has been reviewed by Scheller & Ulvskov, 2010; Fontes & Gilbert, 2010.

5. Extraction and cleanup

The methods used for isolating carbohydrates depend on the carbohydrate type, matrix, and purpose or type of analysis. However, some extraction procedures are commonly used for isolating carbohydrates from other classes of compounds in plants and foods. As an example, foods are usually dried under vacuum to prevent thermal degradation, ground to a fine powder to enhance extraction efficiency, and then remove the fats using appropriate solvent extraction.

A commonly used method for extracting low molecular weight carbohydrates from foods is to boil a sample with a 70-80% alcohol solution (Hall 2003, Asp 1993, Smith 1973.). Monosaccharides and oligosaccharides are soluble in alcohol solutions; however, most proteins, polysaccharides and dietary fiber are insoluble. The soluble components can then be separated from the insoluble components by filtering, soluble portion passes through the filter and the insoluble part retained by the filter. The two fractions can then be dried using lyophilization or nitrogen blow down techniques. In addition, monosaccharides and oligosaccharides and various other small molecules (e.g. organic acids, amino acids) may be present in the alcoholic extract. It is usually necessary to remove those components prior to carrying out a carbohydrate analysis, for example, with clarifying agents or by elution through one or more ion-exchange resins.

Water extracts of many foods contain substances that are colored or produce turbidity, and may interfere with analyses of carbohydrates; as a result, clarifiers may be needed. The most commonly used clarifying agents are heavy metals (e.g. lead acetate) which form insoluble complexes with interfering substances that can't be removed by either filtration or centrifugation. Ion-exchange is another method for removing interfering components prior to analysis. Many monosaccharides and polysaccharides are polar non-charged molecules and can therefore be separated from charged molecules by passing samples through an ion-exchange column. By using a combination of cationic and anionic resins it may be possible to remove most charged contaminants. Non-polar molecules can be removed by eluting through a column with a non-polar or hydrophobic stationary phase. Proteins, amino acids, organic acids, and hydrophobic compounds can be potentially removed from the carbohydrates in this manner prior to analyses.

Before analysis of the carbohydrates, residual alcohol (or other organic solvents) can be removed, if necessary, from the solution by evaporating under nitrogen or under vacuum using a rotary evaporator. For aqueous solutions, the sample can be concentrated using lyophilization.

Solid phase extraction (SPE) has also been reported for the cleanup and quantification of sugars and organic acids in herbal dry extracts. A three step SPE sequence was used for the separation of sugars from the other components. A hydrophobic cartridge was used as the first cartridge followed ion and cation exchange cartridges (Schiller et al., 2002).

6. Analysis

Once the carbohydrate fraction has been isolated from other components of the plant, either the total carbohydrate content can be determined, or individual carbohydrates can be isolated, identified and quantified. The analysis of carbohydrates can be performed using any of several different methods. Two of these techniques include gas chromatography (GC) and liquid chromatography (LC). There are also spectral methods available including nuclear magnetic resonance (NMR), infrared (IR) and Raman spectroscopy. In this review, our focus is on the chromatographic and mass spectrometric methods.

7. Derivatization for GC or GC/MS analyses

The most prevalent method used for analyzing carbohydrates is probably GC and GC coupled with mass spectrometry (MS) due to the high resolution of GC and definitive nature of MS. Since carbohydrates are nonvolatile, it is necessary to hydrolyze the sugars and then derivatize them to increase their volatility so they can elute through a GC column for analysis. Methods involving the formation of methylated glycosides, acetates, acetals, trimethylsilyl ethers, and more volatile alditol acetate derivatives of monosaccharides have been widely used (McInnes et al., 1958; Bishop & Cooper 1960; Bishop 1964; Lehrfeld 1981; Blakeney et al., 1983). More recently, trimethylsilyl (TMS) derivatives of carbohydrates have been used principally due to their relative ease of preparation and increased volatility. (Sweeley et al. 1963; Sullivan & Schewe 1977; Honda et al., 1979; Li et al., 1983; Twilley 1984). Different structural forms of carbohydrates can complicate their chromatograms due to the production of several (as many as 5) peaks for each monosaccharide. Formation of the corresponding oxime TMS-derivative reduces the number of potential peaks (Decker & Schweer 1982; Al-Hazmi & Stauffer 1986; Long & Chism 1987). Dmitriev et al. (1971) prepared the aldononitrile acetate derivatives with the oxime intermediate. Churms (1990) found the derivatization process was not affected by the presence of water in the reaction mixture, helping to minimize processing steps. Methods for the separation of neutral sugars in gums have also been reported using similar methods (Al-Hazmi & Stauffer, 1986).

Silylation is a versatile technique to increase the volatility of various analytes, including carbohydrates, making them amenable to GC and GC/MS analyses. There are several practical considerations that should be addressed prior to derivatization of a sample by this method. One major disadvantage of silylation derivatives is that they are susceptible to hydrolytic attack by any moisture present in the sample, resulting in incomplete silylation. However, the trimethylsilylation of aqueous samples of hydroxyl compounds has been achieved using a large excess of derivatizing reagent (Valdez 1985). Evershetd (1993) discussed another problem associated with silylation of carbohydrates, the existence of multiple reaction products, resulting in complicated chromatograms. The multiple products result from the formation of anomers and interconversion between pyranose and furanose rings. Interconversion of the anomers occurs via the open chain form of the sugar, while mutarotation results from the opening and closing of the ring. The interconversions can be minimized by the use of rapid and mild derivatization conditions. If silylation is the method of choice for derivatization, it may be desirable to protect the keto group of the monosaccharides prior to silylation in order to prevent the formation of enol-TMS ethers. These derivatives are unstable and complicate the analyses by giving rise to multiple products that can't be prepared quantitatively (Halket 1993).

In most instances, the silylating reagent is an adequate solvent. However, sometimes an additional solvent is required in the reaction. The selection of that solvent is critical to the success of the derivatization process. Any active hydrogens, including those present in the solvent, may be silylated. Pyridine has been found to be an ideal solvent for silylation reactions due to the increased solubility of the carbohydrates and their derivatives in that solvent (Evershed 1993). Heating slightly is often utilized to aid in efficient silylation,

One of the earliest reagents used for silylation was hexamethyldisilazane (HMDS). Usually, there is no need for additional solvents when HMDS is used. Recently, Ruiz-Matute et al. (2010) reviewed derivatization techniques of carbohydrates for GC and GC/MS analyses. Included in the discussion were derivatization of common sugars through the formation of ethers and esters, oximes, alditol acetates, aldononitriles, and dithoacetals (Evershed 1993).

Another silylating reagent is trimethylsilylimadazole (TMSI). Garland et al. (2009) analyzed soybean roots for pinitol using GC/MS (see Figs. 1-3). Roots were extracted in methanol and derivatized using TMSI. In this example a DB-5 capillary column was used in the splitless mode. The column eluents were analyzed by a double-focusing, four-sector mass spectrometer in the electron-ionization mode. Accurate mass measurements were also performed to determine the elemental composition of the parent and fragment ions. Under these conditions, a pinitol standard produced a single peak in the total ion chromatogram with a retention time of 9.18 min as shown in Fig. 1. Although several peaks appeared, pinitol's peak at 9.18 min was well-resolved.

The mass spectrum of TMSI-derivatized pinitol in Fig. 2 shows the major ion fragments detected from this, the most common carbohydrate in soybeans (Garland et al, 2009). In this example, the base ion is m/z 260. A comparison of the extracted ion plots of the soybean extract is shown in Fig. 3. A vertical, solid black line was added to each at the retention time of derivatized pinitol as determined from the standard. In the extracted ion plot of the soybean root, Figure 4 shows the total ion chromatogram of a TMSI-derivatized sugar beet extract. In this example no significant peaks appeared at the retention time of pinitol. The sugar beet root extract also showed no substantial peaks with the m/z 260 mass fragment.

The concentration of pinitol in soybean roots was approximately 4% of the soybean root's dry mass using a dry/fresh weight ratio of 54.5 mg DW/g FW (which is similar to 73.6 mg DW/g FW reported for alfalfa by Fougere, *et al.* (1991). The methanol extraction method appears to be effective for removing pinitol from the root tissue of soybean plants. The extent of extraction at the cost of time was encountered as well by Streeter and Strimbu's simultaneous extraction and derivatization method (Streeter & Strimbu 1998). Although they were able to reduce processing time, they were unable to extract as much pinitol from fibrous plant tissues in pyridine in 1 h when compared to complete extraction with ethanol for 24 h before derivatization (Streeter & Strimbu 1998).

Another benefit to using methanol extraction and TMSI derivatization is the relative simplicity of the resulting chromatograms. Eleven peaks were observed in the soybean extract chromatogram in Fig. 1, with pinitol clearly defined near 9.18 min. This compares with only 6 major peaks from sugar beet (Fig. 4) and 10 from snap bean roots (Fig. 5). The simplicity of the chromatograms is an indicator that pinitol and a small amount of other compounds are present in the methanol extract, which reduces the likelihood of coelution or some other interfering matrix effect with pinitol. This also provides support for the possible use of methanol extraction as a first step in the purification of pinitol from soybean root tissue.

The mass spectrum of the derivatized pinitol shown in Figure 2 is very similar to that reported previously (Savidge & Forster, 2001). Identification of pinitol by mass spectrometry is made exceedingly easy by the presence of a high-intensity m/z 260 fragment ion. The fragment ion at m/z 260 appears to be a unique ion associated with pinitol and the other O-methylinositols compared with the other sugars observed using this analytical procedure. This allows for a high probability of quantitative results even in the event of another analyte coeluting with pinitol. The elemental composition obtained from accurate mass measurements for m/z = 260 was determined to be $C_{11}H_{24}O_3Si_2$, which was matched within 4.3 millimass units (mmu).

We have also extracted roots using 80% ethanol rather than methanol. This led to the extraction of a greater variety of inositols and O-methylinositols from several plant roots (unpublished data).

Permethylation is another derivatizing method for the analysis of carbohydrates. The methods using permethylation initially provided relatively long retention times. Some of the reactions to form permethylated derivatives include the use of methyl iodide/silver oxide (Gee & Walker 1962; Walker et al., 1962; Kircher, 1960) methylsulfinylcarbanion/methyl iodide (Hakomori 1964; Corey & Chaykovsky 1962; Moor & Waight 1975), and potassium/liquid ammonia/methyl iodide (Muskat 1934a; Muskat 1934b). Permethylation has also become very popular in the LC/MS analysis of carbohydrates.

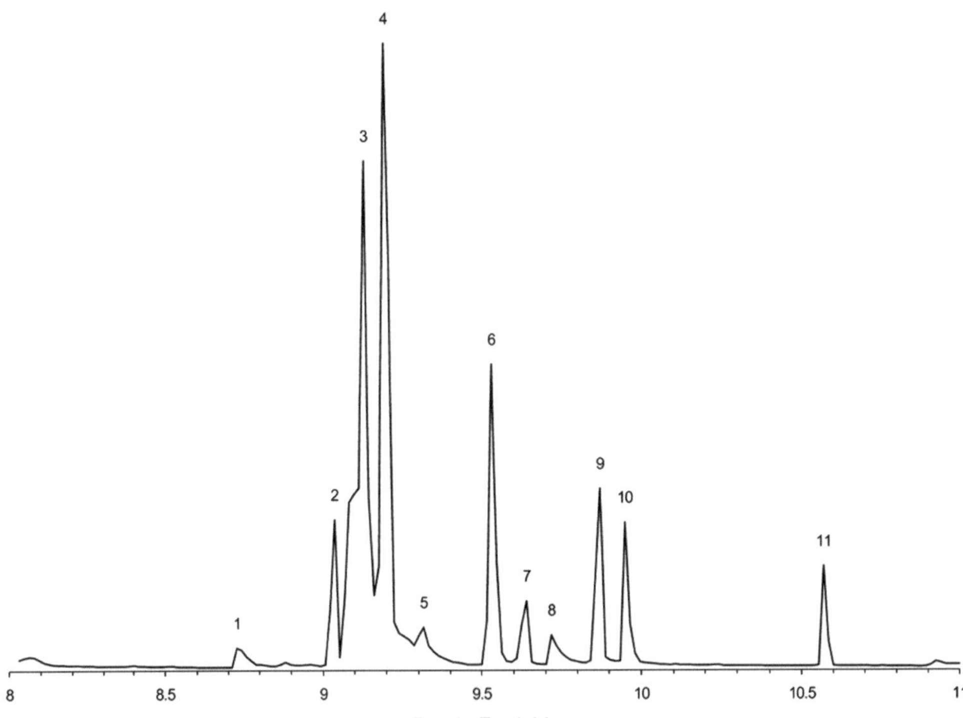

Fig. 1. Total ion chromatogram of an extract of soybean roots. Peak 4 was determined to be pinitol. From Garland, *et al* (2009).

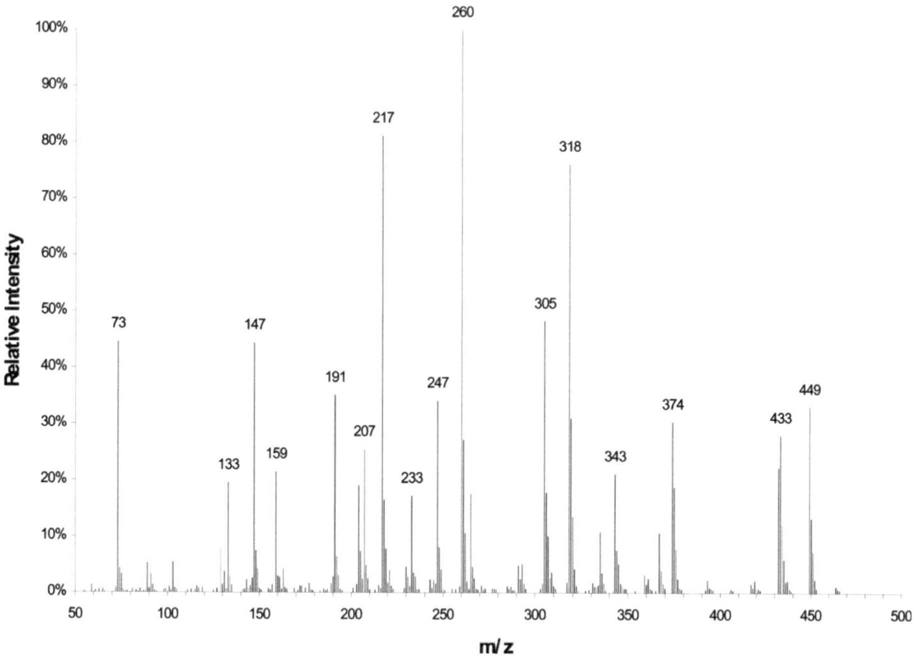

Fig. 2. Mass spectrum of TMSI-derivatized pinitol. From Garland, *et al.* (2009).

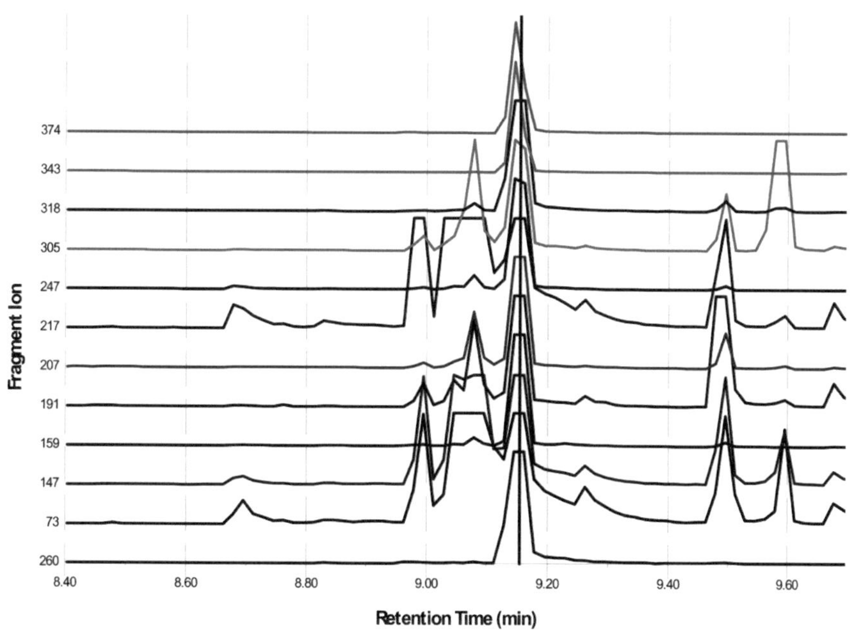

Fig. 3. Extracted ion plot of TMSI-derivatized soybean root extract. The labels on the vertical axis indicate the fragment mass of each extracted ion chromatogram. The chromatograms were spaced for easier representation. All peaks are on the same scale relative to their baselines. A vertical black line was inserted at the retention time of pinitol for reference. From Garland, *et al.* (2009).

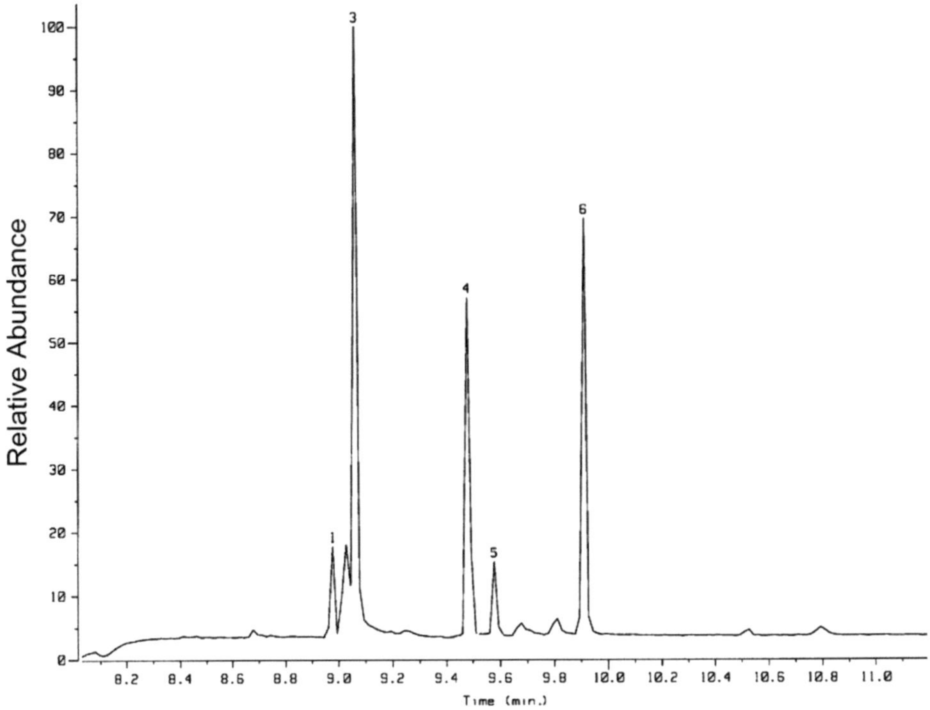

Fig. 4. Total ion chromatogram of derivatized sugar beet extract. Conditions were those of the chromatogram in Figure 1. Pinitol (retention time 9.2 min.) was not detected, as confirmed by MS analysis (Garland, *et al.*, 2009).

8. Other analytical techniques

Another technique for the separation and analysis of carbohydrates is liquid chromatography (LC). The column used in LC to provide the separation depends on whether the carbohydrates have been derivatized or not. Underivatized carbohydrates are commonly separated using ion exchange resins with water as an eluent and refractive index (RI) for detection. Refractive index detectors are, however, typically low in sensitivity, so samples need to be concentrated for quantitative analyses. The concentration of the carbohydrate must be in the percent range, and the RI detector can only be used with isocratic elution (Martens & Frankenberger 1990).

Other alternative detectors including both UV/visible absorbance and fluorescence require either pre-column or pre-detection derivatization of sugars, due to the fact that carbohydrates do not have a chromophore. Evaporative light scattering (ELS) is a detection technique used in high performance chromatography (HPLC) and supercritical fluid chromatography (SFC). It has been used for the analysis of carbohydrates and can act as a qualitative or quantitative detector (Wei & Ding 2000 ; Karlsson et al., 2005). The ELS is limited to solutes of low volatility. With the ELS, the column effluent is passed through a nebulizer and then into a heated drift tube; the solvent is evaporated leaving behind a particulate or aerosol form of the target compound. Light striking the dried particles that exit the drift tube is scattered and the photons are detected by a photodiode or photomultiplier tube at a fixed angle from the incident light. (LaPosse & Herbtreteau 2002).

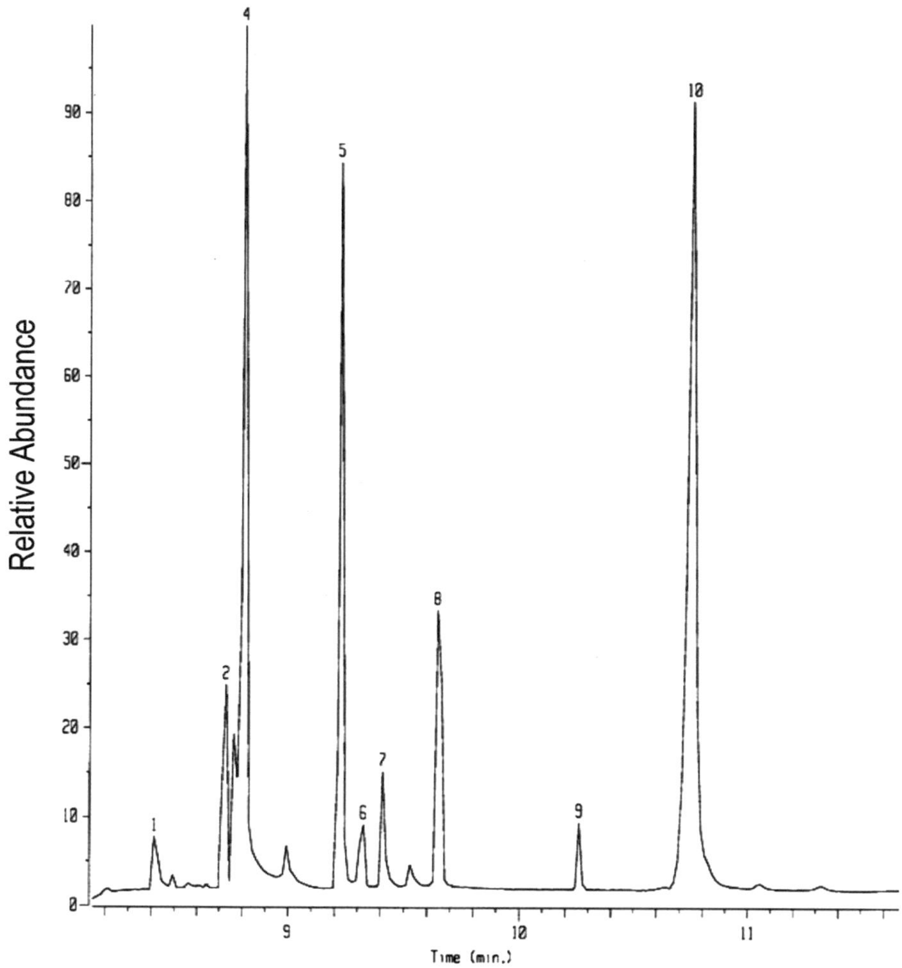

Fig. 5. Total ion chromatogram of derivatized snap bean root extract. Peak 5 was at a similar retention time to that of pinitol in Fig 1 (9.2 min.), but MS analyses were unable to detect pinitol in snap bean root extract (Garland, *et al.*, 2009).

Another detector commonly used is a pulsed amphoteric detector (Lee 1996; Johnson et al., 1993).

One derivatization procedure for carbohydrates to provide a chromophore for LC analysis involves a reaction with p-nitrobenzoyl chloride and pyridine. The reaction replaces the active hydrogens with a nitrobenzoyl group. The method was applicable to mono-, di-, and trisaccharides except fructose (Nachtmann & Budna 1977; Nachtmann 1976). Many of the derivatization reactions for carbohydrates are discussed by Knapp (1979). In addition, other derivatization techniques have been discussed (Meulendijk & Underberg 1990).

Mass spectrometry can also be coupled with LC. Examples are LC/MS and capillary electrophoresis/MS. Many of the LC techniques allow carbohydrates to be analyzed without prior derivatization as is necessary in GC and GC/MS analyses.

It should be noted that there is not one LC column that has been reported to separate every carbohydrate. Togami et al. (1991) discussed the separation of carbohydrates using cation-exchange columns. Richmond et al. (1991) separated carbohydrates in dairy products. Henderson and Berry (2009) have utilized Zorbax columns for the separation of carbohydrates in Stevia sweetener. Romano (2007) discussed carbohydrate analysis in food products emphasizing column chemistries and detection. Several vendors offer LC columns for carbohydrate separation. Wilcox et al. (2001) also discussed several column types used for carbohydrate separation. Hydrophilic interaction chromatography (HILIC) has also been reported as a method for analyzing ionic or polar compounds, particularly biomolecules and drug metabolites (http://www.laboratoryequipment.com/article-is-hilic-in-your-future-ct92.aspx). Simple carbohydrate separations can also be performed on functionalized silica or resin-based columns (http://www.labnews.co.uk/feature_archive. php/4000/5/just-juice).

The separation of mono- and oligosaccharides are also performed using capillary electrophoresis. Different formats are capillary zone electrophoresis (CZE), capillary isoelectric focusing (CIEF), capillary isotachophoresis (CITP), and micellar electrokinetic chromatography (MEKC). These techniques are summarized in a review by Thibault and Honda (2003).

9. Liquid chromatography/mass spectrometry (LC/MS) and other MS techniques

Efficient separation methods such as high performance liquid chromatography (HPLC) and capillary electrophoresis combined with detection methods (e.g. mass spectrometry) that supply structural or compositional information is a preferred tool for the analysis of biomolecules, particularly carbohydrates. Liquid chromatography/mass spectrometry with both electrospray (ESI) and atmospheric pressure ionization (APCI) has spurred a major interest in the analysis of carbohydrates.

In ESI , the liquid containing the analyte(s) of interest is dispersed into a fine aerosol. Because the ion formation involves extensive solvent evaporation, the typical solvents for electrospray ionization are prepared by mixing water with volatile organic compounds (e.g. methanol, acetonitrile). To decrease the initial droplet size, compounds that increase the conductivity (e.g. acetic acid) are customarily added to the solution. Large-flow electrosprays can benefit from additional nebulization by an inert gas such as nitrogen. The aerosol is sampled into the first vacuum stage of a mass spectrometer through a capillary, which can be heated to aid further solvent evaporation from the charged droplets. The ions observed by mass spectrometry may be quasimolecular ions created by the addition of a hydrogen ion and denoted $[M + H]^+$, or of another cation such as sodium ion, $[M + Na]^+$, or the removal of a proton, $[M - H]^-$. Multiply-charged ions such as $[M + nH]^{n+}$ are often observed (Gaskell 1997). As examples, Fountain and Grumbach (2009) used negative ion electrospray mass spectrometry for the analysis of fructose, glucose, sucrose, and lactose. Taormina et al. (2007) and Mauri et al. (2002) used flow injection techniques with mass spectrometry. Fugimoto et al. (2005) used rubidium in the mobile phase as a complexing agent for both nuclear magnetic resonance and electrospray mass spectrometry analysis. Taylor et al. (2005) utilized ESI/MS to study fragmentation patterns of carbohydrates. Schlichtherle-Cerny et al. (2003) utilized a HILIC column coupled with ESI/MS for the analysis of amino acids, peptides, glycoconjugates, and organic acids in foods without prior derivatization.

In APCI, typically the mobile phase containing eluting analyte is heated to relatively high temperatures (above 400 C), sprayed with high flow rates of nitrogen and the entire aerosol cloud is subjected to a corona discharge that creates ions. Often APCI can be performed in a modified ESI source. The ionization occurs in the gas phase, unlike ESI, where the ionization occurs in the liquid phase. A potential advantage of APCI is that it is possible to use a nonpolar solvent as a mobile phase solution, instead of a polar solvent, because the solvent and molecules of interest are converted to a gaseous state before reaching the corona discharge pin. Typically, APCI is a harder ionization technique than ESI, i.e. it generates more fragment ions relative to the parent ion.(Kostianinen et al., 2003). Kumaguai (2001) used atmospheric pressure chemical ionization mass spectrometry for the analysis of sugars and sugar alcohols without derivatization but did use methylene chloride or chloroform that was added post column to increase the sensitivity. The ions detected included $(M+Cl)^-$. Shimadzu application note also used solvent addition post column to improve sensitivity. This application also used APCI in the negative ion mode. Keski-Hynnila et al. (2004) compared APCI, atmospheric pressure photoionization, and electrospray in the analysis of phase II metabolites.

Other types of mass spectrometers used for analysis of carbohydrates include quadrupole time-of-flight (QTOF) mass spectrometers which allow both accurate mass (elemental composition) and MS/MS studies to be performed. Another mass spectrometer very useful for the analysis of carbohydrates is the ion trap (IT) MS. Ion trap technology has been described in (March & Todd 2005a, 2005b), and its major advantage includes the capability of MS^n which can provide additional structural information. Examples of glycoprotein analysis using IT have been described by (Stumpo & Reinhold 2010; Jiao et al., 2010; Reinhold et al., 1990).

Another technique that has been utilized for the analysis of carbohydrates is matrix assisted laser desorption/time-of-flight mass spectrometry (MALDI/TOFMS)(Harvey 1999, 2009)). In MALDI, the sample to be analyzed is mixed with a matrix, which in turns absorbs heat energy from irradiation with a nitrogen laser light. For example, dihydroxybenzoic acid (DHB) or ferulic acid which are commonly used as a matrices have a carboxyl group on a benzene ring. The DHB absorbs the energy and acts as a proton donor (Zenobi & Knochenmuss 1998). Time-of-flight mass spectrometry allows the majority of the ions generated throughout the mass range to be collected by the detector. MALDI has been primarily used to obtain spectra of very large polymers, biomolecules, and a variety of thermally labile materials (Hillenkamp et al., 1991, Nelson et al., 1990. We have also used MALDI/TOF for the analysis of smaller molecules (e.g. <500 amu) (Goheen *et al.*, 1997; Campbell *et al.*, 2001).

10. The inositols

Inositols (Fig. 6) are polyols of cyclohexane with the empirical formula $C_6H_{12}O_6$. There are potentially 9 stereoisomers of inositol but only five are naturally occurring (structure shown below). They are *myo*-inositol, *chiro*-inositol, *scyllo*-inositol, *muco*-inositol, and *neo*-inositol. Of these, *myo*-inositol is the precursor of the other four. *myo*-Inositol is synthesized from glucose.

The synthesis of *myo*-inositol uses the enzyme L-*myo*-inositol 1-phosphate synthase to catalyze the reaction which produces L-*myo*-inositol-1-phosphate from D-glucose 6-phosphate (Hoffmann-Ostenhof and Pittner, 1982). The L-*myo*-inositol-1-phosphate is then dephosphorylated through inositol monophosphate to produce *myo*-inositol (Loewus & Murthy, 2000). The enzyme that catalyzes this step is L-*myo*-inositol 1-phosphate synthase

(Stieglitz et al, 2005). The four other inositol isomers are derived from *myo*-inositol (Loewus and Murthy, 2000). The sequoyitol can then be epimerized to D-pinitol (See Fig. 7) which is demethylated to D-*chiro*-inositol using NADP-specific D-pinitol dehydrogenase (Stieglitz et al, 2005).

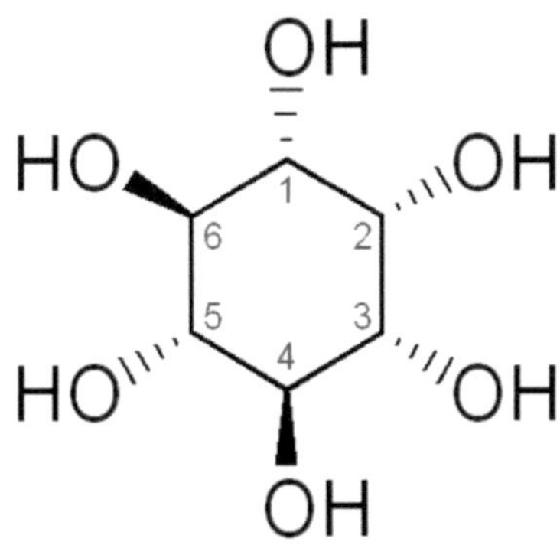

Fig. 6. Inositol.

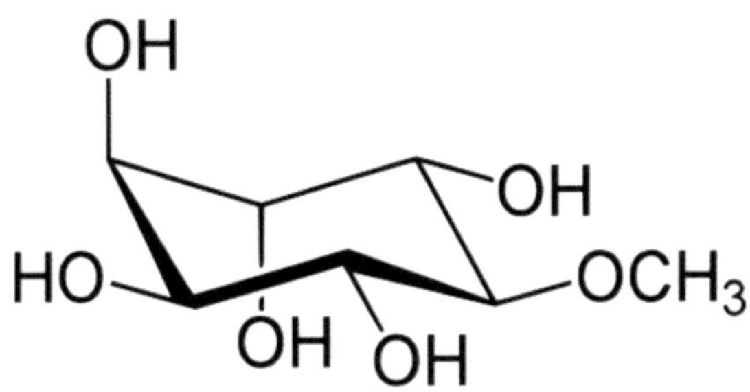

Fig. 7. Pinitol.

In addition to the five stereoisomers of inositol, the *O*-methylinositols can also be synthesized from *myo*-inositol. Of these, ononitol and pinitol are common to soybeans. Ononitol is a precursor to pinitol in soybeans (Loewus and Murthy, 2000; Chiera et al. 2006). Of the *O*-methylinositols, pinitol is most abundant in soybeans.

myo-Inositol is probably the most studied of all the inositols because it is the most commonly available. It has a very important function as it is required in the formation of Lecithin, which protects cells from oxidation and is an important factor in the building of cell membranes. Inositol, also has a metabolic effect in preventing too much fat to be stored in the liver, which is why it is called a lipotropic and is a vital part in maintaining good health.

Inositols have been found in many plants both foodstuff and other plants at varying evolutionary stages (Clements & Darnell, 1980; Chiera *et al.*, 2006; Guo & Oosterhuis, 1997; Henry, 1976; Johansen *et al.*, 1996; Johnson and Sussex, 1995; Johnson & Wang, 1996; Lind *et al.* 1998; Loewus *et al.*, 1984; Manchanda and Garg, 2008; Ogunyemi *et al.*, 1978; Phillips, et al 1982; Sheveleva *et al.*, 1997; Streeter *et al.*, 2001). Different soybean plant parts contain different levels of inositols as do soybean plants in vegetative verses reproductive growth stages (Phillips and Smith, 1974). Comparison of total inositols among plants should be examined carefully because each plant may produce different proportions of the various inositols (Larson & Raboy, 1999). Research with pinitol in soybean documents that this cyclitol is a major consitutent of soybean (Phillips & Smith, 1974; Streeter, 1980; Phillips, et al. 1982; Dougherty & Smith, 1982). Because pinitol diffuses faster than carbohydrates during imbibition, it is theorized that loss of pinitol from soybean seed encourages the growth of Bradyrhizobium (Rhizobium) species in the soil needed for nitrogen fixation (Nordin, 1984). Accumulation of ononitol and pintol in soybean and other plants under drought conditions has been documented (Streeter *et al.*, 2001; Guo & Oosterhuis, 1997; Manchanda & Garg, 2008; Sheveleva, et al, 1997).

Inositols are very important in general plant growth, seed storage, nitrogen fixation and protection of plants during stress. Inositol metabolism and its role in photosynthesis, plant health, and subsequent potential increase in yield is complex but new discoveries in this area may lead to future yield improvements. The role of inositols in nitrogen fixation is also complex and not currently fully understood. Inositols play an important role in phosphorus movement in the environment. Efforts are being made to alter the phytate content of soybean so animals can use the phosphorus and also reduce the amount that is excreted as manure. There are implications here not only for animal health but also for the preservation and sustainability of watersheds.

Phytate, *myo*-inositol hexakisphosphate, is found in almost all plant and animal cells and serves as an important phosphate reserve in plants (Irvine & Schell, 2001). Exposure of soybean cell suspension to *Psuedomonas syringae* pv *glycinea* indicated that whether a virulent or avirulent strain is used, the plant starts defense systems at the expense of housekeeping cell functions (Logemann *et al.*, 1995; Shigaki & Bhattacharyya, 2000). Part of this defense reaction involves cellular cyosolic inositol and the IP3 pathway. This pathway is involved in cell division, growth and elongations and there is evidence that this pathway is inhibited when the plant is exposed to pathogens (Perera *et al.*, 1999; Shigaki & Bhattacharyya, 2000). Selection of plants with reduced phytate levels raised the question of these plants' response to stress in the form of diseases. Murphy et al., 2008 found that disruption of phytate biosynthesis resulted in increased susceptibility in *Arabidopsis thaliana* to virus (potato virus Y), fungal (*Botrytis cinerea*) and bacterial (*Psueodomonas syringae*) diseases. The role of phytate in basal resistance to plant pathogens was previously unknown. Klink et al. (2009) found 1-phosphatidylinositol phosphodiesterase-related genes expressed when soybean plants are exposed to *Heterodera glycines*, a pathogen of soybean. The findings of inositols in plant defense are important findings and the next step is to determine whether the defense reaction is a general reaction or specific to different types of attacks.

Transgenic plants that release extracellular phytase from their roots have a significantly increased ability to acquire phosphorus from inositol phosphates from growth medium; however, there is less evidence that phosphorus nutrition of plants can be improved in plants grown in soil (George, *et al.*, 2004). Phytate and phytic acid represent the major form of phosphorus in animal feed derived from plants. Phosphorus in seeds and tubers is stored primarily as phytate (*myo*-inositol exa*kis*phosphate), which is poorly digested by non-ruminant animals such as swine, poultry and fish (Saghai Maroof *et al.*, 2009; Kim *et al.*, 2006). The lack of the hydrolytic enzymes necessary for phytate to be utilized by these animals requires supplemental phosphate. Plant breeding efforts involve plant selections for improved phosphorus usage by animals and different feed additives resulting in less environmental pollution.

Inositol is synthesized sparingly in the body but is present in many foods. The inositols are essential nutrients for plants (Loewus and Murthy, 2000) and animals (Holub, 1986). Concentrations of the inositols and their metabolites can be much higher in some plant species than in mammalian tissue. For example, in soybeans, the concentration of pinitol alone approaches 30 mg/g (Streeter & Strimbu 1998; Garland *et al.*, 2009) whereas in human blood, the levels of free *myo*-inositol is 3000 times lower (1 mg/100 mL). Levels of pinitol in blood is not widely known, but are anticipated to be orders of magnitude less than *myo*-inositol.

One form of inositol, inositol hexaniacinate, has been used to support circulatory health because it functions like niacin in the body. The major dietary forms of myo-inositol are inositol hexaphosphate or phytic acid, which is widely found in cereals and legumes and associated with dietary fiber, and myo-inositol-containing phospholipids from animal and plant sources.

Inositol is involved in the glucuronic acid and pentose phosphate pathways. Inositol exists as the fiber component phytic acid, which has been investigated for its anti-cancer properties. Inositol is primarily used in the treatment of liver problems, depression, panic disorder, and diabetes (Narayanan, 1987). Used with choline, it also aids in the breakdown of fats, helps in the reduction of blood cholesterol, and helps to prevent thinning hair (Walker, 2010). It promotes the export of fat from the liver. Inositol is required for the proper function of several brain neurotransmitters. Inositol may improve nerve conduction velocities in diabetics with peripheral neuropathy. Inositol may help protect against atherosclerosis and hair loss. There has also been the suggestion that it may help to reverse some nerve damage caused by diabetes (Gregersen et al. 1978; *Ibid*,1983). Inositol has also been tried for other psychological and nerve-related conditions including the treatment of side effects of the medicine lithium. Inositol also has a prominent calming effect on the central nervous system, so it is sometimes helpful to those with insomnia. Inositol may also be involved in depression.

Under pinitol deficiency, detrimental health conditions may exist such as higher blood sugar in diabetics (Geethan and Prince, 2008). Myo-inositol deficiency can lead to depression and other mental disorders (Levine et al, 1995; Benjamin et al, 1995; Fux, et al, 1996). Also, polycystic ovary syndrome (PCOS) has been reported to be related to a deficiency in dietary inositol (Gerli, et al. 2003; Ibid, 2007). Correlations with depression and similar disorders may be related to the abundance of inositol phospholipids in brain and other nervous system tissues. However, the relationship between pinitol and blood sugar levels is more likely correlated with the similarities in structure between the 0-methyl inositol and glucose.

There is no recommended daily allowance for inositol, but the normal human dietary intake is about 1 gram per day. Inositol is available from both plant and animal sources. Natural sources of inositol include soybeans, wheat germ, brewer's yeast, bananas, liver, brown rice, oak flakes, nuts, unrefined molasses, vegetables, and raisins. Most dietary inositol is in the form of phytate, a naturally occurring plant fiber.

Dietary effects of pinitol and ononitol are still in the earlier stages of discovery. It has recently been shown that pinitol lowers blood glucose levels in type II diabetics while significantly decreasing total cholesterol, LDL-cholesterol and the LDL/HDL-cholesterol ratio (Kim et al 2005). The dietary benefits or hazards of the other metabolites of isomers of inositol (other than myo-inositol) are under active investigation.

11. Conclusions and future directions

It is clear from this review that there are many different tools to study the carbohydrates in soybean plants. Results from any of the various analytical methods can be compared as long as they have been tested with adequate standards. The outstanding carbohydrate found in soybeans is pinitol, part of the inositol family. There has been considerable research into the value of the inositols, but most of the emphasis has been on myo-inositol, probably because it is widely available. However, for the soybean industry, it would be valuable to better understand the role of pinitol in health and nutrition. There are good indications that pinitol may have unique nutritional value, and key roles in soybean plant biology. Future directions should include the use of effective analytical methods to perform more research into the roles of pinitol and related inositols in various fields of nutrition, medicine, and plant biology.

12. Acknowledgements

We acknowledge Shaun Garland and Luther McDonald for their technical support.

13. List of abbreviations

GC	gas chromatography
LC	liquid chromatography
MS	mass spectrometry
RI	refractive index
UV	ultraviolet
APCI	atmospheric pressure chemical ionization
IT	ion trap
ELS	evaporative light scattering
ESI	electrospray ionization
SPE	solid phase extraction
MS/MS	mass spectrometry/mass spectrometry
LC/MS	liquid chromatography/mass spectrometry
GC/MS	gas chromatography/mass spectrometry
MALDI/TOFMS	matrix assisted laser desorption/time-of-flight mass spectrometry
QTOF	quadrupole time-of-flight
TMS	trimethylsilyl
TMSI	trimethylsilyl imidazole
HMDS	hexamethyldisilazane

14. References

Anderson, L. & Wolter, K.E. (1966). *Cyclitols in plants: Biochemistry and physiology*. Annual Review Plant Physiology 17:209-222.

Al-Hazmi, M.I. & Stauffer, K.R. (1986). Gas chromatographic determination of hydrolyzed sugars in commercial gums. *Journal of Food Science* 51:1091-1092, 1097.

Anthony, R.M.; Nimmerjahn, F.; Ashline, D. J.; Reinhold, V. N.; Paulson, J. C. & Ravetch, J. V. (2008). *Recapitulation of Intravenous Ig Anti-Inflammatory Activity with a Recombinant IgG Fe.* Science, 320: 373-376.

Asp, N-G. (1993). Nutritional importance and classification of food carbohydrates , In: *Plant Polymeric Carbohydrates*, Meuser, F., Manners, D.J., & Seibel, (Ed.), W. Royal Soc. Chem., Cambridge, U.K., pp. 121-126.

Benjamin, J.; Levine, J.; Fux, M.; Aviv, A. & Belmaker, R.H. (1995). Double-blind, placebo-controlled, crossover trial of inositol treatment for panic disorder. *American Journal of Psychiatry*. 152: 1084-1086.

Berridge, M.J.; Helsop J.P.; Irvine, R.F. & Brown K.D. (1984). Inositol trisphosphate formation and calcium mobilization in Swiss 3T3 cells in response to platelet-derived growth factor. *Biochem. J.* 222: 195-201.

Binder, R. G. & Haddon, W. F. (1984). *Analysis of O-methylinositols by gas-liquid chromatography-mass spectrometry Carbohydr.* Res.129:21-32.

Bisho , C.T. (1964). *Gas-Liquid chromatography of carbohydrate derivatives*. Advances in Carbohydrate Chemistry 19:95-147.

Bishop, C.T. & Cooper, F.P. (1960). Separation of carbohydrate derivatives by gas-liquid partition chromatography. Candian J. Chem., 38:388-398.

Blakeney , A.B.; Harris, P.J.; Henry, J. & Stone, B.A. (1983). A simple and rapid preparation of alditol acetates for monosaccharide analysis. Carbohydrate Research, 113:291-299.

Bohnert, H.J.; Nelson, D.E. & Jensenay, R.G. (1995). The Plant Cell, *Adaptations to Environmental Stresses,* Vol. 7, 1099-1111.

Brockwell, J.; Bottomley, P.J. & Thies, J.E. (1995). Manipulation of rhizobia microflora for improving legume productivity and soil fertility: a critical assessment. *Plant Soil*. 174: 143-180.

Campbell, J. A.; Hess, W.P.; Lohman, J.R. & Goheen, S.C. (2001). Analysis of Hanford-related organics using matrix-assisted laser desorption ionization time-of-flight mass spectrometry. J. Radioanalytical and Nucl. Chem., 250: 247-253.

Campling, J.D. & Nixon, D.A. (1954). The inositol content of foetal blood and foetal fluids. J. Physiol. (I954). I26, 7I-80

Chiera, J.M.; Streeter, J.G. & Finer, J.J. (2006). Ononitol and pinitol production in transgenic soybean containing the inositol methyl transferase gene from *Mesembryanthemum crystallinum*. Plant Sci. 171: 647-654.

Churms, S.C. (1990). Recent developments in the chromatographic analysis of carbohydrates. *Journal of Chromatography*. 500:555-583.

Churms, S.C. (1996). Recent progress in carbohydrate separation by high-performance liquid chromatography based on size exclusion. *Journal of Chromatography*, A., 720:151-166.

Clements, R.S., Jr. & Darnell, R. (1980).myo-Inositol content of common foods: development of a high-*myo*-inositol diet. *J.Am. Clin. Nutr.* 33:1954-1967.

Colowick, S.P. & Kaplan, N.O. (1951). Carbohydrate metabolism. *Annual Review of Biochemistr*, 20:513-558.

Corey, E. J. & Chaykovsky, M. (1962). Methylsulfinylcarbanion. *J. Am. Chem. Soc.* 84: 866-867.

Cote, G.G. & Crain, R.C. (1993). Biochemistry of phosphoinositides. *Annual Review of Plant Physiology*. 44:333-356.

Daughaday, W.H.; Larner, J. & Hartnett, C, (1955). The synthesis of inositol in the immature rat and chick embryo. *J. Biol. Chem.* 212:869-875.

Decker, P. & Schweer, H. (1982). Gas-liquid chromatography on OV-225 of tetroses and aldopentoses as their O-methoxime and O-n-butoxime pertrifluoroacetyl derivatives and of C3-C6 alditol pertrifluoroacetates. *J. of Chromatography*, 236:369-373.

Denison, R.F.; Hunt, S. & Layzell D.B. (1992). Nitrogenase activity, nodule respiration and O_2 permeability following detopping of alfalfa and birdsfoot trefoil. *Plant Physiol.* 98:894-900.

Dmitriev, B.A.; Backinowsky, L.V.; Chizhov, O.S.; Zolotzrev, B.M. & Kochetkow, N.K. (1971). Gas-liquid chromatography and mass spectrometry of aldononitrile acetates and partially methylated aldononitrile acetates. Carbohydrate Research 19:432-435.

Dogruel, D. & Williams, P. (1995). Detection of human IgM at m/z 1 MDa. Rapid Comm Mass Spectrom. 9:625

Eagle, H.; Oyama, V.I.; Levy, M. & Freeman, A.E. (1957) . myo-Inositol as an essential growth factor for normal and malignant human cells in tissue culture. *J. Biol. Chem.* 266: 191-205

Ensminger A.; Ensminger, M.; Konlande, J. & Robson J. (Ed.). (1995). *The Concise Encyclopedia Of Foods & Nutrition*, Boca Raton, London, Tokyo: CRC Press, pp.580-581.

Evershed, R.P. (1993). Advances in Silylation. In: *Handbook of derivatives for chromatography*, K. Blau & J.M. Halket, (Ed.), Chichester, U.K.: Wiley, p.53, p.59.

Farquhar, G.D. & Sharkey, T.D. (1982). Stomatal conductance and photosynthesis. Annu. Rev. Plant Physiol. 33:317-345.

Fontes, C.M.G.A. &. Gilbert, H.J. (2010). Cellulosomes:Highly efficient nanomachines designed to deconstruct plant cell wall complex carbohydrates. Annual Review Biochem. 79:655-681.

Fougere, F.; LeRudulier, D. & Streeter, J.G. (1991). Effects of salt stress on amino acid, organic acid, and carbohydrate composition of roots, bacteroids, and cytosol of alfalfa (*medicago sativa L.).* Plant Physiol. 96: 1228-1236.

Fountain, K.J.; Hudalla, C.; Grumbach, E.S. & McCabe, D. (2009). Analysis of carbohydrates by ultraperformance liquid chromatography and mass spectrometry. LCGC, ASIA Pacific, Volume 12, issue 4.

Fugimoto, T.; Sakari, S.; Tsutsuri, A.; Furihata, K.; Machinami, T. & Tashiro, M. (2005). Observation of 1,6-anhdro-beta-maltose and 1,6-anhdro-beta-D-glucopyranose complexed with rubidium by NMR spectroscopy and electrospray ionization mass spectrometry. Anal. Sci., 21: 1245-1247.

Fux, M.; Levine, J.; Aviv, A. & Belmaker, R.H. (1996). Inositol treatment of obsessive-compulsive disorder. American Journal of Psychiatry, 153:1219-1221.

Garland, S.; S. Goheen, S.; Donald, P.; McDonald, L & Campbell, J. (2009). Application of derivatization gas chromatography/mass spectrometry for the identification and quantitation of pinitol in plant roots. Anal. Letters 42:2096-2105.

Gaskell, S. (1997). Electrospray: Principles and Practices. *Journal of Mass Spectrometry*, 32:677-688.

Gee, M. & Walker H.G. (1962). Gas-liquid chromatography of some methylated mono-, di-, and trisaccharides. Anal. Chem. 34:650-653.

Geethan, P.K. & Prince, P.S. (2008). Antihyperlipidemic effect of D-pinitol on streptozotocin-induced diabetic Wistar rats, J. Biochem. Mol. Toxicol. 22:220-224.

George, T.S.; Richardson, A.E.; Hadobas, P. & Simpson R.J. (2004). Characterization of transgenic *Trifolium subterraneum* L. which expresses *phyA* and releases extracellular phytase: growth and P nutrition in laboratory media and soil. Plant Cell Environ. 27:1351-1361.

George, T.S.; Simpson, R.J.; Hadobas, P.A. & Richardson, A.E. (2005). Expression of a fungal phytase gene in *Nicotiana tabacum* improves phosphorus nutrition of plants grown in amended soils. Plant Biotechnol. J. 3:129-140.

Gerli, S.; Mignosa, M. & Di Renzo, G.C. (2003). Effects of inositol on ovarian function and metabolic factors in women with PCOS: a randomized double blind placebo-controlled trial. European Review for Medical and Pharmacological Sciences. 7:151-159.

Gerli, S.; Papaleo, E.; Ferrari, A. & Di Renzo, G.C. (2007). Randomized, double blind placebo-controlled trial: effects of Myo-inositol on ovarian function and metabolic factors in women with PCOS. European Review for Medical and Pharmacological Sciences, 11(5):347-354.

Goheen, S. C.; Wahl, K.L.; Campbell, J.A. & Hess, W.P.(1997). Mass spectrometry of low molecular mass solids by matrix-assisted laser desorption/ionization. J. Mass Spectr. 32: 820-828.

Gomes, C. I.; Obendorf, R.L. & Horbowicz, M. (2005). myo-Inositol, D-chiro-Inositol, and D-Pinitol synthesis, transport, and galactoside formation in soybean explants. Crop Sci. 45:1312-1319.

Gregersen, G.; Bersting, H.; Theil, P & Servo, C. (1978). Myoinositol and function of peripheral nerves in human diabetes. Acta Neurol. Scand. 58:241-248.

Gregersen, G.; Bartelsen, B. & Harbo, H. (1983). Oral supplementation of myo-Inositol: effects of peripheral nerves in human diabetics and on the concentration in plasma, erythrocytes, urine, and muscle tissue in human diabetics and normals. Acta Neurol. Scand. 67:164-172.

Guo, C. & Oosterhuis, D.M. (1997). Effect of water-deficit stress and genotypes on pinitol occurrence in soybean plants. Environ. Exper. Bot. 37:147-152.

Hakomori, S. (1964). A rapid permethylation of glycolipid, and polysaccharide catalyzed by methylsulfinyl carbanion in dimethyl sulfoxide. J. Biochem 55:205-208.

Halket, J.M. (1993). Derivatives for GC-MS. In: Handbook of derivatives for chromatography, ed. K. Blau and J.M. Halket, Chichester, U.K., Wiley, 304.

Halford, N.G.; Curtis, T.Y.; Muttucumaru, N.; Postles, J. & Mottran, D.S. (2010). Sugars in crop plants. Ann Appl Biol 158:1-25.

Hall, M.B. (2003). Challenges with nonfiber carbohydrate methods. J. Anim. Sci. 81:3226-3232

Halliday, J.W. & Anderson, L. (1955). The synthesis of myo-inositol in the rat. Journal of biological chemistry 217:797-802.

Hanneman, A. R.; Cesar, J.; Ashline, D. E. & Reinhold, V. N. (2006) Isomer and Glycomer Complexities of Core GlcNAcs in *Caenorhabditis e/egans*.Glycobiology, 16: 874 - 890.

Hansen, B. & Ortmeyer, H. (1996). Inositols-Potential roles for insulin action in diabetes: Evidence form insulin-resistant nonhuman primates. Lessons from Animal Diabetes U. Ed. E. Shafrie, pp. 333-348.

Harvey, D.J. (1999). Matrix-assisted laser desorption/ionization mass spectrometry of carbohydrates. Mass Spectrom. Rev. 18:349-450.

Harvey, D.J. (2010). Analysis of carbohydrates and glycoconjugates by matrix-assisted laser desorption/ionization mass spectrometry: an update for 2003-2004. Mass Spectrom. Rev. 28:273-361.

Hasegawa, P.M.; Bressan, R.A.; Zhu, J.K. & Bohnert, H.J. (2000). Plant cellular and molecular responses to high salinity, Annu. Rev. Plant Physiol. Plant Mol. Biol. 51:463–99.

Henderson, J.W., Jr.; & Berry, J. (2007). Isocratic Stevia seetner analysis using selective ZORBAX columns. Agilent Application Note. Food.

Henry, E.W. (1976). Determination of inositol in inbred and hybrid corn (*Zea mays*) seedlings. *Journal of Experimental Botany* 27:259-262.

Hillenkamp, F.; Karas, M.; Beavis, R.C. & Chait, B.T. (1991). Matrix-assisted laser desorption/ionization of biopolymers. *Anal. Chem.* 63:1193A-1203A

Hoffmann-Ostenhof, O. & Pittner, F. (1982). The biosynthesis of myo-inositol and its isomers. Can. J. Chem.60:1863-1871.

Holm, P.J.; Booth, M.H.; Schmidt, T.; Greve, H. & Callesen, H. (1999). Theriogenology 52: 683-700. "High bovine blastocyst development in a static in vitro production system using sofaa medium supplemented with sodium citrate and myo-inositol with or without serum-proteins"

Holub, B. J. (1986). Metabolism and function of myo-inositol phospholipids. Ann. Rev. Nutr. 6: 563-597.

Honda, S.; Kakehi, K. & Okada, K. (1979). A convenient method for the gas chromatographic analysis of hexosamines in the presence of neutral monosaccharides and uronic acids. Journal of Chromatography. 176:367-273.

Horecker, B.L. & Mehler, A.H. (1955). Carbohydrate metabolism. Annual Review of Biochemistry 24:207-274.

Irvine, R.F., & Schell, M.J. (2001). Back in the water: the return of the inositol phosphates. Mol. Cell. Biol. 2:327-338.

Ismail, M.N.; Stone, E.L.; Panico, M.; Lee, S.H.; Luu, Y.; Ramirez, K.; Ho, S.B.; Minory M.; Haslam J.D.; Stuart, M. & Dell, A. (2010). High Sensitivity *O-glycomic Analysis of Mice Deficient in* Core 2131,6-N-*acetylglucosaminyl* transferases. Glycobiology. 21:82-98.

Jaindl, M. & Popp, M. (2006). Cyclitols protect glutamine synthetase and malate dehydrogenase against the induced deactivation and thermal denaturation. Biochem and Biophysical Research Communications 345:761-765.

Jiao, J,; Zhang, H. & Reinhold, V. (2010). High Performance IT-MS" Sequencing of Glycans. Int. J. Mass Spectrometry. 303:109-117.

Johansen, S. L.; A. Sivasothy, A.; Dowd, M.K.; Reilly, P.J. & Hammond, E.G. (1996). Low-molecular weight organic compositions of acid waters from vegetable oil soapstocks. Journal of the American Oil Chemists' Society 73:1275-1286.

Johnson, D.C.; Dobberpuhl, D.; Roberts, R. & Vandeberg, P. (1993). Pulsed amperometric detection of carbohydrates, amines, and sufur species in ion chromatography-the current state of research. *J. of Chromatography* 640:79-96

Johnson, M. D., & I. M. Sussex, I.M. (1995). 1-L-*myo*-inosital 1-phosphate synthase from *Arabidopsis thaliana*. Plant Physio 107:613-619.

Johnson, M.D. & Wang, X. (1996). Differentially expressed forms of 1-L-*myo*-inosital 1-phosphate synthase (EC) in *Phaseolus vulgaris*. *J Biol Chem* 271:17215-17218.

Karlsson, G.; Winge, S, & Sandberg, H. (2005). Separation of monosaccharides by hydrophilic interaction chromatography with evaporative light scattering detection. *J. of Chromatogr.* A, 1092:246-249.

Keski-Hynnila, H.; Kurkela, M.; Elovarra, E.; Antonio, L.; Magdalou, J.; Luukanen, L,; Taskinen, J. & Kostianen, R. Comparison of electrospray, atmospheric pressure chemical ionization, and atomospheric pressure photoionization in the identification of apormophine, dobutamide, and entacapone phase II metabolites in biological samples. *Anal. Chem.* 74:3449.

Kim, J.I.; Kim, J.C.; Kang, M.J.; Lee, M.S.; Kim, J.J. & and Cha, I.J. (2005). Effects of pinitol isolated from soybeans on glycaemic control and cardiovascular risk factors in Korean patients with type II diabetes mellitus: a randomized controlled study.*Eur. J. Clin Nutrition* 59: 456-458.

Kim, T.; Mullaney, E.J.; Porres, J.M.; Roneker, K.R.; Crowe, S.; Rice, S.; Ko, T.; Ullah, A.H.J.; Caly, C.G.; Welch, R & Lei, X.G. (2006). Shifting the pH profile of *Aspergillus niger* PhyA phytase to match the stomach pH enhances its effectiveness as an animal feed additive. *Applied and Environmental Microbiology*. 72:4397-4403.

Kircher, H. W. (1960). Gas-liquid partition chromatography of methylated sugars. *Anal. Chem*. 32: 1103-1106.

Klink, V.P., Hosseini, P.; Matsye, P.; Alkharouf, N.W. & B.F. Matthews, B.F. (2009). A gene expression analysis of synctia laser microdissected from the roots of the *Glycine max* (soybean) genotype PI 548402 (Peking) undergoing a resistant reaction after infection by *Heterodera glycines* (soybean cyst nematode). *Plant Molecular Biology* 71:525-567.

Knapp, D.R. (1979). Carbohydrates In: *Handbook of Analytical Derivatization Reations*. John Wiley and Sons, New York, New York, 1979, Chapter 13, pages 539-598.

Kostianinen, R.; Kotiaho, T.; Kuuranne, T. & Auriola, S. (2003). Liquid chromatography/atmospheric pressure-ionization- mass spectrometry in drug metabolism. *J. Mass Spectrometry* 38:357-372.

Korak, D.A. (1984). Novel biosynthesis of D-pinitol in *Simmondsia chinensis*."*Phytochemistry* 23: 65-66. "

Kumaguai, H. (2001). Application of liquid chromatography/mass spectrometry to the analysis of sugars and sugar-alcohol. Application note. 5988-4236EN. Agilent Technolgies.

Lambers, H., Stulen, I. & Van Der Werf, A. (1996). Carbon use in root respiration as affected by elevated atmospheric CO2. *Plant and Soil*. 187:251-263.

LaFosse, M. & Herbreteau, B. (2002). In: *Carbohydrate Analysis by Modern chromatography and electrophoresis, J. of Chromatography Library,* Z. El Rossi (Ed.), Elesevier Science, Chapter 30, 1101-1134.

Larson, S & Raboy, V. (1999). Linkage mapping of maize and barley *myo*-inositol phosphate synthase DNA sequences: Correspondence with a low phytic acid mutation. *Theoretical Applied Genetics* 99:27-36

Lau, K.; Partridge; E. A.; Grigorian, A.; Silvescu, C. I.; Reinhold, V. N.; Demetriou;Michael; Dennis.James W. (2007). Conditional interpretation of extracellular cues by N-glycan multiplicityand Golgi processing. Cell, 129: 123-134.

Layzell, D.B., Hunt, S.; Moloney, A.H.M.; Fernando, S.M. & Diaz del Castillo. L. (1990). Physiological, metabolic and developmental implications of O2 regulations in legume nodule. Pp 21-32. In: ? P.M. Gresshoff, (Ed.), Nitrogen fixation: Achievements and objectives. Chapman and Hall, New York.

Lee, L.S. & Morris, N.J. (1963). Isolation and identification of pinitol from peanut flour, *Agr. Food Chem.* 11: 321-322

Lee, Y.C. (1996). Carbohydrate analyses with high-performance anion-exchange chromatography. *J. Chromatogr A.* 720:137-149.

Lehrfield, J. (1981). Differential gas-liquid chromatography for the determination of uronic acids in carbohydrate mixtures. *Anal. Biochem.* 115:410-418.

Levine J.; Barak, Y,; Kofman, O.; & Belmaker, R.H. (1995). Follow-up and relapse analysis of an inositol study of depression. *The Israel Journal of Psychiatry and Related Sciences.* 32: 14-21.

Li, B.W.; Schuhmann, P.J.;. &. Holden, J.M.(1983). Determination of sugars in yoghurt by gas-liquid chromatography. Journal of Agricultural and Food Chemistry.l 31:985-989.

Lind, Y.; Engman, J.; L. Jorhem, L. & Wicklund, G.A. (1988). Accumulation of cadmium from wheat bran, sugar-beet fibre, carrots and cadmium chloride in the liver and kidneys of mice. *British Journal of Nutrition* 80:205-211.

Logemann, E.; Wu, S.; Schröder, J.; Schmelzer, E.;. Somssich, I.E. & K. Hahlbrock, K. (1995). Gene activation by UV light, fungal elicitor or fungal infection in *Petroselinum crispum* is correlated with repression of cell cycle-related genes. Plant J. 8:865-876.

Loewus, M. W.; Bedgar, D.L. & Loewus, F.A. (1984.) 1-L-*myo*-Inositol 1-phosphate synthase from pollen of *Lilium longifolium. J Biol Chem* 259: 7644-7647.

Loewus, F.A. & Murthy, P.P.N. (2000). *myo*-Inositol metabolism in plants. *Plant Science* 150:1-19.

Long, A.R. & Chism III, G.W. (1987). A rapid direct extraction-derivatization method for determining sugars in fruit tissue. *Journal of Food Science* 52:150-154.

McInnes, A.G.; Ball, D.H.; Cooper, F.P. & Bishop, C.T. (1958). Separation of carbohydrate derivatives by gas-liquid partition chromatography. *J. of Chromatog* 1:556-557.

Manchanda, G. & Garg, N. (2008). Salinity and its effects on the functional biology of legumes. Acta Physiol. Plant. 30:595-618.

Maniatis, S.; Hui, Z. & Reinhold, V. N. (2010). Rapid De-O-Glycosylation Concomitant with Peptide Labeling Using Microwave Radiation and anAlkyl Amine Base. Anal. Chem.,82(7):2421-2425.

March, R.E. & Todd, J.F.J. (2005a). A historical review of the development of the quadrupole ion trap. In: *Quadrupole Ion Trap Mass Spectrometry*, 2nd edition, John Wiley and Sons, Inc., Hoboken, N.J., Chapter 1.

March, R.E. & Todd, J.F.J. (2005b). Linear quadrupole ion trap mass spectrometer. In: *Quadrupole Ion Trap Mass Spectrometry*, 2nd edition, John Wiley and Sons, Inc, Hoboken, N.J., Chapter 5.

Martens, D.A. & Frankenberger, W.T. (1990). Determination of saccharides by high performance anion-exchange chromatography with pulsed amperometric detection. *Chromatographia* 29:7-12.

Masle, J.; Farquhar, G.D. & Gifford, R.M. (1990). Growth and carbon economy of wheat seedlings as affected by soil resistance to penetration and ambient partial pressure of CO_2. Aust. J. Plant Physiol. 17:465-487.

Mauri, P.; Minoggio, M.; Somonetti, P.; Gardana, C. & Pietta, P. (2002). Analysis of saccharides in beer samples by flow injection with electrospray mass spectrometry. Rapid Commun. Mass Spectrom., 16:743-748

Munnik, T.; Irvine, R.F. & Musgrave, A. (1998). Phospholipid signalling in plants, Biochim. Biophys. Acta 1389 222–272.

Murphy, A.M.; Otto, B.; Brearley, C.A.; Carr, J.P. & Hanke, D.E. (2008). A role for inositol hexakisphosphate in the maintenance of basal resistance to plant pathogens. Plant Journal 56:638-652.

Nachtmann, F. & Budna, K.W. (1977). Sensitive determination of derivatized carbohydrates by high-performance liquid chromatography. J. Chromatography 136:279-287.

Nachtmann, F. (1976). Some properties of 4-nitrobenzoates of saccharides and glycosides; application to high-pressure liquid chromatography. Z. Anal. Chem.282:209-213.

Narayanan, C. (1987). Pinitol-A new anti-diabetic compound from the leaves of *Bouganivillea*. Current Science, 56:139-141.

Nelson, R.W.; Meulenkijk, J.A.P. & Underberg, W.J.W. (1990). Ultraviolet-Visible Derivatization. In: *Detection -Oriented Derivatization Techniques in Liquid Chromatography*, H. Lingeman & W.J.M. Underberg, (Ed.), Marcel Dekker, Inc, New York, New York, pp. 268-270.

Nordin, P. (1984). Preferential leaching of pinitol from soybeans during imbibition. Plant Physiol 76:313-315.

Ochoa, S. & Stern, J.R. (1952). Carbohydrate metabolism. Annual Review of Biochemistry 21:547-602.

Ogunyemi, O.; Pittner, F., & Hoffmann-Ostenhof, O. (1978). Studies on the biosynthesis of cyclitols. XXXVI. Purification of *myo*-inositol-I-phosphate synthase of the duckweed *Lemna gibba* to homogeneity by affinity chromatography on NAD Sepharose. Molecular and catalytic properties of the enzyme. Hoppe-Seyler's Z Physiol Chem 359: 613-616.

Perera, I. Y.; I. Heilmann, &. Boss, W.F. (1999). Transient and sustained increases in inositol 1,4,5-trisphosphate precede the differential growth response in gravistimulated maize pulvini. Proc. Natl. Acad. Sci. USA 96:5838-5843.

Pessarakli, M. (2005). Handbook of photosynthesis M. Pressarakli ed, 2nd edition. Taylor & Francis Group, CRC, Boca Raton, FL.

Phillips, D.V. & A. Smith, A. (1974). Soluble carbohydrates in soybean. Canadian Journal of Botany 52:2447-2452.

Phillips, D.V.; Dougherty, D.E. & Smith, A.E. (1982). Cyclitols in soybean. Journal of Agricultural & Food Chemistry 30:456-458.

Phillips, D.V. & Smith, A. (1974) Influence of sequential prolonged periods of dark and light on pinitol concentration in clover and soybean tissue. Physiol. Plant. 54: 31-33.

Raboy, V. (2007). Seed phosphorus and the development of low-phytate crops. In: Turner, B.L., Richardson, A.E., Mullaney, E.J. (Eds.)., Inositol phosphates linking agriculture and the environment. CABI, Oxfordshire.

Reinhold, V.N.; Ashline, D.J.; Zhang, H. (2010).Unravellng the Structural Details of the Glycoproteome by Ion Trap Mass Spectrometry, *in Practical Aspects of Trapped Ion Mass Spectrometry. Vol.* 4: *Theory and Instrumentation.* Edited by Raymond E. March and John F.J. Todd.Chapter 23,706-736. CCR Press, Boca Raton, FL.

Richmond, M.L.;. Barfuss, D.L.; Harte, B.R.; Gray, J.I. & Stine, C.M. (1982). Separation of carbohydrates in dairy products by high performance liquid chromatography. J. Dairy Sci. 65:1394-1400.

Romano, J. (2007). Carbohydrate analysis: column chemistries and detection. Presented at the Carbohydrates in Feeds Methodology Forum, AOAC 2007 Annual Meeting, Anaheim, CA, September 18, 2007.

Ruiz-Matute, A.I.; Hernandez-Hernandez, O.; Rodriguez-Sanchez, S.; Sanz, M.L. & Martinez-Castro, I. (2010). Derivatization of carbohydrates for GC and GC-MS analyses. J. of Chromatography B Available online 8 December 2010

Saghai Maroof, M.A.; Glover, N.M.; Biyashev, R.M.; Buss, G.R. &. Grabau, E.A. (2009). Genetic basis of the low-phytate trait in the soybean line CX1834. Crop Sci: 49: 69-76.

Sartini, C.M.; Spinabelli U.; C. Riponi, C. & Galassi, S. (2001). Determination of carboxylic acids, carbohydrates, glycerol, ethanol, and 5-HMF in beer by high-performance liquid chromatography and UV-refractive index double detection. J. Chromatogr. Sci., June, 39(6):235-238.

Sanz, M.L., Villamiel, M. & Martinez-Castro, I. (2004). Inositols and carbohydrates in different fresh fruit juices. Food Chemistry 87: 325-328.

Sasaki, K. & Taylor, I.F.P. (1986). *myo*-Inositol synthesis from [1-^{3}H]Glucose in *Phaseolus vulgaris* L. during early stages of germination. Plant Physiology 81:493-496.

Savidge, R. A. & Forster, H. (2001). Coniferyl alcohol metabolism in conifers II. Coniferyl alcohol and dihydroconiferyl alcohol biosynthesis. Phytochem. 57: 1095-1103.

Scheller, H.V. & Ulvskov, P. (2010). Hemicelluloses. Annual Review of Plant Biol. 61:263-289.

Schiller, M. H.; von der Heydt; Marz, F. & Schmidt, P.C. (2002). Quantification of sugars and organic acids in hygroscopic pharmaceutical herbal dry extracts. J. of Chrom A., 968: 101-111.

Schlichtherle-Cerny, H.; Affolter, M. & Cerny, C. (2003). Hydrophilic interaction liquid chromatography coupled to electrospray mass spectrometry of small polar compounds in food analysis. Anal. Chem. , 75:2349-2354.

Scott, R.P.W. (2011). Liquid Chromatography, Chrom Ed. Series, http://www.chromatography-online.org/HPLC/Refractive-Index/rs33.html, accessed March 8, 2011.).

Shah, M.M., Clauss, T.R.; Clauss, S.C.; Campbell, J.A.; Goheen, S.C. & Bennett, K.A. (1998). Determination of Molecular Weight Distribution of Insoluble Carbohydrates from corn fiber using HPLC and Light Scattering Detector. Presented at the Corn Utilization Conference, June 1-3, 1998, St. Louis, Mo.

Sharp, R.E. & Davies, W.J. (1979). Solute regulation and growth by roots and shoots of water-stressed maize plants. Planta 147:43-49.

Sheveleva, E.; Chmara, W.; Bohnert, H.J. & Jensen, R.G. (1997). Increased salt and drought tolerance by D-ononitol production in transgenic *Nicotiana tabacum* L. Plant Physiol. 115:1211-1219.

Shigaki, T. & Bhattacharyya, M.K. (2000). Decreased inositol 1,4,5-trisphosphate content in pathogen-challenged soybean cells. MPMI 13:563-567.

Smith, D. (1973). The nonstructural carbohydrates. In: Chemistry and Biochemistry of herbage, Vol. 1, eds. G.W. Butler, R.W. Bailey, Academic Press, London.

Somerville, C. (2006). Cellulose synthesis in higher plants. Annual Review of Cell and Developmental Biology. 22:53-78.

Stieglitz, K.A.; Yang, H.; Roberts, M.F. & B. Stec, B. (2005). Reaching for mechanistic concensus across life kingdoms: Structure and insights into catalysis of the *myo*-inositol-1-phosphate synthase (mIPS) from Archaeoglobus fulgidus. *Biochemistry* 44: 213-224.

Streeter, J. G. (1980). Carbohydrates in soybean nodules.II. Distribution of compounds in seedlings during the onset of nitrogen fixation. *Plant Physiology*. 66:471-476.

Streeter, J.G. & Strimbu, C. (1998). Simjultaneous extraction and derivatization of carbohydrates from green plant tissues for analysis by gas-liquid chromatograpy. Anal. Biochem. 259:253-257..E,. Strimbu

Streeter, J. G.; Lohnes, D.G. &. Fioritto, R.J. (2001). Patterns of pinitol accumulation in soybean plants and relationships to drought tolerance. *Plant, Cell and Environment* 24:429-438.

Stumpo, K.A.R.& Reinhold, V. (2010). The N-Glycome of Human Plasma. *Journal of Proteome Research*, 9(8):4823-4830.

Sullivan, J.E. & Schewe, L.R. (1977). Pareparation and gas chromatography of highly volatile trifluoroacetylated carbohydrates using N-methylbis[trifluoroacetamide]. *J. of Chromatographic Science* 15:196-197.

Sweeley, C.C.; R. Bentley, R.; Makita, M. & Wells, W.W. (1963). Gas-liquid chromtograpy of trimethylsilyl derivatives of sugars and related substances. J. of Am. Chem. Soc. 85:2497-2507.

Taormina, C.R.; Baca, J.T.; Asher, S.A.; Grabowski, J.J. & Finegold, D.N. (2007). Analysis of tear glucose concentration with electrospray ionization mass spectrometry. *J. Am. Soc. Mass Spectrom.*, 18:32-336.

Tate, R.L. (1995). Soil microbiology (symbiotic nitrogen fixation), pp. 307-333. John Wiley and Sons, Inc., New York, N.Y.

Taylor, V.F.; March, R.E.; Longerich, H.P. & Stadey, C.J. (2005). A mass spectrometric study of glucose, sucrose, and fructose using an inductively coupled plasma and electrospray ionization. Int. *J. Mass Spectrom.*, 243:71-84.

Thibault, P. & Honda, S. (Eds.). (2003). *Capillary electrophoresis of carbohydrates. Methods in Molecular Biology.* Humana Press, Totowa, N.J., 318 pages.

Togami, D.W.; Poulsen,B.J.; Batalao, C.W. & Rolls, W.A. (1991). Separation of carbohydrates and carbohydrate derivatives with cation-exchange columns at high pH. Biotechniques 10(5):650-655.

Twilly , J.W. (1984). The analysis of exudate glant gums In their artistic applications: An interim report. Archaeological Chemistry, Vol. 3, J.Lambert, (Ed.), American Chemical Society, Advances in Chemistry series, no. 205, Washington, D.C., American Chemical Society. 357-394.

Udvardi, M.K. & Day, D.A. (1997). Metabolite transport across symbiotic membranes of legume nodules. Annual Review of Plant Physiology and Plant Molecular Biology 48:493-523.

Vallance, S.L.; Singer, B.W.; Hitchen, S.M. & Townsend, J.H. (1998). The development and initial application of a gas chromatographic method for the characterization of gum media. *Journal of American Institute for Conservation.* 37:294-311.

Vernon, D. M. & Bohnert, H.J. (1992). A novel methyl transferase induced by osmotic stress in the facultative halophyte *Mesembuyanthemum crystalliunum.* EMBO J 11:2077-2085.

Vessey, J.K.; Walsh, K.B. & Layzell, D.B.(1988a). Oxygen limitation of N_2 fixation in stem girdled and nitrate-treated soybean. Physiol Plant. 73:113-121.

Vessey, J.K.; Walsh, K.B. & Layzell, D.B. (1988b). Can a limitation in phloem supply to nodules account for the inhibitory effect of nitrate on nitrogenase activity in soybean? Physiol. Plant. 74:137-146.

Walker, W. H.(2010). Is the "comb over" dying? A mouse model for male patter baldness (Androgenic Alopecia). Endocrinology 151: 1981-1983.

Walker, H. G.; Gee, M. & McCready, R.M. (1962). Complete methylation of reducing carbohydrates. J. Org. Chem. 27: 2100-2102.

Wei, Y. & Ding, M. (2000). Analysis of carbohydrates in drinks by high-performance liquid chromatography with a dynamically modified amino column and evaporative light scattering detection. J. of Chromatography A. 904:113-117.

Weisshaupl, V.; Scheiner, O.; Dworsky, P.; Hoffmann-Ostenhof, O. & Allgem, Z. (1976). Mikrobiol. 16: 81-86.

Wilcox, M.; Hefley, J.R. & Walsh, J.W. (2001). HPLC determination of carbohydrates: A comparison of six different columns with ELS detection. Presented at the Carbohydrate Session 59A, 2001 IFT Annual Meeting,. New Orleans, La.

Williams, J.H.H. & Farrar, J.F. (1990). Control of barley root respiration. Physiolo. Plant. 79:259-266.

Wilson, R.F. (2004). Seed composition Pp 621-677. In H.R. Boerma and J. E. Specht, co-editors, Soybeans: Improvement, production and uses. Third Edition Number 16 Agronomy Monograph. American Society of Agronomy, Inc., Crop Science Society of America, Inc., and Soil Science Society of America, Inc.. Madison, WI, USA.

Zeeman, S.C.; Smith, S.M. & Smith, A.M. (2007). The diurnal metabolism of leaf starch. *Biochemical Journal* 401:13-28.

Zeeman, S.C.; Kossmann, J. & Smith, A.M. (2010). Starch:its metabolism, evolution and biotechnological modification in plants. Annual Review Plant Biology 61:209-234.

Zenobi, R.; & Knochenmuss, R. (1998). Ion formation in MALDI mass spectrometry. Mass Spectrom. Rev. 17:337-366.

Zhu, S.G; Long, S.P. & Ort, D.R. (2010). Improving photosynthetic efficiency for greater yield. Annual Review Plant Biol. 61:235-261.

21

Extraction and Enzymatic Modification of Functional Lipids from Soybean Oil Deodorizer Distillate

Carlos F. Torres, Guzmán Torrelo and Guillermo Reglero
Departamento de Producción y Caracterización de Nuevos Alimentos
Instituto de Investigación en Ciencias de la Alimentación (CIAL), CSIC-UAM
Madrid
Spain

1. Introduction

Crude vegetable oils contain triacylglycerols as major component and various minor components such as diacylglycerols, monoacylglycerols, free fatty acids, phospholipids, tocopherols, sterols, squalene, color pigments, waxes, aldehydes, ketones, triterpene alcohols and metals that may affect the quality of the final product. The minor components are removed partially or entirely by either physical or chemical refining in order to make the vegetable oils suitable for human consumption. Deodorization is the last major processing step in the refining of edible oils. It has the responsibility for removing both the undesirable ingredients occurring in natural fats and oils and those which may be imparted by prior unit processes such as caustic refining, bleaching, hydrogenation, or even storage conditions. It is this unit process that finally establishes the oil characteristics of "flavor and odor," which are those most readily recognized by the consumer (Gavin, 1978).

Deodorizer distillate is a by-product of deodorization, which is the last major step in vegetable oil refining process. It is a complex mixture of free fatty acids, mono-, di- and triacyglycerols, sterols and their esters, tocopherols, hydrocarbons, pesticides, and breakdown products of fatty acids, aldehydes, ketones and acylglycerol species (Ramamurthi & McCurdy, 1993). Deodorizer distillate is an excellent source of valuable compounds such as phytosterols, tocopherols and squalene, which can be recovered and further used as food additives, in pharmaceutical industry and cosmetics. Their commercial value however, is mainly dependent on their tocopherol content (Fernandes & Cabral, 2007). Although in recent years, efforts from industry resulted in a significant number of reports, describing better and improved methods for phytosterol recovery and purification. Such surge is closely related to growing market for phytosterols, particularly given the widespread dissemination of functional foods (Fernandes & Cabral, 2007).

Numerous procedures have been described to isolate bioactive compounds from soybean oil deodorizer distillate to improve the value and the quality of this by-product. All these procedures can be grouped in three generic categories: crystallization and precipitation, chemical and enzymatic modification, and extraction and fractionation.

One important bioactive compound concentrated in intermediate byproducts and waste streams during the refining of soybean oil is squalene. Recently, steroidal hydrocarbons and squalene in soybean oil deodorizer distillate have been isolated and identified. Separation, purification, and chemical characterization of hydrocarbons fraction in soybean oil deodorizer distillate will help researchers finding a better utilization of these byproduct (Gunawan *et al.*, 2008b, Kasim *et al.*, 2009). Both squalene and steroidal hydrocarbons have been purified from soybean oil deodorizer distillate by means of modified soxhlet extractions with hexane to obtain two main fractions: one fraction rich in fatty acid steryl esters and squalene and a second fraction rich in tocopherols, free phytosterols, free fatty acids (FFAs) and acylglycerols. Then hydrocarbons are isolated via silica gel column chromatography. Similarly, other procedures described in the literature for isolation of sterols and tocopherols in soybean oil deodorizer distillate are based on the utilization of organic solvents (Lin *et al.*, 2004).

Recently greener technologies for isolation, purification and fractionation of bioactive compounds from soybean oil deodorizer distillate (SODD) have been developed. These methodologies can work in combination or independently to improve the fractionation of soybean oil deodorizer distillate. Hence, lipase-catalyzed methyl or ethyl esterification of SODD to transform free fatty acids into their corresponding fatty acid methyl or ethyl esters coupled to molecular distillation and/or supercritical fluid extraction has been described as a strategy to improve the separation between tocopherols, sterols and free fatty acids (Torres *et al.*, 2009). Alternatively, enzymatic esterification of the sterols with the fatty acids already present in the deodorizer sludge makes the separation of tocopherols and sterols simpler using short-path distillation or supercritical fluid extraction (Shimada *et al.*, 2000). However, short path distillation (Ito *et al.*, 2006), supercritical fluid extraction (Chang *et al.*, 2000), and enzymatic modifications (Torres *et al.*, 2007) have been also utilized independently on soybean oil deodorizer distillate. Therefore, these three technologies will be analyzed and discussed on the present chapter to evaluate the feasibility of these procedures for the valorization of side-stream products obtained during refining of soybean oil.

In addition, oxidation of sterols during refining steps such as heating, degumming, neutralization, bleaching, and deodorization, and during storage and handling should be also considered (Verleyen *et al.*, 2002b). However, limited information is available on the levels of sterols in the by-product fractions collected from chemical and physical refining processes (Verleyen *et al.*, 2001c) and the biological effects of phytosterol oxidation products on animal and human health need more investigation, as feed quality is crucial for animal health and welfare, and ultimately human health. It has been reported that the formation of sterol oxidation products is affected not only by the chemical nature of the sterols but also by their quantity (Dutta *et al.*, 2006). Positive correlations between total sterols and total phytosterol oxidation products in the by-products collected from refining processes have been found (Ubhayasekera & Dutta, 2009). Therefore, other aspects concerning soybean oil deodorizer distillate such as nonfood applications, direct consumption, and oxidative quality will be also analyzed and discussed in the present chapter.

2. Soybean oil deodorizer distillate production and characterization

The vegetable oils correspond about 70% of demand of natural oils and fatty acids consumed in the world. The soybean oil corresponds from 20 to 30% of the vegetable oils world market (Bockisch, 1998) and its production involves several steps that are necessary to render the soybean oil suitable for human consumption. These production steps have been broadly characterized as 1) soybean preparation, 2) oil extraction, and 3) oil refining.

Soybean preparation generally includes the steps of cleaning, drying, cracking, and dehulling.

Oil extraction basically consists of separating the oil from the remainder of the soybean, known as soybean meal. The great majority of commercial soybean extraction processes use a solvent to separate the oil from the meal. In the solvent extraction process, the beans are flaked to provide a large surface area. A solvent, commonly hexane, is then pumped through the soybean flakes, dissolving the oil in the hexane. The hexane is then separated from the oil and recycled.

The crude oil resulting from the extraction process must then be subjected to additional treatments, collectively called "refining", to remove various materials in order for the oil to be suitable for consumption. These materials include hydratable and non-hydratable phospholipids, free fatty acids, and various color and flavor components.

Crude soybean oil contains phosphorous compounds called hydratable phospholipids, and small amounts of calcium and magnesium that complex with a portion of the phospholipids to form non-hydratable phospholipids. Hydratable phospholipids are normally removed by a process known as "degumming", in which the oil is agitated or otherwise intimately combined with water to precipitate gums from the oil. The gums are then removed by centrifugation.

These precipitated gums can be used as a feed additive, or evaporated to remove moisture, the end product is called lecithin. Lecithin has various end uses such as food emulsifier. The degummed oil is dried under vacuum to remove any water. Removal of non-hydratable phospholipids is considerably more difficult and expensive, requiring further chemical treatment, typically chemical refining, to break the chemical bonds between the calcium or magnesium ions and the phospholipids, followed with extensive bleaching of the oil.

In most processes, free fatty acids are then removed from the oil by a process known as caustic refining, also called chemical or alkali refining, in which the oil is mixed with a caustic material, such as sodium or potassium hydroxide, which undergoes a saponification reaction with the acids, forming soaps that are then removed by centrifugation. In this case, the non-hydratable phosphotide are removed along with the free fatty acids. In addition, a significant quantity of the oil is captured by the soaps, adversely affecting oil yield.

Free fatty acid removal by a process known as physical refining has been used for oils that are low in non-hydratable phospholipids, such as lauric oils, particularly palm oil. In physical refining, the oil is vacuum distilled at high temperatures, e.g., from about 230 °C to about 260 °C to separate more volatile components from the oil. This process is used to remove various flavor components, and will also remove free fatty acids. However, the process has not been viable for removing free fatty acids from oils such as soybean oil, which contains higher levels of non-hydratable phospholipids (more than 20 ppm based of phosphorous content). The high temperatures required for physical refining tend to break down the non-hydratable phospholipids that are present in the soybean oil, producing chemical compounds that cause an unacceptable flavor and color.

Conventional refining processes also involve some bleaching of the soybean oil to remove color pigments (i. e., carotenoids, chlorophyll) that adversely affect the color of the oil. Bleaching process employs the use of adsorbents such as acid-activated clays.

Finally, deodorization is the last process step used to improve the taste, odor, color and stability of the oil by means of removing undesirable substances. The goal of deodorization is to obtain a final product, finished oil that has a bland flavor, a maximum FFA content of 0.05% and a zero peroxide value. All commercial deodorization, whether in continuous, semicontinuous or batch units, is essentially a form of physical distilling by steam, in which the oil is subjected to high temperatures (210 °C - 280 °C) under a high vacuum (1 – 6 mm Hg) for a short period of time, which is sufficient to remove FFA and other volatile flavor-causing compounds. During the process, peroxide decomposition products, color bodies and their decomposition products are eliminated and the content of sterols, sterol esters and tocopherols is also reduced.

The modern commercial deodorizers are equipped with a packed column that has three sections: vapour scrubbing section, stripping section and heat bleaching section.

Bleached oil is pre-heated by outgoing deodorized oil and sprayed into the Deaerator where dissolved air and moisture are reduced to a minimum. The oil is then heated to full temperature by hot deodorized oil in the Deodorizing Economizer and high pressure steam in the Final Heater. A portion of the free fatty acids in the oil will be flashed off as the oil temperature increases.

The hot oil enters the Packed Column, which is filled with special structured packing so that the oil is distributed into a thin film and is evenly agitated by stripping steam flowing counter currently from the bottom of the column. As a result, free fatty acids and other remaining volatile impurities in the oil are evaporated and removed with the steam. The residence time in the column is only a few minutes.

Next, the stripped oil enters the heat bleaching section where it flows through the channels of a series of vertically stacked compartments (trays) while agitated by stripping steam. The prolonged thermal action breaks down color bodies (carotenes) and other heat sensitive compounds are volatilized and removed, or rendered inactive, resulting in a lighter oil color. Also, the amount of remaining free fatty acids in the oil is reduced to an absolute minimum.

The deodorized oil is pre-cooled by deaerated oil and then sprayed into the Post Deodorizer where the final "off-flavor" compounds are removed.

Fatty acids and other materials, evaporated from the oil, are condensed by contact with recycled and cooled distillate in the Vapor Scrubbing section. The soybean oil deodorizer distillate is circulated by the Distillate Pump via the Distillate Cooler where it is cooled by cooling water. Accumulated distillate is discharged from the Scrubber to storage.

SODD is a by-product of deodorization and is a complex mixture of free fatty acids, mono-, di- and triacyglycerols, sterols and their esters, tocopherols, hydrocarbons, pesticides, and breakdown products of fatty acids, aldehydes, ketones and acylglycerol species (Ramamurthi & McCurdy, 1993, Verleyen, 2001c). SODD corresponds between 0.1 and 0.4% of crude soybean oil. Deodorizer distillate is an excellent source of valuable compounds such as phytosterols and tocopherols, corresponding approximately 10% and 20% respectively (Czuppon *et al.*, 2003), which can be recovered and further used as food additives, in pharmaceutical industry and cosmetics (Lin & Koseoglu, 2003). Their commercial value however, is mainly dependent on their tocopherol content, depending on the market demand for this ingredient (Dumont & Narine, 2007).

SODD contains high levels of free fatty acids and acylglycerols (Chu *et al.*, 2002). Fatty acids represent 25-75% and acylglycerols about 3-56% of Vegetal Oil Deodorizer Distillates (VODD), depending on the raw material being refined and on the type and conditions of the refining process (Ramamurthi & McCurdy, 1993). The free fatty acids from deodorizer distillate are mostly used as additives for animal food, fluidizing agents for lecithin or as medium-grade soaps. Such fatty acids also can be used as precursors in a wide variety of molecular synthesis schemes such as the production of dibasic acids of different chain lengths (Gangopadhyay *et al.*, 2007). Alternatively, deodorizer distillate have non-food applications, such as biodiesel or can be used mixed with the fuel oil to fire the steam boilers (Svensson, 1976).

Tocopherols (vitamin E) are natural antioxidants found in vegetable oils and contribute significantly to their oxidative stability. Due to the high tocopherol content of crude oils, they can be stored long periods of time if they are protected from air, moisture and high temperature (Norris, 1979) without any significant deterioration. However, the concentration of natural tocopherols in soybean oil is too high for optimum oxidative stability and flavor, because they can act as pro-oxidants by peroxide formation (Jung & Min, 1990, Warner, 2005).

Moreover, tocopherols exert several beneficial activities, such as protective role of vitamin A, β-carotene and essential fatty acids (Ferrari *et al.*, 1996). Tocopherols also prevent diseases like cancer (Kline *et al.*, 2007), cardiovascular and cataracts (Block & Langseth, 1994, Munteanu & Zingg, 2007, Rimm *et al.*, 1993). They are used in food, cosmetics and pharmaceutical industries (Chu, 2002) and a mixture of α, β, γ and δ isomers containing 60 wt% tocopherols is widely used as additive to many kinds of foods (Shimada, 2000).

On other hand, in recent years a significant number of reports, patents, and scientific publications describing improved methods for phytosterol recovery and purification have been developed. This phenomenon is closely related to the growing market for phytosterols, particularly given the widespread dissemination of functional foods (Fernandes & Cabral, 2007).

Phytosterols are useful hypocholesterolemic agents since a daily intake of 2-3 g lowers LDL cholesterol concentrations by 10-15 % as found in various populations (Kritchevsky & Chen, 2005, Quílez *et al.*, 2003). The proposed mechanism is that plant sterols reduce the micellar solubility of cholesterol and consequently lower intestinal absorption of both exogenous and endogenous cholesterol (de Jong *et al.*, 2003, Trautwein *et al.*, 2003), but also experimental investigations suggest that sterols may act modulating lipid and protein metabolism (Mulligan *et al.*, 2003, Plat & Mensink, 2005). In addition to their cholesterol lowering effect, plant sterols may possess anti-cancer (Awad *et al.*, 2003), antiatherosclerosis (Moghadasian *et al.*, 1999, Moghadasian *et al.*, 1997), anti-inflammation (Bouic, 2001) and antioxidation activities (van Rensburg *et al.*, 2000). Phytosterol compounds exhibit virtually no side effects and they have shown no evidence of *in vitro* mutagenic activity or subchronic toxicity in animals (Rozner, 2006). These compounds have been extensively used as a food ingredient in the functional food industry.

Moreover, phytosterols are valuable precursors in the production of hormones (Donova, 2007). They are used in manufacturing progesterone, corticoids, estrogens, contraceptives, diuretics, male hormones and vitamin D. They are, also, used in cosmetics (Balazs, 1987, Fernandes & Cabral, 2007).

Another important bioactive compound concentrated in intermediate byproducts and waste streams during the refining of soybean oil, is squalene, a hydrocarbon that has been used in applications such as natural moisturizer in cosmetics and biochemical precursor in the synthesis of steroids. Recently, steroidal hydrocarbons and squalene in soybean oil deodorizer distillate have been isolated and identified. Separation, purification, and chemical characterization of hydrocarbons fraction in soybean oil deodorizer distillate will help researchers finding a better utilization of these byproduct (Gunawan, 2008b, Kasim, 2009).

The refining process induces changes in the structure and concentration of tocopherols, sterols (free and bound) and squalene.

Of these various components, most attention is given to the tocopherols. Jung and coworkers (Jung *et al.*, 1989) and Ferrari (Ferrari, 1996) have studied the tocopherol content at all stages of processing for all isomers in the finished oil. The tocopherol content decreases during each step of processing and may be markedly reduced during deodorization, as the tocopherols are volatile under these conditions. The processing removed between 30-60% of tocopherols in crude soybean oil. Even though total tocopherol content decreased during processing, the relative compositions of tocopherols in soybean oils were constant during processing.

Sterol content present in soybean oil also tend to be diminished in processing and the magnitude of such decrease is about the same as the tocopherols (Ferrari, 1996). It has been shown that the absorption of sterols is increased extensively with increased amounts of bleaching clay. The lipid extract from the bleaching clay had high concentrations of sterols in unchanged form.

Squalene content also decreases during processing (Nergiz & Çelikkale, 2010), but not drastically until deodorization, when it is partially volatilized. Total losses during all the stages of refining were found to be 31 % as compared to its content in crude soybean oil.

Numerous procedures have been described to isolate bioactive compounds from soybean oil deodorizer distillate to improve the value and the quality of this by-product. All these procedures can be grouped in four generic categories: classic method such as crystallization and precipitation, molecular distillation, supercritical fluid extraction and chemical and enzymatic modification.

3. Classic methods to obtain functional lipids from SODD

In the past, recovering tocopherols and sterols from deodorizer distillates and related mixtures has been proved to be complicated and expensive. One difficulty associated with isolating one or more distillate fractions enriched in fatty acids, tocopherols, and/or sterols from deodorizer distillates is that the molecular weights and volatilities of sterols are similar to those of tocopherols (Ghosh & Bhattacharyya, 1996). For this reason, it is difficult to recover concentrates of tocopherols and phytosterols with good yield and high quality (Lin, 2002). In addition, in order to separate the squalene present in the distillate, the main challenge is to isolate them from each other, especially in the case of the following pairs of components: tocopherol-squalene, tocopherol-fatty acids, tocopherol-sterol and sterol-squalene.

Another difficulty is that deodorizer distillate can undergo thermal degradation if it is processed for extended periods at the temperatures at which sterols and tocopherols vaporize, such temperature conditions which can cause fatty acids to convert into undesirable trans isomeric forms and may cause the degradation of tocopherols (Chu, 2002).

Classical methods for recovering tocopherols and sterols include solvent extraction, chemical treatment, crystallization, complexation, and molecular distillation (Rohr & Trujillo-Quijano, 2005). The separation process involves a series of chemical and physical techniques which are used alone or in combination. In general, most processes are designed to remove either fatty acids or sterols in the initial step, followed by tocopherol concentration by other methods.

Crystallization has frequently been used to purify sterols from SODD, either following or preceding other separation methods. Brown (Brown & Smith, 1964) reported a phytosterols product prepared by a continuous two-stage liquid-liquid extraction (LLE) with a solvent pair of methanol and hexane, and then followed by crystallization using acetone as a solvent at 4 °C for 24 h. By this approach, 73% sterol concentrate was obtained from SODD containing 6.5% sterol. Sheabar and Neeman (Sheabar & Neeman, 1987) have shown the preparation of a tocopherol concentrate through removal of sterol from SODD by a two-stage crystallization at -20 °C with hexane and acetone as crystallization solvents. Attempts have been made to isolate tocopherols from SODD by supercritical fluid extraction technology with crystallization as pretreatment to first remove sterols (Lee *et al.*, 1991). SODD was esterified with methanol using HCl as catalyst, then a solvent pair of hexane-methanol was used to obtain tocopherols-sterols concentrate from which sterols were recovered by crystallization at -20 °C with acetone as a solvent (Brown & Smith, 1964). The results were similar to those mentioned above. Nevertheless, the information of total yield of sterols was not provided in these publications.

Crystallization seems successful as a simple and efficient process to remove and concentrate sterols and tocopherols from SODD. This process has the advantage of not causing tocopherol oxidation, because the low temperature utilized, and it does not use high pressure. While there is much information in the literature on the recovery of sterols from SODD by crystallization, little attempt is made pertaining to its optimal conditions such as solvent type, crystallization temperature and time. Lin and Koseoglu (Lin & Koseoglu, 2003) have shown crystallization of sterols from SODD without any pretreatment is practical. The best results were achieved by crystallization at -20 °C for 24 h using a solvent mixture of acetone-methanol (4:1, v/v) at a solvent-to-SODD ratio of 3:1 (v/w), followed by centrifugation, filtration, and twice washing of the wet cake. Over 90% of the original tocopherols and squalene, were retained in the filtrate fraction, while 80% of the original sterols were crystallized in the cake fraction. Khatoon and coworkers (Khatoon *et al.*, 2010) developed a method for the preparation of phytosterols from SODD by crystallisation using hexane and water. Direct crystallisation yielded a phytosterol fraction with lower recovery of 13.2–17.8% while treatment with alkali to remove FFA and the glycerides followed by organic solvent extraction yielded unsaponifiable matter containing phytosterols with a recovery of 74.6%. Later the unsaponifiable matter was purified by double crystallisation into a mixture of phytosterols of 87% purity. Moreira and Baltanás (Moreira & Baltanás, 2004) studied the impact of the principal process variables (solvents and cosolvents, cooling rate, crystallization temperature, and ripening time) on the quality and yield of the recovered phytosterols, but in this case by using a sunflower oil deodorizer distillate "enriched" (i.e., preconcentrated). In this study, a sterols recovery as high as 84% (with 36% purity) was achieved by using a single-stage batch crystallization of hexane/ethanol mixture (ratio of 4:1, v/v) at -5 °C.

On the other hand, a modified industrial process was developed by Xu and coworkers (Xu *et al.*, 2005) to recovery and purify valuable compounds from SODD. In this process, tocopherols and fatty acids methyl esters (FAMEs) was obtained from SODD after a process with methyl esterification by sulfuric acid catalyst, transesterification by alkaline catalyst, crystallization of sterols and molecular distillation. The waste residue of SODD was obtained after the molecular distillation and it mainly contains steryl esters, acylglycerols, and hydrocarbons.

In turn, Yang and coworkers (Yang *et al.*, 2009) developed a catalytic and crystallization process to recover phytosterols from waste residue of SODD (WRSODD). A catalyst was employed to decompose WRSODD so as to transform steryl esters into phytosterols. The mixed solvent that generated the best crystallization results was acetone and ethanol (4:1, v/v). The yield and the purity of recovered phytosterols were 22.95 wt. % and 92-97 %, respectively.

Nevertheless, crystallization has the disadvantage of the solvents available at present are not sufficiently selective to obtain, through the current processes, a reasonable separation between the unsaponifiable components and free fatty acids. Due to this, it is often necessary to use more than one solvent, which in turn complicates and increases tremendously the cost of recovery and recycling of these solvent mixtures. Furthermore, solvents or solvent mixtures are used in very large proportions, when compared to the quantity of the material submitted for extraction, and the solvents need additional processes for their removal and/or recycling in the extraction and pre-concentration process of the valuable products. The foregoing reasons make solvent based-processes, expensive, unattractive and less environmentally friendly, resulting in a scarce and expensive final product.

Saponification is also a common practice to concentrate tocopherols and sterols since it produces alkali metal soap which, due to its insolubility in the solvent used in the process, can be separated from the dissolved tocopherols, thereby permitting recovery of the tocopherols in a form relative free from fatty acids and glycerides. The processes themselves are costly, however, and tocopherols are produced in low yield. The sterols are then isolated from the resulting concentrate mixture by crystallization (Brown & Meag, 1963, Kijima *et al.*, 1964, Kim & Rhee, 1982).

Of the saponification processes, the lime saponification process is the most widely used. Hickman, U.S. Patent No. 2.349.270 (Hickman, 1944), discloses that deodorizer distillate can be treated with calcium hydroxide, traditionally called slaked lime, to saponify the fatty acids, followed by extraction of the unsaponifiable fraction (tocopherols and sterols) with acetone, in which the saponification products are insoluble. The extract is then washed and concentrated, as for example by solvent distillation, and then cooled to crystallize sterols which are removed by filtration, leaving a high purity tocopherol fraction. The fatty acid soaps formed by the process can be acidulated and converted into free fatty acids. Andrews, U.S. Patent No. 2.263.550 (Andrews, 1941), discloses saponification of deodorizer distillates with sodium hydroxide, followed by metathesis (a molecular process involving the exchange of bonds between the two reacting chemical species, in this case a ion exchange) with calcium chloride to convert the sodium soaps to calcium soaps (not water soluble), from which the tocopherols and other unsaponifiable matter are then extracted with acetone.

The disadvantage of each of these processes is that the calcium soap is formed in a wide particle size distribution, ranging from fine particles to lumps. The result is a soap mass which is lumpy in form and from which the unsaponificable matter is difficult to extract. To permit the extraction to take place, the soap mass must be ground into particulate form, a process which entails a substantial capital investment. Even then, solvent consumption is high and the recovery of tocopoherols and other useful unsaponificable matter such as sterols is low.

Grinding is avoided in the process disclosed by Brown and coworkers (Brown & Meag, 1963), which uses calcium silicate as a powdering agent in combination with acetone to facilitate the separation of the soluble tocopherols and sterols from the insoluble soap mass. Unfortunately, this process requires a large amount of powdering agent which remains in the soap mass, and the effectiveness of the powdering agent is diminished if the moisture content of the soap mass is too high.

Although saponification is effective to remove free fatty acids and acylglycerols, it involves the use of a large amount of alkali which is harmful to tocopherols, thus leading to low yields. Recently, molecular distillation combined with crystallization was more attractive to separately concentrate tocopherols and sterols (Gapor *et al.*, 1989, Hunt *et al.*, 1997, Kijima, 1964, Kim & Rhee, 1982, Smith Frank, 1967). To increase the separation efficiency, esterification and/or transesterification are usually carried out prior to molecular distillation. Free fatty acids and acylglycerol are converted to fatty acid methyl esters, which are more easily removed by vacuum distillation due to their higher vapor pressure than those of the corresponding free fatty acids and acylglycerol. However, this step made the whole process more complicated and labor-intensive when compared with the saponification process. Another drawback of molecular distillation is that it is energy consuming to maintain high vacuum all of the time during operation. Consequently, the final product is also expensive.

4. Enzymatic modification

Enzymatic reactions are based on the selective biotransformation of determined compounds in order to modify their chemical or physical properties. Hence, the utilization of enzymes, for instance, makes easier the separation of tocopherols from SODD by converting sterols to steryl esters, acylglycerols to free fatty acids and free fatty acids to fatty acid methyl or ethyl esters (FAMEs or FAEEs). Then, it is easier to separate the new product mixture by distillation or supercritical fluid extraction. From published literature, it can be point out that the main difficulties of the enzymatic processes are the numerous parameters involved such as moisture content, enzyme concentration, time, temperature, ratio of the reactants, stability, recovery and reutilization of the enzyme preparation, among others (Ramamurthi *et al.*, 1991), (Ramamurthi & McCurdy, 1993).

The conversion of FFAs to FAMEs or FAEEs is an important step in the concentration and purification of tocopherols. If this step is omitted, the separation of FFA and tocopherols by distillation cannot be achieved due to their similar boiling points (Shimada, 2000). Furthermore, if methanol is used for the biotransformation of FFA to FAMEs, concomitant sterol esterification with fatty acids is inhibited. To avoid this problem, a lipase can be used in a two stages procedure: first to carry out hydrolysis of acylglycerols and then to promote the esterification of sterols with free fatty acids. The different components are then successfully separated by short path distillation or supercritical fluid extraction since their boiling points are now sufficiently different.

In the literature, many enzymatic procedures for the preparation of sterol esters are described, but most of them require organic solvents, water and molecular sieves or other drying agents (Haraldsson, 1992), (Shimada *et al.*, 1999), (Jonzo *et al.*, 1997), (Hedström *et al.*, 1992). Although these strategies gave good conversion rates for the formation of sterol esters, the use of such multiphasic systems may complicate the final purification of the products in the case of larger scale productions. However, the enzymatic preparation of fatty acid esters of sterols, stanols and steroids in high yield by esterification and transesterification of fatty acids and other carboxylic acid esters, in vacuum at moderate temperature using immobilized lipases have been also reported (Weber *et al.*, 2001). In this case neither organic solvent, nor water or any drying reagent such as molecular sieves, are used. This and others studies (Shimada, 2000) showed that in the process of esterification of sterols with free fatty acids, the best results are obtained with *Candida rugosa* lipase and *Pseudomonas* sp. However, enzymatic conversion of FFAs to FAMEs or FAEEs is carry out frequently in the presence of *Candida antactica* lipase or *Alcalygenes* sp. (Torres, 2007), (Nagao *et al.*, 2005).

In the following paragraphs some examples of methodologies using enzymes in the pre-treatment of SODD are described. Most of them will be further developed in following sections:

Shimada and coworkers (Shimada, 2000) converted sterols from SODD to fatty acid sterol esters and completely hydrolyzed acylglycerols by applying lipase reactions (*Candida rugosa* or *Pseudomonas* sp., at 35 °C for 24 h) to the purification of tocopherols and sterols, resulting in an efficient fractionation of tocopherols and sterols as fatty acids steryl esters (FASEs) by short-path distillation. This process included the drawback that FFA and tocopherols were not efficiently fractionated because the boiling points of the two substances were close. This problem could be solved by conversion of the FFA to their corresponding methyl esters. An attempt to develop a reaction system in which the methyl esterification of FFA proceeded simultaneously with the conversion of sterols to FASEs and the hydrolysis of acylglycerols has been also reported (Watanabe *et al.*, 2004).

Nagao and coworkers (Nagao, 2005) and Watanabe and coworkers (Watanabe, 2004) have applied a procedure based on using a lipase to promote the simultaneous esterification of sterols with free fatty acids and hydrolysis of acylglycerols before the esterification of the free fatty acids with methanol. These authors use *Candida rugosa* lipase for the purification of tocopherol in SODD. Watanabe and coworkers reported 80% conversion of the initial sterols to FASEs, complete hydrolysis of the acylglycerols, and a 78% decrease in the initial FFA content by methyl esterification in 40 h. Tocopherols did not change throughout the process. Distillation of the reaction mixture purified tocopherols to 76.4% (recovery, 89.6%) and sterols to 97.2% as FASEs (recovery, 86.3%). Nagao and coworkers reported a more effective sterols esterification, with a degree of esterification reached 95%. The second-step reaction was then conducted at 30 °C for 20 h with *Alcaligenes* sp. lipase. 95% FFAs were converted to FAME, and steryl esters synthesized by the first-step reaction were not reconverted to free sterols. Finally, tocopherols and steryl esters were purified from the reaction mixture by short-path distillation. Tocopherols were purified to 72% (yield, 88%) and steryl esters were purified to 97% (yield, 97%). One of the main disadvantages of this method is that the remaining free fatty acids are not completely separated from the tocopherols.

Lipase-catalyzed esterification of sterols and ethyl esterification simultaneously, are governed by the concentration of water present. The degree to which esterification of sterols occurs relative to ethyl esterification requires to attain a balance not always easy to achieve because the presence of an excess of water favours hydrolysis, whereas esterification predominates when a very limited amount of water is present (Marangoni & Rousseau, 1995). By appropriate choice of reaction conditions, however, it is possible to separate the sterol esterification and ethyl esterification in time or space. It is then possible to optimize each of these reactions independently, thereby minimizing costs or improving the yield of the desired final reaction products.

This is precisely the procedure carried out by Torres and coworkers (Torres, 2007), who proposed a two-step enzymatic procedure to obtain FASEs, tocopherols, and fatty acid ethyl

esters (FAEEs) from SODD. Firstly, SODD was mixed with oleic acid to reduce its melting point and to enhance the free phytosterols esterification. The first enzymatic step (using *Candida rugosa* lipase) allowed the efficient conversion of more than 90% free phytosterols within 5 h. The second one (using Novozym 435) converted more than 95% FFAs in less than 3 h. The final product obtained was used as starting material to purify FASEs, tocopherols, and FAEEs via supercritical CO_2 extraction (Torres, 2009).
Weber and coworkers (Weber *et al.*, 2002) have also reported the use of lipases for the conversion of sterols into steryl esters leading to a higher degree of purity (90%), however the methodology is more complex and involves deacidification, flash chromatography and solvent fractionation.
Another methodology was developed to focus more specifically on the conversion of FFAs into fatty acid butyl esters (FABEs) (Nagesha *et al.*, 2004). Nagesha and coworkers (Nagesha, 2004) used immobilized *Mucor miehei* lipase in supercritical carbon dioxide at high pressure and obtained a maximum recovery of 88% and a FABE purity of 95% from SODD.

5. Molecular distillation

Most of the substances that are present in soybean deodorizer distillate are molecules of high molecular weight and thermally sensitive. These properties hinder the separation or purification of these compounds through traditional methods, because they are decomposed when subjected to high temperatures.
An alternative separation/purification procedure of such products is the use of molecular or short-path distillation. It consists of transferring molecules from the surface of an evaporating liquid to the cooled surface of a condenser through a short path, which is on the order of 2-5 cm. In this process, distillation of heat-sensitive materials is accompanied by only negligible thermal decomposition (Lutisan *et al.*, 2002) because materials, by using high vacuum, are submitted to relatively reduced temperatures, and short residence times (Lutisan, 2002) inside the equipment. Furthermore, this process has advantages over other techniques that use toxic or flammable solvents as the separating agent, avoiding toxicity and environmental problems.
The combination of a small distance between the evaporator and the condenser of only a few centimetres and a high vacuum in the distillation gap, results in a specific mass transfer mechanism with evaporation outputs as high as 20–40 gm^{-2} s^{-1} (Cvengros *et al.*, 2000). Under these conditions (e.g., short residence time and low temperature), distillation of heat-sensitive materials is accomplished without or only negligible thermal decomposition. Therefore, molecular distillation shows potential in the separation, purification and/or concentration of natural products, usually constituted by complex and thermally sensitive molecules such as tocopherols.
In lipid chemistry, it has been used for the purification of monoacylglycerols (Szelag & Zwierzykowski, 1983), recovery of carotenoids from palm oil (Batistella & Wolf Maciel, 1998), fractionation of polyunsatured fatty acids from fish oils (Breivik *et al.*, 1997), recovery of squalene (Sun *et al.*, 1997), and recovery of tocopherols (Batistella *et al.*, 2002), among others.
Normally, SODD have a high content of FFA and acylglycerol. To increase the separation efficiency of the compounds of interest, esterification and/or transesterification reactions are usually carried out prior to molecular distillation. Free fatty acids and acylglycerols are converted to fatty acid methyl esters, which are more easily removed by vacuum distillation

due to their higher vapor pressure than those of the corresponding free fatty acids and acylglycerols. However, this step made the whole process more complicated and labor-intensive when compared with the saponification process. Other problem is that the separation of tocopherols from phytosterols is difficult because they have similar molecular weights, boiling points and vapor pressure, and, consequently, they are distilled together (Ghosh & Bhattacharyya, 1996).

Different processes have been proposed in the literature to eliminate FFA by molecular distillation and purify tocopherols and phytosterols. Most of them include a preliminary chemical or enzymatic treatment step.

Ramamurthi and McCurdy (Ramamurthi & McCurdy, 1993) studied the pretreatment of deodorizer distillate using a lipase-catalyzed esterification reaction to convert FFA into methyl esters, followed by vacuum distillation (1-2 mm Hg) to remove them and concentrate tocopherols and sterols (recoveries were over 90%).

Hirota and coworkers (Hirota *et al.*, 2003) isolated naturally occurring Fatty Acid Steryl Esters (FASEs) from SODD. SODD was firstly subjected to molecular distillation at 250 °C and 0.02 mm Hg to obtain a residue which was rich in DAGs and TAGs, and steryl esters. Enzymatic lipolysis was then conducted to specifically hydrolyze DAGs and TAGs at 35 °C for 24 h, resulting in a mixture from which fatty acid steryl esters were later purified using a two-stage molecular distillation (180 °C and 0.2 mm Hg, and 250 °C and 0.02 mm Hg). The recovery and purity of FASEs were about 87.7% and 97.3 wt%, respectively.

Purification of tocopherols from SODD was carried out by Shimada and coworkers (Shimada, 2000). SODD was distilled using molecular distillation at 250 °C and 0.02 mm Hg and the resulting distillate was used as a starting material. Sterols in SODD were converted to FA sterol esters and acylglycerols were completely hydrolyzed by applying lipase reactions. FASEs were recovered as residue from the reaction mixture via molecular distillation at 250 °C and 0.2 mm Hg. However, the last stage of molecular distillation failed to separate FFAs and tocopherols. A second esterification of free phytosterols was applied at 35 °C for 24 h, followed by another four-stage molecular distillation (160 °C and 0.2 mm Hg, 200 °C and 0.2 mm Hg, 230 °C and 0.04 mm Hg, and 255 °C and 0.03 mm Hg) which yielded tocopherols with purity and recovery of about 65.3 wt% and 54.6%, respectively.

Watanabe and coworkers (Watanabe, 2004) isolated tocopherols and free phytosterols as their esters from SODD tocopherol/sterol concentrate (SODDTSC). SODDTSC was obtained via molecular distillation at 240 °C and 0.02 mm Hg, resulting in a distillate rich in tocopherols and free phytosterols (SODDTSC), and a residue rich in FASEs, DAGs, and TAGs. SODDTSC, which contained also MAGs, DAGs, FFAs, and unidentified hydrocarbons, were then subjected to a two-step in situ enzymatic reaction. SODDTSC were treated with *Candida rugosa* lipase (200 U/g activity) to convert free phytosterols to FASEs, acylglycerols (MAGs and DAGs) to FFAs, and FFAs to FAMEs at 30 °C for 40 h, achieving 80% conversion of the initial sterols to FA steryl esters, complete hydrolysis of the acylglycerols, and a 78% decrease in the initial FFA content by methyl esterification. Tocopherols did not change throughout the process. To enhance degree of steryl and methyl esterification, FASEs and FAMEs enriched in the reaction product were then removed by a two-step molecular distillation. In the first step, FAMEs was removed in the distillate (160 °C and 0.2 mm Hg). In the second step (240 °C, 0.2 mm Hg), FASEs was isolated in the residue and the distillate containing tocopherols, free phytosterols, and FFAs were treated again with lipase. A three-step molecular distillation of the reaction mixture purified tocopherols to 76.4 wt% purity (89.6% recovery) and free phytosterols to 97.2 wt% purity as FASEs (86.3% recovery).

Nagao and coworkers (Nagao, 2005) carried out similar steps of isolation of tocopherols and free phytosterols as FASEs from SODDTSC. SODDTSC were first treated with *Candida rugosa* lipase (250 U/g activity) to convert free phytosterols to FASEs at 40 °C for 24 h, achieving about 95% conversion. Unreacted FFAs contained in the reaction mixture was then converted to FAMEs by *Alcaligenes* sp. lipase at 30 °C for 20 h, achieving 95% conversion. Reaction mixture was then subjected to a four-stage molecular distillation (160 °C and 0.2 mm Hg, 175 °C and 0.2 mm Hg, 230 °C and 0.02 mm Hg, and 240 °C and 0.02 mm Hg, respectively) to isolate tocopherols (72.3 wt% purity, 87.6% recovery) in the third distillate fraction and free phytosterols as FASEs (97 wt% purity, 97% recovery) in the last (fourth) residue.

Jacobs (Jacobs, 2005) proposed a method for recovering tocotrienols from fatty acid distillate FAD, which initially contained 1 wt% tocopherols and 0.3 wt% free phytosterols, by stripping FFAs (condition: 0.5–1.5 mm Hg, 180–240 °C, and 0.5–1.5 min) to form a first stripped product. Short path distillation of the first stripped product gave a first distillate. Saponifying the second stripped product resulted in a saponified product with all FFA converted to FAMEs. A second short path distillation of the saponified product generated a second distillate without FAMEs. Solvent wintering (via filtration) of the second distillate gave a stripped filtrate. The stripped filtrate from the previous step is subjected to a third short path distillation at a temperature of 180° and an absolute pressure of 0.01 mm Hg. The resulting final tocotrienol product, about 1% of the original feed, contains from about 50% tocotrienols, 1% sterols and 49% other unsaponifiables and unknowns. Additionally, the final product contains from about 15% to about 30% tocopherols.

Fizet (Fizet, 1996) esterified free phytosterols from deodorizer distillate with FFAs at either 180 °C for 2.5 h or 250 °C for 1.5 h. Esterification product was then distilled at 120–150 °C and 0.08 mm Hg to obtain a residue containing mostly tocopherols and FASEs and a distillate containing mostly fatty acids. The residue was then distilled again at 200–220 °C and 0.1 mbar to obtain a distillate containing mostly tocopherols and a residue containing mostly FASEs. Tocopherols enriched in distillate were then subjected to an ion exchange chromatography and FASEs enriched in residue were then subjected to an acid-catalyzed transesterification with methanol to produce free phytosterols.

Even so, some authors (Martins *et al.*, 2006) have been trying to achieve an efficient FFA separation from SODD with the lowest loss of tocopherols by molecular distillation, without preliminary steps. This separation is difficult to achieve although is technologically viable at least at lab scale, due to the differences between molecular weights and vapour pressures of FFA (MW 180 g mol^{-1}, VP at 200 °C = 4 mm Hg) and tocopherols (MW 415 g mol^{-1}, VP at 200 °C = 0.15 mm Hg). Martins and coworkers (Martins, 2006) employed molecular distillation at 160 °C, 7.5×10^{-4} mm Hg, and 10.4 g/min feed flow rate to remove FFAs into the distillate fraction and obtain a residue fraction, which contained 6.4 wt% FFAs and 18.3 wt% tocopherols from a SODD feed which contained 57.8 wt% FFAs and 8.97 wt% tocopherols. They succeeded in removing 96.16% FFAs and recovering 81.23% tocopherols. Martins and coworkers (Martins *et al.*, 2005) reported the isolation of tocopherols by first converting acylglycerols in SODD into FFAs through saponification at 65 °C followed by acidulation, and then submitting the unsaponifiable product to five stages of molecular distillation. They succeeded in enriching tocopherols (34.14% purity) by 5.8 times. The major disadvantage in these cases is the residual free fatty acids in the tocopherol mixture.

Among the great variety of processes that have been patented for the purification of the compounds of the SODD, only the processes of esterification of fatty acids and acylglycerols with methanol or ethanol followed by high vacuum distillation, have been developed on a commercial scale for the concentration of tocopherols (Takagi & Kai, 1984), (Su-Min *et al.*, 1992), (Yong-Bo *et al.*, 1994), (Rohr & Trujillo-Quijano, 2002). These processes are the most time efficient and economical methods, however high purity of sterols or tocopherols cannot be achieved due to the similar boiling points of these two substances.

In the case of separation by distillation of unsaponifiable valuable products of SODD subjected to saponification, the difference between the boiling point of volatile products, such as unsaponifiable components, and the boiling point of the sodium and potassium organic acid soaps is so great that separation is theoretically possible at a high level of efficiency. However, a problem related to this separation technique is that the soaps have a very high melting point, close to the decomposition temperature of the sodium or potassium soaps (i.e. the sodium or potassium salts of fatty acids, rosin acids etc), and, when melted, these soaps produce an extremely viscous liquids. These two factors combined make industrial handling difficult. Furthermore, while at the high temperature necessary to maintain their flow, these soaps are in permanent decomposition, compromising the separation output and the quality of the final product, as many of the unsaponifiable valuable products are heat sensitive.

6. Supercritical fluid extraction (SFE)

Although the conventional methods, vacuum and molecular distillation, have been applied to commercial production of tocopherols from SODD, there are some drawbacks such as residual solvents, high temperature, large amounts of energy consumption, high production costs and the unreliable quality of the products that require further developments. Since thermal degradation of tocopherols is commonly caused by processing at high temperatures (de Lucas *et al.*, 2002), new alternative isolation techniques are desired.

Supercritical carbon dioxide extraction is a process where carbon dioxide passes through a mixture of interest at a certain temperature and pressure until it reaches an extractor. This process is used because supercritical carbon dioxide has a low viscosity, a high diffusivity and a low surface tension that provides selective extraction, fractionation and purification, allowing its penetration in micro- and macro-porous materials. Carbon dioxide is the most desirable supercritical fluid solvent for the separation of natural products used in foods and medicines because of its inertness, nontoxicity, low cost, and high volatility. The major advantage of this method is the easy post-reaction separation of the components by depressurization, so resultant extract does not contain solvent residue and hence natural-quality extracts can be obtained. Another advantage is the low temperatures used for the majority of the experimentations because carbon dioxide has a near-ambient critical temperature (31.1 °C), so is suitable for thermolabile natural products.

However, the use of high pressure conditions to concentrate tocopherols makes the system energetically expensive, but the industrial process can be economically viable using conditions of approximately 90 atm and 40 °C (Mendes *et al.*, 2002). At these specific conditions, only fatty acids are separated from tocopherol (Mendes *et al.*, 2005). An increase in pressure and temperature increases the oil extraction and tocopherol recovery, although different pressure–temperature systems need to be used in order to separate the different components (sterols, tocopherols, fatty acids and squalene). It is important to know that recycling the solvent does not endanger the viability of the process. The value of the rate of return on investment and time of return on investment for the process that does not recycle the carbon dioxide is higher than those of recycling the solvent. This is due to the compression cost that represents more than 59% of the total cost of the production (Mendes, 2002).

SODD as such will not be feasible to work with SC-CO_2 for the tocopherol enrichment, owing to its poor SC-CO_2 solubility. So, to concentrate tocopherols from SODD, pre-treatment of the raw material, including the esterification of free fatty acids and the removal of sterols with alcohol recrystallization, is needed to obtain the primary tocopherols concentrate with improve solubility in SC-CO_2 (Shishikura *et al.*, 1988). For that, triglycerides and FFAs which constitute a major component in SODD have to be chemically modified to obtain free fatty acids and then their methyl esters by esterification.

Several researchers have tried to concentrate tocopherols from SODD by supercritical CO_2 (Lee, 1991), (Brunner *et al.*, 1991), (Brunner, 1994b), (Zhao *et al.*, 2000), (Nagesha, G. K. *et al.*, 2003), (Fang *et al.*, 2007), but the operation parameters, especially pressure, differ from author to author. Moreover, in all the cases, tocopherols content of the extract depended on the composition and properties of the natural matrix.

The interest in the tocopherol concentration using supercritical fluid extraction started with Lee and coworkers (Lee, 1991) followed by Brunner and coworkers (Brunner, 1991) and Brunner (Brunner, 1994b). The operational conditions used varied from 35 to 90 °C and from 200 to 400 bar using extractors or countercurrent columns. Lee and coworkers (Lee, 1991) attempted to modify soybean sludge chemically, to improve the solubility in SC-CO_2. A simple batch process was utilized to recover tocopherols at 40% concentration at a pressure of 400 bar from the esterified soybean sludge which initially contained 13-14% tocopherols. The solubility of the esterified soybean sludge in supercritical carbon dioxide was more than 4 to 6 times higher than that of the sterols. Brunner and coworkers (Brunner, 1991) obtained a higher enrichment of tocopherols from SODD using supercritical carbon dioxide as a solvent compared to results obtained by Lee and coworkers This group recovered tocopherols from a model mixture of squalene, tocopherols, and sterols using two continuous countercurrent fractionation columns. Squalene was separated from the model mixture in the first column. Sterols were removed from the bottom of the second column, resulting in 85-95% concentration of tocopherol being obtained at the top of the second column.

These works concluded that the fatty acids are extracted initially and the tocopherols are enriched inside the extractor. The results also indicated that the solubility of the tocopherols is intermediary when compared to the solubilities of squalene and stigmasterol.

Chang and coworkers (Chang, 2000) worked on the separation of several SODD components. Their supercritical fluid extraction apparatus had a separation and an extraction unit. Free fatty acids, squalene and tocopherols were recovered in the extract and the sterols were recovered in the raffinates. The average tocopherol concentration factor was 1.38, which means that the mixture in the extract did not separate. However, the author mentioned that with the increase of CO_2 volume, the separation factor can reach 1.7, but the poor increase in the concentration factor does not justify the raise in gas volume. The following research groups focused on the separation of the problematic pairs by supercritical fluids using synthetic mixtures: (Mendes, 2005), (Mendes *et al.*, 2000), (Wang *et al.*, 2004), (Nagesha, 2004) and (Nagesha, 2003).

There are some interesting relationships and conclusions that can be deduced from the results obtained by Mendes (Mendes, 2005, Mendes, 2000) and Chang (Chang, 2000) regarding the yield and concentration factor of pairs of compounds at different conditions of pressure and temperature. The binary mixture of tocopherol and squalene cannot be separated at low pressure conditions. An acceptable separation needs a raise in pressure to almost 203 bar ((Mendes, 2005) and (Mendes, 2000)). However, a recovery of 90% and a purity of 60% of α-tocopherol has been achieved using a pressure swing adsorption (PSA) device, that is a widely used process in the separation of gas mixtures for air-drying, oxygen and nitrogen separation of air, hydrogen purification, and various other separations. The PSA process is based on the regeneration of adsorber by the difference in adsorbed amounts of gas solute as a function of pressure. In the case of a two-bed process, one bed is in the adsorption step, while the other is simultaneously in the desorption step. The adsorption and desorption steps had pressure conditions of 160 and300 bar, respectively (Wang, 2004).

The ternary mixture of tocopherol, fatty acids and squalene behaved differently from the tocopherol–fatty acid binary mixture. For the same conditions of pressure (160 and 300 bar), the binary mixture had a total separation while the ternary mixture did not achieve any separation. Squalene and stigmasterol mixtures are also very difficult to separate. At low pressure conditions, the yield is less than 10% but at higher pressure, the yield is 76%. It is important to note that low temperatures were used in these studies.

Results of these authors suggest that the supercritical-CO_2 process could be used for the separation of squalene, fatty acids and tocopherols. In order to enhance the squalene, fatty acids and tocopherols separation, deodorizer distillate mixtures should be processed several times in supercritical-CO_2 at different temperature and pressure conditions. However, there is no data reported regarding the total extraction time of the mixture and this makes it impossible to estimate the operational cost. The major advantage of this method is the total removal of free fatty acids from the mixture.

Zhao and coworkers (Zhao, 2000) concentrated tocopherols up to 75% at 120 bar using a fractionation column with a gradient of temperature from 30–80 °C in a pilot plant scale.

Nagesha and coworkers (Nagesha, 2003) performed chemical modification of SODD containing about 2.9 wt% of tocopherols, as well as triglycerides (56 wt%), free fatty acids (25.3 wt%), sterols (7.8 wt%), hydrocarbons (0.6 wt%), and unsaponifiables (6.4 wt%) apart from tocopherols. Chemical modification of SODD included saponification and esterification steps to result in fatty acid methyl esters from free fatty acids, so as to improve the solubility of SODD in SC-CO_2 extraction. Reactions were conducted in dark with continuous flushing of N2 and 1.0 wt% of pyrogallol was added to prevent the oxidation of tocopherols. After chemical modification, esterified SODD contained about 3.7 wt% of tocopherols. Tocopherols concentrates of about 36% was obtained by SC-CO_2 extraction at the pressure 180 bar and temperature 60 °C.

Fang and coworkers (Fang, 2007) carried out a pretreatment of methyl esterification and methanolysis reactions, which converted most of free fatty acids and glycerides to fatty acid methyl esters (FAMEs), respectively, to simplify the composition of SODD and improve his solubility in supercritical CO_2 extraction. The mixture was held at 3 °C in a refrigerator for 12 h, as a result most of sterols were crystallized and removed by filtering under a reduced pressure. Supercritical CO_2 fractionation was employed to concentrate tocopherols from Methyl Ester Oil Deodorizer Distillate (ME-DOD) product, mainly contained FAMEs (65–80 wt.%), tocopherols (10–15 wt.%), and impurities (such as residual sterols, glycerides, squalene, pigments, and long chain paraffins, comprising in total about 10–15 wt.%). The initial pressure, feed location, temperature gradient, and ratio of CO_2 to ME-DOD were optimized for separating FAMEs. For the following tocopherol concentration step, a final pressure of 200 bar resulted in the greatest average tocopherol content (>50%) and tocopherol recovery (about 80%).

The important step in concentrating natural tocopherols from these systems is to remove the FAMEs. FAMEs are important chemical materials in biofuel, metal-cutting oil, and cleaning agent production, as well as in the synthesis of other fatty acid products (Swern, 1986).

Some works on phase equilibrium for the realistic system of modified esterification SODD/supercritical CO_2 (Fang *et al.*, 2005) established that the separation factor [1] between tocopherols and FAMEs was always smaller than unity in the range investigated. This indicates that when supercritical CO_2 is used as the separation solvent, tocopherols, unlike FAMEs, tend to enrich in the liquid phase. In particular, the separation factors at pressures lower than 200 bar were relatively small. At 40 °C, for instance, the separation factor remained lower than 0.2 for all pressures lower than 150 bar. As pressure increased, the separation factor increased greatly, reaching 0.35 at 200 bar. The influence of temperature was contrary to that of pressure, with an increase in temperature leading to a decrease in separation factor. Low pressure and high temperature result in high selectivity, indicated by a low separation factor, which is advantageous in the separation of FAMEs from tocopherols with supercritical CO_2.

King and coworkers (King *et al.*, 1996) combined supercritical fluid extraction (SFE) with supercritical fluid chromatography (SFC) for concentrating tocopherols and the optimized conditions were 250 bar/80 °C for SFE and 250 bar/40 °C for SFC. Approximately 60% of the available tocopherols in soyflakes can be recovered in the SFE step, yielding enrichment factors of 1.83-4.33 for the four tocopherol species found in soybean oil. Additional enrichment of tocopherol species can be realized in the SFC stage, with enrichment factors [2] ranging from 30.8 for delta-tocopherol to 2.41 for beta-tocopherol.

Starting with a feed containing 48.3 wt% tocopherols, Gast and coworkers (Gast *et al.*, 2005) were able to obtain tocopherols with a purity of 94.4 wt% in bottom phase by supercritical CO_2 extraction at 230 bar and 80 °C, with a solvent-to-feed ratio of 110 and a reflux ratio of 4.6. Squalene was completely recovered in top phase.

Torres and coworkers (Torres, 2007) proposed a two-step enzymatic procedure to obtain FASEs, tocopherols, and fatty acid ethyl esters (FAEEs) from SODD, together with minor amounts of squalene, free fatty acids, free sterols and triacylglycerols. The final product obtained was used as starting material to purify FASEs, tocopherols, and FAEEs via supercritical CO_2 extraction The phytosterol esters were then purified from this mixture using supercritical carbon dioxide (Torres, 2009). Experimental extractions were carried out in an isothermal countercurrent column (without reflux), with pressures ranging from 200 to 280 bar, temperatures of 45-55 °C and solvent-to-feed ratios from 15 to 35 kg/kg. Using these extraction conditions, the fatty acid esters were completely extracted and, thus, the fractionation of tocopherols and phytosterol esters was studied. At 250 bar, 55 °C and a

---

[1] The separation factor represents the process selectivity for separating methyl oleate from tocopherol. In detail, a lower value indicates higher selectivity, whereas a higher value indicates that it is more difficult to separate the two compounds under certain conditions. Furthermore, when the separation factor equals unity, the composition in the gas phase is similar to that in the liquid, and the supercritical CO_2 process cannot separate methyl oleate from tocopherol.

[2] The enrichment factors were the ratio of individual tocopherols in extracts versus the same tocopherol content initially found in the soyflakes.

solvent-to-feed ratio of 35, the phytosterol esters were concentrated in the raffinate up to 82.4 wt-% with satisfactory yield (72%).

Other supercritical fluids have been explored but unsuccessfully for the separation of different pairs of components. An attempt at using similar methodology to (Mendes, 2000) and (Mendes, 2005) but using liquid gas petroleum instead of carbon dioxide did not change the poor concentration factor between the critical pairs of components (Buczenko *et al.*, 2003). Buczenko and coworkers (Buczenko, 2003) performed the saponification of the raw material and the extraction of unsaponifiable matter as pre-treatment of VODD.

As discussed above, there are a lot of experimental studies proving the efficiency of the supercritical extraction to concentrate the vitamin E from different raw material or in some cases, from synthetic mixtures representing the deodorizer distillate, but the extraction of sterols using supercritical fluid from the deodorizer distillate was not described in the literature.

On the other hand, due to the low content of squalene in SODD, specific extraction processes of squalene using supercritical fluid from SODD was not described in the literature. Existing studies are models of fractionation of artificial mixtures such as those mentioned above (Chang, 2000). For example, Brunner (Brunner, 1994a) studied the phase equilibrium for recovering α-tocopherol from a mixture of squalene, tocopherol, and campesterol. He concluded that the separation factor for squalene/α-tocopherol varied between a value of 4 at low squalene concentrations (0.5 wt %), to a value of 1 at high squalene concentrations (85 wt %), at pressures ranging from 200 to 300 bar and temperatures ranging from 70 to 100 °C.

Bondioli and coworkers (Bondioli *et al.*, 1993) esterified FFAs into their corresponding glycerides and then applied a supercritical carbon dioxide extraction to produce a squalene-enriched fraction (purity 90.0%, yield 91.1%), but from olive oil deodorizer distillates.

7. Other purification techniques

There has been also limited literature reported on the following alternative methods for purification of tocopherols and sterols from SODD. Hence, Nagesha and coworkers (Nagesha, G. *et al.*, 2003) using nonporous denser polymeric membranes to separate tocopherols from SODD by permeation. The separation in a denser membrane is generally based on a solution-diffusion mechanism. The lower polarity of tocopherols compared to that of free fatty acids appears to have facilitated the preferential permeation of tocopherols through the hydrophobic membrane. Selectivity of the membrane for tocopherols improved with esterified SODD, because the presence of FAME decreased the viscosity of the feed and thereby increased convective flow, which in turn improved permeate flux.

Alternatively, Maza and coworkers (Maza, 1992) concentrated tocopherols and sterols by addition of melted deodorizer distillates to a solution of urea and alcohol which separate fatty acids from the mixture.

In addition, Gunawan and coworkers (Gunawan *et al.*, 2008a) proposed a facile procedure to isolate naturally occurring FASEs from SODD without degradation of FASEs. SODD was first subjected to a modified soxhlet extraction (MSE) to enrich FASEs in a non polar lipid fraction (NPLF). Modified silica gel column chromatography (MSE) was then applied to NPLF to collect FASEs in a third fraction with a purity of 79.99 wt% and a recovery of 97.38%. The third fraction was then subjected to a binary solvent (water/acetone = 20/80, v/v) extraction to purify FASEs to a purity of 86.74 wt% with a total recovery of 85.32%.

These methods do not present a significant advance regarding the most frequently utilized methods and probably their application to production scale would be little profitable. However, they can be used to isolate tocopherols and sterols from SODD at laboratory scale or complement other methods at industrial scale.

8. Degradation and oxidation of functional lipids from SODD

8.1 Influence of refining on phytosterols

During refining of edible fats and oils, the content of total sterols decreases due to degradation and formation of products through isomerization (D5 to D7-sterol), dehydration, polymerization, and formation of hydrocarbons or sterenes and sterol oxidation products (Dutta, 2006). These qualitative and quantitative changes in sterols can be traced in the refined oil and in by-products such as soapstocks and distillate fractions collected after chemical and physical refining processes (Dowd, 1998).

Acid hydrolysis of steryl esters may occur upon bleaching with an acid activated bleaching earth. The slight reduction of the total sterol content is due to the formation of steradienes and disteryl ethers. A gradual reduction in the total sterol content is observed at increasing deodorization temperature due to distillation and steradiene formation. Increasing the temperature from 220 °C to 260 °C resulted in a gradual reduction of the total sterol recovery from 90.4% to 67.7% in physical refining and from 93% to 62.7% in chemical refining. However in physical refining, an increase of 40% in the steryl ester fraction is observed due to an esterification reaction, promoted by high temperature between a sterol and a fatty acid. Due to the absence of free fatty acids in the chemical refining their esterification did not occur (Verleyen *et al.*, 2001b). The influence of refining on free and esterified sterols has been studied by (Verleyen, 2002b).

Phytosterols are progressively lost during refining while continuously altering the ratio of free and esterified sterols (Kochhar, 1983). During chemical neutralization, the free sterol content is significantly reduced especially upon addition of weak caustic solution due to the loss in the soapstock (Gutfinger & Letan, 1974).

Bleaching effects on phytosterols are generally minor and mainly limited to the formation of some nonpolar dehydration products (Ferrari, 1996) and partial hydrolysis of sterol esters (Homberg & Bielefeld, 1982). Steradienes and disteryl ether dehydration products (Figure 1) are formed during bleaching step by the bleaching temperature and the degree of acid activation of the bleaching earth, while during the deodorization, the degree of sterol dehydration is mainly influenced by deodorization temperature giving rise to a concentration of the steradienes in the distillate (Verleyen, 2002b, Verleyen, 2001c). The presence of steradienes can also be used as a marker for the presence of refined oils (Grob *et al.*, 1994).

Whenever applied, hydrogenation has a tremendous effect on sterol structures, including hydrogenation of double bonds, opening of cyclopropane rings, and positional isomerization of side chain unsaturation (Strocchi & Marascio, 1993).

A part of a multinational EU research project (FOOD-CT2004-007020) was to carry out qualitative and quantitative assessment of sterols and sterol oxidation products in samples of by-products from chemical and physical refining of edible fats and oils collected from various locations in Europe. To the best of our knowledge, this is the first report on the contents of oxidized sterols in soapstock and distillate fractions from edible oil refining processes. The levels of sterol oxidation products were higher in acid oil obtained from chemical refining (AOCHE) samples than in acid oil obtained from physical refining (AOPHY) samples, with ranges 0.02–17.0 and 0.01–1.5 mg/100 g, respectively. The lower content of sterol oxidation products in AOPHY samples may be due to the high temperature applied during vacuum distillation accelerating the breakdown and transformation of the sterol oxidation products into other unidentified degradation products. Further formation of sterol oxidation products has been prevented by the high amounts of natural antioxidants in AOPHY distillate (Verleyen, 2001c). Some sterols appeared to be more liable to breakdown than others, e.g. there was a higher content of oxybrassicasterols than the other sterol oxidation products in this study, although the content of brasicasterol in the sample was lower than other sterols. Similar results have been reported previously (Dutta, 2006). This may be due to the structural arrangement in the brassicasterol molecule rendering it more easily oxidized than other sterols. However, systematic studies are required to clarify this phenomenon. Although stigmasterol has a double bond in the side-chain, similar to brassicasterol, the quantities of phytosterol oxidation products or oxyphytosterols observed in this study were quite different. Stigmasterol has an ethyl group at position C24 while brasicasterol has a methyl group, and this difference may affect in the relative rate of formation of their oxidation products (Dutta, 2006). Further studies are needed on this point.

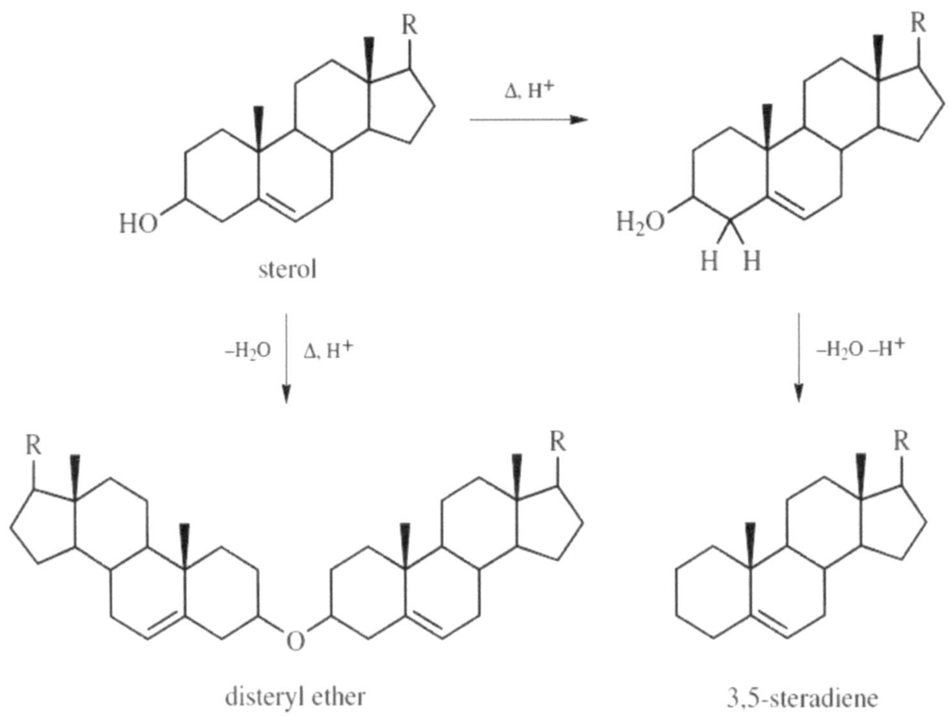

Fig. 1. Reaction products of sterols during refining.

It has been reported that the formation of sterol oxidation products is affected not only by the chemical nature of the sterols but also by their quantity (Dutta, 2006). There were positive correlations between total sterols and total phytosterol oxidation products in the by-products collected from both refining processes.

The biological effects of oxycholesterol have been extensively studied (Bjorkhem *et al.*, 2002); however, the amount of biological research on oxyphytosterols is rather scarce, mostly dated and has never been extensively reviewed before (Francesc, 2004) and (Dieter, 2004).
Most reports available so far on oxyphytosterols cover the methodological aspects of their measurement in foods. The usual perception about oxyphytosterols is that these components present a concern in terms of food quality and health. This perception originates from the parallel that is made between oxycholesterol and oxyphytosterols. Whether oxyphytosterols may indeed play similar and/or different biological roles compared to oxycholesterol has not been elucidated yet.
A review (Hovenkamp *et al.*, 2008) summarise the current knowledge on the possible biological effects of oxyphytosterols and to identify future research needs, which will help in clarifying the possible impact of oxyphytosterols on human health. The review focuses on the more common oxyphytosterols which differ only in a few structural changes from the parent sterol.
Over the last thirty years a diversity of potential biological effects, including modulation of cholesterol homeostasis, anti-inflammatory and anti-tumour activities, as well as lipid-lowering and anti-diabetic properties, have been attributed to specific oxyphytosterols. Although these studies were not all carried out with oxyphytosterols also identified in the human body, these results suggest that oxyphytosterols may have systemic effects in vivo and therefore, the potential to modulate human metabolism.
Despite some putative desirable effects, oxyphytosterols may be perceived as presenting a concern in terms of food quality and health. Indeed, oxyphytosterols have been reported to exert, in vitro, cytotoxic effects comparable to those attributed to oxycholesterol. However, high, non-physiological concentrations of oxyphytosterols were needed to exert adverse effects. In addition, data from one animal study do not support a role of oxyphytosterols in atherosclerosis promotion. However, this aspect deserves more attention in future research. Altogether, the currently available observations do not suggest that oxyphytosterols, in relatively low concentrations such as those reported in human plasma, may exert in vivo deleterious effects similar to those attributed to oxycholesterol. In addition, although probably different in structure than the potentially deleterious ones, some oxyphytosterols may also have the ability to activate transcription factors involved in cholesterol metabolism. Nevertheless, more detailed investigations are needed to evaluate the biological impact of long-term exposure to physiologically relevant concentrations of oxyphytosterols in humans.

8.2 Influence of refining on tocopherols

During deodorization, all tocopherols present in the bleached oil will be partitioned either in the deodorized oil or in the deodorizer distillate. A significant loss in the tocopherol mass balance in the range of 25%-35% was observed originating from technological and/or chemical origin.
The loss of tocopherols can be caused either by a thermal breakdown at temperatures higher than 240 °C, by oxidation reaction or by chemical reaction such as the formation of tocopheryl esters (Verleyen *et al.*, 2001a). Extensive analysis of vegetable oils by HPLC and comparison with synthesized tocopheryl esters did not show any adsorption in the elution region of tocopheryl esters, indicating that esters of tocopherols with fatty acids are not present in crude oils (Verleyen, 2001c). Therefore the stability of tocopherols during

deodorization has been studied under various process conditions. The presence of oxidation products has no influence on the loss of tocopherols during deodorization based on the fact that two successive deodorization steps yielded identical loss of tocopherols.
Experiments using spiked triolein with 2000 ppm of α-tocopherol showed that the addition of tertbutylhydroquinone (TBHQ) as a strong antioxidant reduces the loss of tocopherols with more than 50% in comparison with the reference procedure. α- Tocopherol (2000 ppm) was dissolved in triolein and heated to 254 °C, 5-6 mbar, for 80 min, with no steam injection. 9% of tocopherol loss was observed in the control sample and 3% for the sample with 1500 ppm TBHQ. The more active TBHQ will compete with tocopherols to scavenge radicals and consequently the tocopherol loss in the mass balance is reduced as more natural tocopherols stay in the oil or in the distillate (Verleyen *et al.*, 2002a, Verleyen *et al.*, 2003).
In vegetable oils, the addition of TBHQ from 0 to 1500 ppm establishes a gradual reduction in tocopherol loss from 26.7% to 17.6% while the concentration of tocopherols in the distillate rises from 1.85% to 2.35%. Performing deodorization with nitrogen as stripping agent showed an important reduction in the tocopherol loss (Verleyen, 2002a). In the model study with triolein no reduction of α-tocopherol was observed while using corn oil a reduction of 30%-50% was observed. The highest reduction was detected at severe deodorization conditions (260 °C, 3 mbar) (Verleyen, 2002a). These experiments show that tocopherols are thermally stable compounds and probably the loss of tocopherols is due to oxidation reactions, which leads to compounds such as α- tocopherol dimer quinone, 4α, 5-epoxytocopherolquinone, 7, 8-epoxy tocopherol quinone, tocopherol dimer quinone, tocopherol spirotrimer and ditocopherol ethers (Verleyen, 2001a). These compounds can be found in the finished oil and in the distillate.
In a model experiment using 3500 ppm α- tocopherol in triolein and heating at 240 °C for 90 min at a reduced pressure of 6-7 mbar 4α, 5-epoxytocopherolquinone, 7, 8-epoxy tocopherolquinone and α-tocopherol quinone were identified as oxidation products supporting that the tocopherol loss during deodorization is mainly due to oxidative degradation (Verleyen, 2002a).

9. Nonfood applications

As commented previously, soybean deodorizer distillate represent good source of valuable compounds such as phytosterols, tocopherols and squalene, which can be recovered and further used as food additives, in pharmaceutical industry and cosmetics. Alternatively, deodorizer distillates have nonfood applications, such as biodiesel or can be used mixed with the fuel oil to fire the steam boilers (Svensson, 1976).
Refined vegetable oils are the predominant feedstocks used for the production of biodiesel. However, their relatively high cost renders the resulting fuels unable to compete with petroleum-derived fuel and makes the use of side-stream refining products (soapstock, acid oil and deodorizer distillate) important alternatives as a feedstock for biodiesel production.
Biodiesel is produced from deodorizer distillates by direct esterification of the FFA or by conversion of FFA to acylglycerols prior transesterification (Figure 2). Esterification of the FFA is also performed as a preliminary step in the purification of the tocopherols and sterols in order to reduce their boiling points, thereby facilitating their separation.

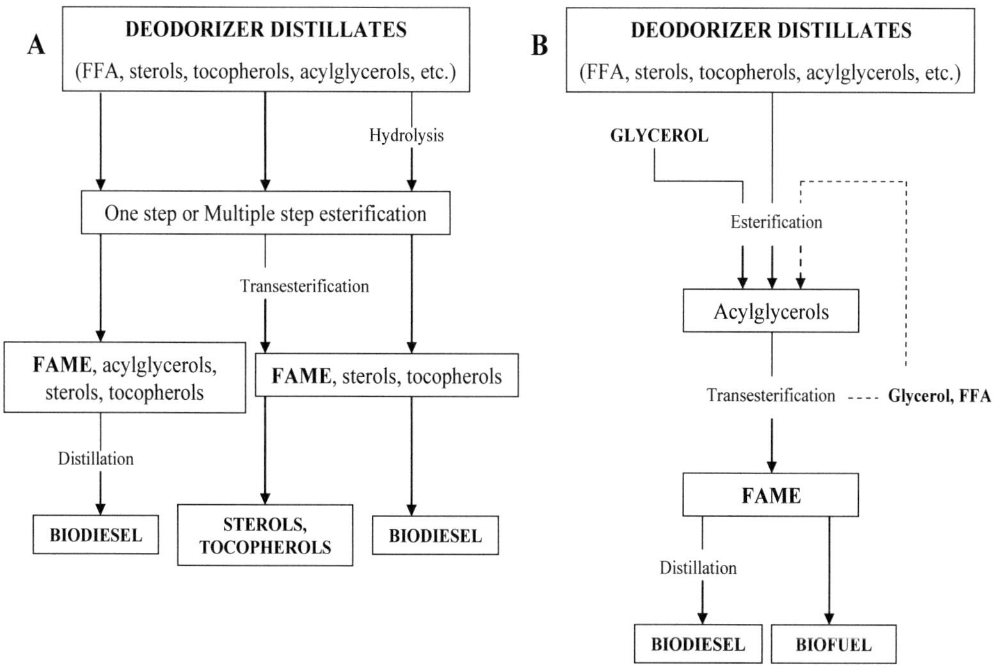

Fig. 2. Production of biodiesel from deodorizer distillates by direct conversion (A) and *via* acylglycerols route (B).

9.1 Production of biodiesel by direct conversion

9.1.1 Chemically catalyzed process

Verhé and coworkers (Verhé *et al.*, 2008) reported a process of converting the deodorizer distillates to biodiesel using methanol in a weight ratio 1:1 and 5 % w/w sulphuric acid as catalyst, at 75 °C for 5 h. Under the mentioned conditions, the FFA have undergone esterification while MAG reacted *via* transesterification, resulting in methyl esters. The crude biodiesel was further washed with 20 % water for 15 min, dried and distilled in order to increase the quality of the methyl esters. The distillation pitch was further processed for the recovery of sterols and tocopherols.

Facioli and Arellano (Facioli & Barrera-Arellano, 2002) described a process to obtain ethyl esters from SODD. SODD contained 47.5 % FFA C18:1, 26.2 % acylglycerols and 26.2 % unsaponifiable matter using concentrated sulphuric acid as catalyst. The optimum conditions found in this study were for EtOH:FFA between 6.4:1 to 11.2:1, H_2SO_4 from 0.9-1.5 % and reaction time from 1.3 h to 2.6 h. Under the described conditions a conversion of 94 % of the fatty acids to ethyl esters was achieved. Tocopherols losses were below 5.5 %. A molar excess of ethanol in relation to SODD:FFA was found to be necessary to obtain the best conversion.

Hammond and coworkers (Hammond & Tong, 2005) described a three-stage acid catalyzed esterification using a molar ratio acid oil:methanol:sulphuric acid of 1:1.3:0.03 for the first stage (25 h). The reaction mixture was centrifuged, the supernatant lipid phase was separated from the sludge (glycerol, water, acid and methanol), and further reacted with methanol and acid, keeping the previous mentioned ratios of unreacted lipid:methanol:sulphuric acid.

It was seen that the reaction proceeded rapidly during the first hour of reaction and then slowed down considerably. In contrast, the second and third stage showed a gradual increase in FAME over time. The maximum FAME conversion obtained for 12 tested acid oils averaged 81%. However, the ester phase could not be increased above 85% even after a fourth-stage reaction or if a base catalyst (sodium methoxide) was used in large excess. If higher amount of methanol was used, the initial reaction tended to go faster, but the reaction reached the plateau in a short time. Furthermore, an increase in the acid catalyst concentration above 1.2 % did not affect the initial reaction rate.

9.1.2 Enzymatically catalyzed process

Several enzymatic methods have been developed for the conversion of fatty acids into FAMEs or FAEEs with positive results. One of the main disadvantages of use of biocatalysts is the high price compared to chemical catalyst, although unfortunately, no rigorous economical viability of these enzymatic procedures has been reported.

Facioli and Arellano (Facioli & Barrera-Arellano, 2001) investigated the enzymatic esterification of the free fatty acids from SODD with ethanol using immobilized fungal lipase (Lipozyme IM) as biocatalyst. SODD contained 47.5 % FFA, 26.22 % neutral oil and 26.23 % unsaponifiable matter. The effect of three independent variables: temperature, enzyme concentration and EtOH:FFA molar ratio on the conversion rate of FFA to ethyl esters was studied. The best conversion (above 88 %) was obtained with EtOH:FFA ratio 1.7-3.2:1, temperature in the range 46.4 °C to 53.6 °C, lipase concentration from 10.7 to 23.0% and the reaction time of 2 h. All three variables had statistically significant effect on the conversion of the FFAs to ethyl esters. During the above mentioned esterification process no tocopherols losses were observed.

The esterification of SODD with butanol using *Mucor miehei* lipase as a biocatalyst and supercritical carbon dioxide (SC-CO_2) has been described by Nagesha and coworkers (Nagesha, 2004). The SODD contained 56.0 % neutral oil, 25.3 % FFA, 7.2 % sterols, 2.9 % tocopherols, 0.6 % hydrocarbons and 0.13 % moisture. It was preliminary filtered in order to remove sediments and sterols and enzymatic hydrolyzed to free fatty acid using immobilized lipase (*Candida rugosa*) in SC-CO_2 reactor unit. The operational conditions were as follows: pressure 160 bar, temperature 45 °C, moisture content 60 % (w/w) and enzyme concentration 200 U/g of SODD. Hydrolyzed SODD containing 87.8 % (w/w) FFA was further esterified for 3h in presence of butanol (1.2 M) using 15 % enzyme (w/w) (*M. miehei*), pressure 120 bar and temperature 35 °C. The maximum yield of 95 % FABE was achieved.

The high content of residual glycerides (3.10 %) present in the final FABE precluded its direct use as biodiesel. However, the process was designed as preliminary step for the purification of tocopherols, since hydrolysis/esterification helps their recovery.

Wang and coworkers (Wang *et al.*, 2006) described a process for simultaneously conversion of FFA (28 %) and acylglycerols (60 %) from SODD to alkyl esters using a mixture of two enzymes (3 % Lipozyme TL IM and 2 % Novozym 435) in the presence of *tert*-butanol as co-solvent. It was found that the negative effects on the enzyme stability caused by the excessive methanol ratio and by-product glycerol could be minimized by using *tert*-butanol. The lipase activity remained stable after 120 cycles. The maximum yield of FAME (84 %) was achieved with an increase of *tert*-butanol content up to 80 % (based on the oil weight). However, a further increase of the solvent resulted in a decrease of the FAME yield which was explained by the dilution effect on reactants.

Du and coworkers (Du *et al.*, 2007) investigated the enzymatic esterification of SODD containing 28 % FFA, 60 % TAG and 6 % tocopherols. The reaction was lipase mediated methanolysis using Novozym 435 as catalyst, at 40 °C in a solvent free medium. The enzyme kept its activity after being reused for 10 cycles, each cycle of 24 h. The highest biodiesel yield of 95 % was achieved by adding 10 fold of 3 Å molecular sieves (based on the maximum water produced from FFA esterification). The investigation of the lipase to methanol tolerance revealed that the lipase could maintain its stability and activity in the presence of even 3 molar concentration of methanol. This tolerance was attributed to the presence of other compounds apart from triglycerides, namely FFA, sterols and tocopherols. A linear relationship between the FFA content and the lipase tolerance to methanol was observed but the presence of sterols and tocopherols showed no effect. The correlation between the initial FFA present in the feedstock and the rate of conversion was confirmed by other authors (Hammond & Tong, 2005).

9.2 Production of biodiesel via acylglycerols route

Another approach reported in the literature consists on esterification of FFA with glycerol to form acylglycerols, followed by conventional transesterification. Synthesis of MAG from deodorizer distillate was mainly studied due to the large number of applications as additives (e.g. enhancing plasticity of fats) in the food, medicine and cosmetic industry. Among synthesized acylglycerols, the monoester has the highest surface activity and therefore, its concentration is very important for direct utilization of the reaction mixture as emulsifier.

Although the use of a large number of different heterogeneous catalysts have been reported in literature, most of the research has been done on the synthetic samples and less on the side stream refining products. Different studies summarized hereunder describe processes for synthesis of acylglycerols as an intermediate step in the production of biodiesel/biofuels. These processes are catalyzed either chemically or enzymatically, or conducted under non-catalytic conditions.

9.2.1 Enzymatically catalyzed process

Tangkam and coworkers (Tangkam *et al.*, 2008) described the enzymatic esterification in a solvent free medium of different deodorizer distillates resulting from the refining of various vegetable oils. A direct esterification of mixed distillates (61 % FFA and 39 % acylglycerols) with glycerol using immobilized lipase B from *Candida Antarctica* (Novozym 435) led to moderate proportions (46 %) of DAG. Application of a two-stage reaction consisting of a hydrolysis step of deodorizer distillate to increase the FFA content followed by esterification with glycerol led to a higher formation (>61 %) of DAG. Furthermore, it was observed that the high initial concentration of free fatty acids in the distillate has a positive influence on the concentration of DAG in the final product (>71 %). This observation is consistent with other literature data (Yamada *et al.*, 1999). Enrichment of DAG in the final products by short-path vacuum distillation led to concentrates containing up to 94 % DAG, ~ 5 % TAG and no unesterified fatty acids and MAG.

9.2.2 Non-catalytic process

Smet (Smet, 2008) described a process for the esterification of fatty acid distillate (93 % FFA) with technical grade glycerol. The reaction was carried out in a high pressure Parr reactor (stirred and thermostated reactor of stainless steal). The following parameters have been checked: temperature, reactor design, agitation speed, molar ratio and influence of the catalyst. The best results were obtained at 200 °C, pressure of 90 mbar and agitation speed of 60 rpm. It was seen that by using a molar ratio 1:1 FFA:glycerol, a total glycerides content of 85.3 % was obtained within 345 min reaction time. The formation of MAG was faster in the first hours and than reached the plateau, while the formation of DAG was slower at the beginning of the reaction and faster at the end. Furthermore, an increase in the molar ratio of 1:2 FFA:glycerol slow down the reaction, the total glycerides content reaching 64.9 % within 345 min reaction time. A molar ratio 2:1 FFA:glycerol gave an increase of the MAG and DAG at the beginning of the reaction, followed by an decrease of MAG after 90 min, the glycerol being completely consumed in within 345 min reaction time. The percent of DAG and TAG increased gradually during reaction, reaching a final yield of 86.2 % of total acylglycerols.

However the FFA content was still high, a distillation step of the residual FFAs and glycerol was necessary in order to increase the purity of the synthesized acylglycerols. The by-products of distillation were further re-used as reaction products in the synthesis of acylglycerols. The novelty of the process consists in synthesizing acylglycerols in a relatively short time (<6 h) in a catalyst free medium.

10. Concluding remarks

Deodorizer distillate is an excellent source of valuable compounds such as phytosterols and tocopherols. Numerous procedures have been described to isolate bioactive compounds from soybean oil deodorizer distillates to improve the value and the quality of this by-product. All these procedures can be grouped in four generic categories: classic methods such as crystallization and precipitation, chemical and enzymatic modification, molecular distillation, and supercritical fluid extraction.

Crystallization seems successful as a simple and efficient process to remove and concentrate sterols and tocopherols from SODD. However solvent based processes are expensive, unattractive and less environmentally friendly, resulting in a scarce and expensive final product.

To increase the separation efficiency of the compounds of interest from SODD, esterification and/or transesterification reactions are usually carried out prior to the purification or fractionation procedure. Hence, the utilization of enzymes, for instance, makes easier the separation of tocopherols from SODD by converting sterols to steryl esters, acylglycerols to free fatty acids and free fatty acids to fatty acid methyl or ethyl esters (FAMEs or FAEEs). Then, it is easier to separate the new product mixture by distillation or supercritical fluid extraction. The main difficulties of the enzymatic processes are the numerous parameters involved such as moisture content, enzyme concentration, time, temperature, ratio of the reactants, stability, recovery and reutilization of the enzyme preparation, among others. However, it is possible to separate the sterol esterification and ethyl esterification in time or space to optimize each of these reactions independently, thereby minimizing costs or improving the yield of the desired final reaction products.

Among the great variety of processes that have been patented for the purification of the compounds of the SODD, only the processes of esterification of fatty acids and acylglycerols with methanol or ethanol followed by high vacuum distillation, have been developed on a commercial scale for the concentration of tocopherols

Regarding supercritical fluid fractionation SODD is not adequate feed material to work with SC-CO_2 for tocopherol enrichment, owing to its poor SC-CO_2 solubility. So, to concentrate tocopherols from SODD, pre-treatment of the raw material is needed to obtain the primary tocopherols concentrate with improve solubility in SC-CO_2.

Alternatively, deodorizer distillates have also non-food applications, such as biodiesel or can be used mixed with the fuel oil to fire the steam boilers. The use of deodorizer distillate instead of refined vegetable oils is an important alternative as a feedstock for biodiesel production.

Two main degradation products from sterols can be observed in SODD, namely dehydration and oxidation products. The degree of sterol dehydration is mainly influenced by deodorization temperature giving rise to a variable concentration of steradienes in the distillate. The content of oxidized sterols in deodorization distillate fractions from edible oil refining processes fluctuates depending on both the temperature applied during vacuum distillation and the breakdown and transformation of oxidized sterols into other unidentified degradation products. Finally, formation of oxidized sterols can be partially prevented by the high amounts of natural antioxidants in acid oil obtained from physical refining distillate. The current knowledge on the possible biological effects of oxyphytosterols is limited and further research to clarify the possible impact of oxyphytosterols on human health is needed.

11. References

Andrews, J. S. (1941). Process for preparation of vitamin e concentrate and antioxidant. U. S. Patent 2.263.550.

Awad, A. B., Roy, R., & Fink, C. S. (2003). Beta-sitosterol, a plant sterol, induces apoptosis and activates key caspases in MDA-MB-231 human breast cancer cells. *Oncol Rep.* Vol. 10, No. 2, pp. 497-500.

Balazs, I. (1987). Refining and use of byproducts from various fats and oils. *Journal of the American Oil Chemists' Society*. Vol. 64, No. 8, pp. 1126-1128.

Batistella, C. B., Moraes, E. B., Maciel Filho, R., & Maciel, M. R. (2002). Molecular distillation: rigorous modeling and simulation for recovering vitamin E from vegetal oils. *Appl. Biochem. Biotechnol.* Vol. 98-100, No. pp. 1187-206.

Batistella, C. B. & Wolf Maciel, M. R. (1998). Recovery of carotenoids from palm oil by molecular distillation. *Computers & Chemical Engineering*. Vol. 22, No. Supplement 1, pp. S53-S60.

Bjorkhem, I., Meaney, S., & Diczfalusy, U. (2002). Oxysterols in human circulation: which role do they have? *Current Opinion in Lipidology*. Vol. 13, No. 3, pp. 247-253.

Block, G. & Langseth, L. (1994). Antioxidant vitamins and disease prevention. *Food Technology*. Vol. 48, No. 7, pp. 80-84.

Bockisch, M. (1998). *Fats and oils handbook*. (AOCS, Champaign, IL.

Bondioli, P., Mariani, C., Lanzani, A., Fedeli, E., & Muller, A. (1993). Squalene recovery from olive oil deodorizer distillates. *Journal of the American Oil Chemists' Society*. Vol. 70, No. 8, pp. 763-766.

Bouic, P. J. D. (2001). The role of phytosterols and phytosterolins in immune modulation: a review of the past 10 years. *Current Opinion in Clinical Nutrition & Metabolic Care*. Vol. 4, No. 6, pp. 471-475.

Breivik, H., Haraldsson, G., & Kristinsson, B. (1997). Preparation of highly purified concentrates of eicosapentaenoic acid and docosahexaenoic acid. *Journal of the American Oil Chemists' Society*. Vol. 74, No. 11, pp. 1425-1429.

Brown, W. & Meag, K. H. (1963). Process for recovery of tocopherols and sterols. U.S. Patent 3,108,120.

Brown, W. & Smith, F. E. (1964). Process for separating tocopherols and sterols from deodorizer sludge and the like. U.S. Patent 3,153,055.

Brunner, G. (1994a). *Gas Extraction*. Springer: New York.

Brunner, G. (1994b). *Gas Extraction-An Introduction to Fundamentals of Supercritical Fluid and the Application to Separation Processes*. Springer, Berlin

Brunner, G., Malchow, T., Stürken, K., & Gottschau, T. (1991). Separation of tocopherols from deodorizer condensates by countercurrent extraction with carbon dioxide. *The Journal of Supercritical Fluids*. Vol. 4, No. 1, pp. 72-80.

Buczenko, G. M., de Oliveira, J. S., & von Meien, O. F. (2003). Extraction of tocopherols from the deodorized distillate of soybean oil with liquefied petroleum gas. *European Journal of Lipid Science and Technology*. Vol. 105, No. 11, pp. 668-671.

Cvengros, J., Lutisan, J., & Micov, M. (2000). Feed temperature influence on the efficiency of a molecular evaporator. *Chemical Engineering Journal*. Vol. 78, No. 1, pp. 61-67.

Czuppon, T., Kemeny, Z., Kovari, E., & Recseg, K. (2003). Process for recovery of plant sterols from by-product of vegetable oil refining. WO2004000979.

Chang, C. J., Chang, Y.-F., Lee, H.-z., Lin, J.-q., & Yang, P.-W. (2000). Supercritical Carbon Dioxide Extraction of High-Value Substances from Soybean Oil Deodorizer Distillate. *Industrial & Engineering Chemistry Research*. Vol. 39, No. 12, pp. 4521-4525.

Chu, B. S., Baharin, B. S., & Quek, S. Y. (2002). Factors affecting pre-concentration of tocopherols and tocotrienols from palm fatty acid distillate by lipase-catalysed hydrolysis. *Food Chemistry*. Vol. 79, No. 1, pp. 55-59.

de Jong, A., Plat, J., & Mensink, R. P. (2003). Metabolic effects of plant sterols and stanols (Review). *The Journal of Nutritional Biochemistry*. Vol. 14, No. 7, pp. 362-369.

de Lucas, A., Martinez de la Ossa, E., Rincón, J., Blanco, M. A., & Gracia, I. (2002). Supercritical fluid extraction of tocopherol concentrates from olive tree leaves. *The Journal of Supercritical Fluids*. Vol. 22, No. 3, pp. 221-228.

Dieter, L. (2004). Sterol autoxidation: from phytosterols to oxyphytosterols. *British Journal of Nutrition*. Vol. 91, No. pp. 3-4.

Donova, M. (2007). Transformation of steroids by actinobacteria: A review. *Applied Biochemistry and Microbiology*. Vol. 43, No. 1, pp. 1-14.

Dowd, M. K. (1998). Gas chromatographic characterization of soapstocks from vegetable oil refining. *Journal of Chromatography A*. Vol. 816, No. 2, pp. 185-193.

Du, W., Wang, L., & Liu, D. (2007). Improved methanol tolerance during Novozym435-mediated methanolysis of SODD for biodiesel production. *Green Chemistry*. Vol. 9, No. 2, pp. 173-176.

Dumont, M.-J. & Narine, S. S. (2007). Soapstock and deodorizer distillates from North American vegetable oils: Review on their characterization, extraction and utilization. *Food Research International*. Vol. 40, No. 8, pp. 957-974.

Dutta, P. C., Przybylski, R., & Eskin, M. (2006). Formation, analysis and health effects of oxidized sterols in frying fat, In: *Deep frying: chemistry, nutrition and practical applications*, M. D. Erickson, Editor, pp. 111-164, AOCS Press, Urbana, IL.

Facioli, N. & Barrera-Arellano, D. (2002). Optimization of direct acid esterification process of soybean oil deodorizer distillate. *Grasas y Aceites*. Vol. 53, No. 2, pp. 206-212.

Facioli, N. L. & Barrera-Arellano, D. (2001). Optimisation of enzymatic esterification of soybean oil deodoriser distillate. *Journal of the Science of Food and Agriculture*. Vol. 81, No. 12, pp. 1193-1198.

Fang, T., Goto, M., Sasaki, M., & Hirose, T. (2005). Phase Equilibria for the Ternary System Methyl Oleate + Tocopherol + Supercritical CO_2. *Journal of Chemical & Engineering Data*. Vol. 50, No. 2, pp. 390-397.

Fang, T., Goto, M., Wang, X., Ding, X., Geng, J., Sasaki, M., & Hirose, T. (2007). Separation of natural tocopherols from soybean oil byproduct with supercritical carbon dioxide. *The Journal of Supercritical Fluids*. Vol. 40, No. 1, pp. 50-58.

Fernandes, P. & Cabral, J. M. S. (2007). Phytosterols: Applications and recovery methods. *Bioresource Technology*. Vol. 98, No. 12, pp. 2335-2350.

Ferrari, R., Schulte, E., Esteves, W., Brühl, L., & Mukherjee, K. (1996). Minor constituents of vegetable oils during industrial processing. *Journal of the American Oil Chemists' Society*. Vol. 73, No. 5, pp. 587-592.

Fizet, C. (1996). Process for tocopherols and sterols from natural sources. U.S. Patent 5,487,817.

Francesc, G. (2004). Phytosterol oxidation products: state of the art. *Reprod. Nutr. Dev.* Vol. 44, No. 6, pp. 597-598.

Gangopadhyay, S., Nandi, S., & Ghosh, S. (2007). Biooxidation of Fatty Acid Distillates to Dibasic Acids by a Mutant of Candida tropicalis. *Journal of Oleo Science*. Vol. 56, No. 1, pp. 13-17.

Gapor, A., Leong, L. W., & Ong, A. (1989). Production ofhigh concentration tocopherols and tocotrienols from palm oil by-products. European Patent Application 0,3 33,472442.

Gast, K., Jungfer, M., Saure, C., & Brunner, G. (2005). Purification of tocochromanols from edible oil. *The Journal of Supercritical Fluids*. Vol. 34, No. 1, pp. 17-25.

Gavin, A. (1978). Edible oil deodorization. *Journal of the American Oil Chemists' Society*. Vol. 55, No. 11, pp. 783-791.

Ghosh, S. & Bhattacharyya, D. (1996). Isolation of tocopherol and sterol concentrate from sunflower oil deodorizer distillate. *Journal of the American Oil Chemists' Society*. Vol. 73, No. 10, pp. 1271-1274.

Grob, K., Biedermann, M., Bronz, M., & Giuffré, A. M. (1994). The Detection of Adulteration with Desterolized Oils. *Lipid / Fett*. Vol. 96, No. 9, pp. 341-345.

Gunawan, S., Ismadji, S., & Ju, Y.-H. (2008a). Design and operation of a modified silica gel column chromatography. *Journal of the Chinese Institute of Chemical Engineers*. Vol. 39, No. 6, pp. 625-633.

Gunawan, S., Kasim, N. S., & Ju, Y.-H. (2008b). Separation and purification of squalene from soybean oil deodorizer distillate. *Separation and Purification Technology*. Vol. 60, No. 2, pp. 128-135.

Gutfinger, T. & Letan, A. (1974). Quantitative changes in some unsaponifiable components of soya bean oil due to refining. *Journal of the Science of Food and Agriculture*. Vol. 25, No. 9, pp. 1143-1147.

Hammond, E. G. & Tong, W. (2005). Method of converting free fatty acids to fatty acid methyl esters with small excess of methanol. U.S. Patent 6965044.

Haraldsson, G. (1992). *The application of lipases in organic synthesis*. (John Wiley,

Hedström, G., Slotte, J. P., Backlund, M., Molander, O., & Rosenholm, J. B. (1992). Lipase-Catalyzed Synthesis and Hydrolysis of Cholesterol Oleate in Aot/Isooctane Microemulsions. *Biocatalysis and Biotransformation*. Vol. 6, No. 4, pp. 281 - 290.

Hickman, K. C. D. (1944). Purification of sludges, scums, and the like to prepare relatively purified tocopherol. U. S. Patent 2.349.270

Hirota, Y., Nagao, T., Watanabe, Y., Suenaga, M., Nakai, S., Kitano, A., Sugihara, A., & Shimada, Y. (2003). Purification of steryl esters from soybean oil deodorizer distillate. *Journal of the American Oil Chemists' Society*. Vol. 80, No. 4, pp. 341-346.

Homberg, E. & Bielefeld, B. (1982). Free and bound sterols in crude and refined palm oils. . *Fette Seifen Anstrichm*. Vol. 84, No. pp. 141-146.

Hovenkamp, E., Demonty, I., Plat, J., Lütjohann, D., Mensink, R. P., & Trautwein, E. A. (2008). Biological effects of oxidized phytosterols: A review of the current knowledge. *Progress in Lipid Research*. Vol. 47, No. 1, pp. 37-49.

Hunt, T. K., Jeromin, L., Johannisbauer, W., Gutsche, B., Jordon, V., & Wogatzki, H. (1997). Recovery of tocopherols. U.S. Patent 5,646,3 1 1.

Ito, V., Martins, P., Batistella, C., Filho, R., & Wolf Maciel, M. (2006). Natural compounds obtained through centrifugal molecular distillation. *Applied Biochemistry and Biotechnology*. Vol. 131, No. 1, pp. 716-726.

Jacobs, L. (2005). Process for the Production of Tocotrienols. U.S. Patent 6,838,104.

Jonzo, M. D., Hiol, A., Druet, D., & Comeau, L. C. (1997). Application of Immobilized Lipase from *Candida rugosa* to Synthesis of Cholesterol Oleate. *Journal of Chemical Technology & Biotechnology*. Vol. 69, No. 4, pp. 463-469.

Jung, M., Yoon, S., & Min, D. (1989). Effects of processing steps on the contents of minor compounds and oxidation of soybean oil. *Journal of the American Oil Chemists' Society*. Vol. 66, No. 1, pp. 118-120.

Jung, M. Y. & Min, D. B. (1990). Effects of α-, γ-, and δ-Tocopherols on Oxidative Stability of Soybean Oil. *Journal of Food Science*. Vol. 55, No. 5, pp. 1464-1465.

Kasim, N. S., Gunawan, S., & Ju, Y.-H. (2009). Isolation and identification of steroidal hydrocarbons in soybean oil deodorizer distillate. *Food Chemistry*. Vol. 117, No. 1, pp. 15-19.

Khatoon, S., Raja Rajan, R., & Gopala Krishna, A. (2010). Physicochemical Characteristics and Composition of Indian Soybean Oil Deodorizer Distillate and the Recovery of Phytosterols. *Journal of the American Oil Chemists' Society*. Vol. 87, No. 3, pp. 321-326.

Kijima, S., Ichikana, N., & Naito, K. (1964). Purification process of tocopherol containing materials. U.S. Patent 3,122,565.

Kim, S. K. & Rhee, J. S. (1982). Isolation and purification of tocopherols and sterols from distillates of soy oil deodorization. *Kor. J. Food Sci. Technol.* Vol. 14, No. pp. 174-178.

King, J. W., Favati, F., & Taylor, S. L. (1996). Production of Tocopherol Concentrates by Supercritical Fluid Extraction and Chromatography. *Separation Science and Technology*. Vol. 31, No. 13, pp. 1843 - 1857.

Kline, K., Lawson, K. A., Yu, W., & Sanders, B. G. (2007). Vitamin E and Cancer, in *Vitamins and Hormones*. pp. 435-461.

Kochhar, S. P. (1983). Influence of processing on sterols of edible vegetable oils. *Progress in Lipid Research*. Vol. 22, No. 3, pp. 161-188.

Kritchevsky, D. & Chen, S. C. (2005). Phytosterols--health benefits and potential concerns: a review. *Nutrition Research*. Vol. 25, No. 5, pp. 413-428.

Lee, H., Chung, B., & Park, Y. (1991). Concentration of tocopherols from soybean sludge by supercritical carbon dioxide. *Journal of the American Oil Chemists' Society*. Vol. 68, No. 8, pp. 571-573.

Lin, K.-M. & Koseoglu, S. S. (2003). Separation of sterols from deodorizer distillate by crystallization. *Journal of Food Lipids*. Vol. 10, No. 2, pp. 107-127.

Lin, K.-M., Zhang, X., & Koseoglu, S. S. (2004). Separation of tocopherol succinates from deodorizer distillate. *Journal of Food Lipids*. Vol. 11, No. 1, pp. 29-43.

Lin, K. M. (2002). National Chung Hsing University, M.S., Texas A& M University. No.

Lutisan, J., Cvengros, J., & Micov, M. (2002). Heat and mass transfer in the evaporating film of a molecular evaporator. *Chemical Engineering Journal*. Vol. 85, No. 2-3, pp. 225-234.

Marangoni, A. G. & Rousseau, D. (1995). Engineering triacylglycerols: The role of interesterification. *Trends in Food Science & Technology*. Vol. 6, No. 10, pp. 329-335.

Martins, P. F., Batistella, C. s. B., Maciel-Filho, R., & Wolf-Maciel, M. R. (2005). Comparison of Two Different Strategies for Tocopherols Enrichment Using a Molecular Distillation Process. *Industrial & Engineering Chemistry Research*. Vol. 45, No. 2, pp. 753-758.

Martins, P. F., Ito, V. M., Batistella, C. B., & Maciel, M. R. W. (2006). Free fatty acid separation from vegetable oil deodorizer distillate using molecular distillation process. *Separation and Purification Technology*. Vol. 48, No. 1, pp. 78-84.

Maza, A. (1992). Process for separating mixed fatty acids from deodorant distillate using urea. U.S. Patent 5078920.

Mendes, M. F., Pessoa, F. L. P., Coelho, G. V., & Uller, A. M. C. (2005). Recovery of the high aggregated compounds present in the deodorizer distillate of the vegetable oils using supercritical fluids. *The Journal of Supercritical Fluids*. Vol. 34, No. 2, pp. 157-162.

Mendes, M. F., Pessoa, F. L. P., & Uller, A. M. C. (2002). An economic evaluation based on an experimental study of the vitamin E concentration present in deodorizer distillate of soybean oil using supercritical CO_2. *The Journal of Supercritical Fluids*. Vol. 23, No. 3, pp. 257-265.

Mendes, M. F., Uller, A. M. C., & Pessoa, F. L. P. (2000). Simulation and thermodynamic modeling of the extraction of tocopherol from a synthetic mixture of tocopherol, squalene and CO_2. *Brazilian Journal of Chemical Engineering*. Vol. 17, No. pp. 761-770.

Moghadasian, M. H., McManus, B. M., Godin, D. V., Rodrigues, B., & Frohlich, J. J. (1999). Proatherogenic and Antiatherogenic Effects of Probucol and Phytosterols in Apolipoprotein E–Deficient Mice : Possible Mechanisms of Action. *Circulation*. Vol. 99, No. 13, pp. 1733-1739.

Moghadasian, M. H., McManus, B. M., Pritchard, P. H., & Frohlich, J. J. (1997). "Tall Oil"–Derived Phytosterols Reduce Atherosclerosis in ApoE-Deficient Mice. *Arterioscler Thromb Vasc Biol*. Vol. 17, No. 1, pp. 119-126.

Moreira, E. & Baltanás, M. (2004). Recovery of phytosterols from sunflower oil deodorizer distillates. *Journal of the American Oil Chemists' Society*. Vol. 81, No. 2, pp. 161-167.

Mulligan, J. D., Flowers, M. T., Tebon, A., Bitgood, J. J., Wellington, C., Hayden, M. R., & Attie, A. D. (2003). ABCA1 Is Essential for Efficient Basolateral Cholesterol Efflux during the Absorption of Dietary Cholesterol in Chickens. *Journal of Biological Chemistry*. Vol. 278, No. 15, pp. 13356-13366.

Munteanu, A. & Zingg, J. M. (2007). Cellular, molecular and clinical aspects of vitamin E on atherosclerosis prevention. *Molecular Aspects of Medicine*. Vol. 28, No. 5-6, pp. 538-590.

Nagao, T., Kobayashi, T., Hirota, Y., Kitano, M., Kishimoto, N., Fujita, T., Watanabe, Y., & Shimada, Y. (2005). Improvement of a process for purification of tocopherols and sterols from soybean oil deodorizer distillate. *Journal of Molecular Catalysis B: Enzymatic*. Vol. 37, No. 1-6, pp. 56-62.

Nagesha, G., Subramanian, R., & Sankar, K. (2003). Processing of tocopherol and FA systems using a nonporous denser polymeric membrane. *Journal of the American Oil Chemists' Society*. Vol. 80, No. 4, pp. 397-402.

Nagesha, G. K., Manohar, B., & Udaya Sankar, K. (2003). Enrichment of tocopherols in modified soy deodorizer distillate using supercritical carbon dioxide extraction. *European Food Research and Technology*. Vol. 217, No. 5, pp. 427-433.

Nagesha, G. K., Manohar, B., & Udaya Sankar, K. (2004). Enzymatic esterification of free fatty acids of hydrolyzed soy deodorizer distillate in supercritical carbon dioxide. *The Journal of Supercritical Fluids*. Vol. 32, No. 1-3, pp. 137-145.

Nergiz, C. & Çelikkale, D. (2010). The effect of consecutive steps of refining on squalene content of vegetable oils. *Journal of Food Science and Technology*. Vol. No. pp. 1-4.

Norris, F. A. (1979). Handling, Storage and Grading of Oils and Oil Bearing Materials, In: *Bailey's Industrial Oil and Fat Products*, D. Swern, Editor, pp. 601-635, Interscience Publishers, New York.

Plat, J. & Mensink, R. P. (2005). Plant Stanol and Sterol Esters in the Control of Blood Cholesterol Levels: Mechanism and Safety Aspects. *The American Journal of Cardiology*. Vol. 96, No. 1, Supplement 1, pp. 15-22.

Quílez, J., García-Lorda, P., & Salas-Salvado, J. (2003). Potential uses and benefits of phytosterols in diet: present situation and future directions. *Clinical nutrition (Edinburgh, Scotland)*. Vol. 22, No. 4, pp. 343-351.

Ramamurthi, S., Bhirud, P. R., & McCurdy, A. R. (1991). Enzymatic methylation of canola oil deodorizer distillate. *JAOCS, Journal of the American Oil Chemists' Society*. Vol. 68, No. (12), pp. 970-975.

Ramamurthi, S. & McCurdy, A. (1993). Enzymatic pretreatment of deodorizer distillate for concentration of sterols and tocopherols. *Journal of the American Oil Chemists' Society*. Vol. 70, No. 3, pp. 287-295.

Rimm, E. B., Stampfer, M. J., Ascherio, A., Giovannucci, E., Colditz, G. A., & Willett, W. C. (1993). Vitamin E Consumption and the Risk of Coronary Heart Disease in Men. *New England Journal of Medicine*. Vol. 328, No. 20, pp. 1450-1456.

Rohr, R. & Trujillo-Quijano, J. A. (2002). Process for Extraction and Concentration of Liposoluble Vitamins, Growth Factors and Animal and Vegetable Hormones from Residues and By-products of Industrialized Animal and Vegetable Products. U.S. Patent No 6,344,573.

Rohr, R. & Trujillo-Quijano, J. A. (2005). Process for separating unsaponifiable valuable products from raw materials. U.S. Patent No. 846941.

Rozner, S., Garti, N. (2006). The activity and absorption relationship of cholesterol and phytosterols. *Colloids and Surfaces A: Physicochemical and Engineering Aspects* Vol. No. pp. 435–456.

Sheabar, F. Z. & Neeman, I. (1987). Concentration of tocopherols from soy oil deodorizer scum. *La Rivista Italiana Delle Sostanze Grasse ZXZV*. Vol. 2, No. pp. 19-222.

Shimada, Y., Hirota, Y., Baba, T., Sugihara, A., Moriyama, S., Tominaga, Y., & Terai, T. (1999). Enzymatic synthesis of steryl esters of polyunsaturated fatty acids. *JAOCS, Journal of the American Oil Chemists' Society*. Vol. 76, No. 6, pp. 713-716.

Shimada, Y., Nakai, S., Suenaga, M., Sugihara, A., Kitano, M., & Tominaga, Y. (2000). Facile purification of tocopherols from soybean oil deodorizer distillate in high yield using lipase. *JAOCS, Journal of the American Oil Chemists' Society*. Vol. 77, No. 10, pp. 1009-1013.

Shishikura, A., Fujimoto, K., Kaneda, T., Arai, K., & Saito, S. (1988). Concentration of Tocopherols from Soybean Sludge by Supercritical Fluid Extraction. *J. Jpn. Oil Chem. Soc.* Vol. 37, No. pp. 8-12.

Smet, P. (2008). Valorisatie van vetzuurdestillaten als biobrandstof door herverestering met glycerol. Master thesis: 1-68.

Smith Frank, E. (1967). Separation of tocopherols and sterols from deodorizer sludge and the like. U. S. Patent 3.335.154.

Strocchi, A. & Marascio, G. (1993). Structural Modifications of 4,4'-Dimethyl Sterols during the Hydrogenation of Edible Vegetable Oils. *Lipid / Fett*. Vol. 95, No. 8, pp. 293-299.

Su-Min, J., Yong-Hun, P., Bong-Hyon, J., Hyon-Ho, S., & Jong-Dok, S. (1992). Method of separation and concentration of sterol and tocopherol from distillated soybean powder. KR Patent No 9,205,695.

Sun, H., Wiesenborn, D., Tostenson, K., Gillespie, J., & Rayas-Duarte, P. (1997). Fractionation of squalene from amaranth seed oil. *Journal of the American Oil Chemists' Society*. Vol. 74, No. 4, pp. 413-418.

Svensson, C. (1976). Use or disposal of by-products and spent material from the vegetable oil processing industry in Europe. *Journal of the American Oil Chemists' Society*. Vol. 53, No. 6, pp. 443-445.

Swern, D. (1986). *Bailey's Industrial Oils and Fats*. (John Wiley & Sons, New York.

Szelag, H. & Zwierzykowski, W. (1983). The Application of Molecular Distillation to Obtain High Concentration of Monoglycerides. *Fette, Seifen, Anstrichmittel*. Vol. 85, No. 11, pp. 443-446.

Takagi, Y. & Kai, Y. (1984). Process for preparation of tocopherol concentrates. U.S. Patent 4454329.

Tangkam, K., Weber, N., & Wiege, B. (2008). Solvent-free lipase-catalyzed preparation of diglycerides from co-products of vegetable oil refining. *Grasas y Aceites*. Vol. 59, No. 3, pp. 245-253.

Torres, C. F., Fornari, T., Torrelo, G., Señoráns, F. J., & Reglero, G. (2009). Production of phytosterol esters from soybean oil deodorizer distillates. *European Journal of Lipid Science and Technology*. Vol. 111, No. 5, pp. 459-463.

Torres, C. F., Torrelo, G., Señorans, F. J., & Reglero, G. (2007). A two steps enzymatic procedure to obtain sterol esters, tocopherols and fatty acid ethyl esters from soybean oil deodorizer distillate. *Process Biochemistry*. Vol. 42, No. 9, pp. 1335-1341.

Trautwein, E. A., Duchateau, G. S. M. J. E., Lin, Y., Mel'nikov, S. M., Molhuizen, H. O. F., & Ntanios, F. Y. (2003). Proposed mechanisms of cholesterol-lowering action of plant sterols. *European Journal of Lipid Science and Technology*. Vol. 105, No. 3-4, pp. 171-185.

Ubhayasekera, S. & Dutta, P. (2009). Sterols and Oxidized Sterols in Feed Ingredients Obtained from Chemical and Physical Refining Processes of Fats and Oils. *Journal of the American Oil Chemists' Society*. Vol. 86, No. 6, pp. 595-604.

van Rensburg, S. J., Daniels, W. M. U., van Zyl, J. M., & Taljaard, J. J. F. (2000). A Comparative Study of the Effects of Cholesterol, Beta-Sitosterol, Beta-Sitosterol Glucoside, Dehydro-epiandrosterone Sulphate and Melatonin on In Vitro Lipid Peroxidation. *Metabolic Brain Disease*. Vol. 15, No. 4, pp. 257-265.

Verhé, R., Van Hoed, V., Echim, C., Stevens, C., De Greyt, W., & Kellens, M. (2008). *Production of Biofuel from Lipids and Alternative Resources*. (John Wiley & Sons, Inc., 9780470385869)

Verleyen, T., Kamal-eldin, A., Dobarganes, C., Verhé, R., Dewettinck, K., & Huyghebaert, A. (2001a). Modelling of alphatocopherol loss and oxidation products formed during thermoxidation in triolein and tripalmitin mixtures. *Lipids in Health and Disease*. Vol. 36, No. 7, pp. 719-726.

Verleyen, T., Kamal-Eldin, A., Mozuraityte, R., Verhé, R., Dewettinck, K., Huyghebaert, A., & De Greyt, W. (2002a). Oxidation at elevated temperatures: competition between α-tocopherol and unsaturated triacylglycerols. *European Journal of Lipid Science and Technology*. Vol. 104, No. 4, pp. 228-233.

Verleyen, T., Sosinska, U., Ioannidou, S., Verhe, R., Dewettinck, K., Huyghebaert, A., & De Greyt, W. (2002b). Influence of the vegetable oil refining process on free and esterified sterols. *Journal of the American Oil Chemists' Society*. Vol. 79, No. 10, pp. 947-953.

Verleyen, T., Verhe, R., Cano, A., Huyghebaert, A., & De Greyt, W. (2001b). Influence of triacylglycerol characteristics on the determination of free fatty acids in vegetable oils by Fourier transform infrared spectroscopy. *J. Amer. Oil Chem. Soc.* Vol. 78, No. 10, pp. 981-984.

Verleyen, T., Verhe, R., Garcia, L., Dewettinck, K., Huyghebaert, A., & De Greyt, W. (2001c). Gas chromatographic characterization of vegetable oil deodorization distillate. *Journal of Chromatography A*. Vol. 921, No. 2, pp. 277-285.

Verleyen, T., Verhé, R., & Kamal-eldin, A. (2003). Competitive oxidation between alfa tocopherol and unsaturated fatty acids under thermoxidation conditions. *Lipid Oxidation Pathways*. Vol. No. pp. 70-84.

Wang, H., Goto, M., Sasaki, M., & Hirose, T. (2004). Separation of α-Tocopherol and Squalene by Pressure Swing Adsorption in Supercritical Carbon Dioxide. *Industrial & Engineering Chemistry Research*. Vol. 43, No. 11, pp. 2753-2758.

Wang, L., Du, W., Liu, D., Li, L., & Dai, N. (2006). Lipase-catalyzed biodiesel production from soybean oil deodorizer distillate with absorbent present in tert-butanol system. *Journal of Molecular Catalysis B: Enzymatic*. Vol. 43, No. 1-4, pp. 29-32.

Warner, K. (2005). Effects on the Flavor and Oxidative Stability of Stripped Soybean and Sunflower Oils with Added Pure Tocopherols. *Journal of Agricultural and Food Chemistry*. Vol. 53, No. 26, pp. 9906-9910.

Watanabe, Y., Nagao, T., Hirota, Y., Kitano, M., & Shimada, Y. (2004). Purification of tocopherols and phytosterols by a two-step in situ enzymatic reaction. *JAOCS, Journal of the American Oil Chemists' Society*. Vol. 81, No. 4, pp. 339-345.

Weber, N., Weitkamp, P., & Mukherjee, K. D. (2001). Fatty acid steryl, stanyl, and steroid esters by esterification and transesterification in vacuo using Candida rugosa lipase as catalyst. *Journal of Agricultural and Food Chemistry*. Vol. 49, No. 1, pp. 67-71.

Weber, N., Weitkamp, P., & Mukherjee, K. D. (2002). Cholesterol-lowering food additives: Lipase-catalysed preparation of phytosterol and phytostanol esters. *Food Research International*. Vol. 35, No. 2-3, pp. 177-181.

Xu, Y., Shi, X., Du, X., Xing, M., Xu, T., Meng, J., & Feng, Z. (2005). A method to extract natural vitamin E from by-product of refined vegetable. CN Patent ZL 200510114851.X.

Yamada, Y., Shimizu, M., Sugiura, M., & Yamada, N. (1999). Process for producing diglycerides. WO 1999/09119.

Yang, H., Yan, F., Wu, D., Huo, M., Li, J., Cao, Y., & Jiang, Y. (2009). Recovery of phytosterols from waste residue of soybean oil deodorizer distillate. *Bioresource Technology*. Vol. 101, No. 5, pp. 1471-1476.

Yong-Bo, H., Kim, N. H., Park, Y. H., Jong, B. H., Shin, H. H., & Son, J. D. (1994). Method of Separation and Purification of Natural Tocopherol for Deodorized Sludge of Soybean oil. KR Patent No 9,402,715.

Zhao, Y., Sheng, G., & Wang, D. (2000). Pilot-scale isolation of tocopherols and phytosterols from soybean sludge in a packed column using supercritical carbon dioxide, *Proceedings of the Fifth International Symposium on Supercritical Fluids*, Atlanta, Georgia, U.S.A.

22

Soybean Phospholipids

Daicheng Liu and Fucui Ma
College of Life Science, Shandong Normal University
China

1. Introduction

As soybean phospholipids are coproducts of soybean oil processing, the production of soybean phospholipids rises with the continuous increase of soybean oil yield. Phospholipids have been already applied widely in such fields as medicine, food, agriculture and industry etc., relating to various aspects of everyone's clothing, food, shelter and transportation. New phospholipids products will constantly sprout in large numbers with the development of science and technology.
The authors describe the structure, composition, physical and chemical properties and applications of soybean phospholipids based on the research data in hand. This chapter is focused on the processings of concentrated soybean phospholipids, powdery soybean phospholipids and modified phospholipids as well as the isolation and purification of phosphatidylcholine (PC), phosphatidylethanolamine (PE), phosphatidylinositol (PI), phosphatidylserine (PS) and phosphatidic acid (PA) in soybean phospholipids. The exploration of technologies for isolating and purifying individual molecular species of a certain phospholipids class is now one of the hot and difficult research issues in the world. The breakthrough in these technologies will enormously improve the great development of medicine (e.g. biomembrane bionics, liposomes and intracellular drug carriers etc.) and chemical industry (e.g. aggregation and dispersion of nano materials) etc.

2. The structure, composition and physical and chemical properties of soybean phospholipids

Food Chemicals Codex (FCC) definites phospholipids as follows: Food grade phospholipids are complex mixtures obtained from soybean and other plants consisting of acetone insolubles (AIs) which are mainly phosphatidylcholine (PC), phosphatidylethanolamine (PE) and phosphatidylinositol (PI).

2.1 Soybean phospholipids structure

Phospholipids mainly include glycerol phosphatides and sphingomyelin. In this chapter, we mainly discuss glycerol phosphatides. The structures of phospholipids are shown in Fig. 1.

2.2 Soybean phospholipids composition

Phospholipids constitute 0.3%-0.6% of soybean seed, or 1.5%-3.0% of crude soybean oil. The phospholipids composition is shown in Table 1. The fatty acid composition of soybean phospholipids is shown in Table 2.

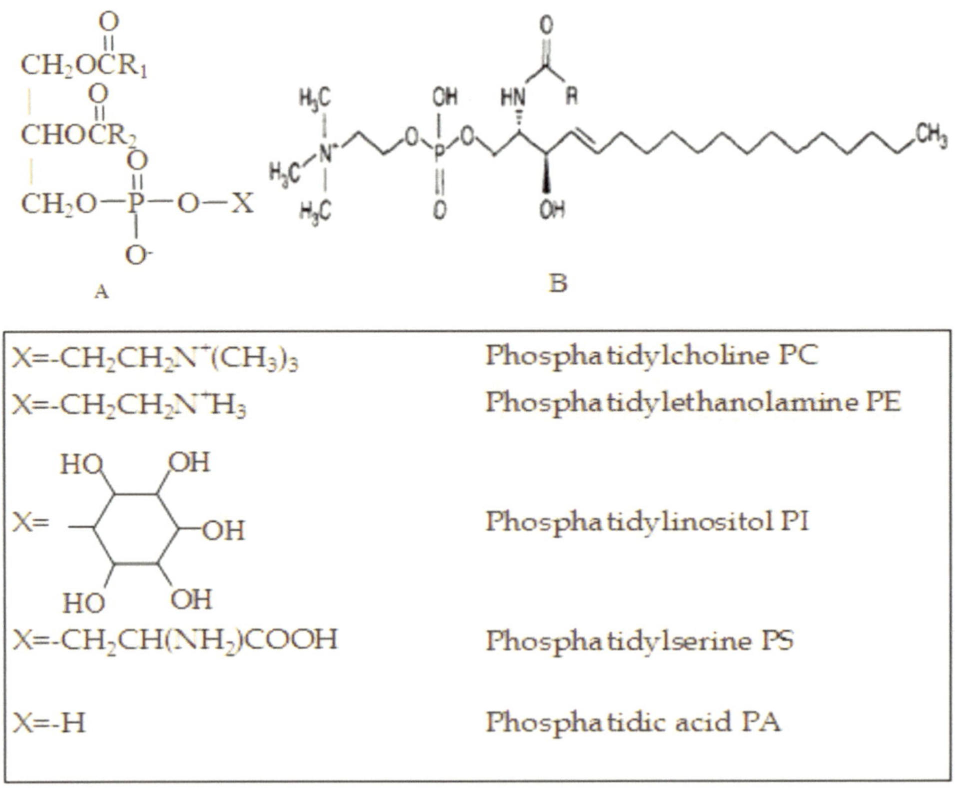

Fig. 1. Structures of phospholipids; A: Glycerol phosphatides structure; B: Sphingomyelin; R1, R2, R: Hydrocarbon chains; Point 'X' is likely composed of structures noted in the box.

Component	Abbreviation	Range(%)		
		Low	Intermediate	High
Phosphatidylcholine	PC	12.0-21.0	29.0-39.0	41.0-46.0
Phosphatidylethanolamine	PE	8.0-9.5	20.0-26.3	31.0-34.0
Phosphatidylinositol	PI	1.7-7.0	13.0-17.5	19.0-21.0
Phosphatidic acid	PA	0.2-1.5	5.0-9.0	14.0
Phosphatidylserine	PS	0.2	5.9-6.3	-
Lysophosphatidylcholine	LPC	1.5	8.5	-
Lysophosphatidylinositol	LPI	0.4-1.8	-	-
Lysophosphatidylserine	LPS	1.0	-	-
Lysophosphatidic acid	LPA	1.0	-	-
Phytoglycolipids		-	14.3-15.4	29.6

Table 1. Composition of Soybean Phospholipids (Szuhaj, 1989)

Fatty acid	Range(%)		
	Low	Intermediate	High
Myristic(C14:0)	0.3-1.9	-	-
Palmitic(C16:0)	11.7-18.9	2.5-26.7	42.7
Palmitoleic(C16:1)	7.0-8.6	-	-
Stearic(C18:0)	3.7-4.3	9.3-11.7	-
Oleic(C18:1)	6.8-9.8	17.0-25.1	39.4
Linoleic(C18:2)	17.1-20.0	37.0-40.0	55.0-60.8
Linolenic(C18:3)	1.6	4.0-6.2	9.2
Arachidic(C20:0)	1.4-2.3	-	-

Table 2. Fatty Acid Composition of Soybean Phospholipids (Szuhaj, 1989)

Soybean phospholipids are by-products of soybean oil refining process. Phospholipids composition can be affected by the oil refining processes and may decrease after frost. The lipase may contribute to phospholipids decrease during storage. Other minor compositions in soybean phospholipids include water, pigment, galactosyl glyceride, glycolipids, carbohydrates, sterols and tocopherol etc.

2.3 Physical and chemical properties

2.3.1 Physical properties

Pure phospholipid is a white solid at room temperature, odorless and colorless. The color of phospholipid may be from light yellow to brown due to refining methods, product categories and storage conditions etc. Non decolored, once decolored and twice decolored are three grades of phospholipid color which is determined by Gardner colorimeter (AOCS official method Td-La-64). The chromaticities are from 9 to 17.

Soybean phospholipids are soluble in aliphatic hydrocarbons, aromatic hydrocarbons and halogenated hydrocarbons solvents, such as ether, benzene, chloroform and petroleum ether etc., and particularly soluble in aliphatic alcohol, for example ethanol. As other non-polar surfactants, soybean phospholipid is insoluble in polar solvent, for example methyl acetate, especially acetone (solubility less than 0.03g/L at 5 degrees Celsius). Phospholipids solubility in methyl acetate and acetone increases when there is a small amount of oil in the phospholipids. PC is soluble in ethanol while PI not. The soluble and insoluble portions of PE in ethanol are about equivalent. The differences of soybean phospholipids solubilities in the above solvents may provide references for isolation, purification and quantification of phospholipids. Soybean phospholipids are soluble in animal fat and vegetable oil, mineral oil and fatty acids, but insoluble in cold animal fat and vegetable oil actually.

The hydrophilic phosphate group and alkaline and hydrophobic hydrocarbon keys in phospholipids molecules help to form a interface between water and oil which lowers the interfacial tension between water and oil and makes them stable colloidal. Soybean phospholipids exist in oil and have obvious hydrophilic colloid property. When mixed with suitable amount of water, phospholipids are isolated from oil. Particularly, in hot alkalescent water (pH>8) the phospholipids are more likely to absorb water and expand and then the colloidal solution is formed. Due to the above property, phospholipids are obtained from crude oil and are widely applied (Lu, 2004).

Phospholipids consist of fluidic and plastic phospholipids. Fluidic phospholipids have the flow property of Newtonian fluid and the fluidity of plastic phospholipids increases with the addition of fatty acids. The viscosity of phospholipids is affected by such factors as AI (acetone insoluble) content, moisture, mineral content, acid value (AV) and various additives for example plant-based surfactant. Generally, high AI or water content results in high viscosity while high AV results in low viscosity. Some bivalent minerals for example Ca^{2+} affect viscosity too.

N-hexane insolubles (HIs) make fluidic phospholipids turbid. The turbidness not only influence the appearance of the products, but also leads to precipitation in long-term storage. The phospholipids also get turbid when the water content is over 1% (Wu, 2001).

2.3.2 Chemical properties

Phospholipids are very unstable when exposed to the air or the sunshine, and color deepening and oxidative rancidity easily happen. However, phospholipids are stable in oil without water. So the oil in concentrated phospholipids can prevent oxidative rancidity and is conductive to the phospholipids storage. Phospholipids are unstable at high temperature. In oil and phospholipids processing, the color of the oil get deeper at 150 degrees Celsius and the odor of phospholipids get worse. Phospholipids decompose at over 150 degrees Celsius.

Hydrolysis of phospholipids occurs upon exposure to strong acid at high temperature. Saponification of phospholipids happens when heated in alkaline ethanol or water solution, and soaps are produced. The salts of phosphoglycerol and inositol phosphate are further heated to be hydrolyzed into glycerol, inositol and phosphoric acid. Free fatty acids and free substances of the above compound are produced after acid hydrolysis or high pressure hydrolysis.

Phospholipids can be hydrolyzed by enzymes. At least four kinds of lipases can cleave ester bonds formed by carboxylic acid and phosphoric acid attached to the glycerol molecule and some of them can only cleave unsaturated fatty acids from phospholipids and can't act on saturated fatty acids. These actions result in production of so called lysophospholipids which have strong effect of hemolysis.

Phospholipids may be modified under certain conditions. The modification reactions include hydroxylation, acetylation, hydrogenation, sulfonation, hydroxyl acetylation and enzymatic modification etc. Modified phospholipids vary in their properties and functions (Lu, 2004).

Phospholipids are regarded as the synergist of antioxidant, and they can synergize or prolong the antioxidation functioin of tocopherol. The synergism of phospholipids differs due to the differences of oil and phospholipids. Mixtures of PE, mixed tocopherol and synthetic antioxidant exhibit the highest antioxidant property (Wu, 2001).

3. Soybean phospholipids processing

3.1 Preparation of concentrated soybean phospholipids

Concentrated soybean phospholipids are products obtained by drying and dehydrating hydrated soybean crude oil foot. Industrial methods preparing concentrated soybean phospholipids include continuous and batch processings.

3.1.1 Continuous processing

The processing steps are as follows: The crude oil is heated to 80 degrees Celsius and then centrifuged and passed through the flowmeter followed by addition of 80 degrees Celsius water of 2% (w/w) of the oil.

Degumming oil and oil foot sediments are produced and they can be separated by centrifugation. The hydrated oil foot should be concentrated immediately to avoid microbial rancidity due to the high moisture and neutral oil content. Oil foot (or mixed with hydrogen peroxide or fluidity agent in advance) is pumped into the agitated film dryer. Phospholipids film is formed under gravity or centrifugal force and the pressure of incoming production materials and flow to the bottom of the vessel while moisture evaporates under high temperature and vacuum conditions. The motionless fluid product is dried at vacuum (726 mm Hg) and 100-110 degrees Celsius for 2 min and then cooled to obtain concentrated soybean phospholipids with less than 1% moisture content. The concentrating procedure should be operated under vacuum as phospholipids are thermosensitive (Ji & Li, 2005).

3.1.2 Batch processing

3.1.2.1 Preheating

Mechanically pressing crude soybean oil is preheated to 80 degrees Celsius after removal of impurities by filtration.

3.1.2.2 Hydration

The amount of water added is determined by phospholipids content in the oil and the changes of phospholipids granules formed during heating and is normally 3.5 times (w/w) the content of phospholipids. The water added is usually boiling or 0.7% hot salt solution is used. The speed of adding water is determined by the water absorption velocity of phospholipids. The faster the latter the faster the former, and vice versa. When adding water, the stirring speed must be fast and is normally 80-100rpm at the beginning and is slowed down 20-30min later when large flocculent phospholipids granules are formed and the stirring is continued for another 20-30min. Then the liquid is left standing still to settle. The supernatant of the upper phase is dehydrated to produce refined oil while the oil foot of the lower phase need to be concentrated to obtain phospholipids products.

3.1.2.3 Concentration

The hydrated phospholipids oil foot is drawn into the concentrating tank by vacuum and subjected to temperature rising and stirring. Vacuum dehydration of phospholipids occurs at about 80 degrees Celsius. When there is slight silk flash while stirring the fluidic phospholipids the moisture content is consistent with the specification. The moisture content is about 5%. Phospholipids after concentration is a brown semisolid and can be used in food, medicine and industry.

3.1.2.4 Decoloration

Decoloration of concentrated phospholipids is needed for preparation of high quality phospholipids. The amount of 30% hydrogen peroxide added to the concentrating tank is 2%-2.5% (w/w) of the concentrated phospholipids. The phospholipids are decolored in the closed tank for 1h at 50 degrees Celsius without vacuum. Then turn on the vacuum pump and heat the mixture to 70 degrees Celsius. Dehydrate until there is no water in the water knock vessel. The decolored phospholipids are light brown.

Mixed fatty acids and mixed fatty acid ethyl ester are added as fluidity agents during the vacuum concentrating procedure to improve the fluidic property of concentrated phospholipids and prevent phospholipids separating with the oil and guarantee the stability of phospholipids products.

The products obtained can flow freely at room temperature. If mixed fatty acids added is inadequate, it will not act as fluidity agent. On the contrary, excess addition of mixed fatty acids may raise the AV of phospholipids and get them rancid. The amount of mixed fatty acids added is usually 2.5%-3% (w/w) of the concentrated phospholipids. The addition of mixed fatty acid ethyl ester does not affect the AV and flavor of phospholipids and can gain high quality products but the cost is high. The amount is 3%-5% (w/w) of the concentrated phospholipids (Ji & Li, 2005).

3.2 Preparation of powdery soybean phospholipids

The applications of concentrated phospholipids are limited due to its high content of neutral oil, fatty acids and other substances and its low purity and off-flavor formation. Refining and purifying processings are needed to consistent the phospholipids products with the high purity and non off-flavor specifications of functional food material.

Methods of producing high purity phospholipids from concentrated phospholipids include solvent extraction, ultrafiltration purification, supercritical carbon dioxide extraction etc. So far, acetone solvent extraction is the most widely used method in industry.

3.2.1 Solvent extraction

3.2.1.1 Preparation of powdery phospholipids of one kind of purity from one kind of materials

The acetone solvent extraction theory is isolating phospholipids from oil by precipitation due to the fact that water and oil is soluble in acetone while phospholipids not.

30% hydrogen peroxide is pumped into the closed agitated container with the amount of 2%-3% (w/w) of the concentrated phospholipids. Concentrated phospholipids are pumped into the above container while stirring with the rotate speed of 30-40rpm. Decoloration occurs after the temperature reaches 60 degrees Celsius with a processing time of 6h. After that, the temperature is raised up to 70-75 degrees Celsius and decolor for 0.5h to decompose residual hydrogen peroxide into water. The decoloring procedure is optional due to the product requirements.

Acetone with purity above 98% is pumped through a flowmeter into the closed agitated container. Concentrated phospholipids with acetone residues of the amount of 1:10 (w/w) are pumped into the above container. Stir for another 20-30min with the speed of 80rpm. After that, the liquid is statically sedimented for 0.5h and the upper acetone extract is discharged. The above procedure is repeated three more times with each time a 5:1 ratio (w/w) of acetone to concentrated phospholipids and prolonging sedimentation time. The total amount of acetone is 25 times (w/w) that of concentrated phospholipids.

Phospholipids settle down at the fourth time is discharged and centrifuged. The diameter of the centrifuge rotor is 800mm and the rotate speed is 1200-1600rpm. Centrifuged phospholipids go directly into the lower closed agitated-container. Acetone of 2 times (w/w) the weight of the concentrated phospholipids is pumped into the same container while stirring (80rpm). The extraction procedure lasts 0.5h and then the liquid is discharged and centrifuged to produce phospholipids with 25%-50% (w/w) acetone. The phospholipids are fed into the double conic dryer with the amount of 1/3-1/2 of the whole dryer volume. The drying parameters are as follows: drying temperature 50-55 degrees Celsius, rotate speed 10rpm, vacuum -0.083--0.09 MPa, time 4-6h. Then light yellow powdery phospholipids without acetone residue are obtained and weighed for packaging.

The above method can be applied to prepare powdery phospholipids from such various raw materials as soybean, rapeseed, peanut and corn etc. as well as concentrated phospholipids prepared from hydrated oil foot and alkalized oil foot. The powdery phospholipids produced have a phospholipids content of 90%-98% due to the quality of the concentrated phospholipids (Liu & Yang et al., 2006; Liu & Feng et al., 2006; Liu, 2007).

3.2.1.2 Preparation of powdery phospholipids of various purities from one kind of materials

In acetone solvent extraction, the phospholipids purity increases with the increase of acetone amount and extraction times. The increase of phospholipids purity results in longer time needed to settle the whole phospholipids in acetone solution.

If the purity of the powdery phospholipids obtained in 3.2.2.1 is 97%-98%, half of the phospholipids will be settled in 0.5-1h in the fourth extraction while the other half in 4h. The upper phospholipids solution of acetone is discharged when the time has passed 2.5-3h and centrifuged and dried. The purity of the phospholipids produced can reach up to over 99%.

Acetone of 2 times the weight of concentrated phospholipids is added into the extraction tank with agitation. The extraction time is 0.5h and static settle time is 1.5-2h. The upper phospholipids solution of acetone is discharged and centrifuged and dried to obtain phospholipids product with purity of over 95%.

The residual liquid is discharged, centrifuged and dried to produce phospholipids product with purity of about 90% (Liu & Ma, 2011).

This method can produce phospholipids products with various purities due to the product purity obtained in 3.2.2.1 and discharging time to meet the market requirements, and make the best use of the materials.

3.2.2 Supercritical carbon dioxide extraction

Extraction temperature, pressure and time are important technological parameters of supercritical carbon dioxide extraction. Extraction yield increases with the increase of one of the parameters in a certain range while the other two conditions remain unchanged. However, there are also problems of increased cost, power consumption and unsafe factors. Generally, the extracting effect is rather good at 50 degrees Celsius and 20MPa for 5h.

Supercritical carbon dioxide extraction used to extract soybean phospholipids has significance for the industrial application and is an applicative technology with wide prospect as it has the advantages of simple, non solvent residue, safe and reliable and high purity product and it consists with the trend of current green chemical technology (Shi, 2005).

3.2.3 Ultrafiltration purification

The crude phospholipids are subjected to derosination and dissolved in solvents and then passed through ultrafiltration film with certain pore size. Components of suitable sizes pass through the membrane and are isolated.

Ultrafiltration lecithin introduced by ADM (Archer Daniels Midland Co.) which has the property of dry, easy to be mixed with other materials, high quality and high purity is produced by removing the glycerides in phospholipids by ultrafiltration. Ultrafiltration lecithin can be precisely quantified and conveniently used. In certain situation which has strict requirements for flavor ultrafiltration lecithin is precious as it has good flavor (Shi, 2005).

3.3 Preparation of modified soybean phospholipids

3.3.1 Chemical modification

3.3.1.1 Hydrogenation

After hydrogenation, the unsaturated double bonds of the phospholipids are saturated to improve the stability, oxidative stability, color and odor of the phospholipids. Hydrogenated phospholipids are mainly used in cosmetics, dyes and lubricants.

Powdery soybean phospholipids are dissolved in the mixture of dichloromethane and ethanol (3:1, v/v) with a 1:6 (w/v) ratio in the stainless steel autoclave. A 5% palladium/carbon catalyst is added into the autoclave followed by leakage checking. Then the air in the autoclave is displaced by hydrogen for several times. The reacting parameters include a temperature of 50 degrees Celsius, a pressure of 0.6MPa, a stirring speed of 300r/min, and a reacting time of 3h under constant temperature and pressure. After reaction, the temperature and pressure are reduced. The catalyst is removed and recycled by centrifuging the reaction product. 30% hydrogen peroxide with the amount of 5% (w/w) is added into the liquid portion to decolor and the solvent is removed by rotate evaporation at 55 degrees Celsius. Light yellow solid hydrogenated soybean phospholipids are obtained after vacuum drying at 70 degrees Celsius for 8h. It may be better to use pure ethanol as solvent than the mixture of dichloromethane and ethanol when hydrogenating phospholipid that is soluble in ethanol such as PC (Huang et al., 2003).

3.3.1.2 Acetylation

PE is transformed into N-acylphosphatidylethanolamine after acetylation, and its 'zwitter ion' structure is modified to obtain improvements in Hydrophile-Lipophile Balance (HLB) value, thermostability, oil in water emulsifying ability and viscosity property. Meanwhile, N-acylphosphatidylethanolamine's large solubility in acetone facilitates isolation and purification of phospholipids. Acetylation with acetic anhydride is used to produce acetylated phospholipids in industry.

Considering acetylated phospholipids are mainly applied in food processings, direct heating (noncatalytic) acetylation process is adopted to produce food grade acetylated soybean phospholipids. Acetic anhydride is added into crude phospholipids with the amount of 1%-4% (v/w) due to the PE content in phospholipids and the amino conversion rate. The process requires temperatures of 60-70 degrees Celsius and stirring reacting time of 1h-1.5h. After acetylation, the mixture is neutralized with sodium hydroxide or potassium hydroxide of certain concentration and then dried under vacuum. The specifications of acetylated phospholipids are: free amino 0.7%-1.7%, pH 6.5-8, and HLB value 5-6 (Xu et al., 2008).

3.3.1.3 Hydroxylation

The hydroxylation theory is that two hydroxyls are introduced into the double bonds of the unsaturated fatty acids of phospholipids molecules, i.e., crude phospholipids react with hydrogen peroxide with the existence of organic weak acid such as lactic acid to hydroxylate the unsaturated bonds in phospholipids and oil. The ethanolamine group of PE is also modified. The obvious hydrophilic property makes modified phospholipids more easily disperse in cold water. The degree of hydroxylation modification is controlled by the amount of hydrogen peroxide added and usually measured by the drop in iodine value (IV). The products with 10%-25% drops in IV have good water dispersibility and oil in water emulsifying property. The emulsifying property decreases and hydrophilic property increases with the increase of drop in IV, but the cost rises too.

Phospholipids hydroxylation processes include such various methods as 'lactic acid + hydrogen peroxide + phospholipids', 'acetic acid + hydrogen peroxide + phospholipids', 'peracetic acid + phospholipids' and 'basic potassium permanganate + phospholipids' etc., which belong to alkyleneortho-dihydroxylation and have various hydroxylation effects. In industrial production, the 'lactic acid + hydrogen peroxide + phospholipids' process is generally adopted as it's a mild method with no problems of the three wastes(waster gas, waster water and industrial residue) and meets the food grade requirements. 75% lactic acid and 30% hydrogen peroxide with the amount of 1%-3% and 5%-15% (v/w), respectively, are added into crude phospholipids. The reaction is carried out at 50-70 degrees Celsius with stirring for 1h-3h. The mixture is neutralized with sodium hydroxide of certain concentration and then dried under reduced pressure until a less than 1% moisture content is reached. The specifications of hydroxylated phospholipids include: drop in IV 10%-25%, pH 6.5-7.5, HLB value 9-10 (Xu et al., 2008).

3.3.1.4 Acetyl-hydroxylation

Acetyl-hydroxylation refers to acetylation of phospholipids followed by hydroxylation, i.e., double modification. Hydroxylation occurs between phospholipids and hydrogen peroxide with the help of acetic acid produced by acetylation. The procedures are as follows: acetic anhydride is added into the crude phospholipids with the amount of 1%-4% (v/w) due to the PE content in phospholipids and the amino conversion rate. The reaction is carried out at 60-70 degrees Celsius for 1h-1.5h with stirring. Then hydrogen peroxide of 5%-15% (v/w) is added. Temperatures of 60-75 degrees Celsius and stirring reacting time of 1h-3h are required. At last the mixture is neutralized with sodium hydroxide of certain concentration and then dried under reduced pressure until reach a less than 1% moisture content. The specifications of acetyl-hydroxylated phospholipids are drop in IV 10%, free amino no more than 1.65%, pH 7-8, HLB value 9-10 (Xu et al., 2008).

3.3.1.5 Hydroxyl-chlorination

100 portions of soybean phospholipids are dissolved in 300 portions of n-hexane. Sodium hypochlorite of 22.5% (w/w) of the total phospholipids is added and the pH is adjusted to 4.5 with acetic acid. The reaction is carried out at 50 degrees Celsius for 1h with stirring. The mixture is washed 3 times with each time 100 portions of water is used. The upper phospholipids solution is evaporated to recycle solvent and obtain hydroxyl-chlorinated soybean phospholipids. The emulsion stability, dispersibility and wettability are improved enormously compared with that of non-modified phospholipids (Xu et al., 2008).

3.3.1.6 Sulfonation

The most likely positions for sulfonation are the double bonds of long chain unsaturated alkanes and α-carbon near ester bonds. When sulfonation of phospholipids including PC, PE, PA and PI etc. happens, the position which is most likely to be introduced with active group is hydroxyl of PI. That is to say, sulfonation occurs on double bonds while esterification (sulfation) occurs on hydroxyls. The double bonds in products will diminish or vanish if sulfonation occurs on double bonds totally or partly. The decrease in unsaturation of sulfonated soybean phospholipids results in the drop of IV. So we can determine whether sulfonation on double bonds happens or not by measuring IV.

There have been a lot of reports on sulfonation and sulfation of phospholipids, but maturer method is sulfur trioxide gas phase continuous film sulfonation which is developed in China. The film sulfonation pipe need heat preservation jacket. The parameters of the sulfonation process are a feed temperature of 40 degrees Celsius, a sulfur trioxide/air flow rate of $1.5m^3/h$ and a protective wind flow rate of $0.25m^3/h$. Continuous sulfonation happens in film sulfonator followed by neutralization with alkali and decoloration with 5%-20% hydrogen peroxide. The sulfonated phospholipids with a 4% sulfur trioxide binding amount and 6-8 pH exhibit such properties as light color, hydrophilic property, emulsifying property and good permeability.

Sulfonation provides soybean phospholipids with special properties and raises the HLB value from 1-2 to 12-16. The physical and chemical properties of phospholipids are improved enormously to facilitate the wide applications of phospholipids as fatliquoring agent, flotation agent and emulsifying agent in leather, pharmacy and farm chemical etc. (Zhang et al., 2004).

3.3.1.7 Alkoxylation

Alkoxylation technology including ethoxylation and propoxylation etc. is a main technology producing nonionic surfactant. It is carried out by addition-condensation reaction of oxirane or epoxypropare with initiators (aliphatic alcohol and nonyl phenol etc.), and the initiator-ethoxylation or initiator-propoxylation products are obtained.

Alkoxylated phospholipids are obtained by addition-condensation reaction of alkoxylated reactant such as oxirane and phospholipids containing hydroxyl such as PE and PI. As hydrophilic oxethyl groups are introduced into the polar end of PE and PI molecules, the HLB value and hydrophilic property are increased and the oil-in-water emulsifying ability improved. As with sulfonation and hydrogenation, the alkoxylation process is very complex and the products are mainly used in non-food industry.

There are not many manufacturers producing this kind of products. R & R551 is the representative ethoxylated soybean phospholipids of ADMC. The ethoxylated phospholipids have a 12.5 HLB value.

According to patents that have been made public and reports related, the soybean phospholipids alkoxylation process is mainly as follows: addition reaction occurs between phospholipids (PE and PI) and alkoxylation reactant (oxirane, epoxypropare and glycidol etc.) and alkoxylated soybean phospholipids are produced. For example, 23.5 pounds of oil-containing soybean phospholipids are dissolved in 15 pounds of dimethylbenzene. 4.5 pounds of oxirane is added. The reaction is carried out at 100 degrees Celsius and 0MPa for 3h followed by removal of solvents. The ethoxylated soybean phospholipids which are resinous, insoluble in dimethylbenzene and soluble in water are obtained (Xu et al., 2008).

3.3.2 Enzymatic modification

Chemical modification of phospholipids improves their emulsifying and hydrophilic properties, but damages natural structure of phospholipids as well. Enzymatic modification exhibits such advantages as no need for purifying the reactant, mild reacting conditions, fast, complete, less by-products, exact action position of enzymes and easy to obtain etc. Phospholipases including phospholipase A_1, A_2, C and D can catalyze various hydrolysis of phospholipids as well as esterification and interesterification reaction with the existence of certain acyl receptor and donor to change or modify the structure of phospholipids which will gain different structures and applications. Phospholipase A_2 and D are used in industries while the other enzymes are on the experimental status (Gu et al., 1999).

3.3.2.1 Phospholipase A_1 and number 1,3-position specific phospholipase

Phospholipase A_2 can specifically hydrolyze the acyl at number Sn-1 position of natural phospholipids. But acyl at number Sn-2 position can be easily transferred onto the thermodynamically stable number Sn-1 position and this results in producing of the same products as with phospholipase A_2. The source of phospholipase A_1 is very limited.
Phospholipase with number 1,3-position specificity can selectively hydrolyze acyl at number Sn-1 position of phospholipids, and can replace phospholipase A_1. Number 1,3-position specific phospholipase can directly catalyze interesterification of phospholipids and fatty acids or oleic acid in organic solvents to produce new phospholipids. For example, Lipozyme IM20 can catalyze the intersterification of PC and fish oil fatty acids with 45% eicosapentaenoic acid (EPA). The parameters are: enzyme amount 1.5% (w/w), the optimal ratio of phospholipids to fatty acids 1:2 and the optimal organic solvent n-hexane. The binding ratio of EPA at number Sn-1 position is 17.7%. Polyunsaturated fatty acids such as EPA and docosahexaenoic acid (DHA) which are good for cardiovascular and cerebrovascular health can be attached to phospholipids and obtain better digestion and absorption properties than that with triglyceride through this kind of reaction (Gu et al., 1999).

3.3.2.2 Phospholipase A_2

Phospholipase A_2 (EC 3.1.1.4) specifically catalyzes the hydrolysis of acyl at number Sn-2 position of phospholipids to produce lysophospholipids and fatty acids. Modified soybean phospholipids exhibit obviously improved hydrophilic and emulsifying properties. They can maintain good emulsifying property under conditions of high or low temperatures or low pH or various salt concentrations. Lysophospholipids are applied in bakery food. They form complexes with amylose and retard aging of breads effectively. Lysophospholipids are two times the price of ordinary phospholipids but they have the advantages of smaller dosage, better effect, oxidative stability and antibiotic property. They are industrially produced in Japan and America. The process is as follows: phospholipids are subjected to moisture content adjustion previously and then added into the solutions with phospholipase A_2 of 0.02% (w/v) and calcium chloride of 0.3% (w/v) with stirring. The temperatures are 50-55 degrees Celsius and the reacting time is 7h-9h. The hydrolyzing degree reaches 35%-40% when the acid value (AV) is in the range of 33-30. The following procedures are required to obtain powdery lysophospholipids: concentration under reduced pressure, pressure filtration, washing with acetone, solvent removal under reduced pressure and vacuum drying (Song et al., 2007).

3.3.2.3 Phospholipase C

Phospholipase C acts on phospholipids to produce diglyceride, phosphoinositide, phosphocholine, phosphoethanolamine and phosphoric acid etc. Diglyceride is a bioactive substance which acts as the second messenger in cell signaling transmission and affects the cell metabolism. There are three kinds of specificities of microbial phospholipase C: the first one specifically hydrolyze PI into diglyceride and cyclic phosphoinositide; the second one specifically hydrolyze sphingomyelin and the third one has relatively wider specificity and takes PC as the optimal substrate (Song et al., 2007).

3.3.2.4 Phospholipase D

Phospholipase D (EC 3.1.4.4) can hydrolyze PC into phosphatidic acid and choline. In microwater system with alcohol, phospholipase D can catalyze transacylation which results in exchange of primary or secondary hydroxyl of some molecules with ethanolamine or choline groups of phospholipids and formation of new phospholipids. This character is called phospholipids' transfer characteristic or base exchange reaction of phospholipase D (Song et al., 2007).

4. Extraction and isolation of soybean phospholipids

4.1 Soybean phosphatidylcholine (SPC)

4.1.1 Organic solvent extraction

Fractions in soybean phospholipids are isolated due to their solubilities' differences in organic solvents. PC exhibits large solubility in lower alcohol (C1-C4) whereas PE and PI have small solubilities. PC- and PI-enriched products can be obtained by their solubilities' difference. When treated with lower alcohol, PC in deoiled phospholipids is soluble in alcohol leaving insoluble matter consisting mainly of PE and PI. The ratio of PC to PE increases from 1:1 (w/w) in raw material to 3:1 (w/w), and even to 12:1 (61% PC, 5% PE).
Better isolation effect on PC can be obtained by isopropanol. Mixtures of high-purity phospholipids and isopropanol with the ratios of 8:157-16:157 (12:157 is optimal) is added into the agitated- and refluxed-closed container. The extraction is conducted in thermostatic water bath or cryohydrate bath with isopropanol volume fraction of 95%-100% (100% is optimal) at -5-15 degrees Celsius (-5 degrees Celsius is optimal) for 5min-11min (5min is optimal). After the extraction, the mixture is filtered, evaporated to remove isopropanol and dried to obtain product with mass fraction of PC increased from 25.6% in raw material to 66.8%. Isopropanol extraction is the most commonly employed step to obtain high PC-containing phospholipids from deoiled phospholipids (An et al, 2001).

4.1.2 Lower alcohol with salt or acid extraction

It's also an effective method to fractionate phospholipids with the property that phospholipids can react with some salts or acids and precipitate. This method is more promising than organic solvent extraction as metal ions or acids can 'recognize' phospholipids molecules more effectively than solvents.
100g of phospholipids containing 45% PC is dissolved in 1L ethanol of 95% before addition of 4.5g of zinc chloride. The light yellow phospholipid-zinc chloride compound precipitate formed is centrifuged, decomposed with 250ml freezed acetone under nitrogen to obtain 36.7g of phospholipids containing 99.6% PC (Ni, 1995).

4.1.3 Supercritical fluid extraction

Supercritical fluid extraction (SFE) is a rapidly developed new technology in recent years. Supercritical fluids most commonly used are carbon dioxide, ammonia, ethylene, propylene and water etc. Carbon dioxide is most frequently used due to its following properties: critical temperature and pressure easy to get, stable chemical properties, non-toxic, odorless, non-corrosiveness and reusable. Supercritical carbon dioxide extraction (SCE) is a method with bright prospect as it can obtain high purity products with no solvent residues and maintain the nutritional and functional properties of the products and need simple process, single equipment and low cost (Guan et al., 2005).
Tekerikler et al. (2001)obtained phospholipids containing 91% PC after SCE of deoiled phospholipids with 10% ethanol as entrainer at 17.2MPa and 60 degrees Celsius. Increasing the pressure to 20.7MPa increased the extraction yield and PC content (95%). Increasing the

temperature to 80 degrees Celsius decreased the extraction yield and PC content which was attributed to decrease of solubility and selectivity of solvents to PC as the solvents density decrease at high temperatures.

4.2 Soybean phosphatidylinositol (SPI)

PI causes concern as it is involved in the transmission of messages in the cell. PI plays and important role in maintaining normal physiological functions of central nervous system, especially in regulating calcium homeostasis. PI on cell membrane can be hydrolyzed by phospholipase C into 1,4,5-triphosphate inositol that goes into intracellular aqueous phase as the second messenger and 1,2-diacylglycerol that stays in the cell wall. These two substances synergetically induce cell reactions such as contraction, secretion, metabolism and proliferation etc.

Soybean phospholipids are rich in PI. PI is a white amorphous solid with its sodium salt a crystal and is wet-sensitive. PI is soluble in water, chloroform and benzene, slightly soluble in methanol, diethyl ether and petroleum ether, insoluble in acetone, ethanol and water. It can be easily oxidized upon exposure to the air (Deng et al., 2003).

4.2.1 Solvent method

The solvent method is conducted to isolate and purify phospholipids and increase the content of a certain constituent with single or mixture of such solvents as methanol, ethanol, isopropyl ketone, acetone, n-hexane and chloroform etc.

The deoiled soybean phospholipids are extracted with appropriate amount of ethanol. The induced ethanol-insoluble phase is vacuum dried to obtain a mixed phospholipids with more PI and less PC which are dissolved in n-hexane before adding 55% ethanol with sodium acetate. The mixture is put into the separating funnel, shaked, allowed to rest and layered. The same procedure is carried out again except that the 55% ethanol is sodium acetate free. The PI obtained is 40%-50% pure. If sodium acetate is replaced by aqueous ammonia with a 8.0 pH and the ethanol concentration increased to 90%, PI of 85% pure can be obtained with the same method.

Purer PI can be obtained by some chemical reaction methods. Soybean phospholipids containing 40% mixed phospholipids are dissolved in such organic solvents as anhydrous pyridine, acetonitrile, dimethyl formamide (DMF) and dimethylsulphoxide (DMSO) etc. Chloride dimethyl tertiary butyl silicon, chloride trimethyl silicon or allyl bromide are added into the mixture to protect hydroxyls of PI by reacting with them. Then PI is isolated from the mixture with solvents such as acetone or ethanol-acetone and hydrolyzed by alkali or acid at room temperature to remove the blocking groups and recover the hydroxyls of PI. PI obtained this way is 98% pure and applied in treating of central nervous system disorder (Deng et al., 2003).

4.2.2 Column chromatography

The phospholipids mixture is dissolved in the mixture of chloroform and methanol in the 1:1 (v/v) ratio before adding aluminium oxide. The eluate contains PC, lysophospholipids, neutral lipids and glycolipids etc. Residues are washed and extracted with the mixture of chloroform, methanol and 1% hydroxyl ammonium acetate in the 1:1:0.3 (v/v/v) ratio and the eluate is loaded on silica column of which the dimension is 30cm. Neutral lipids are eluted with chloroform; glycolipids and PE are sequentially eluted with chloroform and methanol in the 80:20 (v/v) ratio; PE is further removed with chloroform and methanol in the 20:80 (v/v) ratio; PI is finally eluted with chloroform, methanol and 25% ammonia in the 80:20:5 (v/v/v) and 65:25:5 (v/v/v) ratios. The PI-containing fraction is evaporated and dried to obtain PI of no less than 98%-99% pure (Deng et al., 2003).

Column chromatography can obtain high purity PI but the long time needed and the use of complex solvent mixture reduce its feasibility in the commercial world.

4.2.3 Enzymatic method

Phospholipids can be hydrolyzed by such phospholipase as phospholipase A_1, A_2, C and D. When treating the ethanol-treated phospholipids (containing minor PC), phospholipase selectively catalyze hydrolization of PE and PC but not PI. More special, alkaline or acid phospholipase catalyzes hydrolization of PA and some salts produced by PE hydrolization but doesn't act on PI, PC or PE. PI products used in various fields can be obtained by this method and the purity can reach up to 99%. Lypase can be used to purify PI as well, and the purifies are 60%-70% (Deng et al., 2003).

Enzymatic method has wide application prospect as it is simpler, more convenient and environmental prospective compared with solvent method and column chromatography.

4.2.4 Other methods

Ion exchange resin may be applied to isolate PI from phospholipids mixture. The resin adsorbing PI include diethylaminoethylcellulose, diethylaminoethylagarse and quaternaryammoniumethylsephadex etc. (Deng et al., 2003).

4.3 Soybean phosphatidylethanolamine (SPE)

Solid-liquid extraction is performed using powdery soybean phospholipids and ethanol. The parameters are as follows: ratio of phospholipids to ethanol 30g/L, ethanol concentration 95%-100% (100% is optimal), extracting temperatures -15-50 degrees Celsius (-15 degrees Celsius is optimal) and extracting time 5min-11min (8min is optimal). PE content increases from 19.8% in raw material to 62.8% (An et al., 2006).

Phospholipids are dissolved in isopropanol below 65 degrees Celsius to reach a final concentration in the range of 1%-4% (w/v). The mixture is cooled to 26 degrees Celsius, allowed to rest and the precipitate is filtered and dried to obtain PE of 74.7%-79.9% pure (Ni, 1995).

Zhensheng Zhong et al. (2008) removed oil and fatty acids in powdery phospholipids with acetone first, and removed PC with repeated ethanol extraction due to PC has larger solubility in aliphatic alcohol than PE and PI, and finally enriched PE with petroleum ether extraction due to PE is soluble in ether while PI not.

The powdery soybean phospholipids are extracted repeatedly with acetone to remove oil and refined phospholipids containing 35% PE are obtained. PC is removed by repeated absolute ethanol extraction with heating and stirring and alcohol insolubles are obtained. The alcohol insolubles are extracted 3 times with petroleum ether in the 1:3-1:6 (1:4 is optimal, v/v) ratios at 30-60 degrees Celsius (30 degrees Celsius is optimal). PE obtained this way is 93.5% pure and has a yield of 91.9% (Zhong & Wei, 2008).

4.4 Soybean phosphatidylserine (SPS)

Pure PS is a white waxy solid. It's soluble in most of the nonpolar solvents containing little water, insoluble in anhydrous acetone and can be extracted from histiocyte with chloroform

and methanol. When PS is dissolved in water, most of the insoluble lipids form micell while very few form true solution. PS has one positive and two negative charges, resulting in a net negative charge. PS can be hydrolyzed by weak base into metal salts of fatty acids and a remained portion, and by strong alkali into fatty acids, serine and glycerol phosphate. PS is ready to oxidize upon exposure to the air, and the color gets darker from white to yellow and finally black. Natural PS practically isn't affected by alcohol while saturated PS can form interwoven catenulated gel with alcohol and dipalmiloyl-phosphatidylserine can interact with 5% alcohol at room temperature to form regular gel.
PC is dissolved in organic phase while the enzyme and serine are dissolved in aqueous phase. After preheating for a while, the two phases are combined, and reaction occurs at the interface under certain conditions. PS is obtained by isolating and extracting of the organic solvent phase and quantified by thin layer chromatography (TLC). The parameters are as follows: ratio of organic phase to aqueous phase 4:4 (v/v), PC concentration 75mg/ml, reacting temperature 40 degrees Celsius, pH of aqueous phase 4.0 and reacting time 12h. PS yield is 68.9% (Yang, 2010).
Blokland et al. (1999) compared the effects of bovine cortex phosphatidylserine (BCPS), SPS and egg phosphatidylserine (EPS) on cognitive competence of middle aged rats. The dosage given to lab mice was 15mg/kg.d. Changes of emotional behavior and cognitive competence in open field experiment, Morris water maze and two-dimension active avoidance experiment were observed. Arjan Blokland discovered that SPS and BCPS exhibit similar cognitive competence-improving effects which were higher than that of EPS. SPS might be a substitute for BCPS.

4.5 Preparation of phospholipids for injection

1 portion of powdery soybean phospholipids is mixed with 12 portions of distilled water and stirred to form colloidal dispersion liquid in boiling water bath before 1.8 times the weight of raw material of anhydrous sodium sulfate is added. The saturated sodium sulfate solution is discarded after blocky phospholipids are precipitated. Then 5 portions of distilled water and 0.8 portions of anhydrous sodium sulfate are used to repeat the salting out procedure. The salting-out soybean phospholipids are dried at reduced pressure and 70 degrees Celsius in water bath in vacuum drier, transferred into three-mouth bottle followed by addition of 8.7 times the weight of raw material of 95% ethanol and reflux extraction at 80 degrees Celsius in water bath with stirring for 1h. After cooling, the ethanol solution containing phospholipids is poured out and stored in refrigerator overnight to precipitate PC. The ethanol solution containing soybean phospholipids is poured out, heated to about 35 degrees Celsius in water bath before addition of activated aluminum oxide of 0.5 times the weight of raw material, stirred for 1h and filtered. The ratio of powdery phospholipids, ethanol and aluminum oxide is 1:8:0.5 (w/w/w).
The above ethanol solution is poured out followed by addition of activated carbon of 0.22 times the weight of raw material, stirring for 1h and filtration with sintered funnel. The filtrate is transferred into the distillation flask and subjected to reduced pressure distillation at 70 degrees Celsius in water bath under nitrogen to remove ethanol. Diethyl ether of 0.75 times the weight of raw material is added into the distillation flask to dissolve the dried soybean phospholipids. The diethyl ether solution is bottled and the bottle is airtight after filling in nitrogen and stored in refrigerator overnight before ultrafiltration. The diethyl ether is removed by reduced pressure evaporation at 40 degrees Celsius in water bath under nitrogen in evaporator. Anhydrous acetone is added into the glutinous soybean

phospholipids in the evaporator. The mixture is pestled and embathed for several times to remove the residual oil and moisture. Then a powdery parenteral soybean phospholipids product is produced. It exhibits the following characteristics: AV 9.9, IV 91.29, nitrogen content 1.9%, phosphorus content 4.08% and AI 99.3% (PC content is 96.7%) (Shao et al., 2000).

5. Extraction and purification of individual molecular species of soybean phospholipids

Extraction and purification of phosphatidic acid of C_{18} fatty acids

Powdery soybean phospholipids containing 20% PA are extracted with five folds (by weight) of 95% ethanol at 45 degrees Celsius for 2h with stirring at the speed of 100rpm. After centrifugation at 700xg for 10min, the supernatant is discarded. The above extraction procedure is repeated four more times until the ethanol fraction is colorless. The solid fraction is extracted with 5 folds (by weight) of methanol with a stirring speed of 100rpm for 12h at room temperature. The methanol fraction is obtained after centrifugation at 700xg for 10min. The methanol extraction procedure is repeated three times in total. The methanol fractions are combined. If the methanol is removed by evaporation, the solid residue will contain about 50% PA.

The pH of the methanol extract is adjusted to 8-9 by 1mol/L sodium hydroxide and obvious white precipitate is observed. The supernatant is obtained by centrifugation at 700×g for 5min. If the methanol is removed, the solid residue will contain about 70% PA. The pH of the supernatant is adjusted to 5-6 by 1mol/L hydrochloric acid before n-hexane of four times the volume of the methanol solution is added and mixed. The n-hexane phase is obtained after extracting for 15min and being left standing still for 2h. The n-hexane is evaporated at 45 degrees Celsius and the residue is dissolved in methanol of ten times the volume of the n-hexane phase. 60% zinc chloride solution is added into the methanol solution until no more white precipitate is formed. The precipitate is obtained by centrifugation at 700×g for 10min, washed three times with acetone which is removed by filtration and dried under nitrogen gas steam. The solid obtained is PA of C_{18} fatty acids of about 98% pure. The yield is 1/60 of the raw materials (Liu et al., 2008).

6. Applications of phospholipids

Phospholipids are widely applied in pharmaceutical field, food, feed, agriculture, daily chemical industries and other chemical industries.

Pharmaceutical field: Phospholipids exist in all of the biomembranes and play important roles in such various physiological processes as regulating of cell osmosis and membrane enzymes, transmitting of lipoids and sterols and metabolizing of cyanocobalamin, folic acid and methionine etc. Brain tissues contain 25% of phospholipids, and the metabolic abnormality of phospholipids may lead to such diseases as cancer and Alzheimer's Disease etc. High-dosage phospholipids can effectively treat neurological disorders and other diseases of nervous system. In recent years, almost 25% of the non-food application patents of phospholipids are about their applications in pharmaceutical field, especially the applications of liposomes. Phospholipids exhibit huge potential in health care products market.

Food industry: The amount of phospholipids used in food is usually 0.1%-2% of the fat in food. Phospholipids are used in margarines, shortenings, candies, soup bases, pot foods, instant foods (e.g. milk powder), bakery products (e.g. bread, cookie, dessert, biscuit and cracknel) and processed foods of meat and seafood etc. They are also used as coatings of can, soup packaging and casing of meat such as sausages etc.

Feed industry: Phospholipids are applied in animal feed such as milk replacer for calf and feed of cattle, pig, poultry, hairy animals, pets and aquatic animal (e.g. fish and crustaceans) etc. Phospholipids are the essential additives of the eel feed as they decrease the diseases of the eel and improve their growth.

Agriculture: Phospholipids can inhibit the growth of powdery mildew on cucumber, eggplant, green pepper and strawberry. The solution of 0.1% phospholipids-sodium carbonate can effectively inhibit orange green mold, cucumber powdery mildew and rice blights. Phospholipids are used as the coating components after harvesting of fruits and vegetables to improve the storage effect. Phospholipids are additives of pesticide formulae which can improve the adhesivity and permeability of pesticide and reduce their toxicity to plants.

Daily chemical industries: Phospholipids are applied in such cosmetics as moisturizer, facial cleanser, sunblocking cream, soap, bath oil, shampoo, hair care agent, shaving cream, shave clean agent, nail polish, makeup powder, blush, rouge, eye shadow, lipstick and hairspray etc. Applying of phospholipids in detergent can improve the dirt-removing power of anionic detergents.

Other chemical industries: Phospholipids are widely used in various paints, wax, shoes polish, wood preservatives, mold spray, tape coating, printing ink, ink, toner, additives of photographic materials and polyamide coating etc. as well as in papermaking and printing. They are also widely used in cement, pitch, tar shingle, surface sealant of linoleum and putty gum etc.

7. References

An, H.; Wang, M.L. & Cheng, L.B. (2001). Fractionation of Soya Lecithin With High Phosphatidylcholine Content Using Isopropanol. *Fine Chemicals*, Vol.18, No.7, pp. 385-387, ISSN 1003-5214

An, H.; Zhang, L. & He, X.F. (2006). The Fractionation of Phosphatidylethanolamine. *Food Science*, Vol.27, No.12, pp. 343-346, ISSN 1002-6630

Blokland, A.; Honig, W.; Browns, F. & Jolles, J. (1999). Cognition-Enhancing Properties of Subchronic Phosphatidylserine (PS) Treatment in Middle-Aged Rats: Comparison of Bovine Cortex PS with Egg PS and Soybean PS. *Nutrition*, Vol.15, No.10, pp. 778-783, ISSN 0899-9007

Deng, Q.G.; Qi, L. & An, H. (2003). Study On Separating Technique of Soybean Lipositol. *Chemical Engineer*, No.6, pp. 14-16, ISSN 1002-1124

Gu, L.W.; Gao, X.R. & Zhao, J.L. (1999). Recent Advances in Enzymatic Modification of Phospholipids. *China Oils and Fats*, Vol.24, No.6, pp. 60-62, ISSN 1003-7969

Guan, R.L.; Zou, J. & Zhu, H. (2005). Recent Advances in Isolation and Purification of Soybean Lecithin. *China Food Additives*, No.5, pp. 44-48, ISSN 1006-2513

Huang, G.W.; Zhao, J.T. & Gu, K.R. (2003). Hydrogenation of Soybean Powder Lecithin. *China Oils and Fats*, Vol.28, No.2, pp. 51-54, ISSN 1003-7969

Ji, B.P. & Li, B. (2005). *Soybean Production Safety and Quality Control,* Chemical Industry Press, ISBN 7-5025-7036-5, Beijing, China

Liu, D.C. (2007). A Method of Producing Powdery Phospholipids Using Basified Concentrated Phospholipids, Patent number ZL200410024223.8

Liu, D.C.; Feng, J.J.; An, L.G.; Zhu, H.R. & Han, L.Y. (2006). A New Method of Producing Powdery Phospholipids Using Concentrated Phospholipids of Peanut, Soybean or Colza As Raw Materials, Patent number ZL200310105345.5

Liu, D.C. & Ma, F.C. (2011). A Method of Producing Powdery Phospholipids of Various Phospholipids Contents With One Kind of Concentrated Phopholipids, Patent number 201110060102.9

Liu, D.C.; Tao, Y.H.; Sun, F.Z. & Li, Y.L. (2008). Extraction and Purification of Phosphatidic Acid of C_{18} Fatty Acids from Powdered Soybean Phospholipids. *Asian Journal of Chemistry,* Vol.20, No.7, pp. 5595-5600, ISSN 0970-7077

Liu, D.C.; Yang, Y.T. & An, L.G. (2006). A New Technique of Producing Powdery Phospholipids, Patent number ZL200510042316.8

Lu, Q.Y. (2004). *Production Technology of Oil and Fat Chemical Products,* Chemical Industry Press, ISBN 7-5025-5098-4, Beijing, China

Ni, H.Z. (1995). Refining and Isolation of Phospholipids. *Jiangsu Chemical Industry,* Vol.23, No.1, pp. 12-15, ISSN 1002-1116

Shao, X.F.; Zhang, J.L.; Wang, F.L. & Chen, Q.H. (2000). Studying on Injection Phosphatide. *China Oils and Fats,* Vol.25, No.5, pp. 34-35, ISSN 1003-7969

Shi, Y.G. (2005). *Soybean Products Technology* (2nd edition), China Light Industry Press, ISBN 7-5019-4807-0, Beijing, China

Song, Y.Q.; Yu, D.Y.; Luo, S.N. & Wang, S.R. (2007). Modification of Soybean Phospholipids: A Review. *Soybean Bulletin,* Vol.4, pp. 22-29, ISSN 1009-2765

Szuhaj, B.F. (Ed.). (1989). *Lecithins: Sources, Manufacture and Uses (AOCS Monograph),* American Oil Chemists' Society, ISBN 9780935315271, Urbana, USA

Tekerikler, L. (2001). Selective Extraction of Phosphatidylcholine from Lecithin by Supercritical Carbon Dioxide/Ethanol Mixture. *Journal of the American Oil Chemists' Society (JAOCS),*Vol.78, No.2, pp. 115-118, ISSN 0003-021X

Wu, S.M. (2001). *Functional Oils and Fats,* China Light Industry Press, ISBN 7-5017-3049-X, Beijing, China

Xu, Z.S.; Wang, L.J. & Liu, B.Z. (2008). Modification of Soybean Phospholipid and Its Key Technique. *Cereals & Oils,* No.11, pp. 3-9, ISSN 1008-9578

Yang, W.D. (2010). Synthesis of Phosphatidylserine Catalyzed by Phospholipases D. *Modern Food Science and Technology,* Vol.26, No.9, pp. 994-996, ISSN 1673-9078

Zhang, H.F.; Gao, H.; Huang, Y.R.; Chen, J.Y. & Xia, X.Y. (2004). Study of Structures and Performances About the Soybean Lecithin Sulfonate (I)— Structures and Physical Chemistry Performances of Soybean Lecithin Sulfonate. *China Leather,* Vol.33, No.13, pp. 48-51, ISSN 1001-6813

Zhong, Z.S. & Wei, B. (2008). Extraction and Purification of Cephalin from Soybean Phospholipid. *Modern Food Science and Technology,* Vol.24, No.4, pp. 333-335, ISSN 1673-9078

23

Soybean Protein Fibres (SPF)

Tatjana Rijavec and Živa Zupin
University of Ljubljana
Slovenia

1. Introduction

Soybean protein fibres (SPF) are manufactured fibres, produced from regenerated soya *Glycine Max* soybean proteins in combination with synthetic polymer (polyvinyl alcohol) as a predominant component. According to textile fibre labelling (FTC, 2010), textiles from SPF can be marked as azlons from soybean. Azlons are manufactured fibres in which the fibre-forming substance is composed of regenerated naturally occurring proteins (FTC, 2011).
The first commercially successful method for producing regenerated protein fibres was developed by the Italian chemist Antonio Ferretti in 1935 (Ferretti, 1944; White, 2008). In 1936 Snia Viscosa (Milan) started with the production of the world's first commercially produced protein fibres Lanital™ which were made from milk casein (Anon., 1937). Courtaulds in Great Britain (casein fibres Caslen, Fibrolan), Enka in Netherlands, Germany and United States of America (casein fibres Aralac, R-53) soon followed with their commercial productions. Fibres were treated with formaldehyde or aluminium salts, to create cross-links between proteins in the fibre and improve fibre's wet properties. In the year 1945 Snia Viscosa replaced Lanital™ fibres with Merinova™ casein fibres (Fig. 1), which had better properties than Lanital™ fibres.
In the middle of the 20th century and until 1960, vegetable regenerated protein fibres from oilseed peanuts proteins (Ardil fibres, produced by British ICI Company) (Fig. 1) and from corn zein proteins (Vicara fibres produced by American Virginia-Carolina Corporation) were also produced among casein fibres. Fabrics made from regenerated protein fibres were soft, lustrous, resilient, with a good hand and thermal resistance. They were used as a wool or silk substitute by many European fashion designers.
Rapid development of cheaper synthetic fibres with excellent mechanical properties in the early sixties had influence on the commercial production of regenerated protein fibres that was completely discontinued in the middle of the 1960s.
Nowadays, increasing world population need additional quantities of textiles. The world fibre production increases from year to year and in 2010 there was globally produced 78 million tons of fibres, including about one million tons of wool and 0.15 million tons of silk (Kanitkar, 2010). Wool and silk are still very expensive fibres, with selling prices of about 14−23 €/kg and 28−40 €/kg, respectively (Reddy & Yang, 2007).
Today's fibre production strategy is redirected from crude oil to renewable raw materials, eco-friendly and sustainable fibres, that could be biodegraded or recycled. Important raw materials for future textile fibres production could be cheap and worldwide available agricultural by-products, like lignocellulose (from rice straw), wheat gluten (Yang et al., 2006), casein protein from milk after butterfat is removed, zein protein from corn after starch manufacture, and soybean protein after beans are pressed and oil is removed.

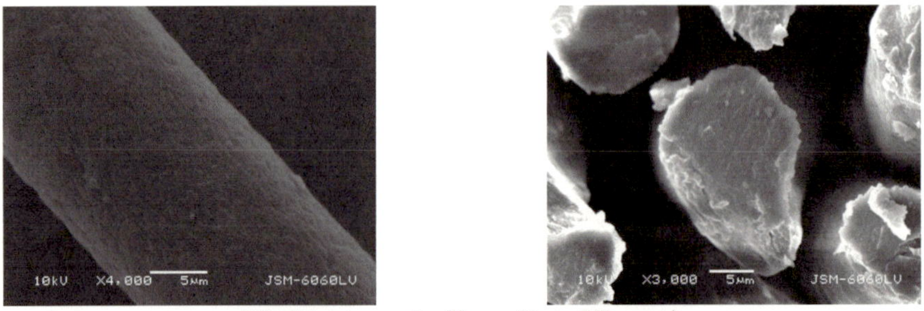

Merinova casein fibres (Snia Viscose)

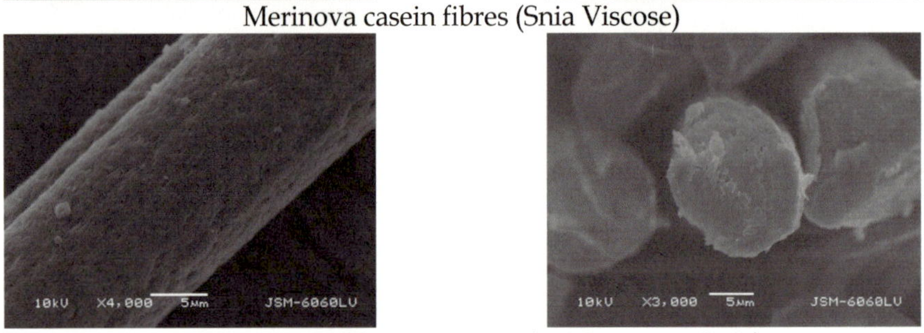

Protilon casein fibres (Les Textiles Nouveaux S. A.)

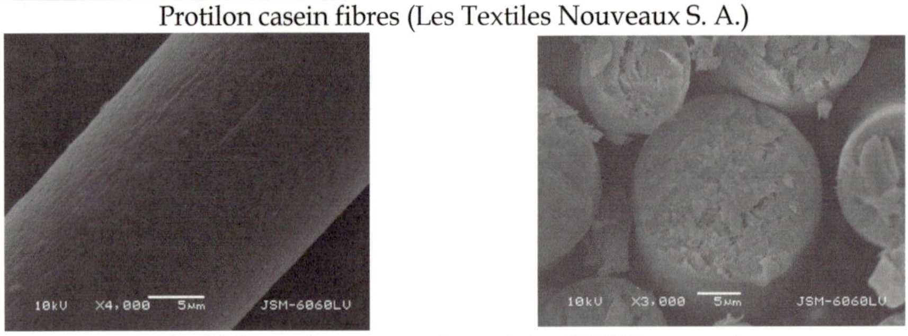

Ardil fibres from peanuts proteins (ICI Company)

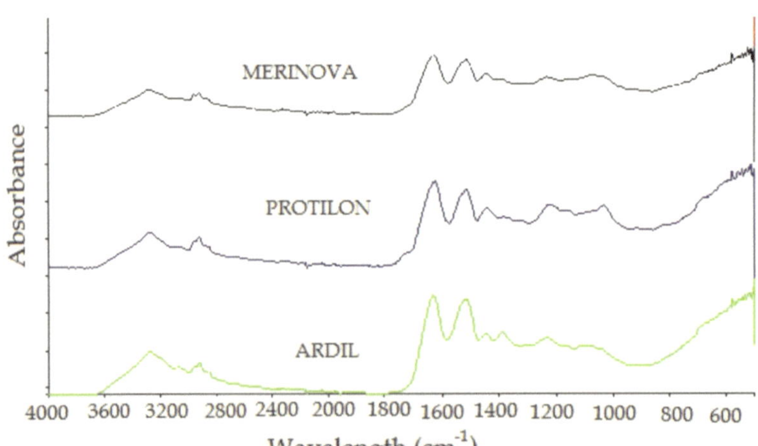

Fig. 1. Scanning electron microscope views and comparative FT-IR/ATR spectra of pure protein fibres with typical absorption peaks at 1658 cm^{-1} (amide I) and 1538 cm^{-1} (amide II).

First protein fibres had low tensile properties, especially in wet state. In order to improve mechanical properties of protein fibres, proteins were combined with synthetic polymers such as acrylonitrile or vinyl alcohol by graft copolymerization or polyblending. First such fibres, made on the patent basis of Morimoto (Morimoto et al., 1962), were produced by the Japanese Toyobo in 1969. The copolymer fibres Chinon® were made from 30% casein and 70% acrylonitrile. Acrylonitrile was grafted on protein with the addition of minor amounts of vinyl or vinylidene chloride for flame retardation. Fibre's density was 1.22 g/cm^3, tensile stress in dry state 3.5-4.5 cN/dtex and moisture regain 4.5–5.5%.
Combining natural proteins and synthetic polymers to get fibres with good moisture absorbency and high tenacity led to new researches in the field of fibres at the beginning of the 21st century.
New fibres from casein proteins have been commercialized as milk protein fibres in China by Shanghai Zhengjia Milkfiber Sci& Tech Co., Ltd., under the brand name ZhengJia®. In the year 2005 a Chinese patent for producing the fibres was granted (Shanghai Z., 2011). Milk fibres are chemical casein acrylic fibres made from graft copolymer of casein and acrylonitrile. Fibres contain about 25–30% of milk proteins and 70–75% of acrylic component. The process is ecological (in 2004 it passed the Oeko-Tex Standard 100 green certification) with no formaldehyde content. Milk fibres with about 55% crystallinity have round cross section with many irregular vertical trenches and pockmarks on the surface (Wang et al., 2009). The fibres with linear density 2.22 dtex have breaking tenacity 2.5 cN/dtex and higher, breaking elongation 35.5%, elastic recovery 76.5%, moisture regain 4–5% and bacteria resistance ≥80% (Shanghai Z., 2011). Milk fibres could be dyed with reactive and acid dyes and after treated with crease-resist finishing and softening agents (Arslan, A., 2007, 2008, 2009).
New viscose filament yarn Lunacel, produced by Kurabo Industries Ltd. (Osaka, Japan), has combined properties of vegetable and animal fibres. The fibres are made from cellulose cotton linter pulp that is cross-linked with water-soluble food protein (Kurabo, 2007).
Using animal proteins as raw material for spinning fibres is very expensive. New soybean protein fibres (SPF) from soybean proteins and polyvinyl alcohol were developed in China by G. Li at Huakang R&D Center (Li, 2003, 2007). The fibres are first manufactured fibres, invented by China. The production process for new fibres was laboratory established in 1993 and commercially promoted in 2000. In 2001 the fibres were standardised and in 2003 they were launched.
The objective of this study was to investigate the contemporary SPF biodegradation in soil at controlled laboratory conditions.

2. Soybean protein fibres

2.1 Fibre forming soybean proteins

Soybeans are very reach with proteins (about 37–42% of dry bean) (Krishnan et al., 2007) in comparison to milk (3.2%), corn (10%) and peanuts (25%). Soybean proteins are used for food and feed and in many industries as adhesives, emulsions, cleansing materials, pharmaceuticals, inks, plastics and also textile fibres. Raw material for spinning textile fibres is obtained from soybean remaining flakes after the extraction of oils and other fatty substances (Li, 2004).
Amino acids content of soybean proteins is given in Fig. 2. Soybean proteins contain 18 different amino acids. There are about 23% of acidic amino acids (glutamic acid and aspartic

amino acid), about 25% of alkaline amino acids (serine, arginine, lysine, tyrosine, threonine, tryptophan) and about 30% of neutral amino acids (leucine, phenylalanine, valine, alanine, isoleucine, proline, glycine). Sulphur containing amino acids are present also in soy proteins: about 1.0% of cysteine and 0.35% of methionine.

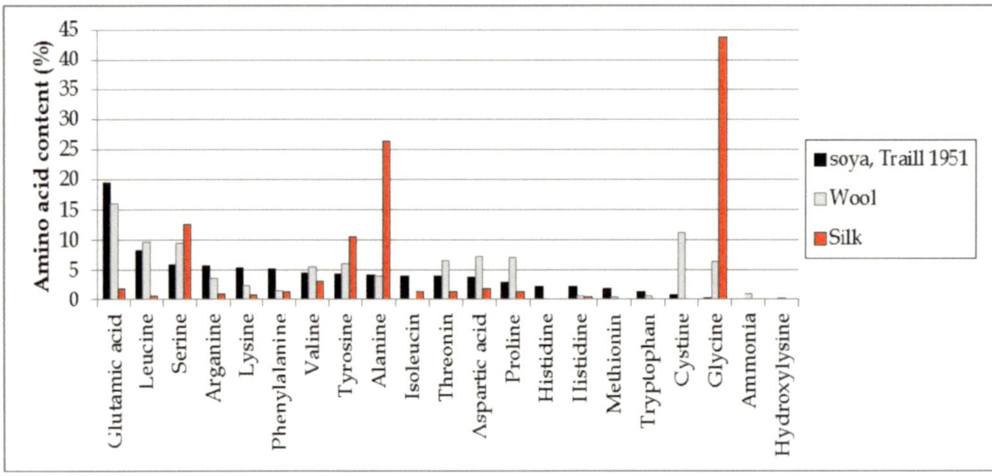

Fig. 2. Amino acids content in soybean proteins, wool keratin and silk fibroin (Brooks, 2005).

Soybean proteins consist of various groups of polypeptides with a broad range of molecular size: about 90% are salt-soluble globulins (soluble in dilute salt solutions) and the remainder is water-soluble albumins (Zhang, 2008). Very important as raw material for producing textile fibres are storage globulins with predominant β-conglycinin (30–50% of the total seed proteins) and glycinin (ca. 30% of the total seed proteins). β-conglycinin is a heterogeneous glycoprotein composed of three subunits (α', α, β) contained asparagine, glutamine, arginine and leucine amino acids. Subunits are non-covalently associated into trimeric proteins by hydrophobic interactions and hydrogen bonding without any disulphide bonds. Glycinin is a large hexamer, composed from acidic and basic polypeptides linked together by disulphide bonds (Zhang, 2008). On the basis of the sedimentation coefficient, a typical ultracentrifuge pattern of soybean proteins has four major fractions: 2S, 7S, 11S, and 15S (Zhang, 2008).

Globular proteins are composed of segments of polypeptides connected with hydrogen bonds, electrostatic interactions, disulphide bonds and hydrophobic interactions. Conformational changes of unfolding globular proteins through denaturation process (Zhang & Zeng, 2008) and reducing the inclination of denaturated proteins to form aggregates are important for spinnability of a spinning dope with proper relative viscosity. It is also important for later drawing of fibres and crystallization of proteins in fibres. Denaturation (Fig. 3) is modification of the secondary, tertiary, and quaternary structure of protein. Exposure of soybean proteins to strong alkali/acids, heat, organic solvents, detergents and urea causes the denaturation of native globular proteins, i.e. converting into unfolded polypeptide chains, which are connected with interchanging of disulphyde bonds. Extruded fibres coagulate in a precipitation acid bath and new disulphide bonds are formed. The structure of soybean proteins and changes at converting globular proteins into fibre forming proteins are given in Fig 3.

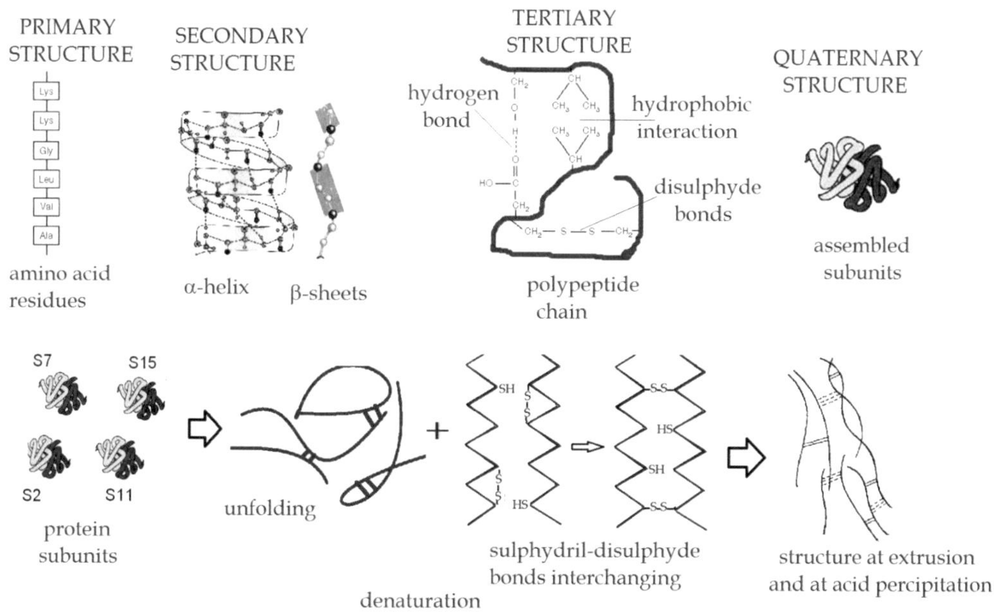

Fig. 3. Protein levels and conversion of globular proteins into fibre forming proteins (Zhang & Zeng, 2008; Kelly & Pressley, 1996).

Oils extraction with solvents used in the mid-twentieth century, was critical for the whole spinning process of soybean fibres, because the chosen temperatures, pH, urea, salts, organic solvents (hexane) and reducing agents influence on the degree of denaturation of proteins, degradation of proteins and changing of proteins colour. Protein degradation is detrimental to the production of high-strength protein fibres. Modern method of modifying soybean globular proteins is biochemical with using enzymes and auxiliary agent (Swicofil, 2011).

2.2 Pure soybean protein fibres from the mid-twentieth century

First researches for developing fibres from soybean proteins were made by the Japanese. In the year 1940 the first US patent was granted to Japanese Toshiji Kajita and Ryohei Inoue (Kajita & Inoue, 1940). The oil-free protein substance was extracted with dilute alkaline solution and precipitated by adding metallic salts. The protein was then washed in water and added by tartaric acid when the precipitate was wet. Then it was again dissolved in alkaline solution to form a spinning dope. Fibres were spun in an acid bath with organic coagulating agent (alcohol, formaldehyde, acetone etc.), where filaments hardened (Kajita & Inoue, 1946). The fibres had natural white to light tan colour. They were crimped, with high resiliency, warmth and soft feel. In comparison to wool they had lower tensile strength, especially in wet state, and lower moisture absorbency.

Patents for spinning fibres from soybean proteins were granted to the American Oscar Huppert from Glidden Company (Huppert, 1945) and Robert A. Boyer from Ford Motor Company (Boyer et al., 1945). In 1939 the American Ford Motor Company produced soybean protein fibres for their car's upholstery and seats fillings. The fibres, which have never been commercialized, had about 80% the strength of wool, higher elongation in dry and wet state than wool and didn't wet so easily as wool or casein fibres (Boyer, 1940).

Chemical and dyeing properties of pure regenerated soybean protein fibres were similar to wool.
Soybean protein fibres were also produced in Japan under the name Silkool (Myers, 1993). In 1939 the fibre production reached about 450–1,200 tons.
Low tensile strength of soybean protein fibres in wet state limited their commercial application. Fibres were used predominantly in blends with wool, cotton or synthetic fibres in woven and knitted fabrics for apparel and in upholstery, also in cars, despite of lower abrasion resistance than wool (Fletcher, 1942). The production of the mid-twentieth soybean protein fibres was ceased at the end of the World War II.

2.3 Researches on soybean protein fibres in the early twenty-first century

Huang et al. (1995) have made experimentally the textile fibres from soybean protein by re-examining the wet spinning method, described in the literature (Croston et al., 1945). The properties of the fibres made by wet spinning method from alkaline solution of soybean protein isolate and coagulated in acid bath were compared with the fibres made by dry spinning method of water solution of soybean protein isolate. Tensile properties of treated fibres were 0.77 cN/dtex at 11% relative humidity (r.h.), 0.75 cN/dtex at 65% r.h. and only 0.08 cN/dtex in wet state. They were lower than those of wool in most conditions (Huang et al., 1995). They found out that dry spinning was a suitable method for spinning soybean protein fibres because of their good solubility in water and glycerol. In the next experimental step they tried to increase tensile properties by decreasing the moisture absorption of soybean protein fibres. They used relatively nonpolar zein proteins (20, 30 and 40%), which were added to soybean protein into the spinning dope. Fibres were made by dry spinning method. The optimum soy protein-zein blended fibre was made from a suspension containing 80% of soybean protein and 20% of zein in glycerol (Zhang et al., 1997), but the tenacity was only 0.20 cN/dtex.
Another idea to improve low tensile strength and decrease shrinkage in boiling water was using water-soluble polymer, such as polyvinyl alcohol (PVA). PVA fibres are produced in similar conditions as phytoprotein fibres. Zhang (Zhang et al., 1999) experimented with bicomponent fibres from soybean protein and PVA. Fibres with a side-by-side configuration were not successful because of splitting of the components. The reason was in too large difference in swelling of the components in water. The next experiment of spinning sheath-core bicomponent fibres, with PVA component in the sheath and soybean proteins in the core, showed brittle core that couldn't be drawn. „The degradation of the soybean protein and the existing microgels in the protein spinning solution were thought to be the causes for the poor fibre drawability" (Zhang et al., 1999).
After ten years of intensive researches the Chinese scientists with Guanqi Li succeeded in producing high-tenacity soybean protein fibres from soybean protein and polyvinyl alcohol (Li, 2007). The process and fibre's properties are presented in section 2.4. Polyvinyl alcohol adds strength and acceptable wearability characteristics to the new SPF.
Biconstituent fibres from a biocompatible soy protein isolate and cellulose were produced experimentally from new aqueous solution NaOH/thiourea/urea. Strong hydrogen bonds between hydroxyl groups of cellulose and amid groups of protein were formed. Fibres with linear density of 6.2 dtex were produced with tensile strength of 1.86 cN/dtex and breaking elongation of 10.3% (Zhang et al., 2009).
High-wet strength fibres containing 5–23% of a soybean protein isolate from oiled soybean cake and 77-95% of polyvinyl alcohol were developed by scientists with Guanqi Li at

Huakang R&D Center in China (Li, 2003, 2007). A soybean protein isolate is treated with an auxiliary agent and biological enzymes to modify the structure of globular proteins. Additives break the disulphide bonds in globular proteins and convert them into linear molecules, which are stable in temperature range 55–90 °C (Mathur & Hira, 2004).

Fibres are wet spun from deaerated spinning dope composed of a soybean protein and polyvinyl alcohol dissolved in distilled water, followed by adding of borax or boric acid and mixing at temperature between 40 and 98 °C. After coagulation in a water bath with salt and alkali, as spun fibres are wet drawn, then dried, pre-heat set, heat-set at 170-185 °C, cooled, winded, stabilised by acetalysing, washed, oiled, crimped and cut into staple fibres. Production process doesn't pollute the environment. Most added agents in the process can be recovered from semi-finished fibres and used again.

The molecules of protein are laterally bonded with molecules of polyvinyl alcohol in the fibres. This enables during additional extension, orientation and crystallisation of proteins in the fibres during drawing. The morphological structure of SPF consists of less oriented sheath and well oriented microfibrilar core. The fibres have about 10% of hydrophilic groups in amorphous regions (Mathur & Hira, 2004).

Properties of soybean protein fibres taken from yarn SoySilk™ and milk protein fibres taken from yarn SilkLatte® are given in Table 1 (Brinsko, K. M., 2010).

Yarns	soybean protein fibres	milk protein fibres
available from	Southwest Trading Company	Southwest Trading Company, Tempe, AZ
cross-section and longitudinal view	bean-shaped with pronounced and elongated micro-pores inclusions	bean-shaped with small micro-pores inclusions
birefringence	0.021–0.027	0.016–0.024
melting point	250–260	235–245
chloroform, AcOH, acetone, DMF	insoluble	insoluble
formic acid	swell	swell
conc. H_2SO_4, conc. HNO_3	partially soluble	soluble, gels
characteristic peaks on FT-IR spectrum	amide I at 1640 cm^{-1} amide II at 1530 cm^{-1}	amide I at 1640 cm^{-1} amide II at 1530 cm^{-1}

Table 1. Properties of soybean protein fibres taken from yarn SoySilk™ and milk protein fibres taken from yarn SilkLatte® (Brinsko, K. M., 2010).

Adding some metallic salts into spinning dope, endows soybean fibre with far-infrared, negative ion and anti-bacterial functions. Only 3% of such fibres added into yarn can give stable and permanent antibacterial effect. Another technology from the same university is adopting $ZnSO_4$ as the dehydrating agent for soybean fibre spinning. In the course of after-processing, $ZnSO_4$ reacts with NaOH, forming ZnOH, which after drying is deoxidized into nanograde ZnO that can form covalent bond with fibre itself, taking a strong screen effect to ultraviolet radiation (Yang, 2011).

2.4 Commercial soybean protein fibres in the early twenty-first century

SPF based on the Li Guanqi patent (Li, 2007) are the first industrially produced fibres from soybean proteins in the world and they are the only soybean protein fibres present on the

market today. These fibres are also the first manufactured fibres, developed by China. The production process of the new SPF was laboratory established in 1993 and commercially promoted in 2000. In 2001 the fibres were standardised and in 2003 launched.
About 1,500 tons of the fibres per year are produced under the brand name Winshow by Shanghai Winshow Soybean Fibre Industry Co., Ltd. Six manufacturing bases were established in four provinces in China for producing SPF (Shanghai, 2011). Zhejiang Jiali Protein Fiber Co., Ltd. is the owner of the soybean protein fiber international intellectual property rights and production line.
The Chinese manufacturer of soybean protein fibres Harvest SPF Textile Co., Ltd. (www.spftex.com) is a Chinese-foreign joint venture co-incorporated by China Harvest International Industry Ltd. and Zhejiang Jiali Protein Fiber Co. Ltd. (Shanghai, 2011). They are specialized in the research and development of new textile fibre raw material application technologies and application of the new-type textile materials from SPF. Fibres and yarns from soybean protein fibres are also available from Swicofil AG Textile Service (Anon, 2011), South West Traiding Company with yarn SoySilk™ (SWTC, 2011).
Since SPF resemble in their softness and shine to silk and cashmere, producers market them as "artificial cashmere", "vegetable cashmere" or "soy silk" fibres to partially decrease needs for natural silk and cashmere fibres. Cashmere goats cause damages to lands, so reducing their number has ecological benefits.
Physical and chemical properties of soybean protein fibres are given in Table 2.

PROPERTIES	SPF	Cotton	Viscose	Silk	Wool
Breaking strength (cN/dtex) in dry state	3.8–4.0	1.9–3.1	1.5–2.0	2.6–3.5	0.9–1.6
Breaking strength (cN/dtex) in wet state	2.5–3.0	2.2–3.1	0.7–1.1	1.9–2.5	0.7–1.3
Breaking elongation (%) in dry state	18–21	7–10	18–24	14–25	25-35
Initial Modulus (kg/mm²)	700–1300	850–1200	850–1150	650–1250	
Loop strength (%)	75–85	70	30–65	60–80	
Knot strength (%)	85	92–100	45–60	80–85	
Moisture regain (%)	8.6	9.0	13.0	11.0	14–16
Density (g/cm³)	1.29	1.50–1.54	1.46–1.52	1.34–1.38	1.33
Heat resistance	Yellowing and tackifing at about 120 °C (Bad)	Becoming brown after long time processing at 150 °C (Excellent)	Strength down after longtime processing at 150 °C (Good)	Keep stable when temperature <=148 °C (Good)	(Good)
Alkali resistance	At general level	Excellent	Excellent	Good	Bad
Acid resistance	Excellent	Bad	Bad	Excellent	Excellent
Ultraviolet resistance	Good	At the general level	Bad	Bad	Bad

Table 2. Comparison of physical and chemical properties of soybean protein fibres (SPF) in comparison to cotton, viscose, silk and wool (Swicofil, 2011).

A raw SPF has light yellow colour, like silk oak. Before dyeing into light colours they should be bleached with hydrogen peroxide or reduction bleached. SPF fibres can be dyed at temperatures lower than 100 °C with weak-acid dyes and substantive dyes for very few colours, because the dyeing fastness is poor. As SPF are less sensitive to high pH, they could be also dyed with reactive dyes (Mathur & Hira, 2004). SPF fibres have good light fastness and good resistance to ultraviolet radiation, which is better than that of cotton, viscose and silk. They are stable to washing even at higher temperatures, but they yellow at dry heat at 120–160°C (Anon., 2003).

Likewise regenerated cellulose bamboo fibres, SPF fibres are promoted on the market as biocompatible and health giving with natural antibacterial properties. The Chinese herbal medicine with sterilising and anti-inflammatory properties can be bonded on side chains of the proteins during the production of SPF (Yi-you, 2004) due to the bacterial resistance of SPF fibres to *Styphalococcus aureuses, coli bacillus* and *Candica albicans* (Swicofil, 2011). Mathur has mentioned that SPF resistance to golden and yellow *Styphalococcus aureuses* is more than 5.8 and hence they are inherently anti-bacterial fibres (Mathur & Hira, 2004).

Beside in yarns from 100% of SPF, the SPF cotton type fibres could be used in yarn mixtures with cotton, polyester, viscose and bamboo viscose. The wool type SPF should be mixed with cashmere (80/20 SPF/cashmere), lyocell, silk or wool (50/50 SPF/wool). Smooth surface of SPF has influence on low spinnability because of low friction coefficient and low cohesion force, and on pilling.

Fabrics with SPF should not be mercerised because SPF are not resistant to strong caustic soda. Woven and knitted fabrics can be used for apparel (personal underwear, T-shirts, pullovers, sweaters, evening dresses, children's clothing and sportswear) and home textiles (towels, bed linen, blankets, bathrobes, pyjamas). Since the fibres have lower abrasion resistance than wool they can be used as upholstery in automobile textiles.

A cloth made of SPF fibres exhibits good wiping properties (Reek, 2008). At least 10% of Winshow SPF of linear density 1.5 dtex and 38 mm length from Shanghai Winshow Soybeanfibre Industry Co., Ltd. of Shanghai, China is used in combination with viscose and/or other textile fibres in thermo-bonded nonwoven fibrous material.

2.5 Biodegradation of contemporary SPF

Biodegradable fibres degrade relatively quickly through biological process, which depends on many factors, such as chemical and morphological structure, temperature, pH, relative humidity and remains of auxiliary agents, which are accumulated (brought) on fibres during manufacturing and are not completely washed after finishing process (Simončič & Tomšič, 2010).

The chemical structure has influence on biodegradability with its hydrophilic nature (wettability), crystallinity of the polymer, chemical linkages in the polymer backbone, pendant groups, end groups and molecular weight distribution. Peptide bonds are susceptible to enzymatic degradation. Additional polymers may (interaction with other polymers) act as barriers to prevent migration of microorganisms, enzymes, moisture or oxygen into the polymer domain of interest (Zee, 2005).

The biodegradation process of proteins is initiated through exposure to water. Long macromolecules under hydrolytic process convert into many small molecules, which are more proper for the metabolism of microorganisms.

The mid-twentieth pure soybean protein fibres mildewed less easily than natural and casein fibres but more easily than synthetic fibres (Fletcher, 1942). Mid-twenty century soybean protein fibres were susceptible to microbiological growth. Casein fibres were readily damaged by mildew, they quickly mildewing especially in damp conditions. Changing protein molecules by chemicals and tanning (hardening) has influence on lower biodegradability of fibres (Wormell, 1954).

Very little data is yet available about biodegradability of contemporary soybean protein fibres. The fibres are promoted as biodegradable fibres in landfill (Mathur & Hira, 2004; Swicofil, 2011). Fibres from water-soluble polyvinyl alcohol are biodegradable in soil. Considering the chemical structure of SPF (Fig. 2), the soybean proteins susceptibility to biodegradation should be similar to wool and not to silk. Wool contains 80% of keratin, the rest are no-keratin proteins. Degradation of wool is mostly caused by fungus and less by bacteria. Ideal conditions for growth of microorganisms on wool fibres are temperature 30°C, relative humidity of 95% and pH from 6,5 to 8,5 (Edwards & Vigo, 2001). In the initial stage, of biodegradation of wool is hard to be noticed. When the growth of microorganisms increases, unpleasant odour appears, coloured spots can be seen on fabrics and tensile strength as result of defibrillation decreases (Edwards & Vigo, 2001, Szostak-Kotowa, 2004).

3. Experimental part

3.1 Materials

Ring spun yarns and twill 2/2 woven fabrics were used in our experiments of biodegradation of contemporary soybean protein fibres:

- 100% soybean protein yarn with linear density of 15 tex (SPF yarn) and 100% cotton yarn with linear density of 19 tex for comparison;
- a fabric with yarn from soybean protein fibres in weft direction and cotton yarn in warp direction (SPF/CO) and a 100% cotton fabric with cotton yarn in warp and weft direction for comparison (CO).

The same cotton yarn with linear density of 28 tex were used for warp and SPF yarn with linear density of 15 tex were used for weft for all woven fabrics. The density of fabrics was 30 ends/cm and 28 picks/cm for SPF/CO and 100% cotton fabrics.

3.2 Methods

3.2.1 Method of controlled biodegradation in soil

A laboratory experiment of biodegradation (Fig. 4) of yarns and woven fabrics was made in accordance with the standardised method SIST EN ISO 11721-1. Commercial humus, rich in microorganisms, was used as a soil. During experiment the soil humidity was 60±5%, which was regularly measured by a hygrometer and maintained by spraying the soil with tap water. The temperature of the soil was 25−30 ^{0}C. Samples of yarns and woven fabrics were buried in the soil for 2, 7, 11, 16 and 21 days. After that, the samples were washed out in tap water, then immersed into ethanol for 30 minutes to stop the activity of microorganisms, and dried in the air.

Tensile properties of samples were measured on dynamometer Instron 5567 in accordance with the standard SIST EN ISO 2062 for yarns and SIST EN ISO 13934 for fabrics. For measuring tensile properties in wet state, the yarns were immersed into distilled water with detergent at room temperature for an hour. Tensile properties of yarns were analysed with the DINARA program (Bukošek, 1988). Tensile properties of fabrics were measured only in weft direction.

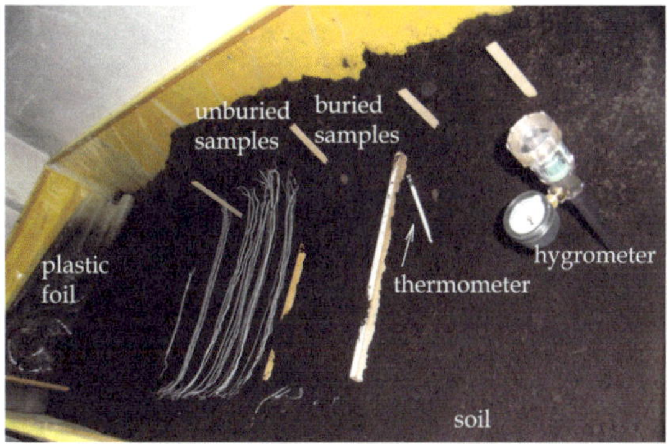

Fig. 4. Experiments of biodegradation were made in a wooden box surrounded with a foil and filled with humus soil.

3.2.2 Other methods

Fourier transform infrared spectra (FTIR/ATR) were obtained on the Spectrum GX (Perkin Elmer) with the Michelson interferometer and Spectrum 5.01 software using 16 scans at a resolution of 4 cm^{-1} in a range of wavenumber from 4000 to 500 cm^{-1}. Microphotographs were made with the Jeol JSM 6060 LV scanning electron microscope and the Nikon SMZ 800 stereomicroscope.

4. Results and discussion

4.1 Fibres properties

The SPF yarn was made from cotton type soybean fibres of 1.27 dtex with an average length of 39.5 cm. Fibres were thermoplastic with melting point at 224°C. Dry fibres absorbed 2.47% of moisture when exposed for 48 hours to the air of relative humidity 50% and temperature of 23 °C.

The cross-section shape of used soybean protein fibres was bean-shaped with diameter of 11-20 μm in longer axle and 6-7 μm in shorter axle (Fig. 5). A very smooth surface of fibres imparted high lustre to fibres. On the longitudinal view irregular grooves and wrinkles can be seen. These grooves can help to transport moisture along fibres. On the optical microscope photograph a nonhomogeneous structure with many voids is seen.

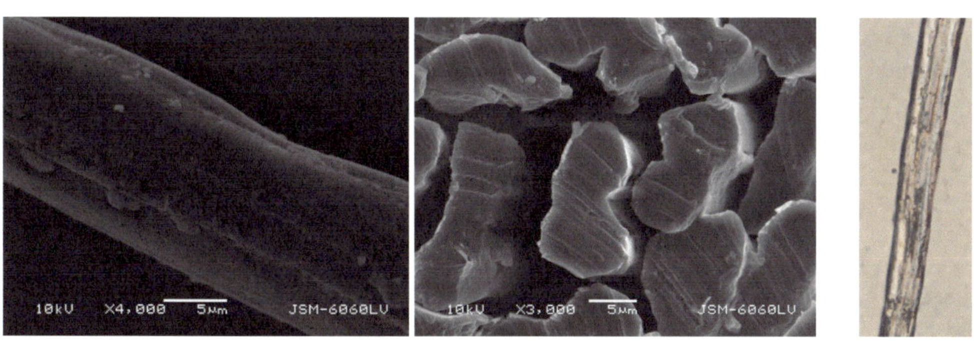

Fig. 5. Scanning electron microscope view of soybean protein fibre: top: longitudinal view at magnification of 4.000 and bottom: cross-section at magnification of 3.000. Right: optical microscope longitudinal view of soybean protein fibre.

Soybean protein fibres are composed of a mixture of two polymers, soybean proteins and polyvinyl alcohol. Protein and polyvinyl alcohol macromolecules are connected by intermolecular interactions like hydrogen bonds (Fig. 3) and van der Waals hydrophilic and hydrophobic forces.

The soybean protein fibres, used in experiment, consisted of polyvinyl alcohol and soybean proteins. The SPF FT-IR spectrum (Fig. 6) has very intensive peaks at 3301 cm^{-1}, which is typical for stretching O-H bonds, and at 1408 cm^{-1} and 1327 cm^{-1}, which corresponds to N-H stretching in amide III. FT-IR absorption spectrum of SPF is different to FT-IR spectrum of PVA fibres at peaks 1644 cm^{-1} and 1535.32 cm^{-1}.

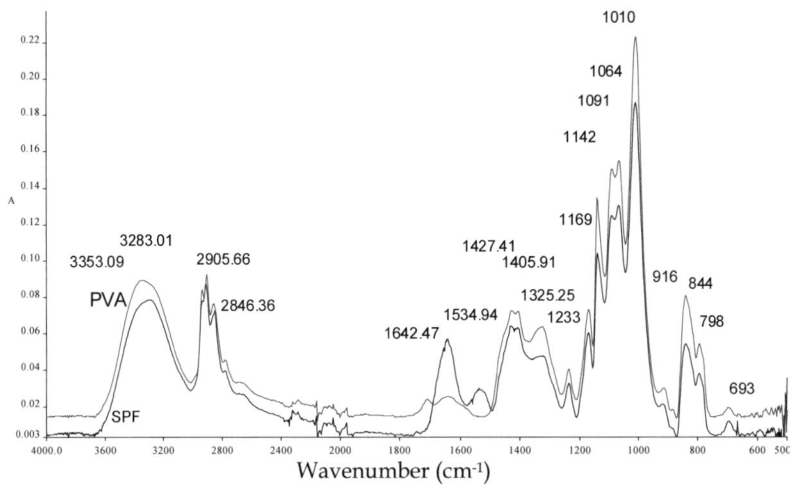

Fig. 6. FT-IR/ATR absorption spectra of soybean protein fibres (SPF) and PVA Kuralon®.

Pure soybean protein isolate has typical infrared absorption bands at 1636-1680 cm^{-1} and 1533-1559 cm^{-1} that are attributable to the -NH- bonds of amide I at 1640 cm^{-1} and at 1550 cm^{-1} for amide II in peptide bonds forming primary backbone of proteins. The absorption peak at 3294 cm^{-1} refers to the hydrogen-bond association between protein chains and moisture in protein. The absorption band at 1241–1472 cm^{-1} is attributable to the (C)O-O and C-N stretching and N-H bending (amide III) vibrations (Su et al., 2008).

At room temperature pure polyvinyl alcohol powder with -OH groups on carbon chains has a typical infrared absorption band at 2918-3565 cm^{-1}, which corresponds to -OH absorption (Su et al., 2008).

4.2 Yarns properties

4.2.1 Tensile properties of SPF yarn in dry and wet state

Water has a significant influence on tensile properties of the SPF yarn (Fig. 7, Tab. 3). After an hour in distilled water, the yarn lost its specific breaking stress for almost one third. Wet yarn had lower modulus then dry yarn in the whole deformation range and attained by 11.4% higher breaking elongation than dry yarn.

4.2.2 Biodegradation of SPF yarns

Biodegradation of SPF yarn was studied after the yarn had been buried for 2, 7, 11, 16 and 21 days in the soil with temperature about 30 °C and 65% relative humidity (Fig. 8). For the

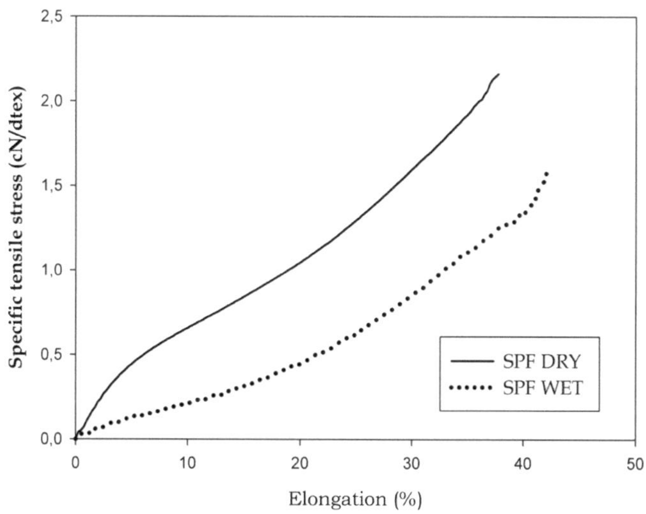

Fig. 7. Stress-elongation curves of dry and wet SPF yarns

Properties of yarns from SPF	DRY	WET	Δ (%)
Specific breaking stress (cN/dtex)	2.16	1.58	-26.9
Breaking elongation (%)	37.75	42.05	+11.4
Initial modulus (GPa)	5.08	3.25	-36.0
Specific work of rupture (mJ/kg)	39.81	22.71	-43.0

Table 3. Tensile properties of dry and wet SPF yarns and relative differences between them (Δ).

purpose of comparison, cotton yarn was buried at the same time in the soil. After 7 days cotton yarn degraded very intensively and only small remains of yarn were left in the soil. Tensile properties of biodegraded cotton yarn could be measured only after 2 days. Biodegradation of cotton showed that microorganisms in the soil were active during the experiment.

The microphotographs of SPF yarns in Fig. 8 show that the quantity of bacteria and fungus, present on the surface of soybean protein fibres, increased with time of biodegradation. After 21 days in the soil, it is hard to say that there are any physical degradations of the fibre's surface because the fibres have natural irregular grooves and wrinkles (0 days).

Soybean protein fibres in comparison to the mid-twentieth century protein fibres (Fig. 1) have essentially smoother surface and relatively lower quantity of surface grooves that could enable bacteria to penetrate into the fibres.

Specific tensile stress-elongation curves of biodegraded cotton yarns (Fig. 9 and 10) show a significant decrease of breaking force and breaking elongation after 2 days, while for the yarns from soya protein fibres, they didn't change essentially.

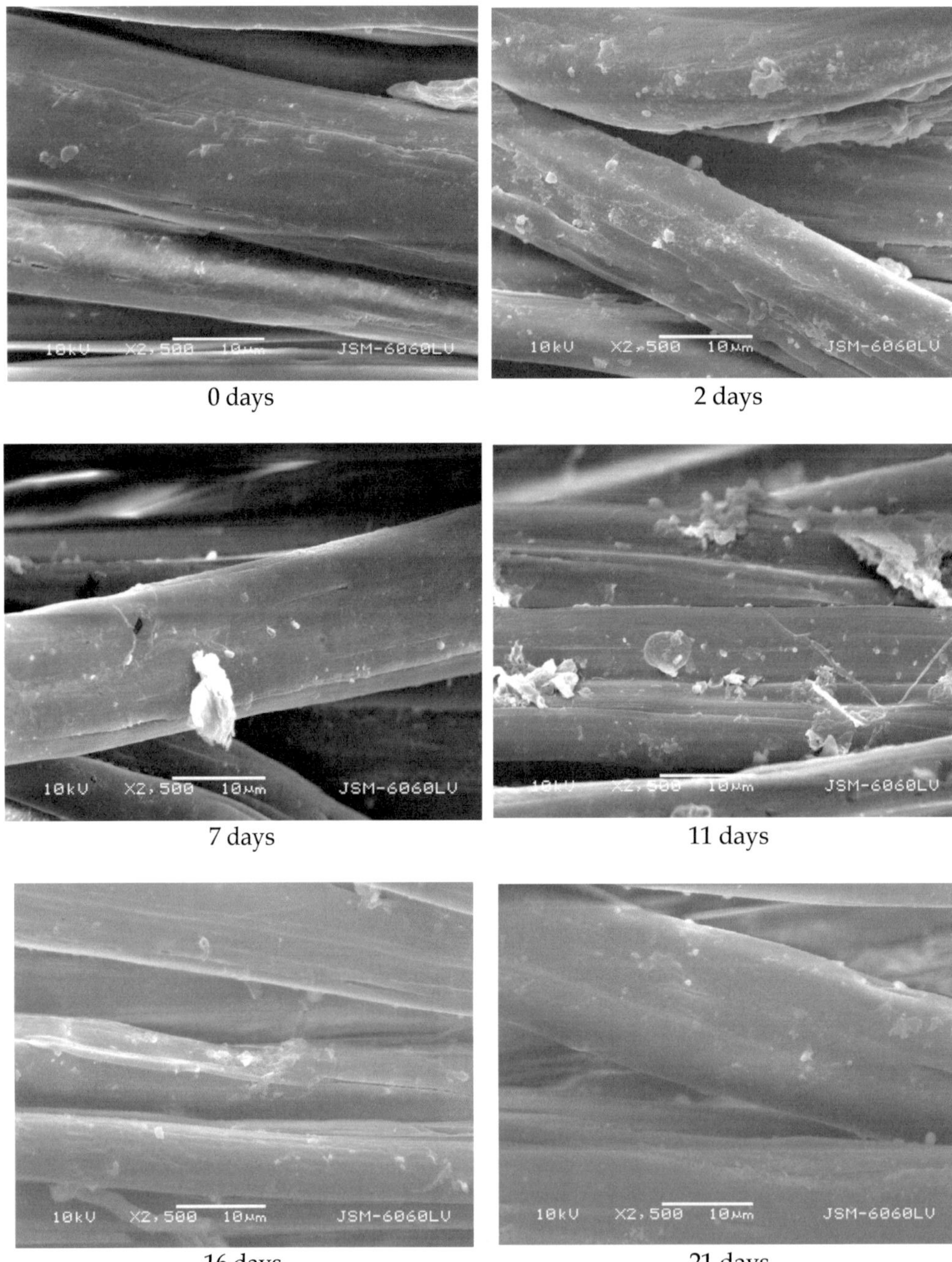

Fig. 8. SEM microphotographs of SPF yarns after having been buried for 0, 2, 7, 11, 16 and 21 days in the soil (at magnification 2500-x).

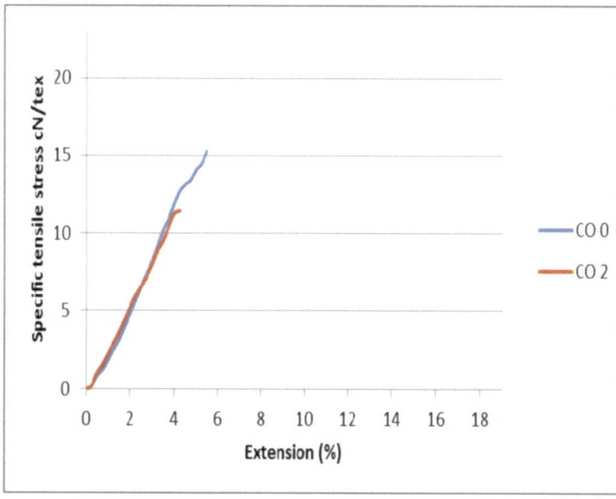

Fig. 9. Stress-elongation curves of cotton yarns after having been buried for 0 (CO 0) and 2 (CO 2) days in the soil.

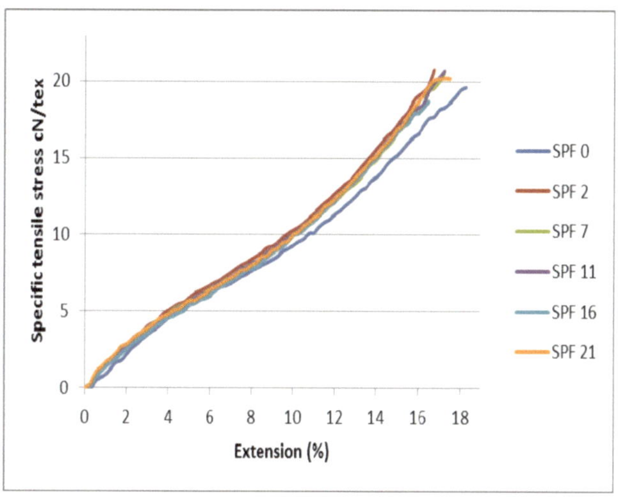

Fig. 10. Stress-elongation curves of cotton yarns after having been buried for 0 (SPF 0) to 21 (SPF 21) days in the soil.

4.3 Biodegradation of fabrics with SPF yarns in weft direction

Woven fabrics were buried at the same time as yarns into the soil at temperature 30°C and relative humidity 65% for 2, 7, 11, 16 and 21 days. Fabric samples (Fig. 11) changed the colour and became browner with many colour spots on the surface, which confirmed the existence of fungus.

Pure cotton fabrics degraded in one weak to such degree that they broke up into pieces when we tried to dig them out of the soil. After 21 days only very little remains were found in the soil. Fabrics with yarns from soybean protein fibres in weft direction were more compact in weft direction than pure cotton fabrics. But in warp direction from cotton yarns the fabrics lost their strength and were easily torn (Fig. 11).

Fig. 12 shows that after 21 days in the soil the cotton cuticle was destroyed. On the soybean protein fibres the quantity of fungus and bacteria increased, but the surface of fibres was not damaged.

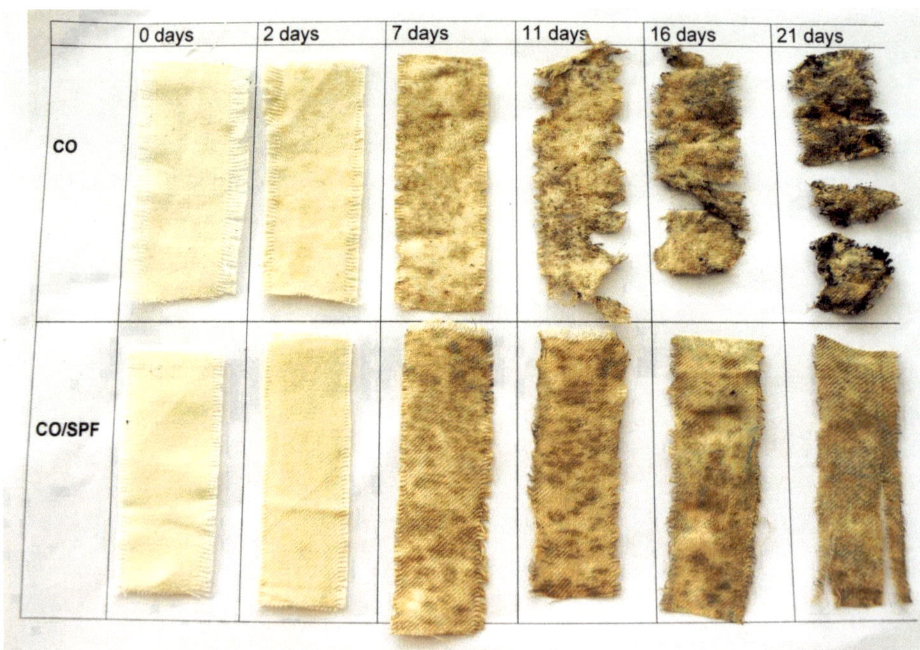

Fig. 11. Fabric samples after having been buried in the soil for 0 to 21 days. (Photo: Marica Starešinič)

The longer was the time of being buried in the soil, the greater was the loss of tensile strength of cotton fabrics: by 12% after 2 days and by 62% after 7 days of being buried in the soil. Breaking elongation decreased also rapidly: from 23% of the unburied fabric to only 8% after 7 days of being buried in the soil (Fig. 13).
Degradation of cotton yarns in warp direction affected tensile strength of SPF/CO fabrics in weft direction (Fig. 14). Fabrics buried in the soil for 7 days lost 12% of their tensile force, but after 21 days of being buried in the soil their breaking force decreased by additional 13% in comparison to unburied fabrics. Breaking elongation in weft direction of SPF/CO fabrics did not change significantly.

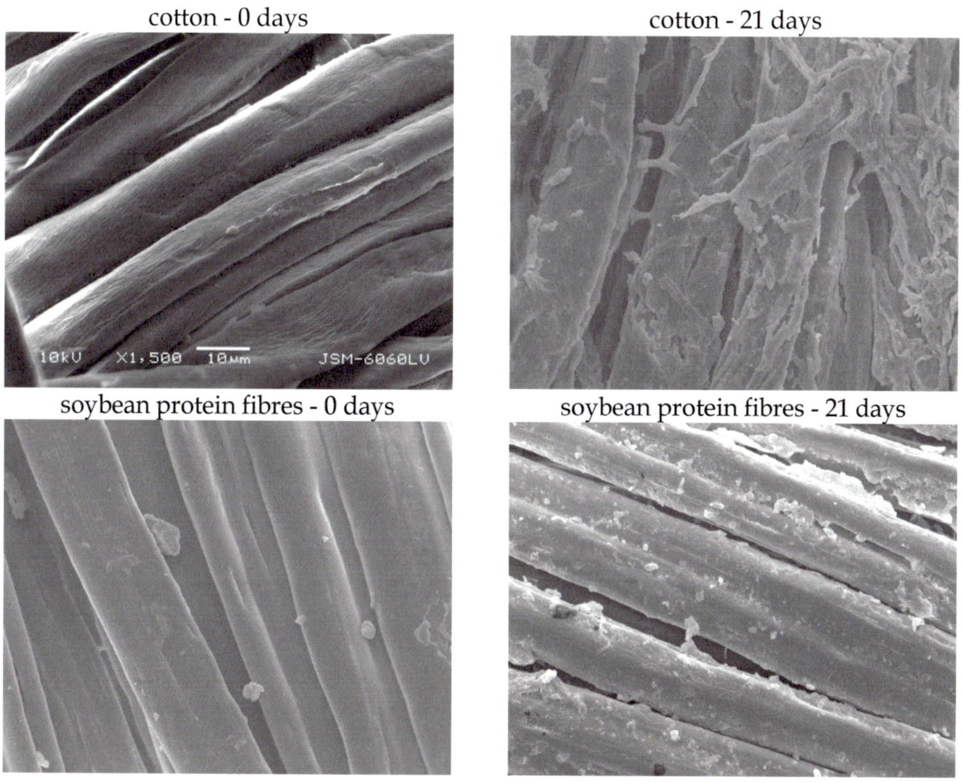

Fig. 12. SEM microphotographs of CO fabrics and SPF/CO woven fabrics (magnification 1500-x).

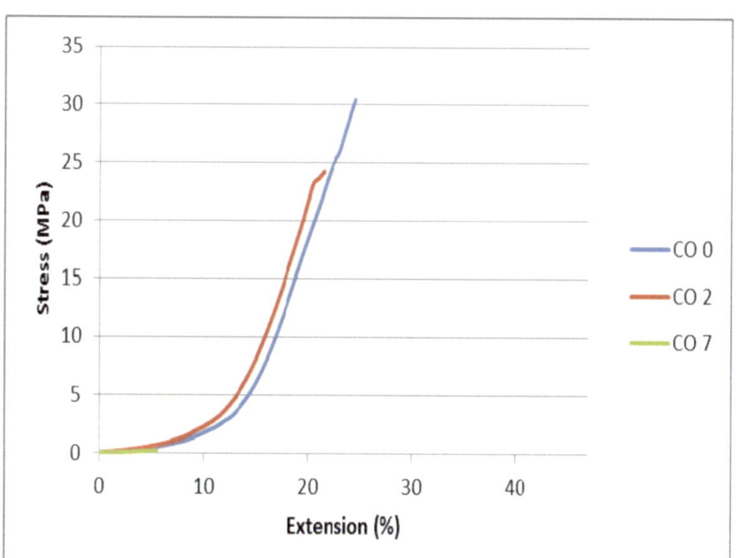

Fig. 13. Stress-elongation curves for cotton woven fabrics.

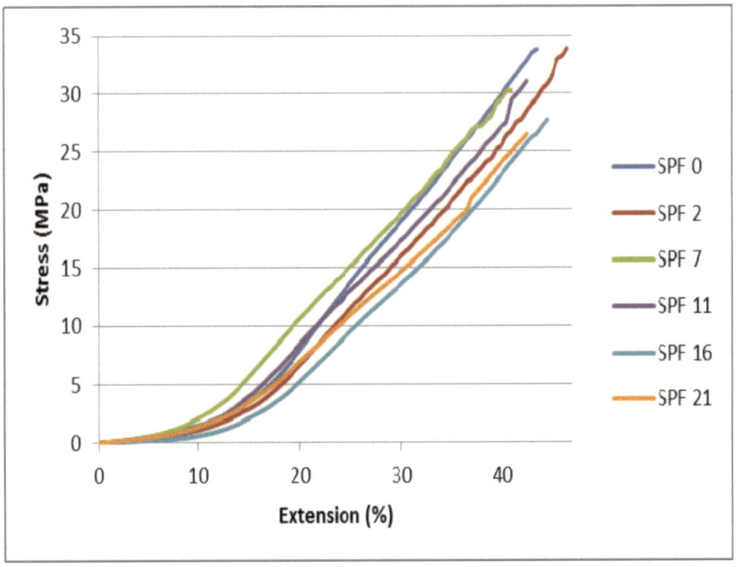

Fig. 14. Stress-elongation curves for woven fabrics with SPF yarn in weft.

5. Conclusion

The mid-twentieth century regenerated soybean protein fibres were made from pure soy proteins treated with formaldehyde or aluminium salts. Because of low tensile strength in wet state they were, like all other mid-twentieth protein fibres, noncompetitive to synthetic fibres in the 1970th.

Due to increasing prices of petroleum and a growing concern about the environmental damage arising from a slow degradation and poor biodegradability of synthetic fibres, researchers began to search for new possibilities of developing fibres from renewable raw materials, also from soybean proteins.

The fact that proteins are renewable and biodegradable materials has attracted considerable attention of many researchers in the area of textile fibres in the last two decades to re-examine the production of fibres from soybean proteins and casein. Soybean proteins have a greater potential for use as textile fibres because of their lower cost than casein proteins derived from milk.

In all experiments made until now, a soybean protein isolate (SPI) has been used, which is a highly purified protein (>90% w/w), obtained after extracting oils and fats from protein cakes. The residues after purification of protein for producing fibres can be also used as foodstuff (Yi-you, 2004).

In the last two decades researches have focused on different spinning methods (Huang et al., 1995), on new fibres from soybean proteins and polyvinyl alcohol (Zhang et. al., 1999, Li, 2007) or zein proteins (Zhang et al., 1997), on new economical biochemical processes that modify physical structure of soya proteins, and on new solvents (Zhang et al., 2009).

The experimental soybean protein fibres were made from two macromolecular components combined together into:

- *biconstituent fibres*, where a spinning dope was prepared from a homogeneous mixture of two solutions - a soybean protein water solution and a water solution of synthetic polymer polyvinyl alcohol or cellulose or zein proteins. Single fibres made from such spinning dopes had homogenous structure.

- *bicomponent fibres*, where the fibre's core was made from a soybean protein and the fibre's sheath from polyvinyl alcohol.

Polyvinyl alcohol was used, because it is a water-soluble polymer, it dissolves at similar conditions as proteins and when added to proteins, it increases the fibre's strength. Polyvinyl alcohol is also biodegradable in the soil (Brooks, 2005).

The combination of cotton yarns and the yarns from soybean proteins in woven fabrics imparts comfort, soft hand and good moisture absorption properties to undergarments, outerwear, infants' wear, towels and beddings. Biodegradation of contemporary soybean protein fibres in early phase, up to 21 days in the soil at 30 °C and 65% relative humidity, is a slow, hardly perceivable process.

6. References

Anon. (2003). Innovations China develops soyabeanf fibre, *Textiles Magazine*, Vol. 30, No. 2, p. 4, ISSN 1367-1308

Anon. (1937). Artificial Wool Production in Italy, *Nature*, Vol. 140, pp. 1090, ISSN 0028-0836

Arslan, A. (2007). Milk protein fibres (Part 1), Chemical Fibre International, Vol. 57, 108-109. ISSN 1434-3584

Arslan, A. (2008). Milk protein fibres (Part II), Melliand International, vol. 14, 20-21. , ISSN 0947-9163

Arslan, A. (2009). Milk protein fibres (Part III), Melliand International, vol. 14 (2009), 92-93. , ISSN 0947-9163

Boyer, R. A. (1940). Soybean protein fibers. Experimental production, *Industrial and Engineering Chemistry*, Vol. 32, p. 1549-1549, April 27, 2011, Available from: <http://pubs.acs.org/doi/abs/10.1021/ie50372a004>

Boyer, R. A., Robinette, C. F. & Atkinson, W. T. (1945). *Artificial fibres and manufacture thereof*, US Patent 2,377,854

Brinsko, K. M. (2010). Optical characterization of some modern "eco-friendly" fibers, *Journal of Forensic Sciences*, Vol. 55, No. 4, pp. 915-923, ISSN 1556-4029, April 15, 2011, Available from: <interscience.wiley.com>

Brooks, M. M. (2005). Soya bean protein fibres - past, present and future, In: *Biodegradable and sustainable fibres*, R. S. Blackburn, pp. 398-440, Woodhead Publishing Series in Textiles, No. 47, ISBN 1-85573-916-X (Woodhead Publishing), Cambridge

Bukošek, V. (1988). *Program »DINARA«: program izračuna in vrednotenja mehanskih in visokoelastičnih lastnosti vlaken iz diagrama specifična napetost /raztezek*, NTF Oddelek za tekstilstvo, Ljubljana

Croston, C. B., Evans, C. D. & Smith, A. K. (1945). Zein fibers - Preparation by wet spinning, *Industrial & engineering chemistry*, Vol. 37, No. 12, pp. 1194-1198, ISSN 1541-5724

Edwards, J.V. & Vigo, T.L. (2001). *Bioactive fibres and polymers*, Americal Chemical Society, pp. 421-430, ISBN 0-8412-3714-X, Washington, DC

Federal Trade Commission [FTC] 2011). *Rules and Regulations Under the Textile Fiber Products Identification Act, 16 CFR Part 303.§303.7(g) Generic names and definitions for manufactured fibers*, USA Federal Trade Commission, Mar. 05, 2011, Available from: <http://www.ftc.gov/os/statutes/textile/rr-textl.htm>

Federal Trade Commission [FTC], (2010). Textile fiber labeling, April 10, 2011, Available from: <http://www.nordstromsupplier.com/NPG/PDFs/Product%20Integrity/Labeling%20Requirements/Textile%20Fiber%20Labeling.pdf>

Ferretti, A. (1944). *Process for manufacturing artificial textile fibers from* casein, US Patent 2,338,917

Fletcher, H. A. (1942). Synthetic Fibers and Textiles, *Kansas Bulletin* 300, p. 8-10

Huang, H. C. , Hammond E. G. , Reitmeier C. A. & Myers D. J. (1995). Properties of fibers produced from soy protein isolate by extrusion and wet-spinning, *Journal of the American Oil Chemists' Society*, Vol. 72, No. 12, p. 1453-1460, ISSN 0003-021X (Print) 1558-9331 (Online)

Huppert, O. (1945). *Process of manufacture of synthetic wool from soya bean protein*, US Patent 2,377,885

Kajita, T. & Inoue, R. (1940). *Process for manufacturing artificial fiber from protein contained in soya bean*, US Patent 2,191,194

Kajita, T. & Inoue, R. (1946). *Process for manufacturing artificial fiber from protein contained in soybean*, US Patent 2,394,309

Kanitkar, U. P. (2010). Man Made Cellulose Fibre Industry over-view. North South Textile Summit, May 2010, Available from: <http://nsts.in/images/UPKANITKAR-GRASIM-Presentation-Man-Made-Fibre-20052010.pdf>

Kelley, J. J. & Pressley, R. (1966). Studies with soybean protein and fiber formation, *Cereal Chemistry*, Vol. 43, pp. 195-206, ISSN 0009-0352

Krishnan, H. B., Natarajan, S. S., Mahmoud, A. A. & Nelson, R. L. (2007). Identification of glycinin and beta-conglycinin subunits that contribute to increased protein content of high-protein soybean lines, *J. Agric. Food Chem.*, Vol. 55, p. 1839-1845, ISSN 1520-5118

Kurabo, 2007. New viscose filament yarn Lunacel, *Chemical Fibers International*, Vol. 57, pp. 315, ISSN 1434-3584

Li, G. (2003). *Phytoprotein synthetic fibre and the method of making the same*, WO/2003/056076

Li, G. (2007). *Phytoprotein synthetic fibre and method of manufacture thereof*, US Patent 7,271,217

Mathur, M. and Hira, M. (2004). Speciality fibres - I: soybean protein fibre, *Man-made Textiles in India*, Vol. 32, No. 10, pp. 365-369, ISSN 0377-7537

Morimoto, S., Ishihara, M., Yamamoto, A., Hamada, K., Imai, K. & Otsuka, M. (1962). *Graft copolymer containing spinnable solution and method for preparing and spinning thereof*, U.S. Pat. 3,104,154

Myers, D. (1993). Past, present and potential uses of soy proteins in non-food industrial applications, *Proceedings of the World Conference on Oilseed Technology and Utilization*, ISBN 0-935315-45-4, Budapest, March 10, 2011, Available from: <http://books.google.si/books?id=Fh2rNOiEIWQC&pg=PA278&lpg=PA278&dq=Past,+present+and+potential+uses+of+soy+proteins+in+non-food+industrial+applications,+Proceedings+of+the+World+Conference+on+Oilseed+Technology+and+Utilization&source=bl&ots=-Z7SEywKa5&sig=Torn8hK4rwCcap0fypN9bYxXDZ0&hl=sl&ei=Gpu9TefjG8TFswawp5z6BQ&sa=X&oi=book_result&ct=result&resnum=1&ved=0CBUQ6AEwAA#v=onepage&q&f=false>

Reddy, N. & Yang, Y. (2007). Novel Protein Fibers from Wheat Gluten, *Biomacromolecules*, Vol. 8, pp. 638-643, ISSN 1525-7797

Reek, A. (2008). *Wipe materials comprising regenerated plant-protein fibres*, PCT/US2009/059406 WIPO

Shanghai Winshow Soybean Fibre Industry. Co., Ltd. (2011). About us, April 10, 2011, Available from: <http://swsficl.en.china.cn>

Shanghai Zhengjia milkfiber sci & tech Co ., ltd., (2011). Zhengjia milk fibre. January 20, 2011, Available from: < http://www.milkfashion.com>

Simončič, B. & Tomšič, B. (2010). *Biorazgradnja tekstilnih vlaken in njihova protimikrobna zaščita*, Naravoslovnotehniška fakulteta, Oddelek za tekstilstvo, ISBN 978-961-6045-81-0, Ljubljana

South West Traiding Company [SWTC], (2011). April 20, 2011, Available from: <www.soysilk.com>

Sue, J.-F., Huang, Z., Yang, C.-M. & Yuan, X.-Y. (2008). Properties of Soy Protein Isolate/Poly(vinyl alcohol) Blend "Green" Films: Compatibility, Mechanical Properties, and Thermal Stability, *Journal of Applied Polymer Science*, Vol. 110, pp. 3706–3716, ISSN 1097-4628

Szostak-Kotowa, J. (2004). Biodegradation of textiles, *International biodeterioration & biodegradation*, Vol. 53, pp. 156-170, ISSN 0964-8305

Swicofil AG, (2011). Soybean protein fibres, April 27, 2011, Available from: <http://www.swicofil.com/soybeanproteinfiber.html>

Wang, N., Ruan, C., Yu, Y., Zheng, Y. & Yu, J. (2009). Composition and structure of acrylonitrile based casein fibers, *Chemical Fibres International*, Vol. 59, No.2, pp. 88-89, ISSN 1434-3584

White, M. (2008). *U.S. Alien Property Custodian patent documents: A legacy prior art collection from World War II – Part 2, statistics*, World Patent Information, Vol. 30, No. 1, pp. 34-42, ISSN 0172-2190

Wormell, R. L. (1954). *New fibres from proteins*, London, Butterworths Scientific Publications, pp. 145

Yang, V. (2011). Soybean fibre application stepping into a new stage in China. April 23, 2011, Available from: < http://www.fibre2fashion.com/industry-article/textile-industry-articles/soybean-fiber-application-stepping-into-a-new-stage-in-china/soybean-fiber-application-stepping-into-a-new-stage-in-china1.asp >

Yang, Y. et al. (2006). *Process for the production of high quality fibers from wheat proteins and products made from wheat protein fibers*, WO 2006/138039

Yi-you, L. (2004). The soybean protein fibre – a healthy & comfortable fibre for the 21st century, *Fibres & Textiles in Eastern Europe*, Vol. 12, No. 2, pp. 8-9, ISSN 1230-3666

Zee, M. van der (2005). Biodegradability of polymers – mechanisms and evaluation methods. In: *Handbook of biodegradable polymers*, Catia Bastioli , pp. 1-32, Rapra Technology, ISBN 978-1-85957-389-1, Shawbury, Shrewsbury, Shropshire

Zhang, M., Reitmeier, C. R., Hammond, E. G., & Myers, D. J. (1997). Production of textile fibers from zein and a soy protein-zein blend. *Cereal Chemistry*, Vol. 74, pp. 594-598, ISSN 0009-0352

Zhang, Y., Ghasemzadeh, S., Kotliar, A. M., Kumar, S., Presnell, S. & Williams, L. D. (1999). Fibers from soybean protein and poly(vinyl alcohol), *Journal of Applied Polymer Science*, Vol. 71, pp. 11-19, ISSN 0021-8995

Zhang, L., & Zeng, M. (2008). Proteins as sources of materials, In: *Monomers, Polymers and Composites from Renewable Resources*, M. N. Belgacem and A. Gandini, pp. 479-493, Elsevier, ISBN 9780080453163, Oxford, Boston

Zhang, L. (2008). *Physicochemical, morphological, and adhesion properties of sodium bisulfite modified soy protein components – B. S. Thesis.* Kansas, Manhattan: Kansas State University. April 20, 2011, Available from: <http://krex.k-state.edu/dspace/bitstream/2097/1707/1/Lu%20Zhang2008.pdf>

Zhang, S., Zhao, S., Tian, Y., Li, F.-X. & Yu, J.-Y. (2009). Preparation of cellulose/soy protein isolate blend biofibers via direct dissolving approach, *Chemical Fibers International*, Vol. 59, No. 2, pp. 106-107. ISSN 1434-3584

24

Characterization of Enzymes Associated with Degradation of Insoluble Fiber of Soybean Curd Residue by *Bacillus subtilis*

Makoto Shoda and Shinji Mizumoto
Chemical Resources Laboratory, Tokyo Institute of Technology, Nagatsuta, Midori-ku, Yokohama, Japan

1. Introduction

Soybean curd residue is a residue of soy milk processing in which most soluble nutrients of soybean are extracted to liquid phase, and thus major carbon sources of the residue are insoluble fibers (O'tool, 1999) which amount to 40.2- 43.6 % on a dry matter basis (Van der Riet et al.,1989). Approximately 700,000 tons of the soybean curd residue were produced annually as a byproduct of *tofu* manufacturing in Japan and most of them is incinerated as an industrial waste. We re-utilized the soybean curd residue as a solid substrate of solid-state fermentation (SSF) using *Bacillus subtilis* (Mizumoto et al., 2006).

The insoluble fibers of soybean consist of cellulose, hemicellulose and lignin. Cellulose is the most abundant biological polymer on earth and is the major constituent of the plant cell wall. This lineal polymer is composed of D-glucose subunits linked by β-1,4 glycosidic bonds forming cellobiose molecules and the long chains are linked together by hydrogen bonds and van der Waals forces (Perez et al., 2002). Hemicellulose is a complex of polymeric carbohydrates which contains xylan, xyloglucan, (heteropolymer of D-xylose and D-glucose), glucomannan (heteropolymer of D-glucose and D-mannose), galactoglucomannan (heteropolymer of D-galactose, D-glucose and D-mannose) and arabinogalactan (heteropolymer of D-galactose, D-glucose and arabinose). Among them, xylan, a complex polysaccharide comprising a backbone of xylose residues linked by β-1,4-glycosidic bonds, is the major component. Xylan is the second most abundant polysaccharide in nature, accounting for approximately one-third of all renewable organic carbon on earth (Collins et al., 2005). Lignin is an amorphous non-water soluble and optically inactive heteropolymer. It consists of phenylpropane units joined together by different types of linkages (Perez et al., 2002) Although lignin is the most abundant polymer in wood fiber along with cellulose, its content in non-wood fiber such as straw, grass and seed hull is low (Sun & Cheng, 2002). The lignin content in the soybean seed coat is reported to be low (Krzyzanowski et al., 2001), and thus it is speculated that the soybean curd residue contains relatively small amount of lignin.

B. subtilis has ability to produce several antibiotics with a variety of structures, especially peptides that are either ribosomally or non-ribosomally synthesized (Leclere et al., 2005; Ongena et al., 2005; Stein, 2005). We previously isolated several strains of *B. subtilis* and the wild strains and their derivatives suppressed 26 types of plant pathogen *in vitro* (Phae et al., 1990) and a fungal disease *in vivo* (Asaka & Shoda, 1996) by producing three lipopeptide antibiotics, iturin A, surfactin and plipastatin (Asaka & Shoda, 1996; Hiraoka et al., 1992; Tsuge et al., 1996, 1999). The suppressive effect of one of the isolates, *B. subtilis* RB14, was mainly associated with the cyclolipopeptide antibiotic iturin A, which contains seven α-amino acids and one β-amino acid. *B. subtilis* RB14-CS, a derivative of the original strain RB14 and a sole producer of iturin A, produced iturin A in SSF using the soybean curd residue 3-fold higher than in submerged fermentation (SmF) (Mizumoto et al., 2006). This suggests that RB14-CS could degrade some kinds of insoluble fibers in soybean curd residue and utilize them as carbon sources during SSF. In this chapter, insoluble fibers in soybean curd residue that RB14-CS could degrade during SSF were clarified and the fiber-degrading enzymes were purified and characterized.

2. Materials and methods

2.1 Strain

B. subtilis RB14-CS which is a spontaneous mutant derived from RB14-C is a single iturin A producer. *B. subtilis* RB14-C is a streptomycin-resistant mutant from a parent strain RB14 and is a co-producer of the antibiotics iturin A and surfactin (Asaka & Shoda, 1996).

2.2 Solid-state fermentation (SSF)

The detail of SSF was described in the previous paper (Mizumoto et al., 2006). The L medium used for the growth of the bacterium contained 10 g of Polypepton (Nippon Pharmaceutical Co., Tokyo, Japan), 5 g of yeast extract and 5 g of NaCl (per liter). One ml of L medium culture broth after 24 h cultivation at 30°C was inoculated into 100 ml of number 3S (no. 3S) medium consisting of 30 g of Polypepton S (Nippon Pharmaceutical Co., Tokyo), 10 g of glucose, 1 g of KH_2PO_4, and 0.5 g of $MgSO_4 \cdot 7H_2O$ (per liter) (pH 6.8), and the culture was incubated at 120 strokes per minute (spm) at 30°C for 24 h in a shaking flask and used as a seed for SSF.

The soybean curd residue was supplied from a *tofu* company in Tokyo and stored at - 20°C. Each of fifteen grams of thawed soybean curd residue was placed in a 100-ml conical flask and autoclaved twice at 120°C for 20 min at an interval of 8-12 h to kill spore-forming microorganisms inhabiting the material. After cooling to room temperature, the following solutions were added as nutrient supplements for every 15 g of soybean curd residue and moisture content was adjusted to 79%: 833 μL of 0.45 g glucose /ml, 75 μL of 1 M KH_2PO_4, 150 μL of 1 M $MgSO_4 \cdot 7H_2O$ and 367 μL of deionized distilled water. Then, 3 mL of an RB14-CS culture grown in no. 3S medium was added to 15 g of soybean curd residue and mixed with a stainless steel spatula. All flasks were incubated statically in a water incubator at 25°C, and at a specified time, one flask was taken and the whole soybean curd residue in a flask was used as a sample for analysis.

2.3 Preparation of samples for acid and neutral detergent fiber analysis

After 5 days of SSF by *B. subtilis* RB14-CS, the whole solid culture was dried by microwave and ground by using a pestle and a mortar. Raw soybean curd residue was used as a control.

2.4 Acid and neutral detergent fiber analysis

2.4.1 Acid detergent fiber

The content of acid detergent fiber, which contains mainly cellulose and lignin, was analyzed in the following manner (Van Soest, 1963). In a 150 mL-flat bottom flask, 0.45 - 0.55 g of ground sample was weighed using micro-balance and 50 mL of acid detergent solution (20 g/L cetyl trimethylammonium bromide in 0.5 M sulfuric acid) was mixed. The flask was placed in an oil bath under the cold water condenser and boiled within 5-10 min. Sample was refluxed for 60 min from onset of boil. After approximately 30 min, the inside of flask was washed with minimal amount of acid detergent solution. After refluxed, sample was filtrated under reduced pressure with a tared Gooch crucible. The crucible was washed twice with hot water, then twice with acetone and was dried at 105°C overnight. After cooled to room temperature in a desiccator, the weight of the crucible was measured.

2.4.2 Neutral detergent fiber

The content of neutral detergent fiber which contained mainly cellulose, lignin, and hemicellulose was analyzed in the following manner (Van Soest, et al., 1991). In a 300 mL-round bottom flask, 0.45 - 0.55 g of ground sample, 50 mL of neutral detergent solution (13.5 g of sodium dodecyl sulfate, 8.38 g of EDTA disodium salt, 3.07 g of $NaB_4O_7 \cdot 10H_2O$, 5.18 g of $Na_2HPO_4 \cdot 12H_2O$ and 4.5 mL of tryethylene glycol per 450 mL) and 0.5 g of sodium sulfite were mixed. The flask was placed in an oil bath under the cold water condenser and boiled for 5 min. After 5 min of boiling, 2 mL of α-amylase solution, which consists of heat-stable α-amylase (Kleistase T10S; Daiwa Kasei, Shiga, Japan) and 50 mM sodium phosphate buffer (pH 6.0) (1:39 [vol/vol]), were mixed. Then, the sample was refluxed for 60 min. After approximately 30 min, the inside of flask was washed down with minimal amount of neutral detergent solution. After refluxed, the sample was filtrated under reduced pressure with a tared Gooch crucible. The crucible was filled with 2 mL of α-amylase solution and hot water, and incubated for at least 2 min. Then, the crucible was washed twice with hot water, and then twice with acetone. The crucible was dried at 105°C overnight. After cooled to room temperature in a desiccator, the weight of the crucible was measured.

2.4.3 Calculation of content of insoluble fibers

As the amount of acid detergent fiber was regarded as total amount of cellulose and lignin, the amount of the neutral detergent fiber minus the amount of acid detergent fiber was regarded as the content of hemicellulose.

2.5 Iturin A production in liquid culture using insoluble fibers

In a 200-mL conical flask, 40 mL of liquid medium consisting of 10 g of fibrous carbon sources, 10 g of Polypepton S, 1 g of KH_2PO_4 and 0.5 g of $MgSO_4 \cdot 7H_2O$ (per liter) (pH 6.8) was prepared. As fibrous carbon sources, xylan (Tokyo Chemical Industry, Tokyo, Japan), avicel, carboxymethyl cellulose, and pectin were used. As a control carbon source, glucose was used. Four hundreds μL of a seeding culture was inoculated into the medium and the flasks were incubated at 30°C at 120 spm.

For measurement of iturin A concentration, 1 mL of culture broth was acidified to pH 2.0 with 12 N HCl. Iturin A was collected by centrifugation at 18,000 ×*g*, at 4°C for 10 min, and extracted with 1 mL of methanol. The extract was injected into a high-performance liquid chromatography (HPLC) with a column (Chromolith Performance RP-18eb 4.6 mm diameter× 100 mm height, Merck, Germany) to determine iturin A concentrations. The HPCL system was operated at a flow rate of 2.0 mL/min with acetonitrile-10 mM ammonium acetate (65:35 [vol/vol]) at a column temperature of 40°C. The elution was monitored at 205 nm by a UV detector (880-UV, Intelligent UV/VIS Detector, Jasco, Tokyo, Japan).

Although iturin A has 8 homologues with different side-chain structures (Asaka & Shoda, 1996), the concentration of iturin A was defined as the total amount of five major homologues. The correlation between the peak heights and the concentration of pure iturin A (Sigma-Aldorich, Tokyo, Japan) was used for quantification. Iturin A concentration was expressed as µg/ g initial wet soybean curd residue.

2.6 Xylanase activity assay

Dinitrosalicylic acid (DNS) solution was prepared in the following manner. Solution A was prepared by mixing 300 mL of 4.5 % NaOH, 880 mL of 1 % 3,5- DNS and 225 g of potassium sodium (+)-tartrate tetrahydrate. For the preparation of solution B, 22 mL of 10 % NaOH and 10 g of phenol was mixed and filled up to 100 mL. To 69 mL of the mixture, 6.9 g of $NaHCO_3$ was added. Solutions A and B were mixed thoroughly and placed at room temperature for 2 days. After filtration, the mixture was used as DNS solution.

Xylanase activity was determined by measuring the amount of reducing sugar released from xylan. One hundred µL of enzyme sample was added to 1 mL of 1 % xylan in 100 mM sodium phosphate buffer (pH 6.5) in a test tube (15 mmΦ × 10.5 cm) and incubated statically at 50°C for 5 min. Two mL of DNS solution was added and cooled immediately in an ice bath. Then the test tubes were boiled for 5 min and cooled in an ice bath. After centrifugation at 18,000×*g* at 4°C for 5 min, absorbance of the supernatant at 540 nm was measured by spectrophotometer (UV2400; Shimadzu, Kyoto, Japan). Xylose was used as the standard. One unit (U) of xylanase activity was defined as the amount of enzyme that liberates 1 µmol of reducing sugars (xylose equivalent) per min.

2.7 Measurement of xylanase activity during SSF

SSF was carried out as described in Section 2.2 without addition of glucose. One gram of solid culture sample and 9 ml of sterile distilled water were mixed in a sterile 18-mm-diameter test tube, the test tube was vortexed thoroughly and shaken at 150 spm for 5 min at room temperature. The suspension was centrifuged at 18,000 ×*g* at 4 °C for 10 min and the supernatant obtained was used for xylanase assay.

2.8 Measurement of concentration of protein

Protein concentrations were determined by the Bradford method (Bradford, 1976) with the Protein Assay Kit II (Bio-Rad, Tokyo, Japan) with bovine serum albumin as the standard protein.

2.9 Purification of xylanase

Solid cultures (90 g) incubated for 5 days in SSF were mixed with 900 mL of distilled water and stirred for 10 min. The suspension was centrifuged at 6,500×*g* at 4°C for 20 min and the supernatant was frozen at -20°C and then thawed. The sample was centrifuged for removal of polysaccharides under the same condition. Ammonium sulfate was added to the

supernatant to 30 % saturation, and the precipitate was removed by centrifugation. Then, ammonium sulfate was added to 70 % saturation. The precipitate was recovered by centrifugation, suspended in 50 mM MES buffer (pH 6.0) and dialyzed overnight against the same buffer. Then the sample was concentrated by ultrafiltration with YM10 (molecular mass cut-off 10 kDa; Advantec, Tokyo, Japan).
The concentrate was applied to a CM-Toyopearl column (1.3 cm Φ×8.3 cm; Tosoh, Tokyo, Japan) pre-equilibrated with buffer A (50 mM MES buffer, pH 6.0), and fractions were eluted with a continuous linear gradient of 0-0.5 M NaCl in buffer A (total volume 120 mL). The flow speed and the volume of one fraction were 4 mL/min and 8 mL, respectively. In this process, xylanase activity was detected in two fractions, one of which was trapped in the column (Fraction I) and the other was not trapped in the column but passed through (Fraction II). These fractions were subjected to further purification processes.
Fraction I was concentrated using Centriprep YM-10 (molecular mass cut-off 10 kDa; Millipore, Tokyo, Japan), diluted with buffer A and applied to a RESOURCES column (0.6 cm Φ×3.0 cm; Pharmacia Biotech, Uppsala, Sweden) pre-equilibrated with buffer A. Fractions were eluted with a continuous linear gradient of 0-0.15 M NaCl in buffer A (total volume 30 mL). The flow speed and the volume of one fraction were 1 mL/min and 1 mL, respectively. The xylanase active fractions were concentrated with Centriprep YM-10 and applied to a Superdex 75 column (1.6Φ×60 cm; Amersham Bioscience, Tokyo, Japan) pre-equilibrated with buffer A containing 0.2 M NaCl. The elution was carried at a flow rate of 1 mL/min and a volume of one fraction was 2 mL.
The pH of the Fraction II was adjusted to 9.5 by adding NaOH and applied to a QAE-Toyopearl (1.6Φ×3.7 cm; Tosoh) pre-equilibrated with buffer B (25 mM piperazine buffer, pH 9.5), and fractions were eluted with a continuous linear gradient of 0-0.5 M NaCl in buffer B (total volume 120 mL). The flow speed and the volume of one fraction were 4 mL/min and 8 mL, respectively. The xylanase active fractions were concentrated with Centriprep YM-10, and fractions were diluted with buffer B and applied to a QAE column. Step elution was performed with 0.07 M NaCl (total elution volume 96 mL). The flow speed and the volume of one fraction were 4 mL/min and 8 mL, respectively.
The xylanase active fractions were supplied to the subsequent Butyl-Toyopearl chromatography. A column of Butyl-Toyopearl (1.6Φ×4.5 cm; Tosoh) pre-equilibrated with 25 mM piperazine buffer containing 1 M ammonium sulfate was used. Ammonium sulfate was added to the active fractions and its concentration was adjusted to 1 M. This solution was then applied to the column and the elution was carried out with a linear gradient of 1-0 M ammonium sulfate in 25 mM Piperazine buffer (total volume 180 mL). The flow speed and the volume of one fraction were 4.5 mL/min and 9 mL, respectively.

2.10 Molecular mass determination

Sodium dodecyl sulfate-polyacrylamide gel electrophoresis (SDS-PAGE) was performed with a 12.5 % gel in accordance with the Laemmli method (Laemmli, 1970). M. W. Marker "Daiichi" II (Daiichi Pure Chemicals, Tokyo, Japan) was used as a molecular mass marker. After electrophoresis, the gel was stained with Coomassie brilliant blue R-250 (CBB).

2.11 N-terminal sequence analysis

SDS-PAGE of xylanases was performed according to the above-described method and then the xylanases on the gel were electroblotted to a commercial membrane (Immobilon-P;

Millipore, Tokyo, Japan) with a horizontal blotting apparatus (ATTO, Tokyo, Japan). For the blotting of pure enzyme of Fraction II, 0.01 % of SDS was added to transfer buffer to improve protein transfer efficacy. Parts of the membrane blotted with xylanases were cut out and then amino acid sequencing analysis was performed with an amino acid sequencing apparatus (PPSQ-21; Shimadzu, Kyoto, Japan) according to the standard method (Edman, 1949).

Searches for homologous amino acid sequences were performed by a *B. subtilis* database BSORF (http://bacillus.genome.jp/) and the nonredundant database at The National Center for Biotechnology Information (http://www.ncbi.nlm.nih.gov/) with the BLASTP.

2.12 pH and temperature profiles and thermostability of xylanases

Xylanase activity was examined in pH range of 3.0 to 11.0. For pH from 3.0 to 4.0, 100 mM sodium citrate buffer was used. For pH from 4.0 to 6.0, 100 mM sodium acetate buffer was used. For pH from 6.0 to 8.0, 100 mM sodium phosphate buffer was used. For pH from 8.0 to 9.0, 100 mM Tris-HCl buffer was used. For pH from 9.0 to 11.0, 100 mM glycine-NaOH buffer was used. To investigate the effect of temperature, the xylanase activity was measured at 20-70°C at pH 6.5. Xylanase thermostability was measured at 50, 55 and 60°C.

2.13 Thin layer chromatography (TLC) analysis of the digestion products

The digestion products of xylan and xylooligosaccharides (Wako Pure Chemical Industries, Osaka, Japan) by xylanase were analyzed by thin layer chromatography (TLC) according to the method previously reported (Kiyohara et al., 2005) with some modifications.

As a substrate solution, 0.5 % xylan or 0.5 % xylooligosaccharides in 100 mM sodium phosphate buffer (pH 6.5) was used. In a test tube (15 mmΦ × 10.5 cm), 0.5 mL of substrate solution and 0.5 mL of enzyme solution containing 0.5 U of xylanase in 100 mM sodium phosphate buffer (pH 6.5) were mixed and the reaction mixture was incubated at 120 spm at 37°C. After 1, 3, and 16 h of incubation, 100 μL of reaction mixture was sampled to microtube, and mixed with 200 μL of ethanol. Then, the mixture was centrifuged at 18,000×*g* for 10 min and the supernatant obtained was evaporated with a centrifugal concentrator (VC-36N; Taitec, Saitama, Japan). The dried material was dissolved in distilled water and spotted on a Silica Gel 60 TLC plate (Merck, Tokyo, Japan), which was then developed with *n*-butanol/acetic acid/ water (10:5:1, by vol.). After development, the TLC plate was sprayed with aniline hydrogen phthalate reagent. The reagent consisted of 0.93 g of aniline, 1.48 g of phthalic anhydride, 84.5 mL of *n*-butanol and 15.5 mL of distilled water (Partridge, 1949), and heated at 100°C to visualize the digestion products.

3. Results

3.1 Degradation of insoluble fibers in soybean curd residue by *B. subtilis* RB14-CS in SSF

To evaluate the ability of *B. subtilis* RB14-CS to degrade insoluble fibers in soybean curd residue, residual fibers after SSF were analyzed by acid and neutral detergent fiber methods. The same analyses were repeated three times. The average values of three samples are shown in Figure 1. After SSF of RB14-CS, no change in content of cellulose and lignin was observed. On the other hand, the content of hemicellulose decreased to 15 % of initial one, indicating that RB14-CS degraded hemicellulose in soybean curd residue.

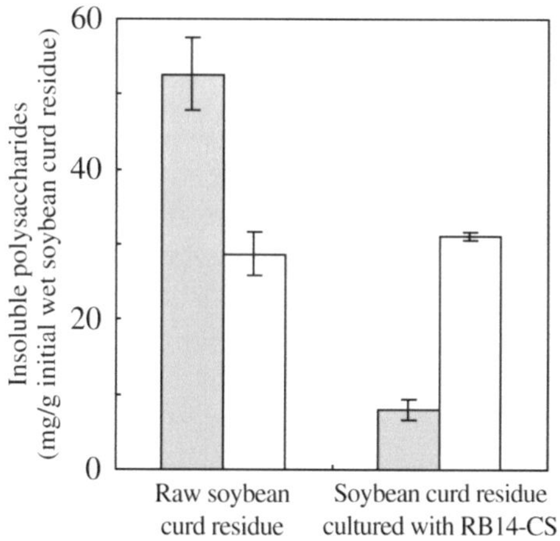

Fig. 1. Analysis of insoluble fiber contents in raw soybean curd residue and soybean curd residue cultured with *B. subtilis* RB14-CS (N=3). Gray bars, hemicellulose; Open bars, cellulose and lignin.

3.2 Iturin A production by *B. subtilis* RB14-CS using insoluble fibers in submerged fermentation

To investigate the effect of insoluble fibers on iturin A production of RB14-CS, each of insoluble fibers was added to a liquid medium as a carbon source and RB14-CS was cultivated in the medium. Results are shown in Figure 2. Xylan exhibited iturin A production at the same level with glucose which has been used as a carbon source for iturin A production in the previous reports (Asaka & Shoda, 1996; Tsuge et al., 2001). Other insoluble fibers, avicel and carboxymethyl cellulose, showed the similar level of iturin A production with control where no additional carbon was added. Pectin, a hardly-soluble or sometimes insoluble fiber which is contained in soybean curd residue (Kasai et al., 2004) did not enhance the iturin A production.

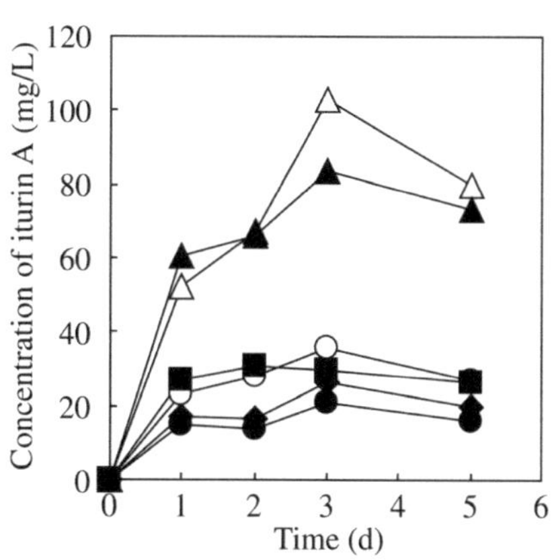

Fig. 2. Iturin A production during submerged fermentation in liquid medium containing fibers (N=3). Symbols: open circles, no additional carbon sources (control 1); open triangles, glucose (control 2); solid circles, pectin; solid triangles, xylan; solid squares, avicel; solid diamonds, carboxymethyl cellulose.

3.3 Xylanase activity of *B. subtilis* RB14-CS during SSF

As RB14-CS degraded xylan, a major hemicellulose in plant cell wall (Beg et al., 2001), in submerged fermentation, xylanase activity was measured during SSF, in which glucose was not added as medium component. Results are shown in Figure 3. The culture of RB14-CS exhibited xylanase activity in SSF. The activity increased after 12 h of incubation, reached the maximum value of approximately 50 U/g wet soybean curd residue at 3 d, and maintained the level during fermentation. When xylanase activity was detected, almost no reducing sugars were detected (data not shown), indicating that RB14-CS immediately utilized the saccharides released from hemicellulose as carbon sources. Changes in cell number and pH were similar to those in SSF of RB14-CS using soybean curd residue previously reported (Mizumoto et al., 2006).

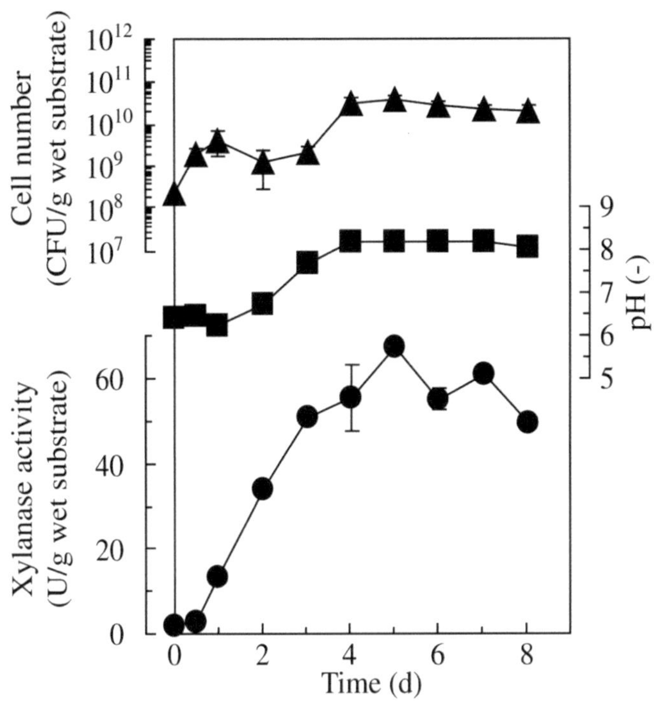

Fig. 3. Xylanase activity of *B. subtilis* RB14-CS during SSF.
Symbols: circles, xylanase activity; squares, pH; triangles, viable cell number.

3.4 Purification of xylanases produced by *B. subtilis* RB14-CS in SSF

Xylanases were purified as described in materials and methods. When the crude enzyme solution was applied to a cation exchange CM-Toyopearl column, xylanase activity was found in both the trapped fraction (Fraction I) and non-trapped fraction (Fraction II). From these fractions, two enzymes were purified and the two enzymes are homogeneous and have different sizes because each single protein band on SDS-PAGE was observed (Figure 4). This indicates that RB14-CS produces two different xylanases. Purified enzymes of Fraction I and II were designated as Xyl-I and Xyl-II, respectively. The molecular masses of the Xyl-I and Xyl-II estimated from SDS-PAGE were 24 and 58 kDa, respectively.

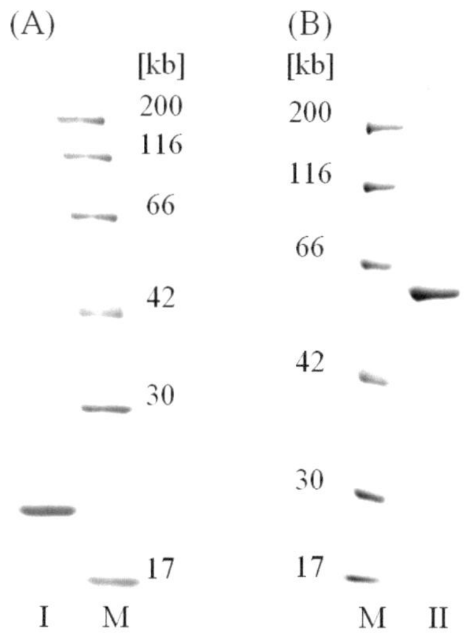

Fig. 4. SDS-PAGE of purified xylanases.(A) Xyl-I, (B) Xyl-II
Lanes: M, molecular mass standards; I, Xyl-I; II, Xyl-II.

3.5 Physicochemical properties of xylanases

Effects of temperature and pH on xylanase activity and thermal stability of the two enzymes are shown in Figure 5. The optimal temperature and optimal pH of Xyl-I were 50-60°C and 6-7, respectively. At 50°C, approximately 30 % of the initial activity of Xyl-I remained after 3 h. At 55 and 60°C, Xyl-I was completely inactivated within 2 and 3 h and the half lives were approximately 18 and 8 min, respectively. The optimal temperature of Xyl-II was 70°C or higher and the optimum pH was 5.5-6. At 50°C, approximately 80 % of the initial activity of Xyl-II remained after 3 h. At 60°C, Xyl-II was inactivated within 3 h and the half life was approximately 40 min.

3.6 Analysis of hydrolytic products

The hydrolysis products released from xylan or xylooligosaccharides by Xyl-I and Xyl-II were analyzed by TLC. From hydrolysis of xylan by both Xyl-I and Xyl-II xylotriose was liberated, but neither xylose nor xylobiose was released. This indicats that these xylanases were not β-D-xylosidase.

3.7 Identification of xylanases by N-terminal sequencing and database matching

The N-terminal sequences of Xyl-I and Xyl-II were determined by automated Edman degradation and compared with databases. Results are summarized in Table 1. Xyl-I displayed 90 % amino acid identity with endo-1,4-β-xylanase (XynA) of *B. subtilis* 168, a standard strain whose complete genome has been sequenced (Kunst et al., 1997). The molecular mass estimated by SDS-PAGE was similar to the database value. Moreover, pI value (9.64) of database was identical to that of purified Xyl-I.

Xyl-II has exactly the same N-terminal sequence as α-amylase (AmyE) secreted by *B. subtilis* X-23 (Ohdan et al., 1999). Actually, Xyl-II exhibited α-amylase activity because reducing sugar was increased when soluble starch was treated with Xyl-II (data not shown). It is

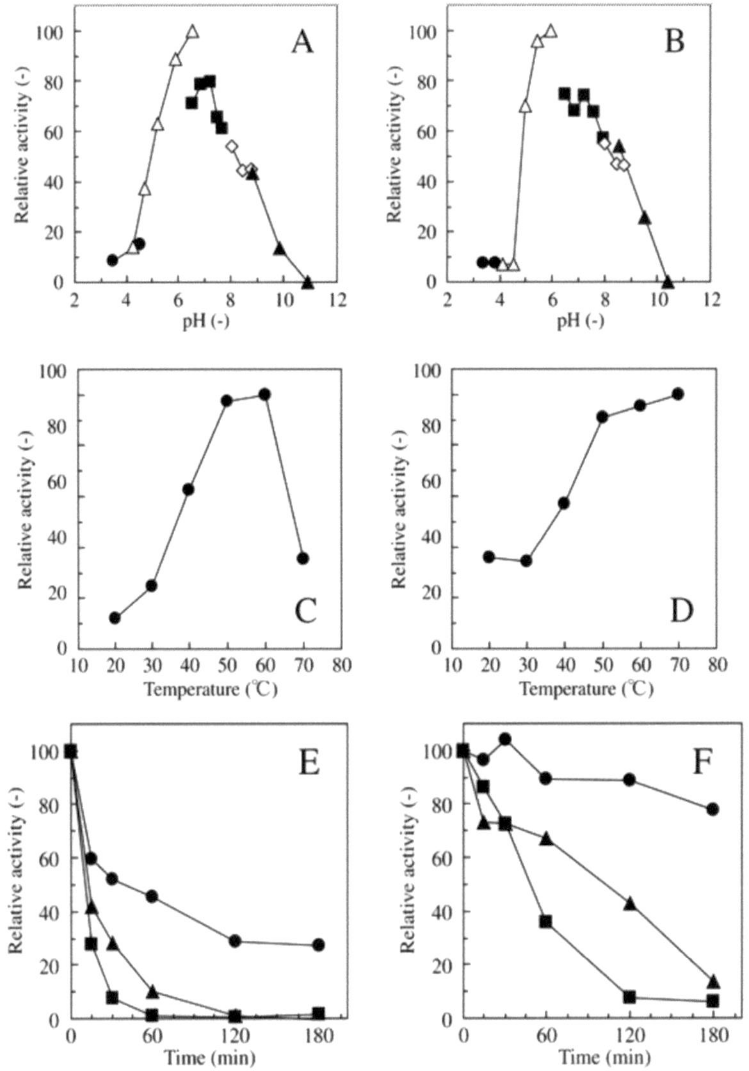

Fig. 5. Effects of pH and temperature on xylanase activities of Xyl-I and Xyl-II. Effects of pH on Xyl-I (A) and -II (B), respectively; Effects of temperature on Xyl-I (C) and -II (D), respectively; Thermal stability of Xyl-I (E) and -II (F), respectively. Symbols in (E) and (F): circles, 50°C; triangles, 55°C; squares, 60°C.

assumed that 45 amino acid residues prior to these sequenced residues deduced from the nucleotide sequence of the *B. subtilis* X-23 are the signal peptide that is removed during the secretion process. Xyl-II also displayed 80 % amino acid identity with α-amylase of *B. subtilis* 168 (Kunst et al., 1997). Although the molecular mass of Xyl-II estimated from SDS-PAGE was different from those in the previous reports, the C-terminal structures of α-amylase of *B. subtilis* were reported to be variable (Ohdan, et al., 1999). The pI value of α-amylase of *B. subtilis* 168 (5.85) was identical with the value of purified Xyl-II. This also reflected in that Xyl-II was trapped in anion exchange chromatography when piperazine buffer of pH 9.5 was used for elution.

	This work		Database					
	Sequence	Size (kDa)	Sequence	Gene	Protein identity	Size (kDa)	pI	References
Xyl-I	AGTDYWQNWT	24	ASTDYWQNWT	*xynA*	endo-1,4-β-xylanase	23	9.64	Kunst et al.
Xyl-II	SVKNGTILHA	58	SVKNGTILHA	*amyE*	α-amylase	47, 67	-	Ohdan et al.
			SIKSGTILHA	*amyE*	α-amylase	73	5.85	Kunst et al.

Table 1. N-terminal amino acid sequences of purified xylanases.

4. Discussion

B. subtilis RB14-CS degraded xylan in soybean curd residue and utilized it as a carbon source during SSF by producing xylanases. Xylanases are produced from xylan by fungi, yeast and bacteria, including *Bacillus* sp. (Beg et al., 2001; Blanco et al.,1995; Gallardo et al., 2004;Heck et al., 2005; Sa-Pereira et al., 2003) and physicochemical properties, structures and specific activities of these xylanases were diverse.

In this study, two xylanase-active enzymes were isolated. One of them (Xyl-I)was endo-1,4-β-xylanase (XynA), which has been found in many strains of *Bacillus* sp.(Blanco et al., 1995; Gallardo et al., 2004; Nishomoto et al., 2002). Characteristics of the Xyl-I obtained in this work are similar to those previously reported in that there is β-D-glucosidase activity and the values of optimum pH and temperature of Xyl-I are similar to those in other xylanases (Table 2). Another xylanase-active enzyme obtained (Xyl-II) was identified as α-amylase. As shown in Table 2, physicochemical properties of Xyl-II except for molecular mass were similar to those reported previously. Distribution of α-amylase is wide from common mesophilic bacteria to hyperthermophilic archaeon *Pyrococcus furiosus* (Jorgensen et al., 1997). Alpha-amylase of *B. subtilis* is used commercially in various categories such as starch hydrolysis in starch liquefaction process and additives to detergents for both washing machines and automated dish-washers because of its high thermo-stable activity (Nielsen & Borchert, 2000). As α-amylase, which catalyzes the hydrolysis and transglycosylation at α-1,4- and α-1,6-glycosidic linkages, it doesn't seem to be responsible for degradation of xylan. However, it has been shown that, due to the heterogeneity and structural complexity of xylan, the complete hydrolysis of xylan requires a large variety of cooperatively acting enzymes; such as endo-1,4-β-D-xylanases, β-D-xylosidase, α-L-arabinofuranosidases, α-D-glucuronidases, acetylxylan esterases, ferulic acid esterases and *p*-coumaric acid esterases (Collins et al., 2005). Thus, α-amylase of RB14-CS which hydrolyzed α-1,4- or 1,6-glucoside linkage in the reagent grade xylan used in this study may act as the cooperatively acting enzymes to release reducing sugars from xylan.

Two enzymes isolated in this work liberated xylooligosaccharides but not xylose from xylan. However, almost no reducing sugars were detected when xylanase activity was detected in SSF. This indicates that RB14-CS degraded xylooligosaccharides into xylose and utilized it as a carbon source. RB14-CS may produce other enzymes such as β-D-xylosidase for this reaction.

In recent years, biomass containing hemicellulose, such as agricultural and forestry residues, waste paper, and industrial wastes, has been recognized as inexpensive and abundantly available sources of sugar (Katahira et al., 2004). Since the production of iturin A by RB14-

CS in soybean curd residue was almost equivalent to that when glucose was used as carbon source, the utilization of soybean curd residue will be one possible nutrient in peptide production.

	This work	References	
		Gallardo et al.	Blanco et al.
Molecular mass	24 kDa	24 kDa	32 kDa
Optimum pH	6-7	6	5.5
Optimum temperature	50-60°C	60°C	50°C
Thermal stability	Decreased to 30 % at 50°C after 3 h. Deactivated within 1 h at 60°C.	Remained stable at 50°C for at least 3 h. Deactivated within 1 h at 60°C.	-

(A) Xyl-I

	This work	Reference	(Ohdan et al.)
		Ba-S	Ba-L
Molecular mass	58 kDa	47 kDa	67 kDa
Optimum pH	5.5-6.0	5.5	5.5
Optimum temperature	70°C	65°C	65°C
Thermal stability	80 % was retained after 3 h at 50°C. Deactivated within 2 h at 60°C.	60 % was retained after 10 min at 65°C	30 % was retained after 10 min at 65°C

(B) Xyl-II

Table 2. Comparison of characteristics of purified xylanases with previous reports.

5. Conclusion

Soybean curd residue which is the residue of *Tofu* production was used for nutrients for production of a lipopeptide antibiotic, iturin A in solid state fermentation(SSF) using *Bacillus subtilis*. As the main carbon sources of soybean curd residue were insoluble fiber, we expected that *B. subtilis* produced the soybean curd residue-degrading enzymes. Among insoluble fibers in soybean curd residue, hemicellulose was mainly degraded by *B. subtilis* during SSF. Xylan, a major hemicellulose in plant cell wall was degraded by *B. subtilis*, and two enzymes which showed xylanase activity were purified and identified as endo-1,4-β-xylanase and α-amylase. As productivity of iturin A in soybean curd residue was almost equivalent to that in glucose medium, this study gave a possible way to use soybean curd residue in higher and economical production of lipopeptides.

6. References

Asaka, O. & Shoda, M. (1996). Biocontrol of *Rhizoctonia solani* damping-off of tomato with *Bacillus subtilis* RB14. *Applied and Environmental Microbiology*, Vol.62, pp. 4081-4085

Beg, Q. K.; Kapoor, M.; Mahajan, L. & Hoondal, G. S. (2001). Microbial xylanases and their industrial applications: a review. *Applied Microbiology and Biotechnology*, Vol.56, pp. 326-338

Blanco, A.; Vidal, T.; Colom, J. F. & Pastor, F.J. I. (1995). Purification and properties of xylanase A from alkali-tolerant *Bacillus* sp. strain BP-23. *Applied and Environmental Microbiology*, Vol.61, pp.4468-4470

Bradford, M. M. (1976). A rapid and sensitive method for the quantification of microgram quantities of protein utilizing the principle of protein-dye binding. *Analytical Biochemistry*, Vol.72, pp. 248-254

Collins, T.; Gerday, C. & Feller, G. (2005). Xylanases, xylanase families and extremophilic xylanases. *FEMS Microbiological Review*, Vol.29, pp. 3-23

Edman, P. (1949). A method for the determination of the amino acid sequence in peptides. *Archives of Biochemistry and Biophysics*, Vo.11, pp. 475-476

Gallardo, O.; Diaz, P. & Javier Pastor, F. I. (2004). Cloning and characterization of xylanase A from the strain *Bacillus* sp. BP-7: Comparison with alkaline pI-low molecular weight xylanases of family 11. *Current Microbiology*, Vol.48, pp.276-279

Heck, J. X.; Flores, S. H.; Hertz, P. F. & Ayub, M. A. Z. (2005). Optimization of cellulase-free xylanase activity produced by *Bacillus coagulans* BL69 in solid-state cultivation. *Process Biochemistry*, Vol.40, pp.107-112

Hiraoka, H.; Ano, T. & Shoda, M.(1992). Characterization of *Bacillus subtilis* RB14, coproducer of peptide antibiotics iturin A and surfactin. *Journal of General and Applied Microbiology*, Vol.38, pp.635-640

Jorgensen, S.; Vorgias, C. E. & Antranikian, G. (1997). Cloning, sequencing, characterization, and expression of an extracellular α-amylase from the hyperthermophilic archaeon *Pyrococcus furiosus* in *Escherichia coli* and *Bacillus subtilis*. *Journal of Biological Chemistry*, Vol.272, pp.16335-16342

Kasai, N.; Murata, A.; Inui, H.; Sakamoto, T. & Kahn, R. I. (2004).Enzymatic high digestion of soybean milk residue (Soybean curd residue). *Journal of Agricultural and Food Chemistry*, Vol.52, pp. 5709-5716

Katahira, S.; Fujita, Y.; Mizuike, A.; Fukuda, H. & Kondo, A. (2004). Construction of a xylan-fermenting yeast strain through codisplay of xylanolytic enzymes on the surface of xylose-utilizing *Saccharomyces cerevisiae* cells. *Applied and Environmental Microbiology*, Vol.70, pp.5407-5414

Kiyohara, M.; Sakaguchi, K.; Yamaguchi, K.; Araki, T.; Nakamura, T. & Ito, M. (2005). Molecular cloning and characterization of a novel β-1,3-xylanase possessing two putative carbohydrate-binding modules from a marine bacterium *Vibrio* sp. strain AX-4. *Biochemical Journal*, Vol.388, pp. 949-957

Krzyzanowski F. C.; Franca-Neto, J. B.; Mandarino, J. M. G. & Kaster, M. (2001). Comparison between two gravimetric methods to determine the lignin content in soybean seed coat. *Seed Science and Technology*, Vol.29, pp. 619-624

Kunst, F.; Ogasawara, N. & 149 other authors (1997). The complete genome sequence of the Gram-positive bacterium *Bacillus subtilis*. *Nature*, Vol.390, pp.249-256

Laemmli, U. K. (1970). Cleavage of structural proteins during the assembly of the head bacteriophage T4. *Nature*, Vol.277, pp. 680-685

Leclere, V.; Bechet, M.; Adam, A.; Guez, J.S.; Wathelet, B.; Ongena, M.; Thonart, P.; Gancel, F.; Chollet-Imbert, M. & Jacques, P. (2005). Mycosubtilin overproduction by *Bacillus subtilis* BBG100 enhances the organism's antagonistic and biocontrol activities. *Applied and Environmental Microbiology*, Vol. 71, pp.4577-4584

Mizumoto, S.; Hirai M. & Shoda, M. (2006). Production of lipopeptide antibiotic iturin A using soybean curd residue cultivated with *Bacillus subtilis* in solid-state fermentation, *Applied Microbiology and Biotechnology*, Vol.72, pp.869-875

Nielsen, J. E. & Borchert, T. V. (2000). Protein engineering of bacterial α-amylases. *Biochimica et Biophysica Acta*, Vol.1543, pp.253-274

Nishimoto, M.; Honda, Y.; Kitaoka, M. & Hayashi, K. (2002). A kinetic study on pH-activity relation ship of XynA from alkaliphilic *Bacillus halodurans* C-125. *Journal of Bioscience and Bioengineering*, Vol.93, pp.428-430

Ohdan, K.; Kuriki, T.; Kaneko, H.; Shimada, J.; Takada, T.; Fujimoto, Z.; Mizuno, H. & Okada, S. (1999). Characteristics of two forms of α-amylases and structural implication. *Applied Environmental Microbiology*, Vol.65, pp. 4652-4658

Ongena, M.; Jacques, P.; Toure, Y.; Destain, J.; Jabrane, A. & Thonart, P. (2005). Involvement of fengycin-type lipopeptides in the multifaceted biocontrol potential of *Bacillus subtilis*. *Applied Microbiology and Biotechnology*, Vol.69, pp.29-38

O'tool, D. K. (1999). Characteristics and use of soybean curd residue, the soybean residue from soy milk production - A review. *Journal of Agricultural and Food Chemistry*, Vol.47, pp.363-371

Partridge, S. M. (1949). Aniline hydrogen phthalate as a spraying reagent for chromatography of sugars. *Nature*, Vol.164, pp. 443

Perez, J.; Munoz-Durado, J.; de la Rubia, T. & Martinez, J. (2002). Biodegradation and biological treatments of cellulose, hemicellulose and lignin: an overview. *International Microbiology*, Vol.5, pp. 53-63

Phae, C. G.; Shoda, M. & Kubota, H. (1990). Suppressive effect of *Bacillus subtilis* and its products to phytopathogenic microorganisms. *Journal of Fermentation and Bioengineering*, Vol. 69, pp.1-7

Sa-Pereira, P.; Paveia, H.; Costa-Ferreira, M. & Aires-Barros, M. R. (2003). A new look at xylanases: an overview of purification strategies. *Molecular Biotechnology*, Vol.24, pp.257-281

Stein, T. (2005). *Bacillus subtilis* antibiotics: structures, syntheses and specific functions. *Molecular Microbiology*, Vol. 56, pp.845-857

Sun, Y. & Cheng, J. (2002). Hydrolysis of lignocellulosic materials for ethanol production: a review. *Bioresources Technology*, Vol.83, pp. 1-11

Tsuge, K.; Ano T. & Shoda, M. (1996). Isolation of a gene essential for biosynthesis of the lipopeptide antibiotics plipastatin B1 and surfactin in *Bacillus subtilis* YB8. *Archives of Microbiology*, Vol. 165, pp. 43-251

Tsuge, K.; Ano, T.; Hirai, M.; Nakamura Y. & Shoda, M. (1999). The genes, *degQ*, *pps* and *lpa-8* (*sfp*) are responsible for conversion of *Bacillus subtilis* strain 168 to plipastatin production. *Antimicrobial Agents and Chemotherapy*, Vol. 43, pp. 2183-2192

Tsuge, K.; Akiyama, T. & Shoda, M. (2001). Cloning, sequencing, and characterization of the iturin A operon. *Journal of Bacteriology*, Vol.183, pp. 6265-6273

Van der Riet, W. B.; Wight, A. W.; Cilliers, J. J. L. & Datel, J. M. (1989). Food chemical investigation of tofu and its byproduct soybean curd residue. *Food Chemistry*, Vol.34, pp.193-202

Van Soest, P. J. (1963) .Use of detergents in the analysis of fibrous feeds. II. A rapid method for the determination of fiber and lignin. *Journal- Association of Official Analytical Chemistry*, Vol.46, pp. 825-835

Van Soest, P. J.; Robertson, J. B. & Lewis, B. A. (1991). Methods for dietary fiber, neutral detergent fiber, and nonstarch polysaccharides in relation to animal nutrition. *Journal of Dairy Science*, Vol. 74, pp. 3583-3597